O. Grasset • M. Blanc • A. Coustenis •
W.B. Durham • H. Hussmann • R.T. Pappalardo •
D. Turrini
Editors

Satellites of the Outer Solar System

Exchange Processes Involving the Interiors

Previously published in *Space Science Reviews* Volume 153, Issues 1–4, 2010

Springer

O. Grasset
LPGNantes Laboratoire
de Planétologie et Géodynamique
Nantes, France

M. Blanc
Centre d'Études Spatiale
des Rayonnements, CESR
Toulouse, France

A. Coustenis
LESIA, Observatoire de Paris-Meudon
Meudon, France

W.B. Durham
Massachusetts Institute of Technology
Cambridge, MA, USA

H. Hussmann
DLR Institute of Planetary Research
Berlin, Germany

R.T. Pappalardo
Jet Propulsion Laboratory
California Institute of Technology
Pasadena, CA, USA

D. Turrini
Institute of Physics of Interplanetary Space
INAF
Rome, Italy

ISBN 978-1-4419-7438-9
Springer New York Dordrecht Heidelberg London

Cover illustration: Evidence of exchange processes in the icy moons: (left panel) Eruption of water vapor and ice particles on Enceladus (credits: NASA/JPL/Space Science Institute); (center panel) Brownish hydrated salt minerals associated with domes and linear ridges on Europa (credits: NASA/JPL/University of Arizona/University of Colorado); (right panel) Giant methane–ethane cloud on the north pole of Titan (credits: NASA/JPL/University of Arizona/University of Nantes).

Printed on acid-free paper

Springer is part of Springer Science+Business Media (www.springer.com)

Contents

Space Sci Rev (2010) 153: 1–3
DOI 10.1007/s11214-010-9679-2

Foreword

Roger-Maurice Bonnet · Michel Blanc

Received: 19 July 2010 / Accepted: 19 July 2010 / Published online: 25 August 2010

"Satellites of the Outer Solar System: Exchange Processes Involving the Interiors" is the second publication of its kind in the Space Science Series of ISSI. It is the result of a successful joint venture between ISSI and Europlanet.

Europlanet started in 2005, the year of the successful landing of the Huygens probe on Titan's surface, as a network of over 110 European and U.S. laboratories deeply involved in the development of planetary sciences and the support to the European programme of space exploration of planets. Since 2005, Europlanet obtained support from the European Commission to strengthen the planetary science community worldwide, and to amplify the scientific output, impact and visibility of the European space programme, essentially ESA's Horizon 2000 and Cosmic Vision programmes. Its first contract with the Commission, as a "Coordination action" extended from 2005 to 2008 and included 7 networking activities, among which the set-up of discipline working groups covering the main areas of planetary sciences. A new contract with the Commission, this time as a full "research infrastructure network", extends Europlanet's activities into the period 2009–2012. With the broad community connection made through its Discipline Working Groups and other activities, Europlanet offers an ideal base from which to identify new fields of research for planetary sciences and to stimulate collaborative work among its member laboratories. For Europlanet, developing collaboration with ISSI in holding workshops and producing books on these new and emerging subjects is both natural and extremely stimulating, considering the high profile, international standing and proven success of ISSI. For ISSI, collaboration with Europlanet offers a very interesting opportunity to extend its successful series of workshops and books within the area of planetary sciences and to deepen its links with this community.

Working from this clear convergence of interests, in 2006 ISSI and Europlanet initiated implementation of a series of joint workshops, soon after the establishment of Europlanet's

R.-M. Bonnet (✉)
International Space Science Institute (ISSSI), Bern, Switzerland
e-mail: rmbonnet@issibern.ch

M. Blanc
Centre Etude Spatiale des Rayonnements (CESR), Toullouse, France

Discipline Working Groups. Each of Europlanet's nine Discipline Working Groups were invited to select one or two promising subjects for ISSI workshops, which were presented and discussed at a w.g. meeting at ESAC, Villafranca, Spain, April 24th to 26th. The Board of Coordinators of Europlanet then selected a shortlist. In the spring of 2007, the Science Committee of ISSI, chaired by Prof. Len Culhane, selected the proposal for a workshop on "Satellites of the Outer Solar System: Exchange Processes Involving the Interiors" formulated by Olivier Grasset, to become the second joint ISSI-Europlanet workshop.

This very timely choice indeed corresponds to a high priority and challenge of present-day space science. The successful survey of the Jupiter system by the Pioneer, Voyager and Galileo spacecraft in the XXth century, followed by the on-going detailed exploration of the Saturn system and its moons by Cassini-Huygens since 2004, have revealed to the science community and to the public at large the fascinating worlds of the moons of the outer planets. We are still going from surprise to surprise as new data unveil new aspects of their so complex characteristics: the discovery of gigantic geysers on Enceladus, and of the complex geological and meteorological activities at the surface of Titan, are the latest examples of a continuing series of breathtaking discoveries!

This book will provide the reader with an up-to-date view of what we know and understand to-day of the moons of outer planets: by their complexity, their internal differentiation, their geological or atmospheric activity, their still poorly known history and evolution, the largest of these moons at least are actual "planetary worlds", with all the complexity this term implies. In particular, and this is the aspect this book has chosen to emphasize, these bodies are complex evolving systems in which the interactions and chemical exchange processes between the different layers—from core to atmosphere or exosphere—play a key role. These interactions extend to the way they are partly driven by interactions with their central planet, via tidal interactions (which for some of them provide a substantial fraction of their internal heat source) and via interactions with their planet's magnetosphere.

One of the most outstanding features of some of these moons, those with a substantial fraction of water ice, is the generation and long-term maintenance of sub-surface water oceans. These oceans are natural candidates to be habitats for life. The question of their habitability, and therefore of the very possibility of the emergence of life in the outer solar system, has received increasing attention in the last decade or so, to the point that it is now one of the highest-priority challenges of astrobiology. If life can emerge there, far from the classical "habitable zone" of stars, then there may be many more places for life in the Universe, where giant planets systems are widespread, than we ever dared to imagine!

For this reason, and many more, understanding better the moons of outer planets, their origin, evolution and potential habitability has become a high priority of planetary exploration. This book builds on our present knowledge of these moons, inherited from telescope observations and past and present space missions, to provide the reader with a comprehensive view of what we understand and don't understand of these fascinating worlds. It also very nicely shows how current progress in this understanding builds on the interplay between three basic ingredients: new space and ground-based data, laboratory measurements of key parameters of the structural materials of these moons, and sophisticated models capable of integrating data and lab measurements into quantitative and predictive representations of planetary bodies, their different layers, and their interactions.

In line with the spirit of ISSI and Europlanet, it is on the basis of this continuing feedback loop between observations, laboratory measurements and modelling that future space missions can be designed and prepared, ultimately to provide a new set of observations that will bring our understanding of planetary moons further. Preparation for the next generation of missions to the outer planets has already begun: just four hundred year after their discovery by Galileo Galilei, ESA and NASA are jointly preparing a new major step in the exploration of the Jovian moons, with the dual spacecraft EJSM-Laplace mission to the Jovian system; further on, the next logical steps will include a new mission to the Saturn system, deepening the discoveries of Cassini-Huygens with an in-depth investigation of Titan, its surface and atmosphere, and, further out, missions to the Ice Giants, Uranus and Neptune. Our exploration of the outer solar system has just started, and will take several decades to accomplish. No doubt it will unveil a fraction of the many mysteries of its fascinating planetary worlds: how they emerged from the cold outskirts of the solar nebula, how they formed complex differentiated bodies interacting with their environments and, possibly, how some of these bodies may have fulfilled the necessary conditions to develop and sustain life! This is, ultimately, the perspective that the reader will discover through the different chapters of this book.

In conclusion, we would like to thank all those whose hard work and dedication made this second joint publication of ISSI and Europlanet possible. Many thanks, first, to the leaderships of ISSI and Europlanet who designed this collaborative project, and to the leaders of the Discipline Working Groups of Europlanet, Norbert Krupp and Ari-Matti Harri. Many thanks to the Science Committee of ISSI, chaired by Prof. Len Culhane, who did the selection and made useful suggestions on the workshop content. Our special gratitude goes to Olivier Grasset, who originally proposed the workshop's topic and masterly led the whole process, from the first Conveners Meeting to book production. The team of conveners and editors did a fantastic job in defining the structure of the book and in managing the writing and the overall review process: as usual, all chapters were carefully reviewed by independent experts to whom we would also like to extend our gratitude.

Last, but not least, our warmest appreciation goes to the wonderful staff of ISSI, Andrea Fischer, ISSI's programme manager Vittorio Manno and his successor Maurizio Falanga, Saliba Saliba, Brigitte Schutte and Jennnifer Zaugg her successor, Irmela Schweizer, Silvia Wenger, and all their colleagues whose kindness and dedication make ISSI such a convivial and effective operation. They are the greatest asset on which this important collaboration between ISSI and Europlanet will continue to develop and flourish in the years to come.

Space Sci Rev (2010) 153: 5–9
DOI 10.1007/s11214-010-9664-9

Preface

Satellites of the Outer Solar System: Exchange Processes Involving the Interiors

O. Grasset · A. Coustenis · W.B. Durham · H. Hussmann · R.T. Pappalardo · S. Sasaki · D. Turrini

Received: 10 June 2010 / Accepted: 10 June 2010 / Published online: 30 June 2010

1 Introduction

Recent space missions to the outer solar system, Galileo (1996–2003) and Cassini-Huygens (2004–today), together with ground observations, have revealed that the moons of the outer solar system are enigmatic objects, introducing extraordinary challenges for geologists, astrobiologists, organic chemists, and planetologists. Chemical exchange exists through the different layers that form their interiors, and also from the interior to the surface. The most convincing evidence is certainly the discovery of water vapour and ice particles emerging from Enceladus's active south polar region (Fig. 1a). Evidence for exchange with a subsurface liquid ocean has also been provided by the inference of hydrated salts on the surfaces of Jupiter's moons, Europa (Fig. 1b) and Ganymede, as well as the detection of sodium salts in particles originating in Enceladus's plumes. Aqueous exchange with the rocky core may also be possible, considering that ^{40}Ar has been observed in the plumes of Enceladus during one flyby of Cassini and in the atmosphere of Titan. The ongoing CH_4 replenishment in Titan's atmosphere is additional striking evidence of exchange processes within the moons (Fig. 1c).

2 Observations and Measurements

Telescopic observations and, in the past few decades, spacecraft observations, from the Galileo and Cassini-Huygens missions, have led to the discovery of many surface materials, from rock-forming minerals to water ice to more exotic volatiles. The tremendous variability of surface compounds is fundamentally related to the exchange processes within the moons (variability of chemical compounds, the degree of activity, and the nature of

O. Grasset (✉)
Laboratoire de Planetologie et Geodynamique, UMR-CNRS 6112, 2 rue de la Houssiniere, 44322 Nantes Cedex 03, France
e-mail: olivier.grasset@univ-nantes.fr

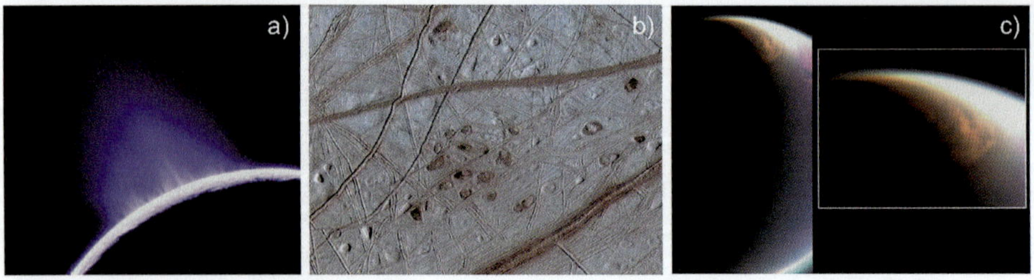

Fig. 1 Evidence of exchange processes in the icy moons: (**a**) Eruption of water vapor and ice particles on Enceladus (credits: NASA/JPL/Space Science Institute); (**b**) Brownish hydrated salt minerals associated with domes and linear ridges on Europa (credits: NASA/JPL/University of Arizona/University of Colorado); (**c**) Giant methane–ethane cloud on the north pole of Titan (credits: NASA/JPL/University of Arizona/University of Nantes)

volcanic/tectonic processes, ...). A comprehensive description of these processes has been provided in the first section of this book.

In the first chapter, Krupp et al. describe the global magnetospheric environments of the gas giants, their global configuration with its large-scale transport processes, but also the local environments, especially the interaction processes between the magnetospheric plasma and the exosphere/atmosphere/magnetosphere of the moons. It also includes the gravity, shape, rotation, astrometric observations and orbital parameters of the icy moons.

Prockter et al. describe the variety of morphological features displayed by the surfaces of the Solar System's icy satellites. Using data from spacecraft missions, the detailed morphology, size, and topography of cryovolcanic, tectonic, aeolian, fluvial, and impact features of large moons and smaller satellites are discussed.

Identifying surface materials and mapping their distributions allows us to constrain interior processes such as cryovolcanism and aqueous geochemistry. Dalton et al. look at all the findings we have so far on the planetary surface composition as derived from remote sensing over the ultraviolet through infrared wavelength ranges. Both ice and non-ice materials are considered, as well as exospheric species. The differences among the various satellites are discussed and brought into context. Other considerations include constraints brought on the interior composition and processes, the transport to the surface and communication with putative oceans.

Coustenis et al. review the characteristics of some satellites with atmospheres like Titan, Enceladus, Io and Europa, based on observations. The authors show that the atmospheres/exospheres of icy satellites greatly vary from one to another in terms of density, composition, structure or steadiness. Looking at satellites around the giant planets, Titan appears as the only icy satellite with a dense nitrogen atmosphere comparable in many ways to that of the Earth's atmosphere. Titan's atmosphere prevents the surface from direct interaction with the plasma environment, but gives rise to terrestrial-like exchanges of energy, matter and momentum. The atmospheres of other satellites are tenuous at best, non existent for most. However, new discoveries by the Cassini-Huygens mission have shown that there is an atmosphere around Enceladus' manifesting itself in large water vapor plumes emanating from surface cracks near the South Pole. The jets extend hundreds of kilometers into space and also include organics and ammonia. Io's SO_2 atmosphere originates from volcanoes that create a torus around the satellite. Europa's tenuous O_2 atmosphere is produced by intense radiation bombardment.

3 Physics and Chemistry of Ices and Organics: Constraints from Laboratory Experiments

Physical and chemical properties of planetary building materials are the medium between the observables described in Sect. 1 and the modeling to be presented in Sect. 3. Thus, laboratory measurements of intrinsic properties and behaviors are key elements in constraining planetary exchange process. Papers in this section describe techniques being developed to permit the measurement of spectral properties of planetary molecules at relevant condition (Dalton), and to determine electrical properties relevant to radar observations (Kofman et al.), thus enabling the interpretation of remotely sensed data.

Critical to our understanding of surface materials are the effects of ionizing radiation, which can sputter molecules from the surface, and can change molecules in atmospheres and surfaces through photolysis and radiolysis, as described in the paper by Cassidy et al. Also described in this chapter are latest results regarding phase relationships among frozen volatile compounds as a function of pressure and temperature, with special emphasis on mixtures of water ice with methane and ammonia (Fortes and Choukroun), which thus constrains the phases present under extreme planetary conditions. Laboratory measurements of rheological properties and thermal transport properties of icy materials, a necessary part of dynamical models, are also detailed (Durham et al.).

4 Modeling and Interpretations

Based on the data from space missions the models describing the interior structure, evolution and processes operating at icy moons have improved considerably. Models of the satellites' interior structures, rotation states and orbital states are required to compute the satellites' internal energy budgets that drive internal activity. In a few cases such activity is ongoing at present. Energy sources that are in some cases linked to the orbital and rotational states by tides and heat transport models are reviewed by Hussmann et al.

Past and recent activity in the icy moons are mainly visible at the surfaces where we find evidence for tectonism and volcanism at specific satellites. However, to distinguish internal processes from the external environment, models of the latter are required. Especially, the interaction with the giant planets' magnetospheres is an important factor. The state and composition of the tenuous atmospheres of the satellites are a result of both internal processes and interaction with the environment. Furthermore, impactors have shaped the satellite surfaces. Models for the impact rates in combination with the measured number density of craters on the surface can allow for the estimate of global and regional surface ages. The external environment including atmosphere-magnetosphere interactions and models for the cratering record are reviewed by Burger et al.

The interface between interior and surface is the lithosphere which is subject to external forcing (e.g., periodic tidal stresses and impact processes). It is also affected by internal processes (e.g., cryo-volcanism and dynamics in the ice shell). In some cases liquid water or volatiles and gases can reach the surface. The activity at Enceladus' south-pole is the most prominent example of such interior-surface communication. The physics of these processes and the conditions and models for individual satellites including Io, Enceladus, Triton and Titan are discussed by Tobie et al.

5 Implications

The state of knowledge about the structure and composition of icy satellite interiors has been significantly extended by combining direct measurements from spacecraft with laboratory experiments, and theoretical modeling. The spacecraft data supplies us with precious information to constrain the models describing satellite evolution, and the possible existence of water reservoirs below the crusts of the icy moons.

The processes governing the formation of the giant planets and the accretion of their gaseous envelopes determine the environments and the initial conditions for satellite formation. Coradini et al. and Mosqueira et al. show how temperature and pressure gradients in circumplanetary disks constrain their chemical evolution and condensation profiles and influence the abundances of elements available to build the satellites. Moreover, the forming planets and their subnebulae interact with the Solar Nebula and the planetesimals that populate the outer Solar System. Gas and planetesimals are captured from the Solar Nebula and spiral toward the growing planet, allowing for different generations of satellites to form and be accreted by the giant planets. In addition, it is shown that irregular satellites did not form where we see them now but originated elsewhere and were captured by the giant planets, introducing exogenous material in their host systems.

Once formed, the satellite systems of the giant planets start their secular evolution to their present state (Schubert et al.). Radiogenic and tidal heating supply the internal energy needed to the larger satellites to differentiate. Dynamical resonances, by exciting the orbital eccentricities of the satellites, interplay with tidal forces and enhance their effects. Internal activity and differentiation shape both the interiors and, through volcanism and tectonism, the surfaces of satellites. In addition, exchange processes take place between regular and irregular satellites or satellites and planetary rings, and high-velocity impacts may expose deeper layers of the satellites. All these processes act to modify both the morphology and the composition of the surfaces of the satellites and must be accounted for in any attempt to interpret the observational data.

The existence of potentially habitable liquid water reservoirs within icy satellites depends on the amount of energy which is available and on the chemistry of the icy materials. Sohl et al. review the chemical evolution of subsurface liquid water oceans, taking into account the processes occuring in aqueous environments and related to material exchange, including processes occuring at the transitions from the liquid water layer to the ice layers above and below, and water-rock exchange when the liquid layer is in contact with the rocky ocean floor.

In the final chapter, Raulin et al. discuss how the exchange processes in the icy moons are comparable to those thought to have played an important role in the emergence of life on Earth. They show how the study of prebiotic processes on and within outer solar system moons is a key goal for exobiology, together with the question of habitability and the search for evidence of past or present life. The question of planetary protection, especially in the case of Europa, is also presented.

6 Conclusion

Despite the large amount of observational evidence that has already been obtained, it is still unclear how the release of internal materials and volatiles onto the surface operates on water-rich bodies. Exchange processes have likely changed as evolution has proceeded, and what we observe today on surfaces and atmospheres is the integrated result of several billion years of complex interactions. In the absence of a precise quantification of these chemical exchange processes, it is also extremely difficult to evaluate the exobiological potential of these water-rich environments. In this volume, a detailed and up-to-date summary is given of the various topics related to the exchange processes in and on the satellites of the outer system.

It is our pleasure to thank all those who have contributed to this volume and to the dedicated workshop. We are grateful to the authors for their contributions, but also to the reviewers who helped us to improve the quality of this book. We also wish to express our warm thanks to the directorate and staff of the International Space Science Institute, for their effort in making the workshop happen and in finalizing the book.

Space Sci Rev (2010) 153: 11–59
DOI 10.1007/s11214-010-9653-z

Environments in the Outer Solar System

N. Krupp · K.K. Khurana · L. Iess · V. Lainey · T.A. Cassidy · M. Burger · C. Sotin · F. Neubauer

Received: 30 September 2009 / Accepted: 2 April 2010 / Published online: 26 May 2010

Abstract The outer planets of our solar system Jupiter, Saturn, Uranus, and Neptune are fascinating objects on their own. Their intrinsic magnetic fields form magnetic environments (so called magnetospheres) in which charged and neutral particles and dust are produced, lost or being transported through the system. These magnetic environments of the gas giants can be envisaged as huge plasma laboratories in space in which electromagnetic waves, current systems, particle transport mechanisms, acceleration processes and other phenomena act and interact with the large number of moons in orbit around those massive planets.

In general it is necessary to describe and study the global environments (magnetospheres) of the gas giants, its global configuration with its large-scale transport processes; and, in combination, to study the local environments of the moons as well, e.g. the interac-

N. Krupp (✉)
Max-Planck-Institut für Sonnensystemforschung, Katlenburg-Lindau, Germany
e-mail: krupp@mps.mpg.de

K.K. Khurana
IGPP, University of California Los Angeles, Los Angeles, CA, USA

L. Iess
Dipartimento di Ingegneria Aerospaziale ed Astronautica, Universita La Sapienza, Rome, Italy

V. Lainey
IMCCE-Observatoire de Paris, Paris, France

T.A. Cassidy
University of Virgina, Charlottesville, VA, USA

M. Burger
Goddard Space Flight Center, Greenbelt, MD, USA

C. Sotin
Jet Propulsion Laboratory, Pasadena, CA, USA

F. Neubauer
University of Cologne, Cologne, Germany

tion processes between the magnetospheric plasma and the exosphere/atmosphere/magnetosphere of the moon acting on time scales of seconds to days. These local exchange processes include also the gravity, shape, rotation, astrometric observations and orbital parameters of the icy moons in those huge systems.

It is the purpose of this chapter of the book to describe the variety of the magnetic environments of the outer planets in a broad overview, globally and locally, and to show that those exchange processes can dramatically influence the surfaces and exospheres/atmospheres of the moons and they can also be used as a tool to study the overall physics of systems as a whole.

Keywords Planetary magnetospheres · Plasma transport · Jupiter · Saturn · Icy moons · Magnetosphere moon interaction

1 Global Magnetic Environment (Magnetosphere) of the Outer Planets

The four outer gas planets Jupiter, Saturn, Uranus and Neptune have strong intrinsic magnetic field and therefore they all have their own intrinsic magnetospheres. These enormous magnetic fields are created in their interior, in a core made out of liquid metallic hydrogen in the case of Jupiter and Saturn, and out of a mixture of molten methane and ammonia ice for Uranus and Neptune. A magnetosphere is formed as a consequence of pressure balance between internal magnetic and particle pressure and external pressure of the solar wind. The magnetopause (boundary where the pressure equilibrium is reached) location is therefore dependent on the strengths of the planetary magnetic fields as well as on the strengths of internal plasma sources. Its nose location towards the sun's direction can extend from a few planetary radii in the case of Mercury's magnetosphere, to hundreds of planetary radii in the case of Jupiter (Fig. 1).

On the anti-sunward facing side of the magnetosphere the magnetic field lines are highly stretched from hundreds to several thousands of planetary radii, forming the so called magnetotails. This means that even the most distant moons of the outer planets traverse through the long magnetotails and interact with magnetospheric plasma as well as with the solar wind when they are outside the magnetosphere in the interplanetary medium.

These magnetospheres are unique entities, huge plasma laboratories in space, providing unique plasma properties under extreme conditions. Gravity and electromagnetic forces form fascinating magnetic environments around those planets in which physical processes take place. In the inner magnetosphere the magnetic field lines are nearly dipolar and high-energy charged particles are trapped forming the so called radiation belts. In the outer part of the magnetospheres the magnetic field gets more and more radially stretched because the oppositely directed fields from the magnetodisc and the magnetotail become comparable to the internal field of the planet (see below).The best studied magnetosphere of all planets is the magnetosphere of the Earth. Therefore it is taken as a reference for the outer planet magnetospheres with which it shares a lot of similarities, but also major differences.

The main differences between the magnetosphere of the Earth and those of the outer gas giants are (1) that the Earth does not have a major internal plasma source, and (2) that the energy driving the system is the planetary rotation for the outer planets while convection plays the major role in the Earth's case.

The outer planets magnetospheres all have major internal plasma sources (moons, rings, planetary atmospheres/ionospheres) while the solar wind, an external plasma source, does not play the same major role as in the case for the Earth's magnetosphere. The dominant

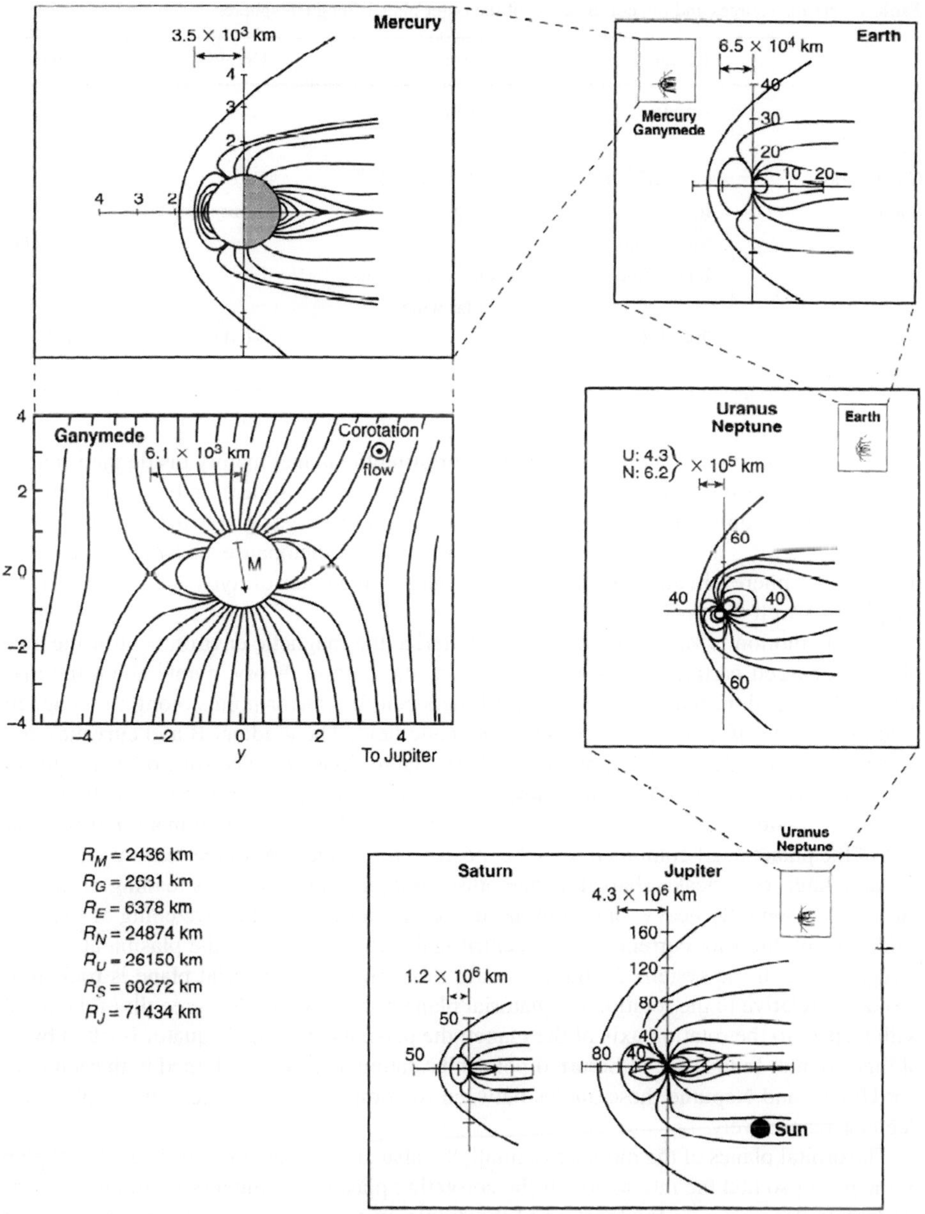

Fig. 1 Size comparison of all known magnetospheres in our solar system (from Williams et al. 1998)

plasma sources and the known composition, density and source strengths are summarized in Table 1.

Some of the sources lie deep within the magnetospheres providing hundreds of kilograms of material in the form of neutral gas, dust and charged particles (especially Io for Jupiter and Enceladus for Saturn).

Table 1 Plasma sources and ion composition of the outer planets magnetospheres

	Jupiter	Saturn	Uranus	Neptune
Density N_{max} [cm^{-3}]	>3000	∼100	∼3	∼2
Composition	O^{n+}, S^{n+}, H^+	O^+, H_2O^+, H^+	H^+	H^+, N^+
Sources	Io Solar wind Ionosphere	Rings Enceladus Other icy moons Solar wind	Ionosphere	Triton Ionosphere
Source-strength [kg/s]	700–1200	∼2–100	∼0.02	∼0.2

All four outer planets harbor icy satellites orbiting the planet mainly inside their magnetosphere, some of them continuously stay outside the magnetosphere in the solar wind. The following table summarizes the number of moons of all four gas giants together with some major icy satellites as well as the nose distance of the magnetopause (indicating the size of the magnetosphere). In addition the spacecraft names and dates of flybys and orbiting times are given as well (Table 2).

Particle motion is very important to characterize the magnetic environment of the outer planets. Charged particles, consistent with adiabatic motion theory, gyrate about the magnetic field direction, bounce back and forth between the mirror points along the magnetic field lines, and drift perpendicular to the magnetic field. The gradient B and curvature drift direction is dependent on the polarity of their charge: ions and electrons drift in opposite directions. The ExB-drift direction, however, is charge independent in the direction of the planetary rotation (corotation). Especially for the outer planets the combination of fast rotation of the planet itself combined with the mostly internal plasma sources ends up in equatorial doughnut- or pancake-shaped magnetodisc configuration. The mass loading of the field lines with essentially heavy ions from these internal sources and the stretching of magnetic field lines by the ring current and the centrifugal force concentrate the plasma at the centrifugal plane of the system. In Jupiter's case this magnetic equatorial plane is rocking up and down relative to the rotational equatorial plane because of a 9.6 degree tilt of the dipole with respect to the rotation axis of the planet; the plasma centrifugal equator is tilted by 2/3 of this amount. In the case of Saturn dipole and rotation axis are co-aligned with each other. For Uranus and Neptune these angles between rotation axis and dipole axis are 59 and 49 degrees, respectively.

The orbital planes of the moons essentially lie also close to the rotational equatorial plane of the planet so that the interaction of the corotating plasma and the moons along their orbit are largely dependent on the location of the dipole-axis relative to the moon. The exchange processes can be significantly different if the moons pass through the center part or through the nearly empty lobe regions of the magnetodisc.

The surrounding neutral and charged particles, with a wide spectrum of energies, directions and ion species and charge states, interact with the moons. On the one hand particles are being implanted on the surfaces of the moons; on the other hand the incoming particle flux can sputter off material from the surfaces. Both processes can cause albedo changes of the surfaces in the images of the bombarded bodies.

The global magnetospheric magnetic field and plasma distribution is disturbed to varying degrees in the vicinity of the icy satellites depending on the mode of interaction. So far such

Table 2 Exploration of the Outer Gas Giants in our solar system, the size of their magnetospheres and the number of known moons and some of the major icy satellites with their distances from the center of the planet in planetary radii

Planet	Spacecraft Flyby/orbiter	Nose distance of magnetopause (in planetary radii)	Number of moons	Major icy moons (distance from the planet in planetary radii)
Jupiter	Pioneer 10 (1974)	60–120 R_J	63	Amalthea (2.54 R_J)
	Pioneer 11 (1974)			Thebe (3.1 R_J)
	Voyager 1 (1979)			
	Voyager 2 (1979)			Io (5.9 R_J)
	Ulysses (1992)			Europa (9.4 R_J)
	Cassini (2001)			Ganymede (14.99 R_J)
	New Horizons (2007)			Callisto (26.37 R_J)
	Galileo (1995–2003)			
	Juno (2016–2018)			
	EJSM (from 2026 on ?)			
Saturn	Pioneer 11 (1979)	22–27 R_S	62	Mimas (3.08 R_S)
	Voyager 1 (1980)			Enceladus (3.95 R_S)
	Voyager 2 (1981)			Tethys (4.88 R_S)
	Cassini (2004–2017)			Dione (6.26 R_S)
				Rhea (8.75 R_S)
				Titan (20.27 R_S)
				Iapetus (59.08 R_S)
				Phoebe (213.54 R_S)
Uranus	Voyager 2 (1986)	25 R_U	27	Miranda (5.08 R_U)
				Ariel (7.48 R_U)
				Umbriel (10.41 R_U)
				Titania (17.05 R_U)
				Oberon (22.79 R_U)
Neptune	Voyager 2 (1989)	26 R_N	13	Triton (14.32 R_N)

$R_J = 71{,}492$ km; $R_S = 60{,}268$ km; $R_U = 24{,}500$ km; $R_N = 25{,}500$ km

disturbances have been observed in detail for the satellites of Jupiter and Saturn only. The three most important factors influencing the exchange processes are:

(1) the existence of an atmosphere;
(2) the existence of large volumes of electrically conducting material like an electrolytically conducting subsurface ocean or a metallic core;
(3) the presence of a permanent magnetic field of dynamo origin or due to natural remnant magnetization.

This will be mentioned later in the chapter when the local environment in the vicinity of the moons will be discussed.

For more details about the topics discussed in this chapter the reader is referred to textbooks like: Uranus by Bergstralh et al., University of Arizona Press, 1991; Neptune and Triton by Cruikshank, University of Arizona Press, 1995; Jupiter, the planet, satellites and magnetosphere by Bagenal et al., Cambridge University Press, 2004; Physics of the Jovian

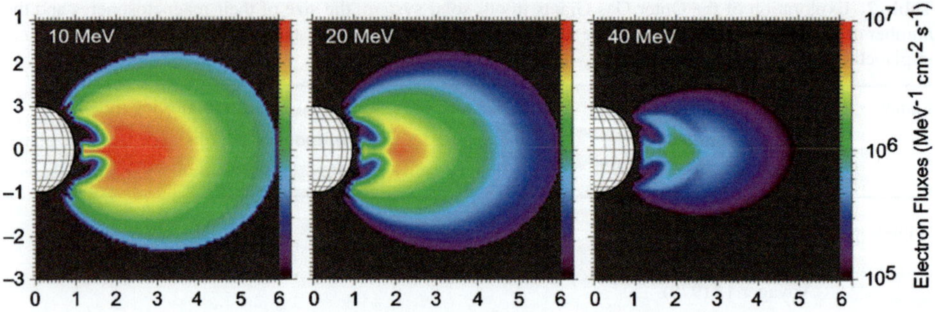

Fig. 2 Meridian cuts of omnidirectional differential electron fluxes for three energies as a function of radial distance, kinetic energy and latitude. From Santos-Costa (2001) and Santos-Costa and Bourdarie (2001) after Bolton et al. (2004)

magnetosphere by Dessler, Cambridge University Press, 1983; Saturn by Gehrels et al., University of Arizona Press, 1984; The Outer Planets and Their Moons by Encrenaz et al. 2005; and recently Saturn from Cassini-Huygens by Dougherty et al., Springer Verlag, 2009.

1.1 Jupiter

The biggest planet in our solar system, Jupiter, has the strongest magnetic field of all planets (first postulated by Burke and Franklin 1955) and therefore also has the biggest magnetosphere, 10 times larger in its dimensions than the solar disk and by far the biggest object inside the heliosphere. Jupiter has 63 known moons today. They orbit the planet at distances between 127.690 km (1.78 R_J) and 30.290.846 km (424 R_J) and interact with different regions of the Jovian magnetosphere.

Starting from the inner magnetosphere we know that the magnetic field is nearly dipolar with closed field lines trapping high-energy electrons and protons in the so called radiation belts. Jupiter's radiation belts are the strongest and most energetic of its kind in the solar system. Energetic particles bounce along the closed field lines and interact with the moons and rings in the inner magnetosphere. In Fig. 2 the fluxes of energetic electrons with energies of 10, 20 and 40 MeV inside the radiation belts are shown in color as a function of distance from the center of the planet.

Inside of about 4 R_J in addition dust is produced when interplanetary impactors collide energetically with the small inner moons Metis, Adrastea, Amalthea and Thebe resulting in a dust cloud and organized in a main ring, an inner halo, and the two fainter and more distant Gossamer rings. Exchange processes between the dust, magnetospheric charged particles, and the moons are important to understand the dust dynamics and some properties of the small moon surfaces. So far only Amalthea has been visited during a targeted flyby of the Galileo spacecraft (see Krüger et al. 2004 and Hamilton and Krüger 2008; Fig. 3).

The moon Io, located at 5.9 R_J, is not only the major source of sulfur and oxygen in the system but also the source of dust streams first detected by the dust detector onboard Ulysses (Grün et al. 1993). Those streams of charged dust particles are transported throughout the magnetosphere and can be detected outside the Jovian magnetosphere. This moon plays an extraordinary role for the global configuration and dynamics of the Jovian magnetosphere. The volcanoes of Io are providing the magnetosphere with 1 ton/s of sulfur dioxide molecules which are being dissociated and are ionized. This material, in the form of oxygen and sulfur ions, is first accelerated from thermal energies to corotating energy forming a torus all the way around Jupiter at the orbit of Io (Fig. 4).

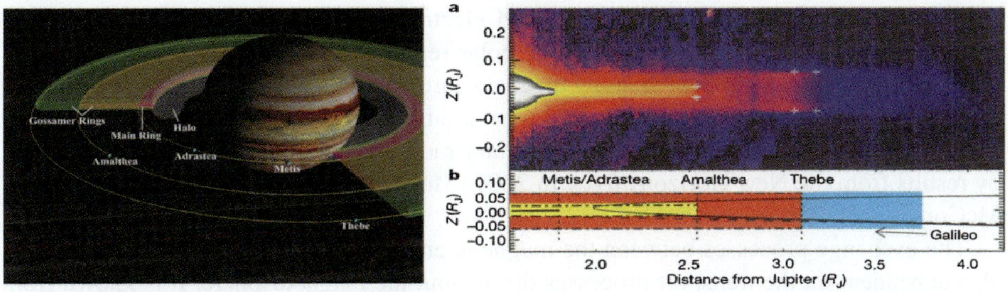

Fig. 3 Dust rings in the inner Jovian magnetosphere and the location of the inner icy satellites (from Hamilton and Krüger 2008)

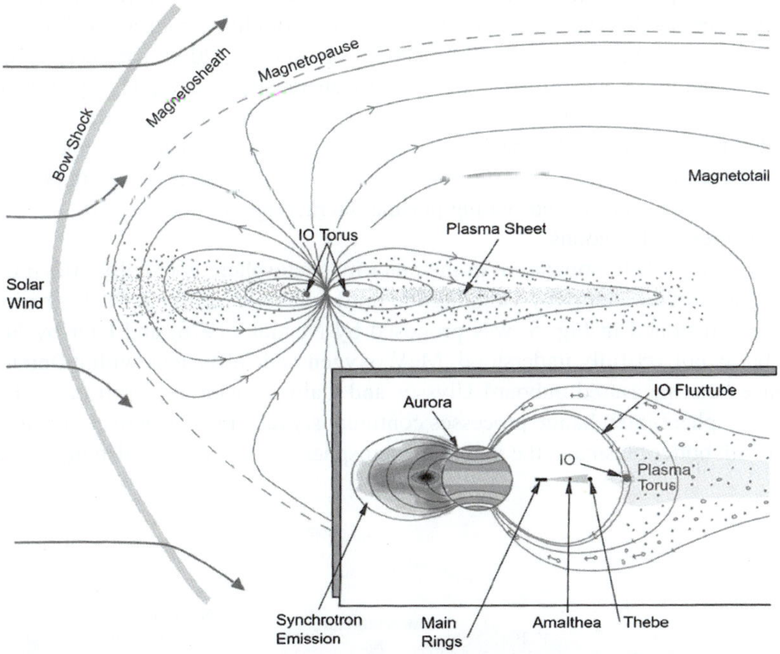

Fig. 4 Structure of Jupiter's inner magnetosphere (after Bolton et al. 2004 and private communication F. Bagenal)

The cold plasma in the torus is transported radially outwards. Due to the centrifugal force and mass loading of the magnetic field lines a magnetodisc or plasmadisc/plasma sheet is formed. The density is highest at the center and falls off as a function of distance from the center. Since the rotation axis and the magnetic dipole axis of Jupiter are tilted against each other by 9.7 degrees the rotating plasmadisc wobbles ± 10 degrees around the equatorial plane of the planet. This means that the number of corotating particles interacting with the moons varies and depends on the location of the magnetodisc.

At distances between 20 and 40–60 R_J (dependent on local time in the Jovian magnetosphere) the plasma in the magnetosphere is not able to fully corotate anymore with the planet. The corotation "breaks down" at this distance. The ionosphere is trying to spin up the magnetosphere to full corotation speed resulting in a field-aligned current system.

This current system accelerates field-aligned electrons from the magnetosphere into the ionosphere/atmosphere of the planet in the polar region. These field-aligned particles impinge into the polar ionosphere creating the main auroral oval. Outside that corotation breakdown region (which is close to Callisto's orbit around the planet the plasma subcorotates. The magnetodisc extends thousands of planetary radii into the magnetotail as documented by results from the New Horizons flyby in 2007 (McNutt et al. 2007; Haggerty et al. 2009; McComas et al. 2007).

The exchange processes between the magnetic environment of Jupiter and the moons is also dependent on the transport processes throughout the magnetosphere. It is known from phase space density calculations that the acceleration of the thermal charged particle population of eV energies up to MeV energies takes place in the outer magnetosphere (Krupp et al. 2004 and references therein). It is assumed that reconnection processes and particle injections might play a major role in this acceleration. It was noted in data from several Galileo instruments that the continuous mass loading from Io exhibits a 2.5–3 day periodicity (Krupp et al. 1998; Woch et al. 1998, 2002; Louarn et al., 2000; Kronberg et al. 2005). In the mass loading of heavy ions from Io the magnetic stress on the highly stretched field lines is too high after about two days so that the magnetic field lines are ripped apart, a plasmoid is released down the magnetotail and the now empty field lines merge and snap back over about a day after which the entire process starts again. It is very important to understand the dynamics of the corotating plasma interacting with the surfaces, exospheres or magnetospheres of the moons.

The charged particle population in the outer magnetosphere is radially diffusing inward again gaining further energy by violating the third adiabatic invariant. Such a recirculation process, as illustrated in Fig. 5, was proposed by Nishida (1976) and later by Fukazawa (2005), but is not yet fully understood. MeV oxygen and sulfur ions with different charge states have been measured onboard Ulysses and Galileo inside the orbit of Io. In the inner magnetosphere interchange processes continuously take place where cold plasma is exchanged with hotter plasma in the outer magnetosphere. Hot plasma is also injected through

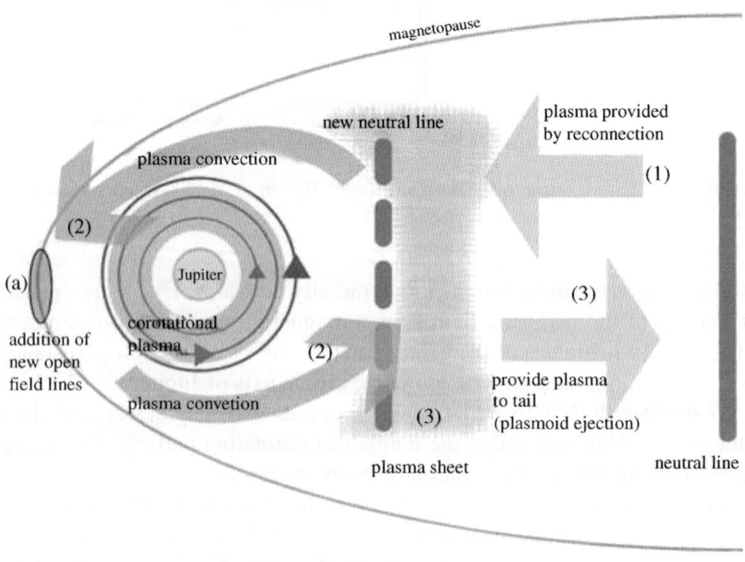

Fig. 5 Recirculation of plasma in the Jovian magnetosphere (Fukazawa et al. 2005)

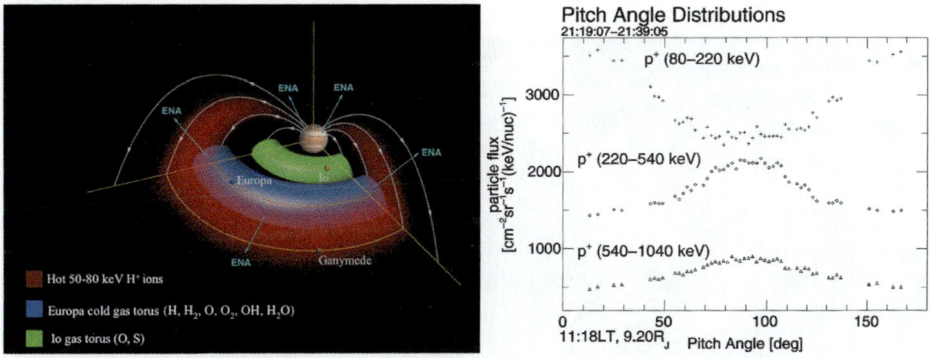

Fig. 6 Sketch of the neutral gas tori around Europa and Io around Jupiter after Mauk et al. (2004) (*left*) and dropout signatures in the energetic proton pitch angle distributions from (*right*) Lagg et al. (2003) from which the neutral gas density has been determined

so called injection events (Mauk et al. 1997) from the outer magnetosphere into the region of the Galilean satellites (Fig. 6).

Another exchange process between a satellite and the surrounding plasma has been discovered from the first energetic neutral atom (ENA) measurements of the Jovian system during the Cassini flyby at Jupiter. ENA measurements confirmed that another neutral gas torus besides the Io torus exists around the orbit of Europa (Mauk et al. 2004). This torus of hot neutral hydrogen atoms is the result of a charge exchange process between cold neutrals lost from Europa and hot ions in the vicinity of the moon creating hot neutrals and cold ions.

The ENAs have been directly detected by the MIMI/INCA instrument on Cassini, a neutral particle camera. The neutral gas density inferred from ENA measurements as well as from dropout signatures in the pitch angle distributions of energetic ions (see Lagg et al. 2003) is about 50 cm^{-3}, the same neutral gas density as of the Io torus. Understanding the process of how the torus is created is one of the questions to be addressed by future missions to Jupiter. It is not fully clear if sputter processes release water molecules from the surface of Europa or if Europa is "active", similar to the Saturnian moon Enceladus (see below).

Finally Ganymede exhibits a very special case since the largest moon in our solar system has its own intrinsic magnetic field, discovered by the Galileo magnetometer (Kivelson et al. 1996). As a consequence Ganymede has its own magnetosphere with a size comparable to Mercury's magnetosphere but embedded in the large Jovian magnetosphere as illustrated in Fig. 7 (Jia et al. 2008, 2009; Burger et al. 2010, this issue).

Plasma transport and acceleration processes between the two magnetospheres are unique in our solar system and are very important to understand the exchange processes in the Jovian system.

1.2 Saturn

Saturn is the second largest planet in our solar system with the second largest magnetosphere (the sizes of the magnetospheres are shown in Fig. 1). With 62 satellites in orbit around the planet, exchange processes on various spatial and temporal scales take place. Some of the moons orbit Saturn outside its magnetosphere and interact with the solar wind plasma. However, when they cross the magnetotail of the planet they interact with magnetospheric plasma resulting in completely different exchange processes.

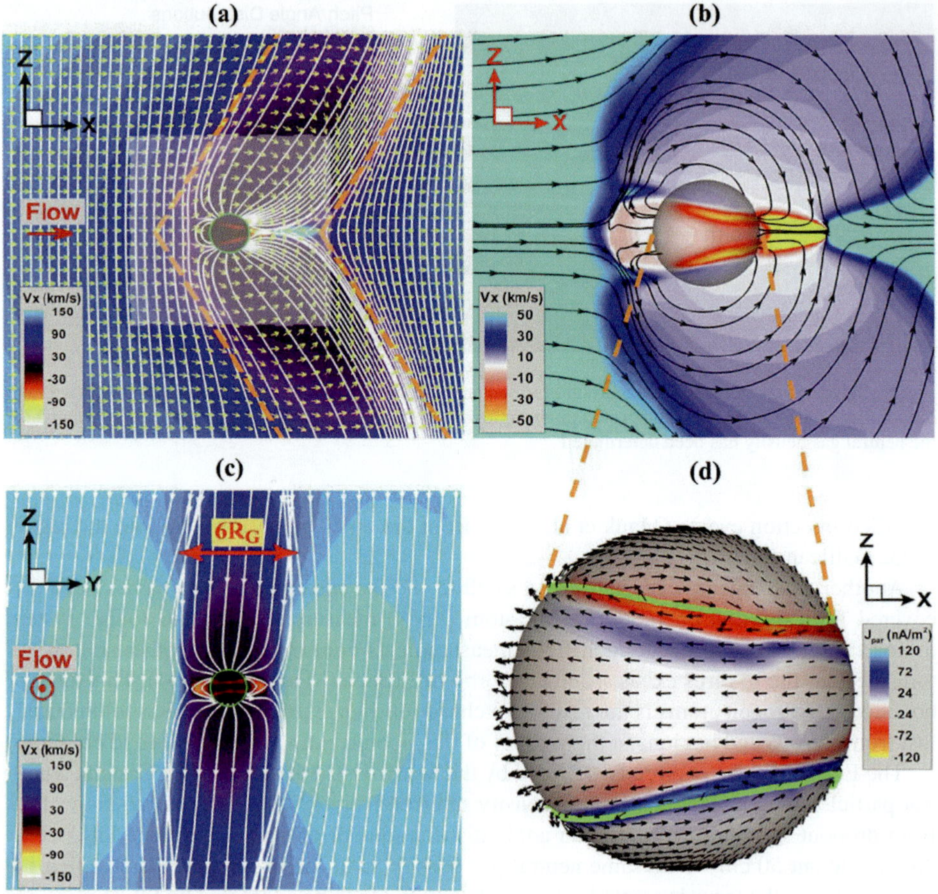

Fig. 7 MHD simulation of Ganymede's magnetosphere after Jia et al. (2009). (**a**) Flows and the projection of field lines (*white solid lines*) in the XZ plane at $Y = 0$. Color represents the Vx contours, and unit flow vectors in yellow show the flow direction. A theoretical prediction of the Alfvén characteristics (*orange dashed lines*) is shown for reference. The projection of the ionospheric flow is also shown as color contours on a circular disk of $r = 1.08$ RG in the center. (**b**) A zoomed-in view of the light area in (**a**). Flow streamlines are superimposed on color contours of Vx. Note that the color bar differs from that in (**a**) and (**b**) in order to illustrate the relatively weak flow within the magnetosphere. (**c**) Same as (**a**) but in the YZ plane at $X = 0$. (**d**) Field-aligned current density along with unit flow vectors shown on a sphere of radius $r = 1.08$ RG

The magnetosphere of Saturn is a mixture of the configuration known from Earth and from Jupiter. On the one hand it is rotationally dominated as in the Jovian case and on the other hand more like the Earth's magnetosphere with respect to the explanation of the aurora as a boundary between open and closed field lines.

The biggest difference between Jupiter and Saturn is the fact that at Saturn the rotation axis is co-aligned with the magnetic dipole axis while they are tilted to each other by 9.7 degrees in the Jovian case. This of course has also impact on the exchange processes between the magnetospheric plasma and the moons. Another significant difference between the two systems is the magnetospheric ion composition. At Jupiter we mentioned already that sulfur and oxygen ions from Io's volcanoes are the dominant species besides protons. At Saturn water group ions like H_2O^+, OH^+, O_2^+, H_2^+, H_3O^+ dominate the Saturnian magnetosphere

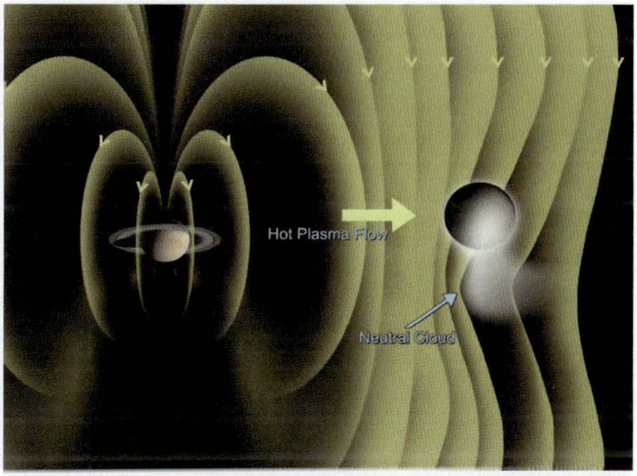

Fig. 8 Effect of the water released from geysers above the south pole from Enceladus on the magnetic environment of the moon (from Dougherty et al. 2006)

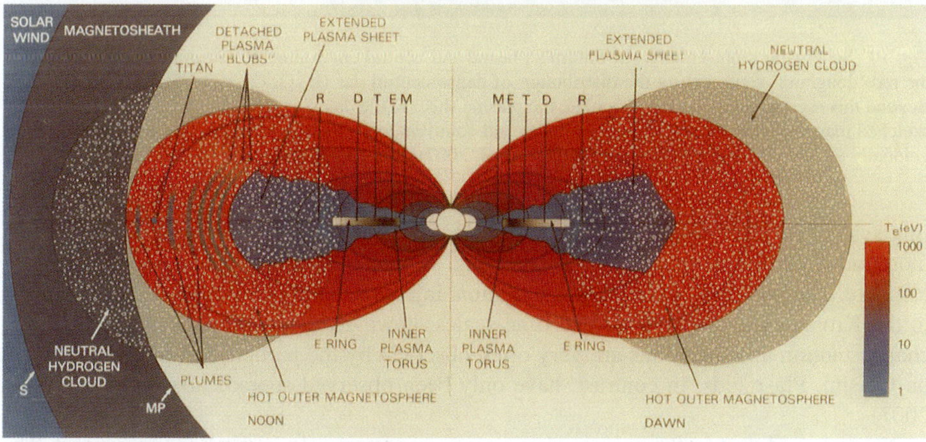

Fig. 9 Magnetospheric regions of Saturn's magnetosphere. Cold regions are colored *blue*, regions of intermediate temperature are *purple* (*blue plus red*), and the hot regions are *red*. The satellite positions of Mimas, Enceladus, Tethys, Dione, and Rhea, the E ring (*gray shaped region*), and the plasma sheet are illustrated (from Sittler et al. 1983)

to a large extent. The main particle source of the magnetosphere is located inside the magnetosphere on the moon Enceladus. Enceladus is special in the sense that it is the source of the neutral cloud and the E-ring in the inner magnetosphere. Enceladus plays the same role as Io does for the Jovian magnetosphere. The constantly released water molecules from the jets (about 100 kg/s) on the south pole of the moon are the source of the water group ions in the magnetosphere (Coustenis et al. 2010, this issue; Tobie et al. 2010, this issue; Dougherty et al. 2006) as sodium, released from the jets points to a possible liquid water ocean underneath the surface (Postberg et al. 2009). The constantly released material has an effect on the magnetic field in the vicinity of the moon above the source region (Fig. 8).

Saturn's magnetosphere has different key regions as illustrated in Fig. 9. Similar to the Jovian case the rotationally dominated magnetosphere is stretched radially outward and an equatorial plasma sheet is formed (Krupp et al. 2009). Between 1 and 9 R_S the Kronian magnetosphere is uniquely structured. It is in that region where 10 icy satellites and the planetary

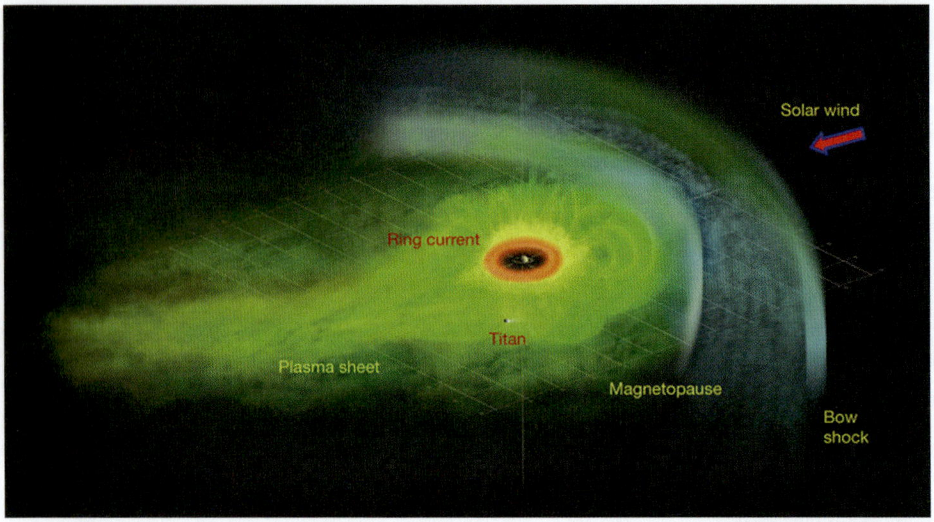

Fig. 10 Schematic representation of Saturn's plasma and neutral environment. Saturn is at the center, with the red 'doughnut' representing the distribution of dense neutral gas (H, O, O_2 and OH) outside the rings. Beyond this region, energetic ions populate the plasma sheet to the day side magnetopause, filling the faintly sketched magnetic flux tubes to higher latitudes and contributing to the ring current. The plasma sheet thins gradually towards the night side (from Krimigis et al. 2007)

ring system exist. At distances of about 10–15 R_S the dipolar field lines are stretched to a more radial configuration due to mass-loading, similar to the situation in Jupiter's magnetosphere. The processes of interchange motion, injections, and plasmoid release are similar in the Kronian magnetosphere. The time scales, however, could be quite different. Interchange motion and injections are very often observed by the particle and fields instrument on Cassini. Plasmoids, in contrast, have only been observed sporadically (Jackman et al. 2007)

Neutrals and dust in the system play a more important role in the Kronian system. The number of neutrals is about 100 times higher than in the Jovian magnetosphere. Schematically this global configuration is shown in Fig. 10 where the plasma and neutral gas environment of Saturn is illustrated based on measurements from the Cassini spacecraft.

The icy satellites Janus, Mimas, Enceladus, Tethys, Dione, and Rhea lie in the regions of the radiation belts while e.g. Iapetus orbits Saturn at about 60 R_S spending most of his time outside the magnetosphere and only inside when crossing the magnetotail. Titan, the only moon in our solar system with an extended atmosphere, orbits Saturn close to the nominal location of the magnetopause. It happens that Titan can be in the magnetosheath or even outside in the solar wind for some time. The result of an exchange or interaction of the Saturnian magnetospheric plasma with the moons is shown in Fig. 11.

The so called macrosignatures of the moons between 2 and 7 R_S are shown as dropouts in energetic particle intensities along the L-shells of the moon's orbits permanently observable at all latitudes and local times. In addition to the macrosignatures the more local interaction with the surrounding plasma is called a microsignature. Microsignatures, however, are count rate decreases strongly dependent on the longitudinal difference from the moon and the signature location of the absorbing body. Those charged particle absorption features in the radiation belts are excellent tracers of magnetospheric parameters. So the exchange processes not only can help to characterize the absorbing body but also the surrounding

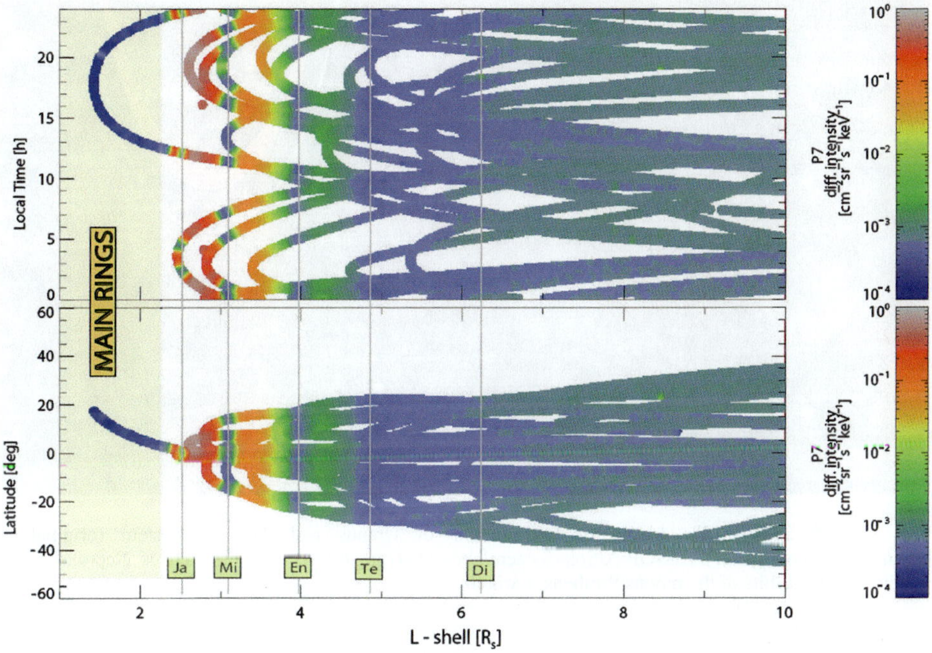

Fig. 11 Macrosignatures of the moons in the inner magnetosphere of Saturn. Color-coded is the intensity of MeV ions plotted as a function of latitude and Local time vs. L-shell (from Roussos 2008)

medium. This absorption effect is much reduced in the Jovian case because of the wobbling of the plasma sheet.

1.3 Uranus and Neptune

Compared to Jupiter and Saturn very little is known about the environment of the other two gas giants in our solar system—Uranus and Neptune. Only the data from the Voyager 2 flyby (1986 and 1989 for Uranus and Neptune, respectively) are available. Uranus has 27 satellites and Neptune harbors 13 moons inside their magnetospheres.

Both of the outermost gas giants have huge ring systems extending over a distance of more than 60,000 km away from the planet (Fig. 12). In the same regions some of the icy moons orbit the planet and exchange processes similar to Jupiter and Saturn are very likely.

The most striking difference between Jupiter and Saturn compared to Uranus and Neptune is the rather extreme tilt angle between the rotation axis and dipole axis at Uranus (59 degrees) and Neptune (47 degrees) as illustrated in Fig. 13.

As a consequence the exchange processes between the planetary magnetosphere and the solar wind are dramatically different and therefore also the interaction with the moons. As an example of charged particle distributions in such an environment Fig. 14 shows the count rates of protons and electrons at various energies as measured aboard the Voyager spacecraft inside the Uranian magnetosphere (for details see Lanzerotti et al. 1987). Those variable count rates are partially caused by the large dipole tilt with respect to the rotation axis and can affect the sputtering rate of the impinging charged particles onto the surfaces of the inner moons.

Fig. 12 *Left* and *middle*: Image of four moons of Uranus and the ring system (original at http://www.eso.org/gallery/v/ESOPIA/SolarSystem/phot-37-07.jpg.html); *right*: Sketch of Neptune's ring system with the orbits of the moons Protheus, Larissa

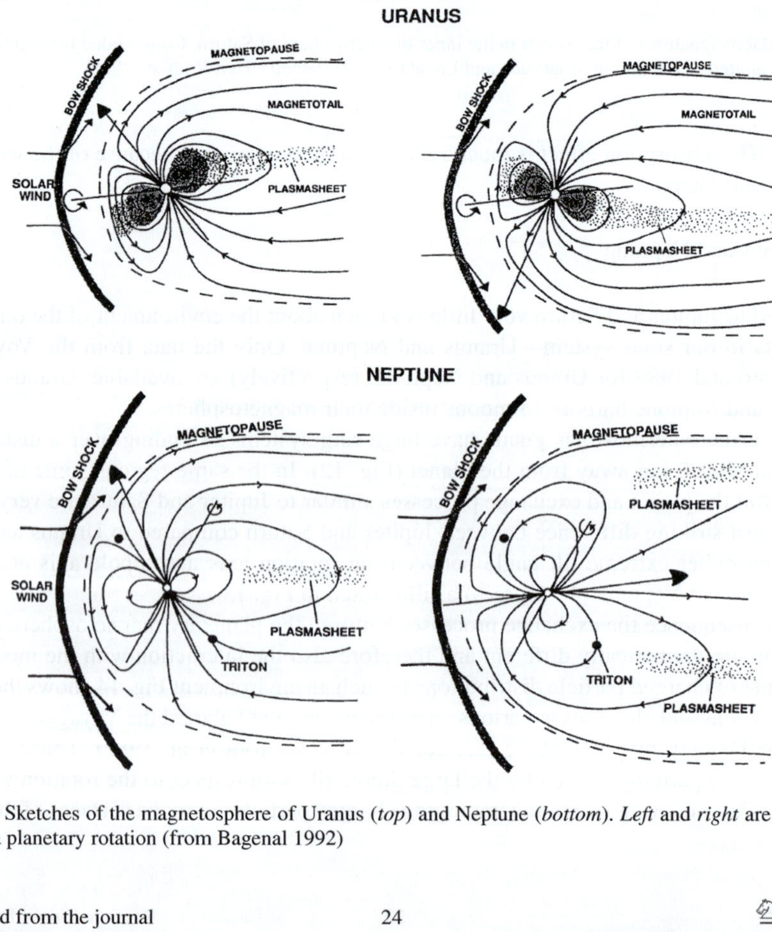

Fig. 13 Sketches of the magnetosphere of Uranus (*top*) and Neptune (*bottom*). *Left* and *right* are separated by half a planetary rotation (from Bagenal 1992)

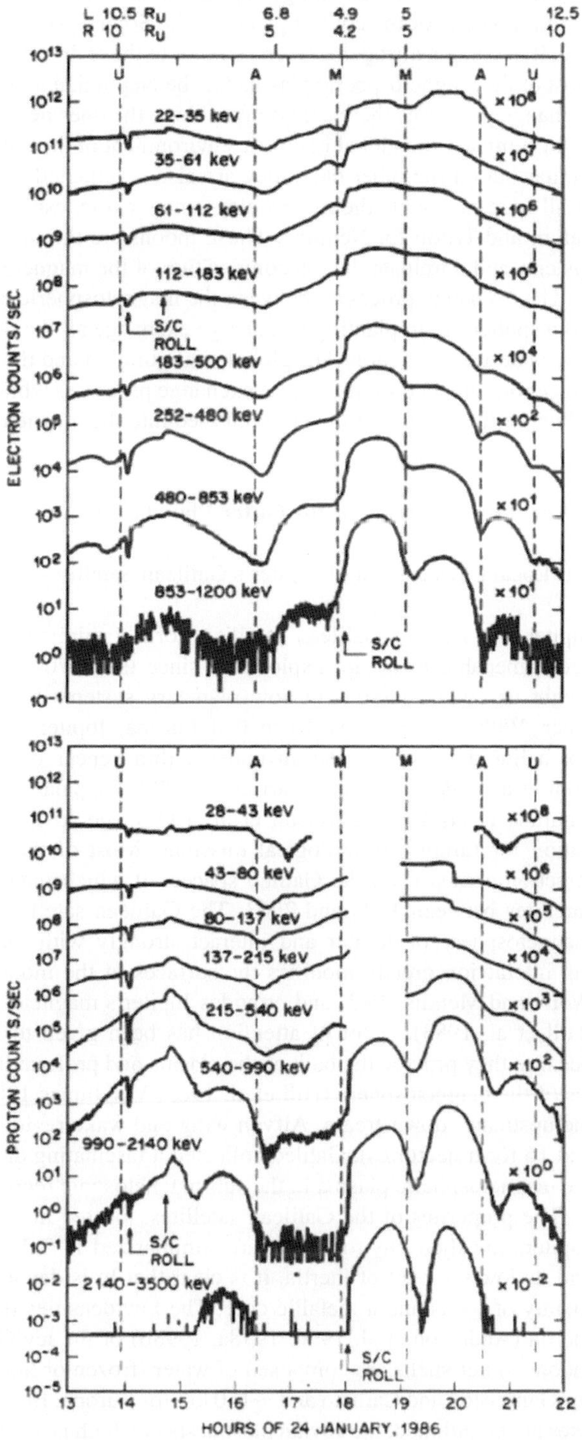

Fig. 14 Protons (*lower*) and electron count rates (*upper*) in the Uranian magnetosphere (from Lanzerotti et al. 1987). *U*, *A*, *M* indicate the location of the minimum L-shells of the moons Umbriel, Ariel, Miranda

One of the most interesting satellites from Uranus and Neptune is the Neptunian moon Triton. It is one of the coldest places in the Solar System but appears to be geologically active. It has active nitrogen geysers, similar to those discovered on Enceladus. Triton provides most of the magnetospheric plasma for the Neptunian magnetosphere. It is expected that the exchange processes there are comparable to the ones near Enceladus.

In summary the global magnetic environment of the outer planets are to some extent very similar but on the other hand they are also significantly different. For the magnetospheres of all four gas giants the moons play a major role, especially Io for Jupiter, Enceladus for Saturn and Triton for Neptune. Those moons are the major internal plasma sources of the systems and dominate the ion composition of the magnetospheres.

The exchange processes between the magnetospheric environment and the moons range from sputtering, implanting and charge exchange all the way to processes related to dust or ring material interaction. The global configuration and the dynamics of the magnetospheres trigger or influence some of those exchange processes while at the same time these dynamics influence the amount of material released into the magnetospheres.

2 Local Environment of the Outer Planets' Moons

2.1 Local Environment of Jupiter's Galilean Satellites

Jupiter and its four Galilean satellites form a miniature solar system next door that has been amenable to in situ exploration since the 1970s. The exploration of this world has taught us crucial lessons in how planetary systems form, evolve and interact with each other. With the recent evidence that Europa, Jupiter's innermost icy satellite, may harbor a liquid subsurface ocean under a thin veneer of ice (Tobie et al. 2010, this issue; Sohl et al. 2010, this issue; Carr et al. 1998; Pappalardo et al. 1998; Geissler et al. 1998; Khurana et al. 1998; Kivelson et al. 1999), the Jovian system is now also a hot destination for future astrobiological missions. Most of the information on moon/plasma interaction comes from the Galileo spacecraft which made 28 close flybys of the Galilean satellites between 1995 and 2003. The Galilean satellites are located in the inner-middle magnetosphere of Jupiter and interact strongly with its field and plasma. The continuing interaction greatly modifies the surfaces of the moons (Spencer and Schneider 1996; Wolfe and Mendis 1983) and provides Jupiter's magnetosphere with the bulk of its plasma (Hill et al. 1983). A lot of attention has been given to Io and its torus, understandably, because they provide the bulk of the plasma and profoundly affect the structure and dynamics of the magnetosphere (Hill et al. 1983; Vasyliunas 1983). By repeatedly flying through the upstream, downstream, Alfvén wing and wake regions of these satellites (see Figs. 15 and 16 for trajectories), Galileo collected a fascinating dataset on how moons interact with the magnetospheric plasma in the sub-magnetosonic regime.

The properties of the Galilean satellites, located in the inner-middle magnetosphere of Jupiter, and their environments are summarized in Table 3. From the high density of Io and its low moment of inertia, it is clear that Io is devoid of any water ice and composed mainly of rocks and a metallic core. The low densities (see Table 3) and low moments of inertia (Anderson et al. 1996, 1998a, 1998b) of the icy Galilean satellites suggest that the moons' outer shells are composed of water (frozen or liquid) with thicknesses of ~400 km for Ganymede and Callisto and ~150 km for Europa. Because Jupiter's magnetic and dipole axes are not aligned, the Jovian plasma sheet which is located in the dipole magnetic equator near these moons, moves up and down relative to these moons at the synodic rotation period

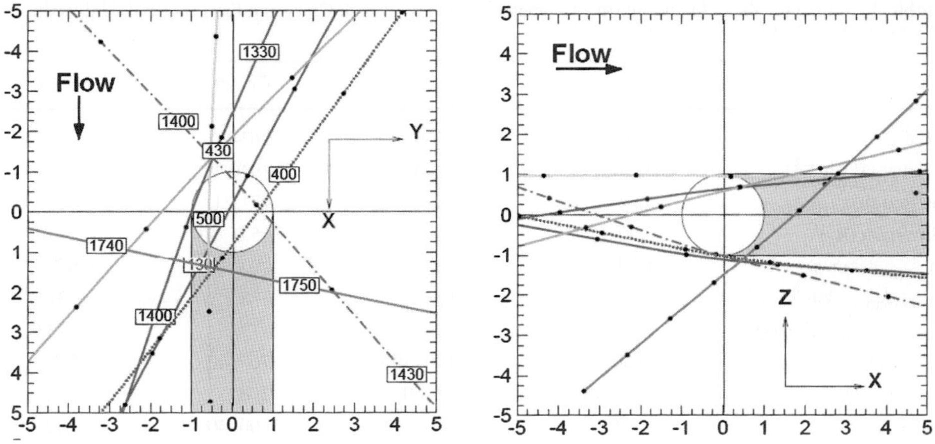

Fig. 15 The trajectories of Galileo during close flybys of Io in the Satellite-φ-B coordinate system. In this coordinate system, the X-axis is parallel to plasma flow direction (system III φ coordinate) and Y-axis is perpendicular to the magnetic field and the X-axis. The left panel show the X–Y projections and the right panels show the XZ projection

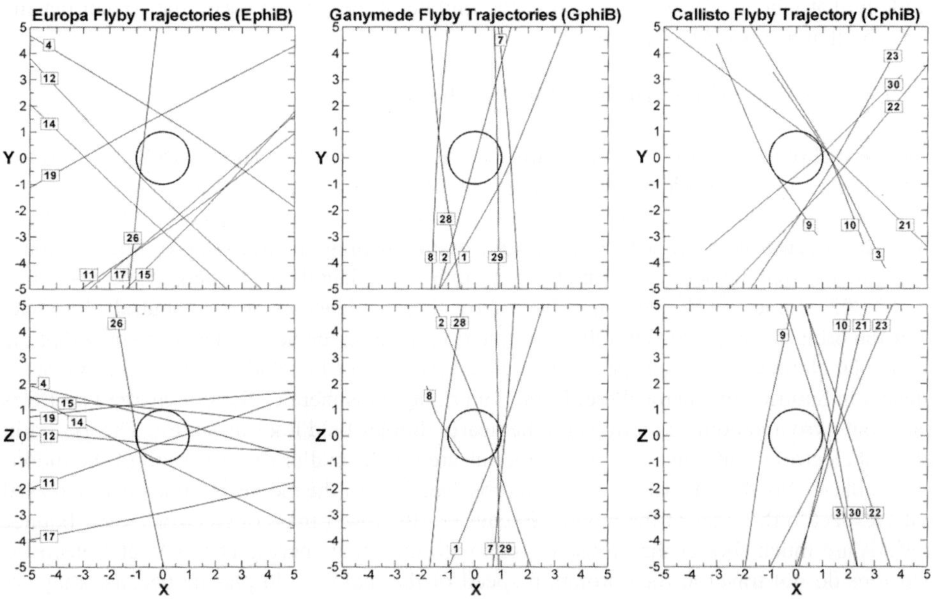

Fig. 16 The trajectories of Galileo during close flybys of the icy Galilean satellites Europa, Ganymede and Callisto in the Satellite-φ-B coordinate system described in the caption of Fig. 15. The upper panels show the X–Y projections and the lower panels show the XZ projections

of Jupiter. Therefore, the plasma density sampled by the moons changes by 50% at Io, by a factor of two at Europa, a factor of four at Ganymede and by a factor of 50 near Callisto over a Jovian rotation. The magnetic field strength also varies near the moons being between 1800 and 2100 near Io, between 400 and 500 nT near Europa, 70 and 140 nT near Ganymede and between 5 and 50 nT near Callisto. Because of the changing field and plasma conditions, the

Table 3 Source for the field and plasma parameters (courtesy: Frank Crary and Fran Bagenal, University of Colorado, Boulder)

Parameter	Io	Europa	Ganymede	Callisto
Radius (km)	1815	1560	2634	2400
Density (kg m^{-3})	3570	2990	1940	1851
Period (h)	42.48	85.22	171.71	400.54
Location (R_J)	5.90	9.4	15.0	26.4
$V_{orbital}$ (km/s)	17.3	13.7	10.9	8.2
$V_{corotation}$ (km/s)	74.1	118	188.2	331.2
$V_{cor.} - V_{orb.}$ (km/s)	57	104.3	177.3	323.0
B_0 (nT)	1800–2100	400–500	70–140	5.5
n_e (cm^{-3})	1500–2200	30–60	2–8	0.02
$T_i + T_e$ (eV)	60–100	200–1100	100–400	100–400
V_A (km/s)	230–330	300–500	150–700	30–2000
M_A	0.17–0.25	0.2–0.3	0.2–0.9	0.1–6.6

Alfvén Mach number of the corotating flow near these moons also changes considerably. It is likely that when Callisto is located in the middle of the plasma sheet, the Jovian plasma flows by it at a super-Alfvénic speed.

2.2 Local Environment of Saturn's Major Satellites

The thermal plasma in Saturn's magnetosphere is effectively corotating with Saturn and has much higher speed than the Keplerian speeds of these moons (see Table 4). The plasma therefore continually overtakes these moons and bombards their surfaces such that the trailing sides receive the highest flux doses. One distinguishing characteristic of Saturn's magnetosphere is that the neutrals are many times more abundant than the charged species (Jurac et al. 2002). Jurac and Richardson (2005) show that the densities of H_2O neutrals dominate near the orbit of Enceladus but OH molecule dominates elsewhere. They also show that the molecular cloud which has its peak density near the orbit of Enceladus is the main source for plasma in Saturn's magnetosphere. In the inner magnetosphere of Saturn, charged particles carry out gyro and bounce motions on the nearly dipolar field lines of Saturn. The gyroradii of most electrons and ions near the icy moons are much smaller than the radii of the moons (see Table 4). So, the effective absorption cross-sections of the moons is somewhat increased with respect to the sizes of the moons. Because of the lower mass of electrons, their bounce periods are much shorter than those of ions with the same energy. The energetic electrons and ions do not travel at the corotation speed of thermal or cold plasma because they are subject to additional drifts in the azimuthal direction from the gradient and curvature of the magnetic field. Ions drift in the direction of the corotation flow and therefore move faster than the corotating cold plasma. Electrons drift in the opposite direction and therefore move slower than the corotation flow. This effect increases the contact time of the electrons with the moons and therefore energetic electrons tend to show very well-developed plasma depletions along the field direction and in the direction of electron wakes near the icy moons. These broad depletions have been given the name micro-signatures. Electrons with an energy of $E = E_{crit}$ (~1 MeV near the orbit of Enceladus) have their drift velocity equal but opposite in sign to that of the corotational flow in the moon's rest reference. Electrons with energies higher then E_{crit}, approach the moon from the leading side and form a wake on the

Table 4 The physical properties of Saturn's large inner icy satellites and their environment (adapted from Khurana et al. 2008)

Satellite	Enceladus	Tethys	Dione	Rhea
Radius	252.1	533.0	561.6	764.4
Density (gm/cm^3)	1.61	0.974	1.48	1.23
Radial distance (Rs)	3.95	4.89	6.26	8.74
$\|B\|$ (nT)	370	167	75	25
v (relative, km/s)	26	34	40	57
n_e (cm^{-3})	40	30	13	2
m_i (AMU)	18	17	17	17
T_e (eV)	10	20	60	100
T_i Heavy ions (eV)	30	50	90	100
V_A (km/s)	301	161	110	93
C_s (km/s)	19	26	37	43
	0.005	0.03	0.14	0.26
M_A	0.09	0.21	0.36	0.61
M_s	1.4	1.3	1.1	1.3
M_{ms}	0.09	0.21	0.34	0.56
f_{ci} thermal ions (Hz)	0.31	0.15	0.066	0.022
f_{ce} thermal electrons (kHz)	10.36	4.68	2.1	0.7
R_g thermal ions (km)	9	25	75	237
R_g thermal electrons (m)	29	90	346	1340

trailing side (see Paranicas et al. 2005). As discussed later, because of the submagnetosonic nature of Saturn's plasma, the absorption cross-section of the moons in the direction of the background magnetic field is highly elongated forming plasma shadow wings in the downstream region. Tethys and Rhea belong to the category of plasma absorbing moons whereas Enceladus and perhaps Dione belong to the category of "conducting" moons because of plasma pick-up in their vicinity (Fig. 17).

The local environment of Rhea is somehow special since Cassini energetic particle absorption signatures revealed evidence of ring-like structures around the moon on bound orbits (Jones et al. 2008). It was concluded that debris is orbiting Rhea with sizes of 10s of cm, most probably remaining from an impact long ago (Fig. 18).

2.3 Local Environment of the Satellites of Uranus and Neptune

The only spacecraft to have visited the Uranian magnetosphere is Voyager 2. As the spacecraft did not make close flybys of any of its major moons, no direct information on their interaction with the magnetosphere of Uranus is available. However, some general observations about the field and plasma conditions in the magnetosphere of Uranus can be made. The equatorial strength of its internal dipole field is 0.23 G at its surface (Connerney et al. 1987). The plasma appears to be derived mostly from the solar wind and the ionosphere of Uranus and consists of mostly H^+ (McNutt et al. 1987). Overall, the plasma densities in the magnetosphere of Uranus are extremely low (<3 cm^{-3} in the plasma sheet, see Fig. 19). Because of the large tilt of the Uranian magnetic dipole with respect to its rotation axis (59°), the moons spend very little time in the magnetic equator of Uranus where the plasma densities are the highest. Thus very week and intermittent moon/plasma interaction is expected.

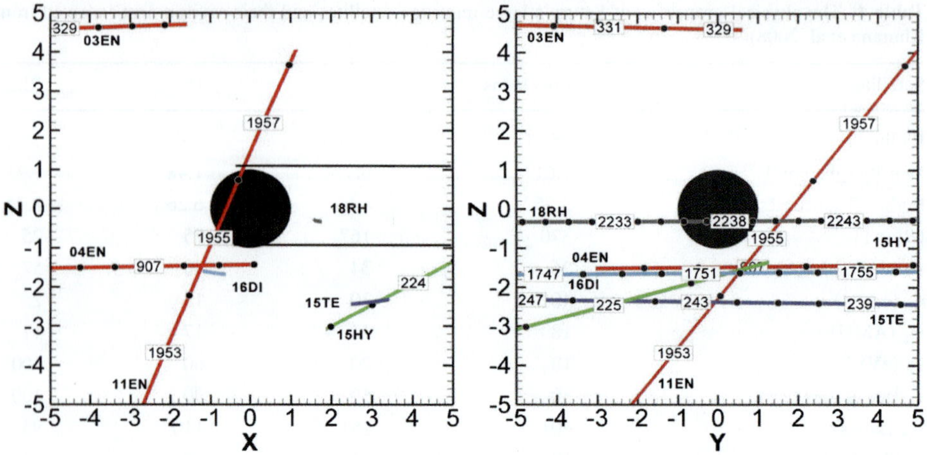

Fig. 17 The trajectory of Cassini near the icy moons in the satellite interaction coordinate system. In this coordinate system, X points in the direction of plasma flow, Y points towards Saturn and Z points along the spin axis of Saturn. Only those portions of the trajectories are shown which satisfied the criterion $-5R_M < X, Y, Z < 5R_M$ where R_M is the radius of the moon. Also drawn is the expected location of cold plasma wake in the X–Z projection. The following color scheme was used for the trajectories: Enceladus = *red*, Dione = *cyan*, Tethys = *blue*, Rhea = *black* and Hyperion = *green*. No identifiable signatures related to the moon have been reported for the Hyperion flyby, presumably because of the large distance of the flyby

This may be the reason why the Uranian moons do not contribute appreciably to the plasma inventory of Uranus's magnetosphere. Lanzerotti et al. (1987) showed that the Uranian magnetosphere does form a stable trapping environment for energetic particles even though the overall fluxes are lower by several orders of magnitude below those found in Jupiter or Saturn's magnetosphere.

Voyager 2 visited Neptune in August 1989 finding it to be magnetized (surface equatorial field strength of 0.14 G). The dipole axis is tilted by 47° with respect to its rotation axis. Because of the large dipole tilt, the plasma sheet appears to bifurcate for part of the Neptunian rotation. The plasma in Neptune's magnetosphere appears to be derived mainly from its moon Triton (N^+) and the ionosphere of Neptune (H^+). Overall, the plasma densities in Neptune's magnetosphere are extremely low ($<0.1\ cm^{-3}$) but surprisingly a heavy ion torus derived from Triton's N_2 atmosphere was discovered (Richardson and McNutt 1990). Richardson (1993) modeled the flux tube volume (NL^2) as a function of radial distance and showed that Triton is most likely the source of both H^+ and N^+ components. Decker and Cheng (1994) modeled the escape of neutral H and N from Triton using a Monte Carlo simulation model and showed that inside the orbit of Triton a neutral torus must exist with H and N densities exceeding 20 particles/cm^3, i.e. two orders of magnitude higher than the ion densities. Paranicas and Cheng (1993) showed that because of the large offset of Neptune's dipole field, on many closed field lines the magnetic field strength profile is monotonic (i.e. devoid of a minimum where particles can get trapped). Because of this unusual field line geometry, in a very large region of the magnetosphere, no stably trapped energetic particles are present (Fig. 20).

2.4 Theory of Plasma/Satellite Interaction

Perhaps, the most important parameter that governs the plasma/satellite interaction is the effective conductivity of the satellite. If the effective conductivity of the satellite is small,

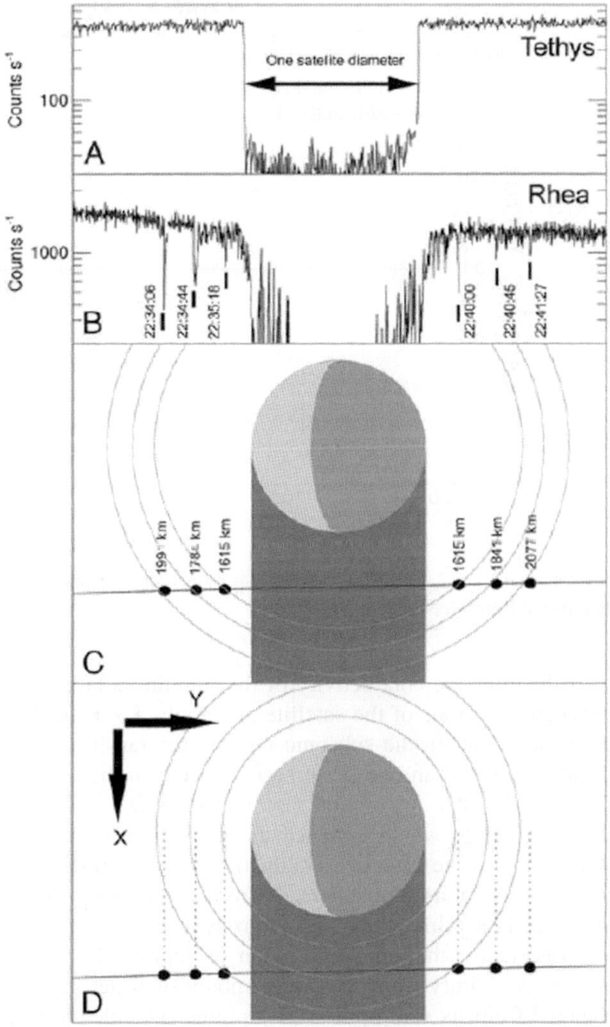

Fig. 18 Evidence of the first ring system around an icy satellite. (**A**) Normal expected count rate of energetic electrons around an icy satellite (here for Tethys), only the moon itself is seen as an absorption in the count rates. (**B**) Same as (**A**) but measured in the vicinity of Rhea: gradual decrease of the fluxes in the vicinity and sharp dropout signatures which could be caused by orbiting material at various bound orbits around the moon (**C**) and (**D**). Picture from Jones et al. (2008)

the satellite does not influence the electric field in its environment, and all of the plasma incident on it, slams into it and is absorbed. An empty wake is formed behind the satellite (Siscoe et al. 1969; Wang and Ness 1972), which is slowly filled by an expansion fan in the downstream plasma. If the flow is supersonic, an umbral region completely devoid of plasma is maintained over a length scale of $R_M M_S$, where R_M is the radius of the moon and M_S is the sonic Mach number. Khurana et al. (2008) have shown that in a sub-magnetosonic plasma, there are many particles which have field-aligned speeds comparable to or faster than the flow speed. Thus as the field lines sweep over the moon, not only those particles which were directly upstream of the moon but even those fast moving particles which were above or below the moon can precipitate on the moon forming absorption shadows defined by the ratio of the field-aligned velocities of the particles and the corotational speed of the plasma. The icy satellites Tethys and Rhea exemplify sub-magnetosonic plasma absorption type of interaction.

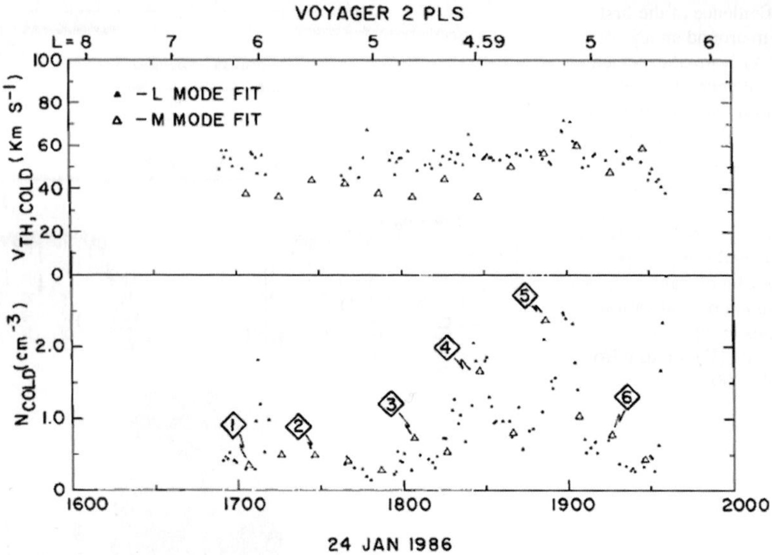

Fig. 19 Plasma density of the main component (cold) of plasma in the Uranian magnetosphere. Figure reproduced from McNutt et al. (1987)

If the effective conductivity of the satellite is large, then the currents flow through the conducting regions of the satellite and apply **J** × **B** forces on the plasma to divert it away from the moon. In the subsonic regime, the far-distance interaction is Alfvenic in nature (Neubauer 1980) and no perturbations occur in the plasma pressure and the field strength. The near-distance interaction also generates fast and slow mode perturbations (Linker et al. 1988, 1991) which enhance the magnetic field upstream and reduce it downstream of the moon. The effective conductivity of the moon can arise either from the Pedersen conductivity of its ionosphere or from the pick-up of the plasma in the vicinity of the moon. Io is a prime example of this type of interaction. Recently it has become evident that Enceladus, the small icy moon of Jupiter also creates an Alfvenic type of interaction in its vicinity from the large amount of water group ions picked up and charged exchanged from its south-pole water plume (Dougherty et al. 2006; Pontius and Hill 2006; Khurana et al. 2007).

2.4.1 Alfvenic Interaction and Plasma Pick-up

Because the corotational velocity of Jovian plasma is much larger than the Keplerian velocities of the Galilean satellites, the plasma continually overtakes the satellites. The conducting satellite extracts some momentum from the flow but most of the plasma is diverted around it. Currents flow in the conducting regions of the satellite (ionosphere and the plasma pick-up region) to make the satellite and the newly picked-up plasma move faster. The currents are eventually closed in Jupiter's ionosphere through a pair of Alfvénic disturbances (called the Alfvén wings) that couple the satellite's environment to the northern and southern ionospheres of Jupiter (see Fig. 14). In the rest frame of plasma, the disturbances propagate along the field but in the rest frame of the moon they propagate along the Alfvén Mach cone (Neubauer 1980). Thus the Alfvén characteristics are given by:

$$\mathbf{V}_{\mathbf{A}}^{\pm} = \mathbf{V}_{\mathbf{0}} \pm \mathbf{B}_{\mathbf{0}}/\sqrt{\mu_0 \rho} \qquad (1)$$

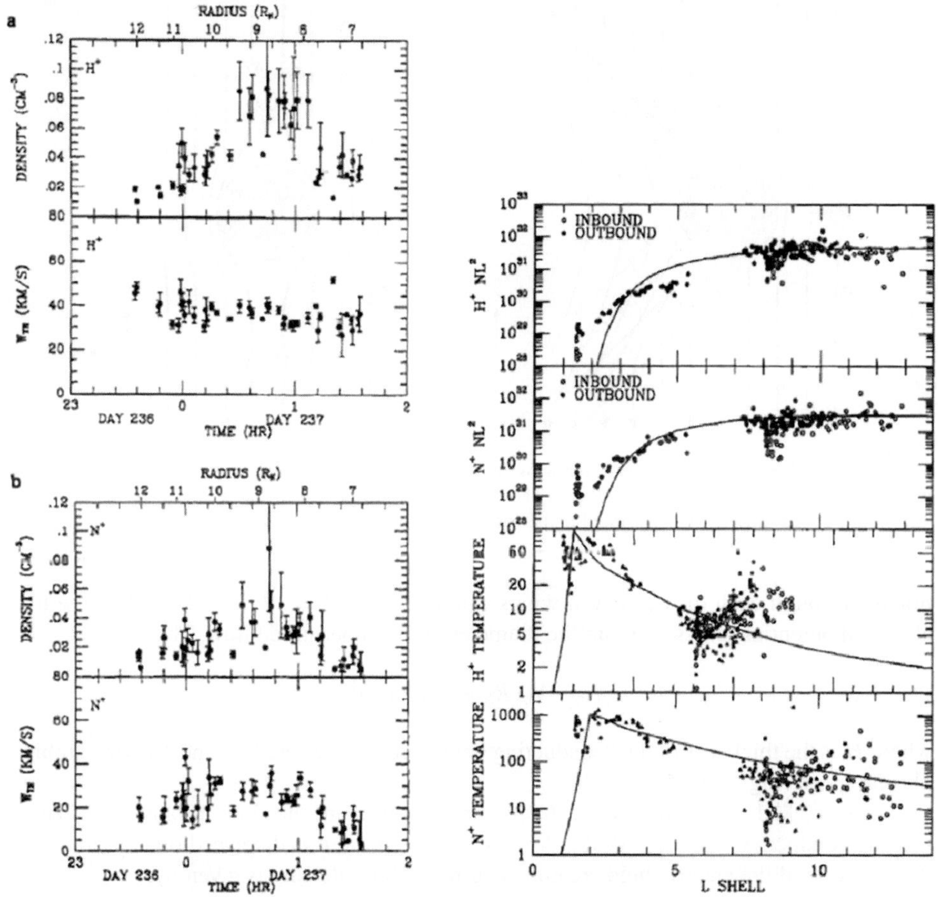

Fig. 20 *Left*, the plasma density and velocity of H^+ and N^+ ions in the magnetosphere of Neptune (Richardson and McNutt 1990) and *right*, the flux tube content and the temperature of plasma in Neptune's magnetosphere (Richardson 1993)

The overall current system enhances the field strength upstream of the satellite (flux pile-up), drapes the field lines over and around it and reduces the field strength in the wake. Neubauer (1980) showed that at distances large from the satellite, the equations of MHD are invariant along the Alfvén wing characteristics and can be combined to yield:

$$\nabla \cdot \mathbf{E} = \mu_0 \mathbf{j} \cdot \mathbf{v}_{\mathrm{A}}^{\pm} \tag{2}$$

Equation (2) admits the solution:

$$j_z = -\Sigma_A \nabla^2 \psi \tag{3}$$

where ψ is the scalar electric potential and $\Sigma_A \approx 1/(\mu_0 V_A)$ is the Alfvén conductance of the medium, $V_A = B/(\mu_0 \rho)^{1/2}$ is the Alfvén velocity and ρ is the background plasma density. The Alfvén conductance may be much smaller than the satellite's effective conductance (Pedersen + pickup) and consequently the current in the Alfvén wing system would be limited by the medium's Alfvén conductance (Neubauer 1980). In this limit, the total current

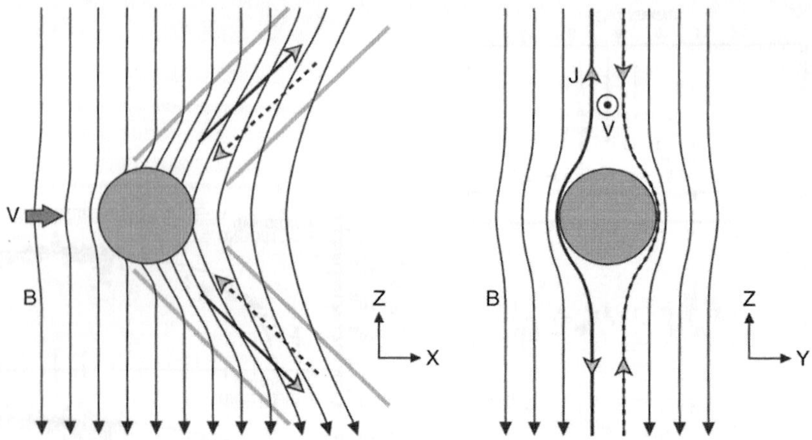

Fig. 21 (*Left*) The geometry of the Alfvén wings in the plane containing the flow direction (X-axis) and the background magnetic field (Z-axis). (*Right*) The Alfvén wings in the YZ plane where Y axis points in the direction of Jupiter and Z-axis is parallel to the background magnetic field

flowing through each of the Alfvén wings cannot exceed $I = 2\Phi\Sigma_A$ where Φ is the total electrical potential across one satellite diameter in the unperturbed flow. Thus

$$I = 4V_0(R_S + H)(\rho\mu_0)^{1/2} \tag{4}$$

where H is the thickness of the "conducting layer" around the satellite, and V_0 is the ambient plasma speed (Fig. 21).

Newly picked up plasma also extracts momentum from the flow through its effective pick-up conductivity (Goertz 1980). As a new ion is born, it is displaced one gyroradius in the direction of the electric field creating a current whose density is given by:

$$\mathbf{J} = \sigma_{\text{pickup}}\mathbf{E} = \tfrac{m_i \dot{n}}{B^2}\mathbf{E} \tag{5}$$

where m_i is the mass of the ion, $\dot{n}$ is the ion pick-up rate, and $\mathbf{E}$ is the background electric field. It is likely that the pick-up conductivity in the close vicinity of Europa and Callisto exceeds the ionospheric Pedersen conductivity and provides the chief closure path for the Alfvén wing currents. Neubauer (1998) has shown how the plasma pick-up and Pedersen conductivities can be combined to obtain generalized solutions for the Alfvén wing model. If the pick-up occurs mainly in a thin shell around Europa, then the current flow paths resulting from Europa's ionospheric conductivity and from the local plasma pick-up would be similar. In such a situation, one can assume that currents flow horizontally through Europa's ionosphere (modeled as a uniform shell). In general, the Pedersen and pick-up current flow paths should be treated separately. The magnetosphere of Ganymede is able to protect a large region of the atmosphere of Ganymede from direct plasma bombardment and pick-up. Therefore, it is likely that plasma pick-up is not a dominant process near Ganymede.

2.4.2 Electromagnetic Induction from Subsurface Oceans

Magnetic field observations from the vicinity of Europa, Callisto and Ganymede reveal that all three moons generate strong electromagnetic induction response to the rotating field of

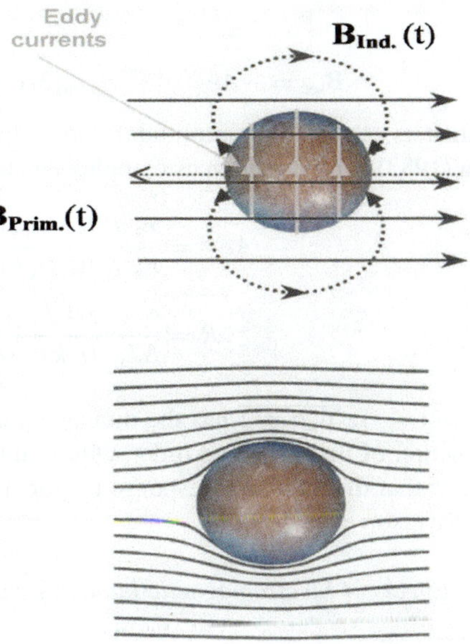

Fig. 22 (*Top*) Time varying primary field (*solid lines*) and induced field (*dotted lines*) generated by a conductor like Europa's ocean and (*bottom*) the sum of primary and induced fields. The total varying field tends to avoid Europa

Jupiter and its magnetosphere (Khurana et al. 1998; Kivelson et al. 1999). It is now well accepted that the induction response arises from subsurface oceans buried at tens to hundreds of km beneath the ice layer (Sohl et al. 2010, this issue). The theory of electromagnetic induction from a global conductor such as an ocean is straight forward. When a conductor is placed in a time-varying primary magnetic field, eddy currents flow on the surface of the conductor in response to the accompanying varying electric field (see Fig. 22, top). These eddy currents generate a secondary or induced field, which tends to reduce the primary field inside the conductor. If the primary field is uniform (i.e. order 1 in external spherical harmonics), and the conductivity distribution has spherical symmetry, the induced field outside the conductor would also be of order 1 in the internal spherical harmonics, i.e. a dipole field. The secondary field may have a phase lag with respect to the primary signal, but its frequency would be identical to that of the primary signal. The sum of primary and secondary fields creates a total field, which avoids the conductor (Fig. 18, bottom).

In order to understand induction from icy satellites, let us model the response of the satellite at any frequency ω by a three-shell model (Zimmer et al. 2000). We assume that the outermost shell of the moon consists of solid ice and has an outer radius r_m equal to that of the moon. This shell is assumed to have zero conductivity. The next shell containing the ocean has an outer radius r_o and conductivity σ. Finally, the innermost shell consisting of silicates is again assumed to have negligible conductivity and radius r_1. As already discussed above, because the primary field is uniform and the conductivity distribution has spherical symmetry, the induced field outside the conductor ($r > r_0$) would be dipolar:

$$\mathbf{B}_{\text{sec}} = \frac{\mu_0}{4\pi}[3(\mathbf{r} \cdot \mathbf{M})\mathbf{r} - r^2\mathbf{M}]/r^5 \tag{6}$$

The induction moment would be opposite in direction to the primary field. The moment $\mathbf{M}$ oscillating at the same frequency ω can therefore be written:

$$\mathbf{M} = -\frac{4\pi}{\mu_0} A e^{i\phi} \mathbf{B}_{\text{prim}} r_{\text{m}}^3/2, \tag{7}$$

and

$$\mathbf{B}_{\text{sec}} = -Ae^{-i(\omega t-\phi)} B_{\text{prim}}[3(\mathbf{r}\cdot\mathbf{e}_0)\mathbf{r} - r^2\mathbf{e}_0] r_{\text{m}}^3/(2r^5) \tag{8}$$

The parameters A (relative amplitude) and ϕ (phase lag) are real numbers, which after Parkinson (1983) can be written as complex equations:

$$Ae^{i\phi} = \frac{r_0}{r_m} \frac{RJ_{5/2}(r_0k) - J_{-5/2}(r_0k)}{RJ_{1/2}(r_0k) - J_{-1/2}(r_0k)} \tag{9}$$

$$R = \frac{r_1 k J_{-5/2}(r_1 k)}{3J_{3/2}(r_1k) - r_1 J_{1/2}(r_1 k)} \tag{10}$$

where $k = (1-i)\sqrt{\mu_0\sigma\omega/2}$ has the dimension of a (complex) wave vector and J_m is the Bessel function of first kind and order. Observations from Europa, Callisto and Ganymede indeed show that the induction response is global and mainly dipolar in nature (Khurana et al. 1998; Kivelson et al. 1999).

2.4.3 The Effects of Electromagnetic Induction and Alfvén Wings on Each Other

Both, electromagnetic induction and the Alfvén wing current system change the geometry and the strength of the magnetic field in the vicinity of the moon. Therefore, they can be expected to influence each other. In particular, the electromagnetic induction tends to re-arrange the instantaneous equatorial magnetic field (which is the time varying component in the moon's frame) so that these field lines avoid the moon (see Fig. 18). The overall magnetic flux penetrating the moon is reduced and the cross-section of the Alfvén wing would become smaller. Neubauer (1999) has predicted that this effect reduces the cross-section of Europa's Alfvén wing by ∼4% and Callisto's Alfvén wing by as much as 40% (see Fig. 22). The Alfvén wing current is also quenched by a similar amount by the reduction in the size of the Alfvén wing. The center of the Alfvén wing is also shifted (see Fig. 23 from observations of the Alfvén wing during the Europa flybys; Fig. 24).

The effect of the Alfvén wing on electromagnetic induction is more subtle. Equation (4) shows that the strength of the Alfvén wing current system changes as the square root of background plasma density. Thus as the moon moves in and out of Jupiter's plasma sheet, the Alfvén wing current would be modulated at the rotation period of Jupiter. Thus, the

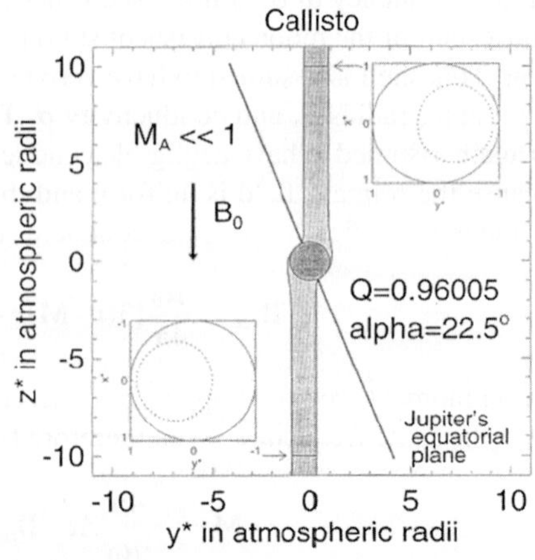

Fig. 23 The expected geometry of the Alfvén wings of Callisto. Notice that the actual cross-sections of the Alfvén wings are substantially smaller than a circle with the radius of Callisto. Figure reproduced from Neubauer (1999)

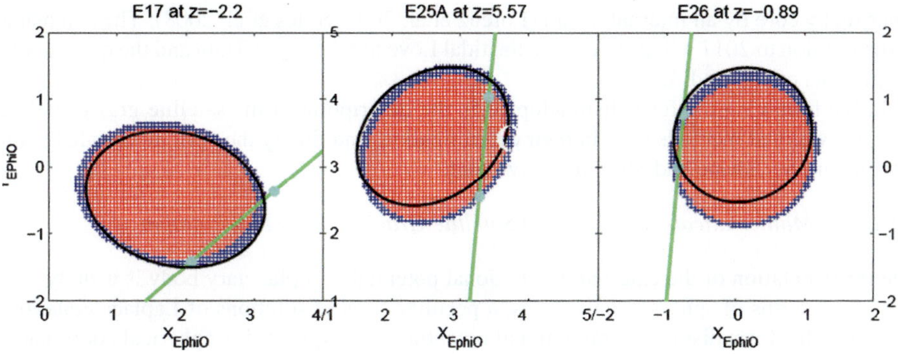

Fig. 24 (*Left*) Cross-section of the Alfvén wing of Europa without induction (ellipse drawn with *thick black line*) and with induction (the central region shaded by *red rectangles*) from three flybys in the EphiO *XY* plane. The larger region shaded with *blue shade* marks the Alfven wing described by the ionosphere. The Alfvén wing is shifted away from Jupiter when induction is present. Two *cyan circles* mark the locations where the Alfvén wing was actually crossed. Figure reproduced from Volwerk et al. (2007)

moon experiences another source of varying field and can be expected to produce an electromagnetic induction response. Neubauer (1999) has shown that the changing magnetic field produced by the Alfvén wing current system is not uniform over the scale size of the moon and therefore would not induce a dipole moment. The leading term of the induced field is quadrupolar, which decays over distance more rapidly than a dipolar field (Fig. 24).

The local environment of the moons can be significantly different from the surrounding plasma parameters. Therefore it is absolutely necessary to have both inputs to address the physics of the exchange processes between the environment and the moons (Cassidy et al. 2010, this issue; Burger et al. 2010, this issue).

2.5 Gravity, Shape, Rotation of the Icy Satellites

In the absence of seismological data, the determination of the interior structure of planetary bodies relies on measurements of magnetic field, gravity, shape and rotation. Indeed, only a handful of missions have so far been targeted on the outer solar system as summarized in Table 2. Flyby spacecraft are of limited use for gravity fields investigations, and even more so when satellites are concerned. Apart from the difficulty in targeting more than one satellite encounter, fast flybys can in general provide only masses (or, more appropriately, the gravitational parameters *GM*). With ancillary knowledge of the volume, obtained from optical observations from ground or onboard cameras, the density is derived.

The Galileo and Cassini missions were designed to carry out in-depth investigations, spending years respectively around Jupiter and Saturn and visiting all of their largest moons. This new exploration phase, although not free from substantial limitations, provided the first clues about the interior structure of solar system satellites. Galileo was able to determine, with the aid of some a priori assumptions, the quadrupole field of Jupiter's Galilean moons Io, Europa, Ganymede and Callisto (Anderson et al. 1998a, 1998b, 2001a, 2001b). Cassini, in orbit around Saturn since 2004, has so far provided a first cut, 3×3 gravity field of Titan (Rappaport et al. 2008b), along with a measurement of Rhea's quadrupole (Iess et al. 2007; Anderson and Schubert 2007; Mackenzie et al. 2008). A major discovery of Cassini is Titan's small deviation from synchronicity, which has been interpreted as an effect of angular momentum exchange between the thick atmosphere and an icy shell, separated from the inner rocky core by an internal ocean (Lorenz et al. 2008; Stiles et al. 2008). The extension of the mission to 2017 will provide also the tidal Love number k_2 of Titan and the quadrupole coefficients of Enceladus.

In the following the formalism adopted in the description of the satellite gravity fields, the observational methods used in their determination, and finally the current knowledge on satellite gravity, shape and rotation is reviewed.

2.5.1 The Mathematical Description of Satellite Gravity, Shape and Rotation

The representation of the external gravitational potential of a planetary body is usually carried out in terms of spherical harmonics, a peculiar class of solutions of Laplace equation. Adopting this formalism, the gravitational potential U is expressed in spherical coordinates as an infinites series of basic functions, the spherical harmonics (see, for example, Bertotti et al. 2003):

$$U(r,\theta,\varphi) = -\frac{GM}{r}\left[1+\sum_{l=2}^{\infty}\sum_{m=0}^{l}\left(\frac{R}{r}\right)^{l}[C_{lm}Y^{c}_{lm}(\theta,\varphi)+S_{lm}Y^{s}_{lm}(\theta,\varphi)]\right] \tag{11}$$

The dipole term ($l = 1$) is missing because the origin of coordinates is chosen in the center of mass. The coefficients C_{lm} and S_{lm} are known as Stokes or harmonic coefficients. Their determination is one of the main goals of planetary geodesy. Expansion (11) depends of course on the mass distribution inside the body, an information of crucial importance in the determination of its interior structure. However, as a consequence of the well known degeneracy of gravity, even a complete knowledge of all harmonics coefficients would not be sufficient to provide the local density. The density distribution can be recovered only if the observational data are combined with models of the interior structure, by means of inversion techniques (see, for example, Palguta et al. 2006). Unfortunately the paucity of observational data severely limits the knowledge of the harmonic coefficients of icy satellites. With a handful of flybys available to date (see Table 5), at most quadrupole terms ($l = 2$) of the expansion have been determined, and in nearly every case the orbital fit required some a priori constraints, as discussed later. However, quadrupole terms are the first and more important correction to spherical symmetry and therefore can be used to put valuable constraints on the interior structure.

The first important information that can be derived from the quadrupole field is the orientation of the principal axes of inertia. Although the moments of inertia cannot be derived from the gravitational quadrupole, the quadrupole tensor **Q** and the inertia tensor **I** share the same eigenvectors. A similarity transformation diagonalizing **Q**, diagonalizes **I** as well. This is a consequence of McCullagh equation, relating **Q** to **I** (Bertotti et al. 2003).

The knowledge of the principal axes of inertia is quite important in satellite gravity and orbital evolution. In the absence of third body perturbations, tidal dissipation locks the satellite to the central planet in such a way that the satellite's long axis (the axis of least moment of inertia) points to the planet at pericenter. Gravity measurements can in principle detect deviations from this condition and therefore single out subtler effects in the spin-orbit coupling.

In general, the quadrupole tensor is built from the degree 2 harmonic coefficients, determined from spacecraft measurements. The relationship between **Q** and the Stokes coefficients is:

$$\mathbf{Q} = MR^2\begin{pmatrix} -\frac{1}{3}C_{20}+2C_{22} & 2S_{22} & C_{21} \\ 2S_{22} & -\frac{1}{3}C_{20}-2C_{22} & S_{21} \\ C_{21} & S_{21} & \frac{2}{3}C_{20} \end{pmatrix} \tag{12}$$

Table 5 The flybys used in the determination of the gravity fields of Jupiter and Saturn. The first column shows the flyby identification code in the nomenclature of the Galileo and Cassini projects. Latitude and longitude of the closest approach point are reported only for close flybys

		Date of closest approach	Altitude (km)	Latitude (°)	Longitude (°)	Tracking band
Io	JOI	7 Dec 1995	897	−9.558	100.941	S/S
	I24	11 Oct 1999	611.2	4.515	224.141	S/S
	I25	26 Nov 1999	300.5	−76.390	44.038	S/S
	I27	22 Feb 2000	197.9	18.536	202.512	S/S
Europa	E4	19 Dec 1996	697	−1.7	36.8	S/S
	E6	20 Feb 1998	591	−17.0	324.7	S/S
	E11	6 Nov 1997	2048	25.7	140.6	S/S
	E12	16 Dec 1997	205	−8.7	225.0	S/S
Ganymede	G1	27 Jun 1996	835	30.62	112	S/S
	G2	6 Sep 1996	260.2	80	123	S/S
Callisto	C3	4 Nov 1996	1118	13	78	S/S
	C9	25 Jun 1997	415	2	259	S/S
	C10	17 Sep 1997	538	5	79	S/S
Enceladus	E03	17 Feb 2005	1511	50	119.1	X/X/Ka
Rhea		26 Nov 2005	1266	−10.2	271.1	X/X/Ka
Titan	T11	27 Feb 2006	1812	−0.2	250.6	X/X/Ka
	T22	28 Dec 2006	1297	40.6	0.0	X/X/Ka
	T33	29 Jun 2007	1933	8.4	63.1	X/Ka
	T45	31 Jul 2008	1614	−43.5	162.7	X/X/Ka
Iapetus		17 Oct 2004	1.11×10^6			X/X
		31 Dec 2004	123363			X/X
Dione		11 Oct 2005	1059	−60.4	92.5	X/X/Ka
Phoebe		11 Jun 2004	2110			X/X
Hyperion		26 Sep 2005	618	−6	232.8	X/X/Ka

When **I** (and therefore **Q**) is diagonal, i.e. when the coordinate axes are parallel to the principal axes of inertia, the McCullagh equation shows that only two Stokes coefficients are different from zero (namely, C_{20} and C_{22}) and their values are

$$C_{20} = \frac{1}{2}\frac{A + B - 2C}{MR^2}, \qquad C_{22} = \frac{1}{4}\frac{B - A}{MR^2} \tag{13}$$

What forces shape the quadrupole field of the largest solar system satellites? The answer is simple only if the body is in a condition of hydrostatic equilibrium, that is if no tangential stresses exist in its interior. Hydrostaticity means that everywhere inside the body the pressure gradient is balanced only by the gradient of the total potential, sum of the gravitational potential of the body, the centrifugal potential and the tidal potential of the central planet

$$\nabla p = \rho \nabla U \tag{14}$$

When this condition is met, the satellite responds like a fluid to external perturbations, namely centrifugal and tidal forces, and its shape is an equipotential surface under the combined effect of its own gravity, tides and rotation. In addition, density and pressure are constant at equipotential surfaces. The important issue here is that the centrifugal potential and tidal potential from the central planet can be also expressed in terms of second degree spherical harmonics (Rappaport et al. 1997), namely

$$\begin{aligned}
U_r &= -\frac{1}{2}\omega^2 r^2 \sin^2\vartheta = -\frac{1}{3}\omega^2 r^2 (1 - P_{20}(\cos\vartheta)) = -\frac{GM}{r}\left[\frac{1}{3}q_r\left(\frac{r}{R}\right)^3\right]\{1 - P_{20}(\cos\vartheta)\} \\
U_t &= \frac{1}{2}\frac{GM_p}{a^3}r^2\left[P_{20}(\cos\vartheta) - \frac{1}{2}P_{22}(\cos\vartheta)\cos 2\phi\right] \\
&= -\frac{GM}{r}\left[\frac{1}{6}q_t\left(\frac{r}{R}\right)^3\right]\left\{P_{20}(\cos\vartheta) - \frac{1}{2}P_{22}(\cos\vartheta)\cos 2\varphi\right\}
\end{aligned} \tag{15}$$

Here r, ϑ and φ are the spherical coordinates (distance from origin, colatitude and longitude), a is the semimajor axis of the orbit, ω the angular velocity, and M_p the mass of the tide-raising planet. The parameters

$$q_r = \frac{\omega^2 R^3}{GM}, \qquad q_t = -3\frac{M_p}{M}\left(\frac{R}{a}\right)^3 \tag{16}$$

are known as the rotational, and tidal parameters. The former is the ratio between the centrifugal and gravity acceleration at the reference radius R (the equatorial radius, for example); the latter is proportional, to the lowest order in (r/a), to the ratio of the tidal and gravity acceleration at the sub-planet point. As all large solar system satellites are tidally locked to their planet and in nearly circular orbit, $q_t = -3q_r$. Table 5 shows the values of the rotational and tidal parameter for the main satellites of Jupiter and Saturn. Satellites are slow rotators. The parameters (16) are therefore small. Under such conditions, it is appropriate to assume that the response of the body to the external forcing potential is linear. Under this assumption, the shape deviates from spherical symmetry and the external potential of the body, described by (11), must be constructed from the same spherical harmonics of the forcing potential. By comparing (11) and (15), we obtain

$$\begin{aligned}
C_{20} &= k_f\left(-\frac{1}{3}q_r + \frac{1}{6}q_t\right) \\
C_{22} &= -k_f\frac{1}{12}q_t
\end{aligned} \tag{17}$$

The scaling factor k_f accounts for the linearity between the forcing potential and the perturbed potential of the body. It is known as the fluid, or secular, Love number. It can be shown (e.g. Bertotti et al. 2003) that for a homogeneous, fluid body its value is 3/2. For an homogeneous elastic body of rigidity (elastic shear modulus) μ its value is

$$k_f = \frac{3}{2}\frac{1}{1 + 19\mu R/2GM\rho} \tag{18}$$

The bulk rigidity for rocky bodies is about 5×10^{10} Pa, while for ice the likely value is 4×10^9 Pa (Hubbard 1984; Yoder 1995). The formula (18) fails for all icy satellites, with the remarkable exception of Rhea, a nearly homogeneous body with a density of 1200 kg/m^3 (therefore made up mostly of ice). Its fluid Love number is close to the one predicted by (18) with a rigidity of 4×10^9 Pa. For the other satellites whose Love number has been determined (Io, Europa, Ganymede and Callisto), the predicted k_f is larger than the actual one even for rigidities in the upper end of the expected range. The preliminary Cassini data indicate that also Titan shows a similar discrepancy. All this should not be too surprising, as the larger icy satellites are not homogeneous, as required to use (18). An important consequence of the hydrostatic equilibrium is that the ratio C_{20}/C_{22} must be exactly $-11/3$ for all tidally locked bodies in circular orbit ($q_t = -3q_r$). The determination of this ratio is therefore quite important. If this ratio is hydrostatic, for example, the moment of inertia (and therefore the degree of concentration toward the core) can be inferred from the Radau-Darwin equation (22). The interpretation of gravity data is much more complex for non-hydrostatic bodies, such as Iapetus.

In many cases gravity data are not available, but the shape (determined with imaging cameras or altimeters) provides useful information, in particular about the hydrostatic behavior of a body. The hydrostatic shape can be determined by requiring that the total potential

$$\begin{aligned} U_g &= U_2 + U_r + U_t \\ &= -\frac{GM}{r}\left\{ \begin{array}{l} 1 + \frac{1}{3}q_r(\frac{r}{R})^3 + [(\frac{R}{r})^2 C_{20} - \frac{1}{3}q_r(\frac{r}{R})^3 + \frac{1}{6}q_t(\frac{r}{R})^3]P_{20}(\cos\vartheta) \\ \quad + [(\frac{R}{r})^2 C_{22} - \frac{1}{12}q_t(\frac{r}{R})^3]P_{22}(\cos\vartheta)\cos 2\varphi \end{array} \right. \end{aligned} \tag{19}$$

is constant at the surface of the body (local topography is neglected). The equipotential surface, which deviates only slightly from a sphere, must be also expressed in terms of spherical harmonics $l = 0$ and $l = 2$. The equation of the hydrostatic ellipsoid obtained from (19) is (Murray and Dermott 1999):

$$r(\vartheta,\varphi) = R\left[1 + \left(C_{20} - \frac{1}{3}q_r + \frac{1}{6}q_t\right)P_{20}(\cos\vartheta) + \left(C_{22} - \frac{1}{12}q_r\right)P_{22}(\cos\vartheta)\cos 2\varphi\right] \tag{20}$$

The shape, as the quadrupole field, is determined by the forcing potential, due to tides and rotation. Rotation alone ($q_t = 0$) would generate an axisymmetric oblate ellipsoid, while the tidal field would change the shape into a prolate ellipsoid, with the long axis toward the planet. The three axes $a > b > c$ of the ellipsoid are easily found from (20), showing that, for an hydrostatic body,

$$\frac{a-c}{b-c} = 4 \tag{21}$$

One has therefore an additional method to assess whether or not a body is in hydrostatic equilibrium. Unfortunately the determination of satellite shapes from camera images is affected by considerable errors, and the departure from hydrostaticity is only detectable in rather extreme cases, such as Iapetus.

Another important consequence of the hydrostatic approximation is the famous Radau-Darwin equation, which relates the moment of inertia C of the body to fluid Love number k_f (and therefore to its quadrupole field):

$$\frac{C}{MR^2} = \frac{2}{3}\left[1 - \frac{2}{5}\left(\frac{4-k_f}{1+k_f}\right)^{1/2}\right] \tag{22}$$

(Cook 1980; Hubbard 1984). Its derivation requires the solution of the hydrostatic equation (22) inside the body. As the moment of inertia factor C/MR^2 indicates the concentration of the body toward its center (it is 0.4 for a homogeneous sphere), it provides one of the most important constraints to geophysical models of the interior. We stress here that its derivation from the gravity field is only possible if the body is in hydrostatic equilibrium, i.e. if $C_{20}/C_{22} = -11/3$ or $(a-c)/(b-c) = 4$. If a body is non-hydrostatic, the moment of inertia can still be determined by combining the quadrupole field and the obliquity (the angle between the orbital and rotational angular momenta), as determined by means of optical or radar measurements. In the absence of external torques, the state of minimum energy of a planet-satellite system is one with a zero obliquity, with coinciding spin and orbit poles. However, due to orbital perturbation, the orbital plane precesses, and the spin axis is left with a small obliquity. The theory requires that the satellite or planet must be in a Cassini state, with the spin axis coplanar with the normal to the orbital plane and the normal to the Laplace plane. As the expected obliquities are small (a few arcminutes), this method has been attempted so far only for Titan (Bills and Nimmo 2008; Lorenz et al. 2008). However, the exchange of angular momentum between the thick atmosphere and the icy shell, decoupled from the core by an internal ocean, may drive Titan away from a Cassini state, preventing the determination of the moment of inertia from obliquity.

2.5.2 The Determination of Satellite Gravity and Rotational State

Gravity fields of solar system satellites are determined by measuring the acceleration they cause on space probes during close flybys. The mathematical formalism is derived from the broad field of estimation and control theory, where the dynamical state of a system is derived from a set of observations which depend on it. Of course the harmonic coefficients are only some of the unknowns of the problem. Other components of the dynamical state include as a minimum the state vectors of the spacecraft and the satellite flown by. The problem of determining the gravity field of a planet or satellite cannot therefore be disentangled from the broader problem of spacecraft navigation and orbit determination.

The main observable quantities used in orbit determination of space probes are range and range rate. In typical operations, an X-band carrier (7.2 GHz), generated from a very stable oscillator (almost invariably a hydrogen maser), is sent from the ground antenna to the spacecraft. An onboard transponder receives the signal and coherently retransmits it back to earth, generally both at X-band (8.4 GHz) and Ka-band (32.5 GHz). The ground receiver regenerates the downlink signal and detects its phase, thus providing an accurate measurement of the range rate. If the carrier is modulated with a sequence of tones or a pseudorandom code, the measurement of the round-trip light time is also possible. Range and range rate are line-of-sight observables, providing information about the spacecraft distance and radial velocity. Accurate angular observables are possible only by observing the spacecraft simultaneously at two or more widely separated antennas, and therefore it is more complex operationally. Conceptually, the method is similar to VLBI, where the direction of arrival of a radio wave is determined by correlating the signal received at two or more antennas (Thorton and Border 2000). However, current accuracies of VLBI are not useful for gravity field determination. State of the art Ka-band radio systems, available at NASA's Deep Space Network (DSN), allow two-way range rate measurements accurate to about 3 μm/s at 1000 s integration times, if suitable transponders are hosted onboard the spacecraft. So far a Ka-band uplink, Ka-band downlink configuration was used only for the Cassini radio science experiments during the cruise phase. Current range systems are based on X-band carriers, accurate to about 1 m two-way. An accurate estimation of gravity fields (and of the other state parameters as well) starts from good data. A Ka-band downlink signal, being higher in frequency, is less affected by noise due to interplanetary plasma turbulence, the main noise source in past S-band gravity experiments such as those carried out by the Galileo mission to Jupiter. At microwave frequencies plasma noise is inversely proportional to the square of the carrier frequency so that, roughly speaking; at Ka-band phase measurements are $(32.5/8.4)^2 \approx 15$ times less noisy than at X-band and $(32.5/2.3)^2 \approx 200$ than at S-band. The use of S-band was therefore a strong limitation to gravity measurements of Galileo at Jupiter (the other being the use of the low gain antenna, which resulted in a large thermal noise). Plasma noise is strongly dependent on the solar elongation angle (or SEP, sun-earth-probe angle). At large solar elongations, plasma density structures ("blobs"), carried along by the solar wind; persist longer in the line of sight (as the solar wind velocity is nearly aligned with the Earth-spacecraft direction) and therefore give rise to a Doppler signal at lower frequency than at small solar elongations. The effect of large structures, more frequent due to the Kolmogorov spectrum of interplanetary plasma turbulence, is therefore pushed out of band at large SEP angle. For probes in the outer solar system, plasma noise has a minimum at solar opposition. Figure 25, adapted from (Asmar et al. 2005), shows the variation of plasma noise as a function of the elongation and tracking frequency.

Range rate is by far the observable of choice in gravity field determination. This is mostly due to the excellent phase stability of current radio links, which allows the measurement of radial accelerations as small as 1.5×10^{-9} m/s^2 over 1000 s. In a flyby the time scale of the velocity change is the ratio b/v between the impact parameter (approximately the distance of closest approach) and the relative velocity.

To fix ideas, the velocity changes (in order of magnitude) due to the monopole and quadrupole component are respectively

$$\Delta v_0 \approx \frac{GM}{b^2}\frac{b}{v} = \frac{GM}{bv}, \qquad \Delta v_2 \approx \frac{GM}{b^2} C_{20} \left(\frac{R}{b}\right)^2 \frac{b}{v} = \frac{GM}{bv} C_{20} \left(\frac{R}{b}\right)^2 \tag{23}$$

For a Titan flyby of Cassini, the time scale is 600 s and the monopole part amounts typically 300 m/s (indeed, due to its large delta-V capability, Titan it the true engine of the mission). The quadrupole contribution, although smaller by a factor $C_{20}(R/b)^2 \approx 10^{-5}$, is still a factor 1000 larger than the measurement noise, so that a good estimation is possible. Of course, flybys with the smaller satellites do not have such a favorable signal-to-noise ratio, but still the harmonic coefficients are accessible to well designed encounters. From the above discussion it is also apparent why current ranging systems are much less valuable: the variation in distance obtained by integrating the range rate is about $(1.5 \times 10^{-6}$ m/s$) \times (1000$ s$) = 1.5$ mm, while the accuracy of a range measurement is only 1 m.

In spite of the accuracy of Doppler measurements, the determination of satellite gravity fields is quite complex. One has to face at least three types of problems. First, only a handful of flybys are available for the largest satellites. This is due to the limitation in the use of the onboard resources, which typically have to be shared between different instruments. Second, the number of parameters to be estimated is quite large. In the estimation of a full quadrupole field (five coefficients), at least 13 additional parameters have to be estimated, namely the state vector of the spacecraft and the satellite (6 + 6 parameters), and the satellite *GM*. In general, one has therefore a solution of rank 18. Not surprisingly, the orbital fit leads to poorly conditioned matrices, whose inversion requires generally some a priori constrains on the parameters. Third, the selection of the free parameters of the dynamical model may be too small for the problem at hand. For example, local, large gravity anomalies, if not

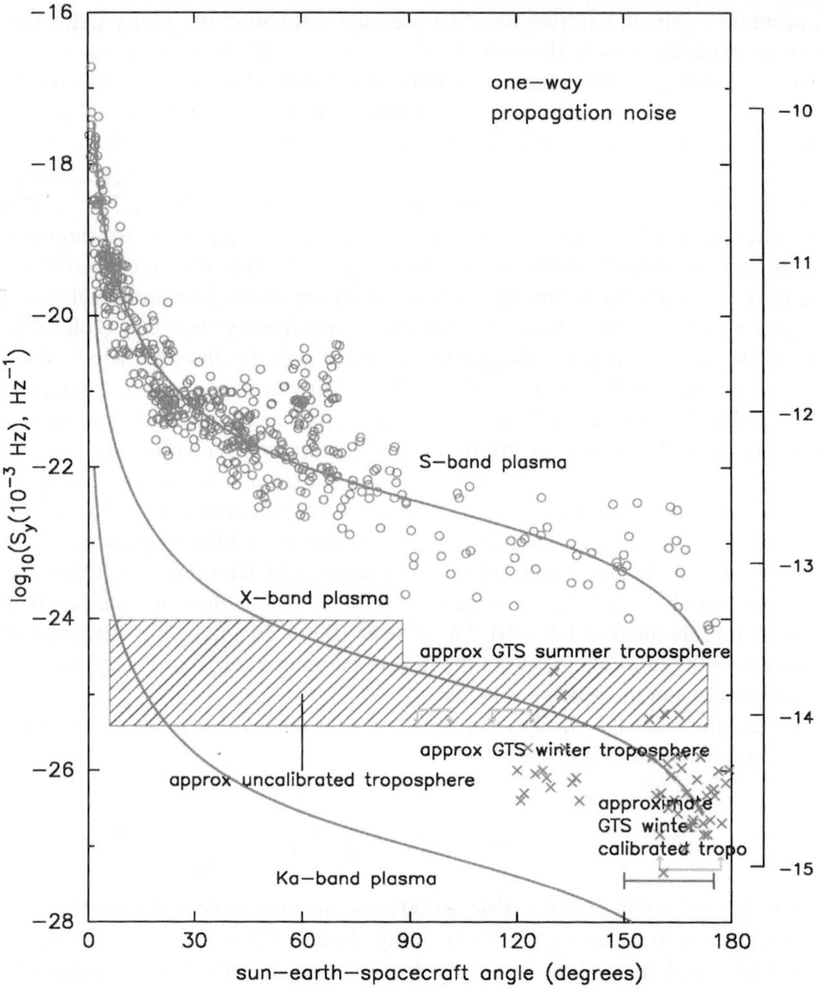

Fig. 25 Plasma noise in Doppler tracking systems used for planetary gravity field determination. The Allan deviation, approximately the rms value of the relative frequency shift, is a measure of the frequency stability of the radio link. An Allan deviation of 10^{-14} corresponds to a one-way range rate of 1.5 μm/s. In Cassini's cruise experiments, carried out at Ka-band (32.5–34 GHz), plasma noise is about a factor 200 lower than at S-band (2.1–2.3 GHz). Troposphere and antenna mechanical noise dominate the error budget at X-band for SEP angles larger than 90 degrees (adapted from Asmar et al. 2005)

included in the model, can bias the estimate of the quadrupole coefficients. On the other hand, increasing the size of the state vector increases the condition number of the matrix to be inverted and weakens the solution. In order to assess if the set of parameters is adequate to the problem at hand, one relies on a sensitivity analysis, where the effect of parameters not included in the state vector but potentially affecting the estimate is evaluated.

We would like to discuss here some consequences of a priori constraints on the solutions. The matter is of particular relevance to the geophysical interpretation, as it may lead to premature or erroneous conclusions. A priori constraints are necessary to improve the condition number of the normal matrix and are always applied when the arc length (the available data span) is not sufficient to remove the rank deficiency. In the estimation of gravity fields, e.g. the harmonic coefficients are generally constrained (with adequate margin) using a Kaula's rule believed to be appropriate to the body. If data from several flybys are available, one can combine the individual solutions into a global solution where the harmonic coefficients are better determined even with looser constraints. Constraints are applied also to the spacecraft and satellite state vector. In this case one makes use of the full covariance matrix, suitably scaled and propagated to the start of the observation interval.

More important for its consequences is the hydrostatic constraint, which has been applied in the analysis of all Galileo flybys. When the data quality is poor, such as in S-band Doppler tracking, the elements of the covariance matrix of the solution may be so large that a geophysical interpretation is difficult. One approach is to reduce the rank of the solution. For example one may limit the estimate to C_{20} and C_{22}, giving up the estimate of the other coefficients. By doing so one makes use of a strong geophysical motivation, namely that tidal locking forces a well defined orientation of the principal axes, with the long axis pointing to the planet at pericenter. If the rotation model of the satellite is constructed in this way, and the satellite is tidally locked and in a Cassini state, then one indeed expects that only C_{20} and C_{22} survive. This assumption, while well motivated on geophysical grounds, may hide important effects and even bias the estimate of the state vector. One may go on and make use of additional geophysical assumptions, such as the hydrostatic constraint $C_{20}/C_{22} = -11/3$, to reduce further the rank of the solution. All these constraints have the effect of reducing the uncertainties in the estimated parameters and improving the conditioning of the normal matrix. Sometimes this is the only available choice to provide a solution sufficiently defined for geophysical interpretation. On the other hand, the geophysicist must be aware that there is no guarantee that the satellite is actually in hydrostatic equilibrium and the theory for hydrostatic bodies (e.g. Radau-Darwin equation) is applicable. When possible, the hydrostaticity of a body must be derived from the data and not imposed a priori.

2.5.3 Gravity Fields of Jupiter and Saturn Satellites

In this section we review the current knowledge of gravity fields and rotational state for the largest satellites of the giant planets. Table 5 reports the Galileo and Cassini flybys used in gravity measurements, while the orbital and physical characteristics of the main satellites are shown in Table 6.

Io. The hot lava flows and its young surface have been studied in depth by the Galileo spacecraft (Prockter et al. 2010, this issue). Gravity measurements have been carried out in four encounters (see Table 5), the first one during Jupiter orbit insertion in 1995, and the remaining three between Oct. 1999 and Feb. 2000. In spite of the large plasma noise (at an average level of 0.1 mm/s), the four encounters allowed a good estimation of C_{20} and C_{22} under the a priori assumption of hydrostatic equilibrium (Table 6). The values obtained by combining orbital solutions from all flybys are $(-1863 \pm 90) \times 10^{-6}$ and $(559 \pm 27) \times 10^{-6}$, where the quoted uncertainties are have been arbitrarily increased by a factor of ten over formal ones (Anderson et al. 2001a, 2001b). The moment of inertia (MoI) factor C/MR^2 of the satellite, inferred from the Radau-Darwin equation (12), was 0.37685 ± 0.00035, a rather large value for a satellite with the largest density in the solar system. (The formal accuracy does not account for limitations in the applicability of the theory, such as deviations from hydrostatic equilibrium.) Combining the MoI with the density, Anderson et al. (2001a, 2001b) develop both a two- and three-layer models of Io's interior. In both cases the existence of a large metallic core was an almost unavoidable conclusion. Acceptable values of its radius are between 550 and 900 km for a Fe–FeS core, or between 350 and 650 km for a Fe core.

Table 6 Mean motion, GM, equatorial radius, orbital semimajor axis, rotational parameter (6), unnormalized Stokes coefficients C_{20} and C_{22}, fluid Love number (from (7)) and moment of inertia factor from Radau-Darwin equation for the main satellites of Jupiter and Saturn. For Enceladus and Dione the quadrupole coefficients have not been determined yet. For Iapetus the value are inferred from a measurement of the shape (Thomas et al. 2006; Bills 2009), where the first value refers to an uncompensated topography, the second to a partial compensation. For the uncertainties in the first five quantities we refer to http://ssd.jpl.nasa.gov/

Satellite	n (deg/day)	GM (km^3/s^2)	R (km)	a (km)	ρ (kg/m^3)	$qr \times 10^6$	$C_{20} \times 10^6$	$C_{22} \times 10^6$	kf	C/MR^2
Io	203.489	5959.9	1821.6	421,800	3529	1710	-1863 ± 90	559 ± 27	1.31	0.378
Europa	101.375	3202.7	1560.8	671,100	3015	498	-435.5 ± 8.2	131.0 ± 2.5	1.06	0.348
Ganymede	50.318	9887.8	2631.2	1,070,400	1943	190	-127.3 ± 2.9	38.18 ± 0.87	0.80	0.311
Callisto	21.571	7179.3	2410.3	1,882,700	1835	37	-32.7 ± 0.8	10.2 ± 0.3	1.10	0.354
Enceladus	262.732	7.2	252.1	238,040	1608	6270				
Dione	132	73	561.7	377,420	1477	1710				
Rhea	79.690	153.9	764.3	527,070	1234	752	-931 ± 12	237.0 ± 4.5	1.26	0.373
Titan	22.577	8978.1	2575.0	1,221,870	1882	39.6	-33.00 ± 0.35	9.98 ± 0.03	1.01	0.341
Iapetus	4.538	120.5	735.6	3,560,840	1084	2.78	$-19000/-1600$	260/22		

Europa. Covered by fractured ice, the second of the Galilean satellites is only slightly smaller than the Earth's Moon, but much brighter (geometric albedo = 0.63). Imaging of the surface has revealed features, such as shallow impact crater, textured surfaces similar to icebergs, and large scale gaps where a young icy crust seems to have formed, which are well explained only by the presence of a global, subsurface ocean (Prockter et al. 2010, this issue; Tobie et al. 2010, this issue). The presence of a global ocean can be detected almost unequivocally by precision measurements of time-variable gravity field (see the discussion on Titan below) and physical librations in longitude. Tides induce a variation in the quadrupole coefficients of order $3ek_2$, between apocenter and pericenter, e being the eccentricity (0.0094) and k_2 the tidal Love number. Unfortunately these variations are too small to be detectable in Galileo data. Forced librations are also driven by the eccentricity of the orbit, but are expected to have an amplitude of only 100 m due to the gravitational coupling with the core. Their measurement by means precision optical imaging would provide the moment of inertia of the shell and the ice thickness, an important quantity. Such measurements could be addressed by the mission to the Jovian system under study by NASA and ESA.

The only available constraints on the deep interior structure of Europa come from gravity measurements. The Galileo spacecraft had as many as seven encounters where Doppler tracking data are available. However, only four of them have been used for the determination of the gravity field (Anderson et al. 1998a, 1998b) with closest approach altitudes between 200 and 2000 km. Unfortunately three of these flybys occurred at small solar elongation angles (25–55 degrees), and were therefore marred by interplanetary plasma noise. Nonetheless the combined solution from the four flybys provided a rather accurate determination of the quadrupole (to about 2%). As in the case of Io, the hydrostatic constraint $J_2 = 11/3C_{22}$ was applied a priori. If Europa is in hydrostatic equilibrium and the quadrupole coefficients are not biased by the constraint, the MoI factor from (12) is 0.346 ± 0.005, indicating a substantial differentiation. The internal structure inferred by Anderson et al. (1998a, 1998b) from the existing constraints entails a metallic core, as large as half the satellite radius, a silicate mantle and a water ice-liquid outer shell. The inferred thickness of the outer shell is 80–170 km. Models without a core are possible, but in this case the interior must be a mixture of metal and rock with a density of $3.8\ \mathrm{g\,cm^{-3}}$.

Ganymede. Ganymede is the largest satellite of the solar system, and the only one endowed with an intrinsic magnetic field. The two Ganymede gravity flybys occurred at altitudes of 835 and 260 km. The latter was low enough to provide for a clear detection of gravity disturbances (up to 1.5 mgal at the flyby altitude) produced by mass anomalies beneath the surface (Anderson et al. 2004; Palguta et al. 2006). The large signature in the Doppler data could be removed by introducing four mass anomalies (two positive and two negative) in the orbital solution, in addition to the usual quadrupole coefficients. In spite of the difficulty in dealing with an intrinsically ill-conditioned problem, Palguta et al. have been able to identify the possible locations of the mass anomalies, once the depth is constrained to the surface or at the rock ice interface at about 800 km. In some cases the location coincided with evident features on Ganymede's surface, such as the Galileo and Marius Regio. The quadrupole coefficients were estimated separately for the two encounters, but no global solution was attempted (Anderson et al. 1996). Rather, geophysical interpretation was obtained from the weighted average of C_{22} obtained from the two encounters $(38.18 \pm 0.87) \times 10^{-6}$, and the assumption of hydrostatic equilibrium. The resulting MoI factor is 0.3105 ± 0.0038, the smallest among the bodies of the solar system and an indication of a completely differentiated interior structure. The suggested interior model entails a large metallic core, with radius between 400 and 1300 km, a silicate mantle and an outer icy shell 800 km thick. The authors conclude that, were not for this icy shell, the two internal layers would otherwise resemble the structure of Io.

Callisto. With a radius of 2410 km, the outer satellite Callisto is only slightly smaller than Ganymede (2631 km), and has nearly the same density (1835 kg/m^3 vs. 1943 kg/m^3). However, the MoI factor is very different (0.354 vs. 0.311). The latter result was obtained using the same approach as for the other three Galilean satellites, namely by fitting Doppler data to a gravity field with the hydrostatic constraint applied a priori, and using Radau-Darwin equation (Anderson et al. 1997, 2001a, 2001b). (See Tables 5 and 6 for further details about the Galileo encounters and the physical characteristics of Callisto.) The MoI factor indicates that Callisto cannot be fully differentiated and that a significant fraction of the satellite (if not all) must be a mixture of ice and rock-metal. The differentiation process may be still going on, as the sinking of silicates and metal through ice can be a very slow process depending on ice viscosity, size and density of particles, and the strength of ice convection. The slowness of the process implies that Callisto is sufficiently rigid to sustain elastic stresses for very long times, and therefore may not be in hydrostatic equilibrium. If so, the MoI factor from Radau-Darwin equation may be an overestimate, and Callisto could be more differentiated, with a lower MoI. Callisto resemblance to Titan would be striking: with nearly identical radius, density and MoI, the question, whether Titan is a Callisto with weather, is surely stimulating (Moore and Pappalardo 2008).

Enceladus, Dione, Hyperion, Iapetus. Little is known about the gravity field of the smaller Saturnian satellites. The Cassini's flybys completed to date allowed only a determination of the mass which, in combination with the volume from optical imaging, provided the density (see Table 6). The general decreasing trend of the density with distance from Saturn is violated only by Titan, a much larger object. Values significantly larger than 1 gm/cm^3 for Enceladus (1608 kg/m^3) and Dione (1477 kg/m^3) indicate a substantial presence of silicates. Before Cassini's arrival at Saturn, ice was believed to be the largely dominant component, and attempts to infer the densities of Enceladus and Dione from the shape determined by Voyager led to a large underestimation (Dermott and Thomas 1990, 1994).

Among the Saturnian satellites only Iapetus has a density in line with the pre-Cassini expectations. However, this moon occupies a special place because of its quite unusual shape. A large fraction of its equator is occupied by a remarkable ridge, which raises as much as 30–50 km above the surface. This still unexplained feature is accompanied by a largely non-hydrostatic shape (Thomas et al. 2006), with a principal axis ratio of 0.95 (11). Obviously, the non-relaxed shape must correspond to an unusual quadrupole and, likely, to a very disturbed higher degree field. Unfortunately nothing is known about Iapetus' gravity besides its monopole, and no further Cassini flybys are planned for the future, so that the solution of this puzzle is probably pushed to the far future.

Hyperion, the largest among Saturn's irregular moons (radii in the range 95–182 km), is known since Voyager flyby to be in chaotic rotation, with a spin vector moving both across the body and the sky. This is not the only peculiar feature of this satellite: its density is the lowest among solar system bodies, a mere 0.544 ± 0.050 g cm^{-3} (Thomas et al. 2007). The determination of the volume, requiring a combination of several Cassini images, was especially complex because of the irregular rotation. The mass was obtained from a dedicated Cassini flyby on Sept. 25, 2006, at a closest approach distance of 618 km. The velocity change of Cassini across the encounter (about 0.1 m/s, much larger than the measurement noise), allowed a very good accuracy in the GM, to about 1%. The uncertainty in the density is almost entirely due to the estimation of the volume.

Rhea. Rhea is one of the two Saturn's satellites whose quadrupole field has been determined (the other being Titan). Three papers have been presented on the topic (Iess et al. 2007; Anderson and Schubert 2007; Mackenzie et al. 2008), with different results. The case is interesting because it shows how critical (and potentially dangerous) the use of a priori constraints in estimation problems could be. Iess et al. (2007) and Mackenzie et al. (2008) did not impose the hydrostatic constrain and fitted the data with three quadrupole coefficients (J_2, C_{22} and S_{22}), in addition to the Cassini and Rhea state vector and Rhea's GM. The two papers used a similar approach, but Mackenzie et al. 2008 relied on a broader data set, leading to a more precise determination of J_{22}. The two results were compatible ($10^6 J_2 = 794.7 \pm 89.2$, $10^6 C_{22} = 235.3 \pm 4.8$ in Iess et al. and $10^6 J_2 = 931.0 \pm 12.0$, $10^6 C_{22} = 237.2 \pm 4.5$ in Mackenzie et al. 2008), the second solution differing only in the estimated (and larger) value of J_2 and a smaller formal uncertainty. The difference in the two estimates of J_{22} is 1.5 times the largest of the two uncertainties. The crucial point is that the solution by Iess et al. 2007 is compatible with hydrostatic equilibrium ($J_{22}/C_{22} = 3.38 \pm 0.12$), while Mackenzie et al. put Rhea in a mildly non-hydrostatic state ($J_{22}/C_{22} = 3.93 \pm 0.03$), with the hydrostatic line at about five sigma. The moment of inertia factor computed using Radau-Darwin equation is $C/MR^2 = 0.3721 \pm 0.0036$, indicating a slight differentiation. Note also that Rhea's measured fluid Love number k_f (1.26, computed from (7)) is very close to the one obtained using (8) for the rigidity recommended by Yoder (1995) for icy bodies ($k_f = 1.30$). On the other hand, the non-hydrostatic solution of Mackenzie et al. (2008) does not allow the use of Radau-Darwin equation to infer the moment of inertia. The interpretation of Anderson and Schubert (2007) differs significantly. By adopting a hydrostatically-constrained solution ($10^6 J_2 = 889 \pm 25$, $10^6 C_{22} = 266.6 \pm 7.5$; Mackenzie et al. 2007), those authors derived a moment of inertia factor of 0.3911 ± 0.0045 and, given also the low density (1233 ± 5 kg/m^3) concluded that Rhea is a homogeneous mix of ice and rock. The use of hydrostatic constraint is not without geophysical justifications. The rigidity of ice is lower than that of rocky bodies, making a hydrostatic structure more likely. The proximity to Saturn exposes Rhea to powerful forces in the despinning and orbit circularization phase, which could probably destroy any internal inhomogeneity. Are these geophysical arguments compelling enough to reduce the quadrupole field to a single parameter? If more flybys were available, there would be no need to bind the solution and an unconstrained estimate would be by far preferred. With just one flyby available, the choice between a hydrostatic and a non-hydrostatic Rhea is a matter of confidence in a specific geophysical model. However, the case of Iapetus, a body with approximately the same mass and density of Rhea (see Table 2) but quite far from hydrostatic equilibrium, suggests caution.

An interesting possibility to reconcile the two approaches relies on the inclusion of topography in the orbital fit. Rhea's images show striking features on the surface of the satellite (Prockter et al. 2010, this issue) some of which may have appreciably affected the estimation process. For example, Cassini's closest approach occurred near Tirawa, a large (300 km) crater at 34N 152E. The basin's mass deficiency, if included in the dynamical model, leads to different values of the quadrupole coefficients. A global estimate of Rhea's topography in terms of spherical harmonics has been already attempted by fitting 22 limb profiles from Cassini's camera (Asmar et al. 2009) and a first evaluation of its effects (at least in the case of no isostatic compensation) shows that it may be sufficient to affect Doppler data. The new quadrupole coefficients would differ from the ones considered in the papers published so far and could well be found in a hydrostatic ratio. If this will turn out to be the case, there are at least two interesting consequences. First, the quadrupole component, shaped by rotation and tides, is a remnant of the initial phases of Rhea's evolution. The eigenvectors of the quadrupole tensor would be oriented at some angles with the current axes of Rhea's body fixed frame. Indeed, a reorientation of the axes would occur after each large impact (Nimmo and Matsuyama 2007). Second, it would be possible to obtain information about the level of isostatic compensation and infer the rigidity of the crust.

Titan. Titan has many unique features among solar system satellites. Second only to Ganymede for its radius and mass, it is embedded in a thick atmosphere dominated by nitrogen, with a surface pressure of about 1.5 bar (Coustenis et al. 2010, this issue). Its bulk composition is roughly 60% rock, a large fraction of water ice (Stevenson 1992; Grasset et al. 2000). SAR imaging of the surface has revealed ample evidence of esogenic processes and features driven by the atmosphere, such as erosion due to wind and precipitation, fluvial channeling and lakes (Prockter et al. 2010, this issue; Elachi et al. 2005). The presence of nitrogen and methane in the atmosphere (a molecule that is photochemically removed over a time scale of a few million years) led Lunine and Stevenson (1987) to propose a four-layer internal structure, made up by a homogeneous rock-ice core, a silicate mantle, a global water-ammonia ocean, and an outer icy shell. Lunine and Stevenson's model was later revisited and modified by several authors (Sohl et al. 2003; Grasset and Sotin 1996; Grasset et al. 2000). Tobie et al. (2005, 2006) propose that the source of the atmospheric methane are episodic outgassing events from methane clathrate hydrates found in the icy shell floating above an ammonia-enriched water ocean. In another model (Fortes et al. 2007), an ammonium hydrosulfate ocean underlying a 70 km thick crust would supply methane via steady outgassing. Although nearly all interior models entail an internal ocean, experimental evidence is still far from complete. The first indication has been provided by measurements of Titan's rotational state and the discovery of a non-synchronous rotation (NSR) as large as 0.36 deg/yr (Lorenz et al. 2008; Stiles et al. 2008). The NSR was measured by Cassini by tracking surface features during the mission, using the onboard synthetic aperture radar (SAR). The determination of the NSR required a precise georeferencing of SAR images and the analysis of the ensuing registration errors, two delicate procedures involving also spacecraft attitude and orbit reconstruction. The NSR was interpreted in terms of exchange of angular momentum between the atmosphere and an icy shell floating above a global internal ocean, as predicted by Titan's global circulation models (Tokano and Neubauer 2005). Were Titan completely solid, the NSR generated by the atmospheric torque would be much smaller. It is remarkable that the predictions of Tokano and Neubauer (NSR $\sim$0.6 deg/yr for a 50 km thick ice shell) are very close to the ones determined initially by Lorenz et al. However, the NSR and the proposed mechanism cannot be easily reconciled with the gravitational coupling between the icy shell and the core. A hydrostatic, non-spherical core would produce a strong quadrupole interaction with the shell, strongly reducing any NSR except for a very low viscosity ice (about 10^{13} Pa s). For ice viscosities of 10^{14} Pa s or larger, NSR may survive only in the presence of a much larger exchange of angular momentum between the atmosphere and the ice lid (Van Hoolst et al. 2009). These theoretical arguments are strong enough to demand a reanalysis of the SAR data, so that, at the time of this writing, the existence of an internal ocean must still be considered an open question. The analysis of time-variations of Titan's quadrupole field is the method of choice to confirm or disprove the existence of an ocean beneath the icy surface. Any internal liquid layer would indeed change dramatically the tidal Love number k_2. This parameter, controlling the elastic response of a body to short term variations of the external tidal field, is about 0.3–0.45 in the presence of an ocean, and about one order of magnitude smaller for a completely solid body. Titan's large orbital eccentricity ($e = 0.03$) induces variations of Saturn's tidal field of order 3e, large enough to be measured by precision Doppler tracking of Cassini. J_2 and C_{22} will show variations of a few percent between pericenter and apocenter in the presence of a liquid layer. This experiment, proposed by Rappaport et al. (1997), is currently under way. However, the four completed gravity flybys of Cassini are not sufficient to determine the Love number with sufficient accuracy, and one has therefore to accumulate more data from additional flybys to provide a reliable value of k_2. A detailed simulation of the experiment and an assessment of the expected sensitivities has been carried out in Rappaport et al. (2008b), indicating that a determination of the tidal Love number is possible to an absolute accuracy of about 0.1, sufficient to discriminate between the ocean/no ocean case with 90–95% confidence level.

3 Astrometric Observations and Orbital Parameters

Astrometry is the discipline that focuses on the position measurements of celestial objects. These observations are used as input of the dynamical models to refine the ephemeris of Solar system objects and provide orbital characteristics (semi-major axis, eccentricity, inclination ...). The classical method uses the star background to calibrate the images, thanks to accurate stellar astrometric catalogues. While the low magnitude of many natural satellites has allowed for old observations to be possible, their proximity to their primary often requires using specific techniques to derive astrometric positions.

3.1 Observations of Events

Observations of events are not strictly speaking astrometric but photometric observations. However, the light flux drop can be used to derive astrometric positions once a model of the event is used.

3.1.1 Eclipses by the Planet

Eclipses of the Galilean moons have been monitored since the middle of the XVIIth century and proved to be useful for seeking long term dynamical effects (Lieske 1987). This kind of observation provides the time of the first speck (at the end of an eclipse), last speck (at the beginning of an eclipse) or mid-time event (Lieske 1986). The accuracy of these observations is measured in seconds of time. Most of the eclipses collected over the last centuries concerned essentially the Galilean moons. This is especially the case of the last third of the XVIIth century and the whole XVIIIth century, as Galilean eclipses were the principal means for the obtention of the longitudes (De Sitter 1931). The global precision of eclipses is about 44.4 s, 42.4 s, 85.8 s and 146.7 s for Io, Europa, Ganymede and Callisto (Lieske 1986), where 1 s of time corresponds to 18 km, 14 km, 11 km and 8 km respectively. Recently, substantial progress has been done to process the eclipse observations (Mallama 2007) and improve their accuracy.

3.1.2 Mutual Events

Mutual events are mutual eclipses or mutual occultations of one satellite by another. Such events occur when the Sun and the Earth cross the equatorial plane of the planet (see Fig. 26).

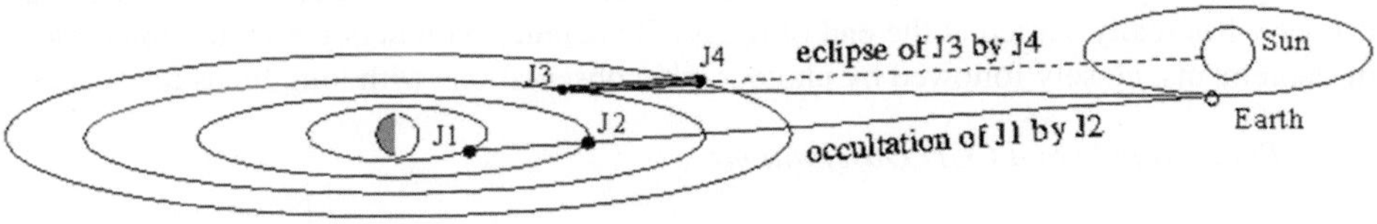

Fig. 26 Definition of the mutual events (copyright: IMCCE)

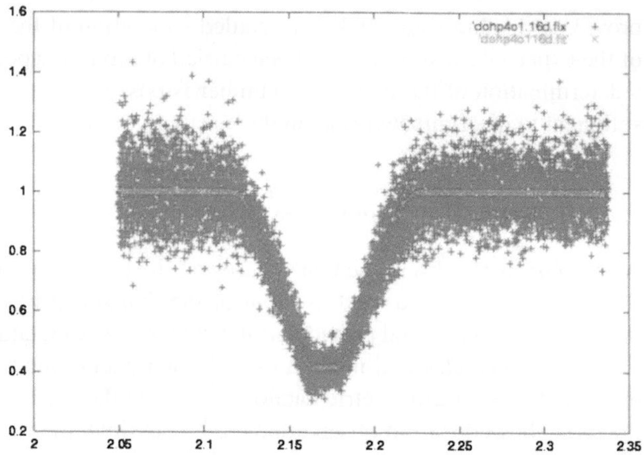

Fig. 27 Fit of a synthetic light curve (x-axis is given in hours and y-axis is the normalized flux). The event shown here is an occultation of Io by Callisto observed at the Haute-Provence Observatory

Hence, each half period of revolution of the planet (6 years, 15 years and 42 years for respectively Jupiter, Saturn and Uranus), international campaign of observations of mutual events can be organized (see Arlot and Sicardy 2008 and references therein). Mutual events can be observed over about 6 months (when the planet is visible) and spend between few minutes to few tens of minutes, most of the time. Once a light flux drop is obtained, one can use a synthetic light curve (see Fig. 27) to fit the mid-time event t_0 and the minimum distance between the two satellites (called impact parameter). Introducing the orbital inclination of both satellites from an ephemeris, the impact parameter can then be shifted to the more usual astrometric coordinates $(\Delta\delta\cos(\delta), \Delta\delta)$ for time t_0.

The great accuracy of this kind of observations stands, first, in the relative independence of the light flux drop to the distance of the satellite to the observer (but it is very dependent on the size of the objects), and second, in the low sensitivity to atmospheric fluctuation compared to astrometric observations from the ground. The usual accuracy of mutual events is about few tens of kilometers for the system of Jupiter, Saturn and Uranus.

3.1.3 Heliometer and Micrometer Observations

Initially developed for measuring the solar diameter, heliometers provide the relative distance (called separation and denoted "s") between two close celestial objects and their relative angle to the North direction (called position angle and denoted "p"). The principle of this instrument stands in the use of two object-glasses of equal focus that can move with one another providing a double image of the observed field. By measuring the shift between the two half-lenses when two celestial objects recover each other from the double image, one can derive their relative distance on the celestial sphere. Micrometer used a different principle based on the pointing of wires successively on two close celestial objects. The mutual distance could then be derived by reading markings on the right hand measuring rod that moved the thimble with an accuracy of few micrometers. For practical use, both instruments where generally calibrated on pairs of standard stars. One can consider that accurate astrometry has really started at the end of the XIXth century with heliometer and micrometer measurements, closely followed by photographic observations with long focus telescopes.

3.1.4 Photographic and CCD Observations

While heliometers and micrometers provided only one coordinate at a time (separation or position angle), photographic observations could provide both right ascension and declina-

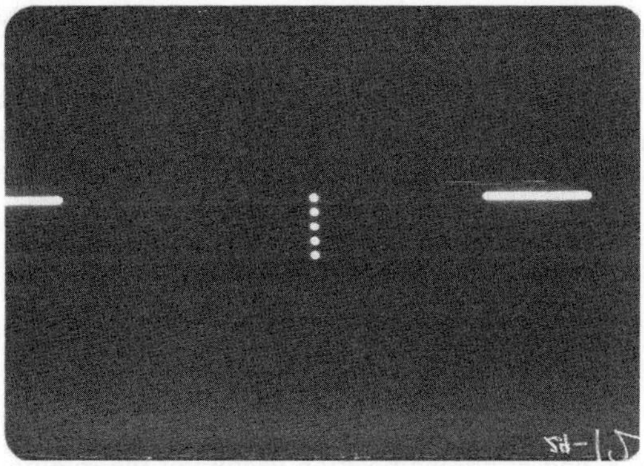

Fig. 28 An example of photographic plate of the Galilean satellites obtained at the long-focus refractor ($f = 9.9$ m) at USNO (Washington DC, USA) by D. Pascu. Five observations of Jupiter and the Galilean satellites (barely visible at each side of the planet on the picture) have been performed including a last one used for applying the trail scale technique

tion simultaneously. Usually the scale of the instrument was deduced as for heliometers and micrometers. Indeed, the low magnitude of the satellites (especially for the Galilean moons) and the small field were a limiting factor to the use of stellar background during the astrometric calibration of the photographic plates. Hence, specific techniques have been tried like the trail method. It consists in stopping the engine of the telescope letting the celestial objects appear like trail on the plate. As the observer knows exactly the time exposure as well as the Earth rotation, he can deduce from the measure of the trail the scale factor along the line of motion (for more information on the astrometric calibration of the observations, one can refer to (Pascu 1994; Peng et al. 2008) and references therein). To save money, more than one observation was also often performed with a photographic plate (see Fig. 28).

More recently, photographic observations have been replaced advantageously by CCD images. Astrometry from ground reaches today an accuracy of 50–100 milli-arcseconds (mas) for most of the main satellites of the giant planets. This is less accurate than mutual events, especially for Saturn's and Uranus' satellites, but these observations can be performed each year. Since the spacecraft era, images from space are also available. While the main source of astrometric error from ground stands in atmospheric fluctuations, astrometric error from spacecraft observations are mainly induced by the spacecraft position and camera pointing uncertainties. In particular, the accuracy of the observations depends critically on the mission and the satellite observed. In the best case, an accuracy at the level of few kilometers is achieved like with ISS instrument onboard of Cassini spacecraft (S. Charnoz 2008, priv. com.).

3.1.5 Fitting the Orbital Parameters

Once astrometric positions are obtained, one can compare the observed positions to the computed ones using a dynamical model of the orbital motions. At a first order of approximation, assuming that the essential difference between an observed and a computed position is the error on the initial conditions in the model, we have[1]

$$\vec{r}_i^{\,O}(t) - \vec{r}_i^{\,C}(t) \approx \sum_{l=1}^{6N} \frac{\partial \vec{r}_i^{\,C}}{\partial c_l}(t) \cdot \Delta c_l \tag{24}$$

[1]For sake of clarity, we compare observed and computed positions as 3-D vectors. In reality, observed positions are mostly provided as angles on celestial sphere and the comparison is done in these variables.

where $\vec{r}_i^O$ and $\vec{r}_i^C$ denote the observed and computed position vector of the satellite P_i, respectively and N is the number of satellites considered in the dynamical model. Assuming that the dynamical equations are integrated in a planetocentric frame with inertial axes, the equation of motion for a satellite P_i can be expressed as (Lainey et al. 2004a)

$$\ddot{\vec{r}}_i = -\frac{G(m_0+m_i)\vec{r}_i}{r_i^3} + \sum_{j=1, j\neq i}^{N} Gm_j\left(\frac{\vec{r}_j - \vec{r}_i}{r_{ij}^3} - \frac{\vec{r}_j}{r_j^3}\right) + G(m_0+m_i)\nabla_i U_{\bar{i}\hat{0}}$$
$$+ \sum_{j=1, j\neq i}^{N} Gm_j \nabla_j U_{\bar{j}\hat{0}} \tag{25}$$

Here, $\vec{r}_i$ and $\vec{r}_j$ are the position vectors of the satellite P_i and a body P_j (another satellite, the Sun, or a planet) with mass m_j, subscript 0 denotes the primary, $U_{\bar{k}\hat{l}}$ is the oblateness gravity field of body P_l at the position of body P_k and $\vec{r}_{kl} = \vec{r}_k - \vec{r}_l$.

Denoting c_l as an unspecified parameter of the model that shall be fitted (e.g. $\vec{r}_j(t_0)$, $d\vec{r}_j/dt(t_0)$), a useful relation is

$$\frac{\partial}{\partial c_l}\left(\frac{d^2\vec{r}_i}{dt^2}\right) = \frac{1}{m_i}\left[\sum_{j=1}^{N}\left(\frac{\partial \vec{F}_i}{\partial \vec{r}_j}\frac{\partial \vec{r}_j}{\partial c_l} + \frac{\partial \vec{F}_i}{\partial \dot{\vec{r}}_j}\frac{\partial \dot{\vec{r}}_j}{\partial c_l}\right) + \frac{\partial \vec{F}_i}{\partial c_l}\right] \tag{26}$$

denoting $\vec{F}_i$ the right hand side of (25) multiplied by m_i. In practice, (25) and (26) are numerically integrated simultaneously and provide for any time t the values of $\vec{r}_i^C(t)$ and $\frac{\partial \vec{r}_i^C}{\partial c_l}(t)$. Then, one can use (24) to derive corrections on the unknowns Δc_l using a least squares procedure.

3.2 Orbital Parameters

We provide in Tables 7–9 below the averaged Keplerian elements of the main satellites of Jupiter, Saturn and Uranus over the period [1920, 2020]. These elements have been computed with the ephemerides L1 (Lainey et al. 2004b) for the Galilean moons, TASS1.6 (Vienne and Duriez 1995) for the Saturnian moons and GUST86 (Laskar and Jacobson 1987) for the Uranian moons. We remind that Keplerian elements can sometimes differ from the geometric elements (see Greenberg 1981). For the main satellites of Jupiter, Saturn and Uranus, only very small differences arise except for Tethys. In particular, the geometric eccentricity of Tethys (that is to be considered in equations of tidal dissipation) is a factor 3.75 smaller than its Keplerian one (Lainey et al. 2008). The secular precession of the ascending nodes Ω and longitude of the periapses ϖ are the term with the highest amplitude in the analytic representation of variable $z = e\exp(-i\varpi)$ and $\zeta = \sin(i/2)\exp(-i\Omega)$.

Table 7 Orbital parameters for the Galilean satellites

Satellite	a (km)	e	i (deg)	$d\Omega/dt$ (deg/day)	$d\varpi/dt$ (deg/day)
Io	421,942	0.0042	0.0383	−0.1326	−0.7395
Europa	671,123	0.0094	0.4646	−0.0326	−0.7395
Ganymede	1,070,400	0.0013	0.2043	−0.0072	0.0073
Callisto	1,882,740	0.0074	0.2277	−0.0018	0.0018

Table 8 Orbital parameters for the main Saturnian satellites. The star (*) for Tethys is to remind that it is the Keplerian eccentricity and not the geometric one

Satellite	a (km)	e	i (deg)	$d\Omega/dt$ (deg/day)	$d\varpi/dt$ (deg/day)
Mimas	186,019	0.0192	1.6223	−0.9995	1.0010
Enceladus	238,409	0.0049	0.0151	−0.4171	0.3380
Tethys	294,973	0.0010*	1.0905	−0.1978	190.6973
Dione	377,647	0.0022	0.0197	−0.0843	0.0843
Rhea	527,228	0.0010	0.3403	−0.0275	0.0014
Titan	1,221,937	0.0289	0.3433	−0.0014	0.0014
Iapetus	3,561,668	0.0283	15.0214	−0.0003	0.0003

Table 9 Orbital parameters for the main Uranian satellites

Satellite	a (km)	e	i (deg)	$d\Omega/dt$ (deg/day)	$d\varpi/dt$ (deg/day)
Miranda	129,859	0.0013	173.6630	0.0556	0.0549
Ariel	190,925	0.0014	179.9059	0.0172	0.0170
Umbriel	265,971	0.0041	179.8505	0.0078	0.0078
Titania	436,253	0.0019	179.8698	0.0050	0.0011
Oberon	583,443	0.0013	179.8750	0.0050	0.0011

3.3 Summary

The understanding of the global magnetic environment of the outer planets as well as the local parameters of the plasma and field environment near the icy satellites, their gravity, shape and rotation are extremely important to understand the exchange processes in the biggest plasma laboratories of our solar system. In-situ measurements from instruments onboard spacecraft dramatically enhanced our knowledge of those small planetary systems. Recent measurements especially of orbiting spacecraft around Jupiter and Saturn have revealed that each new mission to the outer solar system solves questions from previous explorations but also discover new previously unknown phenomena like the Enceladus plume or the evidence of the first ring system around moon (Rhea at Saturn). We know that the magnetic environment of the icy satellites influence the moon's environment, its exosphere and its surface. On the other hand some of the moons influence the entire magnetospheres of the outer planets (Io at Jupiter, Enceladus at Saturn, and Triton at Neptune). In the following chapters the effects of those exchange processes will be discussed in more detail.

Acknowledgements This work was financed through the EC contract No. 001637 (RICA) EURO-PLANET. Work at the Max-Planck-Institut für Sonnensystemforschung has been supported by the German Space Agency DLR through the contracts 50OH0801 and 50OH0802 and by the Max Planck Society. Work in the US has been supported by NASA. V.L. thanks J.-E.Arlot and A.Vienne for fruitful discussions.

References

J.D. Anderson, G. Schubert, Geophys. Res. Lett. **34**, 2 (2007)

J.D. Anderson, E.L. Lau, W.L. Sjogren, G. Schubert, W.B. Moore, Nature **384**, 541–543 (1996)
J.D. Anderson, E.L. Lau, W.L. Sjogren, G. Schubert, W.B. Moore, Nature **387**, 264 (1997)
J.D. Anderson, G. Schubert, R.A. Jacobson, E.L. Lau, W.B. Moore, W.L. Sjogren, Science **281**, 2019–2022 (1998a)
J.D. Anderson, G. Schubert, R.A. Jacobson, E.L. Lau, W.B. Moore, W.L. Sjogren, Science **280**, 1573–1576 (1998b)
J.D. Anderson, R.A. Jacobson, E.L. Lau, J. Geophys. Res. **106**(12), 32963–32969 (2001a)
J.D. Anderson, R.A. Jacobson, E.L. Lau, W.B. Moore, O. Olsen, G. Schubert, P.C. Thomas, in *DPS Meeting 33*, Abstract 35.09. American Astronomical Society (2001b)
J.D. Anderson, R.A. Jacobson, T.P. McElrath, Icarus **153**, 157–161 (2001c)
J.D. Anderson, G. Schubert, R.A. Jacobson, E.L. Lau, W.B. Moore, J.L. Palguta, Science **305**, 989 (2004)
J.E. Arlot, B. Sicardy, Planet. Space Sci. **56**, 1778–1784 (2008)
S.W. Asmar, J.W. Armstrong, L. Iess, P. Tortora, Radio Sci. **40**, RS2005 (2005). doi:10.1029/2004RS003101
S.W. Asmar, F. Nimmo, P.C. Thomas, B.G. Bills, in *XXXX LPSC Meeting*, Abstract 2219 (2009)
H. Backes, F. Neubauer, M.K. Dougherty, N. Achilleos, N. André, C.S. Arridge, C. Bertucci, G.H. Jones, K.K. Khurana, C.T. Russell, A. Wennmacher, Science **308**, 992–995 (2005)
F. Bagenal, Annu. Rev. Earth Planet Sci. **20**, 289–329 (1992)
F. Bagenal, T.E. Dowling, W.B. McKinnon, *Jupiter—The Planet, Satellites and Magnetosphere* (Cambridge University Press, Cambridge, 2004)
J.T. Bergstralh, E.D. Miner, M.S. Matthews, *Uranus* (The University of Arizona Press, Tucson, 1991)
B. Bertotti, P. Farinella, D. Vokrouhlicky, *Physics of the Solar System* (Kluwer, Norwell, 2003)
B.G. Bills, Private communication (2009)
B.G. Bills, F. Nimmo, Icarus **196**, 293 (2008)
S.J. Bolton, R.M. Thorne, S. Bourdarie, I. DePater, B. Mauk, in *Jupiter*, ed. by F. Bagenal, T. Dowling, W. McKinnon (Cambridge University Press, Cambridge, 2004). Chap. 27
M.H. Burger et al., Space Sci. Rev. (2010, this issue). doi:10.1007/s11214-010-8645-z.
B.F. Burke, K.L. Franklin, J. Geophys. Res. **60**, 213 (1955)
T. Cassidy et al., Space Sci. Rev. (2010, this issue). doi:10.1007/s11214-009-9625-3.
M.H. Carr, M.J.S. Belton, C.R. Chapman, M.E. Davies, P. Geissler, R. Greenberg et al., Nature **391**, 363–365 (1998)
S. Charnoz, Private communication (2008)
J.E.P. Connerney, M.H. Acuna, N.F. Ness, J. Geophys. Res. **92**, 15329–15336 (1987)
A.H. Cook, *Interiors of the Planets* (Cambridge Univ. Press, Cambridge, 1980)
A. Coustenis et al., Space Sci. Rev. (2010, this issue). doi:10.1007/s11214-009-9615-5.
D.P. Cruikshank, *Neptune and Triton* (The University of Arizona Press, Tucson, 1995)
W. De Sitter, Mon. Not. R. Astron. Soc. **91**, 706–738 (1931)
R.B. Decker, A.F. Cheng, J. Geophys. Res. Res. **99**, 19027–19045 (1994)
S.F. Dermott, P.C. Thomas, Adv. Space Res. **10**, 165 (1990)
S.F. Dermott, P.C. Thomas, Icarus **109**, 241 (1994)
A.J. Dessler (ed.), *Physics of the Jovian megnetosphere* Cabridge University Press, New York (1983)
M.K. Dougherty, K.K. Khurana, F.M. Neubauer, C.T. Russell, J. Saur, J.S. Leisner, M.E. Burton, Science **311**, 1406–1409 (2006)
M.K. Dougherty, L.W. Esposito, S.M. Krimigis, *Saturn from Cassini-Huygens* (Springer, Berlin, 2009)
C. Elachi et al., Science **308**, 970 (2005)
T. Encrenaz, R. Kallenbach, T.C. Owen, C. Sotin, *The Outer Planets and Their Moons* (Springer, Berlin, 2005)
A.D. Fortes, P.M. Grindrod, S.K. Trickett, L. Vočadlo, Icarus **188**, 139 (2007)
K. Fukazawa, T. Ogino, R.J. Walker, Geophys. Res. Lett. **32**, L03202 (2005). doi:10.1029/2004GL021392
T. Gehrels M.S. Matthews (eds.), *Saturn* The University of Arizona Press, Tucson (1984)
P.E. Geissler, R. Greenberg, G. Hoppa et al., Nature **391**, 368–370 (1998)
C.K. Goertz, J. Geophys. Res. **85**, 2949 (1980)
O. Grasset, C. Sotin, Icarus **123**, 101 (1996)
O. Grasset, C. Sotin, F. Deschamps, Planet. Space Sci. **48**, 617 (2000)
R. Greenberg, Astron. J. **86**, 912–914 (1981)
E. Grün et al., Nature **362**, 428–430 (1993)
D.K. Haggerty, M.E. Hill, R.L. McNutt, C. Paranicas, J. Geophys. Res. 114 (2009) doi:10.1029/2008JA013659
D. Hamilton, H. Krüger, Nature **453**, 72–75 (2008)
T.W. Hill, A.J. Dessler, C.K. Goertz, in *Physics of the Jovian Magnetosphere*, ed. by A.J. Dessler (Cambridge University Press, New York, 1983), pp. 353–394
W.B. Hubbard, *Planetary Interiors* (Van Nostrand–Reinhold, New York, 1984)

L. Iess, N.J. Rappaport, P. Tortora, J.I. Lunine, J.W. Armstrong, S.W. Asmar, L. Somenzi, F. Zingoni, Icarus **190**, 585 (2007)
C.M. Jackman, C.T. Russell, D.J. Southwood, C.S. Arridge, N. Achilleos, M.K. Dougherty, Geophys. Res. Lett., **34** (2007). doi:10.1029/2007GL029764
X. Jia, R.J. Walker, M.G. Kivelson, K.K. Khurana, J.A. Linker, J. Geophys. Res. **113** (2008). doi:10.1029/2007JA012748
X. Jia, R.J. Walker, M.G. Kivelson, K.K. Khurana, J.A. Linker, J. Geophys. Res. **114** (2009). doi:10.1029/2009JA014375
G.H. Jones et al., Science **319**, 1380 (2008)
S. Jurac, J.D. Richardson, J. Geophys. Res. **110**, A09220 (2005)
S. Jurac, M.A. McGrath, R.E. Johnson, J.D. Richardson, V.M. Vasyliunas, A. Eviatar, Geophys. Res. Lett. **29**(24), 2172 (2002). doi:10.1029/2002GL015855
K.K. Khurana, M.G. Kivelson, D.J. Stevenson et al., Nature **395**, 749 (1998)
K.K. Khurana, M.K. Dougherty, C.T. Russell, J.S. Leisner, J. Geophys. Res. **112**, A08203 (2007). doi:10.1029/2006JA012110
K.K. Khurana, C.T. Russell, M.K. Dougherty, Icarus **193**, 465–474 (2008). doi:10.1016/j.icarus.2007.08.005
M.G. Kivelson, K.K. Khurana, C.T. Russell, R.J. Walker, J. Warnecke, F.V. Coroniti, C. Polanskey, D.J. Southwood, G. Schubert, Nature **384**, 537 (1996)
M.G. Kivelson, K.K. Khurana, D.J. Stevenson et al., J. Geophys. Res. **104**, 4609 (1999)
S.M. Krimigis, N. Sergis, D.G. Mitchell, D.C. Hamilton, N. Krupp, Nature **450**, 1050–1053 (2007). doi:10.1038/nature06425
E. Kronberg, J. Woch, N. Krupp, A. Lagg, K.K. Khurana, K.-H. Glassmeier, J. Geophys. Res., 110 doi:10.1029/2004JA010777
H. Krüger, M. Horanyi, A.V. Krivov, A.L. Graps, in *Jupiter*, ed. by F. Bagenal, T. Dowling, W. McKinnon (Cambridge University Press, Cambridge, 2004). Chap. 10
N. Krupp, J. Woch, A. Lagg, B. Wilken, S. Livi, D.J. Williams, Geophys. Res. Lett. **25**, 1249–1252 (1998)
N. Krupp, V.M. Vasyliunas, B.H. Mauk, E.C. Roelof, D.J. Williams, S.M. Krimigis, W.S. Kurth, L.A. Frank, W.R. Paterson, in *Jupiter*, ed. by F. Bagenal, T. Dowling, W. McKinnon (Cambridge University Press, Cambridge, 2004), Chap. 25
N. Krupp, E. Roussos, A. Lagg, J. Woch, A.L. Müller, S.M. Krimigis, D.G. Mitchell, E.C. Roelof, C. Paranicas, J. Carbary, G.H. Jones, D.C. Hamilton, S. Livi, T.P. Armstrong, M.K. Dougherty, Planet. Space Sci. **57**, 1754–1768 (2009)
A. Lagg, N. Krupp, J. Woch, D.J. Williams, Geophys. Res. Lett. **30**(11), 1556 (2003). doi:10.1029/2003GL017214
V. Lainey, L. Duriez, A. Vienne, Astron. Astrophys. **420**, 1171–1183 (2004a)
V. Lainey, J.E. Arlot, A. Vienne, Astron. Astrophys. **427**, 371–376 (2004b)
V. Lainey, L. Beauvalet, A. Vienne, in *DPS Meeting 40*, abstarct 45.01, American Astronomical Society (2008)
L.J. Lanzerotti, W.L. Brown, C.G. Maclennan, A.F. Cheng, S.M. Krimigis, J. Geophys. Res. **92**, 14949–14957 (1987)
J. Laskar, R.A. Jacobson, Astron. Astrophys. **188**, 212–224 (1987)
J.H. Lieske, Astron. Astrophys. **154**(12), 61–76 (1986)
J.H. Lieske, Astron. Astrophys. **176**, 146–158 (1987)
J.A. Linker, M.G. Kivelson, R.J. Walker, Geophys. Res. Lett. **15**, 1311 (1988)
J.A. Linker, M.G. Kivelson, R.J. Walker, J. Geophys. Res. **96**, 21037 (1991)
R.D. Lorenz, B. Stiles, R.L. Kirk, M. Allison, P. Persi del Marmo, L. Iess, J.I. Lunine, S.J. Ostro, S. Hensley, Science **319**, 1649 (2008)
P. Louarn, A. Roux, S. Perraut, W.S. Kurth, D.A. Gurnett, J. Geophys. Res. **105**, 13,073–13,088 (2000)
J.I. Lunine, D.J. Stevenson, Icarus **70**, 61 (1987)
R.A. Mackenzie, P.G. Antreasian, J.J. Bordi, K.E. Criddle, R. Ionasescu, R.A. Jacobson, J.B. Jones, D.W. Parcher, F.J. Pelletier, D.C. Roth, J.R. Stauch, in *AAS/AIAA Space Flight Mechanics Meeting*, Sedona, AZ, January 28–February 1, 2007. Paper AAS 07-248 (2007)
R. Mackenzie, L. Iess, N.J. Rappaport, P. Tortora, Geophys. Res. Lett. **35**, L05204 (2008). doi:10.1029/2007GL032898
A. Mallama, Icarus **192**(2), 576–581 (2007)
B.H. Mauk, D.J. Williams, R.W. McEntire, Geophys. Res. Lett. **24**(23), 2949–2952 (1997)
B.H. Mauk, D.G. Mitchell, R.W. McEntire, C.P. Paranicas, E.C. Roelof, D.J. Williams, S.M. Krimigis, J. Geophys. Res., **109**, A09S12 (2004). doi:10.1029/2003JA010270
D.J. McComas, F. Allegrini, F. Bagenal, F. Crary, R.W. Ebert, H. Elliott, A. Stern, P. Valek, Science **318**, 217–220 (2007). doi:10.1126/science.1147393
R.L. McNutt, R.S. Selesnick, J.D. Richardson, J. Geophys. Res. **92**, 4399–4410 (1987)

R.L. McNutt Jr., D.K. Haggerty, M.E. Hill, S.M. Krimigis, S. Livi, G.C. Ho, R.S. Gurnee, B.H. Mauk, D.G. Mitchell, E.C. Roelof, D.J. McComas, F. Bagenal, H.A. Elliott, L.E. Brown, M. Kusterer, J. Vandegriff, S.A. Stern, H.A. Weaver, J.R. Spencer, J.M. Moore, Science **318**, 220–222 (2007). doi:10.1126/science.1148025
J.M. Moore, R.T. Pappalardo, in *Eos Trans. AGU, 89(53), Fall Meet. Suppl.*, Abstract P11D-06 (2008)
C.D. Murray, S.F. Dermott, *Solar System Dynamics* (Cambridge Univ. Press, Cambridge, 1999)
F.M. Neubauer, J. Geophys. Res. **85**, 1171 (1980)
F.M. Neubauer, J. Geophys. Res. **103**, 19843 (1998)
F.M. Neubauer, J. Geophys. Res. **104**(28), 28671–28684 (1999)
F. Nimmo, I. Matsuyama, Geophys. Res. Lett. **34**, L19203 (2007). doi:10.1029/2007GL030798
A. Nishida, J. Geophys. Res. **81**, 1771–1773 (1976)
J. Palguta, J.D. Anderson, G. Schubert, W.B. Moore, Icarus **180**, 428–441 (2006)
R.T. Pappalardo, J.W. Head, R. Greeley et al., Nature **391**, 365–368 (1998)
C. Paranicas, A.F. Cheng, Geophys. Res. Lett. **20**, 2805–2808 (1993)
C. Paranicas, D.G. Mitchell, S. Livi, S.M. Krimigis, E. Roussos, N. Krupp, J. Woch, A. Lagg, J. Saur, F.S. Turner, Geophys. Res. Lett. **32**, L20101 (2005). doi:10.1029/2005GL024072
W.D. Parkinson, *Introduction to Geomagnetism* (Scottish Academic Press Ltd., Edinburgh, 1983) pp. 308–340
D. Pascu, in *Galactic and Solar System Optical Astrometry, Proceedings of the Royal Greenwich Observatory and the Institute of Astronomy Workshop*, ed. by L.V. Morrison, G.F. Gilmore. Cambridge, June 21–24, 1993 (Cambridge University Press, Cambridge, 1994), p. 304
Q.Y. Peng, A. Vienne, V. Lainey, B. Noyelles, Planet. Space Sci. **56**, 1807–1811 (2008)
D.H. Pontius, T.W. Hill, J. Geophys. Res. **111**, A09214 (2006)
F. Postberg, S. Kempf, J. Schmidt, N. Brilliantov, A. Beinsen, B. Abel, U. Buck, R. Srama, Nature **459**, 1098–1101 (2009). doi:10.1038/nature08046
L.H. Prockter et al., Space Sci. Rev. (2010, this issue). doi:10.1007/s11214-010-9649-8.
N.G. Purves, C.B. Pilcher, Icarus **43**, 51–55 (1980)
N. Rappaport, B. Bertotti, G. Giampieri, J.D. Anderson, Icarus **126**, 313–323 (1997)
N.J. Rappaport, R.A. Jacobson, L. Iess, P. Racioppa, J.W. Armstrong, S.W. Asmar, D.J. Stevenson, P. Tortora, M. Di Benedetto, A. Graziani, R. Meriggiola, in *Eos Trans. AGU, Fall Meeting Suppl.*, Abstract P21A-1343 (2008b)
J.D. Richardson, Geophys. Res. Lett. **20**(14), 1468–1470 (1993)
J.D. Richardson, R.L. McNutt, Geophys. Res. Lett. **17**(10), 1689–1692 (1990)
E. Roussos, Ph.D. thesis, Technical University Braunschweig, Germany (2008)
D. Santos-Costa, Ph.D. thesis (2001)
D. Santos-Costa, S.A. Bourdarie, Planet. Space Sci. **49**, 303–312 (2001)
E.J. Shaya, C.B. Pilcher, Icarus **58**, 74–80 (1984)
G.L. Siscoe, E.F. Lyon, J.H. Binsack, H.S. Bridge, J. Geophys. Res. **74**, 59 (1969)
E.C. Sittler, K.W. Ogilvie, J.D. Scudder, J. Geophys. Res. **88**, 8847–8870 (1983)
F. Sohl, H. Hussmann, B. Schwentker, T. Spohn, R.D. Lorenz, J. Geophys. Res. 108 (E12). CiteID 5130 (2003)
F. Sohl et al., Space Sci. Rev. (2010, this issue). doi:10.1007/s11214-010-964-7.
J. Spencer, N.M. Schneider, Annu. Rev. Earth Planet Sci. **24**, 125–190 (1996)
D.J. Stevenson, in *Proceedings Symposium on Titan, Toulouse, France, 9–12 September 1991* (European Space Agency, Noordwijk, 1992), p. 29
B.W. Stiles, R.L. Kirk, R.D. Lorenz, S. Hensley, E. Lee, S.J. Ostro, M.D. Allison, P.S. Callahan, Y. Gim, L. Iess, P. Persi Del Marmo, G. Hamilton, W.T.K. Johnson, R.D. West, Astron. J. **135**(5), 1669 (2008)
P.C. Thomas, J. Veverka, P. Helfenstein, P. Porco, C.J. Burns, T. Denk, E. Turtle, R.A. Jacobson, and the ISS Science team (2006), in *Shapes of the Saturnian Icy Satellites. XXXVII LPSC Meeting*, Abstract 1639
P.C. Thomas, J.W. Armstrong, S.W. Asmar, J.A. Burns, T. Denk, B. Giese, P. Helfenstein, L. Iess, T.V. Johnson, A. McEwen, L. Nicolaisen, C. Porco, N.J. Rappaport, J. Richardson, L. Somenzi, P. Tortora, E.P. Turtle, J. Veverka, Nature **448**, 50 (2007)
C.L. Thorton, J.S. Border, Descanso series, JPL Publication 00-11 (2000)
G. Tobie, O. Grasset, J.I. Lunine, A. Mocquet, C. Sotin, Icarus **175**, 496 (2005)
G. Tobie, J.I. Lunine, C. Sotin, Nature **440**, 61 (2006). doi:10.1038/nature04497
G. Tobie et al., Space Sci. Rev. (2010, this issue). doi:10.1007/s11214-010-9641-3.
T. Tokano, F. Neubauer, Geophys. Res. Lett. **32**, L24203 (2005). doi:10.1029/2005GL024456
N.A. Tsyganenko, J. Geophys. Res. **100**, 5599 (1995)
T. Van Hoolst, N. Rambaux, Ö. Karatekin, R.-M. Balanda, Icarus **200**, 256 (2009)
V.M. Vasyliunas, in *Physics of the Jovian Magnetosphere*, ed. by A.J. Dessler (Cambridge University Press, New York, 1983), pp. 395–453

A. Vienne, L. Duriez, Astron. Astrophys. **297**, 588–605 (1995)
M. Volwerk, K. Khurana, M. Kivelson, Ann. Geophys. **25**, 905–914 (2007)
Y.C. Wang, N.F. Ness, J. Geophys. Res. **77**, 1109 (1972)
D.J. Williams, B. Mauk, R.W. McEntire, J. Geophys. Res. **103**, 17523 (1998)
J. Woch, N. Krupp, J.A. Lagg, B. Wilken, S. Livi, D.J. Williams, Geophys. Res. Lett. **25**, 1253–1256 (1998)
J. Woch, N. Krupp, A. Lagg, Geophys. Res. Lett. **29**, 1138 (2002). doi:10.1029/2001GL014080
R.S. Wolfe, D.A. Mendis, J. Geophys. Res. **88**, 4749–4769 (1983)
C.F. Yoder, Icarus **115**, 278 (1995)
C. Zimmer, K.K. Khurana, M.G. Kivelson, Icarus **147**, 329 (2000)

Space Sci Rev (2010) 153: 63–111
DOI 10.1007/s11214-010-9649-8

Characteristics of Icy Surfaces

Louise M. Prockter · Rosaly M.C. Lopes · Bernd Giese · Ralf Jaumann · Ralph D. Lorenz · Robert T. Pappalardo · Gerald W. Patterson · Peter C. Thomas · Elizabeth P. Turtle · Roland J. Wagner

Received: 5 June 2009 / Accepted: 24 March 2010 / Published online: 4 June 2010

Abstract The surfaces of the Solar System's icy satellites show an extraordinary variety of morphological features, which bear witness to exchange processes between the surface and subsurface. In this paper we review the characteristics of surface features on the moons of Jupiter, Saturn, Uranus and Neptune. Using data from spacecraft missions, we discuss the detailed morphology, size, and topography of cryovolcanic, tectonic, aeolian, fluvial, and impact features of both large moons and smaller satellites.

Keywords Icy satellites · Europa · Ganymede · Titan · Enceladus · Cryovolcanism

1 Introduction

The icy satellites of the outer solar system show a variety of geologic features on their surfaces, and different compositions and characteristics that help constrain models of how exchange processes are operating. Some satellites are extremely active in terms of exchanges from the interior to the surface and outer space—for example, Enceladus—while others appear to be mostly inactive at present time. This chapter discusses the major satellites

L.M. Prockter (✉) · R.D. Lorenz · G.W. Patterson · E.P. Turtle
Applied Physics Laboratory, Johns Hopkins University, MP3-E169, 11100 Johns Hopkins Road, Laurel, MD 20723, USA
e-mail: Louise.Prockter@jhuapl.edu

R.M.C. Lopes · R.T. Pappalardo
Jet Propulsion Laboratory, California Institute of Technology, M/S 183-601, 4800 Oak Grove Drive, Pasadena, CA 91109, USA

B. Giese · R. Jaumann · R.J. Wagner
German Aerospace Center (DLR), Institute of Planetary Research, Planetary Geodesy, Rutherfordstraße 2, 12489 Berlin, Germany

P.C. Thomas
Cornell University, 422 Space Sciences Building, Ithaca, NY 14853, USA

of Jupiter, Saturn, Uranus and Neptune, emphasizing those where exchange processes are taking place today.

For the Jovian and Saturnian satellites, we have datasets from the Galileo and Cassini-Huygens spacecraft, as well as from Voyager 1 and 2. Both Cassini and the New Horizons spacecraft also obtained some observations of the Jupiter system, although these were distant observations at relatively low spatial resolution. Voyager 2 obtained data on the satellites of Uranus and Neptune.

The discussion below emphasizes how diverse these satellites are, and how exchange processes vary significantly from world to world. The most dramatic interactions are shown by those satellites that exhibit active volcanism, notably Io and Enceladus. Titan is unique because it is the only satellite that has a substantial atmosphere. This atmosphere interacts with the surface in ways not seen on any other icy body but this process is, in many ways, remarkably similar to the surface-atmosphere interaction on Earth and significantly affects exchange processes on both bodies. The smaller Saturnian satellites and the Uranian satellites exhibit a diversity of features on their surfaces demonstrating that even small worlds have dynamic geological histories. Triton's surface shows ample evidence of cryovolcanism, which appears to have ceased in relatively recent geological history. Over the next decade, Pluto's moon Charon will finally be imaged by the New Horizons spacecraft, revealing its secrets for the first time.

In this chapter we summarize the surface features on the icy satellites of the solar system (and include Io for completeness), starting with the Jovian system and working out to the Neptune system. We describe the morphology and topography of cryovolcanic, tectonic, fluvial, aeolian and impact features to lay the foundation for subsequent chapters that describe the proposed formation mechanisms for these features and their implications for exchange processes in the solar system.

2 The Jovian Satellites

2.1 Callisto

Callisto has one of the most heavily cratered surfaces in the Solar System. The dominant landform on Callisto's surface is ancient cratered plains, interspersed with some smaller smooth areas. The surface is close to saturation, in which any new crater is likely to erase an older one (Zahnle et al. 1998). Imaging by Galileo showed that surface degradation is an important process at small scales, manifested by landslides from crater rims and a generally older and degraded appearance compared to similar ancient terrain on Ganymede (Greeley et al. 2000). Callisto's surface shows no clear evidence of communication with the subsurface, although the smooth dark areas appear to mantle and embay the surrounding plains, and have been suggested to result from cryovolcanic resurfacing (Schenk 1995). Early in its history, impact processes formed gigantic basins on Callisto, which would have excavated a considerable amount of material from the subsurface.

The smallest craters on Callisto are simple craters, characterized by raised rims and bowl shape (Moore et al. 2004). The depth-to-diameter ratio is similar to that on Ganymede, about 1/5 (Schenk et al. 2004). Degradation, however, has played a larger role on Callisto than on Ganymede. A great number of small craters were heavily affected by erosion, which produced remnants of rims, rings of isolated massifs, or left only circular depressions filled with dark, smooth material (Moore et al. 1999, 2004; Greeley et al. 2000; Wagner 2007). The simple-to-complex transition diameter on Callisto is the same as on Ganymede, about

2–3 km (Schenk et al. 2004). Also, complex craters including central peaks, central pits, domes and penepalimpsests are remarkably similar to those on Ganymede, including the respective transition diameters (see Sect. 2.2). Therefore, the complex craters on Ganymede represent complex craters on Callisto very well.

Palimpsests have been thought to be more or less absent on Callisto, and those found seem to be mostly degraded (Passey and Shoemaker 1982; Schenk 1995; Greeley et al. 2000; Moore et al. 2004). Further examination of Galileo Solid State Imager (SSI) imagery in context with Voyager data revealed a much greater number of palimpsests on Callisto, though most of them are heavily degraded and difficult to recognize (Wagner 2007).

Callisto features a number of remarkable crater chains (*catenae*), which can reach lengths up to 620 km, such as *Gipul Catena* (Passey and Shoemaker 1982; Schenk 1995; Schenk et al. 2004). Only a small number of such chains are radial to ring basins and therefore secondary in origin (Greeley et al. 2000). Most likely, the majority of crater chains on Callisto and Ganymede were created by cometary bodies fragmented into a string of bolides hitting the surface, similar to the break-up of comet Shoemaker/Levy-9, which impacted Jupiter in July 1994 (McKinnon and Schenk 1995; Greeley et al. 2000).

Unlike Ganymede, Callisto shows a number of comparably well-preserved large multi-ring structures. Typical examples for these structures are Valhalla and Asgard, with maximum diameters of their entire ring systems of ~3800 km (Valhalla) and ~1900 km (Asgard) (Smith et al. 1979a, 1979b; Passey and Shoemaker 1982; Moore et al. 2004). Unlike ring basins on the terrestrial planets, these two basins on Callisto are characterized by distinct structural zones with (a) central bright plains, reminiscent of a palimpsest and up to several hundreds of kilometers in diameter, surrounded (b) by numerous (up to 24) concentric rings of ridges, scarps, or troughs with a more or less constant spacing (Schenk 1995; Greeley et al. 2000) (Valhalla is shown in Fig. 6).

The origin of multi-ring structures on planets and satellites is discussed in several models (e.g., Melosh 1989; Spudis 1993). Of these models, only (a) the Tsunami and (b) the ring tectonic model seem to sufficiently explain the specific morphology of these structures on icy satellites. In the Tsunami model, a massive impact can fluidize the target material and produce wave-like features that are topographically expressed in numerous ridges, scarps, and troughs (Van Dorn 1968; Spudis 1993). The ring tectonic model describes the origin of rings on icy satellites by a massive projectile that impacts into a thin (order of several kilometers to a few tens of kilometers), brittle lithosphere and penetrates into a ductile, possibly in parts fluid, asthenosphere (McKinnon and Melosh 1980). According to this model, the material in the asthenosphere flows gravitationally inward toward the impact center creating extensional stresses in the lithospheric material. From the width and spacing of troughs or furrows, the brittle-to-ductile transition depth can be inferred (McKinnon and Melosh 1980). On Ganymede and Callisto, this transition was, accordingly, at depths of approximately 15 to 20 km at the time of basin formation.

2.2 Ganymede

The surface of Ganymede can generally be divided into two terrain types that exhibit differences in albedo, crater density, and surface morphology. Covering approximately a third of the surface is a lower albedo or dark terrain that is heavily cratered and commonly modified by the presence of large-scale arcuate fracture systems termed furrows. The other two thirds of the surface are covered by vast globe-encircling swaths of a higher relative albedo or light terrain that have a crater density significantly lower than dark material. These swaths are themselves divided into polygons that range in appearance from relatively smooth to heavily modified by linear structures termed grooves. The impact craters that superpose these terrains have interior structures that range from central peaks, to central pits, to central domes with progressively increasing crater diameter (Schenk et al. 2004). The largest impact crater observed on Ganymede is Gilgamesh basin (~590 km in diameter) centered at 57°S, 125°W. Other unique features that superpose these terrains include palimpsests and reticulate terrain.

The topographic database for Ganymede is very limited. The available data suggest that Ganymede's surface is subdued as observed for Europa. Generally, old dark terrain features seem to have more relief than younger, bright terrain features. Similar to Europa, the features with largest relief are domes showing relief of up to 2.5 km on Ganymede (see below). Squyres (1981) detected a dome in Voyager imagery extending over 260 km with a summit elevation 2.5 km above the surrounding plain.

2.2.1 Dark Terrain

Crater density measurements suggest dark terrain represents the oldest preserved surface on Ganymede, though it generally has a lower crater density than the surface of neighboring Callisto (Shoemaker and Wolfe 1982; Neukum 1997; Neukum et al. 1998; Zahnle et al. 2003; Schenk et al. 2004). This terrain is structurally modified by remnants of vast multiringed structures termed furrow systems (Smith et al. 1979a, 1979b), the oldest recognizable structures on the surface of the satellite (Passey and Shoemaker 1982). The majority of the furrows are arranged in sub-concentric sets (Fig. 1), while some isolated furrows have been observed crosscutting others at high angles (Murchie and Head 1988). Because younger swaths of bright terrain disrupt dark terrain, the record of ancient furrow systems

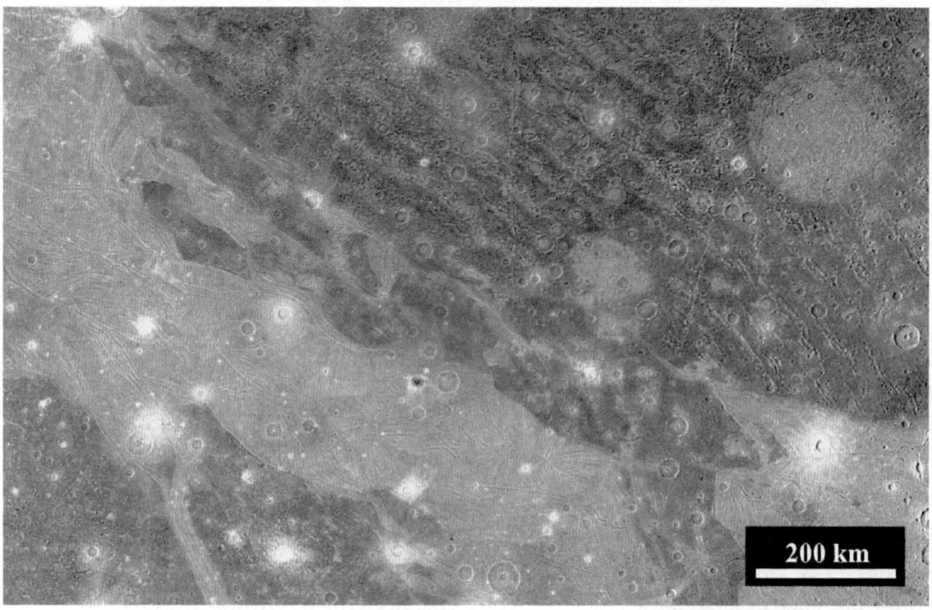

Fig. 1 Typical region on Ganymede, showing dark and light terrains. Linear features trending northwest to southeast are furrows, part of the Lakhmu Fossae system; circular bright feature in the northeast is the palimpsest Memphis Facula

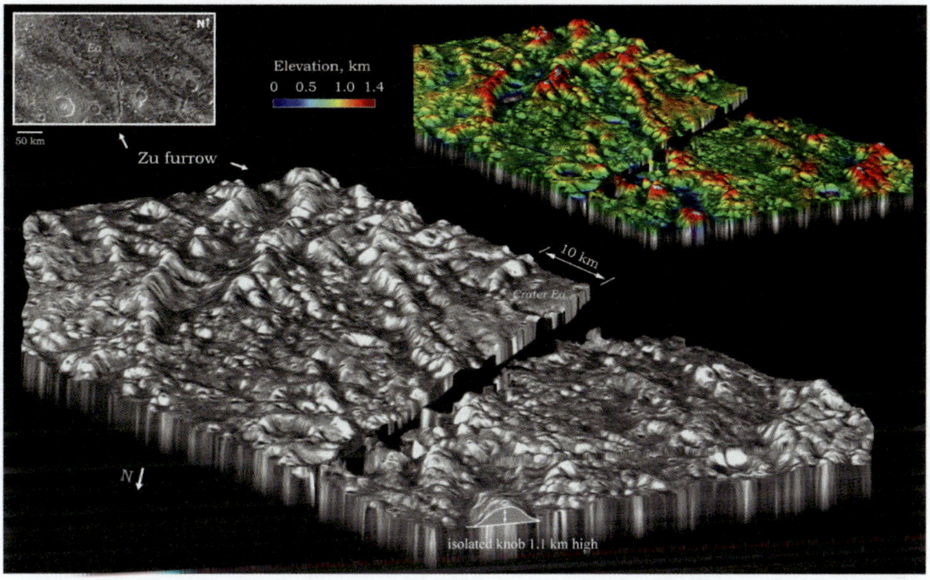

Fig. 2 Perspective view of the prominent Zu furrow located in Ganymede's Galileo Regio. *Arrows* point to the furrow trough between the two raised rims. *Upper left inset* is a cutout of Voyager frame C2063659. The color-coded topographic model was derived from Galileo stereo images with resolutions of ~80 m/pxl (Giese et al. 1998)

is incomplete. When intact, however, the largest known system (e.g., Lakhmu Fossae in Galileo Regio—Fig. 1) would have been hemispherical in scale (Schenk and McKinnon 1987).

Individual furrows are characterized by linear to curvilinear troughs bounded by raised rims, which have a higher relative albedo with respect to the surrounding terrain. They extend from tens to hundreds of kilometers in length and are typically ~6 to 20 km wide, with generally flat or U-shaped floors and sharp raised rims (e.g., Prockter et al. 1998). Spacing between furrows is fairly uniform at ~50 km, although it generally decreases toward the center of a concentric system (Passey and Shoemaker 1982). Topographic models derived from high resolution stereo images within Galileo Regio show that furrow rims can rise a full kilometer above their floor and ~900 m above the level of the surrounding terrain (Prockter et al. 1998) (Fig. 2). This is consistent with estimates from broad-scale shadow measurements of furrow depth (Murchie and Head 1988). On the basis of morphology and planform, along with their similarity to multi-ringed structures on Europa and Callisto, furrows have been interpreted as fault-induced troughs formed in response to large impacts into a relatively thin lithosphere early in Ganymede's history (McKinnon and Melosh 1980).

While geologically uniform at global scales, complex relationships indicating numerous distinct units and heterogeneity in relative albedo within dark terrain have been recognized on local scales (Prockter et al. 1998; Spencer 1987a, 1987b). Observations of cleaner, icy material on sun-facing slopes, along with mass wasted dark material at the base of slopes indicate that the dark terrain represents a relatively thin and dark lag deposit of non-ice material, overlying brighter cleaner icy material. This lag deposit may have been concentrated on the surface by processes such as sublimation, mass wasting, ejecta blanketing, and tectonism (Prockter et al. 1998; Prockter and Pappalardo 2000; Moore et al. 1999; Oberst et al. 1999).

2.2.2 *Light (grooved) terrain*

Crater density measurements suggest that light terrain is younger than dark terrain, with estimates on the initiation of light terrain formation ranging from not long after the formation of dark terrain (Neukum 1997; Neukum et al. 1998) to as little as ~2 Gyr (Shoemaker and Wolfe 1982; Zahnle et al. 1998, 2003). Light terrain forms swaths that crosscut older dark material and contain polygons 10s to 100s of kilometers wide, forming an intricate patchwork across the surface (Fig. 1). These polygons within light terrain are typically characterized by sets of subparallel ridges and troughs, commonly referred to as 'grooves.'

Grooves associated with light terrain exhibit two superposed spacing scales. The larger-scale grooves have a mean spacing of about 8 km, as derived from Fourier analysis of photometric profiles across groove sets in Voyager images at ~1 km/pixel (Grimm and Squyres 1985). These grooves appear in low solar incidence angle images as dark and bright stripes, while in near-terminator images the topography of the grooves is revealed to be broadly U-shaped in cross-section. Galileo spacecraft images of light terrain at better than 100 m/pixel reveal smaller-scale grooves, with a mean spacing of ~1 km (Belton et al. 1996) superposed on the larger grooves. In topographic lows associated with the larger grooves, the small-scale grooves are smaller and more closely spaced, and on topographic highs the small-scale grooves are more widely spaced (Pappalardo et al. 1998). These superposed spacing scales of grooves have been interpreted as resulting from extensional tectonism, likely initiated through necking instabilities and resulting in the development of imbricate normal faulted tilt-blocks (Pappalardo et al. 1998, 2004; Collins et al. 1998; Dombard and McKinnon 2001). In this scenario, light terrain forms from dark terrain with its dark lag deposit being shed via downslope movement and sublimation. This process then exposes the brighter, more ice-rich substrate on the slopes of grooves.

In several available Galileo images, grooves having a morphology resembling graben and intervening flat-topped horst ridges have been observed (Collins et al. 1998; Pappalardo et al. 2004). Rather than downdropping along prominent and distinct bounding faults, offset appears to have occurred by cumulative displacement along several sub-kilometer-scale fractures or faults, which are pervasive across these, grooved terrains (Pappalardo et al. 2004). It has been suggested that cross-cutting relationships between this style of groove formation and that of the more predominately observed normal-faulted tilt block style of grooves indicate the former may have been the preferred morphology of older light terrain (Collins et al. 1998). However, an alternative view utilizing digital elevation models derived from stereo pairs of Voyager and Galileo data of the Sippar Sulcus region suggests the horst-and-graben style of groove formation is topographically lowest and therefore relatively young (Schenk et al. 2001).

The characteristics of grooved terrain in Ganymede's Nicholson Regio are shown in Fig. 3. In the profiled region (from N to S), the inner zone of the band consists of a ~10 km wide swell followed by a pinch, a sawtooth-shaped dark terrain block, a bounding trough, and an elevated dark terrain flank. The swells in the band are high-standing with respect to the surrounding dark terrain (compare this topography also to the elevation model in Giese et al. 2001a), 400 m at highest. The grooved terrain east of Arbela Sulcus provides another example for high standing grooved terrain portions.

Arbela Sulcus, the bright 22 km wide band (Fig. 3), features pronounced smoothness as compared with the adjoining terrains (Head et al. 2002). Apart from an elevated eastern flanking ridge, the eastern half of the band stands by ~100 m lower than the adjoining dark terrain, otherwise the band is about level with it. Bright, smooth terrain exists also in Ganymede's Harpagia Sulcus (Head et al. 2002). Likewise that terrain is, at least locally

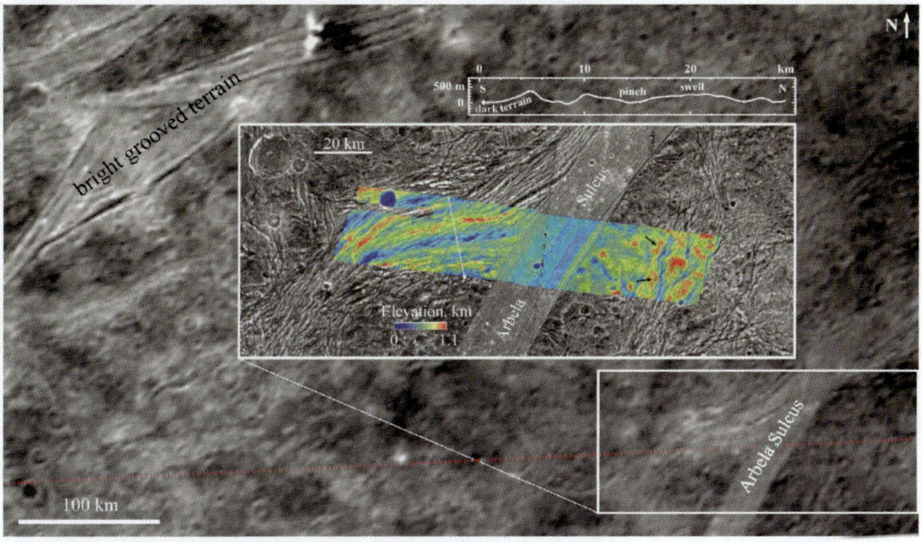

Fig. 3 Cutout of Voyager frame C1640446 (1.4 km/pxl) showing old dark terrain and younger bright terrains in Ganymede's Nicholson Regio. The bright grooved terrain exhibits an undulating topography over 10 km scales (inset with color-coded elevation model), by comparison the bright band "Arbela Sulcus" is smooth. Elevation profile indicates a trough at the boundary to dark terrain and an elevated dark terrain flank. Note also the elevated dark terrain flank at the rift zone east of Arbela Sulcus (*arrows*). The elevation model was derived from 130 m/pxl/40 m/pxl Galileo stereo images (Giese et al. 2001a). Its horizontal/vertical resolution is 600 m/15–30 m

(Giese et al. 2001b), about level with the adjoining dark terrain. In contrast, Schenk et al. (2001) have identified bright, smooth terrains that lie at roughly constant elevations 100 to 1000 meters below the surrounding rougher terrains. However, it has to be noted that this information was inferred from topographic data with resolutions of 7 km horizontally and 0.20–0.24 km vertically, so the errors may be large.

Indirect evidence for volcanic resurfacing in the form of small isolated caldera-like features (Lucchitta 1980; Schenk and Moore 1995; Head et al. 1998; Kay and Head 1999; Spaun et al. 2001) has been identified in association with light material. However, based on evidence gathered from the Galileo mission, cryovolcanism appears to play a relatively minor role in the modification light terrain (Pappalardo et al. 2004).

2.2.3 Reticulate Terrain

Reticulate terrain is distinguished from other terrains by its variable albedo and the presence of distinct grooves with two dominant directions, typically near-orthogonal to each other. This terrain appears to structurally modify both dark and light terrain and is commonly associated with and surrounded by light terrain. Voyager era geological mapping suggested that, based on cross-cutting relationships, reticulate terrains formed shortly before or during the initiation of light terrain formation (Guest et al. 1988; Wilhelms 1997).

2.2.4 Ganymede Craters

The widest range in crater morphology on any planet or satellite is found on Ganymede (Pappalardo et al. 2004). Similar crater forms are found on Callisto as well (see Sect. 2.1),

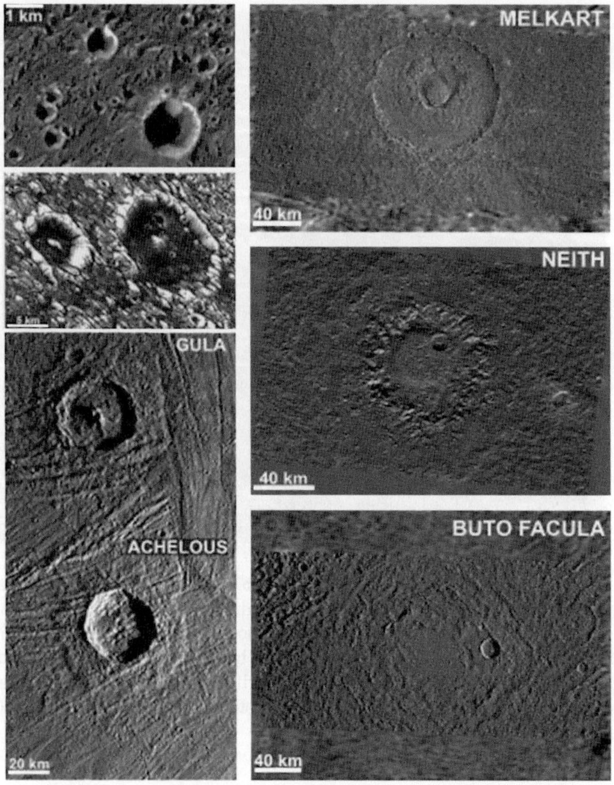

Fig. 4 *Upper left, top two images*: Small (<10 km) simple craters on Ganymede with classic bowl shape (SSI target area 28GSSMOOTH01); below this image are two relatively small complex craters *Chrysor* (*left*) and *Aleyin* (*right*) on Ganymede (target area G1GSMEMPHI01). *Lower left*: *Gula* and *Achelous*: mid-sized complex craters with ejecta pedestals, Achelous also has a bright ray system (SSI target area G7GSACHELS01); *Top right*: *Melkart*: large central pit crater with dome in the pit interior (target area G8GSMELKRT01); *Middle right*: *Neith*: anomalous dome crater or penepalimpsest (target area G7GSNEITH_01); *Lower right*: *Buto Facula*: palimpsest located in dark material of Marius Regio (target area G8GSBUTOFC01). Mosaics in the right column are shown in lower-resolution Voyager context. North is pointing upward except for Neith (North pointing to the right)

but these features are much more degraded than those on Ganymede. Both satellites show crater forms that are completely absent on the terrestrial planets while other forms are reminiscent of craters on terrestrial planets.

The main characteristics of small, simple craters are raised rims and a bowl-shaped interior (e.g., Pike 1977, 1980; Chapman and McKinnon 1986; Melosh 1989; Schenk et al. 2004). The depth-to-diameter relationship of simple craters is similar to the terrestrial planets or the Moon, on the order of 1/5 (e.g., Pike 1980; Schenk et al. 2004). Examples for simple craters on Ganymede are shown in Fig. 4. With increasing diameter, crater forms deviate from a simple bowl shape and become more and more complex. Each planet or satellite has a specific simple-to-complex transition diameter that is approximately inversely proportional to the gravity acceleration of the body (Pike 1980, 1988).

Complex craters are characterized by (1) scalloped rims, (2) wall terraces, (3) flat floors, and (4) interior structures, such as, e.g., central peaks (Pike 1980). On the icy Galilean satellites, the transition diameters from simple to complex forms are significantly lower than those on the terrestrial planets. For Ganymede and Callisto, the transition occurs at 2–3 km diameter, compared to, e.g., the Moon at ~15 km (Schenk et al. 2004). Complex craters on Ganymede and Callisto can be subdivided into four types (Passey and Shoemaker 1982; Chapman and McKinnon 1986; Schenk et al. 2004): (1) craters with central peaks, (2) craters with central pits, (3) craters with central domes, and (4) palimpsests.

Central peak craters at diameters less than 10 km are abundant on Ganymede and also occur at larger sizes. On the rocky bodies, peak rings may form in the center of the crater with increasing diameter (e.g., Grieve and Cintala 1992), but such peak rings are not observed on icy satellites. Instead, at sizes larger than 25–26 km, central peaks are replaced by central pits, rimmed crater-like features in the center of the crater (Passey and Shoemaker 1982; Schenk et al. 2004). Although several models have been proposed, the formation of central pits is not well understood.

Central domes within rimmed central pits occur in craters larger than about 60 km in diameter on both Ganymede and Callisto (Passey and Shoemaker 1982; Moore and Malin 1988; Schenk 1993; Schenk et al. 2004) (Fig. 4). These crater forms range in size from about 50 to 250 km in diameter (Schenk et al. 2004). Several models of formation are discussed for the origin of central domes: (1) refreezing of impact melt (Croft 1983), (2) post-impact diapirism (Moore and Malin 1988), or (3) rapid uplift of ductile material during impact (Schenk 1993).

Impact forms unique to Ganymede and Callisto are termed palimpsests, which were first seen in the Voyager data (Smith et al. 1979a, 1979b; Passey and Shoemaker 1982). These specific forms were interpreted as remnants of former impact craters. Palimpsests are bright, at Voyager resolution more or less smooth, circular or elliptic areas, located predominantly in old, dark terrain, and almost completely devoid of typical crater morphologies such as crater rims (Fig. 1). The size range of palimpsests is from about 50 to 390 km, measured as diameter of the bright circular area (Schenk et al. 2004). Further examination of palimpsests, using, e.g., higher-resolution Galileo images revealed, however, that some of them do show remnants of concentric structures. These palimpsests were classified as *penepalimpsests type I* (Passey and Shoemaker 1982). The origin of palimpsests is still not fully understood, but they likely formed at early times when global heat flow and crustal weakness were still higher (Jones et al. 2003; Schenk et al. 2004).

In some complex craters on Ganymede and Callisto, an outward-facing scarp is observed in the continuous ejecta blanket, located about one crater radius away from the rim. Complex craters with this specific feature were termed *pedestal craters* (Horner and Greeley 1982; Moore et al. 1998, 2001). This feature is similar to rampart craters on Mars and implies the presence of volatiles in the subsurface (Horner and Greeley 1982) (Fig. 4).

Some of the stratigraphically youngest craters on Ganymede have extended ray systems. Bright rays can extend several hundreds or even thousands of kilometers away from the crater (Fig. 5). Investigations using spectral data of the Galileo NIMS instrument revealed a high content of water ice in the crater and its ejecta (Stephan 2007), reflecting a comparably young age of ray craters. In the trailing hemisphere, an unusual number of dark ray craters are observed (Fig. 5). Their low albedo ejecta indicate compositional characteristics of the substrate and/or the projectile (Conca 1981; Schenk and McKinnon 1991). The small

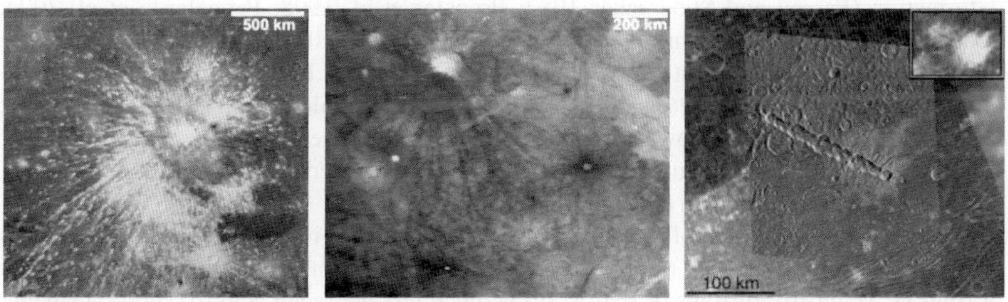

Fig. 5 Special crater forms on Ganymede. *Left*: Bright ray crater *Osiris* (Voyager-2 mosaic); *Middle*: craters with dark rays (Voyager-2 mosaic); *Right*: Crater chain *Enki Catena* (SSI target are G7GSCATENA01 in Voyager-1 context; inserted small image low-resolution Voyager-1 data). North is up in all images

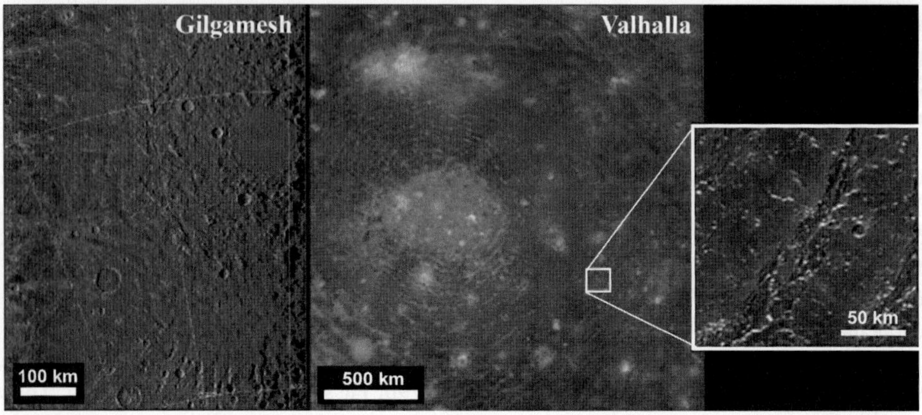

Fig. 6 Examples of multi-ring structures on Ganymede and Callisto. *Left*: *Gilgamesh* on Ganymede with circular central plains, surrounded by a rugged area with concentric ridges and scarps, smooth continuous ejecta, and radial chains of secondary craters (e.g., Schenk et al. 2004), Voyager-2 image mosaic; *Right*: *Valhalla* on Callisto with (a) bright central plains, surrounded by (b) numerous ridges and/or troughs. Concentric troughs are shown in detail on the right. Galileo SSI target area C9CSVALHAL01 (*left panel* in Voyager-1 context). North is pointing upward

number of these impact features could be an effect of rare impact velocities and/or rare composition in members of the impactor family (Conca 1981). An additional possible influence on the darkening of ejecta is an influx of particles from the Jovian magnetosphere (Krupp et al. 2010, this issue; Burger et al. 2010, this issue; Cassidy et al. 2010, this issue).

Crater chains or *catenae* with some range in lengths and crater diameters occur in a number of localities. *Enki Catena* is the largest of such crater chains on Ganymede (Fig. 5). The length of the chain is approximately 160 km, diameters of the 13 craters in the chain range from about 10 to 20 km. The chain transects the bright/dark terrain boundary with a sharp increase in albedo on the bright terrain substrate that is even more visible in lower-resolution, high sun Voyager images (see Fig. 5).

The largest impact features on Ganymede and Callisto are multi-ring basins with a maximum spatial extension up to several thousands of kilometers (Smith et al. 1979a, 1979b; Passey and Shoemaker 1982). On Ganymede, two classes of basins are observable (Schenk et al. 2004): (1) basins reminiscent of lunar basins such as Orientale with a smaller number of rings, and (2) basins quite unlike those on the Moon, featuring numerous rings. On Ganymede, basins of the second class were heavily altered or obliterated by tectonic deformation (Passey and Shoemaker 1982; Prockter et al. 1998; Pappalardo et al. 2004; Schenk et al. 2004). Systems of arcuate, parallel furrows in the dark terrain originate from a basin-forming impact but are only partly preserved because of Ganymede's geologic activity in the past (e.g., Passey and Shoemaker 1982; Prockter et al. 1998; Pappalardo et al. 2004). Furrows are found only in the older, heavily cratered dark terrain of Ganymede. They are 5–10 km wide, several hundred kilometers long (Prockter et al. 1998), and show relief of up to 1 km (Figs. 1, 6).

The more lunar-like type of ring structures on Ganymede is represented by the *Gilgamesh* basin and is characterized by (a) a circular, more or less smooth area, (b) a smaller number of rings (<5), mostly furrows or scarps, (c) is much less degraded, and (d) shows a smaller frequency of superimposed impact craters, therefore represents a relatively young impact structure (Schenk et al. 2004). Gilgamesh is shown in Fig. 6, compared to the multi-ring basin Valhalla on Callisto (bottom).

2.3 Europa

Europa's surface is young by geological standards—only ~50 Ma (Schenk 2005a, 2005b)—and shows evidence of a dynamic history. Many of the landforms have likely communicated with the subsurface, and understanding their geology and topography is key to unraveling Europa's history and global evolution. Although data returned from the Voyager and Galileo spacecraft were limited, much has been learned about the styles of deformation and how they may have changed over time.

2.3.1 Ridges

The most morphologically common features on Europa are linear landforms, the majority of which consist of a central trough flanked by two raised ridges (known simply as a "ridge") (Fig. 7). Ridges are observed over all of Europa's visible surface history (e.g., Figueredo and Greeley 2000, 2004; Kattenhorn 2002). The formation of these features is not well understood, but they may yield information regarding the satellite's geophysical state, including the nature of its tidal processes and the presence and distribution of liquid water. Ridges appear to be morphological features that evolve from cracks caused primarily by diurnal stresses. The influence of diurnal stresses is most significant in the cycloidal shape of some ridges and cracks on Europa's surface, where each arc may result from changing stresses during one diurnal cycle (Hoppa et al. 1999). Exactly how ridge morphologies form from cracks is crucial to understanding their relationship to the subsurface, and whether they bring ocean material up to the surface (see review in Kattenhorn and Hurford 2009).

Europa's ridge features exist in a variety of morphological forms including raised-flank troughs, double ridges, and morphologically complex ridges that consist of a series of anastamosing and inosculating component ridges. It has been suggested that these features may evolve from the simpler forms into more complex configurations, with initial cracks transitioning into raised-flank troughs, which then transition into double ridges, and then into more morphologically complex features (Greenberg et al. 1998;

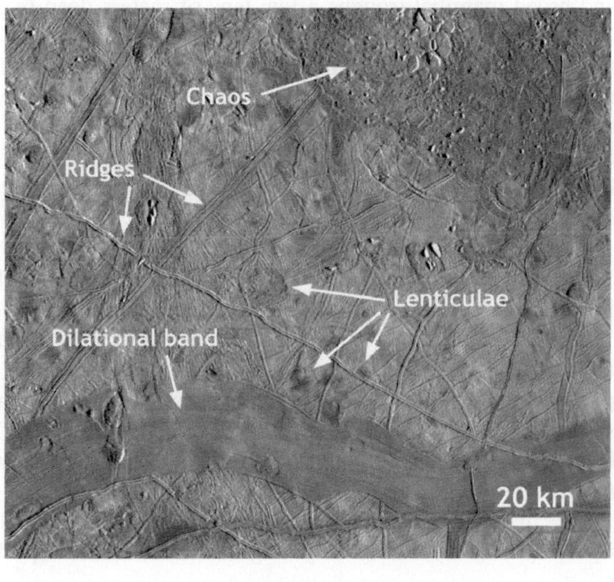

Fig. 7 Typical view of Europa's surface, showing a variety of features including ridges, bands, chaos and lenticulae. From Galileo observation E11REGMAP01 (NASA/JPL)

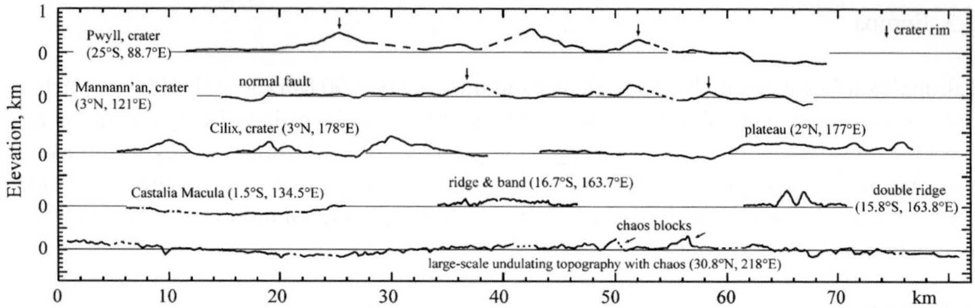

Fig. 8 Topography of landforms on Europa derived by stereo image analysis (Giese et al. 1998) (*dashed lines* are related to gaps in the terrain models). Used Galileo frames—horizontal/vertical resolution of topography: (i) Pwyll: s0383715500, s0383715504, s0426268700—(1500 m/25–50 m). (ii) Mannann'an: s0440948028, s0440948639, s0440948652, s0440948665, s0440948678, s0440948700—(500 m/15–30 m). (iii) Cilix: see Giese et al. (1999)—(500 m/15–30 m). (iv) Castalia Macula: s0440949628, s0383548800, s0383548814—(500 m/10 m). (v) ridge & band and double ridge: see Nimmo et al. (2003)—(150 m/5–10 m) and (400 m/15–30 m), respectively. (vi) chaos: see Nimmo and Giese (2005)—(200 m/5–10 m). Profiles' vertical exaggeration is 4

Head et al. 1999). Double ridges are by far the most common ridge type on Europa and appear to be a fundamental building block that has shaped the landscape we observe today. Ridges and troughs are observed on length scales of up to 1000s of kilometers and can range from linear to cycloidal in planform. Relief of 100 m (Hurford et al. 2005) and more (Nimmo et al. 2003) has been measured at ridges and double ridges. The ridge profiled in Fig. 8 has relief of ∼100 m, and the double ridge is 350 m high.

The most ubiquitous morphology is that of the "double ridge", characterized by the presence of an axial trough flanked by twinned ridge crests (Figs. 7, 8). Double ridges can range in width from ∼200 m to >4 km and have lengths from tens of kilometers to many hundreds of kilometers (Greeley et al. 2004). The morphological characteristics of double ridges are generally uniform over long distances and the ridges themselves can be remarkably straight for significant portions of their length. Some double ridges exhibit subtle and broad marginal troughs a few tens of meters deep (e.g., Pappalardo and Coon 1996; Tufts et al. 2000), and subparallel flanking fractures (presumably tensile) may be found near the outer reaches of these troughs. It has been suggested that these troughs and associated fractures result from either loading of the lithosphere (Tufts et al. 2000; Billings and Kattenhorn 2002) or uplift plausibly due to shear heating and intrusion along the underlying fault (Gaidos and Nimmo 2000). The most complex type of ridge feature consists of a collection of subparallel anastamosing and inosculating single or double ridges (see discussion in Prockter and Patterson 2009).

Some double ridges and ridge complexes are flanked by dark deposits (Lucchitta and Soderblom 1982; Belton et al. 1996) that are generally patchy and diffuse, and may extend for up to 10 km on either side of the feature with which they are associated (Belton et al. 1996). This dark material appears to drape over the surrounding terrain, and is therefore thought to be a surficial deposit (Geissler et al. 1998). In high-resolution images of one prominent ridge, Rhadamanthys Linea, preexisting structural features are clearly visible, suggesting that the deposit is relatively thin (Fagents et al. 2000). Such dark deposits may be continuous along the flanks of a ridge or spaced in discrete subcircular regions along the margins of a ridge (e.g., Rhadamanthys Linea).

The majority of Europa's surface is comprised of relatively old ridged structures so heavily overprinted by numerous younger features that it is prohibitively difficult try to reconstruct their history (Greeley et al. 2004). Ridged plains contain a tremendous diversity of interlaced and interwoven ridges exhibiting a wide range of azimuthal orientations. Fourier analysis of ridges within some of the highest resolution images obtained by the Galileo spacecraft allowed Patel et al. (1999) to recognize seven subtypes of "ridge and trough terrain" within the ridged plains, each with distinct morphological characteristics. These workers were able to derive the approximate spacing of the constituent ridges and troughs within each terrain type, finding a gradation among them that they interpreted to represent multiple formation processes and/or a morphological evolution. Crosscutting relationships and relative albedo suggest that ridged plains are the oldest mappable terrain on Europa (e.g., Greeley et al. 2004). Double ridges and ridge complexes crosscut and are superposed on ridged plains.

The distribution and relative albedo of deposits flanking ridges has been suggested to serve as an indicator of relative age (Belton et al. 1996; Geissler et al. 1998). Geissler et al. (1998) observed that dark flanking deposits are rarely associated with the youngest ridges and concluded that such deposits likely form only after ridge construction has ceased.

While several models have been proposed for ridge formation, it is not yet known whether they represent sites of exchange with the subsurface (Tobie et al. 2010, this issue).

2.3.2 Bands

Another fundamental morphological feature on Europa are termed bands (Fig. 7). The most common types of band are informally known as "pull-apart" bands, or dilational bands (Tufts et al. 2000). In general, the margins of this band type can be reconstructed extremely well (Schenk and McKinnon 1989; Pappalardo and Sullivan 1996; Sullivan et al. 1998) (Fig. 7), indicating that the interiors of pull-apart bands consist of subsurface material that has been emplaced at the surface of Europa. Almost all bands exhibit bilateral symmetry around a central trough oriented parallel to the band margins (Sullivan et al. 1998; Prockter et al. 2002). Opening across dilational bands is primarily extensional (i.e. perpendicular to the band's long axis), although there is commonly also a component of lateral motion resulting in oblique opening (Tufts et al. 2000; Prockter et al. 2002). Maximum band widths are about 30 km, suggesting that some geophysical process limits band formation. In one region of the subjovian hemisphere it has been estimated that over half of the surface is formed from bands alone (Prockter et al. 2002), indicating that the formation of these features clearly represents a significant process by which Europa's crust has been resurfaced (Schenk and McKinnon 1989; Pappalardo and Sullivan 1996).

Bands can be elevated by 100–150 m with respect to the surroundings (Prockter et al. 2002; Nimmo et al. 2003). A convex-shaped 6 km wide and 150 m high-standing band is profiled in Fig. 8. However, although bands tend to stand higher than the surrounding terrain and have distinct boundaries with it, there is no observed spillover of band material or any associated low albedo material. This implies that band material was emplaced in a solid-state rather than as a fluid (Prockter et al. 2002).

The basic process of dilational band formation appears to be consistent across Europa's surface, however, detailed band morphology differs. They are generally classified into two types, known as smooth bands and lineated bands (e.g. Greeley et al. 2000; Prockter et al. 2002; Greeley et al. 2004). Smooth bands appear to have a uniform texture at regional resolutions, but at high-resolution are predominantly comprised of small, closely spaced hummocks with little other original structure. Lineated bands, as their name suggests, exhibit internal lineations oriented subparallel to their central axes. These lineations may either be normal faults, or closely spaced linear ridges and troughs. Many lineated bands appear to be smooth at lower resolutions. Internal band morphology has lead to suggestions that Europa's bands may have formed in an analogous way to terrestrial mid-ocean ridges (Sullivan et al. 1998; Prockter et al. 2002), as they show similar features such as central troughs, small hummocks, and faulting further away from the spreading axis.

A separate band type, bright bands, have internal textures reminiscent of dilational bands, however, they defy reconstruction and have a higher relative albedo than the terrain they disrupt. Formation mechanisms relying on extensional, compressional, and/or strike-slip deformation have all been proposed to explain the unique characteristics of this type of band (see review in Kattenhorn and Hurford 2009).

Geologic mapping has shown that bands formed at a time that was intermediate in Europa's visible history; they postdate the ridged plains, but predate almost all chaos and lenticulae features (see next section) (Greeley et al. 2004; Prockter et al. 2002; Figueredo and Greeley 2004). This implies that they formed at a time when Europa's surface was more mobile, and was able to separate laterally relatively easily. This observation also suggests that Europa's style of surface deformation has changed through time from more lateral deformation (bands) to more vertical deformation (chaos).

2.3.3 Chaos and Lenticulae

Much of Europa's surface has been disrupted into isolated plates of preexisting material with lumpy matrix material between the plates (e.g., Greeley et al. 2004) (Fig. 9). This disrupted terrain covers 20–30% of Europa's surface (Collins and Nimmo 2009), and may take the form of either large disrupted regions (known as "chaos"), or smaller subcircular to elliptical pits, dark spots, domes and microchaos regions (collectively termed "lenticulae") which are ~10–15 km in diameter (Pappalardo et al. 1998). In almost all cases, chaos and lenticulae disrupt other feature types, including bands, and stand very high in the stratigraphic column.

Lenticulae domes can be convex with upwarped but unbroken margins where they meet the plains, while pits are topographically low areas where the surface has downwarped while preserving the preexisting terrain (Pappalardo et al. 1998) (Fig. 9). Many of these features are associated with dark plains material that embays surrounding valleys in the ridged terrain, so was probably relatively fluid when emplaced. Some pits were apparently flooded with dark plains material, forming dark subcircular spots on the surface. Lenticulae, known as "micro-chaos", typically consist of a fine-scale hummocky material, including embedded small plates of preexisting material, commonly with some associated dark plains material. These microchaos regions resemble the larger chaos terrains described below, but at a smaller scale.

Although a range of dome, pit, and spot sizes exists, there is a strong preferred diameter of between ~10 and 15 km (Spaun et al. 1999). This consistency in size and the range in their morphologies suggest that they are genetically related, an inference that is consistent with an origin from convective upwelling of buoyant ice diapirs within Europa's icy shell (see Tobie et al. 2010, this issue). The range of morphologies and levels of degradation that are observed in microchaos regions supports the suggestion that upwelling diapirs may partially melt pockets of briny ice as they rise to the surface, causing disaggregation into matrix material and local flooding by dark, low-viscosity melt.

Regions of chaos are areas in which kilometer-scale blocks of existing ridged plains material have translated and rotated with respect to one another within a mixed-albedo matrix of hummocky material (e.g., Greeley et al. 2004). Chaos regions tend to be defined by an

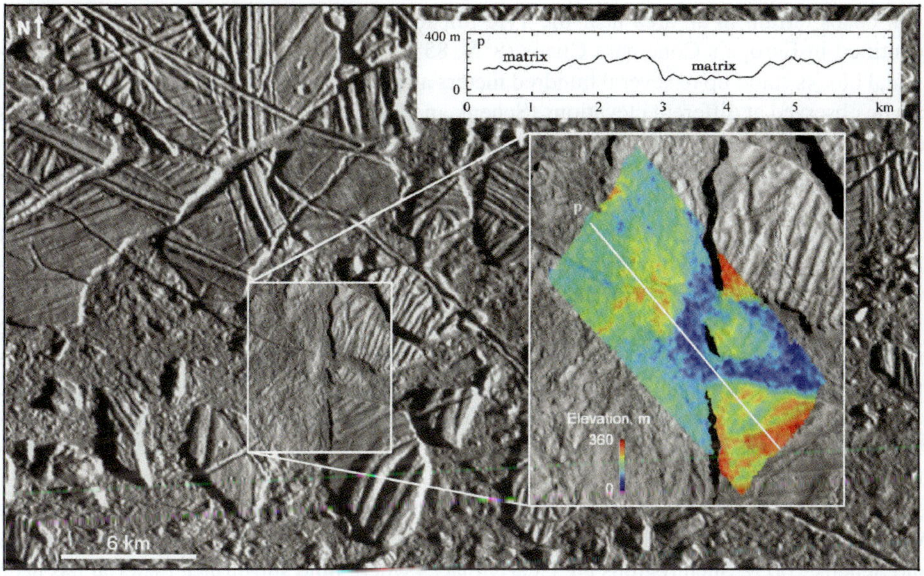

Fig. 9 Color-coded high-resolution (30 m horizontally, 15 m vertically) topographic model of a region in Europa's spectacular Conamara Chaos (8°N, 85°E). This is the best resolution model among all of the stereo models of Europa derived so far. The model features more than 200 m high-standing blocks surrounded by lower-lying matrix material. As compared with the blocks, the matrix material is rough at 200 m scales but smooth and at equipotential level over 1 km scale (profile). The matrix material distributes over two regions lying at different (~70 m) topographic level. The terrain model was derived from Galileo frames s0426271900 and s0426271900 (both 11 m/pxl) using methods of photogrammetry (Giese et al. 1998). Image in the background is a cut out of Galileo frame s0383717513 (60 m/pxl)

inward-facing scarp. The matrix material can be low-lying or high-standing relative to the surrounding plains. In one area, Conamara Chaos, at least 60% of the preexisting terrain has been replaced with or converted into matrix material, and some of the broken plates have been rotated and/or moved by as much as a several kilometers (Spaun et al. 1998). These plates can be reassembled like a jigsaw puzzle, reconstructing portions of preexisting ridges and troughs although much of the original surface has been destroyed.

The proportion of plates to matrix material varies within different chaos areas. Some, such as Conamara chaos, have distinct plates that can be partially reassembled. Others, such as Murias chaos, consist almost exclusively of matrix material at the resolutions available from Galileo images. Other chaos regions, such as Thrace Macula, are predominantly comprised of matrix material, but show faint, yet distinct traces of preexisting structure, as if the preexisting terrain had disaggregated in situ (e.g., Fagents 2003). Thrace also shows evidence of low albedo plains material around its margin, suggesting fluid flow, perhaps of brine-rich material.

In at least one case, Murias Chaos, the matrix material of the chaos region appears to have viscously flowed onto its surroundings (Figueredo et al. 2002). This flow is proposed to have loaded the surrounding crust, resulting in circumferential fractures. Similar troughs have been observed around other chaos regions, and are inferred to result from flexure (Williams and Greeley 1998).

Chaotic terrains and lenticulae are high-standing by 100–200 m above surrounding terrains (Schenk and Pappalardo 2004; Nimmo and Giese 2005). The bottom profile in Fig. 8 indicates long wavelength 50 km-scale topography with amplitude of 150–200 m associated with chaos terrain in an area near the Tyre impact crater. This is similar to domical areas identified in Europa's Conamara Chaos (8°N, 85°E) (Schenk and Pappalardo 2004). Individual blocks stand up to several hundred meters above the smoother matrix material, which can be observed at different elevations (Schenk and Pappalardo 2004) (Fig. 9). Chaos blocks are rotated and tilted (Fig. 8, last profile).

Chaos regions have been interpreted as places where Europa's heat flow has been enhanced, and where local melt-through of ocean water to the surface may have occurred, requiring that Europa has a very thin shell, less than ∼6 km; otherwise, the warm base of the ice shell would flow to maintain its thickness faster than the ice shell could melt from below. An alternative model for chaos formation has been proposed which is analogous as that for lenticulae, where ice diapirs have risen buoyantly through the ice crust, breaking or otherwise interacting with the surface (see discussion in Collins and Nimmo 2009). This mechanism would explain why some chaos areas stand several hundred meters above the surrounding plains, something that is hard to explain if they formed atop liquid water, but feasible if buoyant diapirs rose to their level of neutral buoyancy (see Tobie et al. 2010, this issue).

Smooth plains associated with chaos uplifts have also been observed, such as at Castalia Macula, a ∼30 km-diameter dark patch that lies in a depression up to 350 m deep (Prockter and Schenk 2005). Such regions are clearly places where fluid subsurface material has been emplaced onto the surface.

2.3.4 Europan Craters

Europa's surface is characterized by a comparably low density of impact craters, which implies a relatively young surface age (e.g., Greeley et al. 2004; Schenk et al. 2004). The number of catalogued craters larger than 1 km in diameter is on the order of 150 (Schenk et al. 2004). About 120 of these could be morphologically classified (Moore et al. 1998, 2001). Impact crater depths are strictly limited to values of ∼1 km (Schenk 2002), and larger craters with diameters of ∼20 km are about level with their surroundings (Moore et al. 2001). They possess rim topography and central peak massifs several hundred meters high (Fig. 8). Still larger 100 km craters appear as flat multi-ringed basins with relief <100 m (Moore et al. 2001).

The smallest impact features are simple, bowl-shaped craters. Simple craters have depth-to-diameter ratios slightly lower than, or equal to, those on Ganymede or Callisto, ranging from 0.17 to 0.2 (Schenk et al. 2004) (Fig. 10). In some localities, small craters appear in clusters or chains and are thought to be secondary craters (Bierhaus et al. 2005). The small craters shown in Fig. 10 might originate from the bright ray crater *Pwyll* and are superposed on ice plates in the Conamara Chaos region.

The simple-to-complex transition diameter on Europa is at ∼5 km, somewhat higher than on Ganymede or Callisto but still lower than on silicate surfaces (Schenk et al. 2004). Only 16 complex craters 3–27 km in diameter could be identified in Galileo SSI images (Moore et al. 1998, 2001). A typical example for complex craters is *Cilix*, 18 km in diameter (Figs. 8, 10). The crater somewhat resembles lunar craters and shows terraced walls, a flat floor, a central peak, and continuous ejecta which is mostly dark. A digital elevation model revealed that the flat floor is approximately at the same level as the terrain surrounding the crater, and that the central peak is about 300 m high, comparable to the rim height (Giese et al. 1999).

Pwyll is a complex crater about 24 km in diameter. This crater is characterized by an extensive bright ray system, extending to more than 1000 km from the impact center (e.g., Greeley et al. 2004). Secondary craters from *Pwyll* may be found at great distance (Fig. 10).

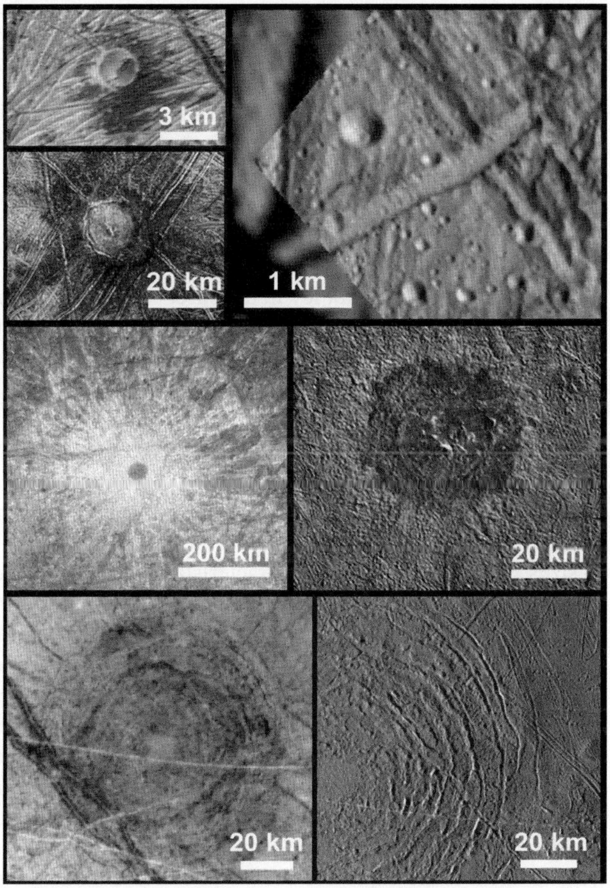

Fig. 10 Crater forms on Europa. *Top left*: simple crater *Mael Duin* (left, SSI target area 12ESWEDGES03) and complex crater *Cilix*, *top right*: cluster of possible *Pwyll* secondary craters in Conamara Chaos (12ESCHAOS_01); *middle left*: bright ray crater *Pwyll*, seen at high sun in lower resolution (E4ESGLOBAL01) and *middle right*: at lower sun in higher resolution (12ESPWYCOL01); *bottom left*: multi-ring structure *Tyre*, at high sun and low resolution (G7ESTYRCOL01) and *bottom right*: at low sun in higher resolution (14ESTYREHR01). *Rectangles* show location of detailed image of *Pwyll* and *Tyre*. North is pointing up in all panels

A digital elevation model was used to show that the crater is very shallow and that the height of the central peak complex is on the order of 100 to 300 m (Giese et al. 1999). Generally, craters on Europa larger than ∼10 km are much shallower than craters of the same size range on Ganymede or Callisto, and crater depth decreases as diameter increases (Moore et al. 2001; Schenk et al. 2004).

Only three craters on Europa are believed to have central domes, such as *Tegid*, a 30-km crater with a distinct ejecta pedestal. Low image resolution and incomplete image coverage impede more complete characterization of these features and identification of others.

The transition from complex craters to multi-ring structures begins at crater diameters larger than ∼30 km on Europa, one order of magnitude lower than on Ganymede or Callisto (Schenk 2002). *Tyre*, shown in Fig. 10 (bottom), and *Callanish* are characterized by similar structural zones such as (a) central plains, (b) concentric, densely spaced ridges and/or troughs, and (c) pits and chains of small craters radial to the center of both impact features which represent secondary crater clusters (Moore et al. 1998, 2001). These secondary crater chains unequivocally confirmed an impact origin of *Tyre* and *Callanish*. Turtle et al. (1999) estimated the nominal crater diameter of these two impact structures at 40–45 km diameter. No craters larger than *Tyre* and *Callanish* have been identified on Europa's surface.

Crater morphologies on Europa reflect the subsurface structure of Europa's icy crust (Schenk and Turtle 2009). Craters such as *Pwyll* formed in a solid icy crust (Turtle and

Pierazzo 2001). The multi-ring structures *Callanish* and *Tyre*, which are stratigraphically older, formed in an icy crust of low viscosity, giving way to ductile flow at shallow depths, with the brittle-to-ductile transition being at depths of ~5–10 km at the time of formation (Moore et al. 2001; Greeley et al. 2004). Whether a liquid water layer was present at a depth of ~10 km at the time when *Tyre* and *Callanish* were formed (Moore et al. 2001; Schenk et al. 2004) is still unknown.

2.4 Io

Although Io's surface is not icy in the strict sense (as it has no water ice), we include it here because of the importance of its exchange processes: Io is the most active satellite known, where material is constantly exchanged from the interior to the surface and out into the Jovian system.

2.4.1 Io Volcanism

The surface of Io (Fig. 11) is dominated by volcanic features and colorful sulfur deposits (see reviews by Lopes and Williams 2005; Lopes 2006). Io's volcanoes cover a wide range of sizes and present varying characteristics such as volcanic style, power output, persistency of activity, and association with plumes. Most of Io's volcanic features are caldera-like volcano-tectonic depressions (paterae) containing lava lakes or flows (Lopes et al. 2004), but there are also small dome and shield type volcanoes and some remarkably large lava flows. The lava flow field from Amirani is ~300 km long, the largest active flow field known in the Solar System. Elsewhere, Io's surface appears smooth except for scarps that cut across the plains, some of which are linear and likely tectonic in origin, while others are irregular and appear to be erosional, sometimes forming a series of mesas or large plateaus. The nature of some of the erosional mechanisms on Io is somewhat puzzling because of the lack of a significant atmosphere or flow of liquid water. McCauley et al. (1979) suggested that the eroding agent is the explosive escape of sulfur dioxide from a subterranean SO_2 "aquifer". The idea of subterranean SO_2 reservoirs gained some support after the analysis of Galileo data of the movement of the Prometheus plume (Kieffer et al. 2000).

Io's surface has four primary color units: red, yellow, white-gray, and black. The red materials are interpreted to be composed of short-chain sulfur molecules (S_3, S_4) resulting

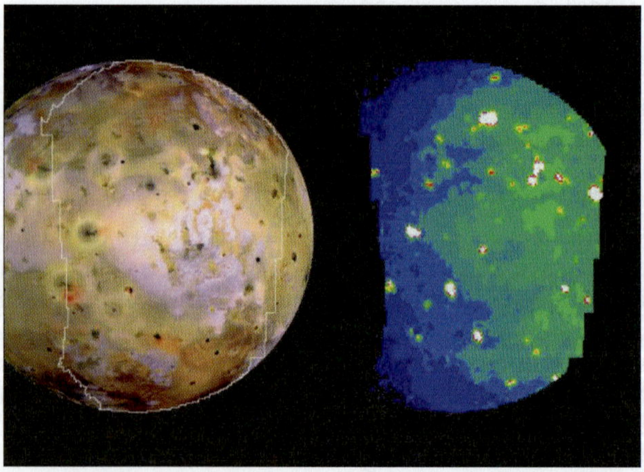

Fig. 11 Io in the visible and infrared. *Left*: Io's colorful surface is shown in this image by Galileo's camera (SSI). Dark features and bright red areas represent active or recently active hot spots. *Right*: Galileo NIMS false-color image of the same view of Io at 4.7 um. Hottest areas are shown in white

from either breakdown of cyclo-S_8 by charged particle irradiation (in polar regions), or by condensation from S_2-rich volcanic gases in the plumes of active vents (in the equatorial rings). The yellow materials (40% of surface) are interpreted to be the sulfur allotrope cyclo-S_8 (Geissler et al. 1999), with or without a covering of SO_2 frosts deposited by plumes, or alternatively polysulfur oxide and S_2O without large quantities of elemental sulfur (Hapke and Graham 1989), that become visible under high-phase illumination (Geissler et al. 2001).

Greenish-yellow patches exist in or near active vents, and are thought to be composed of either sulfur compounds contaminated by iron, or silicates such as olivine or pyroxene with or without sulfur-bearing contaminants. White (to gray) materials (27% of surface) are interpreted to be composed of coarse- to moderate-grained SO_2 resulting from plume fallout and subsequent recrystallization. Black areas (1.4% of surface) mostly correlate with active hotspots and occur as patera floors, lava flow fields, or as dark diffuse materials near or surrounding active vents. These materials are most consistent with Mg-rich orthopyroxene, indicative of silicate lava flows or lava lakes (within paterae) or diffuse silicate pyroclastic deposits near paterae, of mafic to ultramafic composition. Although initial temperature measurements of Io's lavas suggested ultramafic compositions, more recent work (Kezthelyi et al. 2007) argues that measurements of Io's hottest lavas can be explained by superheated basalts. However, interpreting color and color changes on Io's surface is a complex problem, due to variations in terms of image geometry and photometry (Simonelli et al. 1997; Geissler et al. 1999; Geissler 2000). The correlation between color variations and distribution of SO_2 is not straightforward, as shown by the analysis of Galileo NIMS high spatial resolution data (e.g. Lopes-Gautier et al. 2000). Most of Io's surface is covered by SO_2 frost, with the exception of the vicinity of hot spots (e.g. Carlson et al. 1997; Douté et al. 2001). The global distribution of SO_2 on Io's surface is a result of volcanic venting of gaseous SO_2 into the plumes and atmosphere. Other compounds have not been positively identified on Io. For a review of Io's surface composition, see Carlson et al. (2007).

A major question after Voyager was the nature of Io's volcanism—whether sulfur or silicates were predominant. Although temperature measurements from Galileo clearly showed that many hot spots have temperatures far too high for sulfur, the possibility that some sulfur flows occur on the surface cannot be ruled out. The colorful flows around Ra Patera that were argued by Sagan (1979) to be sulfur had been covered over by new eruptions before Galileo's first observations of the area in 1996 (McEwen et al. 1998a). However, other locations may have sulfur flows, such as Emakong patera (Williams et al. 2001) where measured temperatures have been consistently low (Lopes et al. 2001, 2004). An active sulfur eruption may also have been detected by NIMS on the bright flow field Tsui Goab Fluctus in the Culann-Tohil region late in the Galileo Mission (Williams et al. 2004). However, Galileo instruments could not distinguish between sulfur flows or cooled silicates coated by bright sulfurous materials after erupting. One possibility suggested by Greeley et al. (1984) is that rising silicate magma may melt sulfur-rich country rock as it nears the surface, producing 'secondary' sulfur flows (as opposed to 'primary' flows that originate from molten magmas at depth). Another possible magma type on Io is SO_2. High spatial resolution NIMS data provided the opportunity to map SO_2 distribution at local scales (Lopes-Gautier et al. 2000; Douté et al. 2002, 2004). A nearly pure SO_2 region (corresponding to a 90% to 100% concentration of SO_2, Lopes et al. 2001) was found to be topographically confined within the Baldur caldera (see Fig. 12). Smythe et al. (2000) suggested emplacement by liquid flow of SO_2 rising from the subsurface. While liquid would normally boil off when exposed to Io's tenuous atmosphere, estimates by Smythe et al. (2000) suggest that, given sufficiently large quantities, some of the liquid could freeze to form a layer of sulfur dioxide ice inside the

Fig. 12 Io's surface shows a variety of different volcanic features. Insets in this SSI image show, from left, Chaac patera and Baldur patera; Ababinili patera; and Camaxtli patera. Chaac patera shows green deposits on its floor, white SO_2 deposits on the southwest margin, and an active flow (dark) in its southern tip. Next to it is Baldur patera, a nearly pure SO_2 region that is topographically confined and may have formed as a flow. Ababinili patera shows white and orange deposits, probably due to sulfur and SO_2 ice or frost. Camaxtli was observed by Galileo NIMS as an active hot spot, the dark diffuse deposits may be pyroclastics. (*Image montage courtesy of David Williams, Arizona State University; original Galileo image processing by Alfred McEwen, University of Arizona*)

caldera. Another candidate location for an SO_2 flow field, on the southwest part of Tohil Patera, was discussed by Williams et al. (2004). These SO_2 "magmas" are likely coming from subterranean SO_2 "aquifers" or reservoirs, such as discussed by McCauley et al. (1979) and Kieffer et al. (2000).

2.4.2 Io Tectonics

Three percent of Io's surface consists of mountains that are likely tectonic in origin (McEwen et al. 2004; Turtle et al. 2007) and are some of the highest in the solar system. Mountains typically measure tens of kilometers across and may reach heights of ~17 km, although they are ~6 km tall on average. The towering mountains suggest that the lithosphere must be at least a few tens of kilometers thick (Keszthelyi et al. 2004; Jaeger et al. 2003). Most mountains appear to be tectonic in origin: uplifted and sometimes tilted blocks that are bound by steep scarps and, in many cases, fractured. Evidence of mass wasting of loosely consolidated surface materials is found in the form of debris aprons, which surround many mountains. Mountains are somewhat more abundant in two antipodal zones on the satellite that are globally anti-correlated with peaks in the distribution of volcanic centers (Schenk et al. 2001). Intriguingly, at the local scale, individual mountains and paterae are commonly associated, and some mountains show evidence of rifting, perhaps related to uplift (Jaeger et al. 2003).

Although the mountains, in general, are not volcanic structures, they, nonetheless, appear to owe their origins to Io's extreme level of volcanic activity. The rapid resurfacing rate results in a comparable subsidence rate, generating excessive compressional stresses in the lithosphere, which are believed to drive mountain building through large-scale thrust-faulting of crustal blocks (Schenk and Bulmer 1998). Mountain formation may be localized by concentrated stress, such as would result from mantle diapirism (Turtle et al. 2001; Jaeger et al. 2003). This scenario could also explain the local correlation between mountains and volcanic activity, as the release of the compressive stresses may allow the thrust fault to act as conduits by which magma can reach the surface (Turtle et al. 2001; Jaeger et al. 2003). The anti-correlation of mountains and volcanic centers observed globally may be illustrative of the relative roles of stresses due to subsidence and thermal stresses related to long-term, large-scale variations in levels of volcanic activity (McKinnon et al. 2001; Kirchoff and McKinnon 2009). Although their formation mechanisms are not well understood, Io's mountains clearly result from a dynamic environment and likely expose subsurface material.

2.5 Jupiter's Small Satellites

Jupiter's system of satellites includes the four large Galilean satellites, known for 400 years, and more than 50 much smaller objects, mostly discovered in the last decade, but first sampled with the discovery of Amalthea in 1892. The four inner satellites, Metis, Adrastea, Amalthea, and Thebe (Table 1), are regular satellites orbiting at 1.79 to 3.11 Jupiter radii that have been resolved by spacecraft imaging, and are intimately connected to the production and dynamics of the Jovian ring materials (Burns et al. 1999). The outer satellites, with both prograde and retrograde members, orbit as far as 400 Jupiter radii. These objects are mostly a few km in size, inferred from disk-integrated photometry, and likely represent captured objects (Nicholson et al. 2008).

The small inner satellites are dark (albedos <0.1), reddish, and display brighter leading sides. The albedo asymmetry on objects on both sides of the synchronous orbit distance suggests external, non-magnetospheric influence, possibly higher leading side hyper-velocity impact rates (Simonelli et al. 2000). Amalthea has brighter ridges and spots on gravitationally high regions, a relation that suggests some efficacy of downslope motion of regolith. The reddish colors suggest a role for contamination from Io, but this is unlikely to be a simple additive process due to the high-velocity impact regime and magnetospheric charged particle populations (Simonelli et al. 2000). Amalthea and Thebe are resolved sufficiently well to observe large crater densities, consistent with very old, equilibrium densities on other objects (Thomas et al. 1998a).

Table 1 Jupiter satellite sizes. Sizes and densities Metis-Thebe from Thomas et al. (1998a); Io: Thomas et al. (1998b); Europa: Nimmo et al. (2007); Ganymede: Anderson et al. (2001a); Callisto: Anderson et al. (2001b)

Satellite	Mean radius, km	ρ, kg m^{-3}
Metis	21.5	
Adrastea	8.2	
Amalthea	83.5	860 ± 100
Thebe	49.3	
Io	1821.5	3528 ± 3
Europa	1560.8 ± 0.3	3013 ± 4
Ganymede	2631.2 ± 1.7	1942 ± 5
Callisto	2410.3 ± 1.5	1834 ± 3

The Galileo spacecraft made a close pass by Amalthea, and provided a mass from radio tracking data. Combined with the volume derived from Galileo and Voyager images, the mean density of Amalthea is $860 \pm 100\ \mathrm{kg\,m^{-3}}$ (Anderson et al. 2005). This low density suggests either an extremely porous rocky body, or one with a substantial fraction of ice. If it has a substantial amount of ice, then it probably formed much farther from Jupiter than its current orbit (Anderson et al. 2005), which poses interesting dynamical problems. The presence of hydrous minerals (Takato et al. 2004) also suggests formation farther from Jupiter than the current orbit.

3 The Saturnian Satellites

3.1 Titan

Titan is the only satellite in the solar system with a substantial atmosphere, enabling interactions between the surface and atmosphere not seen on other satellites. Titan has a very young surface revealed in detail for the first time by the Cassini spacecraft (radar, visible, and infrared imaging data, acquired by the RADAR, Imaging Science Subsystem (ISS), and Visual and Infrared Mapping Spectrometer (VIMS) instruments, respectively) and the Huygens probe. Optical images from orbit show light and dark regions that generally correlate to radar-dark and radar-bright areas and also light and dark areas in the infrared. The dark, mostly equatorial regions correspond to vast aeolian deposits while the brighter regions correspond well to mountainous and hummocky terrain, including Xanadu, the largest of the bright areas on Titan, originally identified from IR ground-based images and centered near the equator at longitudes 80–140 W.

Titan's geology is, in many ways, remarkably like the Earth's: all major geologic processes appear to be represented, including impact cratering, erosion by wind and liquids, tectonism, and volcanism. Titan's surface has rivers, lakes, and seas of liquid methane, vast fields of dunes, and suffers the effects of wind and rain. Titan's methane cycle is similar to the hydrological cycle on Earth. This methane cycle makes Titan unique among the moons of the outer solar system.

3.1.1 Aeolian Features on Titan

Cassini's initial optical observations in summer 2004 (Porco et al. 2005b) of Titan showed 'streaky' boundaries between light and dark terrain: in particular the eastern boundaries of several features were noted to be diffuse, while the western boundaries were sharp, suggesting possible eastward surface transport, interpreted to be the result of aeolian activity.

RADAR observations in February 2005 (flyby T3), found many radar-dark subparallel features, nicknamed 'cat scratches' which were interpreted (Elachi et al. 2006) as being possibly aeolian in origin, although it was impossible to preclude other processes (Fig. 13). In October of that year, the near-equatorial T8 flyby made the interpretation clear (Lorenz et al. 2006a). First, this flyby covered terrain that was in many places completely covered in the sand material—while on flyby T3, the cat scratches were seen as some dark material deposited above a brighter substrate, on T8 the topographic shape of the sand surface was the major control on the radar appearance. Second, the equatorial ground track of the spacecraft meant that the dunes, oriented E-W, were illuminated broadside-on, giving many opportunities to detect topographic glints. These showed clearly the nature of the dunes as positive relief. Finally, in this dark region the dunes themselves seem to have been larger

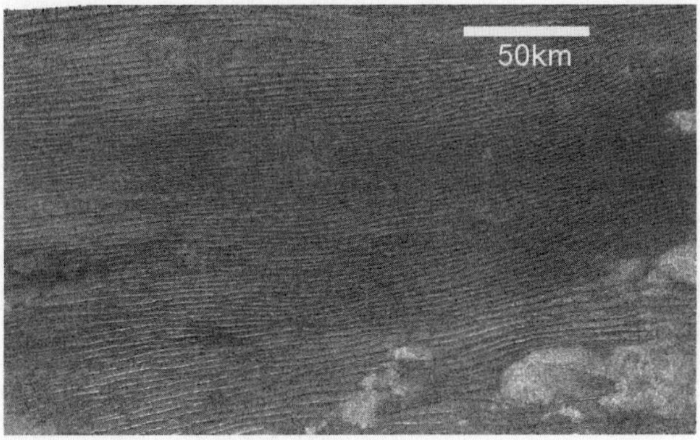

Fig. 13 Section of the T8 swath in Belet (radar illumination from North/*top of image*). Although the dunes are seen as dark and overlying brighter material at lower left, the sharp uprange glints seen throughout, and particularly prominent at lower left, show the dunes to have positive relief of 100–150 m

than in T3. A couple of the dunes observed on T8 were also detected by the near infrared descent imager on the Huygens probe (Lunine et al. 2008).

Radebaugh et al. (2008) present further investigation of the dunes with RADAR data, counting several thousand individual dunes. Some network dunes are noted, suggesting complex wind regimes in specific locations, but the dunes are predominantly linear. A map of dunes observed in the entire prime mission dataset has recently been published (Lorenz and Radebaugh 2009). Dunes have also been detected in a long, thin swath of data acquired by VIMS on the T20 flyby (Barnes et al. 2008). These observations show the interdune areas at this location (~50 W, 10 N in Fensal—see Fig. 14) where dunes had been already identified in T17 RADAR data. The dune material appears spectroscopically to contain less water ice than other units on Titan, suggesting an organic composition. Lorenz and Radebaugh (2009) measured a typical dune spacing of 2 km, and used radarclinometry to determine a dune height in this area of 30–70 m.

The dunes are typically aligned close to the E-W direction, and while in some cases the direction along the dune axis is ambiguous, the overall pattern is overwhelmingly one of net sand transport from W to E. Dunes terminate abruptly at the western edge of obstacles, and pick up gradually thereafter (e.g. Radebaugh et al. 2009). Some obstacles have 'tails' in this downstream direction, which may correlate with the ISS observations of faculae with sharp western margins and diffuse eastern margins (Porco et al. 2005b).

3.1.2 Dune Formation and Composition

In fact, although it now appears that the fraction of Titan's surface covered by dunes is larger than any terrestrial planet, dunes were not expected on Titan (Lorenz et al. 2006a). First, expectations of thermally driven winds near Titan's surface were extremely low, due to the low solar flux, the large column mass of the atmosphere and the small radius of Titan. This kind of energy-flux argument, which correctly predicts windspeeds of a few meters per second on Earth, suggests windspeeds on Titan of only about 1 cm/s. In contrast, despite Titan's thick atmosphere and low gravity, which both favor transport of material by wind, the optimum threshold windspeeds required to move sand (assuming interparticle cohesion similar to terrestrial materials, the easiest particle size to move is 250 microns in diameter) are of the order of 0.5–1 m/s.

The second issue is of sand supply—erosive processes (Lorenz et al. 1995) were expected to be weak due to the weak driving sunlight, slow-falling rain and small temperature variations (no freeze-thaw). Further, the expected presence of liquid hydrocarbons on Titan's

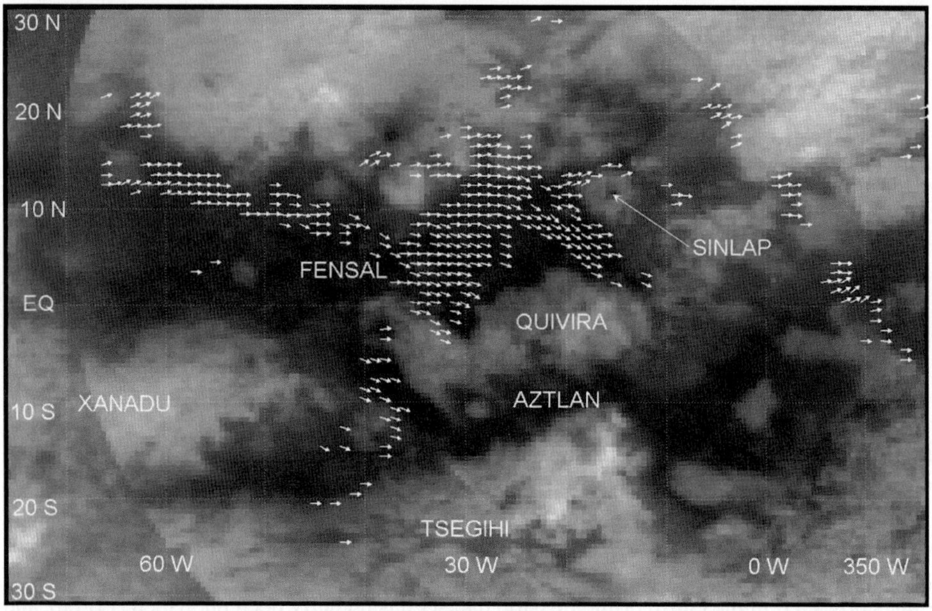

Fig. 14 A map of RADAR -determined dune directions overlain on an ISS basemap (Lorenz and Radebaugh 2009). The RADAR coverage is incomplete, but dunes have been detected on essentially all of the optically dark low-latitude terrain that has been observed so far. Note the generally eastwards transport direction and the diversion of the dunes around the Sinlap ejecta blanket

surface might act as traps for any sand that was formed (for example sand-sized particles in ejecta from impact craters—this is believed to be the principal source of sand on Venus).

Indeed, while fluvial erosion and impact ejecta seem not to be able to provide the large volume of material necessary to form the dunes (Lorenz et al. 2008c have estimated the total sand volume to be between 200 and 800 thousand cubic km of material) a photochemical origin can. Further, the optically dark appearance of the material, and its spectral characteristics, support an organic composition, suggesting that the sand formed from atmospheric haze particles. Conversion of <1 micron haze particles into 250 micron sand grains could occur by sintering over long timescales, or perhaps more likely it may involve cycles of wetting and drying in Titan's lakes.

The distribution of dunes, confined to latitudes equatorward of 30 degrees N and S, seems consistent with meteorological models (e.g. Mitchell 2008) that suggest Titan's tropics should progressively dry out. Second, the predominance of the longitudinal (linear) dune form requires a modestly changing (typically bidirectional) wind regime (e.g. Lancaster 1995). Sources of such a variation include seasonal change (the usual reason for this wind regime on Earth) and possibly the gravitational tide in the atmosphere. Tokano (2008) has explored the near-surface winds in a global circulation model (GCM) and finds that indeed surface winds should occur and not infrequently exceed the saltation threshold of 0.5–1 m/s. (It was noted in Lorenz 1996 that average windspeeds may be of only limited utility in assessing sand transport processes, in that winds above the transport threshold may in fact be quite rare.)

Titan's dunes almost exclusively indicate eastwards sand transport, which is a challenge to atmospheric circulation models which all suggest a predominance of westwards flow at low latitudes. There are regional deviations of up to 45° from east-west (see Fig. 14), typically guided by bright features which appear to present obstacles that range from only a few km across to the 2500-km-scale of Xanadu.

3.1.3 Titan Lakes and Seas

Long before Cassini's investigations of Titan began, it had been predicted that Titan would have lakes or even seas of liquid hydrocarbons on its surface (e.g., Lunine et al. 1983; Lunine 1993). Such reservoirs would be necessary to resupply Titan's atmospheric methane, which would otherwise be broken down over relatively short timescales, $\sim 10^7$–10^8 yrs (e.g., Yung et al. 1984), by photochemical processes that convert the methane into more complex hydrocarbons, creating Titan's thick atmospheric haze and accumulating on the surface in significant quantities over Titan's history (e.g., Strobel 1982; Lorenz and Lunine 2005). Furthermore, clouds inferred to be composed of methane and ethane are frequently detected (e.g., Griffith et al. 1998, 2005, 2006; Porco et al. 2005b; Schaller et al. 2006a, 2006b; Turtle et al. 2009), consistent with an active methane cycle comparable to the hydrologic cycle on Earth.

Early Earth-based observations of Titan's surface showed albedo patterns (e.g., Smith et al. 1996) that were generally interpreted to represent dark hydrocarbon liquids occupying topographically low regions between brighter exposures of presumably water-ice bedrock (e.g., Lorenz and Lunine 2005; Griffith et al. 2003). Cassini-Huygens has revealed Titan's surface in much greater detail, although the compositions of the materials responsible for Titan's albedo variations are complex and still not fully understood even after several years of Cassini observations (Soderblom et al. 2007b). Nonetheless, morphologic interpretations of observations by Cassini's ISS, VIMS, and RADAR experiments and by Huygens' Descent Imaging Spectral Radiometer (DISR), have confirmed that darker regions consist of accumulations of liquid and solid hydrocarbons (e.g., Porco et al. 2005b; Elachi et al. 2005, 2006; Tomasko et al. 2005; Lorenz et al. 2006a; Barnes et al. 2007a; Soderblom et al. 2007a, 2007b; Turtle et al. 2009).

Titan has given away secrets of her hidden surface only gradually. At first the only evidence of surface liquids consisted of widespread and complex systems of channels as well as the rounded cobbles in the plain upon which Huygens landed (e.g., Porco et al. 2005b; Tomasko et al. 2005; Soderblom et al. 2007a; Lorenz et al. 2008a). The frequent occurrence of tropospheric clouds, often exhibiting convective behavior consistent with precipitation, demonstrated a potential source of liquid to carve the observed channels (e.g., Griffith et al. 1998, 2005, 2006; Porco et al. 2005b; Schaller et al. 2006a, 2006b; Turtle et al. 2009). The earliest polar observations were ISS images of the South Pole, which, between the clouds, revealed numerous dark surface features, the largest of which had an outline reminiscent of terrestrial lakes (McEwen et al. 2005), prompting the name Ontario Lacus (Fig. 15). However, the low-resolution, single-wavelength (938-nm) albedo variations did not definitively prove that these dark features were liquid.

It was not until the first polar synthetic aperture radar (SAR) observation (halfway through Cassini's nominal mission) that there was compelling evidence of hydrocarbon liquid on Titan's surface: radar backscatter and radiometric brightness temperatures indicative of exceedingly smooth low-dielectric material such as liquid methane, likely mixed with liquid ethane (e.g., Stofan et al. 2007; Lopes et al. 2007). Cassini RADAR revealed myriad lakes at high northern latitudes (Fig. 15) (e.g., Stofan et al. 2007; Lopes et al. 2007; Hayes et al. 2008). There is significant variation in size and morphology. The former ranges from <1 km^2 to a sea, Kraken Mare, in excess of 400,000 km^2, imaged in its entirety by ISS (Turtle et al. 2009). Morphologically, the lakes are extremely diverse (e.g., Stofan et al. 2007;

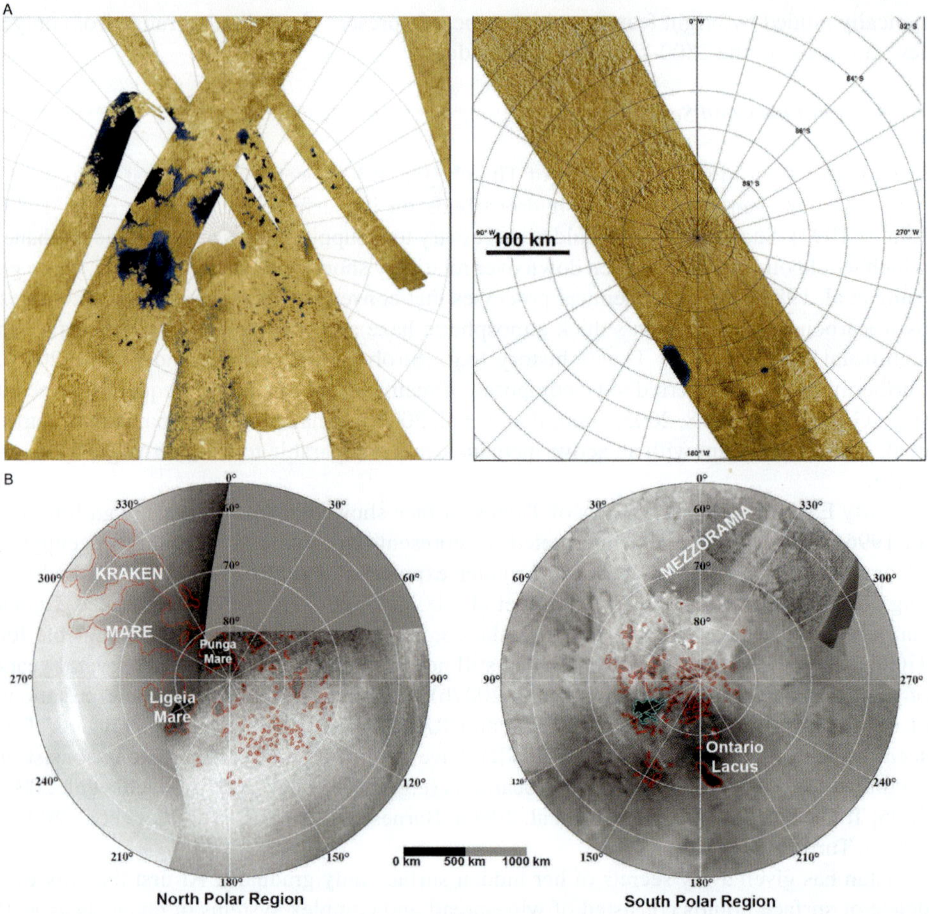

Fig. 15 Cassini RADAR and ISS polar views of Titan. (**A**) False-color RADAR data of Titan's north (*left*) and south (*right*) poles. Areas with radar signals indicative of liquid are blue-black, and radar-bright, presumably solid, surfaces are colored tan. Grid spacing is 10° in longitude for both, but 10° in latitude for the North Pole and only 1° for the South Pole. (**B**) ISS near-infrared mosaics of Titan's North and South Poles (*left* and *right*, respectively). Locations of dark features thought to be lakes or lakebeds are outlined in red, and the blue outline indicates the feature that changed between two ISS observations, interpreted to be the result of flooding after a large storm (Turtle et al. 2009)

Lopes et al. 2007; Hayes et al. 2008): some exhibit rough shorelines with channels leading into them, others resemble flooded river valleys, and still others exist within depressions that have smooth, rounded, steep sides. The first two are consistent with surface runoff, while the last is suggestive of interaction with an underground methane table. In addition to morphologic differences, the lakes vary in radar backscatter from very low, indicative of being liquid-filled, to backscatter as high as the surrounding terrain, interpreted as currently dry lakebeds. Intermediate values of backscatter are seen as well. The distribution of these different types of lakes varies with latitude, with the dark lakes occurring primarily above 65° N latitude and the intermediate and bright cases occurring south of 77° N latitude (Hayes et al. 2008). Intriguingly, although RADAR could detect roughening of the surface by waves, all of the observations of liquid-filled lakes to date indicate they are smooth (Lorenz et al.

2010). This situation may change as seasonal changes bring more dynamic weather to the locations of the lakes.

Further evidence for liquid comes from spectra acquired by VIMS of Ontario Lacus, which are consistent with the presence of liquid ethane (Brown et al. 2008). The VIMS images also reveal details of the shoreline, showing two annuli following the contours of the lake (a bright annulus is also visible in the lower-resolution ISS observations McEwen et al. 2005; Turtle et al. 2009), which Barnes et al. (2009) interpret as evidence of changes in the level of Ontario Lacus over time.

Another line of evidence that could confirm the presence of liquids on Titan's surface would be detection of specular reflections, which indicate smoothness at the scale of the wavelength of the observation. Geometric constraints limit Earth-based and ISS opportunities to observe the specular point to low latitudes, where no such enhancement has been detected at visible or infrared wavelengths (West et al. 2005; Fussner 2006). However, VIMS has recently detected a specular reflection from Kraken Mare at 73°N, 335°W, confirming the presence of material that is smooth at 5 μm (Stephan et al. 2010).

Although ISS images do not discriminate between solid and liquid hydrocarbons, they do provide almost global coverage (Turtle et al. 2009), allowing the interpretations of RADAR and VIMS high-resolution observations to be extrapolated more widely, e.g., revealing the full extent of Kraken Mare (Fig. 15). At high northern latitudes, there is generally good agreement between features that are radar-dark and those that ISS sees as dark in the near-infrared. Intriguingly, only a few of the dark features identified by ISS at high southern latitudes also appear to be radar-dark (Fig. 15). One possibility for these differences is that some of the features ISS sees are dry lakebeds, similar to the bright lakes seen in the northern hemisphere (e.g., Hayes et al. 2008). Another is that changes have occurred on the surface; model precipitation (Tokano et al. 2001; Hueso and Sanchez-Lavega 2006; Barth and Rafkin 2007) and evaporation (Mitri et al. 2007) rates are consistent with such changes given the time that elapsed between the ISS and RADAR observations. Indeed, changes have been documented between repeated observations of the South Pole. An ISS observation in June 2005 showed a large dark area (Fig. 15) that was not present in an observation made by ISS in July 2004. Together with detection of an especially large cloud system in the interim (Schaller et al. 2006a), these observations have been interpreted by Turtle et al. (2009) to represent surface change as a result of flooding by methane rain. More recent observations of Ontario Lacus by both ISS and RADAR suggest shoreline retreat of more than several kilometers in some areas (Turtle et al. 2009; Hayes et al. 2008).

All of the lakes and seas identified to date occur at high latitudes ($\leq \sim 55°$ latitude, e.g., Stofan et al. 2007; Lopes et al. 2007; Hayes et al. 2008; Brown et al. 2008; Turtle et al. 2009). Although channels have been observed globally (e.g., Porco et al. 2005b; Tomasko et al. 2005; Lorenz et al. 2008a) and the Huygens Gas Chromatograph Mass Spectrometer (GCMS) detected liquid methane in the very near subsurface at low latitudes (Niemann et al. 2005; Lorenz et al. 2006b), the long-observed equatorial dark areas appear to be covered by extensive dunes composed of particulate hydrocarbon material (Lorenz et al. 2006a), as discussed previously. This distribution is likely the result of Titan's current climate favoring the accumulation of liquids at the poles. So far, Cassini's observations only extend through Titan's southern autumnal equinox—it will be interesting to follow seasonal changes in weather patterns, especially precipitation at lower latitudes as northern spring progresses. An important constraint implied by the distribution of Titan's lakes and seas is that the amount of liquid they represent does not appear to be sufficient to keep Titan's atmospheric methane resupplied over geologic timescales, suggesting the existence of subsurface or deep interior reservoirs (Lorenz et al. 2008b).

3.1.4 Cryovolcanic Features

The existence of cryovolcanism on Titan was predicted before Cassini (see review in Lopes et al. 2007). Lorenz and Lunine (1996) predicted that cryovolcanic features would be effusive rather than explosive, because the low solubility of methane in water or a water-ammonia solution would make volatile-rich cryomagmas unlikely. Therefore, any flows and domes would be more likely to be present than cinder cones or stratovolcanoes.

Both the RADAR and VIMS instruments have imaged features that have been interpreted to be cryovolcanic, including flow features and caldera-like depressions. Interpretation of edifices that appeared morphologically similar to cryovolcanic domes has been more problematic. Two such features initially suggested to be cryovolcanic, in light of later data, may be tectonic in origin. These are Tortola Facula (Sotin et al. 2005) and Ganesa Macula (Elachi et al. 2005; Lopes et al. 2007). RADAR images of Tortola Facula showed it to be more likely a mountainous or hummocky region, while topography derived from RADAR data (Kirk et al. 2009; Stiles et al. 2009) showed that Ganesa is not a dome as originally inferred. Ganesa's origins are unclear at this point, it may be cryovolcanic or tectonic in origin and has likely suffered considerable erosion and modification. Moore and Pappalardo (2010) go further, suggesting that there is currently no compelling evidence for any cryovolcanic features on Titan.

The features on Titan that to date have been interpreted as cryovolcanic in origin are flow-like features imaged by RADAR and VIMS, several caldera-like depressions and one structure that may be a volcanic cone. The inferred caldera-like features and cones are all associated with lobate deposits that have been interpreted as flow fields (e.g. Lopes et al. 2010). In some cases, both RADAR and VIMS have data over the same region, strengthening the interpretation. Rohe Fluctus (41°W, 47°N) is a region with radar-bright (indicating rough) edges, which is associated with a caldera-like pit initially as a thin tongue (Fig. 16a, 16b), then spreading out. The depression is about 13 km in diameter. The lobate deposit is at least 40 km long. Lopes et al. (2007) used a flow thickness estimate (250–300 m) by Kirk et al. (2005) to make a preliminary rheological analysis of the deposit, inferring an effective yield strength of $\sim 10^4$ Pa, more consistent with ammonia-water-methanol mixtures than ammonia-water mixtures. Because of the large flow thickness and the association with an oval, caldera-like depression, this is one of the best examples of an inferred cryovolcanic flow on Titan.

Ara Fluctus (Fig. 16c, 16d), located at 118 W, 39 N, is an irregularly shaped depression, which has been interpreted as a caldera (Lopes et al. 2007), similar in morphology to a scalloped feature on Ganymede, also interpreted as a cryovolcanic caldera (Schenk et al. 2001; Pappalardo et al. 2004). The Ara Fluctus depression is about 18 km across and oval in shape, with a radar-bright rim around the periphery, indicating rough edges. The associated flow-like deposit is at least 53 km long. Other potential flow features discussed by Lopes et al. (2007) include an unnamed field located at approximately (92 W, 47 N), which is also associated with a bright-rimmed and oval depression interpreted as a caldera. The depression is about 10 km in diameter and appears to be the source of a flow field that heads north-northwest, starting out narrow and spreading out as three main branches, covering an area of at least 1140 km^2 and reaching about 50 km from the source.

A much larger field, Winia Fluctus (Lopes et al. 2007; Mitri et al. 2008), consists of a series of flow-like deposits that may emerge from fissures (Fig. 16e), but unlike the deposit discussed above, the sources for the Winia flows are not clear in RADAR data. Winia (central coordinates $\sim$46 W, 49 N) covers an area of at least 23,700 km^2. Because the northern extent of the deposit has not yet been imaged, the full extent of the associated deposits is not yet

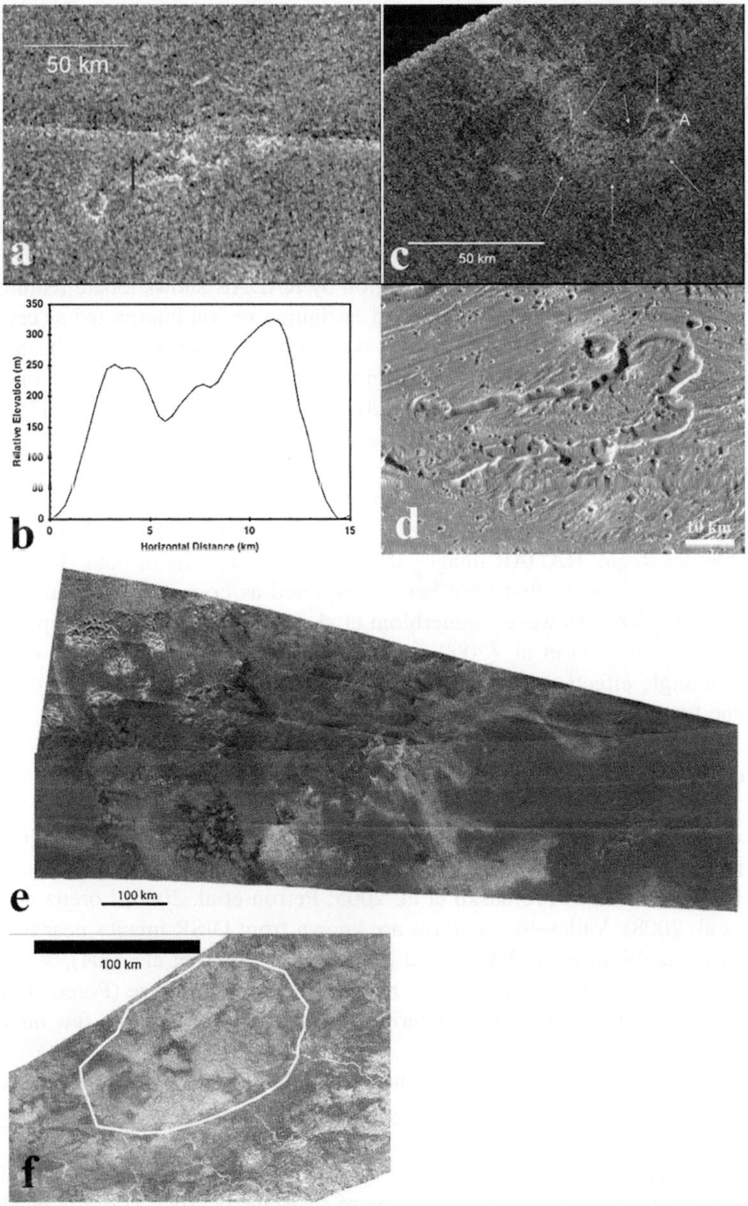

Fig. 16 (**a**) Cassini RADAR image of the Rohe Fluctus depression and inferred flows. Radar illumination is from the bottom of the image. The line indicates the location of a topographic profile across the flow, shown in (**b**), derived using radarclinometry (from Lopes et al. 2007); (**b**) is adapted from Kirk et al. 2005). (**c**) RADAR image of the Ara Fluctus depression (labeled **A**) and associated flow (outlined by arrows). Radar illumination is from the bottom of the image, suggesting that the feature is a depression, perhaps similar to the scalloped depression observed on Ganymede by Galileo, shown in (**d**). (**e**) RADAR image of Winia Fluctus, which has been interpreted to be a flow field. The radar illumination is from the bottom of the image. From Mitri et al. (2008). (**f**) RADAR image of Hotei Arcus. The oval represents the area where changes in brightness may have been detected by the VIMS instrument, including uncertainty in location (from Wall et al. 2009). However, claims of variability have been questioned by Soderblom et al. (2009)

known. Another feature interpreted as cryovolcanic consists of a series of flow fields and a candidate volcanic edifice, located near Sotra facula at 15 S, 40 W (Stofan et al. 2009 LPSC; Lopes et al. 2010).

To date, three possible flows or flow fields have been imaged by both RADAR and VIMS. The first is a small flow at approximately (72 W, 18 N) that appears to be associated with a cone or crater (Lopes et al. 2007). Le Mouelic et al. (2009) used VIMS data to show that this material has a different composition than the surrounding plains. VIMS imaging over the western part of Tui Regio (125 W, 24 S) shows a large ($>3 \times 10^4$ km^2) lobate region that is quite bright at 5 microns and was interpreted by Barnes et al. (2006) to be a field of cryovolcanic flows. Later imaging of this region by RADAR shows lobate features consistent with flows (Stofan et al. 2009). The most intriguing region interpreted as cryovolcanic is Hotei Regio, located at 26 S, 78 W, as this is one of two regions that has been suggested to be currently volcanically active. Hotei Regio has been observed more than once by both VIMS and RADAR (Fig. 16f). Nelson et al. (2009) inferred spectral changes in Hotei Regio, which is unusually bright at 5 microns (Barnes et al. 2005), and also over a second region on the western part of Xanadu (Nelson et al. 2009). VIMS spectra have been interpreted to show the presence of ammonia at Hotei Regio (Nelson et al. 2009); other work (Hayne et al. 2008) proposed that patches of what may be CO_2 ice have been detected in VIMS spectra of Tui and Hotei Regio. RADAR images showed lobate deposits in both Hotei Regio and the western part of Xanadu that have been interpreted as cryovolcanic (Wall et al. 2009; Soderblom et al. 2009). However, Soderblom et al. (2009) infer that the temporal changes in Hotei Regio of Nelson et al. (2009) are instead probably a combination of resolution and emission angle effects related to atmospheric scattering, while Moore and Pappalardo (2010) argue for a fluvial origin.

3.1.5 Fluvial Features

Surface conditions on Titan are different from those on Earth, however, the similarity between terrestrial and Titanian fluvial features, especially at the Huygens site, suggest comparable physical processes (Tomasko et al. 2005; Perron et al. 2006; Lorenz et al. 2008c; Jaumann et al. 2008). Valley-like features are known from DISR images near the Huygens landing site (Soderblom et al. 2007a) and Cassini ISS (Porco et al. 2004), VIMS (Brown et al. 2004) and RADAR (Elachi et al. 2006) observations elsewhere (Porco et al. 2005b). These valleys build dendritic and meandering systems (Fig. 17) up to a few hundred kilometers long.

Valley widths vary substantially from the resolution limit of $\sim$20 m for Huygens-DISR imaging, $\sim$300 m for RADAR, $\sim$500 for VIMS, and several hundred meters for ISS to a few kilometers (Lorenz et al. 2008c). RADAR data indicate very shallow channels with almost no apparent depth suggesting an easily erodible substrate, as well as systems cutting more than 100 m deep into more resistant bedrock (Lorenz et al. 2008c). At least 6 traceable stream orders can be identified (Lorenz et al. 2008c). However, due to the restricted resolutions and taking into account the large widths of the highest order identifiable channel segments, the stream order of the larger systems could be much higher. There is also no widespread evidence for subsurface control on the stream patterns although a few straight segments joining at wider angles (e.g., Tomasko et al. 2005; Soderblom et al. 2007a) may imply some isolated examples of surface/subsurface control but may be also due to surface slope effects (Lorenz et al. 2008c). However, the well-developed dendritic systems and the relative lack of rectangular tectonic control indicate sufficient amounts of fluids to carve the channels at least at the time of their formation. However, due

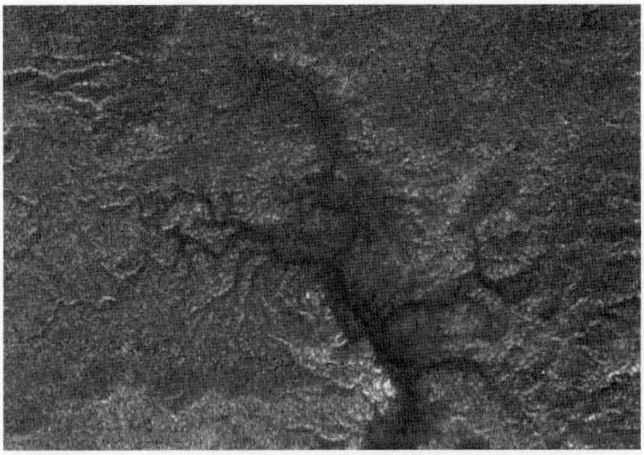

Fig. 17 Dendritic valley system near Titan's North Pole (Cassini RADAR, T28)

to the limited resolution outside the Huygens landing site, only wider valley segments can be identified easily. Nevertheless, the smaller segments at the Huygens landing site justify the assumption that all scales of valley systems might exist on Titan. Traces of fluvial erosion cover the entire surface of Titan and individual channels or networks could be found at the equator, at midlatitudes and at Titan's North Pole (Elachi et al. 2006; Porco et al. 2005b; Lunine and Atreya 2008; Lorenz et al. 2008c; Jaumann et al. 2008) (Fig. 18). On the other hand, there are regions, where channels are almost entirely missing, like the equatorial dune fields.

Titan's surface temperature of 94 K is too cold for liquid water to be stable, and constrains methane and ethane to be the only fluids that remain liquid on the surface. Only methane has a high enough vapor pressure to participate in a hydrological ("methanological") cycle (Lorenz et al. 2008c). Although methane has a short lifetime ($\sim$100 years, Strobel 1982), its high concentration in the atmosphere implies that the atmosphere is recharged by reservoirs on the surface and subsurface (Sotin et al. 2005; Tobie et al. 2006), in which case these fluvial surface features are part of the cycle and should be common on Titan. The viscosity of methane at 95°K is $1.8 \times 10–4$ Pa s, approximately five times lower than that of water at 298°K ($9 \times 10–4$ Pa s) (Hanley et al. 1977; National Institute of Standards and Technology 2005). Thus liquid methane (CH_4) had been suggested to be the major fluid on Titan (Lunine 1993), long before the Huygens results confirmed that parts of Titan's surface have been modified (Tomasko et al. 2005; Soderblom et al. 2007b; Jaumann et al. 2008) by fluvial processes. Based on basic hydraulic considerations for surficial flow, Burr et al. (2006) demonstrate that the required conditions for sediment entrainment and transport by liquid methane on Titan are within a factor of a few (for organic sediment) or several (for water ice) of those required by liquid water flow on Earth or Mars.

Fluvial systems on Earth are characterized by overland flow in the interfluve areas and then channelized flow forming dendritic and branching patterns of surface erosion. The sediments entrained and transported by surficial flow on Earth and Mars are materials derived from the planets' crusts, as well as secondary weathering products. Sediment on Titan would consist of water ice, derived from Titan's water ice outer crust, broken up by impact cratering, mass wasting and fluvial processes (Lorenz and Lunine 1996; Griffith et al. 2003) and organic material, which settles from the atmosphere after formation by photochemical reactions of hydrocarbons (Khare et al. 1978; Tran et al. 2003; Lorenz and Lunine 2005). At Titan's surface temperatures, the brittle behavior of water ice

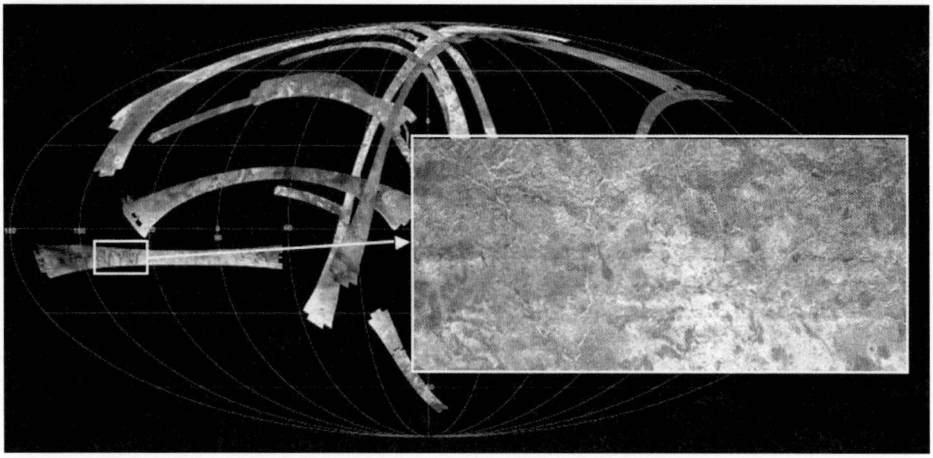

Fig. 18 Global RADAR map of Titan (Mollweide projection). Distribution of fluvial channels marked in *red*. Cassini RADAR: T0A, T03, T07, T08, T13, T16, T17, T18, T19, T21, T23, T25, T28 (after Langhans et al. 2009)

at small scales is similar to that of rocks on Earth, reaching the compressive and tensile strength of granite (Durham et al. 1983; Cuda and Ash 1984), whereas the organic material is weaker and can easily be moved.

Although the necessary flow velocity is generally lower for Titan than for Earth or Mars (Jaumann et al. 2008), the required flow depth for high-density organic sediment on Titan is higher, suggesting that similar overland flow processes occur on Titan as occur on Earth and on Mars (Burr et al. 2006). According to the scaling of sediment transport relations to the conditions on Titan by Burr et al. (2006), frictional shear velocities of 0.5–7 cm/s, unit discharges (mass flow rate per unit area or flow velocity times depth) of 0.005–8 m^2/s equivalent to 0.05–80 m^3/s discharge for 10 m channel width and 5–8,000 m^3/s discharge for 1000 m channel width, respectively (Jaumann et al. 2008) and shallow flow depths of 0.2–6 m at a slope of 0.001 seem to be sufficient to move grains of the size of 0.1 mm–15 cm and indeed rounded material of this size is exposed at the Huygens landing site (Tomasko et al. 2005).

On Earth active channel width and discharge can be correlated empirically (Osterkamp and Hedman 1982). Adapting this correlation to Titan's surface conditions (Jaumann et al. 2008) yields discharges for short recurrence intervals on Titan ranging from <1 m^3/s for the smallest identified channel segments (about 10 m wide) at the Huygens landing site to 1,600 m^3/s for the widest active channel segments (1000 m wide) of valley systems. For longer recurrence intervals these values increase dependent on the channel width to 66 m^3/s up to 7,950 m^3/s for 10 years recurrence and to 178 m^3/s peaking at 13,510 m^3/s for 100 years recurrence, respectively (Jaumann et al. 2008). Inferred discharges of channels on Titan are consistent with the values needed to move sediment as estimated by Burr et al. (2006). Runoff production rates range between 0.1–20 mm/hr for extended systems and peak rates between 60–400 mm/hr for the small steep-sloped systems at the Huygens landing site, with the higher values related to longer recurrence intervals (>100 years) (Jaumann et al. 2008). These rates are consistent with those (0.5–15 mm/hr) needed to move sediment of the size of 1–15 mm (Perron et al. 2006). Modeled methane convective storms on Titan, having rainfall of 110 kg/m^2 within 5–8 hours (20–50 mm/hr) (Hueso and Sanchez-Lavega 2006), are sufficient to trigger such flash floods. Even moderate storms with rainfall of 30 kg/m^2 within 5–10 hours (6–12 mm/hr) will explain the observed run-off rates (Jaumann et al. 2008). On the other hand, precipitation rates as estimated from the current atmospheric conditions might not exceed 1 cm/yr (Lorenz 2000; Rannou et al. 2006; Tokano et al. 2006), comparable to terrestrial deserts. To satisfy the overall energy balance of the hydrological cycle, the heavy rainstorms indicated above must be infrequent—perhaps centuries apart, which is consistent with the estimated run-off production rates (Jaumann et al. 2008). However, we cannot exclude the possibility that cryovolcanic and tectonically induced melting (Sotin et al. 2005) lead to sudden release of large amounts of liquids that could combine with loose surface material to form lahar-like mud flows (Fortes and Grindrod 2006) or hyper-concentrated flows.

3.1.6 Impact Craters

The surface of Titan has been very dynamic due to erosion and deposition and very few impact craters have been identified compared to other bodies. Wood et al. (2010) identified 5 certain impact craters on Titan and 44 additional probable impact craters (based on 22% of the surface imaged by RADAR). The 5 certain impact craters are Menrva (444 km diameter), Sinlap (79 km diameter), Ksa (30 km diameter), Afekan (115 km diameter) and Selk (80 km diameter). These craters have morphologies similar to those of impact craters on other bodies, both rocky and icy. The less certain craters appear to be more eroded, and are often partially covered by dune deposits. Modification may also have occurred by fluvial processes, mass wasting, or submergence in seas. Craters are not distributed uniformly; Xanadu has a crater density 2–9 times greater than other areas on Titan (Wood et al. 2010). There is also a small excess of craters on the leading hemisphere, and craters are deficient in the north polar regions.

The paucity of recognizable craters implies that Titan's surface is quite young, but the model age depends on models of crater production rate. The models of Artemieva and Lunine (2005) and Korycansky and Zahnle (2005) suggestsurface crater retention ages of $\sim$100 My to $\sim$1 Gy, respectively (Lorenz et al. 2007). It is likely that dynamical geological processes have destroyed most of the early history of Titan and multiple processes clearly continue to modify its surface (e.g., Wood et al. 2010; Lopes et al. 2010).

3.1.7 Mountains and Other Tectonic Features

Possible tectonic features on Titan include linear and ridge-like formations that may be chains of hills (Barnes et al. 2007b; Lopes et al. 2010) and dark, sub-parallel and branching lineaments (called virgae) that may represent tectonic fractures later used by liquids as fluvial channels (Porco et al. 2005b; Perry et al. 2007). Linear features are relatively rare on Titan, one of the most evident sets of linear features were observed by VIMS at about 30 S, 330 W (Jaumann et al. 2009). One such feature is about 500 km long and seems to cross both light and dark terrains and was interpreted to be $\sim$1–2 km high (Sotin et al. 2007). Such structures could have been formed by either compression or extension. SAR images showed possible tectonic features: linear chains of radar-bright, topographically high terrain (Radebaugh et al. 2007) and irregularly shaped fragments of radar-bright terrain such as Xanadu (Radebaugh et al. 2010). Lopes et al. (2010) mapped SAR swaths up to T30 and concluded that both types are part of the hummocky and mountainous terrain, the oldest geomorphological unit that can be identified on Titan given the current data. These terrains appear more prevalent or more exposed in the equatorial regions. In many regions, the material is exposed as linear mountain chains, which Radebaugh et al. (2007) suggested were formed by crustal compression tectonics and the up-thrusting of blocks. The driving forces of tectonics on Titan are not well understood, but a possible explanation for contractional tectonism has been proposed by Mitri et al. (2010).

3.2 Enceladus

Despite Enceladus' small size, ~252 km mean radius (Thomas et al. 2007a, 2007b), its surface records a complex history of endogenic geologic activity, which continues to the present. The discovery of active cryovolcanism with plumes emanating from Enceladus' South Pole (e.g., Porco et al. 2006; Spencer et al. 2006) is particularly exciting, but Enceladus, like other icy satellites, has also undergone substantial tectonic resurfacing. The young surface age is demonstrated by Enceladus' extremely high geometric albedo, 1.38 (Verbiscer et al. 2007). The surface is composed almost entirely of water ice, consistent with resurfacing by fallout from the south-polar plumes; spectral variations are indicative of ice particle sizes which range up to ~0.2 mm in the exposed walls of fresh fractures (Porco et al. 2006; Jaumann et al. 2008). Another measure of the youth of Enceladus' surface, and one that has deeper implications for the history of the lithosphere, is the relative lack of impact craters (e.g., Smith et al. 1981, 1982; Porco et al. 2006). The number density of craters varies drastically across the surface, with high northern latitudes being comparatively heavily cratered and the South Polar Terrain (SPT) having no craters >1 km (Porco et al. 2006). The striking heterogeneity of the surface has important implications for Enceladus' internal structure and its evolution.

Not only does the distribution of impact craters vary across the surface of Enceladus, but their morphologies also vary. Craters are concentrated at high northern latitudes as well as in bands on the anti-Saturnian and sub-Saturnian hemispheres (e.g., Bray et al. 2007; Smith et al. 2007; Helfenstein 2010), indicating these areas are older. However, even in such regions, not all of the craters are pristine. In some cases craters are cut by more recent fracturing and in others some of the craters themselves have undergone relaxation by viscous creep, characterized by up-domed floors within generally well-preserved rims (e.g., Parmentier and Head 1981). Figure 19 illustrates craters that have been modified by relaxation as well as subsequent fracturing. As expected, larger craters on Enceladus are more likely to have undergone deformation by viscous creep than smaller craters. In addition, a higher fraction of craters at southern latitudes exhibits characteristics consistent with relaxation than at northern latitudes (Smith et al. 2007; Smith 2008), although there are several examples of northern relaxed craters (e.g., Fig. 19). Viscous creep deformation is especially challenging on such a low-gravity body as Enceladus because the driving force is so small. The fact that it has occurred suggests relatively high thermal gradients (≥20 K/km, Smith 2008) must have existed even in areas that have undergone little tectonic activity.

The tectonic history of Enceladus is quite intricate, with numerous generations of ridges, grooves, and fractures cross-cutting each other (e.g., Fig. 19). The leading and trailing hemispheres exhibit multiple styles of deformation that, especially on the leading hemisphere, vary from region to region; tectonic resurfacing has clearly dominated significant portions of the surface. Yet even amid the incredible geologic complexity of Enceladus' surface, the south-polar region stands out (Fig. 19). The SPT is bounded at ~55°S latitude by a continuous series of arcuate scarps, whose morphologies are consistent with N-S compression (Helfenstein 2010), punctuated by cusps from which bands of extensional fractures stretch northward (Fig. 19). Poleward the terrain is tectonically complex and dominated by four subparallel fractures ~2 km across and more than 100 km long (Fig. 19), commonly called "tiger stripes." Evidence for shear offset, transpression, and transtension has been documented in the SPT (e.g., Helfenstein 2010). Intriguingly, hints of similar morphologies are

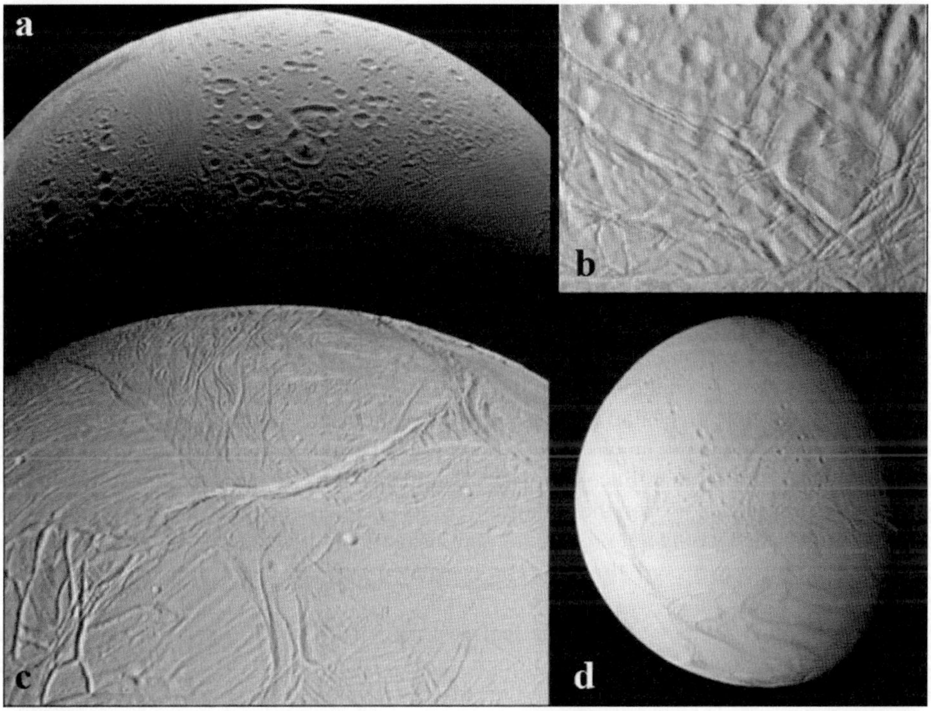

Fig. 19 (**a**) Cassini image (PIA08409) taken above Enceladus' North Pole looking southward over the sub-Saturnian hemisphere. Many of the impact craters, especially the two prominent overlapping craters just above center, have the characteristics of viscous relaxation. Elsewhere, craters have been cut by subsequent tectonic activity. (**b**) Cassini image (PIA06248) of craters cut by multiple sets of fractures. The large crater to the right of center is ~22 km in diameter. (**c**) Mosaic of Enceladus' trailing hemisphere at southern latitudes (south is to the upper right) illustrating a variety of tectonic styles: from ridges at lower left, to individual fractures at lower center and lower right, to a 5-km-wide rift at center, and a network of grooves at top. Pixel scale ranges from ~100–150 meters. (PIA06191). (**d**) False-color view of Enceladus' anti-Saturnian hemisphere illustrating the south-polar terrain (SPT) with the four sub-parallel "tiger stripes" encircled by a cuspate tectonic boundary at 55°S. Enceladus' surface is composed almost entirely of water ice; here, blue colors indicate areas with larger grain size, ~0.2 mm (Jaumann et al. 2008) (PIA06249)

seen at lower latitudes on the trailing hemisphere, suggestive of relict versions of the current SPT (Bray et al. 2006; Helfenstein 2010) and consistent with the hypothesis that the current south-polar location of the tiger stripes is the result of polar wander (Nimmo and Pappalardo 2006).

The cryovolcanic plumes are seen to originate from the tiger stripes (e.g., Porco et al. 2006; Spitale and Porco 2007), and substantial heat is emanating from the surrounding area (Spencer et al. 2006). Indeed, model temperatures of up to 223 K have been derived for the fractures themselves (Abramov and Spencer 2009), considerably higher than the equilibrium temperature of 68 K. High-resolution views reveal morphologic details of the tiger stripes, which are typically ~500 m deep and ~2 km wide with uplifted flanks ~100 m high. Although at moderate resolution (e.g., Fig. 19), Enceladus' surface appears almost blanketed, consistent with deposits from the plumes as well as interaction with the plume-generated E-ring, Cassini's highest-resolution views reveal a rugged surface covered with boulders a few tens of meters across.

3.3 Other Saturnian Satellites

Saturn's system of satellites exhibits somewhat more variety in its dynamical and physical properties than does that of Jupiter. The large satellites comprise the dominant Titan, and six other ellipsoidal ones, including the geologically active satellite Enceladus. Aside from Titan, these moons are smaller than the Galilean satellites, yet have a great variety of surfaces and inferred histories and properties. Small satellites include ring-related objects; Trojan satellites of Tethys and Dione; Hyperion, which orbits between Titan and Iapetus; and a large number of small, irregular satellites. The latter are dominated by Phoebe, mean radius of $\sim$106 km, observed at close range by Cassini just before orbit insertion in 2004. We discuss primarily those objects resolved by spacecraft data. The outer small satellites are summarized in Nicholson et al. (2008).

Mimas is a heavily cratered, apparently internally rigid, and tectonically quiescent object. Its shape is indicative of an approach to hydrostatic relaxation, but suggests about 2 km of supported topography (Thomas et al. 2007a). It had been thought Mimas might be subject to greater tidal heating than Enceladus, thus the greater activity of Enceladus might be regarded as a problem. However, the persistence of its orbital eccentricity indicates a cold and rigid interior, which would not provide for significant tidal dissipation (Barr and McKinnon 2007). Most of the scarps on Mimas may be associated with large impacts and do not require significant internal changes to explain their forms.

Tethys has two primary remarkable features: its density of 980 kg m^{-3} (Thomas et al. 2007a) (Table 2), and the global-scale trough Ithaca Chasma. The very low density requires a small fraction of non-icy materials and greatly restricts many of the possible evolutionary paths dependent upon heating or other effects of significant differentiation. Ithaca Chasma is a 1000 km long terraced trough up to 100 km in width. This tectonic form has been hypothesized to be associated with the 400-km impact basin Odysseus as it is approximately concentric to the basin. However, recent crater counts suggest that the trough is about 4 Gyr in age, and is in fact older than the Odysseus basin (Giese et al. 2007), requiring an origin that is not associated with the impact.

Dione, with a mean radius of 560 km and mean density of 1478 kg m^{-3} (Table 2) has an intermediate rock fraction among the large Saturnian satellites. Its surface has heavily cratered plains, but only one large (400 km) basin, which is younger than many units (Wagner et al. 2008). There appear to have been multiple episodes of tectonism producing a variety of scarps and lineaments at a wide range of scales. Resurfacing of Dione has been shown by the high resolution Cassini images to have been primarily by tectonism rather than by volcanic flows (Wagner et al. 2008).

Rhea is the second largest of Saturn's satellites; its low density of 1200 kg m^{-3} (Thomas et al. 2007a, 2007b) (Table 2) indicates a largely icy interior. The interior configuration of ice and rocky components has been difficult to infer, as the shape is marginally consistent with hydrostatic relaxation. Gravity data are somewhat contradictory, though perhaps consistent with an undifferentiated interior (Anderson and Schubert 2008; Iess et al. 2007). Tectonic structures such as troughs, scarps, and lineaments are not as prominent as on Dione, Enceladus, or Tethys. The surface is heavily cratered and includes two basins 400 and 500 km in diameter (Wagner et al. 2008). Ages of most of the surfaces are probably in excess of 3.6 Gyr (Wagner et al. 2008).

Iapetus, long known for its extreme leading-trailing albedo asymmetry, has proved even more puzzling upon close examination by Cassini. The albedo dichotomy has a complex transition from the dark leading side; the color dichotomy has a smooth transition. These characteristics suggest a role of external impacts coupled with enhanced solar heating in

Table 2 *Saturnian Satellites sizes*. Density values for Mimas—Iapetus are from Thomas et al. (2007a); small satellites are from Porco et al. (2007), except Hyperion, which is from Thomas et al. (2007b)

Satellite	Mean radius, km	ρ, kg m^{-3}
Mimas	198.2 ± 0.5	1150 ± 9
Enceladus	252.1 ± 0.2	1608 ± 5
Tethys	533.0 ± 1.4	973 ± 8
Dione	561.7 ± 0.9	1476 ± 7
Rhea	764.3 ± 2.2	1233 ± 11
Iapetus	735.6 ± 3.0	1083 ± 13
Pan	14.2 ± 1.3	410 ± 150
Daphnis	3.9 ± 0.8	310 ± 200
Atlas	15.1 ± 1.4	460 ± 100
Prometheus	43.1 ± 2.7	470 ± 65
Pandora	40.3 ± 2.2	500 ± 85
Epimetheus	56.7 ± 1.9	690 ± 130
Janus	89.6 ± 2.0	640 ± 64
Methone	1.6 ± 0.6	
Pallene	2.2 ± 0.3	
Telesto	12.4 ± 0.4	–
Calypso	10.6 ± 0.7	–
Polydeuces	1.3 ± 0.4	
Helene	16.5 ± 0.6	
Hyperion	135 ± 4	544 ± 50
Phoebe	106.7 ± 1	1629 ± 50

some depressions in bright areas that lead to runaway thermal segregation of darker surface materials (Denk and Spencer 2008). An additional puzzle emerged in the oblate spheroid shape of Iapetus; if relaxed to today's 79-day spin period, it would deviate from a sphere by only a few meters, yet it has a difference between equatorial and polar radii of $>$33 km (Thomas et al. 2007a). This shape suggests equilibration at a much higher spin rate ($\sim$16 hours), with cooling to a rigid form shortly thereafter. Modeling of heating scenarios (Castillo-Rogez et al. 2007) suggest the formation of Iapetus occurred early enough to incorporate Al^{26} to provide an effective but short thermal pulse.

An additional puzzle is an equatorial ridge spanning more than half the circumference of Iapetus. It reaches heights of more than 20 km above surrounding terrain (Porco et al. 2005a), and is in some places a single ridge, in other parts multiple ridges (Giese et al. 2007). The origin of this ridge remains controversial, as despinning tectonics may not explain tectonic patterns on Iapetus (Singer and McKinnon 2008) and the morphology of the ridge changes significantly along its length. The crater density suggests it is quite old, and high resolution data indicate it is not material piled on from outside.

Hyperion's chaotic rotation derives from its shape and eccentric orbit (Black et al. 1995). Several Cassini flybys have allowed the mass and volume to be determined, and its density of 540 kg m^{-3} is closer to those of the ring-related objects than to other objects measured in the Saturn system. The heavily cratered, sponge-like surface may reflect cratering properties in very porous materials (Thomas et al. 2007b) or a more exotic thermal history. The reddish materials include likely external, carbon rich contaminants (Cruikshank et al. 2007).

Phoebe, with a mean density of 1600 kg m^{-3} (Porco et al. 2005a) and a much lower albedo than the inner icy objects is almost certainly a captured object derived from conditions separate from the Saturn-forming materials (Clark et al. 2005; Cruikshank et al. 2008). It has a heavily cratered surface and is nearly equidimensional in fitted shape. Some bright exposures of ice indicate at least local segregation of ices, but no global layering is evident.

Ring-related small satellites have mean radii less than 100 km, high albedos, and low mean densities (Porco et al. 2007). The shapes of some, especially Pan, and Atlas, suggest influence of ring materials in their accretion (Charnoz et al. 2007; Porco et al. 2007). The larger co-orbitals, Janus and Epimetheus, are also under-dense, ice-rich objects. They are heavily cratered and show some indications of downslope motion of debris. Epimetheus and Pandora exhibit grooves suggesting tidal fracturing (Morrison et al. 2008).

4 Uranian Satellites

The Voyager 2 spacecraft passed closest to the innermost of the five major Uranian satellites, so imaging resolution is best at Miranda, while Titania and Oberon were not imaged sufficiently to resolve tectonic structures. Umbriel appears generally featureless and dark, but careful image processing reveals an ancient tectonic system (Helfenstein et al. 1989). Ariel displays a global-scale structural pattern of tectonic rifts, ridges, and troughs forming an obliquely intersecting pattern across much of the satellite's visible surface (Plescia 1987). Some rift depressions appear to have been resurfaced by a viscous cryolava (Jankowski and Squyres 1988). Abundant extensional tectonic structures on many of the Uranian satellites, as on the Saturnian satellites, are attributed to volume changes linked to thermal evolution (Schubert et al. 1986; Hillier and Squyres 1991).

Despite its small size, Miranda exhibits striking structural geology. Miranda displays three ovoidal regions termed "coronae" that are deformed by subparallel ridges and grooves. An early interpretation of the coronae suggested that they might be folds, which would be consistent with a "sinker" model, in which the satellite was disrupted early in its history by a large impact, with remnants of the proto-satellite's rocky core sinking downward through the reforming satellite and stirring downwelling currents to create the coronae above (Janes and Melosh 1988). However, subsequent analyses showed that the coronae are likely extensional tectonic structures, including tilted blocks and cryovolcanic extrusions, more consistent with a model in which the coronae formed by tectonic deformation above large-scale upwelling plumes (Greenberg et al. 1991; Schenk 1991; Pappalardo et al. 1997). Additional tectonic stress may have resulted from reorientation of Miranda in response to corona formation (Plescia 1988).

5 Neptune's Moon Triton

One of the largest and most intriguing icy satellites is Neptune's moon, Triton. At 2,705 km in diameter, Triton is Neptune's only large moon, and the only one of its moons to be imaged in sufficient detail to discriminate morphological features other than craters. Triton is comprised of rock, water ice, organics and volatile ices such as N_2 and CH_4 (Cruikshank et al. 2000). Voyager 2 explored Triton in 1989, returning images of over half of the surface, which revealed a wealth of geological features on a surface that has few impact craters and appears remarkably young (Fig. 20), and may, like Europa, still be active today (Smith et al. 1979a, 1979b). Modeling of Triton's resurfacing rate based on the paucity of its craters suggests that it may be second only to Io and Europa in level of geological activity in the outer solar system (Stern and McKinnon 2000).

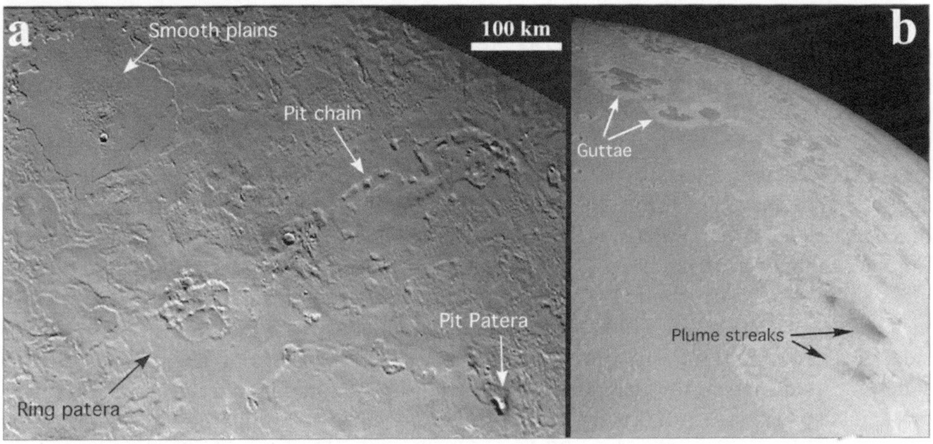

Fig. 20 Some of Triton's volcanic features: (**a**) smooth plains fill shallow circular depressions and form a "ring" patera. Also visible are chains of volcanic pits, and isolated pits, all of which are analogous to volcanic features on terrestrial planets. (**b**) Triton's guttae are giant lobate features with dark cores and bright annuli. They have no obvious analogs elsewhere in the solar system. Dark plume streaks overlie the bright south polar cap. Image: NASA/JPL

Many landforms are interpreted to be cryovolcanic in origin, including eruptive hills, flow lobes, and depressions that resemble terrestrial volcanic calderas (Croft et al. 1995). Different features exhibit superposition relationships hinting at multiple episodes of both explosive and effusive events, and some flows have thicknesses that imply rheologies comparable to terrestrial andesite, dacite, or rhyolite. Some of the landforms look like they were fragmented as they came out of the ground, implying that they may have been emplaced explosively and hence that significant quantities of dissolved gases were present in their source magmas (see Kargel et al. 1999 and references therein).

Triton's cryovolcanic landforms take several forms (Croft et al. 1995). Small cones are common; these are about 7 to 15 km in diameter with summits that generally contain smooth sided pits that can breach the side of the cone. The cones may occur individually or in clusters or chains. They may be analogous to terrestrial cinder cones, and have similar dimensional ratios to cinder cones on Mars and the Moon (although they are much larger), suggesting that they may be formed of particulate material at the angle of repose. Pit paterae are circular to elongate depressions typically 10–20 km in diameter that can have raised rims and are generally located within patches of smooth material 100–200 km in extent. The paterae occur singly or in chains. Ring paterae are much larger, 50–100 km in length, and have an outer rim defined by a ring of coalescing pits. The ring and pit paterae and associated smooth deposits are interpreted to have formed as explosive cryovolcanic craters and cryoclastic deposits. Chains of pit paterae follow tectonic trends in the region and may be analogous to chains of terrestrial explosion and collapse pits in linear volcano-tectonic zones. Some of the most unusual features on Triton's surface are the guttae (Fig. 20), huge dark lobate features each surrounded by a bright aureole of fairly constant width (20–30 km wide). Guttae have smooth surfaces and lobate margins, suggesting they are comprised of extrusive materials that have flowed in a viscous manner. It is inferred that guttae deposits are at least tens of meters thick, implying a relatively high viscosity at the time of emplacement. Suggestions for the origin of the bright aureoles include low viscosity flows, condensed gases, or thermally metamorphosed surface materials.

One unique morphological feature on Triton is the "cantaloupe terrain", containing quasi-circular shallow depressions typically 25–35 km in diameter, with slightly raised rims, giving the uncanny appearance of cantaloupe melon skin. These have been suggested to represent cryovolcanic explosion craters, such as terrestrial maars, on the basis of their similar morphologies. An alternative mode of formation is by diapirism: the organized cellular pattern of the cantaloupe terrain has been proposed to closely resemble terrestrial salt diapirs, and the terrain may have formed due to density inversion in a layered crust composed partly of ice phases other than water ice (Schenk and Jackson 1993). The cavi may be analogous to Europa's lenticulae, which have also been proposed to have a diapiric origin (see Sect. 2.3).

In addition to cryovolcanic flow features, Triton has a significant number of large ridges, measuring ~15–20 km in width, which can be continuous for ~800 km (Fig. 20). Steep-sided in profile and somewhat sinuous, they appear to span a range of ages and degradation states and some appear to stand above the level of the surrounding terrain. A deep continuous axial depression ~5–10 km wide results in a double-ridge morphology, and some ridges are bounded by shallow troughs ~20 km wide. Triton's ridges are similar in morphology to Europa's double ridges, and have been suggested to have a similar formation origin, involving shear along a crack (Prockter and Schenk 2005), or to have formed by the eruption of viscous material into the axes of graben, which then built up as the degree of extension and/or magma availability increased (Croft et al. 1995).

Dark streaks near Triton's bright south polar cap were observed by Voyager 2. What was surprising about these was that they were visible: because of Triton's seasons, nitrogen, methane and carbon dioxide frosts are expected to sublimate from one pole and be cold-trapped at the opposite pole on the cycle of 1 Triton year, equivalent to 165 Earth years. The presence of the dark streaks therefore suggested that they were probably younger than 1 Triton year, and therefore emplaced extremely recently. Stereoscopic images of the surface clearly showed plumes above Triton's illuminated limb, probably composed of fine dark dust and gas (Kirk et al. 1995). The plume columns reached up to 8 km above the surface, and some material fell back to the surface to form the dark streaks. Two models have been proposed for the formation of the plumes; one suggests that the driving mechanism is explosive venting of nitrogen gas pressurized by solar heating (Kirk et al. 1995), while the other suggests that the driving force comes from within Triton. The mechanism by which the plumes form has implications for Triton's current thermal state and evolution.

6 Conclusions

In this chapter we have discussed the wealth of morphological features observed on the surfaces of a number of icy satellites and Io. We have shown that these features exist on a variety of scales ranging from meters to global in extent, and that the way in which they are emplaced varies considerably depending on when and where in the solar system they formed, and under what conditions. Even in recent geological times on some bodies such as Europa, there have been significant changes in the style of morphological features, and hence the type of exchange processes responsible for them. Exchange between the surface and subsurface is still occurring today on satellites such as Enceladus and Io, allowing study of the current state of the interior on those bodies. Understanding how and why material is exchanged between the surface and subsurface of a satellite is crucial to understanding its evolution from formation to the present day. In the chapters that follow, the formation processes for the surface features presented here are evaluated, and their implications for surface-subsurface exchange of material are assessed.

Acknowledgements We thank two anonymous reviewers for their helpful comments, and Olivier Grasset for his tireless efforts in putting together this volume. The portions of this work performed by RTP and RMCL were carried out at the Jet Propulsion Laboratory, California Institute of Technology, under a contract with the National Aeronautics and Space Administration.

References

O. Abramov, J.R. Spencer, Icarus **199**, 189–196 (2009)

J.D. Anderson, G. Schubert, Bull Am. Astron. Soc. **40**, #45.03 (2008)

J.D. Anderson, R.A. Jacobson, E.L. Lau, W.B. Moore, O. Olsen, G. Schubert, P.C. Thomas, Galileo Gravity Science Team, Bull. Am. Astron. Soc. **33**, 1101 (2001a)

J.D. Anderson, R.A. Jacobson, T.P. McElrath, W.B. Moore, G. Schubert, P.C. Thomas, Icarus **153**, 157–161 (2001b)

J.D. Anderson et al., Science **308**, 1291–1293 (2005)

N. Artemieva, J.I. Lunine, Icarus **164**, 471–480 (2005)

J.W. Barnes, R.H. Brown, E.P. Turtle, A.S. McEwen, R.D. Lorenz, M. Janssen, E.L. Schaller, M.E. Brown, B.J. Buratti, C. Sotin, C. Griffith, R. Clark, J. Perry, S. Fussner, J. Barbara, R. West, C. Elachi, A.H. Bouchez, H.G. Roe, K.H. Baines, G. Bellucci, J.-P. Bibring, F. Capaccioni, P. Cerroni, M. Combes, A. Coradini, D.P. Cruikshank, P. Drossart, V. Formisano, R. Jaumann, Y. Langevin, D.L. Matson, T.B. McCord, P.D. Nicholson, B. Sicardy, Science **310**, 92–95 (2005). doi:10.1126/science.1117075

J.W. Barnes, R.H. Brown, J. Radebaugh, B.J. Buratti, C. Sotin, S. Le Mouelic, S. Rodriguez, E.P. Turtle, J. Perry, R. Clark, K.H. Baines, P.D. Nicholson, Geophys. Res. Lett. **33**, L16204 (2006). doi:10.1029/2006GL026843

J.W. Barnes, R.H. Brown, L. Soderblom, B.J. Buratti, C. Sotin, S. Rodriguez, S. Le Mouèlic, K.H. Baines, R. Clark, P. Nicholson, Icarus **186**, 242–258 (2007a). doi:10.1016/j.icarus.2006.08.021

J.W. Barnes, J. Radebaugh, R.H. Brown, S. Wall, L. Soderblom, J. Lunine, D. Burr, C. Sotin, S. Le Mouèlic, B.J. Buratti, R. Clark, K.H. Baines, R. Jaumann, P.D. Nicholson, R.L. Kirk, R. Lopes, R. Lorenz, K. Mitchell, C.A. Wood, Cassini RADAR Team, J. Geophys. Res. **112**(E11), E11006 (2007b). doi:10.1029/2007JE02932

J.W. Barnes et al., Icarus **195**, 400–414 (2008)

J.W. Barnes et al., Icarus **201**, 217–225 (2009)

A.C. Barr, W.B. McKinnon, Geophys. Res. Lett. **34**, L09202 (2007)

E. Barth, S. Rafkin, Geophys. Res. Lett. **34**, L03203 (2007). doi:10.1029/2006GL028652

M.J.S. Belton et al., Science **274**, 377–385 (1996)

E.G. Bierhaus, C.R. Bierhaus, W.J. Merline, Nature **437**, 1125–1127 (2005)

G.J. Black, P.D. Nicholson, P.C. Thomas, Icarus **117**, 149–161 (1995)

V. Bray, E. Turtle, P. Helfenstein, J. Rathbun, J. Perry, C. Porco, Bull. Am. Astron. Soc. **38**, 522 (2006)

V.J. Bray, D.E. Smith, E.P. Turtle, J.E. Perry, J.A. Rathbun, A.N. Barnash, P. Helfenstein, C.C. Porco, in *Lunar Planet. Sci. XXXVIII*, Abst. #1873 (2007)

R.H. Brown, K.H. Baines, G. Bellucci, J.-P. Bibring, B.J. Buratti, E. Bussoletti, F. Capaccioni, P. Cerroni, R.N. Clark, A. Coradini, D.P. Cruikshank, P. Drossart, V. Formisano, R. Jaumann, Y. Langevin, D.L. Matson, T.B. McCord, E. Miller, R.M. Nelson, P.D. Nicholson, B. Sicardy, C. Sotin, Space Sci. Rev. **115**, 111–168 (2004)

R. Brown et al., Nature **454**, 607–610 (2008). doi:10.1038/nature07100

J.A. Burns, M.R. Showalter, D.P. Hamilton, P.D. Nicholson, I. de Pater, M.E. Ockert-Bell, P.C. Thomas, Science **284**, 1146 (1999) 2007

M.H. Burger et al., Space Sci. Rev. (2010, this issue). doi:10.1007/s11214-009-9645-z

D.M. Burr, J.P. Emery, R.D. Lorenz, G.C. Collins, P.A. Carling, Icarus **181**, 235–242 (2006)

R. Carlson, W. Smythe, R. Lopes-Gautier et al., Geophys. Res. Lett. **24**(20), 2474–2482 (1997)

R.W. Carlson, J.S. Kargel, S. Doute, L.A. Soderblom, J.B. Dalton, in *Praxis Publishing Company*, ed. by I. Galileo, R. Lopes, J.R. Spencer (Springer, Berlin, 2007)

T. Cassidy et al., Space Sci. Rev. (2010, this issue). doi:10.1007/s11214-009-9625-3

J.C. Castillo-Rogez, D.L. Matson, C. Sotin, T.V. Johnson, J.I. Lunine, P.C. Thomas, Icarus **190**, 179–202 (2007)

C.R. Chapman, W.B. McKinnon, in *Satellites*, ed. by J.A. Burns, M.S. Matthews (Univ. of Arizona Press, Tucson, 1986), pp. 492–580

S. Charnoz, A. Brahic, P.C. Thomas, C.C. Porco, Science **318**, 1622 (2007)

R.N. Clark et al., Nature **435**, 66–69 (2005)
G. Collins, F. Nimmo, in *Europa*, ed. by R.T. Pappalardo, W.B. McKinnon, K. Khurana (Univ. Arizona Press, Tucson, 2009), pp. 259–282
G.C. Collins, J.W. Head, R.T. Pappalardo, Icarus **135**, 345–359 (1998)
J. Conca, Proc. Lunar. Planet. Sci. **12B**, 1599–1606 (1981)
S.K. Croft, J. Geophys. Res. **88**, B71–B89 (1983)
S.K. Croft et al., Geology of Triton, in *Neptune and Triton* ed. by D.P. Cruikshank, M.S. Matthews (Univ. of Ariz. Press, Tucson, 1995), pp. 879–948
D.P. Cruikshank, B. Schmitt, T.L. Roush, T.C. Owen, E. Quirico, T.R. Geballe, C. de Bergh, M.J. Bartholomew, C.M. Dalle Orek, S. Douté, R. Meier, Icarus **147**, 309–316 (2000)
D.P. Cruikshank et al., Nature **448**, 54–56 (2007)
D.P. Cruikshank et al., Icarus **193**, 334–343 (2008)
V. Cuda, R.L. Ash, Cold Reg. Sci. Technol **9**, 47–52 (1984)
T. Denk, J.R. Spencer, in *AAS/Division for Planetary Sciences Meeting Abstracts* 40, #61.04 (2008)
A.J. Dombard, W.B. McKinnon, Icarus **154**, 321–336 (2001)
S. Douté, R. Lopes-Gautier, R. Carlson, B. Schmitt, L. Soderblom, Galileo NIMS Team, Icarus **149**, 107–132 (2001)
S. Douté, R. Lopes, L.W. Kamp, R. Carlson, Icarus **158**, 460–482 (2002)
S. Douté, R. Lopes, L.W. Kamp, R.W. Carlson, B. Schmitt, Galileo NIMS Team, Icarus **169**(1), 175–196 (2004)
W.B. Durham, H.C. Heard, S.C. Kirby, J. Geophys. Res. **88**, 377–292 (1983)
C. Elachi, M.D. Allison, L. Borgarelli, P. Encrenaz, E. Im, M.A. Janssen, W.T.K. Johnson, R.L. Kirk, R.D. Lorenz, J.I. Lunine, D.O. Muhleman, S.J. Ostro, G. Picardi, F. Posa, C.G. Rapley, L.E. Roth, R. Seu, L.A. Soderblom, S. Vetrella, S.D. Wall, C.A. Wood, H.A. Zebker, Space Sci. Rev. **117**, 71–110 (2005)
C. Elachi, S. Wall, M. Janssen, E. Stofan, R. Lopes, R. Kirk, R. Lorenz, J. Lunine, F. Paganelli, L. Soderblom, C. Wood, L. Wye, H. Zebker, Y. Anderson, S. Ostro, M. Allison, R. Boehmer, P. Callahan, P. Encrenaz, E. Flamini, G. Francescetti, Y. Gim, G. Hamilton, S. Hensley, W. Johnson, K. Kelleher, D. Muhleman, G. Picardi, F. Posa, L. Roth, R. Seu, S. Shaffer, B. Stiles, S. Vetrella, R. West, Nature **441**, 709–713 (2006)
S.A. Fagents, J. Geophys. Res. **108**(E12), 5139–5158 (2003)
S.A. Fagents, R. Greeley, R.J. Sullivan, R.T. Pappalardo, L.M. Prockter, Icarus **144**, 54–88 (2000)
P.H. Figueredo, R. Greeley, J. Geophys. Res. **105**, 22,629–22,646 (2000)
P.H. Figueredo, R. Greeley, Icarus **167**, 287–312 (2004)
P.H. Figueredo, F.C. Chuang, J. Rathbun, R.L. Kirk, R. Greeley, J. Geophys. Res. **107**(E5), 5026 (2002). doi:10.1029/2001JE001591
A.D. Fortes, P.M. Grindrod, Icarus **182**, 550–558 (2006)
S. Fussner, Search for 938 nm Specular Enhancement on Titan. M.S. thesis, Univ. of Arizona, Tucson, AZ (2006)
E.J. Gaidos, F. Nimmo, Nature **405**, 637 (2000)
P. Geissler, in *Encyclopedia of Volcanoes*, ed. by H. Sigurdsson et al. (Academic Press, San Diego, 2000), pp. 785–800
P.E. Geissler, R. Greenberg, G. Hoppa, A. McEwen, R. Tufts, C. Phillips, B. Clark, M. Ockert-Bell, P. Helfenstein, J. Burns, J. Veverka, R. Sullivan, R. Greeley, R.T. Pappalardo, J.W. Head III, M.J.S. Belton, T. Denk, Icarus **135**, 107–126 (1998)
P.E. Geissler, A.S. McEwen, L. Keszthelyi, R. Lopes-Gautier, J. Granahan, D.P. Simonelli, Icarus **140**(2), 265–281 (1999)
P. Geissler, A. McEwen, C. Phillips, D. Simonelli, R. Lopes-Gautier, S. Doute', J. Geophys. Res. **106**(E12), 33,053–33,078 (2001)
B. Giese, J. Oberst, T. Roatsch, G. Neukum, J.W. Head, R.T. Pappalardo, Icarus **135**, 303–316 (1998)
B. Giese, R. Wagner, G. Neukum, J.M. Moore, in *Lunar Planet. Sci. Conf. XXX*, abstract #1709 (1999)
B. Giese, R. Wagner, G. Neukum, R. Pappalordo, J.W. Head, in *Lunar Planet. Sci. Conf. XXXII*, abstract #1743 (2001a)
B. Giese, R. Wagner, G. Neukum, R. Pappalordo, J.W. Head, in *Lunar Planet. Sci. Conf. XXXII*, abstract #1751 (2001b)
B. Giese, R. Wagner, G. Neukum, P. Helfenstein, P.C. Thomas, Geophys. Res. Lett. **34**, 21203 (2007)
R. Greeley, E. Theilig, P. Christensen, Icarus **60**, 189–199 (1984)
R. Greeley, J.E. Klemaszewski, R. Wagner, Galileo Imaging Team, Planet. Space Sci. **48**, 829–853 (2000)

R. Greeley, C.F. Chyba, J.W. Head III, T.B. McCord, W.B. McKinnon, R.T. Pappalardo, P.H. Figueredo, in *Jupiter. The Planet, Satellites, and Magnetosphere*, ed. by F. Bagenal, T.E. Dowling, W.B. McKinnon. Planetary Science, vol. 1 (Cambridge Univ. Press, Cambridge, 2004), pp. 329–362
R. Greenberg, S.K. Croft, D.M. Janes, J.S. Kargel, L.A. Lebofsky, J.I. Lunine, R.L. Marcialis, H.J. Melosh, G.W. Ojakangas, R.G. Strom, in *Uranus*, ed. by Miranda J.T. Bergstralh et al. (U. Arizona Press, Tucson, 1991), pp. 693–735
R. Greenberg, P.E. Geissler, G. Hoppa, B.R. Tufts, D.D. Durda, R. Pappalardo, J.W. Head, R. Greeley, R. Sullivan, M.H. Carr, Icarus **135**, 64–78 (1998)
R.A.F. Grieve, M.J. Cintala, Meteoritics **27**, 526–538 (1992)
C.A. Griffith, T. Owen, G.A. Miller, T. Geballe, Nature **395**, 575–578 (1998)
C.A. Griffith, T. Owen, T.R. Geballe, J. Rayner, P. Rannou, Science **300**, 628–630 (2003)
C. Griffith et al., Science **310**, 474–477 (2005)
C. Griffith et al., Science **313**, 1620–1622 (2006)
R.E. Grimm, S.W. Squyres, J. Geophys. Res. **90**, 2013–2021 (1985)
J.E. Guest, R. Bianchi, R. Greeley, US Geol. Surv. Map I-1934 (1988)
H.J.M. Hanley, W.M. Haynes, R.D. McCarty, J. Phys. Chem. Ref. Data **6**, 597–601 (1977)
B. Hapke, F. Graham, Icarus **79**, 47–55 (1989)
A.G. Hayes, O. Aharonson, D. Stevenson, J. Lunine, R. Lorenz, R. Kirk, B. Stiles, A. Hayes, O. Aharonson, S. Wall, C. Sotin, A. Le Gall, R. Lopes, M. Janssen, Cassini RADAR Team, Joint Analysis of Titan's Surface Using The Cassini Vims And Radar Instruments. Division for Planetary Sciences annual meeting, Ithaca, abstract #34.06 (2008)
P. Hayne, T. McCord, J.-Ph. Combe, J.W. Barnes, G. Hansen, Titan; observational constraints on cryovolcanism. LPSC 39, abstr. 2010 (2008)
J.W. Head, R.T. Pappalardo, J. Kay, G. Collins, L. Prockter, R. Greeley, C. Chapman, M. Carr, M.J.S. Belton, Galileo Imaging Team, in *Lunar and Planetary Science Conference Abstracts* (1998), p. 1666
J.W. Head, R.T. Pappalardo, R. Sullivan, J. Geophys. Res. **104**, 24223–36 (1999)
J. Head et al., Geophys. Res. Lett. **29**(24), 2151 (2002). doi:10.1029/2002GL015961
P. Helfenstein, Nat. Geosci. **3**, 75–76 (2010)
P. Helfenstein, P.C. Thomas, J. Veverka, Nature **338**, 324–326 (1989)
J. Hillier, S.W. Squyres, J. Geophys. Res. **96**, 15,665–15,674 (1991)
G.V. Hoppa, B.R. Tufts, R. Greenberg, P.E. Geissler, Science **285**, 1899–1902 (1999)
V.M. Horner, R. Greeley, Icarus **51**, 549–562 (1982)
R. Hueso, A. Sanchez-Lavega, Nature **442**, 428–431 (2006)
T.A. Hurford, R.A. Beyer, B. Schmidt, B. Preblich, A.R. Sarid, R. Greenberg, Icarus **177**, 380–396 (2005). doi:10.1016/j.icarus.2005.06.019
L. Iess, N.J. Rappaport, P. Tortora, J. Lunine, J.W. Armstrong, S.W. Asmar, L. Somenzi, F. Zingoni, Icarus **190**, 585–593 (2007)
W.L. Jaeger, E.P. Turtle, L.P. Keszthelyi, J. Radebaugh, A.S. McEwen, R.T. Pappalardo, J. Geophys. Res. **108**, 5093 (2003)
D.M. Janes, H.J. Melosh, J. Geophys. Res. **93**, 3127–3143 (1988)
D.J. Jankowski, S.W. Squyres, Science **241**, 1322–1325 (1988)
R. Jaumann, R.H. Brown, K. Stephan, J.W. Barnes, L.A. Soderblom, C. Sotin, S. Le Mouélic, R.N. Clark, J. Soderblom, B.J. Buratti, R. Wagner, T.B. McCord, S. Rodriguez, K.H. Baines, D.P. Cruikshank, P.D. Nicholson, C.A. Griffith, M. Langhans, R.D. Lorenz, Icarus **197**, 526–538 (2008)
R. Jaumann, R. Kirk, R. Lorenz, R. Lopes, E. Stofan, H. Uwe Keller, C. Wood, E. Turtle, C. Sotin, L.A. Soderblom, M. Tomasko, in *Titan after Cassini-Huygens*, ed. by R. Brown, M. Dougherty, L. Esposito, T. Krimigis, J.-P. Lebreton, H. Waite (Springer, Berlin, 2009), pp. 75–140
K.B. Jones, J.W. Head, R.T. Pappalardo, J.M. Moore, Icarus **164**, 188–198 (2003)
J.S. Kargel, P. Delmelle, D.B. Nash, Icarus **124**, 248–280 (1999)
S.A. Kattenhorn, Europa. Icarus **157**, 490–506 (2002)
S.A. Kattenhorn, T. Hurford, in *Europa*, ed. by R.T. Pappalardo, W.B. McKinnon, K. Khurana (Univ. Arizona Press, Tucson, 2009), pp. 199–236
J.E. Kay, J.W. Head, in *Lunar and Planetary Science Conference Abstracts*, 1103 (1999)
L. Keszthelyi, W.L. Jaeger, E.P. Turtle, M. Milazzo, J. Radebaugh, Icarus **169**, 271–286 (2004)
L. Kezthelyi et al.,,Icarus **192**, 491–502 (2007). doi:10.1016/j.icarus.2007.07.008
B.N. Khare, C. Sagan, E.L. Bandurski, B. Nagy, Science **1999**, 1199–1201 (1978)
S.W. Kieffer, R. Lopes-Gautier, A.S. McEwen, L. Keszthelyi, R. Carlson, Science **288**, 1204–1208 (2000)
M. Kirchoff, W. McKinnon, Icarus **201**, 598–614 (2009)
R.L. Kirk et al., in *Neptune and Triton*, ed. by D.P. Cruikshank, M.S. Matthews (Univ. of Ariz. Press, Tucson, 1995), pp. 949–990

R.L. Kirk, P. Callahan, R. Seu, R.D. Lorenz, F. Paganelli, R. Lopes, C. Elachi, Cassini RADAR Science Team, in *Proc. Lunar Planet. Sci. Conf. 36th*, Abstract 2227 (2005)
R.L. Kirk, E. Howington-Kraus, B.L. Redding, T.L. Becker, E.M. Lee, B.W. Stiles, S. Hensley, A.G. Hayes, R.M.C. Lopes, R.D. Lorenz, K.L. Mitchell, J. Radebaugh, F. Paganelli, L.A. Soderblom, E.R. Stofan, C.A. Wood, S.D. Wall, Cassini RADAR Team, Icarus (2009, submitted)
D. Korycansky, K. Zahnle, Planet. Space Sci. **53**, 695–710 (2005)
N. Krupp et al., Space Sci. Rev. (2010, this issue). doi:10.1007/s11214-010-9653-z
N. Lancaster, *Geomorphology of Desert Dunes* (Routeledge, London, 1995)
M. Langhans, R. Jaumann, K. Stephan, R.H. Brown, B.J. Buratti, R. Clark, K.H. Baines, P.D. Nicholson, R.D. Lorenz, in *LPSC 40*, abstract #1681 (2009)
S. Le Mouelic, P. Paillou, M.A. Janssen, J.W. Barnes, S. Rodriguez, C. Sotin, R.H. Brown, K.H. Baines, B.J. Buratti, R.N. Clark, M. Crapeau, P.J. Encrenaz, R. Jaumann, D. Geudtner, F. Paganelli, L. Soderblom, G. Tobie, S. Wall, J. Geophys. Res. **113** (2009). doi:10.1029/2007JE002965
R.M.C. Lopes, in *Encyclopedia of the Solar System*, ed. by L. McFadden, P. Weissman, T. Johnson (Academic Press, San Diego, 2006)
R. Lopes, D. Williams, Io after Galileo. Invited review for Reports on Progress in Physics, Institute of Physics Publishing, 68, pp. 303–340 (2005)
R. Lopes, L.W. Kamp, S. Douté, W.D. Smythe, R.W. Carlson, A.S. McEwen, P.E. Geissler, S.W. Kieffer, F.E. Leader, E. Barbinis, R. Mehlman, M. Segura, J. Shirley, L.A. Soderblom, J. Geophys. Res. **106**(E12), 33,053–33,078 (2001)
R. Lopes, L.W. Kamp, W.D. Smythe, P. Mouginis-Mark, J. Kargel, J. Radebaugh, E.P. Turtle, J. Perry, D.A. Williams, R.W. Carlson, S. Douté, Icarus **169**(1), 140–174 (2004)
R. Lopes et al., EOS **88**, 569–570 (2007)
R.M.C. Lopes, E.R. Stofan, R. Peckyno, J. Radebaugh, K.L. Mitchell, G. Mitri, C.A. Wood, R.L. Kirk, S.D. Wall, J.I. Lunine, A. Hayes, R. Lorenz, T. Farr, L. Wye, J. Craig, R.J. Ollerenshaw, M. Janssen, A. LeGall, F. Paganelli, R. West, B. Stiles, P. Callahan, Y. Anderson, P. Valora, L. Soderblom, Cassini RADAR Team, Icarus **205**, 540–588 (2010). doi:10.1016/j.icarus.2009.08.010
R. Lopes-Gautier, S. Douté, W.D. Smythe, L.W. Kamp, R.W. Carlson, A.G. Davies, F.E. Leader, A.S. McEwen, P.E. Geissler, S.W. Kieffer, L. Keszthelyi, E. Barbinis, R. Mehlman, M. Segura, J. Shirley, L.A. Soderblom, Science **288**, 1201–1204 (2000)
R.D. Lorenz, Planet. Space Sci. **44**(9), 1021–1028 (1996)
R.D. Lorenz, Science **290**, 467–468 (2000)
R.D. Lorenz, J.I. Lunine, Icarus **122**, 79–91 (1996)
R. Lorenz, J. Lunine, Planet. Space Sci. **53**, 557–576 (2005)
R.D. Lorenz, J. Radebaugh, Geophys. Res. Lett. **36**, L03202 (2009). doi:10.1029/2008GL036850
R.D. Lorenz, J.I. Lunine, J.A. Grier, M.A. Fisher, J. Geophys. Res. (Planets) **88**, 26,377–26,386 (1995)
R.D. Lorenz et al., Science **312**, 724–727 (2006a)
R.D. Lorenz, H.B. Niemann, D.N. Harpold, S.H. Way, J.C. Zarnecki, Meteorit. Planet. Sci. **41**, 1705–1714 (2006b)
R.D. Lorenz, C.A. Wood, J.I. Lunine, S.D. Wall, R.M. Lopes, K.L. Mitchell, F. Paganelli, Y.Z. Anderson, L. Wye, C. Tsai, H. Zebker, E.R. Stofan, Geophys. Res. Lett. **34** (2007). doi:10.1029/2006GL028971
R.D. Lorenz, R.M.C. Lopes, F. Paganelli, J.I. Lunine, R.L. Kirk, K.L. Mitchell, L.A. Soderblom, E.R. Stofan, G. Ori, M. Myers, H. Miyamoto, J. Radebaugh, B. Stiles, S.D. Wall, C.A. Wood, Cassini RADAR Team, Planet. Space Sci. (2008a). doi:10.016/j.pss.2008.02.009
R. Lorenz et al., Geophys. Res. Lett. **35** (2008b). doi:10.1029/GL032118
R.D. Lorenz, K.L. Mitchell, R.L. Kirk, A.G. Hayes, H.A. Zebker, P. Paillou, J. Radebaugh, J.I. Lunine, M.A. Janssen, S.D. Wall, R.M. Lopes, B. Stiles, S. Ostro, G. Mitri, E.R. Stofan, Cassini RADAR Team, Geophys. Res. Lett. **35**, L02206 (2008c). doi:10.1029/2007GL032118
R.D. Lorenz, C. Newman, J.I. Lunine, Icarus **207**, 932–937 (2010)
B.K. Lucchitta, Icarus **44**, 481–501 (1980)
B. Lucchitta, L.A. Soderblom, in *Satellites of Jupiter*, ed. by D. Morrison (Univ. of Arizona Press, Tucson, 1982), pp. 521–555
J. Lunine, Rev. Geophys. **31**, 133–150 (1993)
J.I. Lunine, S.K. Atreya, Nature Geosci. **1**, 159–164 (2008)
J.I. Lunine, D.J. Stevenson, Y.L. Yung, Science **222**, 1229–1230 (1983)
J.I. Lunine et al., Icarus **195**, 414–433 (2008)
J.F. McCauley, B.A. Smith, L.A. Soderblom, Nature **280**, 736–738 (1979)
A.S. McEwen, L. Keszthelyi, P. Geissler, D.P. Simonelli, M.H. Carr, T.V. Johnson, K.P. Klaasen, H.H. Breneman, T.J. Jones, J.M. Kaufman, K.P. Magee, D.A. Senske, M.J.S. Belton, G. Schubert, Icarus **135**, 181–219 (1998)

A.S. McEwen, L.P. Keszthelyi, R. Lopes, P.M. Schenk, J.R. Spencer, in *Jupiter. The Planet, Satellites, and Magnetosphere*, ed. by F. Bagenal, T.E. Dowling, W.B. McKinnon. Planetary Science, vol. 1 (Cambridge Univ. Press, Cambridge, 2004), pp. 307–328

A. McEwen, E. Turtle, J. Perry, D. Dawson, S. Fussner, G. Collins, C. Porco, T. Johnson, L. Soderblom, Bull. Am. Astron. Soc. **37**, Abstract 53.04 (2005)

W.B. McKinnon, H.J. Melosh, Icarus **44**, 454–471 (1980)

W.B. McKinnon, P.M. Schenk, Geophys. Res. Lett. **22**, 1829–1832 (1995)

W.B. McKinnon, P.M. Schenk, A.J. Dombard, Geology **29**, 103–106 (2001)

H.J. Melosh, in *Oxford Monogr. on Geol. and Geophys*, vol. 11 (Oxford Univ. Press, New York, 1989), p. 245

J. Mitchell, The drying of Titan's dunes: Titan's methane hydrology and its impact on atmospheric circulation, JGR-Planets (2008)

G. Mitri, A. Showman, J. Lunine, R. Lorenz, Icarus **186**, 385–394 (2007)

G. Mitri, J.I. Lunine, A.P. Showman, R. Lopes, Icarus **196**, 216–224 (2008). doi:10.1016/j.Icarus.2008.02.024

G. Mitri, M.T. Bland, A.P. Showman, J. Radebaugh, R.M.C. Lopes, J. Lunine, R.T. Pappalardo, Cassini RADAR Team, J. Geophys. Res. Planets (2010, submitted)

J.M. Moore, M.C. Malin, Geophys. Res. Lett. **15**, 225–228 (1988)

J.M. Moore, R.T. Pappalardo, Icarus (2010, submitted)

J.M. Moore, E. Asphaug, R.J. Sullivan, J.E. Klemaszewski, K.C. Bender, R. Greeley, P.E. Geissler, A.S. McEwen, E.P. Turtle, C.B. Phillips, B.R. Tufts, J.W. Head, R.T. Pappalardo, K.B. Jones, C.R. Chapman, M.J.S. Belton, R.L. Kirk, D. Morrison, Icarus **135**, 127–145 (1998)

J.M. Moore, E. Asphaug, D. Morrison, J.R. Spencer, C.R. Chapman, B. Bierhaus, R.J. Sullivan, F.C. Chuang, J.E. Klemaszewski, R. Greeley, K.C. Bender, P.E. Geissler, P. Helfenstein, C.B. Pilcher, Icarus **140**, 294–312 (1999)

J.M. Moore, E. Asphaug, M.J.S. Belton, B. Bierhaus, H.H. Breneman, S.M. Brooks, C.R. Chapman, F.C. Chuang, G.C. Collins, B. Giese, R. Greeley, J.W. Head, S. Kadel, K.P. Klaasen, J.E. Klemaszewski, K.P. Magee, J. Moreau, D. Morrison, G. Neukum, R.T. Pappalardo, C.B. Phillips, P.M. Schenk, D.A. Senske, R.J. Sullivan, E.P. Turtle, K.K. Williams, Icarus **151**, 93–111 (2001)

J.M. Moore, C.R. Chapman, E.B. Bierhaus, R. Greeley, F.C. Chuang, J. Klemaszewski, R.N. Clark, J.B. Dalton, C.A. Hibbitts, P.M. Schenk, J.R. Spencer, R. Wagner, in *Jupiter. The Planet, Satellites, and Magnetosphere*, ed. by F. Bagenal, T.E. Dowling, W.B. McKinnon. Planetary Science, vol. 1 (Cambridge Univ. Press, Cambridge, 2004), pp. 397–426

S.J. Morrison, P.C. Thomas, J. Veverka, J.A. Burns, M.S. Tiscareno, C.C. Porco, Bull. Am. Astron. Soc. **40**, 479 (2008)

S.L. Murchie, J.W. Head, J. Geophys. Res. **93**, 8795–8824 (1988)

National Institute of Standards and Technology (NIST), Chemistry Web Book, Standard Reference Data Base, no. 69, June 2005. http:webbook.nist.gov./chemistry

R.M. Nelson, L.W. Kamp, D.L. Matson, P.G.J. Irwin, K.H. Baines, M.D. Boryta, F.E. Leader, R. Jaumann, W.D. Smythe, C. Sotin, R.N. Clark, D.P. Cruikshank, P. Drossart, J.C. Pearl, B.W. Hapke, J. Lunine, M. Combes, G. Bellucci, J.-P. Bibring, F. Capaccioni, P. Cerroni, A. Coradini, V. Formisano, G. Filacchione, R.Y. Langevin, T.B. McCord, V. Mennella, P.D. Nicholson, B. Sicardy, Icarus **199**, 429–441 (2009). doi:10.1016/j.icarus.2008.08.013

G. Neukum, in *The Three Galileos the Man, the Spacecraft, the Telescope*, ed. by C. Barbieri, J. Rahe, T.V. Johnson, A.M. Sohus (Springer, Berlin, 1997), pp. 201–212

G. Neukum, R. Wagner, U. Wolf, B.A. Ivanov, J.W. Head, R.T. Pappalardo, J.E. Klemaszewski, R. Greeley, M.J.S. Belton, Galileo SSI Team, in *Lunar and Planetary Science Conference Abstracts* (1998), p. 1742

P. Nicholson, M. Hedman, R. Clark, M. Showalter, D. Cruikshank, J. Cuzzi, G. Filacchione, F. Capaccioni, P. Cerroni, G. Hansen, Icarus **193**, 182–212 (2008)

H. Niemann et al., Nature **438**, 779–784 (2005)

F. Nimmo, B. Giese, Icarus **177**, 327–340 (2005). doi:10.1016/j.icarus.2004.10.034

F. Nimmo, R.T. Pappalardo, Nature **441**, 614–616 (2006)

F. Nimmo, R.T. Pappalardo, B. Giese, Icarus **166**, 21–32 (2003). doi:10.1016/j.icarus.2003.08.002

F. Nimmo, P.C. Thomas, R.T. Pappalardo, W.B. Moore, Icarus **191**, 183–192 (2007)

J. Oberst, B. Schreiner, B. Giese, G. Neukum, J.W. Head, R.T. Pappalardo, P. Helfenstein, Icarus **140**, 283–293 (1999)

W.R. Osterkamp, E.R. Hedman, US Geol. Surv. Prof. Paper 1242 (1982)

R.T. Pappalardo, M.D. Coon, Lunar Planet Sci. **28**, 997–998 (1996)

R.T. Pappalardo, R.J. Sullivan, Icarus **123**, 557–567 (1996)

R.T. Pappalardo, S.J. Reynolds, R. Greeley, J. Geophys. Res. **102**, 13,369–13,379 (1997)

R.T. Pappalardo, J.W. Head, G.C. Collins, R.L. Kirk, G. Neukum, J. Oberst, B. Giese, R. Greeley, C.R. Chapman, P. Helfenstein, J.M. Moore, A. McEwen, B.R. Tufts, D.A. Senske, H.H. Breneman, K. Klaasen, Icarus **135**, 276–302 (1998)

R.T. Pappalardo, G.C. Collins, J.W. Head III, P. Helfenstein, T.B. McCord, J.M. Moore, L.M. Prockter, P.M. Schenk, J.R. Spencer, in *Jupiter*, ed. by F. Bagenal, T. Dowling, W. McKinnon (Cambridge University Press, Cambridge, 2004), pp. 363–396
E.M. Parmentier, J.W. Head, Icarus **47**, 100–111 (1981)
Q.R. Passey, E.M. Shoemaker, in *Satellites of Jupiter*, ed. by D. Morrison (University of Arizona Press, Tucson, 1982), pp. 379–434
J.G. Patel, R.T. Pappalardo, J.W. Head, G.C. Collins, H. Hiesinger, J. Sun, J. Geophys. Res. **104**(10), 24,057–24,074 (1999)
J.T. Perron, M.P. Lamb, C.D. Koven, I.Y. Fung, E. Yager, M. Ádámkovics, J. Geophys. Res. **111**, E11001 (2006). doi:10.1029/2005JE002602
J.E. Perry, E.P. Turtle, A.S. McEwen, D.D. Dawson, in *38th Lunar and Planetary Science Conference*, Houston, Abstract #2219 (2007)
R.J. Pike, in *Impact and Explosion Cratering*, ed. by D.J. Roddy, R.O. Pepin, R.B. Merrill (Pergamon Press, Oxford, 1977), pp. 489–509
R.J. Pike in *Proc. Lunar Planet. Sci. Conf. XI* (1980), pp. 2159–2189
R.J. Pike, in *Mercury*, ed. by F. Vilas, C.R. Chapman, M.S. Matthews (Univ. of Arizona Press, Tucson, 1988), pp. 165–251
J.B. Plescia, Nature **327**, 201–204 (1987)
J.B. Plescia, Icarus **73**, 442–461 (1988)
C.C. Porco, R.A. West, S. Squyres, A. McEwen, P. Thomas, C.D. Murray, A. Delgenio, A.P. Ingersoll, T.V. Johnson, G. Neukum, J. Veverka, L. Dones, A. Brahic, J.A. Burns, V. Haemmerle, B. Knowles, D. Dawson, T. Roatsch, K. Beurle, W. Owen, Space Sci. Rev. **115**, 363–497 (2004)
C.C. Porco et al., Science **307**, 1237–1242 (2005a)
C.C. Porco, E. Baker, J. Barbara, K. Beurle, A. Brahic, J.A. Burns, S. Charnoz, N. Cooper, D.D. Dawson, A.D. Del Genio, T. Denk, L. Dones, U. Dyudina, M.W. Evans, S. Fussner, B. Giese, K. Grazier, P. Helfenstein, A.P. Ingersol, R.A. Jacobson, T.V. Johnson, A. McEwen, C.D. Murray, G. Neukum, W.M. Owen, J. Perry, T. Roatsch, J. Spitale, S. Squyres, P. Thomas, L. Tiscareno, E.P. Turtle, A.R. Vasavada, J. Veverka, R. Wagner, R. West, Nature **434**, 159–168 (2005b)
C.C. Porco et al., Science **311**, 1393–1401 (2006)
C.C. Porco, P.C. Thomas, J.W. Weiss, D.C. Richardson, Science **318**, 1602 (2007)
L.M. Prockter, R.T. Pappalardo, Science **289**(5481), 941–943 (2000). doi:10.1126/science.289.5481.941
L.M. Prockter, G.W. Patterson, in *Europa*, ed. by R.T. Pappalardo, W.B. McKinnon, K. Khurana (Univ. Arizona Press, Tucson, 2009), pp. 237–258
L. Prockter, P. Schenk, Icarus **177**, 305–326 (2005). doi:10.1016/j.icarus.2005.08.003
L.M. Prockter, J.W. Head, R.T. Pappalardo, D.A. Senske, G. Neukum, R. Wagner, U. Wolf, J.O. Oberst, B. Giese, J.M. Moore, C.R. Chapman, P. Helfenstein, R. Greeley, H.H. Breneman, M.J.S. Belton, Icarus **135**, 317–344 (1998)
L.M. Prockter, J.W. Head III, R.T. Pappalardo, J.G. Patel, R.J. Sullivan, A.E. Clifton, B. Giese, R. Wagner, G. Neukum, J. Geophys. Res. **107**, E5 (2002). doi:10.1029/2000JE001458
J. Radebaugh, R. Lorenz, R. Kirk, J. Lunine, E. Stofan, R. Lopes, S. Wall, Cassini Radar Team, Icarus **192**, 77–91 (2007). doi:10.1016/j.icarus.2007.06.020
J. Radebaugh, R. Lorenz, J. Lunine, S. Wall, G. Boubin, E. Reffet, R. Kirk, R. Lopes, E. Stofan, L. Soderblom, M. Allison, Icarus **194**, 690–703 (2008)
J. Radebaugh, R. Lorenz, T. Farr, P. Paillou, C. Savage, C. Spencer, Linear dunes on Titan and Earth: Initial remote sensing. Geomorphology (2009, in press)
J. Radebaugh, R.D. Lorenz, S.D. Wall, R.L. Kirk, C.A. Wood, J.I. Lunine, E.R. Stofan, R.M.C. Lopes, P. Valora, T.G. Farr, A.G. Hayes, B. Stiles, G. Mitri, H. Zebker, M. Janssen, L. Wye, A. LeGall, K.L. Mitchell, Cassini Radar Team, Icarus (2010, submitted)
P. Rannou, F. Montmessin, F. Hourdin, S. Lebennois, Science **311**, 201–205 (2006)
C. Sagan, Nature **280**, 750–753 (1979)
E. Schaller, M. Brown, H. Roe, A. Bouchez, Icarus **182**, 224–229 (2006a)
E. Schaller, M. Brown, H. Roe, A. Bouchez, C. Trujillo, Icarus **184**, 517–523 (2006b)
P.M. Schenk, J. Geophys. Res. **96**, 1887–1906 (1991)
P.M. Schenk, J. Geophys. Res. **98**(E4), 7475–7498 (1993)
P.M. Schenk, J. Geophys. Res. **100**, 19,023–19,040 (1995)
P.M. Schenk, Nature **417**, 419–421 (2002)
P.M. Schenk, in *Lunar Planet. Sci. Conf. XXXVI*, abstract #2081 (2005a)
P.M. Schenk, in *Lunar Planet. Sci. Conf. XXXVI*, abstract #2321 (2005b)
P.M. Schenk, M.H. Bulmer, Science **279**, 1514–1517 (1998)
P. Schenk, M.P.A. Jackson, Geology **21**, 299–302 (1993)
P.M. Schenk, W.B. McKinnon, Icarus **72**, 209–234 (1987)

P.M. Schenk, W.B. McKinnon, Icarus **79**, 75–100 (1989)

P.M. Schenk, W.B. McKinnon, Icarus **89**, 318–346 (1991)

P.M. Schenk, J.M. Moore, J. Geophys. Res. **100**, 19,009–19,022 (1995)

P.M. Schenk, R.T. Pappalardo, Geophys. Res. Lett. **31**, L16703 (2004). doi:10.1029/2004GL019978

P.M. Schenk, E.P. Turtle, in *Europa*, ed. by R.T. Pappalardo, W.B. McKinnon, K. Khurana (Univ. Arizona Press, Tucson, 2009), pp. 181–199

P.M. Schenk, W.B. McKinnon, D. Gwynn, J.M. Moore, Nature **410**, 57–60 (2001)

P.M. Schenk, C.R. Chapman, K. Zahnle, J.M. Moore, in *Jupiter*, ed. by F. Bagenal, T. Dowling, W. McKinnon (Cambridge University Press, Cambridge, 2004), pp. 427–456

G. Schubert, T. Spohn, R.T. Reynolds, in *Satellites*, ed. by J.A. Burns, M.S. Matthews (U. Arizona Press, Tucson, 1986), pp. 224–292

E.M. Shoemaker, R.F. Wolfe, Creating timescales for the Galilean satellites, in *The Satellites of Jupiter*, ed. by D. Morrison (Univ. of Arizona Press, Tucson, 1982), pp. 277–339

E.M. Shoemaker, B.K. Lucchitta, D.E. Wilhelms, J.B. Plescia, S.W. Squyres, in *The Satellites of Jupiter*, ed. by D. Morrison (Univ. of Arizona Press, Tucson, 1982), pp. 435–520

D.P. Simonelli, J. Veverka, A.S. McEwen, Geophys. Res. Lett. **24**, 2475–2478 (1997)

D.P. Simonelli, L. Rossier, P.C. Thomas, J. Veverka, J.A. Burns, M.J.S. Belton, Icarus **147**, 353–365 (2000)

K.N. Singer, W.B. McKinnon, in *Lunar and Planetary Institute Conference*, Abstracts 39, 2415 (2008)

B.A. Smith, L.A. Soderblom, T.V. Johnson, A.P. Ingersoll, S.A. Collins, E.M. Shoemaker, G.E. Hunt, H. Masursky, M.H. Carr, M.E. Davies, A.F. Cook, J.M. Boyce, T. Owen, G.E. Danielson, C. Sagan, R.F. Beebe, J. Veverka, J.F. McCauley, R.G. Strom, D. Morrison, G.A. Briggs, V.E. Suomi, Science **204**, 951–957 (1979a)

B.A. Smith, L.A. Soderblom, R. Beebe, J. Boyce, G. Briggs, M. Carr, S.A. Collins, A.F. Cook II, G.E. Danielson, M.E. Davies, G.E. Hunt, A.P. Ingersoll, T.V. Johnson, H. Masursky, J.F. McCauley, D. Morrison, T. Owen, C. Sagan, E.M. Shoemaker, R.G. Strom, V.E. Suomi, J. Veverka, Science **206**, 927–950 (1979b)

B.A. Smith et al., Science **212**, 163–191 (1981)

B.A. Smith et al., Science **215**, 504–537 (1982)

P. Smith, M. Lemmon, R. Lorenz, L. Sromovksy, J. Caldwell, M. Allison, Icarus **119**, 336–349 (1996)

D.E. Smith, Viscous relaxation of craters on Enceladus. Masters Thesis, The University of Arizona (2008). Publication Number: AAT 1453010. Source: MAI 47/01, Feb. 2009

D.E. Smith, E.P. Turtle, H.J. Melosh, V.J. Bray, in *Lunar Planet. Sci. XXXVIII*, Abst. #2237 (2007)

W.D. Smythe, R. Lopes-Gautier, S. Doute', S.W. Kieffer, R.W. Carlson, L. Kamp, F.E. Leader, in *Bull. American Astron. Soc., 32nd DPS meeting*, vol. 32(3), p. 1047 (2000)

L. Soderblom, M. Tomasko, B. Archinal, T. Becker, M. Bushroe, D. Cook, L. Doose, L. Galuszka, T. Hare, A. Howington-Kraus, E. Karkoschka, R. Kirk, J. Lunine, E. McFarlane, B. Redding, B. Rizk, M. Rosiek, C. See, P. Smith, Planet. Space Sci. **55**, 2015–2024 (2007a). doi:10.1016/j.pss.2007.04.015

L. Soderblom et al., Planet. Space Sci. **55**, 2025–2036 (2007b)

L.A. Soderblom, R.H. Brown, J.M. Soderblom, J.W. Barnes, R.L. Kirk, C. Sotin, R. Jaumann, D.J. Mackinnon, D.W. Mackowski, K.H. Baines, B.J. Buratti, R.N. Clark, P.D. Nicholson, Icarus **204**, 610–618 (2009)

C. Sotin, R. Jaumann, B.J. Buratti, R.B Brown, R.N. Clark, L.A. Soderblom, K.H. Baines, G. Bellucci, J.-P. Bibring, F. Capaccioni, P. Cerroni, A. Coradini, D.P. Cruikshank, P. Drossart, V. Formisano, Y. Langevin, D.L. Matson, T.B. McCord, R.M. Nelson, P.D. Nicholson, B. Sicardy, L. Le Mouélic, S. Rodriguez, K. Stephan, C.K. Scholz, Nature **435**, 786–789 (2005)

C. Sotin, S. Le Mouelic, R.H. Brown, J. Barnes, L. Soderblom, R. Jaumann, B.J. Buratti, R.N. Clark, K.H. Baines, R.M. Nelson, P. Nicholson, Cassini VIMS Team, in *38th Lunar and Planetary Science Conference*, Houston Abstract #2444 (2007)

N.A. Spaun, J.W. Head, G.C. Collins, L.M. Prockter, R.T. Pappalardo, Geophys. Res. Lett. **25**, 4277–4280 (1998)

N.A. Spaun, L.M. Prockter, R.T. Pappalardo, J.W. Head, G.C. Collins, A. Antman, R. Greeley, in *Lunar Planet. Sci. XXX*, abstract #1847 (1999)

N.A. Spaun, J.W. Head, R.T. Pappalardo, The Galileo SSI Team, in *Lunar and Planetary Science Conference Abstracts* (2001), p. 1448

J.R. Spencer, The Surfaces of Europa, Ganymede, and Callisto: An Investigation Using Voyager IRIS Thermal Infrared Spectra. Ph.D. thesis, University of Arizona (1987a)

J.R. Spencer, Icarus **69**, 297–313 (1987b)

J.R. Spencer, J.C. Pearl, M. Segura, F.M. Flasar, A. Mamoutkine, P. Romani, B.J. Buratti, A.R. Hendrix, L.J. Spilker, R.M.C. Lopes, Science **311**, 1401–1405 (2006)

J.N. Spitale, C.C. Porco, Nature **449**, 695–697 (2007)

P.G. Spudis, *The Geology of Multi-Ring Basins: The Moon and Other Planets*. Cambridge Planet. Sci. Ser., vol. 8 (Cambridge Univ. Press, Cambridge, 1993), p. 263
S.W. Squyres, Icarus **46**, 156–168 (1981)
K. Stephan, Chemisch-physikalische Zusammensetzung der Ganymed-Oberfläche—Zusammenhänge mit geologischen Strukturen und deren Gestaltungsprozessen. Ph.D. Dissertation (in German), Freie Universität Berlin, Germany (2007). http://www.diss.fu-berlin.de/2006/343
K. Stephan, R. Jaumann, R.H. Brown, J.M. Soderblom, L.A. Soderblom, J.W. Barnes, C. Sotin, C.A. Griffith, R.L. Kirk, K.H. Baines, B.J. Buratti, R.N. Clark, D.M. Lytle, R.M. Nelson, P.D. Nicholson, Geophys. Res. Lett. **37** (2010). doi:10.1029/2009GL042312
S.A. Stern, W.B. McKinnon, Astron. J. **119**, 945–952 (2000)
B.W. Stiles, S. Hensley, Y. Gim, D.M. Bates, R.L. Kirk, A. Hayes, J. Radebaugh, R.D. Lorenz, K.L. Mitchell, P.S. Callahan, H. Zebker, W.T.K. Johnson, S.D. Wall, C.A. Wood, M. Janssen, F. Pelletier, R.D. West, C. Veeramacheneni, Cassini RADAR Team, Icarus **202**, 584–598 (2009)
E. Stofan et al., Nature **445**, 61–64 (2007)
E.R. Stofan, T. Farr, R. Kirk, R.M.C. Lopes, R. Lorenz, J.I. Lunine, K.L. Mitchell, P. Paillou, J. Radebaugh, S.W. Wall, C.A. Wood, Cassini Radar Team, in *Lunar Planet. Sci. Conf. 40th*, abstract #1043 (2009)
D.F. Strobel, Planet. Space Sci. **30**, 839–848 (1982)
R. Sullivan, R. Greeley, K. Homan, J. Klemaszewski, M.J.S. Belton, M.H. Carr, C.R. Chapman, R. Tufts, J.W. Head III, R. Pappalardo, J. Moore, P. Thomas, Galileo Imaging Team, Nature **391**, 371–373 (1998)
N. Takato, S.J. Bus, H. Terada, T.S. Pyo, N. Kobayashi, Science **306**, 2224–2227 (2004)
P.C. Thomas, J.A. Burns, L. Rossier, D. Simonelli, J. Veverka, C.R. Chapman, K. Klaasen, T.V. Johnson, M.J.S. Belton, Icarus **135**, 360–371 (1998a)
P.C. Thomas, M.E. Davies, T.R. Colvin, J. Oberst, P. Schuster, G. Neukum, M.H. Carr, A. McEwen, G. Schubert, M.J.S. Belton, Icarus **135**, 175–180 (1998b)
P.C. Thomas et al., Icarus **190**, 573–584 (2007a)
P.C. Thomas et al., Nature **448**, 50–56 (2007b)
G. Tobie, J.I. Lunine, C. Sotin, Nature **440**, 61–64 (2006)
G. Tobie et al., Space Sci. Rev. (2010, this issue). doi:10.1007/s11214-010-9641-3
T. Tokano, Icarus **194**, 243–262 (2008)
T. Tokano, G. Molina-Cuberos, H. Lammer, W. Stumptner, Planet. Space Sci. **49**, 539–560 (2001)
T. Tokano, C.P. McKay, F.M. Neubauer, S.K. Atreya, F. Ferri, M. Fulchignoni, H.B. Niemann, Nature **442**, 432–435 (2006)
M.G. Tomasko, B. Archinal, T. Becker, B. Bézard, M. Bushroe, M. Combes, D. Cook, A. Coustenis, C. de Bergh, L.E. Dafoe, L. Doose, S. Douté, A. Eibl, S. Engel, F. Gliem, B. Grieger, K. Holso, A. Howington-Kraus, E. Karkoschka, H.U. Keller, R. Kirk, R. Kramm, M. Küppers, E. Lellouch, M. Lemmon, J. Lunine, E. McFarlane, J. Moores, M. Prout, B. Rizk, M. Rosiek, P. Rüffer, S.E. Schröder, B. Schmitt, P. Smith, L. Soderblom, N. Thomas, R. West, Nature **438**(8), 765–778 (2005). doi:10.1038/nature04126
B.N. Tran, J.C. Jones, J.P. Ferris, P.D. Persans, J.J. Chera, Icarus **165**, 379–390 (2003)
B.R. Tufts, R. Greenberg, G. Hoppa, P. Geissler, Icarus **146**, 75–97 (2000)
E.P. Turtle, E. Pierazzo, Science **294**, 1326–1328 (2001)
E.P. Turtle, C.B. Phillips, G.C. Collins, A.S. McEwen, J.M. Moore, R.T. Pappalardo, P.M. Schenk, Galileo SSI Team, in *Lunar Planet. Sci. Conf. XXX*, abstr. No. 1882 [CD-Rom] (1999)
E.P. Turtle, W.L. Jaeger, L.P. Keszthelyi, A.S. McEwen, M. Milazzo, J. Moore, C.B. Phillips, J. Radebaugh, D. Simonelli, F. Chuang, P. Schuster, Galileo SSI Team, J. Geophys. Res. **106**, 33175–33200 (2001)
E.P. Turtle, W.L. Jaeger, P.M. Schenk, in *Io After Galileo: A New View of Jupiter's Volcanic Moon*, ed. by R.M.C. Lopes, J.R. Spencer (Springer, Berlin, 2007), pp. 109–131
E.P. Turtle, J.E. Perry, A.S. McEwen, A.D. DelGenio, J. Barbara, R.A. West, D.D. Dawson, C.C. Porco, Geophys. Res. Lett. **36**, L02204 (2009). doi:10.1029/2008GL036186
W.G. Van Dorn, Nature **220**, 1102–1107 (1968)
A. Verbiscer, R. French, M. Showalter, P. Helfenstein, Science **315**, 815 (2007)
R. Wagner, Investigations of chronostratigraphy, impact chronology and geological evolution of the Jovian satellite Callisto, based on Galileo SSI camera data. Ph.D. Dissertation (in German), Freie Universität Berlin, Germany (2007). http://www.diss.fu-berlin.de/2007/806
R.J. Wagner, G. Neukum, B. Giese, T. Roatsch, T. Denk, U. Wolf, C.C. Porco, in *Lunar and Planetary Institute Conference*, Abstracts 39, 1930 (2008)
S.D. Wall, R.M. Lopes, E.R. Stofan, C.A. Wood, J.L. Radebaugh, B.W. Stiles, R.M. Nelson, L.W. Kamp, M.A. Janssen, R.L. Lorenz, J.I. Lunine, T.G. Farr, G. Mitri, P. Paillou, F. Paganelli, K.L. Mitchell, Geophys. Res. Lett. **36**, L04203 (2009). doi:1029/2008GL036415
R. West, M. Brown, S. Salinas, A. Bouchez, H. Roe, Nature **436**, 670–672 (2005)
D.E. Wilhelms, US Geol. Surv. Map I-2422 (1997)

K.K. Williams, R. Greeley, Geophys. Res. Lett. **25**, 4273–4276 (1998)
D.A. Williams, R. Greeley, R. Lopes-Gautier, S. Douté, J. Geophys. Res. **106**(E12), 33,161–33,174 (2001)
D.A. Williams, E.P. Turtle, L.P. Kezthelyi, W.L. Jaeger, J. Radebaugh, M.P. Milazzo, A.S. McEwen, J.M. Moore, R.M.C. Lopes, R. Greeley, Icarus **169**(1), 80–97 (2004)
C.A. Wood, R. Lorenz, R. Kirk, R. Lopes, K. Mitchell, E. Stofan, Icarus, 334–344 (2010). doi:10.1016/j.icarus.2009.08.21
Y. Yung, M. Allen, J. Pinto, Appl. J. Suppl **55**, 465–506 (1984)
K. Zahnle, L. Dones, H.F. Levison, Icarus **136**, 202–222 (1998)
K. Zahnle, P. Schenk, H. Levison, L. Dones, Icarus **163**, 263–289 (2003)

Space Sci Rev (2010) 153: 113–154
DOI 10.1007/s11214-010-9665-8

Chemical Composition of Icy Satellite Surfaces

J.B. Dalton · D.P. Cruikshank · K. Stephan · T.B. McCord · A. Coustenis · R.W. Carlson · A. Coradini

Received: 29 July 2009 / Accepted: 14 June 2010 / Published online: 28 August 2010

Abstract Much of our knowledge of planetary surface composition is derived from remote sensing over the ultraviolet through infrared wavelength ranges. Telescopic observations and, in the past few decades, spacecraft mission observations have led to the discovery of many surface materials, from rock-forming minerals to water ice to exotic volatiles and organic compounds. Identifying surface materials and mapping their distributions allows us to constrain interior processes such as cryovolcanism and aqueous geochemistry.

The recent progress in understanding of icy satellite surface composition has been aided by the evolving capabilities of spacecraft missions, advances in detector technology, and laboratory studies of candidate surface compounds. Pioneers 10 and 11, Voyagers I and II, Galileo, Cassini and the New Horizons mission have all made significant contributions. Dalton (Space Sci. Rev., 2010, this issue) summarizes the major constituents found or inferred to exist on the surfaces of the icy satellites (cf. Table 1 from Dalton, Space Sci. Rev., 2010, this issue), and the spectral coverage and resolution of many of the spacecraft instruments that have revolutionized our understanding (cf. Table 2 from Dalton, Space Sci. Rev.,

J.B. Dalton (✉) · R.W. Carlson
Jet Propulsion Laboratory (JPL), California Institute of Technology, 4800 Oak Grove Drive, Pasadena, CA 91109, USA
e-mail: dalton@jpl.nasa.gov

D.P. Cruikshank
NASA Ames Research Center, Moffett Field, CA 94035-1000, USA

K. Stephan
German Aerospace Center (DLR), Rutherfordstrasse 2, 12489 Berlin, Germany

T.B. McCord
Bear Fight Institute, 22 Fiddler's Road, Winthrop, WA 98862, USA

A. Coustenis
Laboratoire d'études spatiales et d'instrumentation en astrophysique (LESIA), Observatoire de Paris-Meudon, 5, pl. Jules Janssen, 92195 Meudon, France

A. Coradini
Istituto di Fisica della Spazio Interplanetario (IFSI), Via Fosso del Cavaliere 100, 00133 Rome, Italy

2010, this issue). While much has been gained from these missions, telescopic observations also continue to provide important constraints on surface compositions, especially for those bodies that have not yet been visited by spacecraft, such as Kuiper Belt Objects (KBOs), trans-Neptunian Objects (TNOs), Centaurs, the classical planet Pluto and its moon, Charon.

In this chapter, we will discuss the major satellites of the outer solar system, the materials believed to make up their surfaces, and the history of some of these discoveries. Formation scenarios and subsequent evolution will be described, with particular attention to the processes that drive surface chemistry and exchange with interiors. Major similarities and differences between the satellites are discussed, with an eye toward elucidating processes operating throughout the outer solar system. Finally we discuss the outermost satellites and other bodies, and summarize knowledge of their composition. Much of this review is likely to change in the near future with ongoing and planned outer planet missions, adding to the sense of excitement and discovery associated with our exploration of our planetary neighborhood.

Keywords Composition · Icy satellites · Infrared spectroscopy

1 The Galilean Satellites of Jupiter

1.1 Surface Composition of the Icy Galilean Satellites

Io and the icy Galilean satellites Europa, Ganymede and Callisto have as many similarities as differences. They are often considered to lie along a continuum, both in terms of surface activity and composition. This ranges from the highly energetic and active innermost Galilean satellite, Io, to the outermost and less highly differentiated Callisto. Yet all spring from a common origin and undergo similar processes—at varying scales. The vigorous surface activity of Io is powered by the same tidal forces that maintain Europa's—and presumably Ganymede's and Callisto's—oceans. Charged particles trapped in Jupiter's magnetosphere are implanted in the surfaces of all four Galilean satellites. Material ejected from Io arrives at Europa, where it has been incorporated into the interior. Chemical processing, whether internally or externally driven, creates a number of compounds that have been detected and their distributions mapped using observations from recent spacecraft missions. From these observations we can extrapolate, with some confidence, the chemical and physical processes dominating the interiors.

1.1.1 Water Ice

Water is the dominant component of Europa, Ganymede and Callisto but not of Io, as evidenced by their bulk densities (Table 1). Each object has undergone considerable internal thermal evolution (e.g. Schubert et al. 2010), and, therefore, it is not surprising that the surfaces of Europa and Ganymede have a major ice component. Callisto is somewhat less evolved and so it has less ice and more hydrated/hydroxylated silicates on its surface. Io, being closest to Jupiter, has experienced the most heating and thus has not only differentiated but has lost most of any original water.

The possibility of water ice on these objects was first described in the literature by Kuiper (1957), Harris (1961), Moroz (1965) using spectrophotometry with ground-based telescopes. Spectroscopic observations proving the presence of water ice were first published by Johnson and McCord (1971), Pilcher et al. (1972), Fink et al. (1973) and Fink and Larson (1975), with an early review by Johnson and Pilcher (1977). Ground-based telescopic

Table 1 Physical parameters of the Galilean satellites

Satellite	Diameter km	Mass 10^{20} kg	Orbital radius R_J	Orbital radius 10^3 km	Orbital period days	Visual geometric Albedo	Inclin- ation	Eccen- tricity	Bulk density g/cm^3
Io	3643.2 ± 1	893.54	5.89	421.8	1.769	0.62	0.036	0.0041	3.528 ± 0.006
Europa	3121.6 ± 1	480.17	9.38	671.1	3.551	0.68	0.469	0.0094	3.014 ± 0.005
Ganymede	5262.4 ± 3.4	1482.4	14.97	1,070.4	7.155	0.44	0.170	0.0011	1.942 ± 0.005
Callisto	4820.6 ± 3.0	1076.3	26.33	1,882.7	16.69	0.19	0.187	0.0074	1.834 ± 0.004

observations continued with improved spectra showing more clearly the various water ice absorptions in the near-infrared part of the spectrum. The general conclusions were that water ice is a major constituent on the surfaces of Europa and Ganymede and less so on Callisto.

These surfaces were observed using spatially-resolved spectroscopy for the first time from the Galileo Mission's satellite fly-bys while in Jupiter orbit. Water ice on the Galilean satellites has been studied using the Galileo Near-Infrared Mapping Spectrometer (NIMS) data (Carlson et al. 1992). The spectrum of water ice is highly responsive to influences such as temperature, pressure, porosity, crystallinity, structural damage and formation history (Pauling 1935; Fink and Sill 1982; Grundy and Schmitt 1998). Two spectral features in particular are believed highly diagnostic of crystalline versus amorphous water ice. The first is a narrow temperature-sensitive band found near 1.65 μm in crystalline water ice that is not observed in amorphous ice. The second feature is the surface Fresnel reflectance peak near 3.1 μm which is considered to be diagnostic of the lattice order in the top few micrometers of the surface. Amorphous ice is crystallized by heating, while crystalline ice is amorphized through disruption by particle irradiation. Hansen and McCord (2004) report that the lattice structure of water ice on the three icy satellites varies from crystalline to amorphous, and provide analyses of the distributions of these two forms. The surface ice is predominantly amorphous on Europa and predominantly crystalline on Callisto, while both types of ice are found on Ganymede. The distribution of Ganymede ice properties shows a broad global pattern of more amorphous ice in the high-latitude Jovian-facing hemisphere and in the low-latitude trailing hemisphere. The ice at ~1 mm depth on all three satellites is predominantly crystalline. The occurrence of crystalline and amorphous ice suggests a balance between the disruption and crystallization, with Callisto dominated by thermal crystallization and Europa by radiative disruption, and with nearly equal rates between the two processes at Ganymede. The broad, global patterns of crystalline and amorphous surface ice on Ganymede are consistent with the expected distribution of energetic magnetospheric ions and neutrals, but the detailed distribution implies additional control by geologic or other factors. These Galileo measurements, although spatially resolving the surface far beyond Earth-based telescopes, do not have the resolution required to relate the ice properties to detailed geology. This must await the next planned campaign of NASA and ESA Jupiter System missions.

1.1.2 Hydrated Materials

Water absorption features in the dark and disrupted surface regions on Europa and, to a lesser extent, Ganymede, appear asymmetric and distorted, evidence of shifted vibrational transition frequencies. This would be expected if the water were bound to some host molecule. Many hydrated materials exhibit such spectral behavior. Water molecules, attached

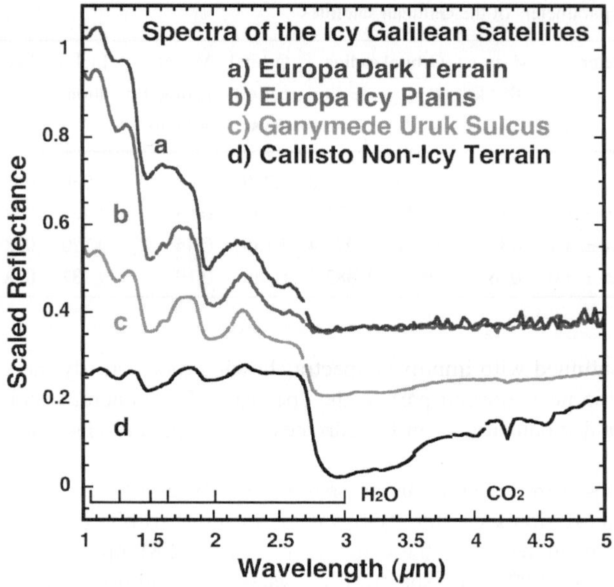

Fig. 1 Galileo NIMS spectra of the icy Galilean satellites. The primary H_2O absorption features at ~1.0, ~1.25, ~1.5, ~2.0 and ~3.0 μm are evident on all four satellites. However they are much weaker on Callisto, indicating the dominance of other components such as silicates, CO_2 and possible hydrocarbons. The 1.65-μm crystalline water ice feature is apparent on all three satellites, but is faint in the Europa dark terrain spectra, which also exhibit strongly asymmetric absorption feature shapes at 1.5 and 2.0 μm representative of bound water such as hydrated salts and sulfuric acid hydrate. The Europa and Ganymede spectra have been offset for clarity, though the Europa spectra are offset by the same amount to highlight the near-saturation at 3.0 μm. The signal is extremely low longward of 3.0 μm, and dominated by radiation spikes, demonstrating the challenges inherent in observing Europa at these longer wavelengths

to a host molecule by weak hydrogen bonds, vibrate at a number of frequencies close to those observed in water ice (Salisbury and Hunt 1971; Crowley 1991). Examination of several compounds in the laboratory has demonstrated that sulfate hydrates have absorption features with similar spectral shapes, center frequency positions, relative strengths and full widths at half maximum (FWHM) to those observed in the spectra of dark material. In particular, varieties of $MgSO_4 \cdot nH_2O$, $Na_2SO_4 \cdot nH_2O$ and $H_2SO_4 \cdot nH_2O$ (where n indicates the number of waters of hydration) provide exceptionally good matches (Dalton et al. 2005). Sulfuric acid may be brought up from the subsurface, or created by radiolysis of water ice with implanted S or endogenic sulfurous salts; or perhaps some combination of these. The hydrated sulfate salts are predicted on the basis of geochemical arguments (Kargel 1991, 2001) and are more difficult to produce through radiolysis.

All of the sulfate hydrates exhibit similar spectral features. This is because the absorptions arise from vibrations of the solvated water molecules, not from the host sulfate complex. However the electromagnetic influence of the host constrains the configuration of the whole molecule, affecting the vibrational frequencies and producing detailed spectral structure that can be highly diagnostic of the host compound and the level of hydration (Dalton 2003). Note that no single compound yet studied has perfectly matched the Europa observations when considered by itself. Other compounds, not yet studied, or a mixture of compounds, may be present on the surface. Alteration of spectral properties by radiation is a process known to occur on Galilean satellites (Nash and Fanale 1977) but requiring further

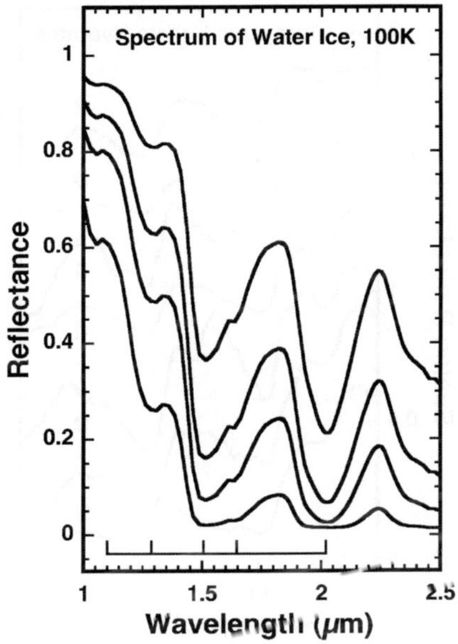

Fig. 2 Spectra of water ice at 100 K. Synthetic spectra, generated from optical constants of cryogenic water ice (Grundy and Schmitt 1998) demonstrate the changes in band depth and shape with grain size. Grain diameters, from top to bottom, are 25 μm, 100 μm, 250 μm, and 1 mm. The primary water ice absorption features are indicated by the bars along the bottom. No offsets or scaling have been applied. The smaller grain sizes have higher reflectance continuum levels due to the higher influence of volume scattering. As grain size increases, optical path length increases, enhancing absorption and reducing reflectance. Note the saturation of the 2.0-μm absorption feature at large grain sizes: no information about the surface is retrievable because no photons are returned from within the surface

study (Johnson et al. 2004; Carlson et al. 2009) before quantitative models can be developed incorporating these effects into interpretations of spectral observations of icy satellites.

Infrared spectra of water, and consequently of hydrates, are highly sensitive to temperature (Pauling 1935; Ockman 1958; Hobbs 1974). At the surface temperatures of icy Galilean satellites (typically within 80–150 K) the infrared spectra of hydrated materials are markedly different than they are at room temperature or even at 273 K. Recent work using cryogenic laboratory spectra of hydrates indicates that the best spectral matches to Europa observations are achieved using mixtures of sulfuric acid hydrate, sodium and magnesium sulfate hydrates, and water ice of grain sizes varying from 50 to 250 microns (Dalton 2007b; Dalton et al. 2009; Shirley et al. 2010).

The composition of the dark surface terrain on Europa may not be identical everywhere (Hibbitts and Hansen 2007; Shirley et al. 2010). This may be due to variations in relative proportions of hydrates in the surface. Furthermore, the spectral features indicating hydrates are correlated with the dark, reddish terrain units, most of which indicate disruption or heating of the surface from below. Even if the water of hydration spectral features can be explained by sulfate hydrates, there must be another component to explain the reddish tint since all of the sulfate hydrates are white or transparent at visible wavelengths. Sulfur allotropes from the radiolysis of sulfates can explain the association of hydrate with dark material produced through Europa's radiolytic sulfur cycle (Carlson et al. 1999b, 2002, 2005).

Ganymede's surface also contains hydrated material, suggested to be chemically similar to Europa's hydrates and derived from a briny ocean within Ganymede (McCord et al.

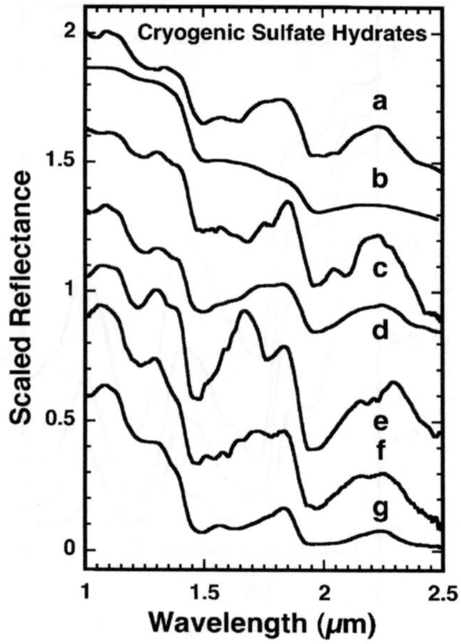

Fig. 3 Spectra of hydrated sulfate salts and sulfuric acid hydrate. (**a**) $MgSO_4$ Brine; (**b**) Sulfuric Acid Hydrate, $H_2SO_4 \cdot 8H_2O$; (**c**) Bloedite, $Na_2Mg(SO_4)_2 \cdot 4H_2O$; (**d**) Magnesium Sulfate Undecahydrate, $MgSO_4 \cdot 11H_2O$ (Meridianiite); (**e**) Mirabilite, $Na_2SO_4 \cdot 10H_2O$; (**f**) Hexahydrite, $MgSO_4 \cdot 6H_2O$; and (**g**) Epsomite, $MgSO_4 \cdot 7H_2O$. All are measured at 100 K, except for Bloedite and Hexahydrite, 120 K, and Sulfuric Acid Hydrate, measured at 140 K. The major absorption features near 1.25, 1.5, and 2.0 μm arise from the bound water of hydration in the crystal lattice, while the exact shape and position of the features (and the fine structure within them) are determined by the interaction of the host sulfate molecule with the waters of hydration. Spectra (**c**) and (**f**) from Dalton (2003); (**b**) from Carlson et al. (1999b); remainder from Dalton et al. (2005)

2001). The dark material on Ganymede could contain sulfur if the hydrated compounds on Ganymede are sulfate salts. A second hydrated species associated with dark crater rays was noted and may be due to contamination or dehydration by impacts (Hibbitts and Hansen 2007). Impacts can also contribute dark material to all of the satellites, but Europa's active surface, with a crater age of about 50 My (Zahnle et al. 2003), must be subducting this infall material and producing less contaminated ice compared to Ganymede and Callisto.

1.1.3 Other Species

In addition to water ice and hydrate, there are other species that are known or thought to exist on the icy satellites; we group these into oxygen, carbon, sulfur, and alkali compounds.

Molecular oxygen is known to exist on all three icy Galilean satellites (Spencer et al. 1995; Calvin et al. 1996; Calvin and Spencer 1997; Spencer and Calvin 2002) and is a radiolytic product of water (Johnson et al. 2004, 2005) that is trapped within defects in the ice (Johnson and Jesser 1997). Irradiation of oxygen-rich ice surfaces can produce ozone (Teolis et al. 2006) and this species has been observed on Ganymede (Noll et al. 1996a, 1996c). Radiolysis of water ice also produces hydrogen peroxide, first found on Europa using infrared and ultraviolet spectra (Carlson et al. 1999a) and later inferred from ultraviolet measurements to exist on Ganymede and Callisto (Hendrix et al. 1999; Lopes et al. 2008).

Ultraviolet signatures of sulfur dioxide are seen in spectra of Europa and Callisto (Lane et al. 1981; Noll et al. 1995, 1997a, 1997b; Lane and Domingue 1997). Europa's SO_2 exists on the trailing side, consistent with implantation of sulfur from the Io plasma torus (Eviatar et al. 1981; Lane et al. 1981) while Callisto's SO_2 was suggested to arise from a neutral sulfur wind (Lane and Domingue 1997). Infrared signatures from Europa, Ganymede, and Callisto have been interpreted as arising from SO_2 (McCord et al. 1997, 1998a; Hibbitts et al. 2000; Hansen and McCord 2008) but carbonates are also a possibility (Johnson et al. 2004). Implanted sulfur from the low-energy plasma torus, present in the form of polymeric sulfur, is likely the cause of the reddish color of Europa's trailing hemisphere (McEwen 1986; Carlson et al. 2009) and high-energy sulfur ions can impact both hemispheres, producing the sulfur signature observed from the leading hemisphere (Spencer et al. 1995). If endogenic sulfates are emplaced at the surface, they could also be sources of sulfur. A band at 3.88 μm in NIMS spectra of Ganymede and Callisto has been interpreted as the S–H stretching transition (McCord et al. 1997, 1998a). H_2S is a possible absorber, but is unstable at Galilean satellite temperatures (Schmitt and Rodriguez 2003). Carbonic acid (H_2CO_3), produced by the radiolysis of CO_2 in H_2O ice (Moore et al. 1991), has also been suggested to explain this feature (Hage et al. 1998).

Carbon dioxide has been observed on the icy Galilean satellites (Carlson et al. 1996; McCord et al. 1997, 1998a; Smythe et al. 1998; Hibbitts et al. 2000, 2003; Hansen and McCord 2008) as well as other outer solar system objects (see below). The source of this ubiquitous molecule is unknown; it could arise from internal degassing as suggested by (Moore et al. 1999) for Callisto, or the CO_2 could be a radiolytic and photolytic product of carbonaceous material in water ice. CO_2 is not expected to occur as pure CO_2 ice due to its high vapor pressure, but rather as molecules trapped in ice or adsorbed on minerals (Carlson and Hand 2006; Hibbitts and Szanyi 2007). Related products, such as carbonic acid and carbonates, may also manifest themselves (see above). A feature at 4.57 μm in NIMS spectra of Ganymede and Callisto has been interpreted as a nitrile compound (–C≡N) (McCord et al. 1997, 1998a) similar to features formed in HCN polymers and tholins (Khare et al. 1984, 1994; Imanaka et al. 2004; Quirico et al. 2008). Hydrocarbons may be present on the Galilean satellites as suggested by the possible presence of symmetric and asymmetric stretch bands of alkanes ($-CH_2-$) in NIMS spectra of Callisto (McCord et al. 1997, 1998a). The dark terrain of Ganymede and Callisto seems to correlate with increased CO_2 absorption band strength (Hibbitts et al. 2000, 2003; Cruikshank et al. 2009) despite the high reflectance of CO_2 ice in the visible. Organic materials or dust from meteoritic infall have been suggested as darkening agents; another possibility is that these may be lag deposits left behind by sublimation degradation in these landscapes (Moore et al. 2004).

1.1.4 Metals

Sodium, potassium, and chlorine are sputtered from the surface of Io, and subsequently escape, becoming ionized and entrained in the plasma torus, and then radially diffuse outward. A fraction of these ions can strike the outer icy satellites, be sputtered, and form tenuous neutral bound atmospheres and exospheres around the icy moons. Endogenic salts from the surface (McCord et al. 1998b) can also contribute to these atmospheres. Europa has a Na and K atmosphere (Brown and Chaffee 1974; Brown and Hill 1996; Brown 2001) and models (Johnson 2000; Johnson et al. 2002; Leblanc et al. 2002, 2005) suggest that much of it may arise from an endogenic component if there is significant diversion of plasma flow around Europa. However the amount of diversion is currently unknown and definitive conclusions cannot yet be drawn (Carlson et al. 2009). A limit on Ganymede's Na atmosphere is placed at $\sim 1/13$ of Europa's (Brown 1997).

1.2 Differences Among the Icy Galilean Satellites

Changes in the spectral properties of Europa, Ganymede and Callisto can be viewed as the result of competing influences that vary with radial distance from Jupiter. Two processes dominate these interactions. First, energetic particles trapped in the Jovian magnetosphere impact the icy surfaces and cause radiolysis, sputtering and implantation of ions (Cooper et al. 2001; Johnson et al. 2004). Secondly, thermal annealing and re-crystallization of the ice during the diurnal temperature cycle counteracts the radiolytic disruption. The incident radiation flux and the surface temperatures change appreciably with distance from Jupiter. The radiation flux decreases by a factor of about $\sim$300 between Europa and Callisto, whereas the surface temperature increases from a maximum temperature of 132 K at Europa (Spencer et al. 1999) to 158 K measured at Callisto (Moore et al. 2004) resulting in differences in the specific thermal crystallization rate by at least five orders of magnitude among the three icy satellites.

Radiolytic disruption of the crystalline structure of surface ice is an important physical modification process on Europa (Paranicas et al. 2001; Johnson et al. 2004; Carlson et al. 2009) that leads to conversion of crystalline to amorphous H_2O ice, forming a top surface layer of amorphous water ice all over the satellite (Hansen and McCord 2004). Thermal annealing and re-crystallization is less operative on Europa because of its relatively cold surface temperatures (Spencer et al. 1999).

The intrinsic magnetic field of Ganymede diverts much of the local plasma flux toward high latitudes. This is supported by the fact that amorphous ice dominates Ganymede's poles (Hansen and McCord 2004) where particles can easily reach the surface. In contrast, the equatorial region of Ganymede is partly shielded from radiation by the intrinsic magnetic field. Only plasma particles $\geq$10 keV penetrate Ganymede's shield. Here, the crystallization timescale seems very similar to the radiolytic disruption timescale resulting in an equilibrium distribution of amorphous and crystalline ice on a global scale. Geologic surface processes at local scales may also influence details of the distribution (Hansen and McCord 2004).

Although the flux of energetic particles is lowest at Callisto, as the furthest large satellite from Jupiter, its surface is fully exposed to this flux unlike in the case of Ganymede. Callisto's total flux of energetic ($\geq$20 keV) particles still reaches a level similar to that at the equatorial region of Ganymede (Cooper et al. 2001). However, as Callisto is the warmest of the icy Galilean satellites, thermal annealing dominates, resulting in a surface completely covered by crystalline ice particles (Hansen and McCord 2004).

1.3 Io

The innermost Galilean satellite Io is strongly heated by tidal effects (Peale et al. 1979), making it the most volcanically active body in the solar system with vigorous internal convection that has dehydrated the moon (Keszthelyi and McEwen 1997; McEwen et al. 1998b; Keszthelyi et al. 2004). Io's volcanic plumes and lava flows very likely arise from silicate magmas containing other compounds as minor constituents. The colorful surface is coated with sulfurous volatiles—sulfur dioxide and elemental sulfur—and other trace species, hiding the silicates and precluding their unequivocal identification. Below we discuss the observed sulfur compounds, the silicates, and known and possible minor species. Io's surface composition was recently reviewed (Carlson et al. 2007); here we include subsequently reported work on Io's sulfur dioxide (SO_2) distribution and the possible presence of disulfur monoxide (S_2O) and polysulfur oxide (PSO).

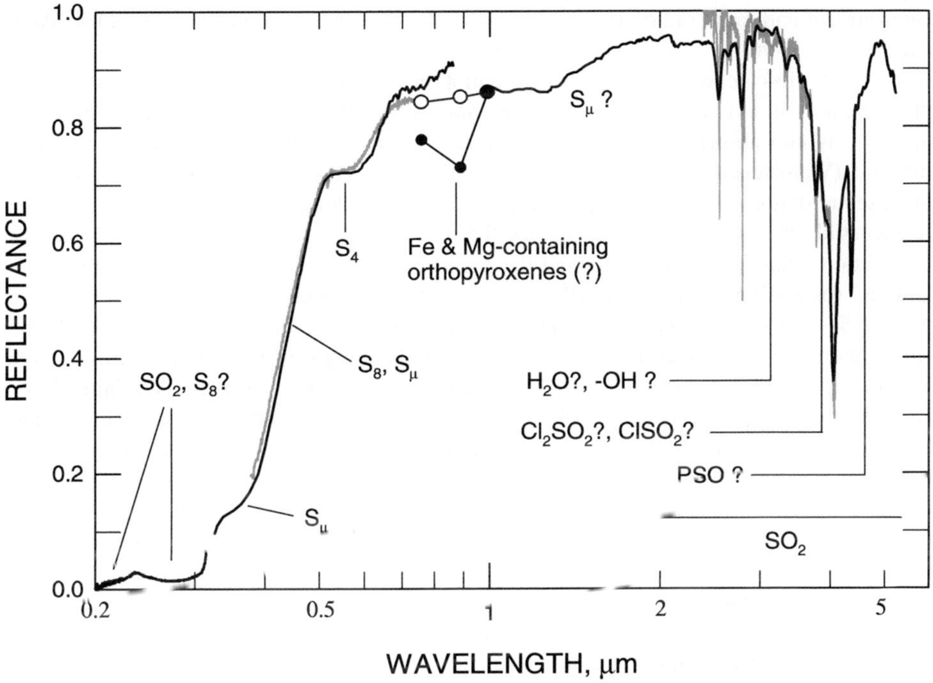

Fig. 4 Composite reflectance spectrum of Io. Measurements in the 0.2 to 0.310-μm region are from Jessup et al. (2002); those from 0.330 μm to 0.860 μm (*black line*) and from 0.380 μm to 0.78 μm (*gray line*) are from Nelson and Hapke (1978) and Spencer et al. (1995), respectively. White and dark areas on Io (*open* and *filled circles*, respectively) are from Geissler et al. (1999). Modest resolution near-infrared measurements (*black line*) are from Carlson et al. (1997) while the higher-resolution spectrum (*gray line*, 2–5 μm) is from Schmitt and Rodriguez (2003). Absorption positions and shapes of known and possible species are shown, including polymeric sulfur (S_μ) and polysulfur oxide (PSO) (Baklouti et al. 2008). Adapted from Carlson et al. (2007)

Most of the absorption features characterizing Io's spectrum can be attributed to SO_2 frost (Fanale et al. 1979; Hapke 1979; Smythe et al. 1979; Schmitt et al. 1994; Nash and Betts 1995; Carlson et al. 1997) (Fig. 4). The global distribution of SO_2 as well as the spatial variations in its grain size on Io's surface has been studied in a variety of ways. Derived maps of Carlson et al. (1997) and Laver and de Pater (2008, 2009) show that optically thick and relatively coarse-grained SO_2 deposits are concentrated close to Io's equator and confirm earlier measurements of McEwen et al. (1988), who studied Io's surface using ultraviolet and visible Voyager images. These equatorial deposits are strongly associated with active volcanoes and probably result from volcanic venting of gaseous SO_2 into the plumes and atmosphere and subsequent deposition onto colder surfaces as frost. In contrast, extensive fine-grained SO_2 frost has been found in medium and higher latitudes (Carlson et al. 1997; Douté et al. 2001; Laver and de Pater 2008, 2009). SO_2 frost easily sublimates during the day due to relatively high surface temperatures (Sandford and Allamandola 1993). Although most of it probably re-condenses during the nighttime, some of it migrates toward higher latitudes and colder regions (Matson and Nash 1983). These SO_2 deposits appear transparent under normal-incidence illumination and therefore are thought to be optically thin (Geissler et al. 2001). They remain as the originally condensed smaller grains because of the colder environment and rapid radiolytic destruction of the SO_2 grains at northern latitudes (Wong and Johnson 1996).

Io's full disk reflectance spectrum cannot be explained by SO_2 alone. It requires the addition of a material such as elemental sulfur that absorbs ultraviolet and blue light but is non-absorbing and spectrally bland, devoid of spectral features, in the near-infrared (Wamsteker et al. 1973; Fanale et al. 1982; Nash et al. 1986). The detection of gaseous diatomic sulfur (S_2) in the plume of the volcano Pele (Spencer et al. 2000; Jessup et al. 2007) serves as evidence for the presence of elemental sulfur on Io's surface. In addition, many plume deposits and the polar regions appear bright red at visible wavelengths, suggesting the presence of tetrasulfur (S_4) produced by polymerization of condensed S_2-rich volcanic gases (in the equatorial regions) or due to breakdown of higher sulfur allotropes by charged particle irradiation (in polar regions) (Wong and Johnson 1996; Spencer et al. 1997; Geissler et al. 1999; Spencer et al. 2000; Lopes and Williams 2005). By analogy with Pele and an association with vent structures, other red deposits indicate eruptions that have happened recently. In addition, reflectance spectra of Io (Fig. 4) show a feature at 0.53 μm attributable to S_4 which gradually polymerizes into the more stable cyclo-octal S_8, polymeric sulfur, and possibly other allotropes (Moses and Nash 1991; Carlson et al. 2007; Baklouti et al. 2008), turning the color of the deposits from red to pale yellow that characterizes up to 40% of Io's intensely colored surface (Spencer et al. 1997; McEwen et al. 1998b; Geissler et al. 1999). Some red deposits show enhanced concentrations of SO_2 (Lopes-Gautier et al. 2000), which is possibly co-deposited with S_2 from SO_2+ S_2-rich volcanic gases (Spencer et al. 2000).

Io's bulk density and moment of inertia as well as the chemistry of the solar system indicate that the satellite's mantle and lithosphere are composed of silicates. Spectroscopy of the surface is diagnostic of volatile compounds and therefore more indicative of Io's volatile reservoir than of the bulk composition. However, a few relatively small areas (1.4% of the surface) were found to be free of sulfurous compounds. They occur in the vicinity of volcanic hot spots where dark recent lavas are exposed (McEwen et al. 1998a, 1998b; Lopes and Williams 2005). These areas are characterized by very high surface temperatures causing rapid vaporization of any SO_2 deposits (Carlson et al. 1997). Temperatures up to 1825 K have been estimated for actively erupting lavas from Galileo Solid-State Imager (SSI) and Near-Infrared Mapping Spectrometer (NIMS) measurements (McEwen et al. 1998a; Davies et al. 2001; Lopes et al. 2001) as well as from Cassini Imaging Science Subsystem (ISS) measurements (Radebaugh et al. 2004). Such temperatures are much higher than the melting point of sulfur and higher even than the melting point of common basalts, requiring either superheating of the magma or (more likely) high-melting point silicates with ultramafic compositions rich in magnesium.

Despite the evidence for silicate volcanism, no spectral features definitively attributable to silicates have been seen, indicating very efficient coating of silicates with more volatile materials. Only the Galileo SSI instrument has fully resolved the dark lavas, detecting an absorption feature at 0.89 μm (Fig. 4) (Geissler et al. 1999). This spectral signature is consistent with Fe-containing Mg-rich orthopyroxenes common in terrestrial mafic and ultramafic lavas (McEwen et al. 1998a; Geissler et al. 1999). Nearly all of the low-albedo lavas observed through the SSI near-infrared filters exhibit the 0.89 μm Fe^{2+} absorption, suggesting that lavas with orthopyroxenes are common, and the entire crust may have a mafic or ultramafic composition. However, orthopyroxenes are expected to show additional weaker absorptions near 2 μm (Adams 1974), which have not been observed in NIMS spectra, perhaps because of limited spatial resolution when NIMS had its full spectral capability (Geissler et al. 2004).

Laboratory near-infrared spectra of sulfur trioxide, SO_3, were compared to ground-based spectra of Io by Nelson and Smythe (1986) who found no evidence for SO_3 on Io. However, Khanna et al. (1995) investigated mid-infrared spectra of several forms of sulfur trioxide and noted the possible presence of SO_3 features in Io's thermal emission spectrum. Disulfur monoxide, S_2O, and polysulfur oxides, suggested to be present on Io by Hapke (1989), were investigated spectroscopically by Baklouti et al. (2008) and compared with Io spectra. They were unable establish the presence of S_2O on Io due to interference by strong neighboring SO_2 bands but suggested the possibility of a broad absorption feature due to polysulfur oxides in the 4.5 μm region (Fig. 4).

Nash and Betts (1995) proposed Na_2S to be responsible for the spectral feature reaching from 0.55 to 0.7 μm. Na, K, and Cl, as well as sulfur and oxygen compounds are seen escaping from Io and populating the neutral and plasma torus (McGrath et al. 2004). The microwave detection of NaCl in plumes (Lellouch et al. 2003) suggests that the torus species originated as salts of NaCl and KCl present in the magma. Alternatively, the chlorine and metals may be bound in other minerals such as Cl_2SO_2 (Schmitt and Rodriguez 2003).

Io is thought to have lost nearly all its hydrogen (Zolotov and Fegley 1999); however, the detection of hydrogen pickup ions by Galileo's plasma analyser (Frank and Paterson 1999) led to the interpretation that Io may provide a flux of hydrogen from the surface to the atmosphere as hydrogen escapes rapidly to space. A possible hydrogen source, hydrogen sulfide, was suggested by Nash and Howell (1989) and Salama et al. (1990) but Schmitt and Rodriguez (2003) placed an upper limit of 10^{-4} for the mixing ratio of H_2S in SO_2 on Io. Possible H_2O absorptions at 2.97 and 3.15 μm were reported by Salama et al. (1990, 1994). The 3.15-μm absorption was confirmed in the NIMS by Carlson et al. (1997) and interpreted as a result of the O–H stretch transition in hydrated minerals, hydroxides, or possibly water. It is also possible that the Jovian magnetosphere, and not Io, is the source of the hydrogen detected by Galileo. It is likely that a new mission to the Jovian system will be necessary before any of these suggested compounds are confirmed to exist on Io.

Other materials of so-far unknown composition are found in regions having negative spectral slopes from 0.76 to 0.97 μm (Geissler et al. 1999; Lopes et al. 2004; McEwen et al. 2004) that are not characteristic of pure sulfur or SO_2 compounds. These regions correspond to greenish areas in Galileo SSI images and are restricted to dark patches (probably crusted surfaces of recent lava flows and lava lakes), and red material deposited on recent lavas has later turned green (Keszthelyi et al. 2001), so the green color may be due to contamination or alteration of the red material by warm silicate lavas (Kargel et al. 1999; Phillips 2000).

Io's surface composition is revealed in studies of the material ejected from Io through sputtering (Johnson 1990) and by impacts (Alvarellos et al. 2008). Sputtering of solid SO_2 can eject SO_2, SO, and SO_3 (Boring et al. 1984) from Io, and Na, K, and Cl will be produced by sputtering of the parent salts. Sputtering of exposed silicates will also contribute products. These atoms and molecules, and their ionization and dissociation products, will be present in Io's atmosphere and as escaping atoms and molecules, forming neutral and plasma tori at Io's orbit (Thomas et al. 2004). Na, K, and Cl are observed as neutrals while S^{n+}, O^{n+}, Na^+, and Cl^+ are found in the plasma torus (Hall et al. 1994; Kuppers and Schneider 2000; Feaga et al. 2004; Thomas et al. 2004). Upper limits for many species are given by Na et al. (1998). Impact ejecta can escape Io and introduce spall fragments and vaporized material into the torus region. The neutral atoms, ions, and ejecta can be sources of material for Europa and other satellites (Alvarellos et al. 2008; Carlson et al. 2009).

1.4 Exchange of Material Between the Jovian Satellites

The primary link between the Jovian satellites is the massive Jovian magnetosphere. Its co-rotating plasma particles continuously bombard the surfaces of the satellites. Sputtered surface atoms form tenuous atmospheres and orbital torii from which material can be exchanged with neighboring satellites. Io orbits Jupiter deep within its magnetosphere at about 6 R_J (Joy et al. 2002; cf. Table 1). This results in significant interaction between Io and the magnetic field. Io's active volcanoes eject approximately 1 ton/s of material (Hill et al. 1983), producing Io's tenuous atmosphere. The interaction of the atmosphere with magnetospheric charged particles produces clouds of neutrals (primarily oxygen and sulfur with a few percent of sodium, potassium, and chlorine) that surround the satellite and move around its orbit (Brown 1974; Smyth and McElroy 1977; Mendillo et al. 1990; Lellouch 1996; Summers and Strobel 1996). Some of the O and S neutral atoms become charged by processes of UV photo-ionization, electron impact ionization and charge exchange (Kivelson et al. 2004) forming a dense plasma (up to 4000 electron/cm^3) called the Io plasma torus, a ring of sulfur and oxygen ions transported around Jupiter by its magnetic field (Bridge et al. 1979; Broadfoot et al. 1979). Once these ions are picked up by the Jovian magnetic field they slowly diffuse throughout Jupiter's magnetosphere, dominating its plasma population (Russel et al. 2005). All of the solid bodies in the Jovian system are impacted by these ions. Charge-exchange processes within the plasma produce fast neutrals that leave the system, unless a further collision occurs. Io's atmosphere is partially ionized by these interactions, driving chemical alteration of the surface and leading to compositional, spectral and optical modifications. Plasma is re-supplied to the torus by sputtering and other loss processes that deplete the surface and atmosphere.

Energetic ions incident on the icy surface of Europa, orbiting Jupiter at a distance of 9.4 R_J (Table 1), eject H_2O molecules as well as hydrogen and oxygen. Photolysis, driven by ultraviolet radiation, and radiolysis driven by magnetospheric charged particles, dissociate water molecules to produce hydrogen and oxygen (Johnson et al. 1982). The hydrogen escapes because of Europa's weak gravity, but the heavier O_2 remains gravitationally bound, forming a tenuous atmosphere (Burger and Johnson 2004; Smyth and Marconi 2006). This atmosphere interacts with the thermal plasma in Jupiter's magnetosphere. Observations from the Hubble Space Telescope (HST; Hall et al. 1995) and Cassini (Hansen et al. 2005) have detected atomic oxygen emissions believed to result from dissociation of O_2 by electrons. Charge exchange with magnetospheric ions and electron-impact ionization removes atoms and molecules from the atmosphere and exosphere, contributing fresh ions to the Jovian plasma (Saur et al. 1998; Shematovich et al. 2005) and creating a torus similar to that of Io that was detected by the Cassini spacecraft as it flew by Jupiter in 2000 (Mauk et al. 2003; Ip et al. 2000).

Along with H_2O, trace species are expected to be sputtered from the surface. Of these, sodium is by far the best-observed and modeled component of Europa's atmosphere (Leblanc et al. 2005; Cassidy et al. 2008). Additionally, an even smaller potassium atmosphere around Europa has been reported by Brown (2001). SO_2 and Cl have also been identified via pickup ions (Volwerk et al. 2001), possibly released by sputtering of the surface material by magnetospheric ions. Initially, these compounds were considered to be released by Io's volcanoes, after which they are ionized in the magnetosphere and implanted into Europa's surface ice, possibly forming hydrated salts like $Na_2SO_4 \cdot 10H_2O$ (mirabilite) and/or $H_2SO_4 \cdot 8H_2O$ (sulfuric acid octahydrate). In contrast, studies by Johnson (2000), Johnson et al. (2002), and Leblanc et al. (2002) favor an endogenic source related to a salty ocean below Europa's crust. Recent efforts (Dalton 2007b; Dalton et al. 2009; Shirley et al. 2010) suggest an intermediate compromise wherein implanted materials may have been transported into Europa's interior over geologic time, processed, and erupted to the surface as hydrated sulfates, including salts and acids. These deposits would subsequently undergo processing driven by the radiolytic flux to produce the present balance of water ice, hydrated sulfate salts, hydrated sulfuric acid, and newly implanted ionic species.

Even though Ganymede and Callisto do not seem to maintain torii near their orbits in the more distant regions of the Jovian system ($\sim$15 and 26 R_J, Table 1), they still exhibit thin atmospheres of their own. Ganymede is known to have a tenuous atmosphere of molecular oxygen (Hall et al. 1998) and atomic hydrogen (Barth et al. 1997). Charged particle measurements indicate that there is an outflow of protons, implying ongoing gas production (Frank et al. 1997). Images of Ganymede obtained with the Hubble Space Telescope Imaging Spectrograph (STIS) show clearly that oxygen emissions are confined to polar regions above latitudes of 40°, and appear to vary in time and relative intensities between northern and southern hemispheres. This is consistent with current models of Ganymede's magnetic field. Plasma bombardment of Ganymede's surface is also supported by the detection of ozone, O_2, and SO_2 embedded in its surface material (Spencer et al. 1995; Calvin et al. 1996; Noll et al. 1995, 1996c; McCord et al. 1998a; Hibbitts 2001).

A thin CO_2 atmosphere on Callisto, up to an altitude of 100 km above the surface, was detected by Carlson et al. (1999a) based on Galileo NIMS observations. The Galileo radio-occultation experiment (Kliore et al. 2002) detected an ionosphere, suggesting the existence of an atmosphere composed mainly of molecular oxygen. Atmospheric CO_2 should be lost easily due to ionization by solar ultraviolet radiation. These ions are then swept away by Jupiter's magnetic field, indicating a steady flux of carbon dioxide from Callisto's surface into the atmosphere (Carlson et al. 1999a).

Europa, Ganymede and Callisto all exhibit spectral evidence for CO_2 trapped in their surface material. This CO_2 may have been supplied by carbonaceous chondrites, released from clathrates, or produced through radiolysis. Although the source of this CO_2 is not yet clear (McCord et al. 1997, 1998a; Hibbitts et al. 2000, 2003; Dalton 2000, 2007b; Hansen and McCord 2004) the CO_2 concentrated on the trailing hemisphere of Callisto at least is consistent with radiolytic production.

2 Saturnian Satellites

2.1 The Surface Composition of Icy Saturnian Satellites

The exact number of Saturnian satellites is difficult to determine when one also takes into account the major and minor satellite fragments in between the rings. In any event, Saturn has more than sixty satellites currently reported, ranging in size from objects of a few kilometers in diameter to Titan, larger than the planet Mercury. Of these, the icy satellites comprise a suite of different and unique bodies spanning a large range of sizes, the largest one being Rhea with a radius of 1530 km (Table 2). The major satellites of Saturn have low eccentricities, elliptical equatorial orbits and tidal effects due to their proximity to Saturn that have led, over time, to the synchronization of their motions. Other general characteristics are resonances among different satellites, which have had a fundamental role in shaping the overall thermal history of the icy satellites. The resonances between "shepherd satellites" and rings lead to the formation of the ring divisions and radially confine the ring materials within them (Goldreich and Tremaine 1979, 1982). Gordon et al. (2006) have compiled an atlas of satellite mean motion resonance locations and amplitudes within the rings of Saturn.

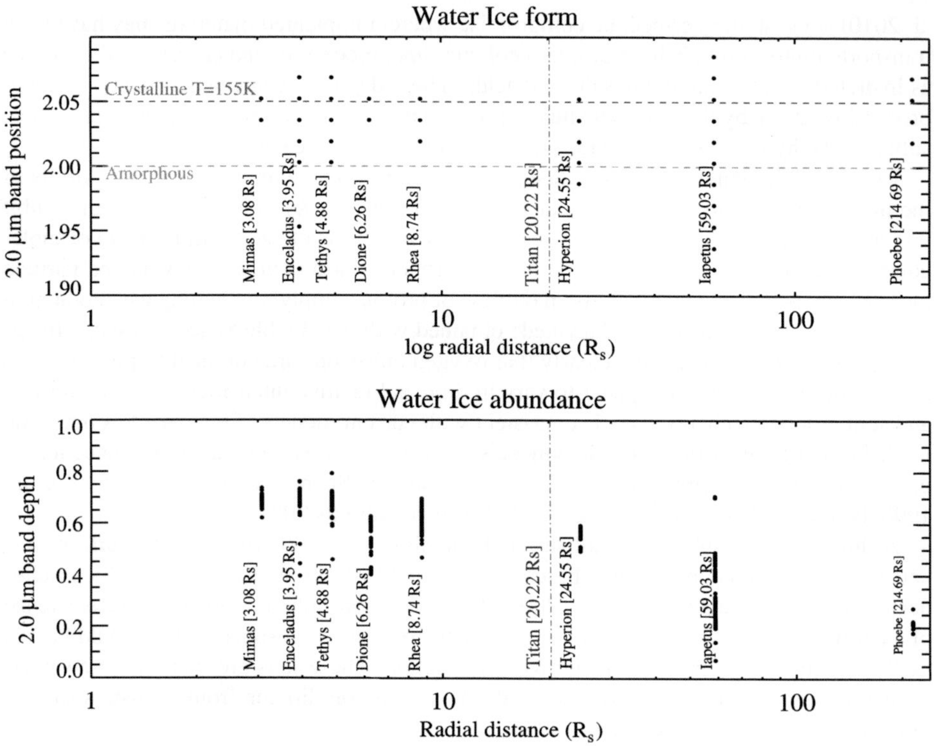

Fig. 5 Radial variation of the water ice 2.0-μm band depth through the Saturnian system of satellites. The higher band depth corresponds to a richer abundance of water ice on the surfaces

This has contributed much to our understanding of the interactions between the satellites and the rings.

Almost all of the satellites have surfaces rich in impact craters of different sizes, dating from early in their histories (Schenk and Moore 2007). Observational data on bulk density (about 1 g/cm^3), albedo and diagnostic molecular bands present on the surface, as seen by Cassini Visual and Infrared Mapping Spectrometer (VIMS) and Ultraviolet Imaging Spectrograph (UVIS), indicates that most of the satellites are composed primarily of water ice (Filacchione et al. 2006). This ice, even if extremely cold and rigid, doesn't need a significant amount of energy to melt or deform: this explains the complex geology observed on the satellites. In some specific locations, however, carbon compounds such as carbon dioxide and methane ice are also present as well as ammonia nitrate and organic materials (Dalton 2007a). Moreover VIMS data allow the identification of carbon dioxide and organics (C–H and possibly C–N) on Phoebe (Clark et al. 2005; Buratti et al. 2008; Coradini et al. 2008), Iapetus (Cruikshank et al. 2008) and Hyperion (Cruikshank et al. 2007) as well as the nature of the chromophores contaminating the surface of Dione (Clark et al. 2008).

The high average albedo of the satellites, and the large distance from the Sun, leads to very low average surface temperatures (about 70 K). However, low-albedo ($\leq$0.1) material is found on many satellites: its distribution in the Saturnian satellite system is of special interest because of its juxtaposition with high-albedo ice (Clark et al. 2005; Buratti et al. 2008; Coradini et al. 2008; Cruikshank et al. 2008). These authors, examining Cassini VIMS data,

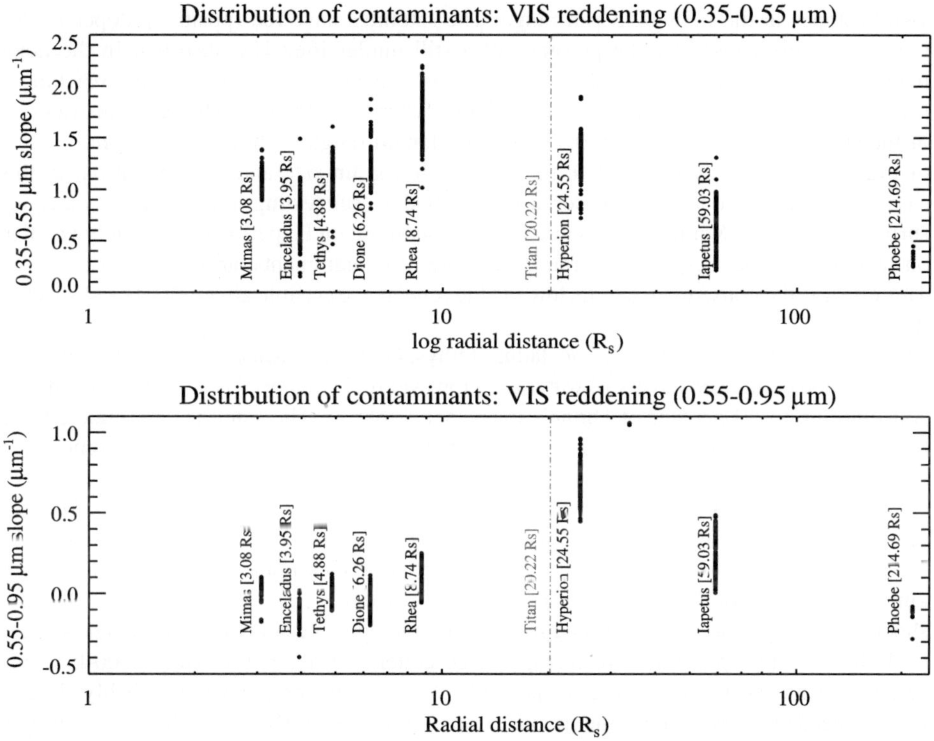

Fig. 6 Radial distribution of non-ice contaminants across the Saturnian system as evidenced by the 0.35–0.55 μm slope (*top panel*) and 0.55–0.95 μm slope (*bottom panel*)

and in particular the spectral region 3.0–3.6 μm have identified a broad band centered at 3.29 μm, that was identified as the vibrational C–H stretching mode in polycyclic aromatic hydrocarbon (PAH) molecules. This material has been interpreted as carbonaceous material with a kerogen- or coal-like structure, similar to that in carbonaceous meteorites. This seems to be an important proof of the link between the "dark matter" and the contaminants coming from sources external to the Saturn System. In order to highlight these differences among the Saturnian icy satellites we discuss below some diagnostic spectral indicators as derived from a statistical analysis of more than 1400 full-disk observations made by Cassini-VIMS during the nominal mission (Filacchione et al. 2007, 2010).

The variation in the 2.0-μm water ice absorption feature band depth with distance from Saturn (Fig. 5) illustrates the radial distribution of water ice abundance. The inner satellites Mimas, Enceladus and Tethys have the deepest absorption features, indicative of the higher abundance of surface water ice. Moving outwards, Dione, having a lower band depth, is slightly more contaminated with respect to its neighbors Tethys and Rhea. The outer moons Hyperion, Iapetus and Phoebe show a linear reduction of the surface water ice fraction. On Iapetus two different clusters of points are apparent: the higher values are measured on the trailing hemisphere (bright regions) while the lower values correspond to the leading hemisphere (dark regions).

The radial distribution of the "contaminants" across the Saturnian system is measured through the visible spectral slopes (e.g. the angular coefficient of the best fit line to the spectrum normalized at 0.55 μm). The top panel of Fig. 6 shows the distribution of the visible slopes in the 0.35–0.55 μm range. The satellites have red (positive) slopes in this range, probably caused by the presence of a still unidentified UV absorber. In the inner region the minimum is observed on the less contaminated object (Enceladus) reaching a maximum on Rhea and finally decreasing from Hyperion to Phoebe. The radial distribution of the spectral slope in the 0.55–0.95 μm wavelength range is shown in the bottom panel of Fig. 6: this quantity is sensitive to both regolith grain size and contamination: in this case the negative values are measured on Enceladus (suggesting a nearly pure water ice surface), neutral on Mimas, Tethys and Dione, maximum on Hyperion and strongly positive on Iapetus. Phoebe, despite the large percentage of contaminants, appears almost neutral. This effect is the consequence of its low visible reflectance (visible albedo 0.04) and spectral flatness.

The inner satellites—Mimas, Enceladus, Tethys, Dione, and Rhea—are roughly spherical in shape, and composed mostly of water ice. However rocky material is also present and may constitute up to 40 percent of Dione's mass (see Table 2). The satellites are cratered and resurfaced, showing great variability, both on the surface of each moon and in comparison with each other.

Mimas is the largest satellite among those closest to Saturn and the smallest of the icy satellites to have a roughly spherical shape, with a large crater, Herschel, that has a diameter equal to one third of the satellite itself, with depths of 10 km and a central peak rising 6 km above the base of the crater. The impact that formed this crater almost disrupted Mimas: shock waves from the impact created fractures on the opposite side of the body. The surface of Tethys has two prominent features, a large crater and a 100-km-wide, 2000-km long valley, Ithaca Chasma. The bodies responsible for such large craters were probably close to the size needed to break apart the moons (Lopes et al. 2008). Dione and Rhea have bright, wispy streaks on their already highly reflective surfaces.

Tethys, Dione and Rhea are twice the size of Mimas and Enceladus and their surfaces are unevenly cratered. Tethys and Rhea are composed almost entirely of ice, while Dione with its rocky core is the densest Saturnian moon after Titan. The lack of complex rings or central peaks that usually characterize the craters on the Moon and Mercury is perhaps attributable to regional relaxation of impact craters (Schenk and Moore 2007). In contrast to Rhea and Mimas, the surface of Dione, while still heavily cratered, exhibits evidence for substantial resurfacing and internal activity. Both moons were found to support smooth, planar regions suggesting that icy material has oozed from the interior to the surface. Dione and Rhea have similar distributions of albedo, which is related to a wide variety of landscapes. Both are characterized by clear divisions between the leading and trailing hemispheres. The leading hemispheres are uniformly bright, while the trailing hemispheres display a distinctive appearance with networks of bright and subtle stripes on a dark background that overlaps with craters. Cassini observations have established that these lineaments created by tectonic fractures are rich in water ice.

Several small moons have been discovered immediately outside the A ring and close to the F and G rings. Possibly four so-called Trojan or co-orbital satellites of Tethys and one of Dione have also been discovered. Recent observations by Cassini ISS (Porco et al. 2007) and theoretical modeling of the accretion processes (Charnoz et al. 2007) have shown that the shape of Atlas is similar to a "flying saucer", with a prominent equatorial ridge of material accreted from the nearby rings.

The outer satellites, Hyperion and Iapetus, also consist mainly of water ice. Iapetus has a very dark region in contrast to the rest of its surface, which is mostly bright. This dark region and the rotation of the satellite are the cause of the variations in brightness that were noticed by J.D. Cassini in 1671. Iapetus is characterized by the presence of an approximately 20-km

Table 2 Physical parameters of the major Saturnian satellites

Satellite	Diameter km	Mass 10^{20} kg	Orbital radius R_S	Orbital radius 10^3 km	Orbital period days	Visual geometric albedo	Inclination	Eccentricity	Bulk density g/cm^3
Mimas	390	0.375	3.07	185.2	0.94242	0.5	1.53	0.0202	1.20
Enceladus	500	0.65	3.94	238.02	1.37021	1.0	0.02	0.0045	1.610
Tethys	1060	6.27	4.88	294.66	1.88780	0.9	1.86	0.0000	0.9735
Dione	1120	11.0	6.26	377.40	2.73691	0.7	0.02	0.0022	1.5
Rhea	1530	23.1	8.74	527.04	4.51750	0.7	0.35	0.0010	1.23
Titan	5150	1345.5	20.22	1221.8	15.94542	0.22	0.33	0.0292	1.88
Hyperion	410 × 260 × 220	0.2	24.58	1481.1	21.27660	0.3	0.43	0.1042	0.54
Iapetus	1472	15.9	59.09	3561.3	79.33018	0.05/0.5	14.72	0.0283	1.083
Phoebe	230 × 220 × 210	0.072	214.9	12952	550.48	0.08	175.3	0.1633	2

wide, 13-km high topographic ridge. Along its roughly 1300 km length this ridge remains almost exactly parallel to the geographic equator to within a couple of degrees. It has a complex geologic structure, with signs of impact erosion, but the physical origin of the ridge has yet to be explained.

Phoebe, the farthest large satellite, moves in a retrograde orbit (in the opposite direction of the orbits of most of the other satellites) with a high inclination relative to Saturn's equator. In regards to composition, it may be the most diverse body in the solar system, with spectral evidence for water and carbon dioxide ices, organic volatiles, alkanes, nitriles, aromatic hydrocarbons, iron-bearing minerals, olivines, pyroxenes, and phyllosilicates (Clark et al. 2005). The wealth of volatiles on Phoebe argues for an origin in the cold outer reaches of the solar nebula rather than the drier inner regions where the terrestrial planets and asteroid belt formed. Phoebe may be a KBO-like body captured by Saturn's gravitational field (Dalton et al. 2005; Coradini et al. 2008).

Voyager and Cassini/ISS images as well as reflectance spectra of Saturn's major satellites show that a coloring agent is present on their surfaces. Dark material in the Saturnian system was first inferred by Cassini (1676), and verified by Murphy et al. (1972) and Zellner (1972). Since then the nature of the low-albedo material has been studied by numerous authors favoring carbon-rich material such as complex organic polymers or minerals as the major compounds (Cruikshank et al. 1983; Vilas et al. 1996; Jarvis et al. 2000; Owen et al. 2001; Buratti et al. 2002; Vilas et al. 2004).

Recent measurements by Cassini/VIMS confirm a gradient in the distribution of low-albedo surface material from the inner to outer regions of the Saturn system. While inner objects (e.g. Mimas, Enceladus, Tethys) are dominated by bright H_2O ice, high concentrations of low-albedo compounds are detected on the three outermost satellites Hyperion, Iapetus and Phoebe. The dark hemispheres of Iapetus, Phoebe and Hyperion display spectra markedly different from pure H_2O ice (Cruikshank et al. 2005, 2007, 2008; Clark et al. 2005, 2008, 2009). The inner Saturnian satellites Dione and Rhea also show evidence for non-ice contaminants on their surfaces (Clark et al. 2008; Stephan et al. 2009, 2010).

The source of the low-albedo material in the Saturn system is still unclear. The low mean densities of the Saturnian satellites indicate that ice is the principal component of their interiors. Both the pattern of emplacement of the low-albedo material of Iapetus and its thickness of a few tens of cm derived from Cassini/RADAR measurements (Black et al.

2004) suggest that it is a superficial, geologically young deposit lying on bedrock of H_2O ice. Several researchers (e.g., Clark et al. 2005) suggest a source outside of the Saturn system, perhaps of Kuiper belt or cometary origin. Owen et al. (2001) suggest that an impact on Titan created a spray of dark, organic-rich material, part of which was swept up by Iapetus' leading hemisphere, and part of which was transported to the inner system to accrete onto the surfaces of the inner icy moons. However, concentrations of low-albedo material on the inner satellites Dione and Rhea occur primarily on their trailing hemispheres, suggesting bombardment by energetic particles from Saturn's co-rotating magnetosphere (Clark et al. 2008; Stephan et al. 2009, 2010). This is also supported by local effects. Crater and cliff walls facing away from the trailing hemisphere are shielded from implantation and expose clean ice. On the other hand, slopes that face trailing direction show a higher abundance of the low material (Clark et al. 2008; Stephan et al. 2010).

The interaction between the satellites' surfaces and Saturn's magnetosphere is also considered responsible for the presence of CO_2 in their surface material. Based on the differences in the position of the CO_2 band in the spectra of Phoebe and Iapetus, Hyperion, and Dione, Cruikshank et al. (2009) suggest that CO_2 on Iapetus, Hyperion and Dione results at least in part from particle or UV irradiation of native H_2O and carbon derived from external sources via either implantation or accretion, or redistribution of native CO_2 from the interior. Phoebe's CO_2 is probably native to the body as part of the initial inventory of condensates and now exposed on the surface.

2.2 Iapetus

Since the discovery of Iapetus by Cassini (1676), it has been known for its variable brightness as seen from Earth over the ~80-day orbital period. This variation arises from the irregular distribution of albedo across the surface; the synchronism with the orbital period establishes that the rotation and revolution are tidally locked. The leading hemisphere has a geometric albedo of ~0.07, while the albedo of the trailing hemisphere is ~0.6, indicating a significant difference in the compositions of the materials covering those hemispheres. Water ice is clearly identified in the near-IR spectrum of the trailing hemisphere, thus accounting for the high albedo, while details of the composition of the low-albedo material on the leading hemisphere are still not entirely clear. The low mean density of Iapetus (Table 2) is a strong indication that ice is the principal component of the interior.

The red color (in the 0.3–2 μm wavelength range) of the low albedo material suggests that complex organic solids are at least one component (e.g., Owen et al. 2001), though Clark et al. (2009) propose that nano-particulate metallic Fe and Fe oxides are a significant contributor to the color and albedo. A spectral absorption band detected by the Cassini VIMS instrument at 3.28 μm matches well with the C–H stretching mode in aromatic hydrocarbons, and there is evidence of C–H absorption bands arising from aliphatic hydrocarbons as well (Cruikshank et al. 2008).

Cassini VIMS also detected the 4.27-μm asymmetric O–C–O stretching mode in CO_2 on Iapetus, seen primarily but not exclusively in the low-albedo surface regions (Buratti et al. 2005; Clark et al. 2009). The central wavelength and the profile of this CO_2 spectral band are variable over the surface, indicating that the CO_2 molecules are complexed in one or more ways with other surface components, such as H_2O and minerals (Clark et al. 2009; Cruikshank et al. 2009). In this context, "complexed" means that the nearest neighbor of a CO_2 molecule is another material, thus giving rise to different energy states for the stretching modes. The complexing may be in the form of clathrate structures in which CO_2 molecules are caged in H_2O, or the CO_2 molecules may be chemisorbed on grains of minerals or macromolecular carbonaceous materials (Hibbitts and Szanyi 2007). A weak absorption band at 0.67 μm was attributed to Fe-bearing minerals (Vilas et al. 1996), but Owen et al. (2001) could not find the additional absorption bands predicted for these materials. A few other weak absorption bands in the spectrum are unidentified.

2.3 Rhea and Dione

Orbiting Saturn in the outer parts of the E-ring, Rhea and Dione have surfaces dominated by H_2O though some regions on their surfaces exhibit distinct contamination with low-albedo material. In contrast to Iapetus these contaminants occur primarily on their trailing hemispheres (Clark et al. 2008; Stephan et al. 2009, 2010).

Again, Clark et al. (2008) point out that nano-phase hematite (with trace amounts of CO_2 and possibly NH_3) can explain Dione non-ice characteristics in the VIS and NIR as well as the UV absorber seen by Cassini UVIS (Hendrix et al. 2009). They also suggest that the low-albedo material throughout the Saturnian system exhibits similar spectral characteristics. The UV reflectance of Rhea was also found to be consistent with the presence of an organic solid intimately mixed with H_2O ice (Cruikshank and Khare 2000; Cruikshank et al. 2005).

CO_2 was detected on Dione (Clark et al. 2008) with the band broadened and slightly shifted to shorter wavelengths as observed on Iapetus and Hyperion (Cruikshank et al. 2009). CO_2 has not been detected on Rhea so far, probably due to the spectral dominance of H_2O (Stephan et al. 2009). Other trace materials produced and trapped within the surface ice of Dione and Rhea such as O_3 (Noll et al. 1997b) have yet to be confirmed by the Cassini instruments.

2.4 Mimas and Tethys

Mimas, with a mean radius of ~197 km is the smallest and the innermost of the classical Saturnian satellites, orbiting just outside of the A ring. It is heavily cratered and is best known for Herschel, the large impact crater on its leading hemisphere. Surface properties of Mimas were found to be quite uniform with color variations less than 10% (Buratti et al. 1990; Verbiscer and Veverka 1992; Filacchione et al. 2010). The photometric uniformity of Mimas for the most part extends to its spectral appearance as well. However, the H_2O absorptions on the leading hemisphere are slightly deeper than on the trailing hemisphere (Emery et al. 2005) similar to the trend seen on Tethys, Dione and Rhea as described by Clark et al. (1984) and Stephan et al. (2009, 2010). There is no indication of any species other than H_2O in the spectra of Mimas so far (Emery et al. 2005) which suggests that the observed variations could be caused by slightly larger particles on Mimas' leading hemisphere.

Tethys is significantly larger than Mimas and its inner neighbor Enceladus, with a radius of 530 km. Like Mimas, Tethys is heavily cratered with the large (400 km) Odysseus as its most prominent impact feature, but a very large fracture, Ithaca Chasma, extends about three-quarters of the way around a great circle roughly centered on Odysseus. In contrast to Dione and Rhea there are no dark/bright differences between cratered and younger tectonically modified regions on Tethys (Stephan et al. 2009, 2010). The leading hemisphere is about 10 to 15% brighter than the trailing hemisphere (Buratti et al. 1990, 1998), which correlates with deeper H_2O absorptions on the leading than on the trailing hemisphere. This is consistent with fewer non-ice contaminants on the leading hemisphere as observed on Dione and Rhea (Clark et al. 1984, 2008; Buratti et al. 1998; Stephan et al. 2009, 2010), although there is no evidence for any other component aside from H_2O thus far (Emery et al. 2005).

2.5 Phoebe and Hyperion

Phoebe and Hyperion each have particularly unusual characteristics that distinguish them from the other Saturnian satellites, all of which are unique in their own ways.

Phoebe is an irregular-shaped body with a mean radius of ~107 km, characterized by a heavily cratered surface, with overlapping craters of varying sizes. This morphology suggests an old surface. There are apparently many craters smaller than 1 km, indicating that projectiles probably smaller than 100 meters once bombarded Phoebe. All images show evidence for an ice-rich body coated with a thin layer of dark material. Its revolution around Saturn is retrograde, and its rotation period is ~9.28 hours; it is generally thought to be a captured object, perhaps originating in the Kuiper Belt. The spectrum of Phoebe is rich with absorption bands of H_2O ice, CO_2, probable ferrous-iron-bearing minerals, organic solids, phyllosilicates, and other unidentified materials (Clark et al. 2005; Coradini et al. 2008). Early reports of C–N bearing molecules require revision in the light of new lab data.

Despite the overall low albedo of Phoebe (~0.09), H_2O ice occurs everywhere on the surface, although with variations in spectral band strength. The CO_2 band at 4.27 μm is different in central wavelength and profile from CO_2 on the other satellites, and is indistinguishable from pure CO_2 ice and a type 2 CO_2 clathrate in H_2O from the laboratory (Blake et al. 1991). Cruikshank et al. (2009) suggest that it is native CO_2 that occurs as a part of the bulk composition of the satellite. The high vapor pressure of CO_2 at the temperature of Phoebe suggests that it is gradually being lost to space. Weak absorption bands in the region 3.2–3.5 μm, and near 4.4 μm, similar to those on Iapetus, are attributed to organic molecules that are probably a component of the low-albedo materials (Coradini et al. 2008) mixed with H_2O and CO_2 ices and distributed unevenly across Phoebe's surface.

Correlations among the abundances of the identified and unidentified surface components of Phoebe are complex, but in general terms there is a strong correlation of the band strength of CO_2 and aromatic hydrocarbons with exposed H_2O ice, while there is an anti-correlation of the unidentified 2.42-μm absorption band with CO_2 (Coradini et al. 2008).

Hyperion is also an irregular-shaped body, with a mean radius of 135 km. It is in a prograde orbit around Saturn, but its rotation is chaotic, with no single stable period. Its surface structures are unique among all the small bodies in the solar system observed at high spatial resolution (Thomas et al. 2007). Hyperion's surface is completely covered by subcircular depressions resembling craters yet distinctly different in form from those on other heavily cratered small bodies. The morphology of many of the cup-shaped depressions suggests an impact origin with extensive subsequent alteration by sublimation and mass wasting. Thomas et al. (2007) attribute the peculiar crater-like morphologies to their formation by impact into a highly porous body; the mean density of Hyperion is astonishingly low at $0.54 \pm 0.05\ \mathrm{g\,cm^{-3}}$. Many, but not all, of the depressions have deposits of low-albedo material on their floors; small white-rimmed craters are abundant on the low-albedo surfaces. Very small but distinct patches of low-albedo material are scattered sparsely in the high-albedo, icy regions. Geomorphological considerations suggest that the low-albedo materials are lag deposits left behind as the surrounding impure ice sublimes away. The white rims on the small craters and other landforms on the low-albedo materials of Hyperion are probably made of H_2O and CO_2 that has moved from the interior or other locations and deposited on these locally cold surfaces.

The reflectance spectrum of Hyperion shows strong absorption bands of crystalline H_2O ice and CO_2, with weak indications of the C–H stretching bands of hydrocarbons in the region 3.2–3.5 μm. The enigmatic and unidentified 2.42-μm band is also prominent. As with

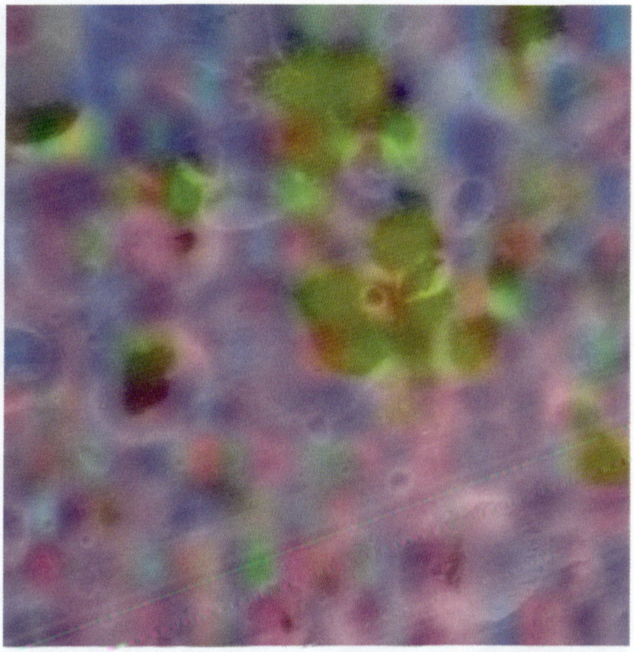

Fig. 7 Compositional map of part of the surface of Hyperion. *Blue* indicates water ice absorption band depth, *red* indicates CO_2 4.27-μm band depth, and *yellow* denotes the depth of an as-yet unassigned feature at 2.42 μm, possibly a nitrile. Note the correspondence between yellow and the dark subcircular deposits. CO_2 is correlated with some but not all impacts, and is also distributed throughout the background ice. Adapted from Cruikshank et al. (2007)

Phoebe, the relationships of these absorption features with one another across the satellite's surface are complex, as shown in the compositional map (Fig. 7) of one region studied by Cruikshank et al. (2007). H_2O absorption is clearly weaker in the low-albedo material, while the 2.42-μm and CO_2 bands are stronger in those regions.

The CO_2 band on Hyperion is shifted to shorter wavelength by ~0.0145 μm compared to the laboratory spectrum of pure CO_2 ice, and its profile (full-width half-maximum) is nearly twice as broad. The shift and broadening are indicative of the complexing of CO_2 molecules at the molecular level with other molecules of volatile and/or refractory materials, as noted above for Iapetus (Cruikshank et al. 2009). The CO_2 on Hyperion may arise from implantation of C atoms from the space environment, from native CO_2 molecules that are mobilized and recaptured on the surface in a complexing arrangement, or created one-by-one *in situ* by irradiation effects on H_2O ice in the presence of carbon-bearing material(s).

2.6 Enceladus

The icy Saturnian satellite that shows the greatest evidence for resurfacing and internal activity is the intriguing moon Enceladus. It is embedded in the E-ring, the largest planetary ring in the solar system. Even before Cassini, Enceladus was proposed to be the source of the E-Ring. Since 2005 active venting has been detected on Enceladus, and this has been considered responsible for the E-Ring replenishment. Enceladus' surface is covered by a complex pattern of interlaced smooth, icy textures, extremely bright and with the smooth appearance of newly formed units. Even the most heavily cratered regions on Enceladus show fewer craters than the other Saturnian satellites. Enceladus also shows many surface cracks and ridges. Since bodies as small as Enceladus should have cooled off very rapidly after their formation, it is still unclear how such recent resurfacing could have taken place. Orbital resonance with Dione is one of the possible sources of energy leading to the intense resurfacing activity of Enceladus.

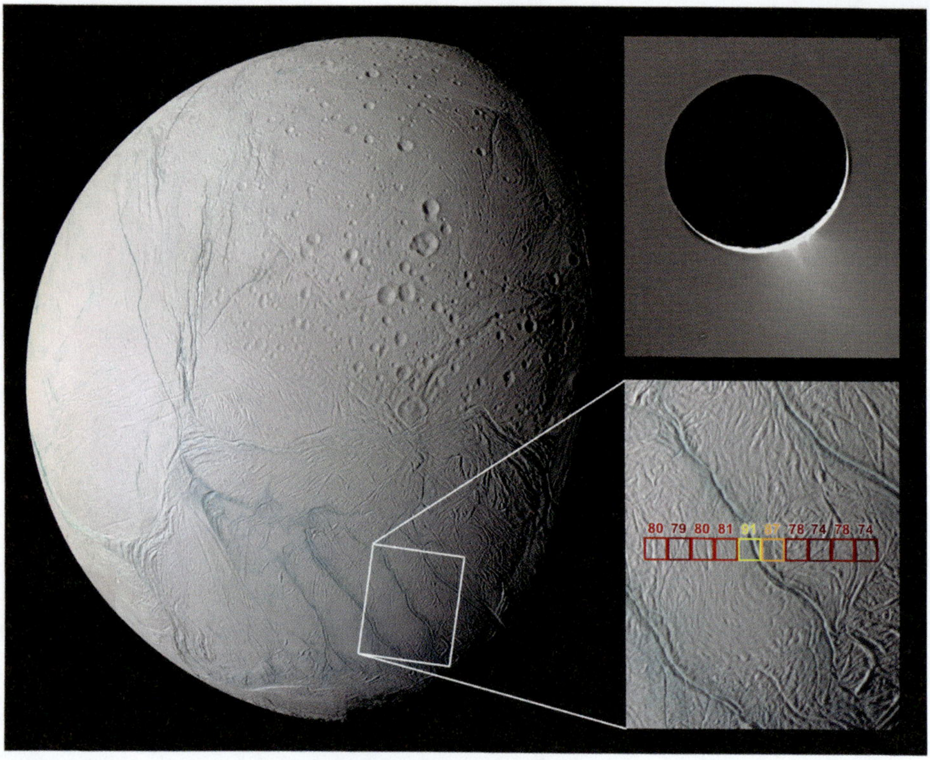

Fig. 8 The surface of Enceladus as observed by the Cassini ISS camera with the south polar region showing distinctive pattern of sub-parallel and ridged tectonic structures informally called "tiger stripes" (see *lower right*) from which geysers of icy particles are ejected into space (see *upper right*). *Inset* at *lower right* shows temperatures determined from the CIRS instrument, demonstrating elevated temperatures across the stripes

Enceladus is the only icy world in the solar system, besides Neptune's moon Triton, to show current geological activity. Because it is so small (500 km, Table 2), the discovery of large plumes by the Cassini Dual Technique Magnetometer (MAG) (Dougherty et al. 2006) currently emanating from surface cracks near the south pole of Enceladus (Fig. 8) venting water vapor, fine icy particles and organics, implies a strong and continuous underground source. Cassini detected an extended atmosphere around it, believed to be composed of ionized water vapor, carbon dioxide, methane and other organic molecules. For a detailed description of the atmosphere around Enceladus see Coustenis et al. (2010, this issue).

Whether the plume emissions arise directly from chambers of pressurized liquid water, evaporation from warm ice, or the explosive release of gases from clathrate ices is currently unknown (Porco et al. 2005, 2006; Kieffer et al. 2006). The origin and evolution of the internal heating source are a major puzzle. Cassini observations during its extended mission may resolve this issue, particularly if elements such as sodium (soluble in liquid water and not ice) are identified in or near the plumes. A hint that sodium might in fact be found eventually in the Enceladus plumes comes from its detection in E-ring particles by the Cosmic Dust Analyzer (Postberg et al. 2006). A direct detection in the plumes would provide strong evidence that sodium originated in the subsurface, and would further support the presence of a subcrustal liquid water layer.

Enceladus' plumes provide us with an opportunity to study phenomena that have been important at some time in the past throughout the outer solar system, when tidal effects and/or higher radiogenic heat fluxes may have powered geysering, melting and aqueous chemistry in a number of icy bodies. They allow us to obtain samples directly from the interior and provide information about the initial composition of the Saturnian nebular material from which Enceladus and Titan accreted. The primary activity after formation was differentiation and the subsequent resurfacing with very clean ice. Enceladus has the highest reflectivity (albedo) of any natural satellite. There have been multiple episodes of resurfacing, as shown by the large range and variation of crater densities across Enceladus' surface. Heavily cratered units in the north are perhaps billions of years old whereas the youngest, the south-polar region, is probably less than ten million years old. The interior became hot, 500 to 800 K, as inferred from the presence of molecular nitrogen in the plume (Matson et al. 2007) and further supported by the discovery of sodium-rich grains in the E-ring (Postberg et al. 2006). These are ideal conditions for the synthesis of complex organic molecules and may also have occurred in Titan. There have been a number of models proposed for structure and heat production close to the surface (e.g., Kieffer et al. 2006; Nimmo and Pappalardo 2006; Nimmo et al. 2007; Collins and Goodman 2007; Tobie et al. 2008; and others).

2.7 Titan

Titan's bulk density of $\sim$1.88 g cm^{-3} (Jacobson et al. 2006) implies a volatile fraction of $\sim$50%, corresponding to a layer of H_2O from the surface to $\sim$700 km depth. Detailed thermal evolution models of Titan's interior suggest a liquid water layer $\sim$300 km thick exists at $\sim$50 km depth below an upper Ice I layer; the lower $\sim$350 km of H_2O consists of high pressure ice phases (Grasset et al. 2000; Tobie et al. 2005). Water ice and other volatiles are therefore expected to be major components of Titan's surface material. Photochemical models, coupled with Cassini observations of Titan's atmosphere, suggest a steady flux of hydrocarbons and nitriles (i.e. "tholin") at the surface (Sagan and Thompson 1984; Wilson and Atreya 2003; Bellucci et al. 2009). Surface reservoirs of liquid hydrocarbons (i.e. liquid ethane and methane "lakes"), are reported in Titan's polar regions (Stofan et al. 2007; Brown et al. 2008).

2.7.1 Titan's Atmospheric Composition and Processes Involving Exchanges with the Surface

A thorough review of the composition of Titan's atmosphere including the results of measurements by the Cassini instruments is provided by Strobel et al. (2009) and Coustenis et al. (2010, this issue). We briefly summarize the major aspects here. The primary constituents of Titan's dense atmosphere, N_2, CH_4, H_2, and the inert, noble gases are known from ground-based observations (Kuiper 1944; Trafton 1972), Voyager measurements (Broadfoot et al. 1981; Hanel 1981; Strobel and Shemansky 1982; Strobel et al. 1993; Samuelson et al. 1997; Vervack et al. 2004: Yelle et al. 2008; Waite et al. 2005) and theoretical modeling (Lewis 1971). N_2 is the most abundant species (97%). Chemical analysis from the Huygens Gas Chromatograph/Mass Spectrometer (GCMS) revealed that CH_4, the next most abundant molecule, increases in abundance from 1.4% in the troposphere to 5% at Titan's surface. In addition, the atmosphere contains small amounts of more complex hydrocarbons, nitrogen-bearing organics and molecules containing oxygen as minor constituents formed mainly by reactions of solar ultraviolet radiation and/or charged particles impinging on this atmosphere (Lara et al. 1994; Lara 1996) resulting in Titan's well-known orange-colored atmospheric haze (Clarke and Ferris 1997; Coll et al. 1999).

In the photochemical reactions anticipated in Titan's atmosphere, photons are able to remove a hydrogen atom from CH_4 to produce the methyl radical CH_3^-. H and H_2 produced from the breakup of CH_4 may enable the chemical evolution of complex organic compounds but easily escape to space. A wide variety of hydrocarbons are produced, many of which have been detected spectroscopically by both spacecraft and Earth-based observations. Ethane (C_2H_6) is the most abundant gas product followed by acetylene (C_2H_2), propane (C_3H_8), ethylene (C_2H_4), methyl acetylene (CH_2C_2H), diacetylene (C_4H_2) and benzene (C_6H_6), which was first detected by ISO measurements (Coustenis et al. 2003) and since confirmed in the neutral atmosphere with the Cassini Composite InfraRed Spectrometer (CIRS) (Flasar et al. 2005) and in the ionosphere by the INMS instrument (Vuitton et al. 2008).

Magnetospheric electrons from Saturn convert N_2 to N atoms that are then free to react with hydrocarbons to produce nitriles (Clarke and Ferris 1997; Clarke et al. 2000; Tran et al. 2003a, 2003b). Voyager/IRIS and Cassini/CIRS detected three nitriles: hydrogen cyanide (HCN), cyanoacetylene (HC_3N) and cyanogen (C_2N_2) in the thermal infrared spectrum of Titan (for a review of these detections and the abundances see Coustenis and Taylor (2008) with HCN being the simplest and most abundant nitrile. Acetonitrile (CH_3CN) has only been observed from Earth (Marten et al. 2002).

Although Titan's atmosphere is a reducing environment, it contains oxygen compounds. Only CO_2 (Samuelson et al. 1983; de Kok et al. 2007; Coustenis et al. 2007), CO (Lutz et al. 1983; Noll et al. 1996b; Hörst et al. 2008) and water vapor (Coustenis et al. 1998; Bjoraker et al. 2008) have been detected so far.

The interaction between the atmosphere and surface produces lakes, dunes, fluvial systems, and other morphological features that Titan shares with Earth. Seasonal changes in Titan's atmosphere are evident in the landscape: the lakes observed in the North may be a seasonal phenomenon, dependent on the atmospheric structure above (temperature, composition and aerosol content). The methane cycle on Titan is closely reflected on the ground (as in the channels near the Huygens landing site) and can be constrained by studying the surface.

The paucity of surface craters indicates a young surface and is likely to be linked to atmospheric phenomena. Morphological differences between Titan's few craters and those on other icy satellites may be attributable to atmospheric or subsurface volatiles.

2.7.2 Early Observations and General Properties of Titan's Surface

Titan's surface properties can be observed remotely in the near infrared through microwave by two primary techniques: (1) measurement of near-IR reflectance of solar radiation, and (2) radar system observations of reflected and emitted radiation, usually in the microwave region. Due to absorption mainly by atmospheric CH_4, surface reflectivity in the near IR ($\sim$1–5 μm) can only be measured within a few (7–10) narrow methane "windows" (Griffith et al. 1991). Scattering by high altitude haze particles contributes to the diffuse field, with the strongest effect at shorter wavelengths (Rages and Pollack 1980).

Radiative transfer modeling of these atmospheric effects, combined with ground-based, full disk spectroscopic observations in the near IR methane windows suggest surface reflectivity consistent with water ice, contaminated with a spectrally neutral material—possibly "tholin" (Griffith 1993; Griffith et al. 1991, 2003; Coustenis et al. 1995, 2006). Longitudinal variations were also observed in these ground based studies (Lemmon et al. 1993; Negrão et al. 2006).

Disk-resolved images of Titan's surface were obtained in the methane windows by the HST and ground-based adaptive optics (Smith et al. 1996; Coustenis et al. 2005) showing dark and bright areas. The latter authors suggested, in addition to water ice and tholins, the presence of methane ice to explain the spectral behavior of the images, but even after further differential spectroscopy of the images was applied (Hirtzig et al. 2007), the evidence remained inconclusive, except to support the presence of water ice. Laboratory characterization of volatile ices under relevant conditions (Dalton 2010, this issue) may resolve this issue.

Ground based radar observations suggest that Titan's surface dielectric constant is consistent with water ice but not with extensive liquid hydrocarbons (Grossman and Muhleman 1992), while Cassini radar observations show considerable geographic variations in microwave (~2.2 cm) optical properties (Wye et al. 2007). Measured microwave emissivities (~0.9) and dielectric constants ($\varepsilon \simeq 1.5$–2.5) are consistent with highly porous or fractured water ice (possibly contaminated with ammonia), solid CO_2, and/or solid and liquid hydrocarbons (but not silicates or complex organics, which have $\varepsilon > 4.0$) at a surface temperature of ~90 K. Scattering models indicate these observations probe the upper ~20 m of Titan's surface (Zebker et al. 2008).

2.7.3 Cassini Observations of Titan's Surface

Observations of the Titan surface by the Cassini ISS generally confirmed the ground-based adaptive optics and HST results, finding dramatic changes in surface albedo within the same regions. Bright and dark areas were found to share Titan's landscape (Porco et al. 2005).

Fluvial modification of the surface was very evident at the Huygens landing site (Fig. 9). Steeply incised channels a few kilometers long and ~30 m across have been observed in the bright highland by the Descent Imager Spectral radiometer (DISR) instrument on the Huygens probe. Models of sediment transport suggest these can be formed in methane rainstorms (Perron et al. 2006). The knee-high vista from the probe after landing showed rounded cobbles characteristic of tumbling in a low-viscosity fluid.

Near-infrared reflectance measurements taken by DISR suggest the presence of water ice and organics, as well as a 5% methane abundance at the surface (Schröder and Keller 2008; Tomasko et al. 2005, Fig. 10). The GCMS instrument on Huygens detected surface constituents (Fig. 11) released upon heating by contact with and radiation from the probe, including ethane, cyanogens, benzene, and carbon dioxide (Niemann et al. 2005). The Huygens landing site is located in an infrared-dark region VIMS observes to be consistent with water ice and CO_2 ice, with trace organics (McCord et al. 2008; Rodriguez et al. 2006; Soderblom et al. 2007). Accelerometer data onboard Huygens indicates the material at the impact site is compressible and probably highly porous (Bettanini et al. 2008).

Observations by Cassini VIMS—combined with RADAR and DISR measurements—confirm earlier results, and also show compositional differences between surface geologic units (McCord et al. 2006; 2008, Soderblom et al. 2007). Titan's surface can be divided into two main units: infrared (and radar) "bright" and "dark". Dark material is largely confined to low latitudes within ±30°, while bright material incised by river channels dominates the mid-latitudes. Lakes of liquid hydrocarbons are reported only in the polar regions (>60°, Hayes et al. 2008). Liquid ethane has been reported in Ontario Lacus (Brown et al. 2008). Although methane opacity in the window regions remains uncertain, the VIMS observations

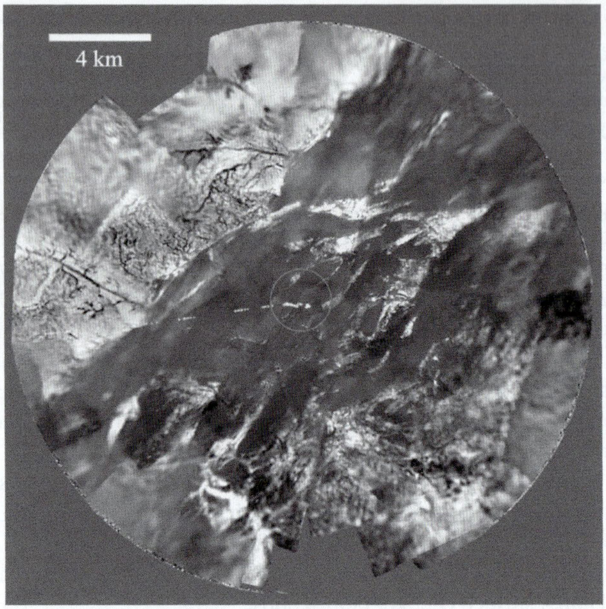

Fig. 9 DISR panorama mosaic of the Huygens landing site taken at altitudes between 8 and 17 km (Tomasko et al. 2005). Dark fluvial channels scar the bright elevated terrain as they flow into the lower, darker terrain

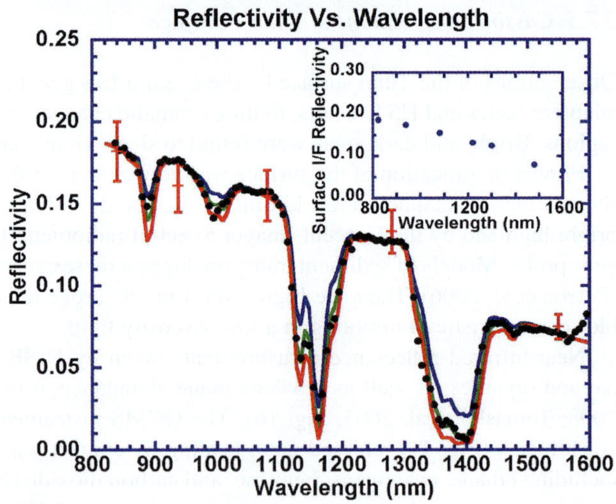

Fig. 10 Titan's reflectivity spectrum, as recorded by DISR, showing that the optimized modeled surface abundance of methane is close to 5% (*green line*). Lamp only downwardloking spectrum from altitude of 21 m (*black* data *points*). The *red line* with three-sigma error bars indicate absolute reflectivity in methane windows estimated from infrared measurements. This spectrum is compared to three models: 3% (*blue*), 5% (*green*), and 7% (*red*) methane mode fractions. Also shown in the *inset* is the blue slope of the surface reflectivity at the Huygens site (Tomasko et al. 2005, Fig. 14)

suggest nearly spectrally pure water ice is exposed within the dark regions—a "bedrock" underlying the more spectrally neutral (possibly hydrocarbon) dunes (Soderblom et al. 2007; McCord et al. 2008).

Titan's infrared- and radar-bright regions have haze-corrected reflectance spectra consistent with CO_2 ice, possibly mixed with water ice and tholin (McCord et al. 2008). Earth-based and Earth-orbital telescopic spectra as well as VIMS data have led to the suggestion that CO_2 could explain the spectral slope found in Titan's spectra in the 3 to 5-μm range. The presence of CO_2 on the surface was supported by the increase in partial pressure of CO_2 as the Huygens Probe Gas Chromatograph/Mass Spectrometer (GCMS) inlet temperature rose after landing and by its suggested compatibility with Infrared Space Observatory

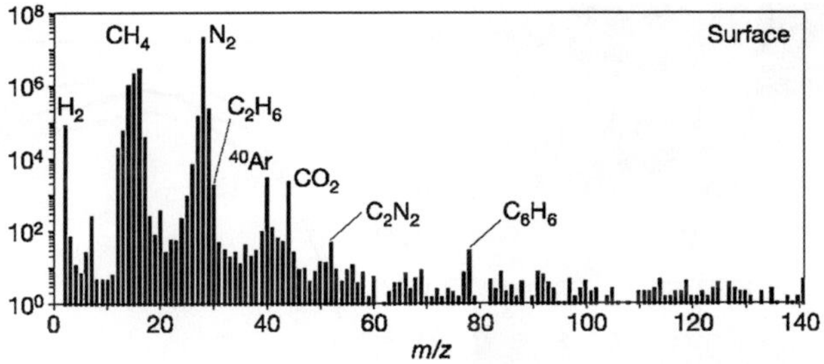

Fig. 11 Mass spectra from Huygens/GCMS at Titan's surface (Niemann et al. 2005). These sample-averaged spectra show the ion count rates per second (c/s) as a function of unit charge (m/z). Several components are detected. A surge of methane and ethane was detected shortly after the probe's impact, indicating an evaporation of volatiles due to heating of the surface by the probe. Furthermore, the presence of argon may indicate some geological activity on Titan that would transport Ar from the interior through liquid then solid layers, providing a means of atmospheric methane replenishment

(ISO) 3-mm observations (Coustenis et al. 2006). Two of Titan's putative cryovolcanoes, Tui Regio (Barnes et al. 2006) and Hotei Regio (Nelson et al. 2009), have reflectivities consistent with relative depletion in water ice. Enrichment in ammonia (Nelson et al. 2009) or carbon dioxide (McCord et al. 2008; Fig. 12) associatedwith outgassing or cryovolcanism have been suggested to explain this relative depletion. However, the presence of either NH_3 or CO_2 as a major ice constituent on Titan's surface remains to be confirmed.

Cyanoacetylene observed in Titan's atmosphere has been suggested as an alternate source for the feature near 4.92 microns in the VIMS Tui Regio spectrum. Various aromatic and aliphatic hydrocarbons, in particular benzene, have been suggested to explain the VIMS data (Clark et al. 2006), but the features observed have mostly low signal-to-noise ratios making it difficult to identify them. Other surface absorption features in the spectra could be compatible with low-molecular-weight alkanes, such as methane and ethane, in agreement with GCMS observations detecting methane evaporation, exhausted from the subsurface as the soil was heated post-landing. Debate continues over the existence of any of these constituents, with the possible exception of benzene, which may be the source for an absorption feature centered near 5.05 mm. Also, if these regions are indeed cryovolcanoes, some signatures of components from the interior water-ammonia liquid ocean would be expected.

2.7.4 The Liquid Reservoir on Titan's Surface

Radar imagery has revealed channels on much larger scales than those seen by Huygens. Radar-bright channels (probably cobbled streambeds) have been observed at low and mid-latitudes (Lorenz et al. 2007), while channels incised to depths of several hundred meters are seen elsewhere, and at high latitudes radar-dark, meandering channels are seen that suggest a lower-energy environment where deposition of fine-grained sediment occurs (Elachi et al. 2005). Whether these larger channels—some of which exceed a kilometer in width—and the large-scale flow features near the landing site (Soderblom et al. 2007) would require a different climate regime to be formed remains to be determined. The fate of the rivers is also a topic of debate.

The Cassini orbiter RADAR imaged a variety of very dark features of different sizes that have been interpreted to be liquid-filled (most probably with ethane) basins—"lakes" or "seas" (Stofan et al. 2007) confined to the region poleward of 55°N. To date some 655 such features have been identified and mapped over seven Titan flybys (Hayes et al. 2008)

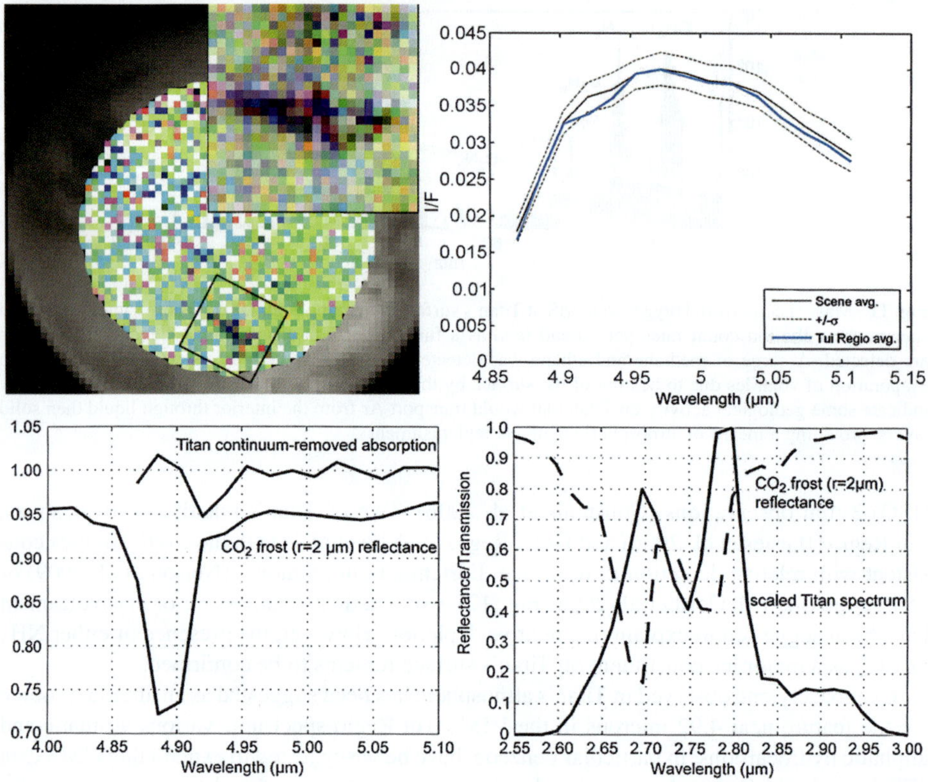

Fig. 12 Possible identification of CO_2 on Titan in Tui Regio. The anti-Saturnian hemisphere is shown, with the Tui Regio bright feature at center-left. The central colored disk in upper left shows deviations from the whole-scene average spectrum, at three channels (*red* = 4.94 μm, *green* = 4.92 μm, *blue* = 4.90 μm). The inset shows the same region, from another dataset. Tui Regio is clearly visible as a dark patch, indicating lower I/F in all three channels, particularly 4.92 μm. *Upper right*: Average (scaled) 5-μm spectra are shown for the portion of Tui Regio appearing *dark blue* in the *inset* of *upper left image*. *Dashed black lines* indicate one standard deviation from the scene average. *Lower two panels*: A comparison of the laboratory spectrum of CO_2 frost and the VIMS spectrum of Tui Regio is shown for two different cases: (*left*) the Tui Regio region continuum-removed spectrum with the 4.92-μm absorption shown in upper right panel and (*right*) the reflectance in the 2.8-μm methane window (*solid line*) compared to a synthetic spectrum of carbon dioxide (*dashed line*) calculated using a grain size of 2 μm. Adapted from McCord et al. (2008)

(Fig. 13). Mapping by Hayes et al. (2008) indicates that above 65°N the dark lakes occupy 15% of the terrain imaged so far. Below 70°N, the radar-dark lakes are replaced by similar features exhibiting less brightness contrast with their surroundings.

Circumstantial support for the hypothesis that the dark lakes are filled with liquid has been advanced (Stofan et al. 2007) based on several arguments, supported by the detection of ethane (Brown et al. 2008). Arguments for the existence of liquid in the lakes include the fact that their reflectivities indicate a smooth surface on the scales of the 2.16-cm wavelength of the radar system. A calm liquid surface or a recently frozen smooth solid surface would produce this result. Also, radiometry measuring the natural thermal emission at the 2.16 cm wavelength of the Cassini RADAR indicates that the dark lakes emit more thermal energy than the surroundings—consistent with hydrocarbons and inconsistent with a smooth surface of water-ice or ammonia-ice (Paganelli et al. 2007), assuming the exposed surrounding

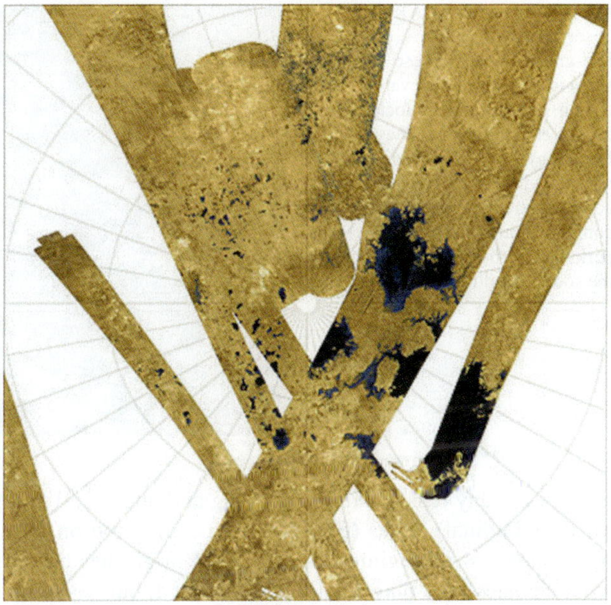

Fig. 13 Cassini flybys of the high northern latitudes of Titan allowed the RADAR instrument to image dark features that have been interpreted to be liquid-filled basins—"lakes". The features range in size from less than 10 km^2 to at least 100,000 km^2

crustal material is water-ice. Finally, the morphology of the boundaries between the largest of the dark lakes and the surroundings resembles a terrain flooded by liquid (Fig. 13), with the dark material appearing to flood valleys between hilly terrain and in some cases occupying networks of channels that feed into or out of the lakes.

Global circulation models predict precipitation of methane onto both or at least the winter pole (Rannou et al. 2006) and a decrease in surface temperatures poleward (Flasar et al. 2005). Currently the northern pole is approaching spring equinox in an annual cycle that is 29.5 years in length. Thus, the latitudinal restriction on the occurrence of the dark lakes is consistent with these models. Furthermore, the presence of channels suggests these lakes periodically empty and are then subjected to channel formation through flow of methane from the surroundings.

Cassini data on the lakes coverage suggest that roughly 0.6% of Titan's surface is potentially covered by liquid methane and ethane if the remaining un-imaged parts contain no lakes. Mitri et al. (2007) modeled evaporation off of high latitude lake surfaces and found that this amount of surface coverage, coupled with advective rates consistent with plausible wind speeds of 0.1–1 m/s (Tomasko et al. 2005), is sufficient to maintain the methane global mixing ratio on Titan at its present value. However, for an average lake depth of 20 m the reservoir of methane in the lakes is between $1/30$ and $1/3$ the methane atmospheric inventory (Lorenz et al. 2008), insufficient to account for the additional methane required to enhance the methane mixing ration in the equatorial atmosphere and permit the convectively- triggered rainstorms required to form the dendritic features at the Huygens landing site (Fig. 9). Either the lakes are on average at least an order of magnitude deeper than the minimum inferred from the radar absorption lengths, or additional methane is presence in subterranean porous or fractured media. Alternatively, the dendritic features might be a relic of a wetter recent past.

Aeolian activity on Titan has proven to be one of the major forces at work at low latitudes. Almost half the terrain within 30° of the equator is covered in dark (presumably organic-rich) streaks or dunes. In a few of the best-imaged regions, these have proven to

Fig. 14 Cassini/RADAR observations of Titan's fine-grained dunes. These dunes seem to flow around elevated terrain, like the bright blocks in the lower right corner. They are formed by a specific wind regime, with a west-to-east main component, and a north-south component alternating periodically to the north and to the south. Credit: NASA/JPL

be dunes, many tens of kilometers long and about 150 m high. Almost all appear to be linear (longitudinal) dunes (Fig. 14), a type common on Earth, where they form typically in bidirectional wind regimes. A tidal wind origin has been proposed for Titan, but seasonal wind changes may play a role. It is assumed, but has not been shown, that these dunes are presently actively maintained and shaped by the winds; they are certainly young relative to nearby geologic features.

The two satellites around Saturn with atmospheres are Titan and Enceladus and their comparative study brings insights on how these atmospheres influence the surfaces.

Indeed, studying geologic processes on sister moons like Titan and Enceladus, of similar ice-rock ratio but three orders of magnitude difference in mass, provides unique insights into how sources of heating manifest themselves in surface activity and tectonics on rock-ice worlds.

Enceladus' plumes provide an analogy that may help to explain the oxygen chemistry in Titan's upper atmosphere. The dominant oxygen-bearing gas in Titan's atmosphere is CO, which at 50 parts per million is likely derived from the methane-dominated hydrocarbon chemistry and an external source of oxygen (e.g., Wilson and Atreya 2004). Although water from micrometeoroids contributes to Titan's atmosphere (Coustenis et al. 1998) this source is not sufficient by itself to explain simultaneously the abundances of CO and CO_2 (Hörst et al. 2008). However, Cassini/ Cassini Plasma Spectrometer (CAPS) observations revealed O^+ ions being supplied to Titan's atmosphere (Hartle et al. 2006). Chemical modeling shows that a combination of atomic oxygen sources and water (or OH) are required to explain the origins of both atmospheric CO and CO_2 without invoking outgassing of CO from the interior. There may still be contributions from internally derived CO_2, but these do not reach the upper atmosphere given the very low tropopause temperature (Samuelson et al. 1983). Cassini observations of the Enceladus plumes indicate that they are the dominant source of O^+, O_2^+, and OH, derived from water coming out of the vents (Waite et al. 2006). The water-ice dominated plumes coming out of Enceladus' interior, magnetospheric processes that transform some of this material into oxygen atoms and ions, and the oxygen chemistry producing CO and CO_2 in Titan's atmosphere are linked.

2.7.5 Constraints on the Interior Composition from Surface Composition

The internal structure of planetary satellites can be inferred mainly by means of geophysical inversions of the gravity field of the satellites, as described in Hussmann et al. (2010, this issue) and Burger et al. (2010, this issue). Several other, complementary approaches may also be used to provide a more complete picture.

The topography exhibited on the surface of icy satellites can also provide important clues to the internal structural and thermal conditions of those bodies (Moore et al. 2007). Even if the surface temperatures of the satellites are very cold, the water ice may undergo substantial deformations over the lifetime of the solar system, provided that the interior is substantially hot. This is not immediately related to the size of the bodies, as expected, but to specific processes that are acting on the satellites. For example, Iapetus, despite its large size (1490-km in diameter), appears to have undergone very little viscous relaxation. On the contrary, Enceladus is characterized by abundant evidence of these processes.

The potential for complex internal evolution is directly related to the sources of energy that are available over the thermal history of the satellite as well as to its overall composition. The relative importance of general composition decreases with the availability of more intense energy sources. For the Saturnian satellites, available sources of energy include radioactive elements, tidal interactions between different satellites, and the original accretional heating. None of these sources alone is sufficient to explain the presence of oceans or layers of liquid and possibly salty water. Usually they have to be considered together and it is important as well to take into account the overall composition of the material. In particular the presence of materials able to reduce the melting point of ice are generally assumed (Multhaup and Spohn 2007) despite the lack of direct detections. The melting temperature of ice will be significantly reduced by the presence of small amounts of salts and/or incorporated volatiles such as methane and ammonia that are highly abundant in the outer solar system.

In general, large icy bodies such as the icy Galilean satellites, Titan and Triton are more likely to possess sub-surface oceans because of the slower cooling rate and the more intense radiogenic heating caused by their larger rock mass fractions, as compared to smaller icy bodies. However, depending on the amount of volatiles incorporated into the icy component during accretion, internal oceans cannot be ruled out for the largest of the medium-sized satellites of Saturn and Uranus and the biggest Transneptunian objects (TNOs), provided they are differentiated into a rock core and a water ice/liquid shell. The suggestion of an ocean within Callisto (Stevenson 2003) argues for at least partial differentiation.

Multhaup and Spohn (2007) computed the thermal evolution of icy satellites taking into account the effects of convection in a stagnant lid regime. They compared the temperatures in the interiors with the NH_3–H_2O solidus, the liquidus temperatures for 5 and 15 weight-% ammonia, and the ice melting temperature. Their models suggest large scale melting of ice I in Dione and Rhea only, for these amounts of ammonia. Similar models, when applied to Mimas and Tethys indicate partial melting of the ice if the satellites are composed of ice and ammonia. The presence of oceans requires that small amounts of ammonia (or another compound with similar properties) are available.

The detection of ammonia and/or ammonia hydrates on the surfaces would represent a chemical expression of internal activity. Unfortunately ammonia is difficult to identify by its weak infrared signature. Since it is easily photoionized it may not remain on icy satellite surfaces for appreciable lengths of time. Recently Waite et al. (2009) have shown that ammonia ice has been found in the plumes emitted by Enceladus. The amount of ammonia is estimated to be less than 1%. This is the first firm identification of ammonia on a satellite.

Another indicator of internal differentiation could be the presence of salts on the surface. These salts have been identified on the surface of Europa (McCord et al. 1998b, 1999), and they were considered a strong indication of the surface expression of the presence of an ocean. Heavily hydrated salt minerals, such as magnesium and sodium sulfates, sodium carbonate and their mixtures, have been identified on Europa, along with sulfuric acid hydrate (Dalton 2007b). These salts were considered the product of aqueous geochemistry involving sulfur-bearing minerals in contact with water, possibly at the ocean-rock interface (Kargel 1991, 2001). Therefore they are strong indicators of the past or present existence of a subsurface ocean. Micron and submicron grains of salts have also been found in the E-ring and are associated with emissions from Enceladus (Postberg et al. 2009). These authors claim that, due to the abundance of sodium salts, such grains could be produced by interaction with liquid water. The abundances identified in these grains are compatible with the anticipated interaction between the liquid ocean and a silicate core. Taken together, these observations indicate a plausible relationship between the surface chemistry of icy bodies and their internal structures.

3 Transneptunian Objects, KBOs, Pluto/Charon, Triton, and the Uranian Satellites

3.1 Kuiper Belt Objects

Several of the brightest transneptunian objects (TNOs), including the subset known as Kuiper Belt objects (KBOs), have yielded some of their compositional secrets to visible and near-infrared spectroscopy in the region 0.3–2.5 μm, as recently reviewed by Barucci et al. (2008) and Barkume et al. (2008). Lying beyond the orbit of Neptune, these objects are very faint and require the world's largest telescopes for spectroscopic studies. Visible region spectroscopy reveals a wide diversity in spectral slope ranging from neutral to very red. The red slope is thought to arise in most cases from the presence of complex macromolecular carbonaceous (organic) material. Broad but shallow absorption bands sometimes seen in this region may result from phyllosilicates and other aqueously altered minerals, but confirmation of their presence has been challenging.

TNOs are not only diverse in color, but in spectral characteristics. Many show no detectable spectral absorption bands, while others are replete with diagnostic bands throughout the visible and near-infrared wavelength regions. Despite the intrinsic difficulties in observing very faint objects, spectroscopy of TNOs is a quickly advancing subject area. In a sample of 45 of the brightest KBOs and Centaurs, the majority show features of H_2O ice within an otherwise featureless continuum. Little correlation of the spectral properties to the visible colors or albedo has been found, supporting the view that this population is quite heterogeneous (Barkume et al. 2008).

All of the discrete spectral bands found to date arise from frozen surface volatiles. Ices found on TNOs and related objects (e.g., the Centaurs) include H_2O, CH_4, C_2H_6, and CH_3OH, with possible NH_3. All of these ices have diagnostic absorption bands in the region 1.4–2.5 μm (e.g., Schmitt et al. 1998; Grundy and Schmitt 1998; Quirico et al. 1999; Douté et al. 1999; Brown and Calvin 2000; Brown et al. 2007a, 2007b). At the low temperatures ($<$40 K) characteristic of bodies beyond the orbit of Neptune, even the highly volatile N_2 and CO molecules are frozen. The presence of crystalline N_2(in the hexagonal β-phase) can be determined directly from its weak absorption band at 2.15 μm from spectral observations of sufficient quality (Grundy et al. 1993). In some cases when CH_4 is also present (e.g., Quaoar and 136199 (2003 UB_{313}) Eris; Licandro et al. 2006a, Schaller and Brown 2007) the presence of N_2 can also be established from the prominent shift of the CH_4 bands to shorter wavelengths (Quirico and Schmitt 1997; Grundy et al. 2002). This "matrix shift" occurs when CH_4 is dissolved in frozen N_2, and was first seen in the spectrum of Triton (Cruikshank et al. 1993). This has been established in laboratory studies (Quirico and Schmitt 1997). Among transneptunian objects, the matrix shift of CH_4 in N_2 is probably most prominent in the spectrum of 136472 (2005 FY_9) Makemake (Licandro et al. 2006b). Pluto, on the other hand shows CH_4 ice bands representing both the pure state (not shifted) as well as CH_4 dissolved in N_2 (Douté et al. 1999).

Nitrogen ice has the interesting property of metamorphosing into large (several cm), polycrystalline slabs, thus allowing long photon path lengths that give rise to the observed spectral absorption bands, which are not apparent from small particles in a scattering geometry. In addition, methane probably sinters at very low temperature ($\sim$30 K) into large slabs with cm-size crystals (Eluszkiewicz et al. 2007). This behavior may explain the very long path lengths in solid CH_4 required to produce the exceptionally strong bands observed in 136472 (2005 FY_9) Makemake noted above.

While the carbon oxides are found on Triton (CO, CO_2) and Pluto (CO only), their detection in the accessible wavelength regions requires spectra of high signal precision and moderately high spectral resolution. Neither molecule has yet been reported on a TNO or a Centaur object (on Centaurs, CO will be quickly lost by evaporation; it may be one of the volatiles that powers cometary activity in some of these objects).

Although spectroscopy beyond 2.5 μm has not yet been performed, photometric measurements of some TNOs and Centaurs at 3.6 and 4.6 μm have been made with the Spitzer Space Telescope. These two photometric points, when joined with spectra and photometry at shorter wavelengths, can help constrain scattering models of the surface compositions of TNOs, as demonstrated by Emery et al. (2006) and Dalle Ore et al. (2009), and serve as a guide to the presence of molecules (e.g., N_2 and CO_2) that are not directly detectable with presently available telescopes and spectrometers.

3.2 Pluto and Charon

The surface of the classical planet Pluto (radius 1150 km) shows clear spectral signatures of N_2, CH_4, and CO ices (Owen et al. 1993; Douté et al. 1999). Methane appears in two forms, pure CH_4 and CH_4 dissolved in N_2 (Douté et al. 1999), discernable by the wavelength-shifted and non-shifted components of the CH_4 bands. Carbon dioxide is notably absent in sufficient quantity to produce the absorption bands at 2 μm that are readily seen on Triton. The distribution of CH_4 on Pluto's surface is not uniform, as can be seen by the variable strength of the absorption bands as the planet rotates, bringing new longitudes into view. Grundy and Buie (2001) found that the N_2, CO, and the weak CH_4 band strengths are not directly correlated with the rotational light curve, while the strong CH_4 bands are correlated. Grundy and Buie (2002) identified three different terrains on Pluto; a N_2-rich, a CH_4-rich, and a volatile-depleted terrain, each having different continuum slopes. They found only weak evidence for any H_2O ice, concluding that H_2O is much less prominent on Pluto than on Triton. The red continuum slope of the globally averaged spectrum of Pluto can be modeled satisfactorily with tholins, and while tholins are a plausible constituent of the planet's surface, the standard spectral modeling caveat must be considered, that there may be other, equally plausible materials that provide the same spectral behavior, and so the modeling match does not necessarily constitute an identification of these complex organics (Cruikshank et al. 2005).

Olkin et al. (2007) and Protopapa et al. (2008) extended the spectroscopy of Pluto to longer wavelengths, but did not clearly identify any new molecular constituents. Protopapa et al. found a strong absorption band at 4.7 μm, possibly indicating the presence of C–N bearing molecules, although no specific identification has yet been made. Reports of C_2H_6 ice on Pluto have not yet been confirmed.

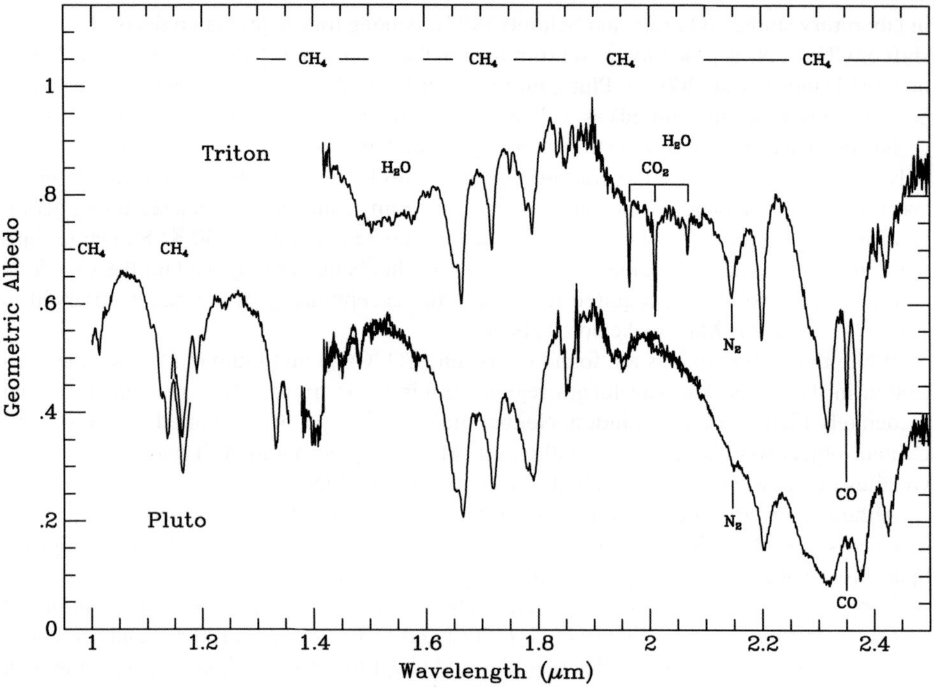

Fig. 15 These spectra of Triton and Pluto (1–2.5 μm) show the large number of absorption bands for the several ices composing the surfaces of these two bodies. These are all overtone or combination bands; the fundamentals for each molecule occur at longer wavelengths. The N_2 band in both spectra indicates the presence of the hexagonal β-phase. Small shifts in the CH_4 band centers, compared to their wavelengths in the laboratory, show that some (or most) of the CH_4 is dissolved in the N2 matrix. The Triton spectrum is adapted from Quirico et al. (1999), and the Pluto spectrum is adapted from Douté et al. (1999)

The light from Pluto's largest satellite Charon (radius 625 km) can be isolated from that of Pluto by advanced optical techniques with some large telescopes. The spectrum of Charon shows distinct absorption bands of crystalline H_2O ice in the region 1–2.5 μm, and a nearly neutral continuum at shorter wavelengths. The presence of hydrated ammonia (Brown and Calvin 2000), or possibly the NH_4^+ ion, is strongly suggested by an absorption band at 2.21 μm (Cook et al. 2007, 2009). Because of the instability of ammonia compounds, the presence of this band may suggest internal activity in Charon that replenishes NH_3 to the surface.

3.3 Triton

Neptune's largest satellite (radius 1353 km) has a rich near-infrared spectrum showing absorption bands of five ices: N_2, CH_4, CO, CO_2, and H_2O, all of which were discovered from ground-based spectroscopy.

Burgdorf et al. (2010) have presented evidence for the detection of the HCN ice band at 4.76 μm. HCN is an expected product of the photolytic processing of nitrogen and carbon-bearing molecules, and represents an intermediate step in the production of more complex materials of astrobiological interest. In addition, it is possible that the polymer of HCN can account for the low-albedo particulate material emanating from the active geyser-like vents revealed in Voyager 2 images of Triton's surface.

Quirico et al. (1999) provide the most thorough analysis of Triton's spectrum. The surface is clearly geologically young and dominated by volatile materials, as seen in images from the Voyager 2 spacecraft, which flew by Neptune and Triton in 1989 (Smith et al. 1989). Its high disk-averaged albedo (~0.8) is consistent with the presence of fresh ices. Triton undergoes an extreme seasonal cycle that forces an exchange of volatile material (principally N_2 and CO) from the frozen state on the surface to a tenuous and variable atmosphere, followed by recondensation. This process also redistributes the ices in different latitude zones over the complex seasonal cycle. Rapid changes in the color of Triton have been observed (Brown et al. 1995; Hicks and Buratti 2004), indicating redistribution of volatile materials on the surface as a consequence of pulses in the sublimation and recondensation cycles.

3.4 Uranian Satellites

The five largest Uranian satellites are (largest to smallest, with radii given in km) Titania (790), Oberon (760), Umbriel (585), Ariel (580), and Miranda (237). All five show near-infrared spectral absorption bands of H_2O ice in the crystalline state (e.g., Cruikshank et al. 1998; Bauer et al. 2002; Grundy et al. 2006). Grundy et al. (2006) have found the trio of CO_2 ice bands at 2 μm in the spectra of Ariel, Umbriel, and Titania, but not Oberon, and conclude that these represent pure CO_2 ice rather than a clathrate or a molecular complex. Notably absent is the forbidden band of CO_2 in H_2O that occurs at 2.134 μm (Bernstein et al. 2005). Grundy et al. (2006) find that the CO_2 band strength declines with increasing distance from Uranus. H_2O ice bands are strongest on the leading hemispheres of Ariel, Umbriel, and Titania, but weakest on the trailing hemisphere of Oberon. CO_2 absorption is strongest on the trailing hemispheres of the three satellites on which it has been found.

Grundy et al. (2006) considered possible sources of the CO_2 they found on three of the Uranian satellites. The coverage of the surfaces by imaging from Voyager showed geologically young regions that indicate varying degrees of resurfacing by fluids from the interior, especially in the cases of Miranda and Ariel (Croft and Soderblom 1991; Kargel 2004). While CO_2 gas might have emerged from the interior, the geographical distribution found by Grundy et al. does not strongly support that interpretation. Perhaps a more likely source is the *in situ* synthesis of CO_2 molecules by ion irradiation of H_2O ice in an intimate mixture with carbonaceous particles, as demonstrated in laboratory experiments by Mennella et al. (2004) and Gomis and Strazzulla (2005). At the same time, CO_2 molecules are readily destroyed by ion bombardment, and it is not clear that the balance of production and destruction results in a net accumulation of sufficient quantities of the CO_2 ice to yield the spectral bands that are detected. In addition, the expected sublimation and recondensation of CO_2 molecules, driven by sputtering and insolation patterns, complicates the picture even further. Grundy et al. (2006) conclude by noting that CO_2 is concentrated on the trailing hemispheres.

Bauer et al. (2002) observed the 1.65-μm component of the H_2O ice band complex on Miranda, demonstrating the presence of the crystalline phase. The also found a weak absorption band near 2.2 μm, which they tentatively attributed to ammonia hydrate. This band and its identification need confirmation; the presence of ammonia in any form would be of great importance because of its short lifetime against destruction by ultraviolet light and other energy sources and the consequent implications of its recent emplacement on the surface.

4 Summary

The icy satellites of the outer Solar system present a variety of chemical compositions arising from the interplay of both endogenic and exogenic processes. Initial complements of ices, minerals, and elements provided during formation have been subjected to aqueous alteration, extremes of pressure and temperature, irradiation by energetic charged particles and photons, and geologic processes ranging from cyclical to cataclysmic in nature. Additional material has been gained via cometary and meteoritic infall, transfer of material between satellites, and magnetospheric interactions. Material processed at the surface may be transported to the deep interiors, where it undergoes further processing; likewise, materials produced in the interior may be brought to the surface by cryovolcanic and tectonic forces. As spacecraft observations and laboratory measurements begin to reveal additional details of surface composition, we can now decipher the complex interactions of these various influences, thereby elucidating the thermal, physical and chemical histories of these enigmatic worlds.

Acknowledgements A portion of this research was carried out at the Jet Propulsion Laboratory, California Institute of Technology, under a contract with the National Aeronautics and Space Administration and funded through the internal Research and Technology Development program. The authors acknowledge the support of the NASA Outer Planets and Cassini Data Analysis programs, the European Space Agency (ESA), the German Aerospace Center (DLR), and the Italian Istituto di Fisica della Spazio Interplanetario (IFSI).

References

J.B. Adams, J. Geophys. Res. **79**, 4829 (1974)
J.L. Alvarellos et al., Icarus **194**, 636 (2008)
D. Baklouti, B. Schmitt, O. Brissaud, Icarus **194**, 647 (2008)
K.M. Barkume, M.E. Brown, E.L. Schaller, Astron. J. **135**, 55 (2008)
J.W. Barnes et al., Geophys. Res. Lett. **33**, 16204 (2006)
C.A. Barth et al., Geophys. Res. Lett. **24**, 2147 (1997)
M.A. Barucci et al., Astron. Astrophys. **479**, L13 (2008)
J.M. Bauer et al., Icarus **158**, 178 (2002)
A. Bellucci et al., Icarus **201**, 198 (2009)
M.P. Bernstein, D.P. Cruikshank, S.A. Sandford, Icarus **179**, 527 (2005)
C. Bettanini, M. Zaccariotto, F. Angrilli, Planet. Space Sci. **56**, 715 (2008)
G. Bjoraker et al., Bull. Am. Astron. Soc. **40**, 448 (2008)
G.J. Black et al., Science **304**, 553 (2004)
D.F. Blake et al., Science **254**, 548 (1991)
J.W. Boring et al., Nucl. Instrum. Methods **1**, 321 (1984)
H.S. Bridge et al., Science **204**, 987 (1979)
A.L. Broadfoot et al., Science **204**, 979 (1979)
A.L. Broadfoot et al., Science **212**, 206 (1981)
R.A. Brown, *Exploration of the Planetary System* (Reidel, Dordrecht, 1974). 527 pp
M.E. Brown, Icarus **126**, 236 (1997)
M.E. Brown, Icarus **151**, 190 (2001)
M.E. Brown, W.M. Calvin, Science **287**, 107 (2000)
R.A. Brown, F.H. Chaffee, Astrophys. J. **187**, L125 (1974)
M.E. Brown, R.E. Hill, Nature **380**, 229 (1996)
R.H. Brown et al., in *Neptune and Triton*, ed. by D.P. Cruikshank (Univ. Arizona Press, Tucson, 1995), p. 991
M.E. Brown et al., Astron. J. **133**, 284 (2007a)
M.E. Brown et al., Nature **446**, 294 (2007b)
R.H. Brown et al., Nature **454**, 607 (2008)
B.J. Buratti, J. Mosher, T.V. Johnson, Icarus **87**, 339 (1990)
B.J. Buratti et al., Icarus **136**, 223 (1998)
B.J. Buratti et al., Icarus **155**, 375 (2002)
B.J. Buratti et al., Astrophys. J. **622**, L149 (2005)
B.J. Buratti et al., Icarus **193**, 309 (2008)
M. Burgdorf et al., Astrophys. J. Lett. (2010). doi:10:1088/2041-8205/718/2/L53
M.H. Burger, R.E. Johnson, Icarus **171**, 557 (2004)
M.H. Burger et al., Space Sci. Rev. (2010, this issue). doi:10.1007/s11214-010-9645-z

W.M. Calvin, R.E. Johnson, J.R. Spencer, Geophys. Res. Lett. **23**, 673 (1996)
W.M. Calvin, J.R. Spencer, Icarus **130**, 505 (1997)
T.A. Cassidy et al., J. Geophys. Res. **113**, E02005 (2008)
J.D. Cassini, Philos. Trans. R. Soc. **12**, 831 (1676)
R.W. Carlson, K.P. Hand, Bull. Am. Astron. Soc. **38**, 540 (2006)
R.W. Carlson et al., Space Sci. Rev. **60**, 457 (1992)
R.W. Carlson et al., Science **274**, 385 (1996)
R.W. Carlson et al., Geophys. Res. Lett. **24**, 2479 (1997)
R.W. Carlson et al., Science **283**, 2062 (1999a)
R.W. Carlson, R.E. Johnson, M.S. Anderson, Science **286**, 97 (1999b)
R.W. Carlson et al., Icarus **157**, 456 (2002)
R.W. Carlson et al., Icarus **177**, 461 (2005)
R.W. Carlson et al., in *Io after Galileo*, ed. by R.M.C. Lopes, J.R. Spencer (Springer-Praxis, Chichester, 2007), p. 193
R.W. Carlson et al., in *Europa*, ed. by R.T. Pappalardo, W.B. McKinnon, K.K. Khurana (Univ. Arizona Press, Tucson, 2009)
S. Charnoz et al., Science **318**, 1622 (2007)
R.N. Clark et al., Icarus **58**, 265 (1984)
R.N. Clark et al., Nature **435**, 66 (2005)
R.N. Clark et al., Bull. Am. Astron. Soc. **38**, 574 (2006)
R.N. Clark et al., Icarus **193**, 372 (2008)
R.N. Clark et al., J. Geophys. Res. **114**, E03001 (2009). doi:10.1029/2008JE003150
D.W. Clarke, J.P. Ferris, Icarus **127**, 158 (1997)
D.W. Clarke, J.C. Joseph, J.P. Ferris, Icarus **147**, 282 (2000)
P. Coll et al., Planet. Space Sci. **47**, 1331 (1999)
G.C. Collins, J.C. Goodman, Icarus **189**, 72 (2007)
J.C. Cook et al., Astrophys. J. **663**, 1406 (2007)
J.C. Cook et al., LPSC **XL**, 2222 (2009)
J.F. Cooper et al., Icarus **149**, 133 (2001)
A. Coradini et al., Icarus **193**, 233 (2008)
A. Coustenis, F.W. Taylor, *Titan: Exploring an Earthlike World*, 2nd edn. (World Scientific Press, Singapore, 2008)
A. Coustenis et al., Icarus **118**, 87 (1995)
A. Coustenis et al., Astron. Astrophys. **336**, L85 (1998)
A. Coustenis et al., Icarus **161**, 383 (2003)
A. Coustenis et al., Icarus **177**, 89 (2005)
A. Coustenis et al., Icarus **180**, 176 (2006)
A. Coustenis et al., Icarus **189**, 35 (2007)
A. Coustenis et al., Icarus **207**, 461 (2010)
A. Coustenis et al., Space Sci. Rev. (2010, this issue). doi:10.1007/s11214-009-9615-5
S.K. Croft, L. Soderblom, in *Uranus*, ed. by J. Bergstralh (Univ. Arizona Press, Tucson, 1991), p. 561
J.K. Crowley, J. Geophys. Res. **96**, 16231 (1991)
D.P. Cruikshank, B.N. Khare, in *A New Era of Bioastronomy*, ed. by G. Lemarchand, K. Meech. ASP Conf. Ser., vol. 213 (2000), p. 253
D.P. Cruikshank et al., Icarus **53**, 90 (1983)
D.P. Cruikshank et al., Science **261**, 742 (1993)
D.P. Cruikshank et al., in *Solar System Ices*, ed. by B. Schmitt, C. deBergh, M. Festou (Kluwer Academic, Dordrecht, 1998), p. 579
D.P. Cruikshank et al., Icarus **175**, 268 (2005)
D.P. Cruikshank et al., Nature **448**, 54 (2007)
D.P. Cruikshank et al., Icarus **193**, 334 (2008)
D.P. Cruikshank et al., Icarus **206**, 561 (2009)
C. Dalle Ore et al., Astron. Astrophys. **501**, 349 (2009)
J.B. Dalton, Astrobiology **3**, 771 (2003)
J.B. Dalton, LPSC **XXXVII**, 2415 (2007a)
J.B. Dalton, Geophys. Res. Lett. **34**, L21205 (2007b)
J.B. Dalton, Space Sci. Rev. (2010, this issue). doi:10.1007/s11214-010-9658-7
J.B. Dalton et al., Icarus **177**, 472 (2005)
J.B. Dalton et al., LPSC **XL**, 2511 (2009)
A.G. Davies et al., J. Geophys. Res. **106**, 33079 (2001)
R. de Kok et al., Icarus **186**, 354 (2007)

M.K. Dougherty et al., Science **311**, 1406 (2006)
S. Douté et al., Icarus **142**, 421 (1999)
S. Douté et al., Icarus **149**, 107 (2001)
C. Elachi et al., Science **308**, 970 (2005)
J. Eluszkiewicz et al., J. Geophys. Res. **112**, 06003 (2007)
J.P. Emery et al., Astron. & Astrophys. **435**, 353 (2005)
J.P. Emery, D.P. Cruikshank, J. van Cleve, Icarus **182**, 496 (2006)
A. Eviatar et al., Icarus **47**, 75 (1981)
F.P. Fanale et al., Nature **280**, 761 (1979)
F.P. Fanale et al., in *Satellites of Jupiter*, ed. by D. Morrison (Univ. Arizona Press, Tuscon, 1982), p. 756
L.M. Feaga et al., Astrophys. J. **610**, 1191 (2004)
G. Filacchione, et al., Lunar Planet. Sci. Conf. **37**, 1271 (2006)
G. Filacchione et al., Icarus **186**, 259 (2007)
G. Filacchione et al., Icarus **206**, 507 (2010)
U. Fink, N.H. Dekkers, H.P. Larson, Astrophys. J. **179**, L155 (1973)
U. Fink, H.P. Larson, Icarus **24**, 411 (1975)
U. Fink, G.T. Sill, in *Comets*, ed. by L. Wilkening (Univ. Arizona Press, Tucson, 1982), p. 164
F.M. Flasar et al., Science **308**, 975 (2005)
L.A. Frank et al., Geophys. Res. Lett. **24**, 2159 (1997)
L.A. Frank, W.R. Paterson, Geophys. Res. Lett. **104**, 21345 (1999)
P.E. Geissler et al., Icarus **140**, 265 (1999)
P. Geissler et al., J. Geophys. Res. **106**, 33253 (2001)
P. Geissler et al., Icarus **169**, 29 (2004)
P. Goldreich, S. Tremaine, Nature **277**, 97 (1979)
P. Goldreich, S. Tremaine, Annu. Rev. Astron. Astrophys. **20**, 249 (1982)
S.H. Gordon et al., Meteorit. Planet. Sci. **41**, 5213 (2006)
O. Gomis, G. Strazzulla, Icarus **177**, 570 (2005)
O. Grasset, C. Sotin, F. Deschamps, Planet. Space Sci. **48**, 617 (2000)
C.A. Griffith, Nature **364**, 511 (1993)
C.A. Griffith, T. Owen, R. Wagener, Icarus **93**, 362 (1991)
C.A. Griffith et al., Science **300**, 628 (2003)
A.W. Grossman, D.O. Muhleman, Bull. Am. Astron. Soc. **24**, 954 (1992)
W.M. Grundy, M.W. Buie, Icarus **153**, 248 (2001)
W.M. Grundy, M.W. Buie, Icarus **157**, 128 (2002)
W.M. Grundy, B. Schmitt, J. Geophys. Res. **103**, 25809 (1998)
W.M. Grundy et al., Icarus **105**, 254 (1993)
W.M. Grundy et al., Icarus **155**, 486 (2002)
W.M. Grundy et al., Icarus **184**, 543 (2006)
W. Hage et al., Science **279**, 1332 (1998)
D.T. Hall et al., Astrophys. J. **426**, L51 (1994)
D.T. Hall et al., Nature **373**, 677 (1995)
D.T. Hall et al., Astrophys. J. **499**, 475 (1998)
R.A. Hanel, Science **212**, 192 (1981)
G.B. Hansen, T.B. McCord, Proc. J. Geophys. Res. **109**, E01012 (2004)
G.B. Hansen, T.B. McCord, Geophys. Res. Lett. **35**, L01202 (2008)
C.J. Hansen, D.E. Shemansky, A.R. Hendrix, Icarus **176**, 305 (2005)
B. Hapke, Geophys. Res. Lett. **6**, 799 (1979)
B. Hapke, Icarus **79**, 56 (1989)
D.L. Harris, in *Planets and Satellites*, ed. by G.P. Kuiper, B.M. Middlehurst (The University of Chicago Press, Chicago, 1961), p. 272
R.E. Hartle et al., Geophys. Res. Lett. **33**, L08201 (2006)
A. Hayes et al., Geophys. Res. Lett. **35**, L09204 (2008). doi:10.1029/2008GL033409
A.R. Hendrix et al., LPSC XXX **964**, 2043 (1999)
A.R. Hendrix et al., Icarus (2009, submitted)
M.D. Hicks, B.J. Buratti, Icarus **171**, 210 (2004)
C.A. Hibbitts, Dissertation, Univ. of Hawaii, AAT 3017400, 2001, 249 pp
C.A. Hibbitts, G.B. Hansen, Bull. Am. Astron. Soc. **39**, 428 (2007)
C.A. Hibbitts, J. Szanyi, Icarus **191**, 371 (2007)
C.A. Hibbitts, T.B. McCord, G.B. Hansen, J. Geophys. Res. **105**, 22541 (2000)
C.A. Hibbitts et al., J. Geophys. Res. **108**, 5036 (2003)

T.W. Hill, A.J. Dessler, C.K. Goertz, in *Physics of the Jovian Magnetosphere*, ed. by A.J. Dessler (Cambridge Univ. Press, Cambridge, 1983), p. 353

M. Hirtzig et al., J. Geophys. Res. **112** (2007)

P.V. Hobbs, *Ice Physics* (Clarendon Press, Oxford, 1974)

S. Hörst, V. Vuitton, R.V. Yelle, R.V. J. Geophys. Res. **113**, E10006 (2008)

Hussmann et al., Space Sci. Rev. (2010, this issue). doi:10.1007/s11214-010-9636-0

H. Imanaka et al., Icarus **168**, 344 (2004)

W.H. Ip et al., Adv. Space Res. **26**, 1649 (2000)

R.A. Jacobson et al., Astron. J. **132**, 2520 (2006)

K.S. Jarvis et al., Icarus **146**, 125 (2000)

K.L. Jessup et al., Bull. Am. Astron. Soc. **34**, 913 (2002)

K.L. Jessup, J. Spencer, R. Yelle, Icarus **192**, 24 (2007)

R.E. Johnson, in *Energetic Charged-Particle Interactions with Atmospheres and Surfaces* (Springer, Berlin, 1990)

R.E. Johnson, W.A. Jesser, Astrophys. J. **480**, L79 (1997)

R.E. Johnson, Icarus **143**, 429 (2000)

R.E. Johnson, T.B. McCord, Astrophys. J. **169**, 589 (1971)

T.V. Johnson, C.B. Pilcher, in *Planetary Satellites*, ed. by J.A. Burns (Univ. Arizona Press, Tucson, 1977), p. 232

R.E. Johnson, L.J. Lanzerotti, W.L. Brown, Nucl. Instrum. Methods **198**, 147 (1982)

R.E. Johnson et al., Icarus **156**, 136 (2002)

R.E. Johnson et al., in *Jupiter*, ed. by F. Bagenal, W. McKinnon (Cambridge University Press, Cambridge, 2004)

R.E. Johnson et al., J. Chem. Phys. **123**, 184715 (2005)

S.P. Joy et al., J. Geophys. Res. **107**, 17 (2002)

J.S. Kargel, Icarus **94**, 368 (1991)

J.S. Kargel, in *11th Annual Goldschmidt Conf.* (2001), p. 3147

J.S. Kargel, Earth, Moon & Planets **67**, 101 (2004)

J.S. Kargel, P. Delmelle, D.B. Nash, Icarus **142**, 249 (1999)

L. Keszthelyi, A. McEwen, Icarus **130**, 437 (1997)

L. Keszthelyi et al., Geophys. Res. **106**, 33025 (2001)

L. Keszthelyi et al., Icarus **169**, 271 (2004)

R.K. Khanna, J.C. Pearl, R. Dahmani, Icarus **115**, 250 (1995)

B.N. Khare et al., Icarus **60**, 127 (1984)

B.N. Khare et al., Can. J. Chem. **72**, 678 (1994)

M.G. Kivelson et al., in *Jupiter—The Planet, Satellites and Magnetosphere*, vol. 6118, ed. by F. Bagenal, T. Dowling, W. McKinnon (Cambridge Univ. Press, Cambridge, 2004), p. 513

A.J. Kliore et al., J. Geophys. Res. **107**, 1407 (2002)

G.P. Kuiper, Astrophys. J. **100**, 378 (1944)

S.W. Kieffer et al., Science **314**, 1764 (2006)

G.P. Kuiper, Astron. J. **62**, 245 (1957)

M. Kuppers, N.M. Schneider, Geophys. Res. Lett. **27**, 513 (2000)

A.L. Lane, D.L. Domingue, Geophys. Res. Lett. **24**, 1143 (1997)

A.L. Lane, R.M. Nelson, D.L. Matson, Nature **292**, 38 (1981)

L.M. Lara, R.D. Lorenz, R. Rodrigo, Planet. Space Sci. **42**, 5 (1994)

L.M. Lara, J. Geophys. Res. **101**, 261 (1996)

C. Laver, I. de Pater, Icarus **195**, 752 (2008)

C. Laver, I. de Pater, Icarus **201**, 172 (2009)

F. Leblanc, R.E. Johnson, M.E. Brown, Icarus **159**, 132 (2002)

F. Leblanc et al., Icarus **178**, 367 (2005)

E. Lellouch, in *The Collision of Comet Shoemaker-Levy 9 and Jupiter*, ed. by K.S. Noll, H.P. Weaver, P.D. Feldman (Cambridge University Press, Cambridge, 1996)

E. Lellouch et al., Nature **421**, 45 (2003)

M.T. Lemmon, E. Karkoschka, M. Tomasko, Icarus **103**, 329 (1993)

J.S. Lewis, Icarus **15**, 174 (1971)

J. Licandro et al., Astron. Astrophys. **458**, L5 (2006a)

J. Licandro et al., Astron. Astrophys. **445**, L35 (2006b)

R.M.C. Lopes, D.A. Williams, Rep. Prog. Phys. **68**, 303 (2005)

R.M.C. Lopes et al., J. Geophys. Res. **106**, 33053 (2001)

R.M.C. Lopes et al., Icarus **169**, 140 (2004)

R. Lopes-Gautier et al., Science **288**, 1201 (2000)

R.M.C. Lopes, B.J. Buratti, A.R. Hendrix, Icarus **193**, 305 (2008)
R.D. Lorenz et al., Geophys. Res. Lett. **34**, 07204 (2007)
R.D. Lorenz et al., Geophys. Res. Lett. **35**, 02206 (2008)
B.L. Lutz, C. De Bergh, T. Owen, Science **220**, 1374 (1983)
A. Marten et al., Icarus **158**, 532 (2002)
D.L. Matson, D.B. Nash, J. Geophys. Res. **88**, 4771 (1983)
D.L. Matson et al., Icarus **187**, 569 (2007)
B.H. Mauk et al., Nature **421**, 920 (2003)
T.B. McCord et al., Science **278**, 271 (1997)
T.B. McCord et al., J. Geophys. Res. **103**, 8603 (1998a)
T.B. McCord et al., Science **280**, 1242 (1998b)
T.B. McCord, G.B. Hansen, C.A. Hibbitts, J. Geophys. Res. **104**, 11827 (1999)
T.B. McCord, G.B. Hansen, C.A. Hibbitts, Science **292**, 1523 (2001)
T.B. McCord et al., Planet. Space Sci. **54**, 1524 (2006)
T.B. McCord et al., Icarus **194**, 212 (2008)
A.S. McEwen, J. Geophys. Res. **91**, 8077 (1986)
A.S. McEwen et al., Icarus **75**, 450 (1988)
A.S. McEwen et al., Icarus **135**, 181 (1998a)
A.S. McEwen et al., Science **281**, 87 (1998b)
A.S. McEwen et al., in *Jupiter: The Planet, Satellites, and Magnetosphere*, ed. by F. Bagenal, T.E. Dowling, W.B. McKinnon (Cambridge Univ. Press, Cambridge, 2004), p. 307
M.A. McGrath et al., in *Jupiter*, ed. by F. Bagenal, T.E. Dowling, W.B. McKinnon (Cambridge Univ. Press, Cambridge, 2004), p. 457
M. Mendillo et al., Nature **348**, 312 (1990)
V. Mennella, M.E. Palumbo, G.A. Baratta, Astrophys. J. **615**, 1073 (2004)
G. Mitri et al., Icarus **186**, 385 (2007)
J.M. Moore, R. Khanna, B. Donn, J. Geophys. Res. **96**, 17541 (1991)
J.M. Moore et al., Icarus **140**, 294 (1999)
J.M. Moore et al., in *Jupiter*, ed. by F. Bagenal, T.E. Dowling, W.B. McKinnon (Cambridge Univ. Press, Cambridge, 2004), p. 397
J.M. Moore et al., LPSC XXXVIII **1338**, 2136 (2007)
V.I. Moroz, Astron. Z. **42**, 1287 (1965)
J.I. Moses, D.B. Nash, Icarus **89**, 277 (1991)
K. Multhaup, T. Spohn, Icarus **186**, 420 (2007)
R.E. Murphy, D.P. Cruikshank, D. Morrison, Astrophys. J. Lett. **177**, L93 (1972)
C.Y. Na et al., Icarus **131**, 449 (1998)
D.B. Nash, B.H. Betts, Icarus **117**, 402 (1995)
D.B. Nash, F.P. Fanale, Icarus **31**, 40 (1977)
D.B. Nash, R.R. Howell, Science **244**, 454 (1989)
D.B. Nash et al., in *Satellites*, ed. by J.A. Burns, M.S. Matthews (Univ. Arizona Press, Tucson, 1986), p. 629
A. Negrão et al., Planet. Space Sci. **54**, 1225 (2006)
R.M. Nelson, B.W. Hapke, Icarus **36**, 304 (1978)
R.M. Nelson, W.D. Smythe, Icarus **66**, 181 (1986)
R.M. Nelson et al., Icarus **199**, 429 (2009)
H.B. Niemann et al., Nature **438**, 779 (2005)
F. Nimmo, R.T. Pappalardo, Nature **441**, 614 (2006)
F. Nimmo et al., Nature **447**, 289 (2007)
K.S. Noll, H.A. Weaver, A.M. Gonnella, J. Geophys. Res. **100**, 19057 (1995)
K.S. Noll et al., Science **273**, 341 (1996a)
K.S. Noll et al., Icarus **124**, 625 (1996b)
K.S. Noll, H.A. Weaver, P.D. Feldman, IAU Colloq. **156**, 213 (1996c)
K.S. Noll et al., Geophys. Res. Lett. **24**, 1139 (1997a)
K.S. Noll et al., Nature **388**, 45 (1997b)
N. Ockman, Adv. Phys. **7**, 199 (1958)
C.B. Olkin et al., Astron. J. **133**, 420 (2007)
T.C. Owen et al., Science **261**, 745 (1993)
T.C. Owen et al., Icarus **149**, 160 (2001)
F. Paganelli et al., Icarus **191**, 211 (2007)
C. Paranicas, R.W. Carlson, R.E. Johnson, Geophys. Res. Lett. **28**, 673 (2001)
L. Pauling, J. Am. Chem. Soc. **57**, 2680–2684 (1935)
S.J. Peale, P. Cassen, R.T. Reynolds, Science **203**, 892 (1979)

J.T. Perron et al., J. Geophys. Res. **111**, E11001 (2006)
C.B. Phillips, PhD Thesis. Univ. Arizona, Tucson, 2000
C.B. Pilcher, S.T. Ridgeway, T.B. McCord, Science **178**, 1087 (1972)
C.C. Porco et al., Nature **434**, 159 (2005)
C.C. Porco et al., Science **311**, 1393 (2006)
C.C. Porco et al., Science **318**, 1602 (2007)
F. Postberg et al., Icarus **183**, 122 (2006)
F. Postberg et al., Nature **459**, 1098 (2009)
S. Protopapa et al., Astron. Astrophys. **490**, 365 (2008)
E. Quirico, B. Schmitt, Icarus **127**, 354 (1997)
E. Quirico et al., Icarus **139**, 159 (1999)
E. Quirico et al., Icarus **198**, 218 (2008)
J. Radebaugh et al., Icarus **169**, 65 (2004)
K. Rages, J.B. Pollack, Icarus **41**, 119 (1980)
P. Rannou et al., Science **311**, 201 (2006)
S. Rodriguez et al., Planet. Space Sci. **54**, 1510 (2006)
C.T. Russel, M.G. Kivelson, K.K. Khurana, Planet. Space Sci. **53**, 937 (2005)
C. Sagan, W.R. Thompson, Icarus **59**, 133 (1984)
F. Salama et al., Icarus **83**, 66 (1990)
F. Salama et al., Icarus **107**, 413 (1994)
J.W. Salisbury, G.R. Hunt, Suppl. to Advances in the Astronautical Sciences (1971), p. 35
R.E. Samuelson et al., J. Geophys. Res. **88**, 8709 (1983)
R.E. Samuelson, N.R. Nath, A. Borysow, Planet. Space Sci. **45**, 959 (1997)
S.A. Sandford, L.J. Allamandola, Icarus **106**, 478 (1993)
J. Saur, D.F. Strobel, F.M. Neubauer, J. Geophys. Res. **103**, 19947 (1998)
F.L. Schaller, M.E. Brown, Astrophys. J. **670**, 49 (2007)
P.M. Schenk, J.M. Moore, LPSC XXXVIII **1338**, 2305 (2007)
B. Schmitt, S. Rodriguez, J. Geophys. Res. **108**, 5104 (2003)
B. Schmitt et al., Icarus **111**, 79 (1994)
B. Schmitt, C. de Bergh, M. Festou, in *Solar System Ices*, ed. by B. Schmitt, C. de Bergh, M. Festou (Kluwer Academic, Dordrecht, 1998), p. 312
S.E. Schröder, H.U. Keller, Planet. Space Sci. **56**, 753 (2008)
G. Schubert et al., Space Sci. Rev. (2010). doi:10.1007/s11214-010-9635-1
V.I. Shematovich et al., Icarus **173**, 480 (2005)
J.H. Shirley et al., LPI **41**, 2395 (2010)
B.A. Smith et al., Science **246**, 1422 (1989)
P.H. Smith et al., Icarus **119**, 336 (1996)
W.H. Smyth, M.L. Marconi, Icarus **181**, 510 (2006)
W.H. Smyth, M.B. McElroy, Planet. Space Sci. **25**, 415 (1977)
W.D. Smythe, R.M. Nelson, D.B. Nash, Nature **280**, 766 (1979)
W.D. Smythe et al., XXIX LPSC, Vol. CD Lunar and Planetary Institute, Houston, Texas, 1998
L.A. Soderblom et al., Planet. Space Sci. **55**, 2015 (2007)
J.R. Spencer, W.M. Calvin, Astron. J. **124**, 3400 (2002)
J.R. Spencer, W.M. Calvin, M.J. Person, J. Geophys. Res.—Planets **100**, 19049 (1995)
J.R. Spencer et al., Icarus **127**, 221 (1997)
J.R. Spencer et al., Science **284**, 1514 (1999)
J.R. Spencer et al., Science **288**, 1208 (2000)
K. Stephan et al., LPSC **XXXX**, 1377 (2009)
K. Stephan et al., Icarus **206**, 631 (2010)
D.V. Stevenson, EGS-AGU-EUG Joint Assembly (2003), p. 3289
E.R. Stofan et al., Nature **445**, 61 (2007)
D.F. Strobel, D.E. Shemansky, J. Geophys. Res. **87**, 1361 (1982)
D.F. Strobel et al., Icarus **103**, 333 (1993)
D.F. Strobel et al., in *Titan after Cassini-Huygens, Atmospheric Structure and Composition*, ed. by R.H. Waite (Springer, Berlin, 2009), p. 235
S.E. Summers, D.F. Strobel, Icarus **120**, 290 (1996)
B.D. Teolis et al., Astrophys. J. **644**, L141 (2006)
N. Thomas et al., in *Jupiter: The Planet, Satellites, and Magnetosphere*, ed. by F. Bagenal, T.E. Dowling, W. McKinnon (Cambridge Univ. Press, Cambridge, 2004), p. 561
P.C. Thomas et al., Icarus **190**, 573 (2007)
G. Tobie et al., Icarus **175**, 496 (2005)

G. Tobie, O. Čadek, C. Sotin, Icarus **196**, 642 (2008)
M.G. Tomasko et al., Nature **438**, 765 (2005)
L.M. Trafton, Astrophys. J. **175**, 285 (1972)
B.N. Tran, J.P. Ferris, J.J. Chera, Icarus **162**, 114 (2003a)
B.N. Tran et al., Icarus **165**, 379 (2003b)
A.J. Verbiscer, J. Veverka, Icarus **99**, 63 (1992)
R.J. Vervack, B.R. Sandel, D.F. Strobel, Icarus **170**, 91 (2004)
F. Vilas et al., Icarus **124**, 262 (1996)
F. Vilas et al., Icarus **170**, 125 (2004)
M. Volwerk, M.G. Kivelson, K.K. Khurana, J. Geophys. Res. **106**, 26033 (2001)
V. Vuitton, R.V. Yelle, J. Cui, J. Geophys. Res. **113**, E05007 (2008)
J.H. Waite et al., Science **308**, 986 (2005)
J.H. Waite et al., Science **311**, 1419 (2006)
J.H. Waite et al., Nature **460**, 487 (2009)
W. Wamsteker, R.L. Kroes, J.A. Fountain, Icarus **23**, 417 (1973)
E.H. Wilson, S.K. Atreya, Planet. Space Sci. **51**, 1017 (2003)
E.H. Wilson, S.K. Atreya, J. Geophys. Res. **109**, E06002 (2004)
M.C. Wong, R.E. Johnson, J. Geophys. Res.—Planets **101**, 23255 (1996)
L.C. Wye et al., Icarus **188**, 367 (2007)
R.V. Yelle, J. Cui, I.C.F. Müller-Wodarg, J. Geophys. Res. **113**, E10003 (2008)
K. Zahnle et al., Icarus **163**, 263 (2003)
H.A. Zebker et al., Icarus **194**, 704 (2008)
B. Zellner, Astron. J. **174**, L107 (1972)
M.Y. Zolotov, B. Fegley, Jr., Icarus **141**, 40 (1999)

Space Sci Rev (2010) 153: 155–184
DOI 10.1007/s11214-009-9615-5

Atmospheric/Exospheric Characteristics of Icy Satellites

A. Coustenis · T. Tokano · M.H. Burger · T.A. Cassidy · R.M. Lopes · R.D. Lorenz · K.D. Retherford · G. Schubert

Received: 8 April 2009 / Accepted: 19 November 2009 / Published online: 16 February 2010

Abstract The atmospheres/exospheres of icy satellites greatly vary from one to the next in terms of density, composition, structure or steadiness. Titan is the only icy satellite with a dense atmosphere comparable in many ways to that of the Earth's atmosphere. Titan's atmosphere prevents the surface from direct interaction with the plasma environment, but gives rise to Earth-like exchanges of energy, matter and momentum. The atmospheres of other satellites are tenuous. Enceladus' atmosphere manifests itself in a large water vapor plume emanating from surface cracks near the south pole. Io's SO_2 atmosphere originates from volcanoes. Europa's tenuous O_2 atmosphere is produced by intense radiation bombardment. This chapter reviews the characteristics of the atmospheres of Titan, Enceladus, Io and Europa based on observations.

A. Coustenis (✉)
LESIA, Observatoire de Paris-Meudon, 5, pl. Jules Janssen, 92195 Meudon, France
e-mail: athena.coustenis@obspm.fr

T. Tokano
Institut für Geophysik und Meteorologie, Universität zu Köln, Köln, Germany

M.H. Burger
University of Maryland and NASA Goddard Space Flight Center, Greenbelt, MD 20771, USA

T.A. Cassidy
Astronomy Department, University of Virginia, Charlottesville, USA

R.M. Lopes
Jet Propulsion Laboratory, Pasadena, USA

R.D. Lorenz
Applied Physics Laboratory, Johns Hopkins University, Laurel, USA

K.D. Retherford
Southwest Research Institute, San Antonio, USA

G. Schubert
Earth and Space Sciences, University of California at Los Angeles, Los Angeles, USA

Keywords Atmospheres · Exospheres · Icy satellites · Titan · Io · Enceladus · Europa

1 Titan's Atmosphere

1.1 Introduction

Titan has a remarkable atmosphere that we discuss in particularly within the context of surface–atmosphere–interior interactions and external exchange processes. Titan is the only icy satellite covered by a dense atmosphere, with a surface pressure of ~1.5 bars. This dense atmosphere gives rise to a diversity of interactions with the surface, which are unique among the icy satellites and which closely resemble those at the Earth's surface. An important feature for the evolution and the interaction with the surface is Titan's climate. The few known isotopic ratios in Titan's atmosphere attest to resupply and loss processes, with different results for the nitrogen and carbon reservoirs. Carbon can exist on Titan in its three forms (gas, liquid and ice). It has already been detected on Titan in the form of methane greenhouse gas and surface liquid deposits and therefore makes an appropriate analog for the Earth's water. The main photochemical sub-product of methane is ethane.

Far from the Sun, methane then plays the active role on Titan that water plays on Earth, acting as a condensable greenhouse gas, forming clouds and rain, and pooling on the surface as lakes (and perhaps as "acquifers" underneath the surface). Titan's icy surface is shaped not only by impact craters and tectonics, but also by volcanism in which the lava is liquid water quickly transformed into ice after ejection ("cryovolcanism"), by rivers of liquid methane, and by tidally and thermally driven winds that sculpt drifts of aromatic organics into long linear dunes. Channels likely carved by liquid methane and/or ethane, lakes and seas of these materials, vast equatorial dune fields of complex organics (produced high in the atmosphere) and shaped by wind, intriguing hints of volcanic flows of aqueous materials across an icy crust, and a dearth of impact craters suggest a world with a balance of geologic and atmospheric processes that is the solar system's best analog to Earth (Coustenis and Taylor 2008).

Most prominent is the existence of a hydrological cycle involving methane clouds, rain and at least transient rivers. The possibility of such a cycle had been noted already since Voyager, as the proximity of Titan's surface conditions to the methane triple point was recognized, and the first evidence of clouds emerged in spectroscopic data (Griffith et al. 1998), reported since 1995, and in Keck Telescope images (Brown et al. 2002). Another analogy with the Earth relates to the seasonal phenomena observed for instance in the polar stratosphere. Titan was observed by Voyager to have a UV-dark 'polar hood,' a dark haze cap over the winter northern pole. This cap was observed in high-phase-angle images to stand above the main haze deck, and connect with the detached haze layer. While connected to the detached haze at high altitude, the region is dynamically isolated by the circumpolar vortex. On Earth, the corresponding circumpolar winds isolate the winter stratosphere from the rest of the atmosphere: the catalytic surfaces of polar stratospheric clouds that form in the winter night cause the destruction of ozone whose concentration becomes locally depleted—the ozone hole.

A key characteristic of Titan is its massive inventory of organic chemicals. As noted previously, the Voyager-1 Titan flyby in the early 1980s verified the presence of methane in a thick background atmosphere of nitrogen. Even more interesting was the detection of a host of complex hydrocarbons and nitriles that resulted from the photolysis and energetic particle bombardment of the atmosphere and the thick organic haze that both scattered and absorbed visible and infrared photons, thereby playing an important role in determining atmospheric thermal structure. Measurements throughout the atmosphere, both remotely and in situ, have thus indicated the presence of numerous hydrocarbon and nitrile gases, as well as a complex layering of organic aerosols that persists all the way down to the surface of the moon (Tomasko et al. 2005; Israël et al. 2005; Coustenis et al. 2007), although their molecular composition remains to be determined.

The abundance of methane and its organic products in the atmosphere, seas and dunes exceeds by more than an order of magnitude the carbon inventory in the Earth's ocean, biosphere and fossil fuel reservoirs (Lorenz et al. 2008a). Indeed, the measured value of the irreversible conversion of the methane in the atmosphere into higher-order organic/nitrile compounds that eventually end up deposited on the surface of Titan (10–100 Myr) is close to that of our own planet, indicating that methane is resupplied and converted at a rate that prevents the build-up of the heavier isotope over time as is the case of nitrogen (Atreya et al. 2006).

Water (and tentatively carbon dioxide) ice has also been reported to exist currently on the surface of Titan (e.g., Coustenis et al. 1995, 2006; Griffith et al. 2003; McCord et al. 2008). Transient episodes of melting of the water ice by either geologic activity or impacts would expose organics to aqueous alteration, as well as contact with carbon dioxide, leading potentially to reaction pathways that mimic those that occurred on the pre-biotic Earth. No other place in the solar system has this type of ongoing chemistry. Titan is an organic paradise that is certain to tell us much about the chemical evolution that may lead to life.

1.2 Titan's Upper Atmosphere

Titan's upper atmosphere (above 400 km in altitude) plays a key role for the chemistry and physics of the entire atmosphere. It is the region where solar EUV radiation is absorbed, dissociating molecular nitrogen and methane and initiating a chain of highly complex photochemistry that affects all layers. Figure 1 shows the three main layers of Titan's atmosphere, and its illumination by solar UV (blue rays), solar visible light (yellow rays) and solar infrared light (red rays). Titan's upper atmosphere interacts with Saturn's magnetosphere, and an induced magnetosphere is formed wherein energy and particles from Saturn's magnetosphere are deposited, breaking and ionizing molecular nitrogen and further contributing to the complex atmospheric chemistry. Here atmospheric escape processes also occur, directly affecting atmospheric evolution, including ion escape processes.

The energy input from Saturn's magnetospheric interaction shows a strong interaction with Titan's upper atmosphere and the corresponding induced magnetosphere with draped magnetic field lines. The keV to 10s of keV O^+ ions and energetic H^+ ions $E > 50$ keV can penetrate across field lines and deposit their energy directly into Titan's atmosphere and lower ionosphere to altitudes <800 km (Cravens et al. 2008) resulting in heating and escape of suprathermal nitrogen and methane atoms and molecules from Titan's upper atmosphere (Michael and Johnson 2005). Energetic neutral atoms (ENAs) are also emitted via charge transfer collisions between magnetospheric energetic ions and Titan's atmosphere (Garnier et al. 2008, 2009). The magnetospheric electrons are magnetically tied to the field lines but can drift to lower altitudes via gradient and curvature drift mechanisms (Hartle et al. 1982). In addition to magnetospheric electrons, photoelectrons from the ionizing solar UV impacting Titan's upper atmosphere produce a non-thermal electron gas within Titan's ionosphere which can then heat thermal electrons to $T_e \sim 1000$ K (Wahlund et al. 2005) and correspondingly reduce positive ion recombination rates. The hot electrons can also produce a polarization electric field ($T_e \gg T_{ION} \sim 180$ K) which will pull ions out of the ionosphere and produce an ionospheric wind as shown in the form of CH_5^+ and $C_2H_5^+$ ions.

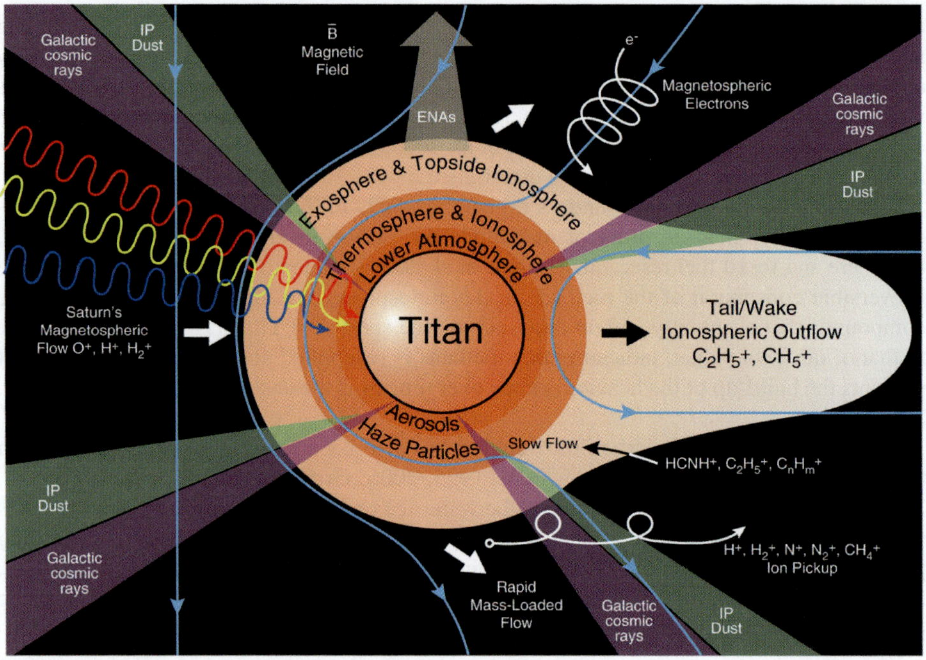

Fig. 1 Energy deposition in Titan's upper atmosphere (adapted by E. Sittler from the figure in Waite et al. 2004)

Also shown in Fig. 1 are pickup ions (PUI) from the plasma interaction with Titan's exosphere which slows the external flow at high altitudes of >1400 km and leads to atmospheric escape of keV methane and dinitrogen PUIs (Hartle et al. 2006a, 2006b), as well as H^+ and $H_2{}^+$ PUIs at even higher altitudes where the H and H_2 corona can extend to Hill sphere distances of about 55,000 km. It is only near the equinox that Titan spends most of its time in Saturn's current sheet (and thus has a strong interaction with the Saturnian plasma) while it remains above or below it during solstice. The escape of methane from Titan's exosphere will translate to a methane torus in Saturn's outer magnetosphere with dimensions ~20 Saturn radii. Figure 1 shows a more isotropic energy input to deeper altitudes from Galactic Cosmic Rays (GCR) with penetration below the 100 km altitude (Fulchignoni et al. 2005) and Interplanetary Dust (IP) which can penetrate down to altitudes lower than 740 km (Molina-Cuberos et al. 2001). The IP dust flux will tend to peak on the side facing Titan's orbital motion around Saturn. In both cases ionization layers are produced deep into Titan's atmosphere where further chemistry can be driven.

Additional growth of haze particles could occur in these layers and the free electrons can attach themselves to the aerosols producing negative heavy ions (Borucki and Whitten 2008). The negative heavy ions formed in Titan's upper atmosphere (Coates et al. 2007) will fall to lower altitudes and result in a loss of ionospheric ions which must continuously be replaced by the ionosphere (Sittler et al. 2008). These heavy ions are now believed to be the primary source for the aerosols in Titan's lower atmosphere which then fall to the surface, where unknown organic and nitrile chemistry can occur. If the negative ions observed by Coates et al. (2007) are fullerenes (C_{60}), then magnetospheric keV oxygen ions from Enceladus can become trapped inside the fullerene cages and be a source of free oxygen in the aerosols and thus the surface chemistry (Sittler et al. 2008).

It is thus essential to determine the ion and neutral composition of the ionosphere with a mass range and resolution allowing the detection and characterization of a very wide range of compounds. However, the altitude range from ~400 km to 950 km of Titan's atmosphere is unfortunately below the reach of Cassini in situ measurements and above the region of intense remote sensing and most Huygens measurements, and its exploration shall therefore have to await a future space mission.

The Cassini observations have nevertheless revealed that Saturn's magnetosphere is deformed into a magnetodisk configuration at Titan orbital distances for Saturn local times (SLT) dusk to midnight to dawn LT while more dipolar on the dayside near noon LT. This deformation is ultimately driven by the large amounts of water vapor dumped into Saturn's magnetosphere by Enceladus originating in a hot interior (Matson et al. 2007) and possibly other icy moons. As shown in Sittler et al. (2008), the heavy ions will be confined within a few degrees of the magnetodisk current sheet and light ions will dominate at higher magnetic latitudes, while for dipolar field lines heavy ions are more likely to be encountered by Titan. When heavy ions dominate the external flow, the energy input to the magnetosphere can be orders of magnitude greater than that when the field is more disk-like and light ions such as H^+ and $H_2{}^+$ dominate the ion composition. In some cases, when Titan encounters the magnetodisk current sheet plasma, the heavy ions will be highly abundant and a very high-energy encounter can occur such as the T5 Cassini encounter (Sittler et al. 2008).

1.3 The Organic Inventory from the Ionosphere to the Surface

In the current picture of Titan's organic chemistry, and already since the Voyager days, it was known that the chemical evolution of the main atmospheric constituents—dinitrogen and methane—produces complex refractory organics which accumulate on the surface, together with condensed volatile organic compounds such as HCN and benzene. The second most abundant constituent, methane, is dissociated irreversibly to produce hydrocarbons (e.g., C_2H_2, C_2H_4, C_2H_6 and C_3H_8) and nitriles, (e.g., HCN, HC_3N), from the coupled nitrogen chemistry. The Cassini–Huygens era of investigation has furthered our understanding of Titan as the largest abiotic organic factory in the solar system and we now know a little more about the processes involved in Titan's organic chemistry.

The structure of the upper atmosphere of Titan was defined by the Cassini Ultraviolet Imaging Spectrometer (UVIS), which observed the extinction by the atmosphere of Titan of two stars during the second Titan flyby (Shemansky et al. 2005). A mesopause was inferred at 615 km with a temperature minimum of 114 K. Six species were identified and measured: methane, acetylene, ethylene, ethane, diacetylene, and hydrogen cyanide at altitude ranges from 450 to 1600 km. The higher order hydrocarbons and hydrogen cyanide peak sharply in abundance and are undetectable below altitudes ranging from 600 to 750 km, leaving methane as the only identifiable carbonaceous molecule in this experiment below 600 km.

Indirect but solid evidence for high complex organic species in the ionosphere was found by the Cassini Ion Neutral Mass Spectrometer (INMS). The analysis of the ionospheric INMS measurements taken during the closest Cassini flybys of Titan shows the presence of many complex organic species, in spite of the very high altitudes (1100–1300 km) (Waite et al. 2007). Extrapolation of Cassini INMS (limited to mass up to 100 daltons) and Plasma Spectrometer (CAPS) data, strongly suggests that high-molecular-weight species (up to several thousands daltons) may be present in the ionosphere (Fig. 2). These findings revolutionize our understanding of the organic processes occurring in Titan's atmosphere, with a strong implication that ionospheric chemistry plays an important role in the formation of complex

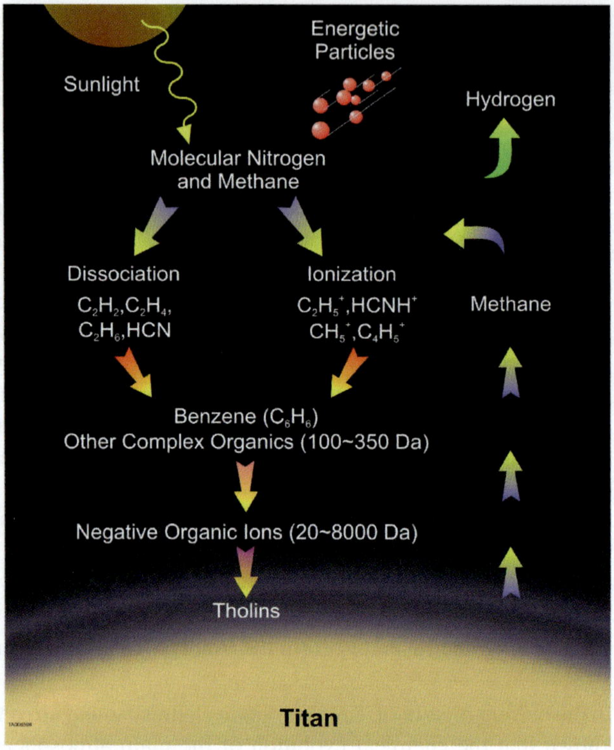

Fig. 2 Organic inventory on Titan from the upper atmosphere to the surface (Waite et al. 2007)

organic compounds in Titan's environment, something that was not foreseen (Waite et al. 2007). These organics are certainly the precursors of the hydrocarbons and nitriles found in the stratosphere, which form aggregates and condensate eventually out on the surface.

The Composite Infrared Spectrometer (CIRS) onboard Cassini has detected these organics in Titan's stratosphere (between 150 and 450 km roughly) and determined their spatial and vertical distributions (Flasar et al. 2005; Coustenis et al. 2007; Vinatier et al. 2007; Teanby et al. 2007). Comparisons with previous Voyager and ISO (Infrared Space Observatory) results (Coustenis et al. 2003) have not yet pointed to any significant temporal variations of these species, but the seasons of these measurements were very similar. Some studies point to a possible trend for an increasing enhancement in the composition of the northern pole with time since the beginning of the Cassini mission (Teanby et al. 2007) and it will be interesting to see if the enhancements observed by Voyager/IRIS in 1980 are confirmed towards the end of the extended mission when the season will be again spring equinox.

The spatial distribution of organic gases can bear on seasonally varying meridional circulations in the middle atmosphere. The break-up of CH_4 and N_2 from photo- and electron-impact dissociation tends to be high in Titan's atmosphere, and the abundances of most of the resulting organic compounds increase with altitude. Figure 3 illustrates the meridional variation of several organic compounds derived from observations by Voyager IRIS (Infrared Interferometer Spectrometer) and Cassini CIRS. In both cases there is an enhancement of several species in the north polar region. The enhancements observed by Voyager in the northern spring are often much larger (by factors of 1.5–2) than those observed by Cassini in northern winter. Since their abundances increase with altitude, this is consistent with sub-

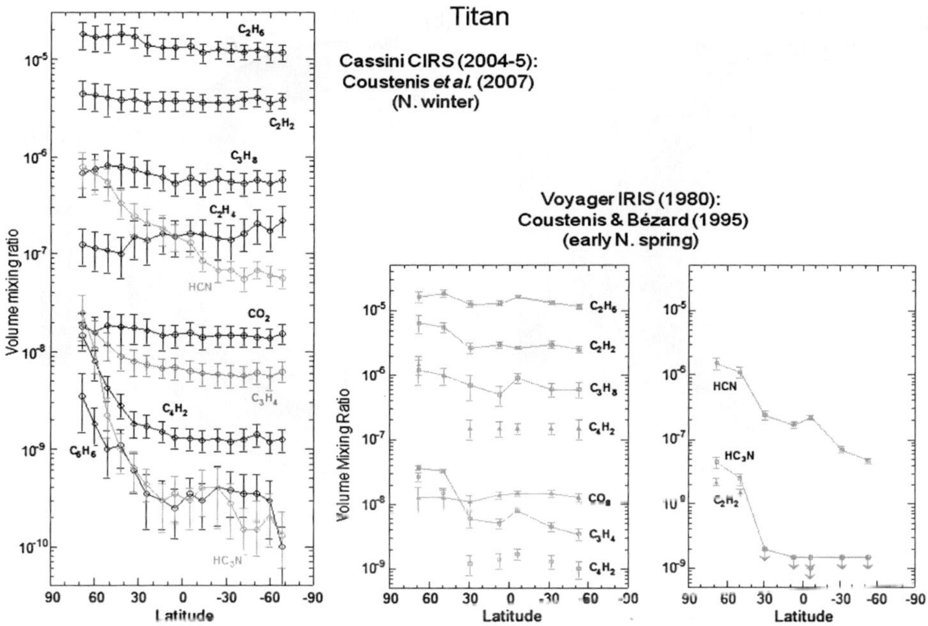

Fig. 3 Meridional distribution of several organic compounds from Voyager (Coustenis and Bézard 1995) and Cassini (Coustenis et al. 2007) thermal-infrared spectra

sidence at high northern latitudes that persists from winter until early spring, probably within the polar vortex that was noted earlier. This is not unexpected and can be interpreted as a general cross-equatorial circulation in the middle atmosphere, with ascent over the summer pole and descent over the winter pole within the strong circumpolar vortex, similarly to what is found on Earth (Flasar et al. 2005).

Figure 4 depicts a meridional cross section of Titan's middle atmosphere, illustrating the warm temperatures and enhanced cyanoacetylene mixing ratio over the winter north pole, consistent with descent over the pole.

The N_2-CH_4 byproducts in Titan's atmosphere create aggregates which form the haze layers observed (the main haze layer is located at about 80–100 km altitude) and eventually end up as sediments on the surface, where they accumulate at a rate of roughly 0.5 km in 4.5 Gyr. Since no large source to resupply methane was detected by Cassini, cryovolcanic outgassing has been hypothesized (Sotin et al. 2005), yet over what timescales and through which internal processes remains unknown. Tobie et al. (2006), for instance, propose an episodic outgassing. Cassini–Huygens also found that the balance of geologic processes on Titan—impacts, tectonics, fluvial, Aeolian—is somewhat similar to the Earth's, more so than in the case of Venus or Mars. Titan may well be the best analog to an active terrestrial planet in the sense of our home planet, albeit with different working materials.

Thus, it appears that Titan is a chemical factory in which the formation of complex positive and negative ions is initiated in the high thermosphere as a consequence of magnetospheric-ionospheric-atmospheric interaction involving solar EUV, UV radiation, energetic ions and electrons. It should also be noted that many of the neutral constituents predicted by models and laboratory measurements and still listed on CIRS "shopping list" (Flasar et al. 2004) remain to be detected.

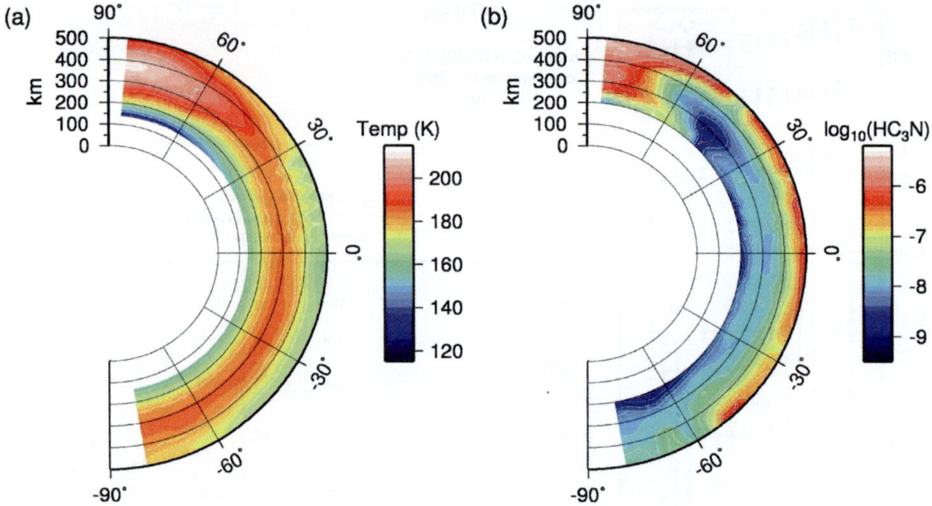

Fig. 4 Meridional cross section of middle-atmospheric temperatures and cyanoacetylene abundance (mixing ratio) retrieved from Cassini CIRS spectra obtained in early northern winter. After Teanby et al. (2007)

Radar observations support the idea that the ultimate fate of this aerosol precipitation is the generation of expansive organic dunes that lie in an equatorial belt on the surface (Lorenz et al. 2006; Radebaugh et al. 2008; Lorenz and Radebaugh 2009).

1.4 The Dynamical Atmosphere and Its Effects on the Surface

The thermal structure of Titan's atmosphere bears significant similarities with the Earth's in that it is subdivided into the troposphere, stratosphere, mesosphere and thermosphere. The vertical structure of the troposphere and stratosphere near the equator was retrieved by the radio-occultation experiment of the Voyager mission despite some sensitivity to the then unknown tropospheric composition (Lindal et al. 1983). This measurement was confirmed by in situ measurements of pressure and temperature by the Huygens Atmospheric Structure Instrument (HASI) (Fulchignoni et al. 2005). The equatorial surface temperature of 93–94 K is only ~3 K above the triple point of methane (major condensable species in the troposphere), but nitrogen is dissolvable in liquid methane, so liquid condensates may exist down to ~80 K, the temperature near 15 km altitude. The lowest temperature (70 K) on Titan is found at the tropopause (40 km). Most stratospheric species condense at their respective cold trap, which is located above the tropopause.

The tropopause temperature is only 7 K above the freezing point of nitrogen, the main constituent of the atmosphere. Therefore, Titan owes its climate stability to a tropospheric temperature that does not undergo a large latitudinal and seasonal variation. Otherwise a substantial fraction of the atmosphere would condense at the winter pole as on Mars or even collapse altogether (Lorenz et al. 1997). At present the combined effect of greenhouse warming and atmospheric heat transport ensures the persistence of the dense atmosphere. While the radiative response time in the upper stratosphere is much shorter than the length of a Titan year, it increases to several Titan years in the lower troposphere (Smith et al. 1981; Flasar et al. 1981). This results in the upper stratosphere (and mesosphere) exhibiting large seasonal and latitudinal variations in temperature (several tens of degrees Kelvin). Temperatures retrieved from the Cassini CIRS during the northern winter indicate large variations

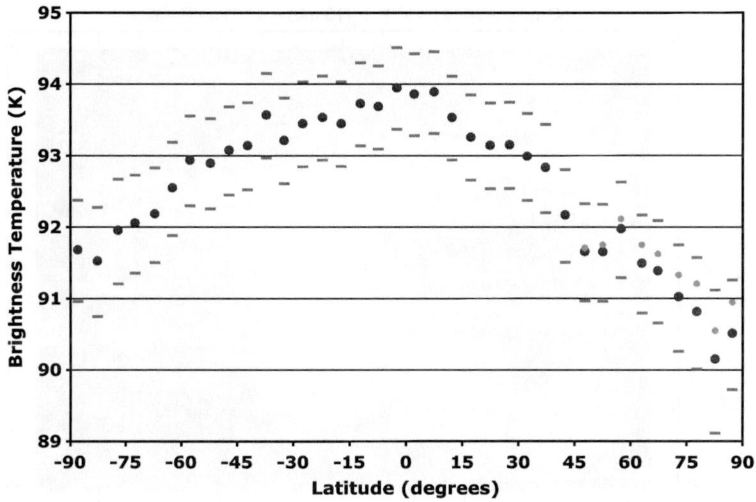

Fig. 5 Latitudinal profile of the surface brightness temperature at 530 cm^{-1} in northern winter measured by the Cassini CIRS (Jennings et al. 2009)

with latitude in the stratosphere with a pronounced north–south asymmetry, with the winter north pole being colder below the 0.1-mbar level (Fig. 4). Seasonal temperature changes mute as one moves downward to the troposphere. There is no observational evidence that the winter polar tropopause temperature drops to the nitrogen freezing point or that polar caps of nitrogen frost form on the surface.

The surface temperature retrieved from the Cassini thermal spectra decreases by 2 to 3 K from the equator to the poles, with a colder winter pole (Jennings et al. 2009) (Fig. 5). This latitudinal profile is more asymmetric about the equator than the profile acquired during the Voyager mission shortly after the equinox (Flasar et al. 1981). The difference between the two profiles is likely to reflect seasonal variation of the surface temperature given the small thermal inertia of the surface compared to that of the tropospheric air (Tokano 2005). Because of sensible heat flux from the surface, the near-surface atmospheric temperature is likely to vary with season despite a sluggish radiative response.

The Huygens probe found a weakly convective planetary boundary layer (PBL) with a thickness of 300 m as evidenced by the vertical profile of potential temperature (Tokano et al. 2006a). The observed PBL structure indicates weak turbulence and a tiny upward turbulent heat flux at the Huygens landing site, and the diurnal variation was inferred to be almost negligible.

The wind system is relevant for the transport of matter, momentum and energy in the atmosphere and on the surface. Before the Cassini–Huygens mission, several lines of indirect evidence had indicated the presence of Venus-like ubiquitous superrotation with wind speeds exceeding 100 $m\,s^{-1}$ in the stratosphere (Flasar and Conrath 1990; Hubbard et al. 1993; Kostiuk et al. 2005). The in situ measurement of the wind speed near the equator by the Huygens Doppler Wind Experiment (DWE) eventually confirmed the presence of superrotation at altitudes higher than 100 km (Bird et al. 2005). At the same time, this measurement also showed that the wind drastically weakens below the main haze layer. Most striking is the layer of very weak winds near ~75 km, in which the wind is even much weaker than further below at the tropopause (40 km). Huygens also found that the wind direction reverses at two altitudes near the surface (Folkner et al. 2006;

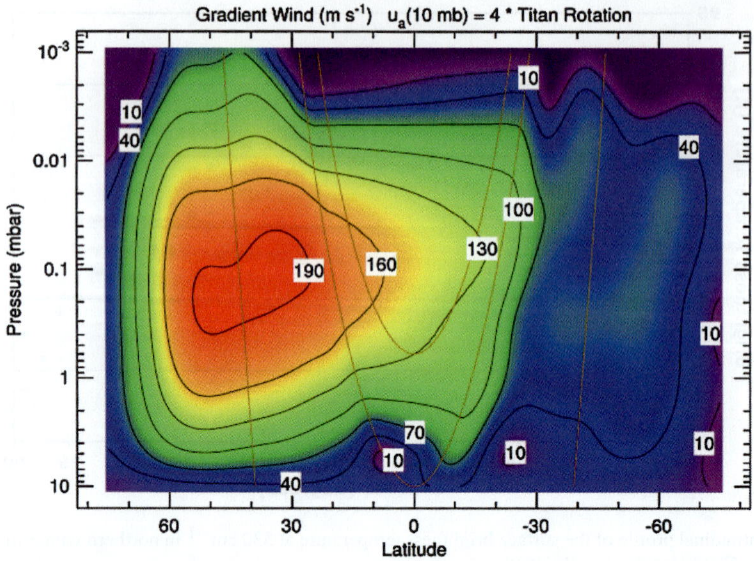

Fig. 6 Zonal winds calculated from the temperatures through the gradient wind equation. Speed contours are in $\mathrm{m\,s^{-1}}$ (Achterberg et al. 2008)

Karkoschka et al. 2007). This may be caused by a combination of seasonal surface temperature variation, sensible heat flux across the surface and coupling between temperature and wind via the thermal wind relation (Tokano 2007), but the mechanism of eastward surface wind near the equator is hard to understand with current insights of meteorology.

The meridional circulation is inherently more difficult to measure than the faster zonal wind, but it is quite important for transport processes. Figure 6 shows results obtained on the zonal winds as inferred from CIRS data (Achterberg et al. 2008). Thermally forced meridional circulation (Rannou et al. 2004) and tidal winds (Tokano and Neubauer 2002) are supposed to transport chemical species and haze to latitudes distant from their respective formation regions. This would eventually affect the deposition of organic material on the surface including the material the sand dunes are made of or species dissolved in the lakes. The latitudinal and seasonal variation of the haze distribution as a result of meridional circulation directly affects the solar radiation budget of the surface, with a strong feedback to the meridional circulation, temperature and zonal wind direction. The most direct confirmation of the presence of meridional circulation came from the Huygens probe, which also drifted in a meridional direction at $\sim 1\ \mathrm{m\,s^{-1}}$ or less (Karkoschka et al. 2007). During the descent the drift direction changed several times below 15 km altitude. Measurements of the radiation balance in the atmosphere by Cassini–Huygens showed that the equatorial region has a surplus of solar heating rate over the thermal cooling rate (Tomasko et al. 2008). This global heat imbalance as a result of differential heating has to be equilibrated by heat transport of a Hadley-type meridional circulation. The tidal wind cannot be unambiguously identified in the one in situ wind profile at the Huygens entry site (Tokano 2007).

While the Huygens wind data represent a snapshot along the Huygens descent trajectory, information on the long-term behavior of surface wind can be gathered from observations of aeolian features. A vast portion of the equatorial region is covered by parallel sand dunes, which have an eastward orientation, with a few exceptions (Lorenz et al. 2006; Lorenz and Radebaugh 2009; Radebaugh et al. 2008, Fig. 7). These dunes have been identified as lon-

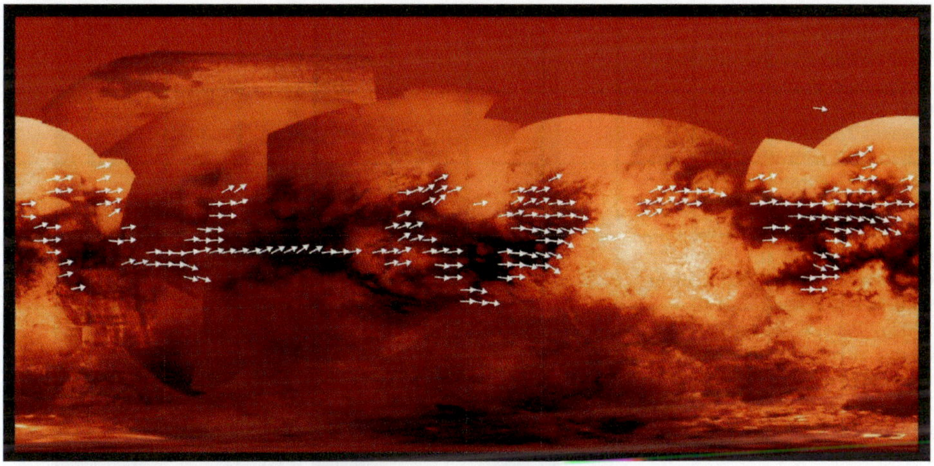

Fig. 7 Orientation of all dunes measured on Titan by the Cassini radar (Lorenz and Radebaugh 2009). The background is a map assembled from images of the Cassini's imaging science subsystem (ISS). The *arrows* indicate the direction in which sand is expected to be transported along the dunes observed in Titan radar data. Many of the equatorial dark areas without arrows may have dunes which have not yet been imaged with radar. The dune orientations represent only the net effect of winds. It could be that sand transport only occurs on rare occasions, and winds from different directions can combine to yield the observed dune orientations

gitudinal dunes, and point to an average eastward surface wind accompanied by a temporal variation in wind direction.

1.5 Haze and Methane Hydrological Cycle

It has been known since the Voyager days that Titan's atmosphere exhibits a hemispheric (north–south) asymmetry (NSA) that varies with season. The winter hemisphere is aerosol-enriched in the stratosphere because the polar night facilitates the condensation of larger aerosol particles and downward transport of aerosol particles is forced by the descending branch of the large-scale circulation in that region (Rannou et al. 2004). In the visible, aerosols are dark, while they are bright in the infrared (Ramirez et al. 2002).

At the time of the Voyager encounter (spring equinox), Titan's northern hemisphere was dark in the visible (Fig. 8-A1). This corresponds to one Titan season after the one of the arrival of Cassini–Huygens, and the observations showed a dark southern limb in the visible (and a bright northern limb in the infrared). The NSA asymmetry was therefore inverted at the time of the first adaptive optics and HST images (Smith et al. 1996; Combes et al. 1997), during Titan's southern winter, with the famous "Titan's smile" (bright southern pole) visible on all data. From HST data in the visible, Lorenz et al. (1999, 2001) showed that the asymmetry inversion was already occurring in the late 1990s. From adaptive optics images in the infrared, Ádámkovics et al. (2006) and Hirtzig et al. (2006) showed that the reversal was completed by 2002. The hemispheric asymmetry and its seasonal reversal are largely explained by global transport of aerosols across the equator (Lorenz et al. 1999). The spatial distribution of stratospheric haze, which primarily changes with season, is predicted to have a large impact on the atmospheric circulation down to the surface and the surface temperature by affecting the radiation balance (Lebonnois et al. 2009).

Besides Earth, Titan is the only other body in the solar system to sustain an active hydrological cycle with liquid methane as a working fluid, as revealed by the Cassini–Huygens mission. Titan's methanological cycle is evidenced by the detection of various

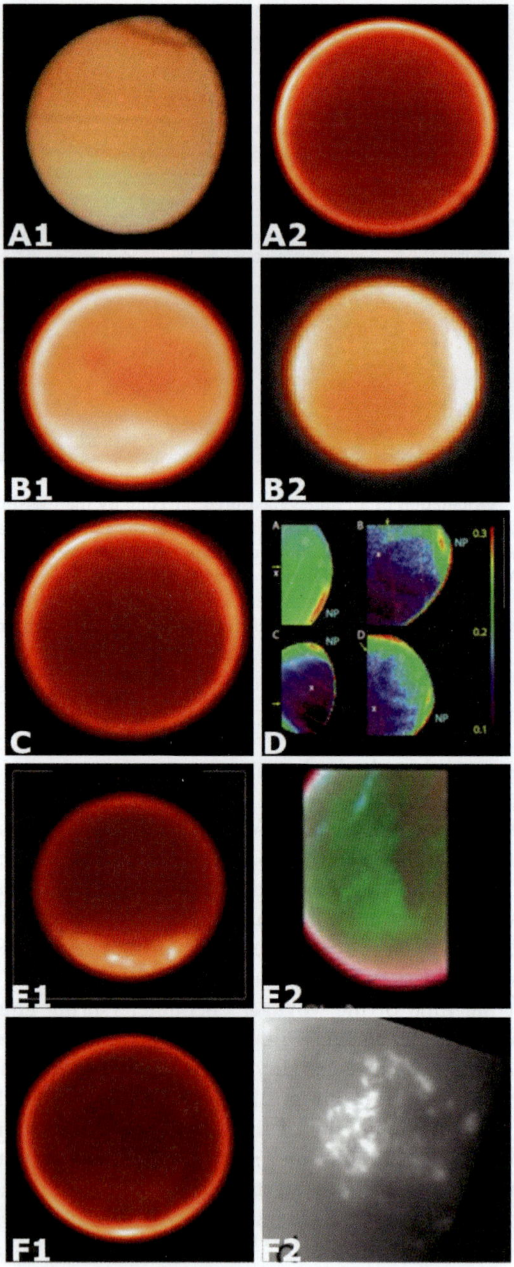

Fig. 8 Images sampling different Titan atmospheric features. Panels **A** illustrate the north–south asymmetry (**A1**: 1982 in the visible with Voyager 2, Smith et al. 1982; **A2**: in 2002 with VLT/NACO at 1.64 μm, Gendron et al. 2004). Panels **B** display two views of the southern "tropopause cirrus" (**B1**: VLT in 2002 at 1.04 μm, Gendron et al. 2004; **B2**: Keck in 2001 at 1.702 μm, Roe et al. 2002a). Panel **C** shows the diurnal effect through a bright morning limb (VLT in 2005 at 2.17 μm, Coustenis et al. 2001; Hirtzig et al. 2006). Images in panel **D** show the stratospheric "ethane cloud" detected by Cassini/VIMS at the northern limb of Titan (Griffith et al. 2006). Panels **E1** and **E2** show the mid-latitude methane clouds unresolved from Earth (**E1**, 2001 Keck data at 2.108 μm, Roe et al. 2002b) and resolved with Cassini/VIMS (**E2**, 2004, near 2 μm, Griffith et al. 2006). Similarly, panels **F1** and **F2** show the cloud systems at the 20 pole (**F1**, 2002 VLT data at 2.12 μm) and with Cassini/ISS (**F2**, in 2004, 928 nm, Porco et al. 2005). From Hirtzig et al. (2009)

clouds, fluvial channels, lakes/seas. While for nearly two decades after Voyager it was debated whether or not Titan might support clouds or precipitation (e.g., Hunten et al. 1984), observations of variable cloud systems—first from ground-based (Griffith et al. 1998; Brown et al. 2002) and afterwards with Cassini images and spectra (Porco et al. 2005; Brown et al. 2006, Fig. 8, Panels C–F)—have demonstrated that convection, condensation, and—presumably—precipitation are indeed prevalent across Titan. Understanding this

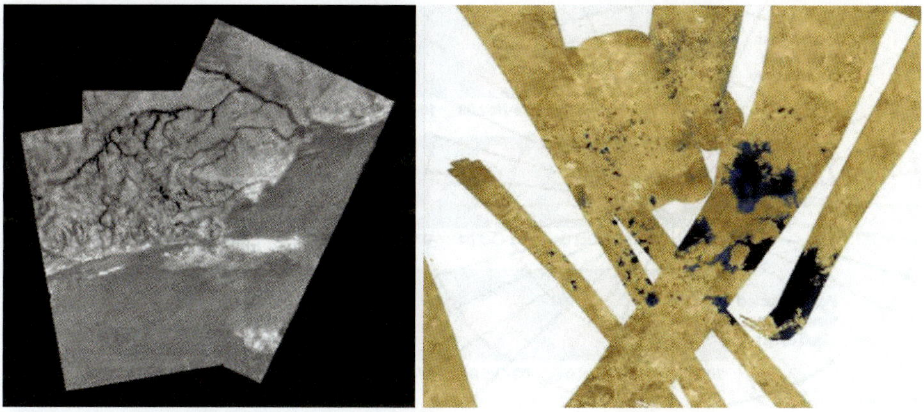

Fig. 9 *Left*: fluvial features at the Huygens landing site imaged by the DISR camera. The thickest channels are ~10 m across. *Right*: mosaic of the northern hemisphere lakes (*dark areas*). (NASA/JPL, USGS)

methanological cycle holds the key to understanding much of the diversity of features on the surface of Titan.

The first irrefutable evidence of the methanological cycle between the atmosphere and surface was provided by the first Cassini and Huygens images of Titan's surface (Tomasko et al. 2005; Porco et al. 2005), which showed rounded pebbles, fluvial channels and shorelines (Fig. 9a). This was followed by the detection of lakes/seas near the north pole (Stofan et al. 2007; Hayes et al. 2008; Lorenz et al. 2008a) and in the south pole (Brown et al. 2008; Turtle et al. 2009) by three imaging instruments onboard Cassini (Fig. 9b). Liquid ethane was spectroscopically detected in Ontario Lacus, one of the southern lakes, and liquid methane is likely to coexist albeit not identifiable in the spectra (Brown et al. 2008).

One hypothesis connecting the lakes and atmosphere put forth is the presence of a seasonal cycle of methane evaporation, global pole-to-pole transport of methane and methane precipitation on lakes at the opposite pole (Mitri et al. 2007; Lunine and Atreya 2008). Precipitation that would be intense enough to pond the lakes has not been directly measured so far. However, there is some circumstantial evidence for a link between the lakes and methanological cycle. Repeated south polar imaging revealed differences in the appearance of Ontario Lacus that is consistent with ponding of hydrocarbon liquids on the surface due to precipitation from a large storm (Turtle et al. 2009). In the opposite, north polar region, several clouds were observed east of the major seas (Brown et al. 2009). They have been described as Titan's analogs of lake-effect clouds because they appear at the same latitudes as the largest seas and the global climate models do not predict such clouds, and were explained by the heat flux from a warm sea surface that triggered moist convection. It is yet unknown whether the lake-effect clouds produce surface precipitation. The Cassini radar performed atmospheric backscatter observation of the north polar region in May 2007 analogously to terrestrial weather radar (Lorenz et al. 2008b). The result was negative in the sense that substantial precipitation was ruled out during the short measurement and a very low upper limit for precipitation was placed, allowing at most extremely sparse or extremely small droplets (drizzle) or both.

Considering the large radiative time constant in the troposphere and the relatively fast response of the surface temperature to seasonal forcing, it is reasonable to assume that the surface directly or indirectly affects the evolution of the cloud systems, although the relationship between season and cloud occurrence is not straightforward. Cloud tracking from

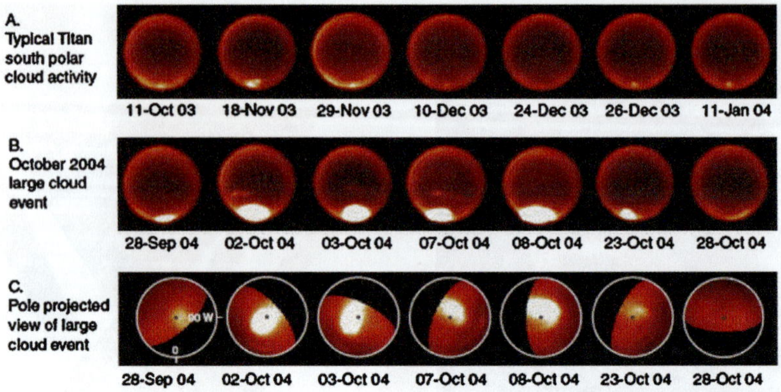

Fig. 10 A comparison between the typical Titan cloud activity around solstice and the rare large cloud outbursts in October 2004 (Schaller et al. 2006a)

ground-based telescopes (Roe et al. 2005; Bouchez and Brown 2005; Schaller et al. 2006a, 2006b; Hirtzig et al. 2006) and the Cassini ISS and VIMS (Porco et al. 2005; Brown et al. 2006, Fig. 10) shows a large variety of clouds in Titan's atmosphere at different locations over the disk.

The largest and most persistent cloud system has been observed during southern summer. A sequence of ISS images (Porco et al. 2005, Fig. 8) shows that the cloud system rapidly evolves during the few hours spanning the sequence. The system was also detected by VIMS (Brown et al. 2006). Schaller et al. (2006b) found that the south polar cloud system, which was visible in almost every single image since 2001, suddenly disappeared in late 2004. The disappearance coincided with the time period in which the south pole had abruptly become no longer the latitude of maximum insolation. In this season the surface temperature at the south pole is predicted to drop (Tokano 2005). However, the surface brightness temperature measured by the Cassini CIRS between 2004 and 2008 shows that the south pole was 2 K colder than the equator (Jennings et al. 2009, Fig. 5). Since it is unknown whether the south polar surface temperature was warmer before 2004, it is hard to confirm or reject the hypothesis of insolation-driven cloud formation.

The temperature lapse rate and methane humidity measured by Huygens is insufficient to cause convective rainstorms (Tokano et al. 2006b), and a higher temperature or a higher humidity would be required to initiate them (Hueso and Sánchez-Lavega 2006; Barth and Rafkin 2007).

The formation of another type of clouds has been more tightly related to surface features. Roe et al. (2005) detected longitudinally extended cloud streaks that were concentrated in a band near 40°S. Continued long-term monitoring showed that these clouds were preferentially forming over a single longitude on the surface of Titan, indicative of some sort of geographic control such as volcanic vent, rather than a seasonal effect.

In contrast to high and middle latitudes, no optically thick clouds have been observed so far in the equatorial region. However, even the equatorial region is not entirely dry despite the presence of sand dunes, which are suggestive of a dried surface. The Huygens landing site, some 20 km distant form the closest dunes, is inferred to consist of a damp surface with at least liquid methane underneath a thin crust (Niemann et al. 2005). The Huygens imager identified an optically thin cloud/aerosol layer near 21 km altitude that was distinct from the background haze (Tomasko et al. 2005). In this region the relative humidity with respect to solid methane was close to 100%, suggesting that the observed opacity source was a cloud composed of solid methane (Tokano et al. 2006b). The Huygens probe measured high-frequency motions exactly in this altitude region, suggestive of cloud turbulence (Lorenz et al. 2007). Further below, the relative humidity was close to 100% with respect to a binary liquid methane–nitrogen mixture and the methane mixing ratio was uniform with altitude, indicative of weak drizzle that reaches the surface (Tokano et al. 2006b). Ground-based telescopic observations one month after the Huygens descent confirmed the presence of an additional tropospheric opacity over the Xanadu region near the equator that is best explained by light scattering by methane drizzle (Ádámkovics et al. 2007). On the other hand, Kim et al. (2008) do not find evidence for such drizzle by an independent telescopic observation. Also, the images of Huygens taken after the landing do not show evidence of raindrops on the surface (Karkoschka and Tomasko 2009). Obviously, the question of whether falling precipitation evaporates during the descent or reaches the surface is of utmost importance for the atmosphere–surface coupling. The fate of falling raindrops may depend on various factors such as the relative humidity at which condensation sets in (Tokano et al. 2001), composition of the raindrops (Barth and Toon 2006) or ethane humidity in the troposphere (Graves et al. 2008), none of which is well known from observations. Nevertheless, there is consensus among observers that rainfall sufficient to carve the fluvial channels near the Huygens site did not occur during the last few years.

The coexistence of sand dunes (Lorenz et al. 2006) and fluvial channels (Tomasko et al. 2005; Soderblom et al. 2007) in the equatorial region cannot be readily reconciled with a climate model that predicts a repeatable annual cycle (Rannou et al. 2006; Mitchell et al. 2006) because the dunes and fluvial channels are suggestive of a dry and wet climate, respectively. Therefore, some workers suggest that the equatorial fluvial channels may not result from recent seasonal rainstorms, but from past wet climate or past stochastic rainfall events (Lunine and Atreya 2008; Griffith et al. 2008). However, while the coexistence of fluvial channels and dunes is paradoxical from a simplistic physical standpoint, such juxtapositions are not infrequently seen in terrestrial desert areas. The very complexity of the Titan landscape, and not only the landforms themselves, is strikingly Earth-like.

1.6 Origin and Evolution of the Titan Atmosphere and Its Relation to the Surface and Interior

Information on the origin and evolution of Titan's atmosphere is becoming available through the scarce accurate measurements of the different volatile compounds present in today's atmosphere and onto the surface, including measurements of the noble gas concentrations and their isotopic abundances, as well as the nitrogen and carbon stable isotope ratios. Cassini–Huygens has provided some important information in this regard. The abundance of the radioactively derived ^{40}Ar indicates that only a few percent of the total volatile inventory has been outgassed from the interior (Niemann et al. 2005). The relatively low abundance of the primordial ^{36}Ar isotope suggests that nitrogen was not delivered during Titan's initial formation as molecular nitrogen, but more likely as ammonia that underwent subsequent chemical conversion into N_2—the predominant constituent of Titan's present-day atmosphere. Furthermore, the enrichment of ^{15}N in N_2 to that of ^{14}N relative to a terrestrial reference suggests that, as at Mars, Titan has lost most of its nitrogen over the course of its evolution (Waite et al. 2007). This is substantiated by the measurement of isotopic separation in the upper atmosphere measured by the Cassini INMS and the escape of methane and hydrogen inferred from the altitude structure of these species in Titan's upper atmosphere (see Sect. 1.3). Moreover, the non-detection of neon, krypton and xenon by Huygens raises fundamental questions about Titan's origin and evolution: have these compounds never been incorporated in Titan's building blocks, or have they been lost or recycled and hidden at the surface and in the interior since Titan's formation? The accurate measurements of the abundances of these noble gases and of their isotopic ratios will provide important clues about the origin and evolution of Titan, and about the overall role of escape, chemical conversion, outgassing and recycling in the evolution of Titan's atmosphere. In particular, the detection of radiogenically-derived isotopes of neon, xenon and krypton will constrain the evolution of the interior and the outgassing history of Titan. All these questions must await new surface analysis techniques such as noble gas enrichment cells, which were not present on the Huygens GCMS (Gas Chromatograph Mass Spectrometer).

The deposition of CH_4–N_2 photochemical materials, including liquid ethane, leads to the creation of lakes and seas of liquid hydrocarbons on Titan's surface in the high latitude regions, making a uniquely evocative landscape. Titan is about to enter a period of rapid seasonal change as the Sun crosses the equator in 2009 into the northern hemisphere. Model predictions based on previous observations suggest that the meridional circulation will eventually reverse to conduct that excess haze from the north back to the south. The north UV polar hood will disappear by about 2013 with consequent changes in the polar clouds and the vortex, which may disappear or persist. In the latter half of the decade 2010–2020, vigorous convective cloud activity (and likely precipitation) will likely occur over the northern polar regions, now known to host many hydrocarbon lakes.

Predictions of how low- and mid-latitude tropospheric clouds will change are less certain: it will likely depend on the extent to which clouds are controlled by insolation-driven upwelling and to what extent by methane availability on the surface—the latter being rather poorly constrained. It does seem likely that northern mid-latitude cloud activity will increase.

Titan is also an analog, albeit with different working materials, of the future state of the Earth if surface conditions preclude stable equatorial/mid-latitude oceans.

The evolution of the solar luminosity makes it almost unavoidable that, in $\sim$0.5–1 gigayear from now, the tropopause temperature of the Earth will be elevated to the point that loss of the Earth's oceans by evaporation and then stratospheric photolysis of the water vapor will occur over a timescale of the order of a hundred million years. From that point on, surface liquid water will be limited to the poles, will have a short photochemical lifetime, and if persistent would be resupplied by mantle/deep crustal water. At the equator, occasional intense rainstorms might scour out fluvial features, but the principal sedimentary deposits will be sand dunes.

The closest analog in our solar system to such a future Earth is Titan. Studying the recent evolution of Titan and how the various geologic and meteorological processes interact, we have a peek at the Earth's future "simulated" on a planetary scale by different materials.

2 Enceladus

The Cassini spacecraft's close flybys of Saturn's icy moon Enceladus have revealed that the moon has a significant dynamic atmosphere. When Cassini had its first encounter with Enceladus on February 17, 2005 at an altitude of 1167 km, the magnetometer instrument saw a striking signature in the magnetic field. On March 9 of the same year, Cassini approached to within 500 km of Enceladus' surface and obtained additional evidence.

The observations showed a bending of the magnetic field, with the magnetospheric plasma being slowed and deflected by the moon. In addition, magnetic field oscillations were observed. These are caused when electrically charged (or ionized) molecules interact

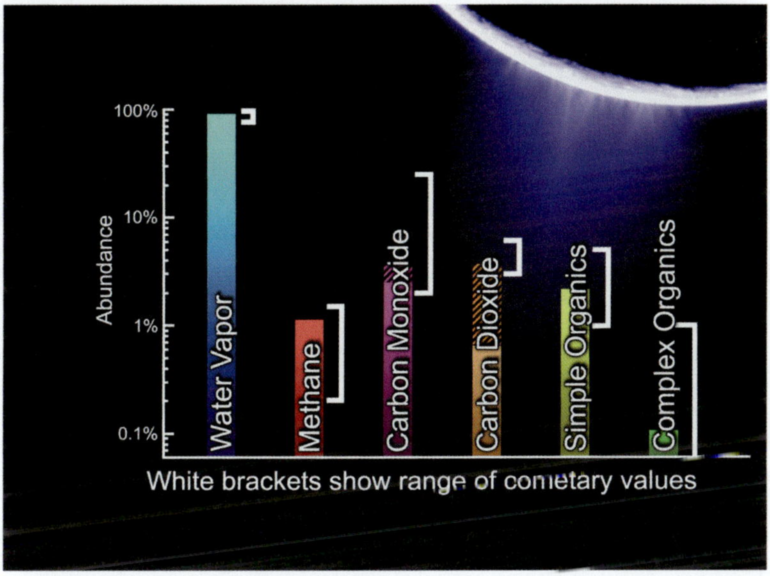

Fig. 11 Schematic based on the Cassini INMS data showing the classes of compounds detected coming out of the Enceladus plumes by the Cassini INMS instrument, with a comparison (*top*) to values measured in comets. Background ISS image shows the plumes at the south pole of Enceladus (Porco et al. 2006)

with the magnetic field by spiraling around the field line. This interaction creates characteristic oscillations in the magnetic field at frequencies that can be used to identify the molecule. The observations from the Enceladus flybys are believed to be due to ionized water vapor. This is the first time since Cassini arrived in orbit around Saturn in 2004 that an atmosphere has been detected around a kronian moon, other than Titan (Fig. 11). Enceladus is a relatively small moon, 500 km in diameter. The amount of gravity it exerts is not enough to hold an atmosphere for very long. Therefore, at Enceladus, a strong continuous source is required to maintain the atmosphere.

The discovery by the Cassini magnetometer of a large plume currently emanating from surface cracks near the south pole of Enceladus (Dougherty et al. 2006), venting water vapor, fine icy particles (or dust) and organics implies an underground heat source. It is the only icy world in the solar system, with Io, proven to have current geological activity (excluding Triton's geysers, likely solar-driven). Enceladus has a strong astrobiological potential and offers a unique opportunity to sample fresh material originating in its interior.

From Cassini measurements we know today that the ejecta form an atmosphere around Enceladus which contains water, carbon dioxide, methane and other organic molecules, and—as mentioned in Sect. 1.3—the plumes are also the source of the E ring. Indeed, ions from the geysers are driven out into the magnetosphere of Saturn and into the E ring. Fine icy particles escaping from the jets also make their way into the ring, confirming Enceladus as the main source for this ring.

Because Enceladus is the source of the Saturnian E-ring (Fig. 12), as well as the extensive neutral O and OH clouds that fill the middle Saturnian magnetosphere, the moon plays a pivotal role in the Saturnian system. Cassini observations of the Enceladus plumes indicate that they are the dominant source of O^+, O^{2+}, and OH, derived from water coming out of the vents (Waite et al. 2006, Fig. 11). The water-ice dominated plumes coming out of Enceladus' interior, magnetospheric processes that transform some of this material into oxygen

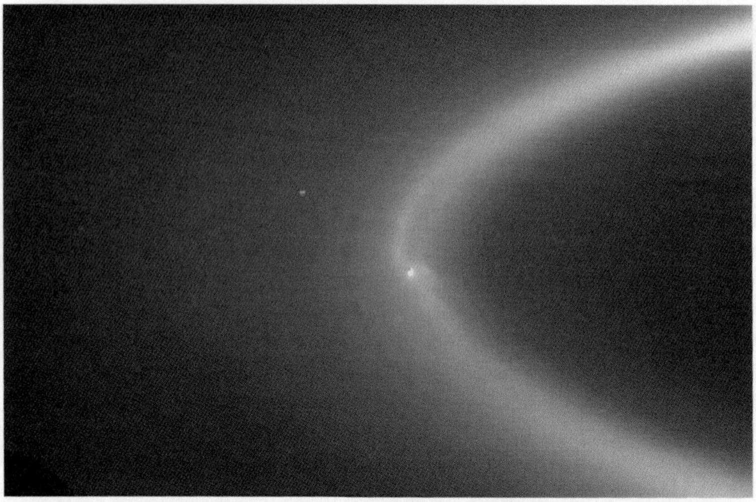

Fig. 12 Cassini image showing Enceladus (near center), throwing off material (wings on either side of moon) that resupplies the tenuous E-ring within which it is embedded. ISS image (NASA/Space Science Inst.)

atoms and ions, and the oxygen chemistry producing CO and CO_2 in Titan's atmosphere are coupled. In addition to the geysers of water ice and vapor that contain complex organic compounds, the Cassini spacecraft has recently found what may be the strongest evidence yet that Enceladus has an ocean beneath its icy surface. It comes from the detection of sodium in Saturn's E ring (Postberg et al. 2009), the band of ice particles believed to be fed by the moon. If the finding is confirmed, it would suggest that the moon may be a promising place to search for signs of life. More recently, the INMS team has provided conclusive evidence for the presence of ammonia (which readily dissolves in water and acts to keep water liquid at lower temperatures than it would otherwise) and argon (which may have been transported to the plume by liquid water coursing through minerals in the interior) in Enceladus' plumes (Waite et al. 2009). These two elements point to the presence of a liquid water in the interior of the satellite.

Whether the plume emissions arise directly from chambers of pressurized liquid water or evaporation from warm ice, or the explosive release of gases from clathrate ices, is currently unknown (Porco et al. 2005, 2006; Kieffer et al. 2006). The origin and time history of the internal heating source are a major puzzle. Cassini observations, including extended mission observations, will allow direct plume sampling as well as further remote sensing observations of the south polar regions and high phase plume imaging.

3 Characteristics of Io's Atmosphere

The atmosphere of Io is unique in terms of its production, composition, and dynamics. The ultimate sources of gas are the volcanoes which cover the surface, venting volatiles from the interior. However, the density and distribution of gas are determined by a balancing act between volcanic supply and the distribution of frosts across the surface. Io's atmosphere also plays the dominant role in supplying plasma to Jupiter's magnetosphere, with the effect that the atmosphere must be continuously replenished.

In this section, we first briefly review the atmospheric observations through the Galileo era (see recent reviews by McGrath et al. 2004 and Lellouch et al. 2007 for more detailed discussions). We then present more recent advances on the spatial distribution of gas from Earth-based observations and the New Horizons flyby of Jupiter, and place them in the context of understanding the balance between the exchanges with the satellite interior (i.e., a volcanically supplied atmosphere), and the exchanges with the surface (frost-buffered atmospheric models). The atmosphere–surface coupling is explored more fully in Tobie et al. (2010, this issue). We also describe atmospheric observations indicative of energy exchange with Jupiter's magnetosphere which can be used to understand the composition and evolution of Io's rapidly changing atmosphere.

The first definitive observation of an atmosphere at Io was made during the Pioneer 10 flyby of Jupiter by means of a radio occultation experiment indicating an ionosphere around the satellite. The peak measured electron density of $\sim 6 \times 10^4$ cm^{-3} suggested an atmosphere with surface density $\sim 10^{10}$–10^{12} cm^{-3} (Kliore et al. 1974, 1975). This observation gave no indications of the atmospheric composition. The first identified species was sodium, observed from the Earth in an extended corona around Io and partial torus orbiting Jupiter (Brown 1974; Macy and Trafton 1975). This discovery was followed by the detection of potassium (Trafton 1975) with a similar distribution. Na and K are minor species in the atmosphere, but the ease of observing them from the ground at optical wavelengths has made them essential tools for understanding atmospheric escape from Io and the supply of mass to the magnetosphere (see Burger et al. 2010, this issue).

The dominant species in Io's atmosphere is SO_2, which was first observed at 7.3 μm by Voyager 1 (Pearl et al. 1979). Voyager also made the first observations of Io's volcanoes (Morabito et al. 1979) and SO_2 surface frosts (Fanale et al. 1979; Smythe et al. 1979), providing the basis for atmospheric source models. Unfortunately, SO_2 was not detected again for over a decade, when Lellouch et al. (1990) successfully observed SO_2 at millimeter wavelengths. SO_2 has also been detected in the UV by the HST (e.g., Balester et al. 1994; McGrath et al. 2000; Feldman et al. 2000) Galileo (Hendrix et al. 1999), and in the infrared from the ground (Spencer et al. 2005).

Improvements in Earth-based observing tools and techniques in recent years have permitted mapping the global distribution of SO_2 gas at Io—revealing spatial inhomogeneities in the atmosphere—and of SO_2 frost deposits on the surface. These have shed some light on the balance between the volcanic and surface frost supplies to the atmosphere. A multi-year observational campaign to observe gaseous SO_2 in the mid-infrared by Spencer et al. (2005) revealed a persistent hemispheric asymmetry in the absorption band depths between the Jupiter-facing (sub-Jovian) and anti-Jupiter facing hemispheres, suggesting that the SO_2 density is an order of magnitude greater on the side facing away from Jupiter. This asymmetry was confirmed with maps of the dayside SO_2 distribution produced from measurements of reflected Ly-α from Io's surface, which is attenuated by SO_2 in the atmosphere (Feldman et al. 2000; Feaga et al. 2009, Fig. 13). Using HST/STIS spectra spanning 4 years, Feaga et al. (2009) were able to characterize the mean dayside SO_2 distribution as (1) approximately stable on a global scale with the largest column densities in the equatorial region ~ 6–8×10^{16} cm^{-2}, (2) temporally variable, but on the order of 1–2×10^{16} cm^{-2}, (3) the column density at the anti-Jovian point is $\sim 3\times$ larger than at the sub-Jovian point, consistent with, although less extreme than, the asymmetry measured by Spencer et al. (2005), and (4) the latitudinal extent of SO_2 over the anti-Jovian hemisphere is greater than over the sub-Jovian hemisphere.

The large number of observations in recent years, and especially the ability for long-term observing campaigns with the same instruments, have indicated that both the volcanic

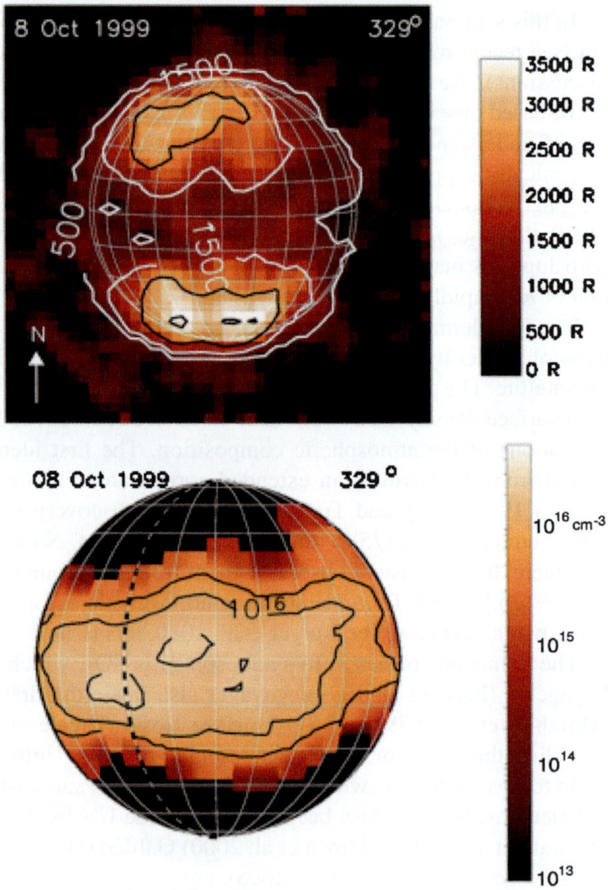

Fig. 13 (**a**) HST/STIS observation of Ly-α1216 Å reflected from Io's surface and attenuated by SO_2 in the atmosphere. *Dark regions* indicated a greater absorbing column of SO_2. (**b**) SO_2 column abundance inferred from the observation in (**a**). From Feaga et al. (2009)

supply and frost-buffered sources are important for maintaining Io's atmosphere. Spencer et al. (2005) and Feaga et al. (2009) describe aspects of their results which support both models. The importance of SO_2 frost in determining the atmospheric distribution is suggested through correlations between longitudinal variations in gas abundance and surface frost distributions inferred by Johnson and Pilcher (1977), McEwen et al. (1988), and Douté et al. (2001). Measurements made by Galileo's Near-Infrared Mapping Spectrometer of surface SO_2 frost (Douté et al. 2001) suggested that SO_2-rich regions (relatively fresh deposits of small grain sizes) likely represent condensation areas of SO_2 gas migrating from the volcanic plumes. In contrast, there was no correlation in either the IR or UV observations between the measured atmospheric columns and heliocentric distance, implying that annual surface temperature variations have little effect on the gas abundance. A test of the relative importance of volcanic supply and sublimation sources was carried out by Retherford et al. (2007) using New Horizons observations of Io's FUV emissions while Io was eclipsed by Jupiter. By measuring the decrease in intensity of Io's auroral emissions (which are indicative of plasma and neutral densities, but do not require direct sunlight), as Io entered Jupiter's shadow, and the subsequent post-eclipse brightening, they concluded that only 1–3% of the dayside atmosphere is supplied directly from volcanoes.

While SO_2 is currently understood to be the primary constituent of Io's atmosphere (McGrath et al. 2004), S_2 is now known to be prominent (8–30%) in the large Pele plume

(Spencer et al. 2000). Jessup et al. (2007) provide evidence that other plumes are similarly enhanced in S_2. Detailed plume chemistry models reported by Moses et al. (2002a) suggest that the relative abundance of atomic sulfur from such S_2-rich plumes supplying the Io plasma torus and neutral clouds is larger relative to atomic oxygen despite the relatively short lifetime of S_2 by photodissociation or loss to the surface. Fegley and Zolotov (2000) describe the subsurface thermochemical reactions that fractionate trace constituents Na, K and Cl, which Moses et al. (2002b) describe further in plume photochemistry models. Lellouch et al. (2003) discovered atmospheric NaCl, which is also reported to be the main constituent in escaping dust particles (Postberg et al. 2006). Since NaCl is not volatile at Io surface temperatures, it is thought to be directly volcanic. SO has also been observed on the dayside at millimeter and infrared wavelengths (Lellouch et al. 1996; de Pater et al. 2002) with abundance ∼5–10% of SO_2. Io's nightside cannot be observed from Earth, but modeling suggests that SO_2 condenses on the surface while non-condensable species such as SO and O_2 remain, with abundances comparable, or possibly greater, than SO_2 (Wong and Smyth 2000).

Observations of atomic species in the atmosphere are especially sensitive to the escaping components. Oxygen and sulfur have both been observed in extensive neutral tori around Jupiter (Brown 1981; Durrance et al. 1983, 1995). This is a complement to the sodium cloud which had been mapped earlier (e.g., Macy and Trafton 1975). The neutral oxygen and sulfur clouds are much more extensive than the sodium cloud due to the facts that (a) Na is a minor species compared with O and S, and (b) sodium is more easily ionized by the plasma torus and therefore does not spread as far from Io. The Na abundance ratio is estimated at only a few percent (Thomas et al. 2004) while the O and S are produced by the dissociation of abundant volcanic molecules such as SO_2 and S_2. The distribution of material in the exosphere has been mapped in emission by HST (O and S by Wolven et al. 2001, Na by Mendillo et al. 2007) and in absorption (Na, using mutual eclipses between the satellites, by Schneider et al. 1991a and Burger et al. 2001). These data have revealed significant differences in the distributions of the major (O and S) and minor (Na) species escaping from Io that stem both from the loss mechanisms from the atmosphere (dissociation versus sputtering) and the emission mechanisms (electron-impact versus resonant scattering) for the major and minor species (see Burger et al. 2010, this issue).

Observations of sodium have revealed several features associated with atmospheric escape from Io (Fig. 14). The exosphere is formed from sodium sputtered from the surface and atmosphere by Io plasma torus ions, and contains a mix of bound and escaping atoms on ballistic orbits. Those that escape form a partial torus around Jupiter extending ahead of Io in its orbit. The characteristic banana shape of the cloud does not result from variations in the loss of sodium from the exobase, but from radial variations in the plasma torus which quickly ionize material drifting radially outward from Jupiter (Smyth and Combi 1988). The material interior to Io's orbit, which has a Keplerian orbital velocity larger than Io's, moves ahead of Io forming the predominantly leading cloud; the trailing cloud does not form since the sodium moving slower from Io was ionized by the plasma. Neutralization of Na^+ pick-up ions created in Io's atmosphere is revealed through a jet of sodium escaping from the near surface–atmosphere (Burger et al. 1999; Wilson et al. 2002). The escape of molecular sodium-bearing ions from Io is revealed through the "stream" feature, a long narrow cloud of neutral Na formed from the dissociative recombination of ions in the Io plasma torus (Schneider et al. 1991b). All of these processes combine to form a cloud of sodium extending over hundreds of Jupiter radii (Mendillo et al. 1990). Variability in the shape and brightness of this cloud over such large distance scales has been attributed to the changing

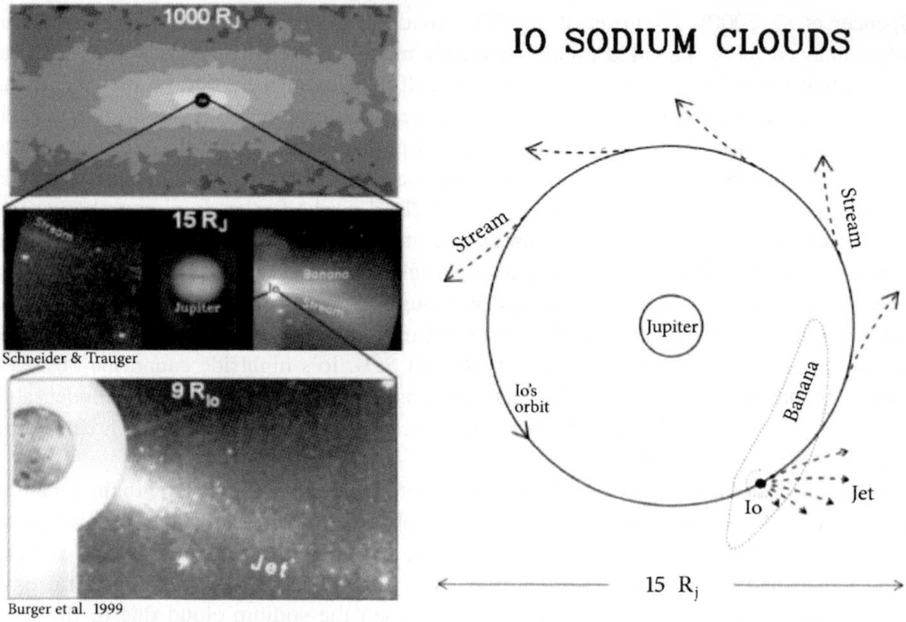

Fig. 14 Io's sodium clouds on three distance scales. (**a**) Large-scale sodium cloud. (**b**) Image of sodium features observed on the scale of Io's orbit, revealing the "banana" formed from sputtering, and the "stream" formed from the recombination of Na-bearing molecular ions in the plasma torus. (**c**) Sodium "jet" observed by Galileo. (**d**) Sketch of the sodium features viewed from above Io's orbit. From Thomas et al. (2004)

contributions of the different escape processes from Io and in variations in the overall escape rate of sodium from the atmosphere (Wilson et al. 2002).

The interaction between Jupiter's magnetosphere and Io's atmosphere is observed through a series of auroral emissions excited by Io plasma torus ions flowing through the atmosphere. Observations of Io while in Jupiter eclipse by Galileo (Geissler et al. 1999, 2001), made through broadband filters, show multiple emissions with distinct spatial distributions: blue glows were associated with volcanic plumes, red glows were seen at the polar limb, and there was a faint green emission across the disk. Ground-based spectroscopy of Io in eclipse identified oxygen and sodium contributing to the red and green emissions, respectively (Bouchez et al. 2000). The blue glows are likely due to SO_2 in volcanic plumes (Geissler et al. 1999, 2004).

HST/STIS observations in the FUV have identified several emissions caused by the magnetospheric interaction. The brightest of these emanate from localized regions near Io's equator, dubbed the "equatorial spots," and contain emissions due to oxygen and sulfur (Roessler et al. 1999; Retherford et al. 2000). The locations and brightness of these spots were seen to vary with Io's System III (magnetic) longitude due to the oscillation of the Jovian magnetic field at Io. Because Jupiter's dipole field is tilted relative to its rotation axis, the magnetic field experienced at Io changes its orientation with a 10-hour period, and Io passes through the center of the plasma torus twice during each field rotation. The locations of the spots at Io were found to almost track the orientation of the field line through Io, and the spots were brightest when Io was in the densest part of the plasma. In addition, Retherford et al. (2000) found that the anti-Jovian spot is consistently brighter than the sub-

Jovian spot. This asymmetry is diagnostic of the electrodynamic interaction with the Jovian magnetosphere occurring in Io's ionosphere (Saur et al. 2000, 2004).

During the Cassini flyby of Jupiter in 2000–2001, ISS observed three complete eclipses of Io by Jupiter (Porco et al. 2003). As with the UV observations from HST, the brightest emissions were seen in two equatorial spots near the sub-Jovian and anti-Jovian points which responded to the magnetic field line orientation at Io (Geissler et al. 2004). These authors report equatorial spot emission from SO_2, O, S, Na, K, and likely S_2.

4 Europa's and Other Tenuous Atmospheres

Neutrals ionized near Io are trapped ("picked up") by Jupiter's rapidly rotating magnetosphere at the rate of about 1000 kg/s. These ions diffuse outward through the magnetosphere where some hit Europa. Most of these ions hit Europa's trailing hemisphere (trailing with respect to Europa's orbital motion) since the magnetosphere overtakes Europa's orbital motion (e.g., Pospieszalska and Johnson 1989). The result is a deposition of sulfur and other ions that leaves a noticeable hemispheric dichotomy. Figure 15 shows the reddish mixture of various sulfur compounds in visible wavelengths.

The hemispheric dichotomy was first observed in the UV from Earth orbit: Lane et al. (1981) and follow-up work (e.g., Noll et al. 1995) found evidence of SO_2 on the trailing hemisphere consistent with the implantation of S ions. The discovery by the Galileo orbiter, in the IR, of an unidentified sulfate hydrate partly supported this *exogenic* view: the sulfate hydrates were concentrated on the trailing hemisphere. The presence of SO_2, a sulfate hydrate and a reddish visible element, possibly S_8, was consistent with the presence of a "radiolytic sulfur cycle" (Carlson et al. 1999b) in which the intense radiation bombardment, in a process known as radiolysis, cycles the sulfur-H_2O mixture through various chemical forms. This exogenic paradigm was further supported by a striking correspondence between high energy electron flux, Europa's largest source of ionizing energy, and hydrated sulfate (Paranicas et al. 2001).

The picture is complicated by higher-resolution views of the surface (e.g., Fig. 16), which show that the dark, reddish and sulfur-rich material also correlates with geological features. The correlation is especially interesting if the sulfate hydrates are salts from Europa's subsurface ocean (McCord et al. 1999).

The issue has been investigated by observing Europa's tenuous atmosphere. Unlike the atmospheres of Titan, Io and Enceladus, Europa's atmosphere is produced by intense radiation bombardment (though occasional venting, never detected, cannot be ruled out). The surface is exposed to unusually intense fluxes of ions and electrons whose energies range up to hundreds of MeV (the plasma implanted into the trailing hemisphere is due to the dominant "cold" component of the plasma, hundreds of eV). Following the detection of this radiation environment by the Pioneer mission flybys, laboratory experiments showed that

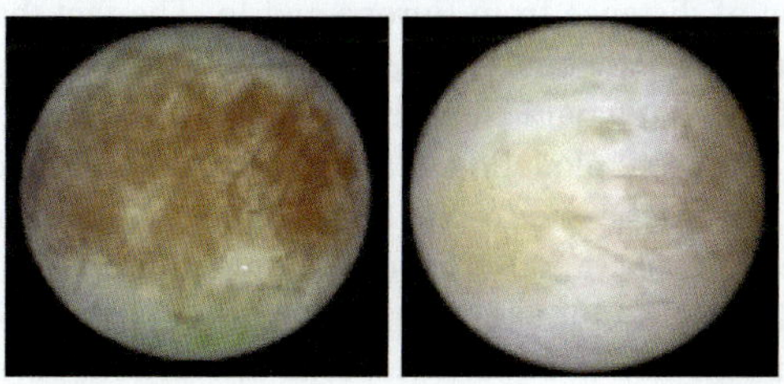

Fig. 15 Europa's trailing (*left*) and leading (*right*) hemispheres in approximate natural color (JPL photojournal). Jupiter's sulfurous plasma is largely responsible for the dark, reddish stain on the trailing hemisphere

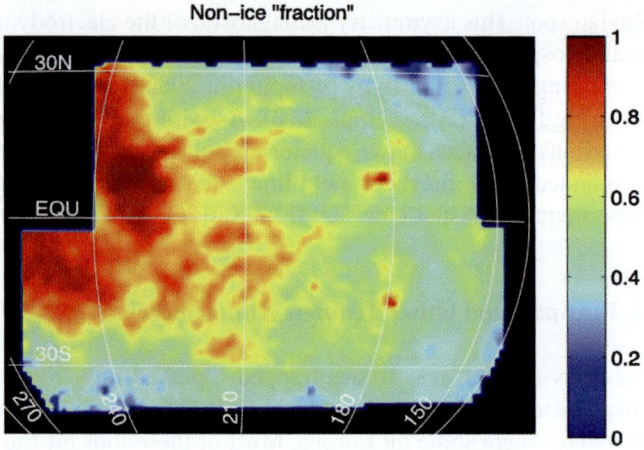

Fig. 16 A map of the fraction of hydrated sulfate fraction in Europa's surface (normalized). It shows a correlation both with plasma bombardment, the abundance of sulfates on the trailing hemisphere (*left*), but also with geologic features, the fine features toward the center of the map which correspond to geological features such as linea and chaos. From McCord et al. (1999)

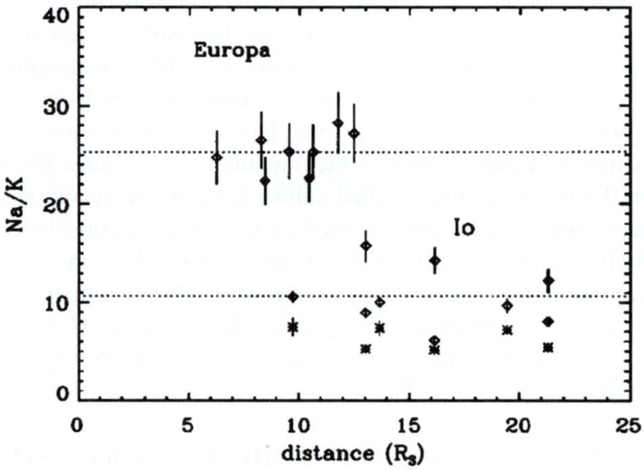

Fig. 17 Na/K ratios in the extended atmospheres of Europa and Io

water ice is particularly sensitive to the effects of radiation: Brown et al. (1978) found that ion irradiation was surprisingly effective at eroding ("sputtering") water ice. "Heavy" energetic ions are particularly effective, such as S and O ions from Io that have been accelerated to large energies.

The ions erode the surface, launching surface molecules into ballistic trajectories and forming a tenuous "atmosphere." Two components of the atmosphere are especially relevant to the exogenic/endogenic question posed above: Na and K. Earth-based observations by Brown (2001) of the two alkalis are shown in Fig. 17. These species are particularly easy to observe by their fluorescence in sunlight. The figure shows the Na/K abundance ratio as a function of distance from Europa and Io. The fact that the ratios are different suggests that Europa's alkali are endogenic, and do not derive from Iogenic implanted plasma. Fractionation effects that would change that conclusion cannot be ruled out (Carlson et al. 2010; this issue), but recent modeling (Cassidy et al. 2009) found that the ejecta from Europa's surface *should*, in fact, be one-to-one representative of the surface composition.

The atmospheric observations, along with the surface spectra, seem to indicate that Europa's Na and K ultimately derive from Europa's interior, and not from Io's atmosphere. This reaffirms the notion of surface/ocean interaction, but does not, however, rule out a large exogenic effect. The radiolytic sulfur cycle, for instance, may operate on both sulfate salts and Iogenic sulfur, and the hemispheric dichotomy is well explained by plasma bombardment.

Another component of the atmosphere with interesting implications for the subsurface ocean is Europa's thickest atmospheric component: O_2 (Hall et al. 1995). The O_2 is produced by the same radiation that erodes the surface (see chapter on radiolysis). Laboratory experiments suggest that O_2 should only be sputtered at about 1/10th the rate of the dominant surface component H_2O, but O_2 is thought to be the dominant atmospheric component because it does not condense onto the surface as H_2O does. Nonetheless, O_2 is seen in Europa's surface—inferred to be solid state via a visible light absorption indicating two touching O_2 molecules (Spencer and Calvin 2002; Hand et al. 2006). The evidence of exchange seen above is especially interesting in light of the presence of trapped O_2—it indicates that the radiolytically-produced O_2 may be supplied to the subsurface ocean, where it could be a source of energy for life (Chyba 2000).

Other satellites to exhibit very tenuous atmospheres in the Jovian system are Ganymede and Callisto. Ganymede is the largest satellite in the solar system and the only one known to possess a magnetosphere, likely created through convection within the liquid iron core (Schubert et al. 1996; Kivelson et al. 2002). This weak magnetosphere is buried within Jupiter's much larger magnetic field and connected to it through open field lines. In spite of negative results by Voyager occultations (Broadfoot et al. 1981), the HST observed auroral atomic oxygen emissions in the far-ultraviolet (Hall et al. 1998; Feldman et al. 2000). The line ratio of these emissions gives evidence for a neutral O_2 atmosphere around Ganymede. While the surface number density is more difficult to constrain than for Europa, owing to a lack of electron temperature measurements along auroral field lines, O_2 column densities of $(0.3–5) \times 10^{14}$ cm^2 are reasonable (Feldman et al. 2000).

Ganymede's thin oxygen atmosphere, similar to Europa, is probably created through sputtering from the magnetosphere. The composition of the atmosphere includes O, O_2, and possibly ozone (O_3) (Hall et al. 1998). Additional evidence of the oxygen atmosphere comes from spectral detection of gases trapped in the ice at the surface of Ganymede. Observations with the Hubble Space Telescope have also allowed for the detection of ozone bands on the trailing hemisphere of Ganymede (Noll et al. 1996). The inferred O_3 column abundance of ozone, 4.5×10^{16} cm^{-2}, can be produced by ion impacts or by photochemical equilibrium with previously detected O_2. An estimated number density ratio of $[O_3]/[O_2] \times 10^{-4}$ to 10^{-3} requires an atmospheric density orders of magnitude higher than upper limits from spacecraft occultation experiments. Furthermore, the modeling work of Marconi (2007) showed that near the subsolar region the atmosphere would be dominated by H_2O whereas away from this region O_2 would be the dominant constituent of the atmosphere.

Apparently, this oxygen atmosphere is trapped in Ganymede's surface ice, an inference consistent with the shift and broadening of the band compared with the gas-phase O_3 band. In 1997 spectroscopic analysis revealed the dimer (or diatomic) absorption features of the molecular oxygen. Such absorption can arise only if the oxygen is in a dense phase. The best candidate is the molecular oxygen trapped in ice. The depth of the dimer absorption bands depends on latitude and longitude, rather than on surface albedo—they tend to decrease with increasing latitude on Ganymede, while the O_3 shows an opposite effect (Calvin and Spencer 1997). Laboratory work has found that O_2 would not cluster and bubble but dissolve in ice at Ganymede's relatively warm surface temperature of 100 K (Vidal et al. 1997). Atomic hydrogen is a minor atmospheric constituent on Ganymede, but hydrogen atoms were observed as far as 3000 km from the surface of the moon. Their density on the surface is about 1.5×10^4 cm^{-3} (Barth et al. 1997).

Whether the satellite has an ionosphere to correspond to its atmosphere is unresolved (Eviatar et al. 2001, and references therein). These molecules are also quickly ionized by ultraviolet radiation and charged particles in the vicinity. The surface may also evaporate, or 'sublimate' when the sun is shining, the way the surface of a comet does. Because there is a magnetosphere, the charged particles in the atmosphere of Ganymede are not transported out of the environment the way they are at Europa.

A thin atmosphere of CO_2 (Carlson et al. 1999a), and perhaps O_2 (Liang et al. 2005), as well as an ionosphere (Kliore et al. 2002) were also detected around Callisto by the Near-Infrared Mapping Spectrometer (NIMS) during the 1999 Galileo flybys. The surface pressure is estimated to be 7.5×10^{-12} bar and particle density 4×10^8 cm^{-3}. Because such a thin atmosphere would be lost in only about 4 days, it must be constantly replenished, possibly by slow sublimation of carbon dioxide ice from the satellite's icy crust (Carlson et al. 1999a) which would be compatible with the sublimation hypothesis for the formation of the bright surface knobs.

Callisto's ionosphere was first detected during Galileo's flybys (Kliore et al. 2002). Its high electron density of 7–17 $\times 10^4$ cm^{-3} cannot be explained by the photoionization of the atmospheric carbon dioxide alone. Hence, it was suggested that the atmosphere of Callisto is actually dominated by molecular oxygen (Liang et al. 2005). However, oxygen has not yet been directly detected in the atmosphere of Callisto. Observations with the Hubble Space Telescope (HST) placed an upper limit on its possible concentration in the atmosphere, based on lack of detection, which is still compatible with the ionospheric measurements (Strobel et al. 2002). At the same time HST was able to detect condensed oxygen trapped on the surface of Callisto (Spencer and Calvin 2002).

Even though Ganymede and Callisto have thin atmospheres, there does not seem to be a Ganymede or Callisto torus, the way there is around Europa and Io.

Further out, around Neptune, revolves another satellite with a unique atmosphere, Triton. Triton's nitrogen and methane atmosphere, revealed by the Voyager-2 spacecraft in 1989 (Smith et al. 1989), has a surface pressure of 14 mbar, a surface temperature of about 35.6 K, and it extends 950 km above the surface. The atmospheric temperature profile shows a troposphere (tropopause? The troposphere cannot be associated with a single altitude) at around 8 km in altitude due to turbulence at the surface, but no stratosphere (McKinnon and Kirk 2007). Well-structured thermosphere (where heat is transported by conduction), ionosphere and exosphere layers exist (Lellouch et al. 1992). The exosphere temperature reaches 95 K. Recent HST and occultation observations have shown that the atmosphere is actually even denser than suggested by Voyager 2, and that the temperature is increasing (Buratti et al. 1999).

The lower atmosphere of Triton contains a variety of condensates. Most of the atmosphere contains a diffuse haze which probably consists of hydrocarbons and nitriles produced by the photolysis of N_2 and CH_4 (McKinnon and Kirk 2007). Discrete clouds can be distinguished at the limbs, most probably consisting of condensed nitrogen. The plumes (Hansen et al. 1990), seen by Voyager-2 near 50°S and rising to altitudes as high as 8 km (named Mahilani and Hili), could be purely atmospheric phenomena with the same composition as the rest of the atmosphere, with nitrogen rather methane ascending and condensing (McKinnon and Kirk 2007).

Thus, in spite of many similarities and interactions, the icy satellites are independent bodies with different exchange processes that need to be further investigated.

References

R.K. Achterberg et al., Icarus **194**, 263 (2008)

M. Ádámkovics et al., J. Geophys. Res. **111**, E07S06 (2006)
M. Ádámkovics, M.H. Wong, C. Laver, I. de Pater, Science **318**, 962 (2007)
S.K. Atreya et al., Planet. Space Sci. **54**, 1177 (2006)
G.E. Balester et al., Icarus **111**, 2 (1994)
E.L. Barth, S.C.R. Rafkin, Geophys. Res. Lett. **34**, L03203 (2007)
E.L. Barth, O.B. Toon, Icarus **182**, 230 (2006)
C.A. Barth et al., Geophys. Res. Let. **24**, 2147 (1997)
M.K. Bird et al., Nature **438**, 800 (2005)
W.J. Borucki, R.C. Whitten, Planet. Space Sci. **56**, 19 (2008)
A.H. Bouchez, M.E. Brown, Astrophys. J. **618**, L53 (2005)
A.H. Bouchez, M.E. Brown, N.M. Schneider, Icarus **148**, 316 (2000)
A.L. Broadfoot et al., Science **86**, 8259–8284 (1981)
M.E. Brown, Icarus **151**, 190 (2001)
M.E. Brown, A.H. Bouchez, C.A. Griffith, Nature **420**, 795 (2002)
M.E. Brown et al., Geophys. Res. Lett. **36**, L01103 (2009)
R.A. Brown, in *Exploration of the Planetary System*, ed. by R.A. Brown (Reidel, Dordrecht, 1974), p. 527
R.A. Brown, Astrophys. J. **244**, 1072 (1981)
R.H. Brown et al., Astron. Astrophys. **446**, 707 (2006)
R.H. Brown et al., Nature **454**, 607 (2008)
W.L. Brown, L.J. Lanzerotti, J.M. Poate, W.M. Augustyniak, Phys. Rev. Lett. **40**, 1027 (1978)
B.J. Buratti, M.D. Hicks, R.L. Newburn, Jr., Nature **397**(6716), 219 (1999-01-21). doi:10.1038/16615
M.H. Burger, N.M. Schneider, J.K. Wilson, Geophys. Res. Lett. **26**, 3333 (1999)
M.H. Burger et al., Astrophys. J. **563**, 1063 (2001)
M.H. Burger et al., Space Sci. Rev. (2010, this issue)
W.M. Calvin, J.R. Spencer, Icarus **130**, 505–516 (1997)
R.W. Carlson et al., Science **283**, 820–821 (1999a)
R.W. Carlson, R.E. Johnson, M.S. Anderson, Science **286**, 97 (1999b)
R.W. Carlson et al., Space Sci. Rev. (2010, this issue)
T.A. Cassidy, R.E. Johnson, O.J. Tucker, Icarus (2009, in press)
C.F. Chyba, Nature **403**, 381 (2000)
A.J. Coates et al., Geophys. Res. Lett. **34**, L22103 (2007)
M. Combes et al., Icarus **129**, 482 (1997)
A. Coustenis, B. Bézard, Icarus **115**, 126 (1995)
A. Coustenis, F.W. Taylor, *Titan: Exploring an Earthlike World*, 2nd edn. (World Scientific, Singapore, 2008)
A. Coustenis et al., Icarus **118**, 96 (1995)
A. Coustenis et al., Icarus **154**, 501 (2001)
A. Coustenis et al., Icarus **161**, 383 (2003)
A. Coustenis et al., Icarus **180**, 176 (2006)
A. Coustenis et al., Icarus **189**, 35 (2007)
T.E. Cravens et al., Geophys. Res. Lett. **35**, L03103 (2008)
I. de Pater et al., Icarus **156**, 296 (2002)
M.K. Dougherty et al., Science **311**, 1406 (2006)
S. Douté et al., Icarus **149**, 107 (2001)
S.T. Durrance, P.D. Feldman, H.A. Weaver, Astrophys. J. **267**, L125 (1983)
S.T. Durrance et al., Astrophys. J. **447**, 408 (1995)
E. Eviatar et al., Astrophys. J. **555**, 1013 (2001)
F.P. Fanale, R.H. Brown, D.P. Cruikshank, R.N. Clarke, Nature **280**, 761 (1979)
L.M. Feaga, M.A. McGrath, P.D. Feldman, Icarus (2009). doi:10.1016/j.icarus.2009.01.029
B. Fegley, M.Y. Zolotov, Icarus **148**, 193 (2000)
P.A. Feldman et al., Geophys. Res. Lett. **27**, 1787 (2000)
F.M. Flasar, B.J. Conrath, Icarus **85**, 346 (1990)
F.M. Flasar, R.E. Samuelson, B.J. Conrath, Nature **292**, 693 (1981)
F.M. Flasar et al., Space Sci. Rev. **115**, 169 (2004)
F.M. Flasar et al., Science **308**, 975 (2005)
W.M. Folkner et al., J. Geophys. Res. **111**, E07S02 (2006)
M. Fulchignoni et al., Nature **438**, 785 (2005)
P. Garnier et al., J. Geophys. Res. **113**, A10216 (2008)
P. Garnier et al., Icarus (2009, in press)
P.E. Geissler et al., Science **285**, 870 (1999)
P.E. Geissler et al., J. Geophys. Res. **106**, 26137 (2001)
P.E. Geissler et al., Icarus **172**, 127 (2004)

E. Gendron et al., Astron. Astrophys. **417**, L21 (2004)
S.D.B. Graves et al., Planet. Space Sci. **56**, 346 (2008)
C.A. Griffith, T. Owen, G.A. Miller, T. Geballe, Nature **395**, 575 (1998)
C.A. Griffith et al., Science **300**, 628 (2003)
C.A. Griffith et al., Science **313**, 1620 (2006)
C.A. Griffith, C.P. McKay, F. Ferri, Astrophys. J. **687**, L41 (2008)
D.T. Hall et al., Nature **373**, 677 (1995)
D.T. Hall et al., Astroph. J. **499**, 475–481 (1998)
K.P. Hand et al., Astrobiology **6**, 463 (2006)
C.J. Hansen et al., Bull. Am. Astron. Soc. **22**, 1070 (1990) [Abstract]
R.E. Hartle et al., J. Geophys. Res. **87**, 13 (1982)
R.E. Hartle et al., Geophys. Res. Lett. **33**, L08201 (2006a)
R.E. Hartle et al., Planet. Space Sci. **54**, 1211 (2006b)
A. Hayes et al., Geophys. Res. Lett. **35**, L09204 (2008)
A.R. Hendrix, C.A. Barth, C.W. Hord, J. Geophys. Res. **104**, 11817 (1999)
M. Hirtzig et al., Astron. Astrophys. **456**, 761 (2006)
M. Hirtzig et al., Astron. Astrophys. Rev. (2009, in press)
W.B. Hubbard et al., Astron. Astrophys. **269**, 541 (1993)
R. Hueso, A. Sánchez-Lavega, Nature **442**, 428 (2006)
D.M. Hunten et al., in *Saturn*, ed. by T. Gehrels, M.S. Matthews (Univ. Arizona Press, Tucson, 1984), p. 671
G. Israël et al., Nature **438**, 796 (2005)
D.E. Jennings et al., Astrophys. J. **691**, L103 (2009)
K.L. Jessup, J. Spencer, R. Yelle, Icarus **192**, 24 (2007)
T.V. Johnson, C.B. Pilcher, in *Planetary Satellites*, ed. by J. Burns (Univ. Arizona Press, Tucson, 1977), p. 232
E. Karkoschka, M.G. Tomasko, Icarus **199**, 442 (2009)
E. Karkoschka et al., Planet. Space Sci. **55**, 1896 (2007)
S.W. Kieffer et al., Science **314**, 1764 (2006)
S.J. Kim, L.M. Trafton, T.R. Geballe, Astrophys. J. **679**, L53 (2008)
M.G. Kivelson et al., Icarus **157**, 507–522 (2002)
A. Kliore et al., Science **183**, 323 (1974)
A. Kliore et al., Icarus **24**, 407 (1975)
A.J. Kliore et al., J. Geophys. Res. **107**, 1407 (2002)
T. Kostiuk et al., Geophys. Res. Lett. **32**, L22205 (2005)
A.L. Lane, R.M. Nelson, D.L. Matson, Nature **292**, 38 (1981)
S. Lebonnois, P. Rannou, F. Hourdin, Philos. Trans. R. Soc. A **367**, 665 (2009)
E. Lellouch, M. Blanc, J. Oukbir, P.-Y. Longaretti, Adv. Space Res. **12**, 113–121 (1992)
E. Lellouch et al., Astrophys. J. **459**, L107 (1996)
E. Lellouch, G. Paubert, J.I. Moses, N.M. Schneider, D.F. Strobel, Nature **421**, 45 (2003)
E. Lellouch, M.A. McGrath, K.L. Jessup, in *Io After Galileo*, ed. by R.M.C. Lopes, J.R. Spencer (Praxis Publishing, Chichester, 2007), p. 231
M.C. Liang et al., J. Geophys. Res. **110**, E02003 (2005)
G.F. Lindal et al., Icarus **53**, 348 (1983)
R.D. Lorenz, J. Radebaugh, The Cassini RADAR Team, Geophys. Res. Lett. **36**, L03202 (2009)
R.D. Lorenz, C.P. McKay, J.I. Lunine, Science **275**, 642 (1997)
R.D. Lorenz, M.T. Lemmon, P.H. Smith, G.W. Lockwood, Icarus **142**, 391 (1999)
R.D. Lorenz, E.F. Young, M.T. Lemmon, Geophys. Res. Lett. **28**, 4453 (2001)
R.D. Lorenz et al., Science **312**, 724 (2006)
R.D. Lorenz et al., Planet. Space Sci. **55**, 1936 (2007)
R.D. Lorenz et al., Geophys. Res. Lett. **35**, L02206 (2008a)
R.D. Lorenz et al., Icarus **195**, 812 (2008b)
J.I. Lunine, S.K. Atreya, Nature Geosci. **1**, 159 (2008)
W.W. Macy, L.M. Trafton, Icarus **25**, 432 (1975)
M.L. Marconi, Icarus (2007)
D.L. Matson et al., Icarus **187**, 569 (2007)
T.B. McCord et al., J. Geophys. Res. **104**, 11827 (1999)
T.B. McCord et al., Icarus **194**, 212 (2008)
A.S. McEwen, L.A. Soderblom, T.V. Johnson, D.L. Matson, Icarus **75**, 450 (1988)
M.A. McGrath et al., Icarus **146**, 476 (2000)
M.A. McGrath et al., in *Jupiter: The Planet, Satellites and Magnetosphere*, ed. by F. Bagenal, T.E. Dowling, W.B. McKinnon (Cambridge University Press, Cambridge, 2004), pp. 457–483. ISBN 0-521-81808-7

W.B. McKinnon, R.L. Kirk, in *Encyclopedia of the Solar System* (2nd edn.), ed. by L.-A. McFadden, P.R. Weissman, T.V. Johnson (Academic Press, San Diego, 2007), p. 483
M. Mendillo, J. Baumgardner, B. Flynn, W.J. Hughes, Nature **348**, 312 (1990)
M. Mendillo et al., Nature **448**, 330 (2007)
J.L. Mitchell, R.T. Pierrehumbert, D.M.W. Frierson, R. Caballero, Proc. Natl. Acad. Sci. **103**, 18421 (2006)
G. Mitri, A.P. Showman, J.I. Lunine, R.D. Lorenz, Icarus **186**, 385 (2007)
M. Michael, R.E. Johnson, Planet. Space Sci. **53**, 1510 (2005)
G.J. Molina-Cuberos et al., Planet. Space Sci. **49**, 143 (2001)
L.A. Morabito, S.P. Synnott, P.N. Kupferman, S.A. Collins, Science **204**, 972 (1979)
J.I. Moses, M.Y. Zolotov, B. Fegley, Icarus **156**, 76 (2002a)
J.I. Moses, M.Y. Zolotov, B. Fegley, Icarus **156**, 107 (2002b)
H.B. Niemann et al., Nature **438**, 779 (2005)
K.S. Noll, H.A. Weaver, A.M. Gonnella, J. Geophys. Res. **100**, 19057 (1995)
K.S. Noll et al., Sciente **273**, 341–343 (1996)
C. Paranicas, R.W. Carlson, R.E. Johnson, Geophys. Res. Lett. **28**, 673 (2001)
J.R. Pearl et al., Nature **280**, 755 (1979)
C.C. Porco et al., Science **299**, 1541 (2003)
C.C. Porco et al., Nature **434**, 159 (2005)
C.C. Porco et al., Science **311**, 5766 (2006)
M.K. Pospieszalska, R.E. Johnson, Icarus **78**, 1 (1989)
F. Postberg et al., Icarus **183**, 122 (2006)
F. Postberg et al., Nature **459**, 1098–1101 (2009)
J. Radebaugh et al., Icarus **194**, 690 (2008)
S.I. Ramirez et al., Icarus **156**, 515 (2002)
P. Rannou, F. Hourdin, C.P. McKay, D. Luz, Icarus **170**, 443 (2004)
P. Rannou, F. Montmessin, F. Hourdin, S. Lebonnois, Science **311**, 201 (2006)
K.D. Retherford et al., J. Geophys. Res. **105**, 27157 (2000)
K.D. Retherford et al., Science **318**, 237 (2007)
F.L. Roessler et al., Science **283**, 353 (1999)
H.G. Roe et al., Icarus **157**, 254 (2002a)
H.G. Roe et al., Astrophys. J. **581**, 1399 (2002b)
H.G. Roe et al., Science **310**, 477 (2005)
J. Saur, F.M. Neubauer, D.F. Strobel, M.E. Summers, Geophys. Res. Lett. **27**, 2893 (2000)
J. Saur et al., in *Jupiter: The Planet, Satellites, and Magnetosphere*, ed. by F. Bagenal, T. Dowling, W. McKinnon (Cambridge University Press, Cambridge, 2004), p. 537
E.L. Schaller et al., Icarus **182**, 224 (2006a)
E.L. Schaller et al., Icarus **184**, 517 (2006b)
N.M. Schneider et al., Science **253**, 1394 (1991a)
N.M. Schneider, D.M. Hunten, W.K. Wells, A.B. Schultz, U. Fink, Astrophys. J. **368**, 298 (1991b)
G. Schubert, K. Zhang, M.G. Kivelson, J.D. Anderson, Nature **384**, 544 (1996)
D.E. Shemansky et al., Science **308**, 978 (2005)
E.C. Sittler et al., Planet. Space Sci. **56**, 3 (2008)
B.A. Smith et al., Science **212**, 163 (1981)
B.A. Smith et al., Science **215**, 504 (1982)
B.A. Smith, L.A. Soderblom et al., Science **246**, 1422 (1989)
P.H. Smith et al., Icarus **119**, 336 (1996)
W.H. Smyth, M.R. Combi, Astrophys. J. **328**, 888 (1988)
W.D. Smythe, R.M. Nelson, D.B. Nash, Nature **280**, 766 (1979)
L.A. Soderblom et al., Planet. Space Sci. **55**, 2015 (2007)
C. Sotin et al., Nature **435**, 786 (2005)
J.R. Spencer, W.M. Calvin, Astron. J. **124**, 3400–3403 (2002)
J.R. Spencer et al., Science **288**, 1208 (2000)
J.R. Spencer et al., Icarus **176**, 283 (2005)
E.R. Stofan et al., Nature **445**, 61 (2007)
D.F. Strobel et al., Astrophys. J. **581**, L51–L54 (2002)
N.A. Teanby et al., Icarus **186**, 364 (2007)
N. Thomas, F. Bagenal, T.W. Hill, J.K. Wilson, in *Jupiter: The Planet, Satellites, and Magnetosphere*, ed. by F. Bagenal, T. Dowling, W. McKinnon (Cambridge University Press, Cambridge, 2004), p. 561
G. Tobie, J.I. Lunine, C. Sotin, Nature **440**, 61 (2006)
G. Tobie et al., Space Sci. Rev. (2010, this issue)
T. Tokano, Icarus **173**, 222 (2005)

T. Tokano, Planet. Space Sci. **55**, 1990 (2007)
T. Tokano, F.M. Neubauer, Icarus **158**, 499 (2002)
T. Tokano, F.M. Neubauer, M. Laube, C.P. McKay, Icarus **153**, 130 (2001)
T. Tokano et al., J. Geophys. Res. **111**, E08007 (2006a)
T. Tokano et al., Nature **442**, 432 (2006b)
M.G. Tomasko et al., Nature **438**, 765 (2005)
M.G. Tomasko et al., Planet. Space Sci. **56**, 648 (2008)
L. Trafton, Nature **258**, 690 (1975)
E.P. Turtle et al., Geophys. Res. Lett. **36**, L02204 (2009)
R.A. Vidal et al., Science **276**, 1839–1842 (1997)
S. Vinatier et al., Icarus **188**, 120–138 (2007)
J.E. Wahlund et al., Science **308**, 986 (2005)
J.H. Waite et al., Bull. Am. Astron. Soc. **36**, 1068 (2004) [Abstract]
J.H. Waite et al., Science **311**, 1419 (2006)
J.H. Waite et al., Science **316**, 870 (2007)
J.H. Waite et al., Science (2009, submitted)
J.K. Wilson et al., Icarus **157**, 476 (2002)
B.C. Wolven et al., J. Geophys. Res. **106**, 26155 (2001)
M.C. Wong, W.H. Smyth, Icarus **146**, 60 (2000)

Space Sci Rev (2010) 153: 185–218
DOI 10.1007/s11214-010-9633-3

Phase Behaviour of Ices and Hydrates

A. Dominic Fortes · Mathieu Choukroun

Received: 5 May 2009 / Accepted: 17 January 2010 / Published online: 2 March 2010

Abstract The primary volatile 'rock-forming' minerals in the icy satellites of the outer solar system include water-ice and various hydrated crystals of methane and ammonia. The rich polymorphism of these substances as a function of pressure and temperature are described in this chapter. This polymorphism has a fundamental influence on the exchange of mass and energy between the core and the surface of icy satellites. We describe the current state-of-the-art in our understanding of the high pressure phase behaviour and the measurements of thermoelastic and transport properties of these substances. In addition we describe the structures and properties of hydrated phases of methanol, sulfuric acid, and various sulfate salts.

Keywords Water ice · Methane clathrate · Ammonia · Methanol · Sulfuric acid · Sulfates · Polymorphism · Phase relations

1 Introduction

The four thousand or so mineral species—primarily silicates—extant on the Earth, and likely to dominate the other terrestrial planets of the inner solar system, represent a relatively small fraction, by mass, of the solar system's mineral wealth. The major rock-forming minerals in the outer solar system are water ice, hydrates of volatile compounds (such as methane and ammonia), and hydrates of the most abundant soluble inorganic substances (such as sulfates). As on the rocky planets, geological activity has acted to process these basic substances into a range of other mineral species. For example, methane outgassed from clathrate hydrates in the interior of Saturn's largest moon, Titan, is likely to be the source of gaseous

A.D. Fortes (✉)
Centre for Planetary Sciences at UCL/Birkbeck, Gower Street, London, WC1E 6BT, UK
e-mail: andrew.fortes@ucl.ac.uk

M. Choukroun
NASA Jet Propulsion Laboratory, California Institute of Technology, Pasadena, CA 91109, USA
e-mail: mathieu.choukroun@jpl.nasa.gov

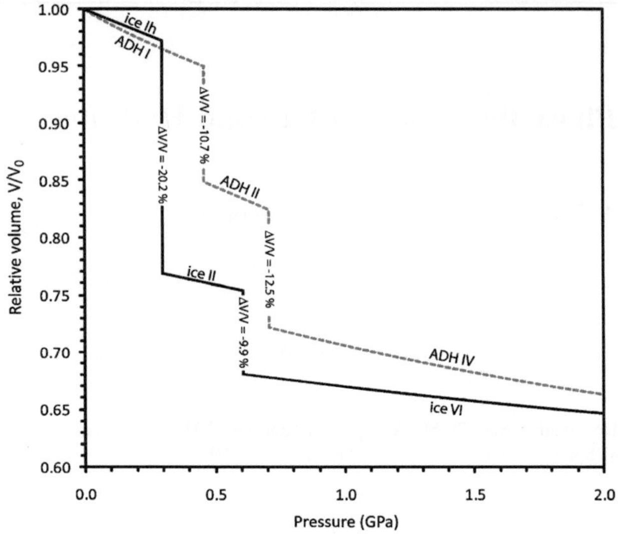

Fig. 1 Variation of density in ice and ammonia dihydrate as a function of pressure (0–2 GPa) at temperatures of 185 ± 10 K. The very large jumps in density correspond to solid–solid phase transitions to denser crystal structures, labeled, ice II, ice VI, etc. Data for both ice and ammonia dihydrate obtained using neutron powder diffraction on deuterated samples (Fortes et al. 2009a)

methane in the satellite's atmosphere (Lunine and Stevenson 1987; Loveday et al. 2001a; Tobie et al. 2006; Lunine et al. 2009), where it is destroyed by electromagnetic and particle radiation, the fragments recombining to form other hydrocarbons—acetylene (C_2H_2) being the most abundant solid by-product—which accumulate as sedimentary minerals on the surface.

The evolution of all planetary bodies consists of an ongoing process of exchange, involving both matter and energy. Primarily, this consists of the generation of radiogenic heat in a planetary interior and its loss to outer space, either by conduction or convection through a solid medium and radiation from a planetary surface (Hussmann et al. 2010, this issue; Tobie et al. 2010, this issue). The stresses imposed on the surface are usually manifested as extensional or compressional geology; for example, rifts, thrusts or folds. Often, there is partial melting of material, which by virtue of being buoyant then advects heat into the surface environment (as an igneous intrusion or eruption). In other cases, matter and energy are delivered from endogenic sources, as in meteorite impacts, or through the interplay of gravitational fields.

The transport of heat from core to surface, and any resultant melting, depends intimately upon the physical properties of the constituent materials. Exchange of energy from endogenic sources is often strongly dependent upon the rheology—and hence the microscopic crystalline structure and the diffusion of defects through that structure—of the near surface materials (Durham et al. 2010 this issue).

In the case of icy satellites, we are concerned primarily with the mineral physics of water ice, but other solids are implicated. These materials are all considerably more compressible than silicates, by virtue of the comparatively weak hydrogen bond (compare the energy in the O–H$\cdots$O hydrogen bond, $\sim$20 kJ mol^{-1}, with that in a typical covalent bond, several hundred kJ mol^{-1}) (Jeffrey 2003). Moreover, the weak hydrogen bond provides nature with many different ways of connecting water molecules that differ almost insignificantly in energy; in other words, there exist a multitude of crystal structures which are thermodynamically stable in a relatively narrow range of variables (e.g., pressure and temperature). These phase changes also result in major discontinuities (e.g., Fig. 1) in thermoelastic properties (thermal expansion, compressibility), and in transport properties (e.g., thermal conductivity, rheology), all of which are critical to understanding and modelling the evolution of planetary bodies.

Melting behaviour is also of central concern to understanding volcanism on icy bodies (cryovolcanism). Water ices and clathrates are the most refractory 'icy' minerals in the outer solar system, the latter requiring very large thermal anomalies to dissociate. Once melting of water ice occurs, the density contrast between the melt and the host 'rock' is quite small (order 10s $\text{kg}\,\text{m}^{-3}$) and positive, meaning that melt will prefer to propagate *downwards* into the mantle rather than rise to the surface. Moreover, there is a barrier in the form of the negative pressure melting curve of low-pressure ice Ih (see Sect. 2); the consequence of this, should a cryomagma be able to rise towards the surface, is that it will progressively freeze as it is decompressed. Contrast this with silicate systems, which have large negative density contrasts with the host rock ($\Delta\rho \approx 500$–$1000\ \text{kg}\,\text{m}^{-3}$), resulting in large positive (upward) buoyancy forces, and also experience greater degrees of partial melting when decompressed. A cosmically abundant molecule such as ammonia can lower the melting temperature of the binary mixture by up to 100 K (see Sect. 4) bringing partial melting of icy bodies within the realm of their energy budgets, and hence plausibility. The resultant melts (in the ammonia–water system) can be buoyant under certain conditions with respect to their surroundings and the liquid may be extruded at the surface, or be intruded as high-level plutons. In sufficiently large icy bodies, the presence of ammonia (or other solutes) may allow melt to persist as a global subsurface ocean throughout geological time (Sohl et al. 2010, this issue).

The objective of this contribution is to describe the structures of the major rock-forming minerals, water ice, methane clathrates, and ammonia hydrates, discussing their polymorphism and physical properties and also concentrating on describing the crystal structures in mineralogical terms, since many share familiar structural characteristics with silicate minerals. In addition, information on a number of other aqueous systems is presented, including methanol, sulfuric acid, and a number of potentially important sulfates.

This is by no means an exhaustive study of ice or related outer solar system solids; many other excellent texts cover material that is beyond the scope of this brief review, including the books on ice physics by Whalley et al. (1973), Hobbs (1974) and Petrenko and Whitworth (1999), the two large compilations of papers relating to solar system ices (Klinger et al. 1985 and Schmitt et al. 1998), the text on gas hydrates by Sloan and Koh (2007), and the collection of (largely) Russian work on the physics of cryocrystals (Manzhelii et al. 1996). As a final note prior to embarking on this review, we mention the difficulty in using the word 'ice' where—amongst planetary scientists—it is taken to mean 'water ice' *as well as* being used to describe a vast collection of other volatile solids. Whalley (1985) favoured the term 'glacial' to refer to solids other than H_2O, though this, sadly, has not caught on. We endeavour to avoid using 'ice' where we do not mean H_2O, but there is no escaping using 'icy' as a broader term to signify planetary bodies in the outer solar system.

2 Water Ice

The largest icy moons (e.g., Ganymede, Callisto, and Titan) probably have ice mantles over 800 km deep; the pressure at the base of such a layer may be as high as 1.5 GPa, depending on the density of the rocky core (e.g., Schubert et al. 1986). In terms of planetary geophysics, it is therefore obvious that the outermost layer of ice must consist of concentric shells of high-pressure ice polymorphs. However, were these bodies to have large hydrous cores, then the ice shell is necessarily thinner, and the pressure at the core-mantle interface may be 0.5–0.7 GPa. Clearly, if giant icy objects such as ocean-planets (e.g. Léger et al. 2004)

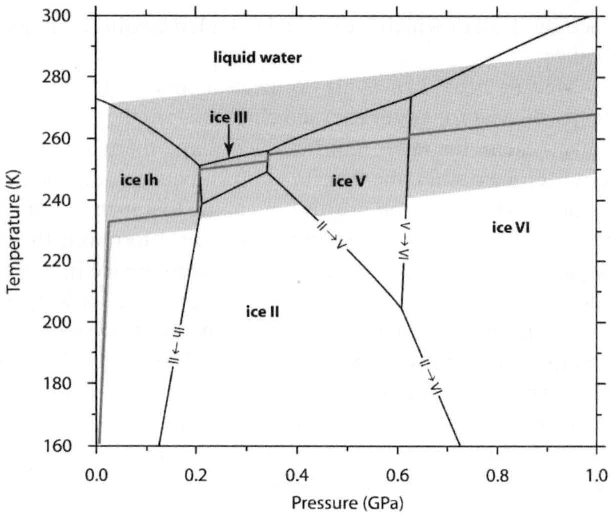

Fig. 2 Equilibrium phase diagram of water ice in the range 0–1 GPa (after Bridgman 1912, 1935, 1937; (Mishima 2000); Nishibata 1972). The *solid red line* depicts a possible geotherm inside a large icy satellite (such as Ganymede); note the steep conductive profile in the lowest-pressure region of the ice Ih region, and the shallower convective (adiabatic) profile at higher pressure. The *wider shaded region* shows the range of plausible 'hot' and 'cold' profiles. Observe that the warmest profile intersects the liquid field over a large range of pressures (= depths), resulting in a deep subsurface ocean underneath a crust of ice Ih

exist in other extra-solar systems, then the pressures encountered in the ice layers will be considerably greater, perhaps in the region of tens of GPa. The phase diagram of water ice as a function of pressure and temperature is amongst the most complex known. Figure 2 depicts the stable phases in the 0–1 GPa regime. It is clear from this phase diagram, that a cold mantle ought to be comprised of concentric shells of ice Ih, ice II (or III if warm enough), ice V, and ice VI, the latter being the thickest shell. In a warmer case, some high-pressure polymorphs may be missing in the internal structure because the negative dT/dP melting curve of ice Ih permits the possibility of having liquid water beneath an ice Ih layer. This phase layering is critical because each ice polymorph has different physical properties, density, viscosity, and thermal conductivity to name a few. The effect of having ice shells with different properties is to alter the pattern of convection, and thus the transport of heat, from core to surface (Hussmann et al. 2010, this issue). Since it is largely this heat transport that drives surface geology, it is obvious that we can better interpret the geology we see in spacecraft imagery if we can constrain the internal structure and dynamics via the physics of the constituent minerals.

2.1 Structures

In the solid state, water molecules adopt a tetrahedral coordination. Each molecule donates two hydrogen bonds to near neighbours, and accepts two hydrogen bonds from other neighbours. A given water molecule may therefore exist in one of six discrete orientations with respect to its four nearest neighbours (see Hobbs 1974 and references therein). It is useful at this stage to define orientationally ordered and disordered structures. In the case of an ordered structure, a given water molecule in a particular crystallographic unit-cell, shares the same orientation as that in the next unit-cell, and this common orientation persists over the long range (1000s of angstroms). In orientationally disordered structures, a given water molecule may be oriented differently to the molecule in the next unit-cell with the same fractional atomic coordinates. As a result there is no long-range order that may be sensed by coherent diffraction of X-ray or neutron beams, the technique of choice for determining the atomic structure of matter, and instead one senses a time and space-averaged structure. In the ice Ih structure this average is constrained by the symmetry elements of the unit-cell to be two half-occupied hydrogen sites along each O–O vector, but in other forms of ice,

Table 1 The ordered and disordered phases of water ice. See text for description of structures

Disordered phase	(space group)	↔	Ordered phase	(space group)
ice Ih	$P6_3/mmc$	↔	ice XI	$Cmc2_1$
ice Ic	Fd3m			
no disordered form			ice II	$R\bar{3}$
ice III	$P4_22_12$	↔	ice IX	$P4_22_12$
ice IV	R3m		*no ordered form*	
ice V	A2/a	↔	ice XIII	$P2_1/a$
ice VI	$P4_2/nmc$	↔	ice XV	$P\bar{1}$
ice VII	Pm3n	↔	ice VIII	$I4_1/amd$
n/a			ice X	Pm3n
ice XII	$I\bar{4}2d$	↔	ice XIV	$P2_12_12_1$

much more complex ordering schemes are possible. Generally speaking, the low temperature forms of ice contain orientationally ordered water molecules, but become disordered at higher temperatures (Table 1). By and large, the ordered forms are extremely difficult to make in the laboratory, since they form at low temperatures where the diffusion timescales necessary for molecular re-orientation become very long (Suga 1997; Salzmann et al. 2006, 2009).

There are presently fifteen known crystalline polymorphs of water ice; importantly, many of these are pairs of ordered and disordered structures (albeit sometimes with reduced symmetry), and some—such as ices Ih and Ic—are closely related polytypes. Hence, the phases listed in Table 1 can be reduced to eight basic structures.

2.1.1 Low-Pressure Structures, Ices Ih and Ic

The low-pressure form of water ice (called ice Ih) consists of water molecules bound together into hexagonal rings; these rings are joined together to form sheets, and these sheets are stacked in such a way that every fourth sheet lies directly over the first (in other words the stacking is AB-AB-AB, or hexagonal close packing). The crystal is hexagonal (hence the subscript 'h' after ice I), crystallizing in space-group $P6_3/mmc$ (e.g., Kuhs and Lehmann 1983). This structure is an analogue of the silica polymorph tridymite. It is also possible to pack the hexagonal layers so that the repeat is ABC-ABC-ABC (cubic close packing) and this results in a crystal of cubic symmetry—space-group Fd3m—being formed (called ice Ic; König 1943; Arnold et al. 1968), which strictly is a polytype of ice Ih. This structure is an analogue of the silica polymorph cristobalite. At temperatures below ~70 K, the proton disordered ice Ih transforms to a proton ordered form with orthorhombic symmetry, which is referred to as ice XI (Leadbetter et al. 1985; Howe and Whitworth 1989; Line and Whitworth 1996; Fukazawa et al. 2002). Although it is normally only possible to form this ordered phase of ice in the laboratory by doping the structure with defects (typically be adding KOH), in the natural environment where eons are available to effect the phase change, then we expect any icy surface which is below 70 K (e.g., Pluto and Triton) to exist as ice XI rather than ice Ih (e.g., Fukazawa et al. 2006).

2.1.2 High-Pressure Structures

The response of the ice structure to pressure is to preserve at all costs the tetrahedral coordination of the water molecule with a minimum of 'strain' (in other words O–H–O angles as close to tetrahedral, 109.47°, and hydrogen bonds as straight as possible) whilst achieving the maximum densification. This is accomplished by forming novel structures containing a range of polygonal rings of water molecules (from 4- to 8-sided), and by allowing H-bonds to pass through some of these rings.

In ice II, six-sided rings are preserved but instead of forming sheets, these become stacked to form six-sided tubes; the tubes are packed side by side, with hydrogen bonds linking them together (Lobban et al. 2002; Fortes et al. 2005). These result in a considerably denser crystal ($\Delta V/V \approx -20\%$ with respect to ice Ih [Fig. 1], which is an enormous density contrast compared to silicate minerals), and highly strained bonding (O–H–O angles from 80–130°) whilst preserving the H_2O coordination. In ice III (and its proton-ordered equivalent, ice IX; La Placa et al. 1973), the structure is composed of ribbons of pentagonal H_2O rings, cross-linked by 7-sided rings (Lobban et al. 2000). Ice III is a topological equivalent to the SiO_2 polymorph keatite. The wholly metastable phase ice IV returns to the motif found in ice Ih, being composed of sheets of H_2O hexagons. However, the structure differs in that it effectively has an extra sheet interposed into the stacking sequence, with the interlayer hydrogen bonding passing through the centre of the rings in the interposed sheet (Engelhardt and Kamb 1981). Ice V (and its proton ordered equivalent, ice XIII; Salzmann et al. 2006) exists under the same P, T conditions as ice IV, but achieves its densification in a quite different manner. In the ice V structure, we find a mixture of 4-, 5-, 6-, and 8-sided rings, with hydrogen bonds penetrating the 8-sided ring. Ice V has the most complex structure of any ice polymorph, with 28 symmetry inequivalent molecules in the monoclinic unit cell (Lobban et al. 2000). Nonetheless, the ribbon-like pattern of square rings in the structure is almost identical to the structural motif found in the high-pressure SiO_2 polymorph coesite.

Ice VI takes the concept of hydrogen-bond interpenetration to the extreme, being comprised of two distinct edingtonite-like frameworks (of 4- and 8-sided rings) one of which is inserted entirely into the interstices of the other (Kamb 1965; Kuhs et al. 1984). This so-called self-clathrate structure has hydrogen bonds passing through each 8-sided ring, and exhibits the greatest degree of molecular strain (O–H–O bond angles as small as 77°). The structure is nonetheless comparatively robust, existing over a pressure range of 0.6–2.1 GPa: contrast this with most of the other polymorphs <2 GPa, which have much narrower stability fields. The most recently discovered phase, ice XV, is an antiferroelectrically ordered variety of ice VI with triclinic symmetry (Salzmann et al. 2009).

Ice XII is the only ice polymorph with an entirely novel structure discovered within the last forty years (Lobban et al. 1998). This phase appears to be entirely metastable (like ice IV), and has been formed under a range of conditions within the stability fields of ices V and VI (Loerting et al. 2002; Johari 2003). The structure is tetragonal, being characterized by a five-sided double-helix arrangement of water molecules aligned along the crystallographic c-axis. Ice XII is proton disordered; a proton ordered form dubbed ice XIV has been observed in doped samples at low temperatures (Salzmann et al. 2006).

Above 2.1 GPa, the phase diagram simplifies considerably. The high-T phase, ice VII replicates the interpenetration observed in ice VI using the ice Ic framework, achieving a very compact body-centred-cubic (bcc) structure (Kuhs et al. 1984). A slight distortion of this framework caused by proton ordering below $\sim$280 K yields the tetragonal ice VIII structure (Kuhs et al. 1984), and the shift of the hydrogen atoms to positions half-way between neighbouring oxygens (so-called H-bond symmetrization) above $\sim$60 GPa generates the ice X structure, which is also body-centred-cubic (e.g., Loubeyre et al. 1999).

Ice polymorphs Ih, II, III, V and VI are the most relevant to the interiors of outer solar system icy bodies, occurring in the P–T range 0–2 GPa and 100–350 K. The equation of state of ice Ih has been extensively studied as a function of both pressure and temperature;

an excellent recent parameterization is given by Feistel and Wagner (2006). The equation of state of ice II has been studied computationally (Fortes et al. 2003b) and experimentally (Fortes et al. 2005). Less is known about the thermal expansion of ices III, V and VI. The small stability fields of ices III and V makes experimental determination of their thermal expansion difficult; some neutron diffraction data are available but only at two state points for ice III and three state points for ice V (Lobban et al. 2000). For ice VI, neutron diffraction (Kuhs et al. 1984), ultrasound (Shaw 1986; Gagnon et al. 1990) and light scattering data (Tulk et al. 1997) are similarly quite sparse. Based on available experimental data, equations of state have been developed for some phases of the H_2O phase diagram by Choukroun and Grasset (2007).

At ambient pressure, the specific heat capacity of ice Ih (Giauque and Stout 1936; Smith et al. 2007) has been determined from 0.5–270 K and the thermal conductivity has been measured over a similar range (Slack 1980; Andersson et al. 1980). Measurements of the thermal conductivity for the high-pressure ice phases (e.g. Ross et al. 1976, 1977), are summarized in a recent review by Andersson and Inaba (2005). In general, the P–T range studied was $100 < T < 300$ K and $P < 2.4$ GPa Both the brittle and ductile rheology of ice Ih is extremely well known under conditions extant in terrestrial ice sheets and glaciers. The ductile rheology of ices II, III, V, and VI are relatively well known at strain rates available under laboratory conditions (see Durham et al. 2010, this issue). Published values are expected to represent upper boundaries on the strength of these materials since planetary strain rates are typically many orders of magnitude smaller than those measured in the laboratory.

2.2 Planetary Implications

As described earlier, the polymorphism of water ice results in the formation of concentric shells in the outermost layers of a differentiated icy satellite consisting of increasingly dense phases of ice. Both the thermoelastic properties and the bulk transport properties of each phase affect the way in which the body loses radiogenic heat from its interior, and the way in which tidal heat is frictionally generated and dissipated. Upon phase transitions, latent heat must be brought to or removed from the system as well, which modifies the energy balance and affects heat transfer. All of the ice phases except ice Ih exhibit a positive dependence of thermal conductivity on pressure, though displaying only very small variations over their fields of stability. Thermal conductivities vary from a low of 1 $\mathrm{W\,m^{-1}\,K^{-1}}$ in ice III to a maximum of $\sim$6 $\mathrm{W\,m^{-1}\,K^{-1}}$ in ice VIII. Of particular planetological interest is the large thermal conductivity contrast between ices Ih and II and between ices Ih and III. The very much steeper conductive geotherm dictated by the low thermal conductivities of ices II and III suggests the possibility that layers within an icy mantle could exceed their melting points, producing cryovolcanic liquids. Steeper geotherms will also make shells of ices II and III more susceptible to convection, which would, conversely, have the effect of lowering temperatures below the liquidus. However, this depends very much on the rheological properties of the ice phases in question.

Of note is the extreme weakness of ice III, having a solid-state viscosity several orders of magnitude smaller than that of the other ice phases, whereas the viscosity contrast between the other phases is relatively small across the pressure ranges within which they are stable. All of the ice phases exhibit a marked dependence of viscosity on temperature that complicates the development of convection models. Determining whether or not a given heat flux is sufficient to make the ice mantles of these bodies unstable to convection requires calculation of the Rayleigh number ($\mathrm{Ra} = g\alpha\rho D^3 \Delta T/\kappa\nu$), where g is the acceleration due to gravity, α is the volume thermal expansion coefficient, ρ is the density, D is the depth of the layer, ΔT

Table 2 Thermodynamic character of selected ice phase transitions, indicating their effect on the propagation of convective flow

	$\Delta\rho$ (kg m^{-3})	$\mathrm{d}P/\mathrm{d}T$ (MPa K^{-1})	Rb/Ra	$\rho_0 g D$ (GPa)	γ Rb/Ra
ice Ih—II	220	0.81	≈35	0.1–0.2	10
ice Ih—III	185	−0.47	≈30	0.3	−2
ice II—V	65	−0.67	≈10	0.3–0.4	−1

the temperature contrast across that layer, κ is the thermal diffusivity, and ν is the kinematic viscosity. Rayleigh numbers greater than $\sim 10^3$ typically indicate that a layer is unstable to convection. For many larger icy bodies, such as Ganymede, Callisto, and Titan, Ra is of the order of 5×10^8, indicating that radiogenic heat is transported by vigorous convection. In the convecting layer, the temperature profile is adiabatic—and may yield ΔT of only 20 K over 1000 km, with much of the temperature drop occurring in conductive boundary layers at the top and bottom of the convecting shell. Calculations suggest that effective Rayleigh numbers $>10^8$ are required for whole mantle convection, and $>10^5$ for convection in the uppermost ice Ih layer. Only Titan, Ganymede, and Callisto are currently likely to have Ra $> 10^8$, and Triton and Pluto have Ra $> 10^5$. Mid-sized icy bodies could have experienced convection earlier in their histories when radiogenic heat flows were larger, but small bodies such as Mimas and Miranda are unlikely ever to have experienced a period of convection.

Changes in buoyancy associated with phase transitions add yet another layer of complexity to convection in icy mantles. The relative density contrasts of water ice polymorphs are much larger than those found in terrestrial mantle silicate phases and so it is instructive to calculate the effect that these phase changes could have on the pattern of convection. We can assess the influence of a phase change on convective motion from the buoyancy Rayleigh number ($\mathrm{Rb} = \Delta\rho g D^3/\kappa\nu$) and the normalized Clapeyron slope, $\gamma = (\mathrm{d}P/\mathrm{d}T)(\Delta T \rho_0 g D)$. The normalized Clapeyron slope is a measure of the extent to which the phase boundary is deflected from its equilibrium position and thus contributes to excess buoyancy. The phase buoyancy parameter (simply $\gamma \cdot \mathrm{Rb/Ra}$) is calculated for some relevant phase changes in Table 2 (see McKinnon 1998).

As a guide to the physical meaning of these values, the endothermic phase change in the terrestrial mantle from the γ-spinel structure to the silicate perovskite at depths of $\sim$670 km yields $\gamma\,\mathrm{Rb/Ra} \approx -0.1$, a value which is thought to be mildly inhibitive to convection and perhaps leads to 'leaky' or intermittent convection between the upper and lower mantle (e.g., Tackley 1995). Clearly, the values derived for the ice Ih–III and ice II–V phase transitions can be taken to suggest that these boundaries are strongly inhibitive to convective motions, whereas the ice Ih–II phase change will enhance convective motion. In relatively small icy bodies such as Triton and Pluto, where only ices Ih and II are expected to be stable, whole mantle convection is favoured (if convection occurs at all). However, in larger bodies containing higher pressure ice V and VI, convection should occur in discrete shells separated by conductive boundary layers. The implication of this result is that ΔT will be much larger and this again raises the possibility that temperatures could exceed the melting point (either at the thermal minimum near the ice Ih–III-liquid triple point, or in a boundary layer between the ice V and ice VI shells).

3 Methane–water (Clathrate Hydrates)

Mineralogists define two broad classes of microporous crystalline materials; the first is characterized by inter-connected voids (zeolites) and the second by unconnected voids (clathrates). On Earth, these are typically composed of a silica framework, the voids and channels being occupied by a range of ions or molecules, but water-based analogues of some silicate clathrates occur, these being referred to commonly as clathrate hydrates. The framework species is referred to as the 'host' and the species which occupy the various cavities are the 'guests'. For the most part, the host structure is only thermodynamically stable when the guest molecules are present; when they are removed, the host structure collapses.

We saw in Sect. 2 that water molecules tend to form ring structures, typically preferring 5- or 6-sided rings, and forming sheets which stack to form a three dimensional crystal structure. In the presence of a suitable guest species ("hydrate-formers"), under the right thermodynamic conditions, rings of water molecules can instead join together to form a variety of polyhedral cages. The simplest is a dodecahedron composed of twelve pentagonal rings (the shorthand description for this unit therefore being 5^{12}), but there are several other possible polyhedra constructed from 4-, 5-, and 6-sided rings. In the clathrate hydrate crystal, these polyhedra are bonded together to create a regular three-dimensional array. There are a limited number of ways of doing this, as described shortly, to form four basic clathrate hydrate structures.

Guest molecules occupy the polyhedral cages. Very small molecules, such as N_2 or O_2 can easily occupy the smaller cages, whereas larger clathrate guest formers (such as propane) are restricted to larger cages. Hence, the type of guest molecule available selects for the type of clathrate structure formed on the basis of available cage dimensions. So, for example, methane may form a so-called structure I clathrate (see Sect. 3.1.1 below), but in the presence of a small amount of ethane, a structure II clathrate (see Sect. 3.1.2) is formed. The guest molecules do not themselves form chemical bonds to the host structure, instead 'rattling around' inside the cages. A few comparatively rare species, such as certain alcohols (e.g. methanol), and organics (e.g., hexamethylenetetramine), do bond to the cage walls, and these are dubbed semi-clathrates. The lack of chemical bonding means there is no requirement to satisfy any stoichiometry, and so clathrate hydrates can vary in composition. This makes them somewhat difficult to deal with experimentally, since composition provides an extra variable which is not always simple to control or measure.

3.1 Low-Pressure Structures

3.1.1 Structure I (sI) Clathrate

This is a cubic crystal (space group $Pm3n$, typical cell parameter, $a \approx 12$ Å, and $V \approx 1700$ Å^3) built up from a three-dimensional network of tetrakaidecahedra (T-type cavities, $5^{12}6^2$), the interstices of which are pentagonal dodecahedra (D-type cavities, 5^{12}). The composition may be represented as $2D{\cdot}6T{\cdot}46H_2O$, resulting in an idealized stoichiometry—when all cages are filled—of $46/8 = 5.75 : 1$. Gases that form sI clathrates have a molecular diameter of 4.3 to 5.3 Å. Typically, CH_4, CO_2, H_2S, and Xe, are sI-hydrate-formers (e.g. Sloan and Koh 2007).

3.1.2 Structure II (sII)

This crystal is also cubic ($Fd3m$, in the centrosymmetric setting, typical cell parameter, $a \approx 17$ Å, and $V \approx 4900$ Å^3); the basic building block consists of hexagonal rings of large hexakaidecahedra (H; $5^{12}6^4$), surrounding a smaller dodecahedron (D, 5^{12}), arranged into sheets parallel to {110}. These sheets are packed vertically (i.e., perpendicular to 110), with an A-B-C–A-B-C repeat. The composition may be represented as $8H \cdot 16D \cdot 136H_2O$, resulting in an idealized stoichiometry—when all cages are filled—of $136/24 = 5.667 : 1$. Two types of gases form sII clathrates: (i) those with a molecular diameter smaller than 4.3 Å but larger than 3.8 Å (Ar, Kr, N_2, O_2); and (ii) those with a molecular diameter larger than 6 Å (ethane, propane, iso-butane), which, in this case, only occupy the large cages $5^{12}6^4$, with an idealized stoichiometry of 17:1.

3.1.3 Structure H (sH)

This crystal is hexagonal (P6/*mmm*, typical cell parameters, $a \approx 12.3$ Å, $c \approx 9.9$ Å, and $V \approx 1300$ Å^3) and is constructed from alternating layers of two types of dodecahedra; the first (D, 5^{12}) is familiar from both the sI and sII clathrates; the second type of dodecahedron (D′, $4^35^66^3$) is more complex. The large interstices between the dodecahedra define a much larger icosahedral cage (E, $5^{12}6^8$). The composition may be represented as $E \cdot 2D' \cdot 3D \cdot 34H_2O$, resulting in an idealized stoichiometry—when all cages are filled—of $34/6 = 3.5 : 1$. Until the advent of studies above 1 GPa (see Sect. 3.2 below), sH clathrates were thought only to exist when a small molecule such as methane is associated with a molecule larger than 7 Å, such as *n*-butane or higher hydrocarbons. Then, the small molecule occupies the 5^{12} and $4^35^66^3$ cages, while the large molecule is restricted to the largest $5^{12}6^8$ cages (Ripmeester and Ratcliffe 1990). As described shortly, recent studies have found that sH is stable over a large pressure range; provided the temperature profile inside a large icy body is sufficiently cool, sH may well be abundant.

3.2 High-Pressure Structures

3.2.1 Structure T (sT)

This crystal is tetragonal ($P4_2/mnm$, typical cell parameters, $a \approx 6.4$ Å, $c \approx 10.6$ Å, and $V \approx 430$ Å^3) and comprises a single cage type, a tetrakaidecahedron (denoted T′, $4^25^86^4$); the structure may be viewed as a space-filling packing of these polyhedra, or alternatively as puckered pentagonal sheets, stacked A-B–A-B, and related to one another by a mirror plane. The bonding between the sheets forms the square and hexagonal rings of the T′-cages. The composition of sT-clathrate may be represented as $T' \cdot 12H_2O$; in the case of argon, the T′ cages contain four Ar atoms, resulting in an ideal stoichiometry of $12/4 = 3 : 1$.

3.2.2 Structure O (sO)—A Form of 'Stuffed' Ice

In addition to the traditional cage-type clathrates, there are hydrate compounds which are referred to as 'stuffed' ices (Loveday et al. 2001b). A number of molecules, such as H_2, He, and Ne, are sufficiently small that they are able to penetrate the interstices of ice phases Ih, Ic, and II without greatly distorting the ice structure (Londono et al. 1992; Lobban et al. 2002). Alternatively, they may be described as *dissolving* in the ice structures. These structures can be thought of as analogous to those stuffed silica structures like the feldspathoid leucite. The presence of small atoms in the cavities of ices Ih and Ic, and in the hexagonal tubes of ice II has a marked effect on the elastic properties of the structures; simply put, they become much stiffer. In addition, the phase behaviour is altered; formation of He-clathrate prevents ices III and V from forming, the ice II-like hydrate being the liquidus phase above ~200 MPa.

Methane clathrate has been found to form a stuffed ice structure at high pressures. Having a body-centred-orthorhombic structure (hence, structure 'O') with a stoichiometry of 2:1 (Loveday et al. 2001b; Hirai et al. 2001), MH-sO is known to be remarkably stable, persisting up to 86 GPa (Machida et al. 2006).

3.3 High-Pressure Behaviour

Methane clathrate, sI, is not stable at ambient conditions; at 273 K, a pressure of 2.6 MPa is necessary to stabilize the structure, and at room temperature this increases to ~100 MPa. MH-sI is only stable at ambient pressure below 193 K (Falabella 1975). Prior to the mid-1990s it was generally believed that MH-sI would dissociate into ice VI plus liquid methane at pressures in excess of ~1 GPa (e.g. Lunine and Stevenson 1987). Experimental studies upon a range of clathrate hydrate at pressure approaching 100 GPa have now revealed a series of polymorphic phase transitions (Hirai et al. 2000, 2001, 2003, 2004; Loveday et al. 2001a, 2001b, 2003; Shimizu et al. 2002; Machida et al. 2006; Ogienko et al. 2006, Choukroun et al. 2007; Loveday and Nelmes 2008). Neutron and X-ray diffraction experiments, and calorimetric and spectroscopic measurements do not necessarily tell the same story concerning the sequence of high-pressure phase transitions, and disagreements remain. The X-ray and neutron diffraction studies, based on the variations in unit-cell dimensions, suggest that sI clathrates *might* experience the following series of transitions (dashed line marked (1) in Fig. 3):

$$\text{sI} \rightarrow \text{sII?} \rightarrow \text{sH} \rightarrow \text{sO} \quad \text{(Fig. 3)}$$

where the question mark indicates uncertainty concerning the behaviour at relatively high temperatures, nearer dissociation; at room temperature, sI transforms directly to sH (Loveday et al. 2003), as discussed in detail below. Clathrates which crystallize in sII at low pressures may experience the following series of transitions (see e.g., Dyadin et al. 1997; Loveday et al. 2003; Manakov et al. 2004; Hirai et al. 2004):

$$\text{sII} \rightarrow \text{sH} \rightarrow \text{sT} \rightarrow \text{sO}$$

It is noteworthy that studies using Raman spectroscopy (Hirai et al. 2001; Shimizu et al. 2002; Kumazaki et al. 2004; Choukroun et al. 2007) suggest a different evolution for sI clathrate hydrates such as methane. These studies suggest that sI clathrate hydrates transform directly into sH clathrates around 1 GPa, based on a reported phase transition. Such a high-pressure phase diagram would suggest a different dissociation curve, which is shown as path 2 in Fig. 3. One might note that, despite a slight curvature of the dissociation data from Dyadin et al. (1997), the difference in temperature between both curves is on the maximum order of 2 K. No uncertainty on temperature measurement is found in Dyadin et al. (1997), we could thus not assess whether the discrepancy is real or related to the accuracy of the measurements. As to a sII methane clathrate, generation of such a metastable structure has been reported at low pressures (below 100 MPa), in regions of the samples that are very rich in methane (Schicks and Ripmeester 2004; Choukroun et al. 2007). It has been shown that this structure later converts back into sI, but its metastability may allow sII to persist in the samples up to at least 500 MPa (Choukroun et al. 2007).

In Raman spectra, the types of cages and occupancies can be inferred from the signature of entrapped methane (e.g. Sum et al. 1997). This property is unique to methane clathrates, because the radius of the CH_4 molecule is so close to that of the small cages of sI clathrates that different levels of compression exist between methane molecules in small and in large

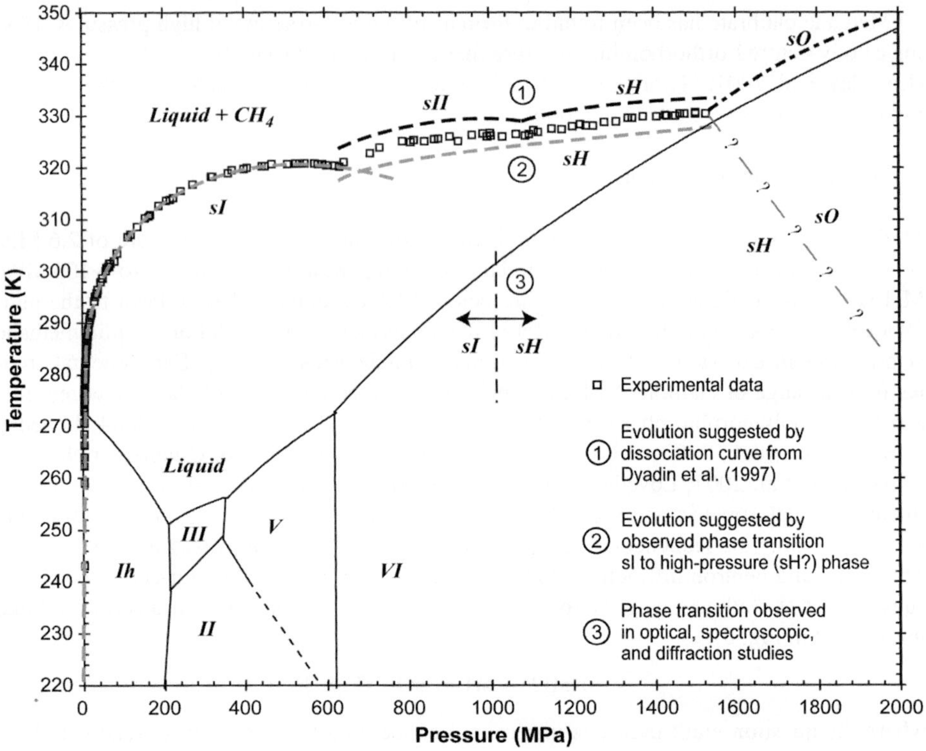

Fig. 3 The *P–T* phase diagram for methane clathrate hydrate, compared to that of H_2O (*solid lines*, from Chizhov 1993, and Choukroun and Grasset 2007). *Dashed lines* correspond to the dissociation curve of methane clathrate hydrates. Above 0.6 GPa, two possible paths have been identified, numbered 1 and 2; the corresponding curves have been shifted in temperature for clarity. Above 1.5 GPa, the dissociation curve of "stuffed-ice" sO is speculative based upon that of Ar clathrate hydrate sO polymorph (Manakov et al. 2004). The transition with number 3 indicates the phase transition form sI to sH reported by Loveday et al. (2001a, 2001b), Hirai et al. (2001), Shimizu et al. (2002), Kumazaki et al. (2004), and Choukroun et al. (2007). See text for details

cages. These induce a change in bond length at equilibrium, hence a change in vibrational energy and frequency, as can be detected by Raman spectroscopy. The high-pressure Raman signature of methane clathrates (see Shimizu et al. 2002; Kumazaki et al. 2004; Choukroun et al. 2007) is subject to abrupt changes in peak positions upon phase transitions. The peak positions are very similar between sI and metastable sII methane clathrate from 1–500 MPa (Choukroun et al. 2007), and the signature of methane trapped in large cages remains at 2905 cm^{-1}. In the phase formed by compression of sI, the signature of methane in small cages remains at the same position, whereas a jump is observed in that of CH_4 in large cages (Shimizu et al. 2002; Choukroun et al. 2007). This is hard to reconcile with the sII clathrate structure. The high-pressure structure may have 5^{12} small cages, but the large cages must be much closer in size to the small 5^{12}, rather like the $4^3 5^6 6^3$ dodecahedral cages in sH hydrate. So, at present, it seems that the transition form sI to sH at ~1 GPa, then to the stuffed ice or "filled-ice" structure sO at ~2 GPa, (path 2/phase transition 3 in Fig. 3), is the favoured evolution. Yet, it must be noted that the dissociation data obtained by Dyadin et al. (1997) is the only dataset at temperatures higher than 320 K. This means that, if a high-pressure sII phase did exist over a very narrow temperature range, as is the case of ice III, the P-T-t path followed by Loveday et al. (2001b), Hirai et al. (2001), Shimizu et al. (2002), or Choukroun et al. (2007) would not have allowed its formation. Further high-pressure studies in the ranges of temperature 300–320 K and pressure 600–900 MPa would clarify which structures do exist.

The experiments described above have yielded values for the incompressibility of MH-sI (~7.1 GPa at 277 K) in agreement with sound speed measurements (Waite et al. 2000), and good values for the incompressibility of MH-sO. The number of observations of MH-sII are too small (and the densities do not agree well between data sets) for the bulk modulus to be measured.

The thermal expansivity of MH-sI is known from 2 K to 200 K from experiments and lattice dynamical calculations (Shpakov et al. 1998; Gutt et al. 2000; Ogienko et al. 2006). The volume coefficient of thermal expansion at 200 K is $\sim 300 \times 10^{-6}\ K^{-1}$, which is approximately three times the expansivity of water ice Ih at the same temperature. The thermal expansivities of the other phases are unmeasured.

Thermal conductivity of MH-sI was measured by Krivchikov et al. (2005), confirming the earlier studies suggesting a glassy behavior; the thermal conductivity is extremely low, and comparable to that of amorphous water ice. Preliminary studies of the rheology of MH-sI were published by Durham et al. (2003), showing that this material is ~10 times stronger than water ice. The consequence of these last two studies is that a layer of clathrate in a planetary crust will be more insulating and more resistant to convective overturn than a similar layer of water ice.

3.4 Planetary Implications

Clathrate hydrates are abundant on Earth (e.g., Buffet 2000). These are dominantly clathrates of methane (produced either biogenically in sediments or abiogenically by serpentinization of oceanic crust), which occur on the ocean floors, and in permafrost terrains. Clathrates of air (N_2, O_2) also form in the terrestrial ice sheets (e.g. Chazallon and Kuhs 2002), and it is likely that CO_2 clathrates occur on Mars (e.g. Kargel and Lunine 1998, and references therein). However, it is in the outer solar system where these minerals may be particularly important.

During planetary accretion, clathrates may have trapped large quantities of gas directly into protoplanetary condensates, particularly at and beyond the orbit of Saturn (e.g. Lunine and Stevenson 1985; Gautier and Hersant 2005; Alibert and Mousis 2007). Prior to the high-pressure studies described earlier, it was believed that methane clathrates which had been accreted into Titan were dissociated, and virtually all of the methane (save for some trapped in the colder inner core) was out-gassed. The cooling of a primordial, dense, hot atmosphere in contact with an ocean of water (± ammonia?), saturated with methane gas, may have led to the formation of a crust of methane clathrate upon cooling (e.g. Lunine et al. 1989), but the general belief was that much of the methane reservoir would exist as a surface ocean many kilometers deep. Once it was established that methane clathrate could persist to high pressure in sH and sO phases, then it was proposed that Titan's primordial methane reservoir may not have been outgassed early in the satellite's history, being preserved to buffer the present atmospheric loss (Loveday et al. 2001a).

Given that the density of methane clathrate sI is lower than that of ice Ih, and all but the least dense aqueous solutions, a primordial clathrate-rich crust may persist for billions of years, providing a mechanically-resistant, low-conductivity shell (e.g. Lunine et al. 2009, and references therein). The effect of this clathrate shell would be: (i) to decrease in the cooling rate of the interior; (ii) to delay of the crystallization of ice Ih from the internal liquid layer until ~4 Gyr; and (iii) form a huge reservoir of methane, the dissociation of which may be able to sustain the continued presence of CH_4 in Titan's atmosphere for long periods of time (e.g. Tobie et al. 2006). Exactly how this de-gassing occurs is the subject of debate. The high-pressure arm on the phase diagram in Fig. 3 (representing the equilibrium between CH_4-clathrate + CH_4 gas and water + CH_4 gas) rises very steeply. As shown in Fig. 3, this dissociation curve lies far above the melting curve of pure water ice; hence clathrate hydrates are the most refractory of outer solar system minerals. Causing them to dissociate therefore requires either a great deal of heat, a reduction of pressure, or the presence of some material component which reduces the dissociation temperature. Choukroun et al. (2008, 2010) have argued that ammonia is just such an agent, allowing clathrates to be devolatilized by much more modest thermal anomalies, as may occur in cryomagmatic systems. Fortes et al. (2007b) have instead proposed that cryomagmas can entrain clathrate crystals and carry them towards the surface as xenoliths (equivalent to a reduction in hydrostatic pressure), a process which is commonplace in terrestrial magmatic systems. In this case, the clathrate xenoliths dissociate at some distance below the surface and give up their gas to the magma, forming bubbles and ultimately drive an explosive eruption. This model has also been applied to explain the geysers at the south pole of Enceladus (Fortes 2007). Both models have their strengths and weaknesses; the very specific scenario described in Fortes et al. (2007b) has a problem with the buoyancy of the proposed aqueous cryomagma (although the xenolith dissociation aspect of the model works equally well in any lower density aqueous system—see Fortes 2007), whilst Choukroun et al. (2008, 2010) require locally high amounts of ammonia within Titan's icy crust. Further experimental data and geophysical models are necessary to improve our current understanding of such processes.

4 Aqueous Ammonia

The water–ammonia system (Fig. 4) has long been of interest to planetary scientists, since astronomical observations and cosmochemical models indicate that ammonia may be a significant component of the outer solar system volatile inventory (e.g., Lewis 1971, 1972; Lewis and Prinn 1980; Prinn and Fegley 1981). Although there is some difference of opinion regarding the ammonia yield from various condensation models (Mousis et al. 2002), and indeed the fate of that ammonia once accreted into large bodies such as Titan (Sohl et al. 2003; Tobie et al. 2005; Fortes et al. 2007b; Grindrod et al. 2008; Choukroun et al. 2010), it remains amongst the most plausible planetary 'anti-freeze' agents, and its physical properties under the appropriate conditions (roughly 0–5 GPa, 100–300 K) must be known in order for it to be accommodated in planetary models. The pressure melting curve, and the expected polymorphism of the stoichiometric ammonia hydrates have implications for the internal structure of large icy moons like Titan, leading to phase layering and the possible persistence of deep subsurface oceans (Grasset and Sotin 1996; Grasset et al. 2000; Sohl et al. 2003; Grasset and Pargamin 2005; Tobie et al. 2005), the latter being sites of high astrobiological potential (Fortes 2000; Simakov 2001; Raulin 2008). Aqueous ammonia is also a candidate substance involved in cryomagmatism on Titan (Kargel 1992; Lopes et al. 2007), and again the melting behaviour, and densities of liquids and solids, in the ammonia–water system must be known to model properly the partial melting and propagation of magma through a planetary crust. Lately, experimental studies of the water–ammonia–methane system have begun (e.g., Kurnosov et al. 2006; Choukroun et al. 2008, 2010; Dong et al. 2009), and proper interpretation of the ternary system must rely on an accurate description of the binary water–ammonia system at high pressures.

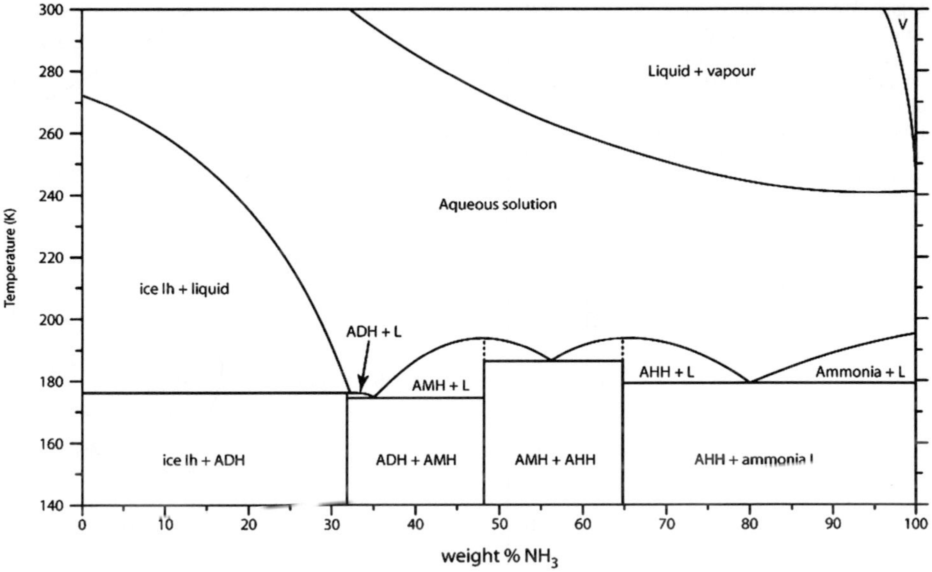

Fig. 4 The binary water–ammonia system at atmospheric pressure. ADH = ammonia dihydrate, AMH = ammonia monohydrate, and AHH = ammonia hemihydrate. Redrawn from data in Pickering (1893a), Rupert (1909, 1910), Smits and Postma (1910, 1911, 1914), Postma (1920), Elliot (1924), Clifford and Hunter (1933), Mironov (1955), Rollet and Vuillard (1956), Van Kasteren (1973), Kargel (1990), Yarger et al. (1993)

The behaviour of ammonia dihydrate and ammonia monohydrate at high pressure have been investigated extensively over the past twenty-five years; for the most part these were Raman and optical studies focused on establishing the high-pressure melting curves (Johnson et al. 1984, 1985; Johnson and Nicol 1987; Croft et al. 1988; Cynn et al. 1989; Boone and Nicol 1991; Kargel 1990; Hogenboom et al. 1989, 1994; Grasset and Sotin 1996; Leliwa-Kopystyński et al. 2002; Mousis et al. 2002; Grasset and Pargamin 2005) but also preliminary rheological studies (Durham et al. 1993). Despite the observation that ADH crystals grown under high-pressure do not exhibit the cubic symmetry of the low-pressure phase (Boone 1989), the general conclusion of the early diamond anvil cell studies was that neither ADH nor AMH exhibited evidence of high-pressure polymorphism (Boone 1989; Koumvakalis 1988). However, dilatometric studies subsequently established the existence of a new high-pressure polymorph in each system, ADH II stable above ~400 MPa at 170 K, and AMH II stable above 340 MPa at 195 K (Kargel and Hogenboom 1995; Hogenboom et al. 1995, 1997). Shortly afterwards, neutron powder diffraction patterns of both phases were reported (Nelmes and Loveday 1998; Nelmes et al. 1999) from studies using a gas pressure cell on the POLARIS diffractometer at ISIS. AMH has since been shown to possess a number of high-pressure polymorphs (Loveday and Nelmes 1999, 2004; Fortes et al. 2009b, 2009c).

Several recent computational and experimental studies of ammonia hydrates have been carried out, including a number of high-pressure studies of ammonia dihydrate (Fortes et al. 2003a, 2003b, 2007a; Fortes 2004). Most recently, the high-pressure neutron diffraction studies of ADH were summarized by Fortes et al. (2007a), and new experimental analysis presented in Fortes et al. (2009a) and in Loveday et al. (2009).

Phase relations in the binary system NH_3–H_2O (Figs. 4, 5) were investigated at ambient pressure by Pickering (1893a), Rupert (1909, 1910), Smits and Postma (1910, 1911, 1914),

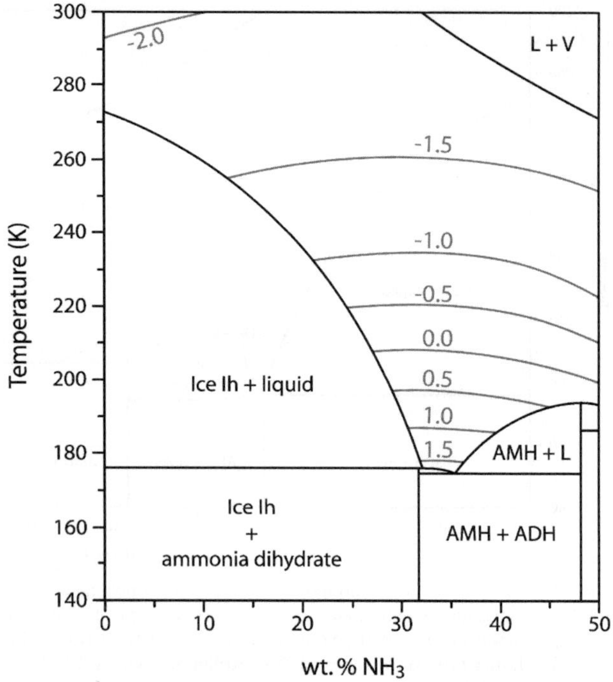

Fig. 5 Water-rich part of the ammonia–water system at atmospheric pressure, showing the increase in viscosity ($\log_{10}$ cP) of the liquid phase as a function of temperature. Drawn using data presented in Kargel et al. (1991)

Postma (1920), Elliot (1924), Clifford and Hunter (1933), Mironov (1955), Rollet and Vuillard (1956), Chan and Giauque (1964), Van Kasteren (1973), Kargel (1990), and Yarger et al. (1993). The system is characterised by three stoichiometric hydrates (in addition to the end members, water ice and solid ammonia), these are ammonia dihydrate (ADH), $NH_3{\cdot}2H_2O$, ammonia monohydrate (AMH), $NH_3{\cdot}H_2O$, and ammonia hemihydrate (AHH), $NH_3{\cdot}\frac{1}{2}H_2O$. Whilst the existence of the latter two hydrates was established very early, the high viscosity (see Fig. 5) and consequent vitrification of the liquid phase near the ice-ADH peritectic and AMH-ADH eutectic meant that the dihydrate was not identified until 1956 (Rollet and Vuillard 1956). The probable low abundance of ammonia relative to water (likely < 15 wt%) means that it is the properties and behaviour of the dihydrate with which we need to be concerned.

4.1 Ammonia Dihydrate I

The phase of ADH which crystallizes at atmospheric pressure (ADH I) melts incongruently at 176.13 K (Chan and Giauque 1964). The first structural study combined X-ray diffraction and infrared spectroscopy, concluding that ADH I was probably orientationally ordered in space group $P2_12_12_1$ (Bertie and Shehata 1984), in agreement with the low temperature heat capacity measurements (Chan and Giauque 1964), which indicated ordering of the hydrogen bonds in the low temperature limit. However, neutron powder diffraction showed that ADH I is cubic, space group $P2_13$ with $Z = 4$ (Loveday and Nelmes 2000, 2004). The three-fold site symmetry of the water molecule rules out an ordered structure under this space group, and so the water molecules in ADH I must be orientationally disordered. As in water ice, this disorder is represented crystallographically by partially occupied hydrogen sites. The zeolite-like heavy atom structure of ADH I is dominated by a strongly hydrogen bonded framework of water molecules which has broad channels along the *a*-, *b*-, and *c*-axes. Ammonia molecules are sited in these channels.

Fortes et al. (2003b) derived ferroelectric monoclinic and antiferroelectric orthorhombic ordered structures and carried out a first principles computational study. The energetically favoured orthorhombic structure was not observed in a subsequent neutron powder diffraction study (Fortes et al. 2003a) but, as with the ordered variants of many water-ice polymorphs, the addition of a dopant might cause the ordering transition to occur on laboratory timescales.

The melting curve of ADH I is known to be quite flat as a function of pressure. A polynomial fit through the data of Hogenboom et al. (1997) in the range $0 < P < 315$ MPa yields an extrapolated $T_m \approx 172(1)$ K at 460 MPa, depending on the order of polynomial fitted (compare 176.09 K at atmospheric pressure). Fortes et al. (2007a, 2009a) observed melting of deuterated ADH I on a metastable extension of this curve between 175–179 K at 460 MPa (bearing in mind that the deuterated system appears to be shifted to higher temperatures by ~3 K). These authors also measured the equation of state along the 175 K isotherm, from 41–450 MPa (curve in Fig. 1), reporting a bulk modulus of 7.5 GPa.

4.2 Ammonia Dihydrate II

ADH II has been formed repeatedly by compression of ADH I to ~450 MPa at 175 K. Neutron powder diffraction data has been used to establish that this phase is monoclinic (space-group $P2_1/n$) with $Z = 4$. It has also been observed that a distorted variant of this structure forms when the phase boundary is crossed very rapidly. The atomic structure of these phases has not been solved yet, although they are likely to differ only in hydrogen bond ordering. This phase has been observed over only a narrow range of pressures (440–550 MPa) so it is difficult to precisely determine its compressibility; initial measurements indicate a bulk modulus the same as the low-pressure phase. The thermal expansion has been measured, and found to be identical (within errors) to that of ADH I at the same temperature (Fortes et al. 2009a).

Hogenboom et al. (1997) observed the ADH I–II transition at temperatures from 150–175 K. However, the slightly negative sign of their T, P slope is consistent with similar work on the water ice phase diagram (Gromnitskaya et al. 2001) wherein increasingly sluggish kinetics below 200 K push observed phase boundaries to higher pressures; in other words, the observed phase boundary is probably not the *equilibrium* phase boundary. By analogy with ice, where the only significant entropy differences between phases are due to orientational ordering, we would suspect that the ADH I–II slope should be vertical if ADH II is orientationally disordered, and have a positive T, P slope if it is ordered.

ADH II appears not to be the liquidus phase over the pressure range in which it is observed to be stable. Warming of ADH II results in a reaction to form the high-pressure polymorph of ammonia *monohydrate*, AMH II, plus water ice II. Figure 6 shows the interpretation of the phase transitions observed to occur upon warming of an ADH + ice mixture at 550 MPa. Breakdown to a mixture of ice II + AMH II occurs at 188–190 K, followed by partial melting to ice II + liquid at 196 K. The breakdown of ADH II is probably a kinetically controlled process, limited by molecular diffusion; hence the observation merely indicates the temperature at which this process runs to completion on the timescale of our diffraction pattern integrations. Partial melting occurs at a simple binary eutectic in the system water–ammonia monohydrate. The shape of the ice liquidus has been inferred by fitting a quadratic expression to three points, these being the melting point of pure D_2O ice V at 550 MPa (271.06 K: Bridgman 1935), observation of the ice II liquidus at ~210 K, 30 wt.%

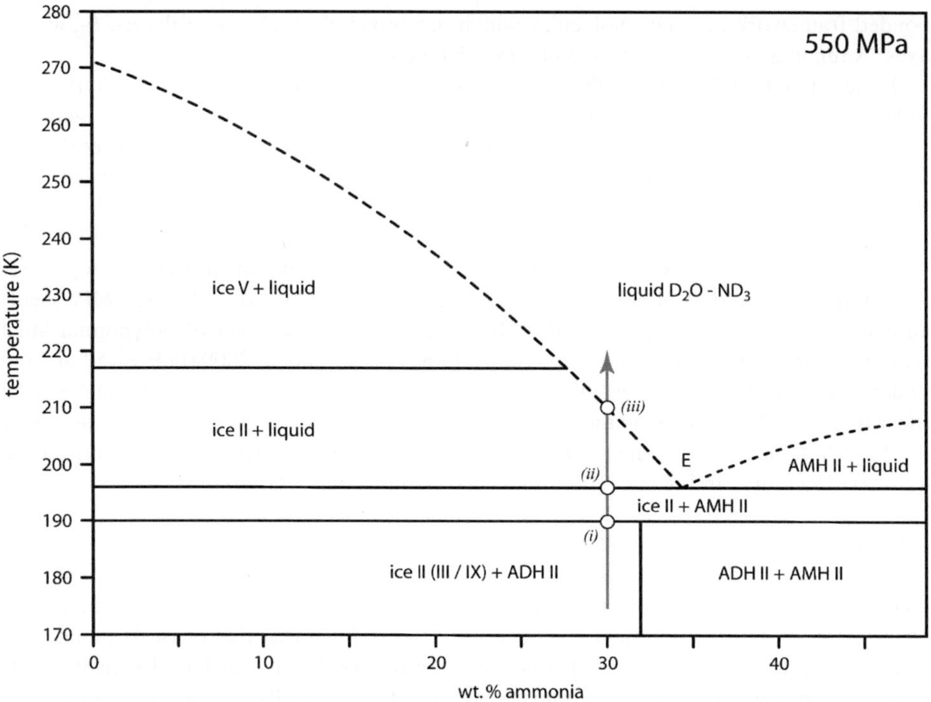

Fig. 6 Derived T–X slice through the P–T–X phase diagram at 550 MPa (Fortes et al. 2009a), showing phase boundaries consistent with the sequence of transitions observed on warming at the this pressure (*red line*), beginning with disproportionation of ADH (+ minor accessory ice) to ice II + AMH II at around 190 K (*i*), partial melting to ice II + liquid at 196 K (*ii*), and complete melting at ~210 K (*iii*)

ammonia (determined from the abundances of ADH and ice in the specimen) and an eutectic at 196 K, the composition of which was varied to give the closest agreement with the observed liquidus curvature in the data of Hogenboom et al. (1997) at 300 MPa. The resulting ice II (ice V above 217 K) liquidus is defined as $T_m = -0.0344X^2 - 1.0021X + 271.06$, where X is the wt.% ammonia concentration. N.B., the liquidus in the protonated system will be 1–2% lower in temperature. The eutectic composition is quite narrowly constrained by the plausible range of liquidus curvatures to be within ±1 wt.% of 34.4 wt.% ammonia.

N.B. This mixture of AMH II + ice II had previously (Fortes et al. 2007a) been interpreted as a single phase of ammonia dihydrate (ADH III), but this is clearly not the case, and ADH III is—temporarily—discredited.

4.3 Ammonia Dihydrate IV

Above 600 MPa, a new polymorph of ADH appears, which *is* the liquidus phase. This phase, ADH IV, appears to be the 'ice VI' of the ammonia dihydrate system, being stable over a very large pressure, and probably comprising a robust interpenetrating framework. ADH IV crystallizes in the orthorhombic system, and whilst the structure has not been solved, the thermoelastic properties have been determined (Fortes et al. 2009a). At temperatures below ~170 K, ADH IV persists to ~6 GPa, whereupon a body-centred-cubic phase forms (Fortes 2004; Fortes et al. 2007a; Loveday et al. 2009). This phase is orientationally and compositionally disordered, probably forming a solid solution series between the equivalent

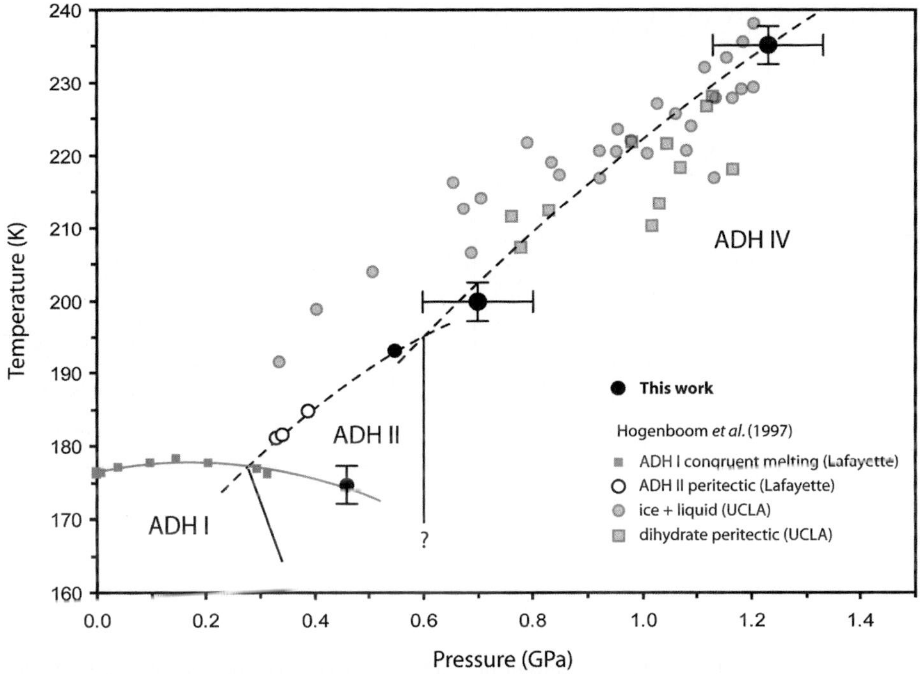

Fig. 7 Pressure dependence of the ADH curve, showing data published by Hogenboom et al. (1997) and the various observations of melting reported by Fortes et al. (2009a) (all reduced by 3 K, the likely difference between protonated and deuterated systems). The grey ADH I congruent melting line is a quadratic fit through the data of Hogenboom et al. (1997) and the metastable melting point at 460 MPa (experiment 1). The *dashed* ADH II and IV melting *lines* are rough guides to the eye. The ADH IV data includes the confirmed partial melting at 1.23 GPa as well as the observation of pressure freezing at ~0.70 GPa

BCC phase of ammonia monohydrate (Loveday et al. 1999), and ice VII. At higher temperatures however, ADH IV becomes unstable at lower and lower pressures, breaking down (as does ADH II) to a mixture of high-pressure ammonia monohydrate + ice. This dehydration reaction was observed in diamond anvil cell experiments by Boone (1989) and confirmed in high-pressure neutron diffraction studies by Fortes (2004). Boone (1989) reported optical studies of ammonia dihydrate up to ~7 GPa, including Raman spectra and images of crystals grown from the melt. No phase transitions were documented between ambient pressure and ~2.5 GPa, although it now appears there are several. Boone also attempted to acquire X-ray diffraction patterns at high-pressure but was not successful (S.C. Boone, personal communication). Interestingly, the optical observations of Boone's high-pressure crystals agree well with the indexing of the unit-cell of ADH IV from neutron powder diffraction patterns.

Hogenboom et al. (1997) published a melting curve up to ~1.2 GPa, yielding $T_m \approx 197$ K at 550 MPa and (extrapolated slightly) $T_m \approx 231$ K at 1.23 GPa (Fig. 7). Their curve was fitted on the assumption that a single liquidus phase exists over this pressure range, which we now know to be incorrect. It is plausible that the melting curve is inflected at ~600 MPa (above which ADH IV is stable) such that $T_m \approx 193$ K (protonated) at 550 MPa agrees quantitatively with our observation of the partial melting of AMH II + ice II (Fig. 6). Given the uncertainties in pressure calibration, the observation of ADH IV melting at 1.23 GPa agrees well with the measurements of Hogenboom et al. (1997).

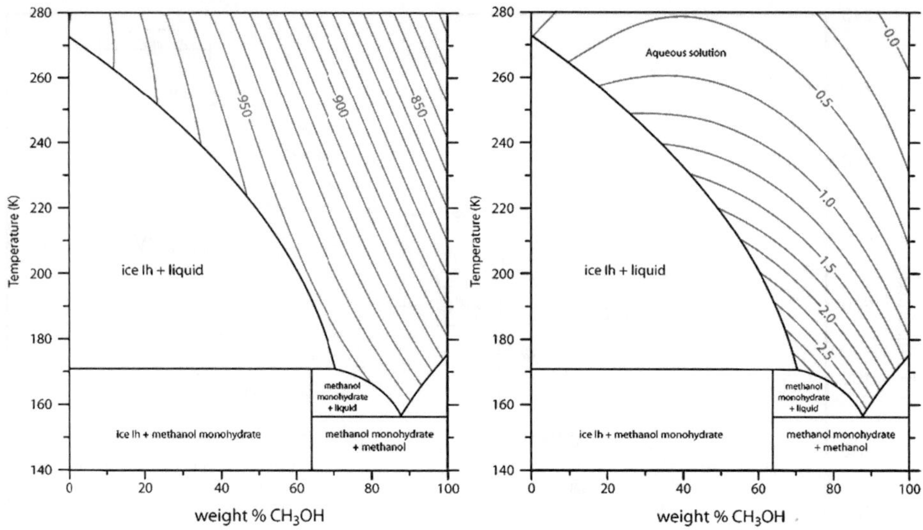

Fig. 8 The binary system methanol–water at atmospheric pressure, showing (*left*) the variation of density ($kg\,m^{-3}$) and (*right*) viscosity ($\log_{10}$ PaS), in the liquid phase (plotted from data in Yergovich et al. 1971)

The high-pressure instability of ammonia dihydrate may have consequences for the history of ammonia in Titan's deep interior, prior to core overturn, and may well be important in understanding the dynamics of exchange processes in larger extrasolar icy planets.

5 Aqueous Methanol

5.1 The Methanol–Water Binary System

The water–methanol system (Fig. 8) is of importance in biological and industrial chemistry (as an inhibitor of clathrate formation in gas pipelines; e.g., Sloan and Koh 2007), and is also of interest to cosmochemists studying ice grain reactions in the interstellar medium and the source of carbon in comets and other primitive solar nebula materials (e.g., Bernstein et al. 1995; Allamandola et al. 1999; Cruickshank 2005). Moreover, methanol may be an important constituent of aqueous cryovolcanic solutions (along with ammonia) on the icy moons of our solar system (Kargel 1992), although this interpretation remains highly controversial. Phase relations in the methanol–water system have been studied extensively (De Forcrand 1892; Pickering 1893b; Godefroy and Varenne 1904; Jones and Getman 1905; Baume and Borovsky 1914; Pushin and Glagoleva 1922; Benjamin 1932; Ewert 1937; Conrad et al. 1940; Ross 1954; Vuillard and Sanchez 1961; Miller and Carpenter 1964; Ott et al. 1979) revealing the existence of an incongruently melting intermediate compound, methanol monohydrate. The high liquidus viscosity in the vicinity of the eutectic meant that the stoichiometric monohydrate was not identified initially, being discovered by Vuillard and Sanchez (1961). A report that a hemihydrate exists (Franks and Ives 1966), and that its structure was solved by Pachler and Von Stackelberg (1963), are erroneous; the citation refers to pentamethylethanol hemihydrate $(CH_3)_5CH_3OH{\cdot}H_2O$. Suggestions that a dihydrate exists (e.g., Takamuku et al. 2005) are incorrect.

Mixtures of methanol monohydrate and water ice were investigated by X-ray powder diffraction (Nakayama et al. 1997; Yamamoto et al. 2000) and Nuclear Magnetic Resonance spectroscopy (Nakayama et al. 1997) which indicated a tetragonal unit cell with dimensions $a = 4.660(1)$ Å and $c = 13.813(5)$ Å at 80 K and two symmetry-independent methanol molecules. The NMR spectra exhibit a central line component above 140 K due to the onset of dynamic reorientation of water molecules in the structure. The structure of perdeuterated methanol monohydrate was established from neutron powder diffraction at 160 K (Fortes 2006). The crystal is actually orthorhombic, space-group $Cmc2_1$ $(Z = 4)$ $\rho_{calc} = 1265.55(2)$ $kg\,m^{-3}$ at 4.2 K, but coincidentally passes through a state of metric tetragonality at ~80 K, the temperature of the earlier X-ray study, by virtue of its anisotropic thermal expansion. The structure consists of water–water chains, linked by an ordered hydrogen bond, extending along the c-axis, which cross link methanol–water chains with disordered hydrogen bonds along the a-axis. These perpendicular water–water and methanol–water chains form sheets which are stacked parallel to the b-axis and which interact only through weaker contacts between the methyl hydrogens and hydroxyl oxygens of neighboring sheets. The volume thermal expansion of methanol monohydrate is exceptionally large $(5.27 \times 10^{-4}\ K^{-1}$ at 160 K, the greatest part of which is along the b-axis), an expansivity five times larger than water ice at the same temperature, a value more characteristic of liquids or liquid crystals. The very weak interaction between adjacent sheets, and the considerable thermal motion of the methyl groups, is likely to be responsible for the large thermal expansion.

It is probable that, under hydrostatic compression, methanol monohydrate will prove to be extremely compressible along the b-axis. If this is the case, then only a relatively modest pressure may be necessary to destabilize the structure (by driving the methyl groups too close together) with respect to a possible clathrate structure; in the ethanol-water system, a metastable semi-clathrate structure ($EtOH{\cdot}4\frac{3}{4}H_2O$) is found to become stable at pressures of just 250 bar. A type II clathrate of methanol ($CH_3OH{\cdot}17H_2O$) apparently forms when the mixture is vapour deposited under vacuum at very low temperatures (Blake et al. 1991; Williams and Devlin 1997; Murthy 1999; Bobev and Tait 2004).

In ammonia-bearing methanol–water mixtures, stoichiometric methanol ammoniates may crystallize (Kargel 1992); both methanol monoammoniate and methanol hemiammoniate are orthorhombic (Fortes et al. 2009d, 2010), and exhibit much more isotropic thermal expansion than methanol monohydrate.

6 Aqueous Salt Solutions

Hydrated sulfates are predicted to be major rock-forming minerals in the interiors of the icy Galilean moons of Jupiter, the result of leaching of chondritic sulfates from their cores during differentiation (Kargel 1990, 1991). This hypothesis is supported by the observation of a spectral signature attributable to highly hydrated sulfate salts on the surfaces of Europa and Ganymede by the Galileo spacecraft's Near Infrared Mapping Spectrometer (McCord et al. 1999, 2001; Dalton 2003; Dalton et al. 2005). Since these hydrated sulfate salts may experience hydrostatic pressures up to 1.5 GPa in the interiors of the largest icy moons, there has been interest in measuring their solubilities and densities (described below), as well as studying the rheology of eutectic salt + ice mixtures (e.g., McCarthy et al. 2007).

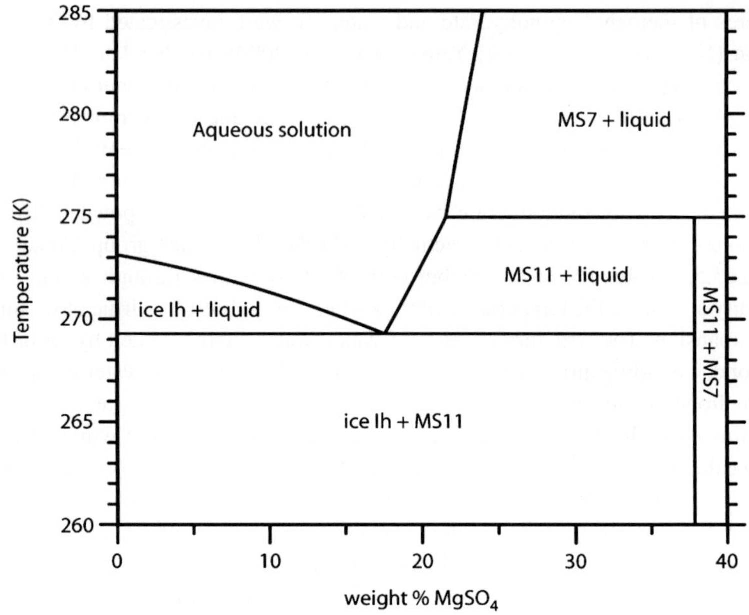

Fig. 9 The water-rich end of the binary system $MgSO_4$–water at atmospheric pressure; MS7 = $MgSO_4 \cdot 7H_2O$ (epsomite), and MS11 = $MgSO_4 \cdot 11H_2O$ (meridianiite)

6.1 $MgSO_4$–Water Binary System

This binary system, shown in Fig. 9 contains two hydrates in the temperature regime of interest to icy moons. Magnesium sulfate undecahydrate—$MgSO_4 \cdot 11H_2O$—is the most water-rich hydrate at ambient pressure; the synthetic substance is referred to as Fritzsche's salt, after its discoverer (Fritzsche 1837), and the naturally occurring substance is the mineral meridianiite (Peterson et al. 2007). For the sake of brevity, we refer to the title substance throughout as MS11, and the other hydrates in the $MgSO_4$–H_2O system as MSx, where x is the number of water molecules bound into the structure (e.g., MS7 for $MgSO_4 \cdot 7H_2O$, epsomite). The phase relations of this substance were first established by Frederick Cottrell circa 1900 (cited in Van't Hoff et al. 1901): as depicted in Fig. 9, MS11 melts incongruently at 1.8°C (274.95 K) to MS7 and a peritectic liquid containing 21.6 wt.% $MgSO_4$. The eutectic between MS11 and ice Ih is at −3.9°C (269.25 K), 17.5 wt.% $MgSO_4$. Thus the stability field is rather small and, with a few exceptions (Cardoso 1930; Polo et al. 1971; Himawan 2002; Gärtner et al. 2005), little work has been done in this narrow region of the phase diagram. Indeed, until very recently, the higher hydrate was believed to be a dodecahydrate, referred to widely as 'MS12' in consequence. However, Peterson and Wang (2006) solved the structure from single crystal X-ray diffraction data and determined that the unit cell contains only eleven water molecules. MS11 has been identified as a naturally occurring terrestrial mineral on frozen brine ponds in British Columbia, Canada (Peterson et al. 2007), and it is probably relatively common in some periglacial environments. MS7 (epsomite, $MgSO_4 \cdot 7H_2O$), is a widespread evaporite mineral on Earth, being the stable phase in contact with a saturated solution at temperatures from 275–320 K.

At the pressures relevant to the interiors of icy satellites, little work has been done on either MS7 or MS11. The earliest high-pressure work on epsomite was carried out by

Bridgman (1948a, 1948b) upon polycrystalline and single crystal epsomite at pressures up to 4 GPa in a piston-cylinder apparatus. Bridgman observed a sluggish phase transition between 1.0–1.5 GPa, and suggested that there were two further sluggish transitions at ~2.5 GPa. The piston compression method was subsequently employed by Livshits et al. (1963) in work on crystalline hydrates of magnesium sulfate. These authors claimed to have observed the following series of phase transitions; I $\leftrightarrow$ II (~0.45 GPa); II $\leftrightarrow$ III (~1.2 GPa); III $\leftrightarrow$ IV (~1.6 GPa); and IV $\leftrightarrow$ V (~2.5 GPa). Of these, the II $\leftrightarrow$ III transition may correspond to Bridgmans' first phase change, and the IV $\leftrightarrow$ V transition to one of Bridgmans' 2.5 GPa phase changes. Livshits et al. (1963) also reported weak evidence for breaks in their pressure–volume curves at ~0.2 GPa and at 0.7–0.8 GPa.

More recently, a number of groups have studied melting relations in the H_2O–$MgSO_4$ system (Hogenboom et al. 1995, to 400 MPa; Grasset et al. 2001a, 2001b, to ~2 GPa; Nakamura 2003, to ~5 GPa) and in more complex ternary systems considered applicable to Europa's subsurface ocean and icy crust; e.g., H_2O–$MgSO_4$–H_2SO_4 (Hogenboom et al. 2001). Of these studies, Grasset et al. (2001a) noted, by visual observation, a probable phase transition in epsomite at 600 MPa. *In situ* neutron powder diffraction studies were begun by Fortes et al. (2006b, 2006c), and this work identified a series of polymorphic phase transitions in the 1–2.5 GPa region followed by a probable dehydration reaction above 3 GPa (all at room temperature). Gromnitskaya et al. (2010) have since reported piston-cylinder measurements of the density and ultrasonic wave velocities through epsomite as a function of pressure in the range $0 < P < 2.8$ GPa, at temperatures from 77–295 K. This work has identified a similar series of phase transitions to those reported by Bridgman (1948b).

In the case of MS11, there is only the ambient temperature neutron diffraction study (of the deuterated phase) by Fortes et al. (2008a). These authors have carried out some difficult high-pressure studies (unpublished), and have concluded that MS11 undergoes a dehydration reaction to a lower hydrate + ice VI when compressed above ~1 GPa at 270 K. The volume change associated with this reaction is of the order of a few percent, supporting the earlier hypotheses that Ganymede's global expansion and subsequent extensional tectonism may be explained by the dehydration of MS11 (see Day et al. 2002; Hogenboom et al. 2002).

6.2 Na_2SO_4–Water Binary System

Sodium sulfate decahydrate—$Na_2SO_4{\cdot}10H_2O$—is the stable phase in contact with an equilibrium mixture of Na_2SO_4 and H_2O at room temperature and pressure (Fig. 10); the synthetic substance is referred to as Glauber's salt, and the naturally occurring substance is the mineral mirabilite. The eutectic between mirabilite and ice Ih is at 271.85 K, 4.15 wt.% Na_2SO_4, and the solubility increases rapidly with temperature up to the peritectic at 305.534 K, 33.2 wt.% Na_2SO_4 where mirabilite undergoes incongruent melting to anhydrous sodium sulfate (orthorhombic phase V, Fddd, thenardite). This is the highest dissociation temperature amongst the isostructural decahydrates of Na_2SO_4, Na_2SeO_4, Na_2WO_4, Na_2VO_4, or Na_2MoO_4. Above this point, the solubility is retrograde, reaching a minimum near 400 K. A metastable phase, $Na_2SO_4{\cdot}7H_2O$, is known at room pressure and temperature (Hamilton and Hall 2008). The metastable heptahydrate–ice Ih eutectic is at 269.60 K, 12.8 wt.% Na_2SO_4, and the incongruent melting point of this phase is at 296.615 K.

Since these hydrated sulfate salts may experience hydrostatic pressures up to 1.5 GPa in the interiors of the largest icy moons, there has been interest recently in measuring the solubility and density of mirabilite at non-ambient conditions (e.g., Hogenboom et al. 1999; Dougherty et al. 2006) to extend earlier high-pressure investigations into the pressure dependence of the ice-mirabilite eutectic (Tammann 1903; Block 1913;

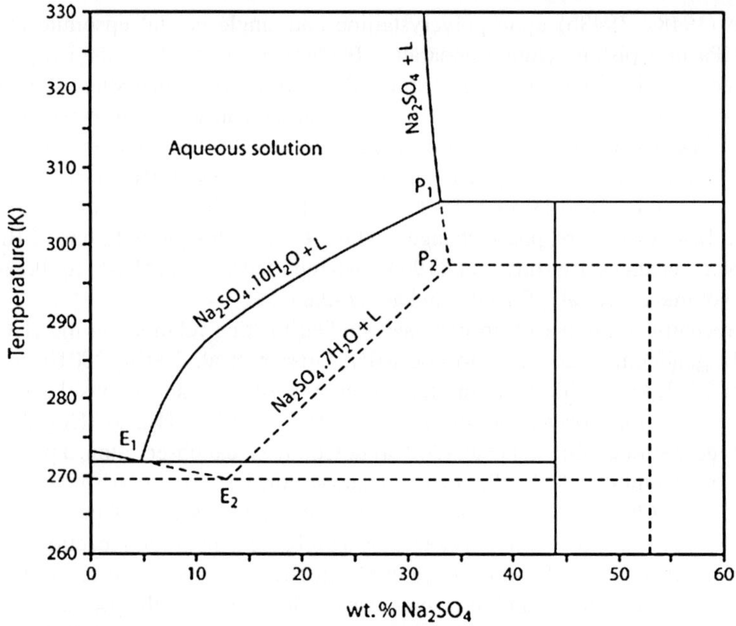

Fig. 10 *T–X* phase diagram of the binary system sodium sulfate–water at room pressure showing stable phase boundaries (*solid*) and metastable phase boundaries (*dashed*). E_1 and E_2 are the mirabilite–ice and the $Na_2SO_4{\cdot}7H_2O$–ice eutectics, respectively. P_1 and P_2 are the peritectics $Na_2SO_4{\cdot}10H_2O(s) \leftrightarrow Na_2SO_4(s) + \text{liquid (L)}$ and $Na_2SO_4{\cdot}7H_2O(s) \leftrightarrow Na_2SO_4(s) + L$ respectively. The *solid vertical line* at 44.09 wt.% Na_2SO_4 corresponds to the composition of mirabilite, and the *dashed vertical line* at 52.97 wt.% corresponds to the heptahydrate. Redrawn after Negi and Anand (1985) with solubility data tabulated in Garrett (2001; Table 7.1)

Geller 1924; Tamman 1929; Tanaka et al. 1992; Kryukov and Manikhin 1960). These various experiments covered the range from 0–800 MPa, and 263–343 K; only Hogenboom et al. (1999) observed any evidence of another solid phase, which they identified tentatively as the heptahydrate, although Geller (1924) reported a branch in the peritectic melting point above 500 MPa. It has since been established using the diamond anvil cell that the phase which crystallizes at 1.54 GPa is an octahydrate (Oswald et al. 2008). There are clearly complex changes in hydration state as well (probably) as conventional polymorphism in this system, which require further investigation. Neutron powder diffraction has been used to study the behavior of mirabilite at atmospheric pressure, as a function of temperature (Brand et al. 2009), and at pressures up to 500 MPa. *Ab initio* calculations (Brand et al. 2010) also reveal evidence of polymorphism at multi-GPa pressures brought about by changes in the hydrogen bonding scheme.

6.3 $(NH_4)_2SO_4$–H_2O Binary System

Ammonium sulfate occurs naturally on Earth as the mineral mascagnite, typically associated with guano deposits. Ammonium sulfate has also been proposed as a major rock-forming mineral in Titan's interior, formed by the interaction of primordial aqueous ammonia with sulfate-bearing solutions leached from the rocky core (Fortes et al. 2007b; Grindrod et al. 2008). It could be argued that this model results from an extreme scenario, where complete leaching of all silicates occurred during accretion and/or differentiation, and leached sulfates

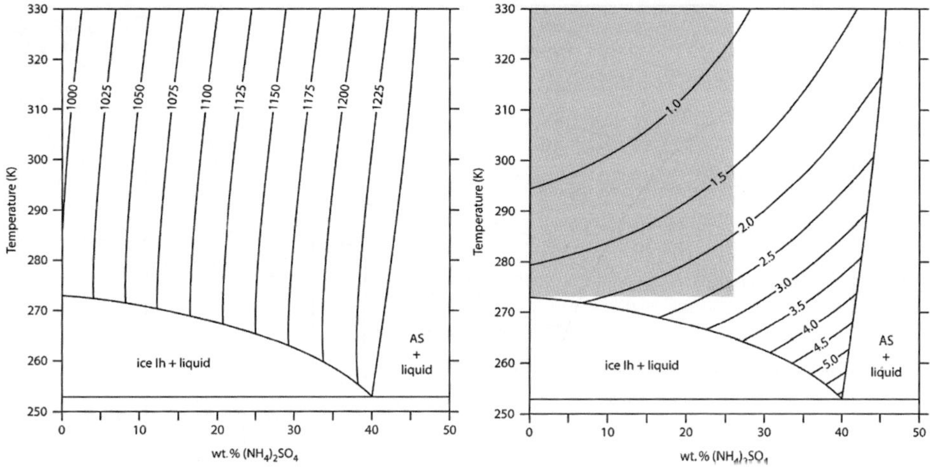

Fig. 11 The water-rich half of the binary $(NH_4)_2SO_4$–H_2O system at atmospheric pressure, redrawn after Xu et al. (1998). The liquidus field on the left is contoured to show the density (kg m^{-3}) of the aqueous solution (Novotný and Söhnel 1988). The liquidus field on the right is contoured to show the viscosity (mPa sec) of the aqueous solution, derived by fitting a polynomial function to data in Sprung (1876) over the range 0–25.5 wt%, 0–60°C (this region, with good data coverage, is shaded grey) and to data in the CRC Handbook (Lide 2006) over the range 0—saturation at 25°C

subsequently reacted with all available putative ammonia to form ammonium sulfates. Yet, the presence of this compound to some extent within Titan cannot be ruled out, and its potential implications in Titan's dynamics remain worthy of exploration.

Ammonium sulfate is well known for the first-order phase transition which occurs upon cooling through ~224 K (at room pressure) from a para-electric crystal (space group P*nam*, $Z = 4$) to a ferro-electric crystal (space-group P$na2_1$, $Z = 4$) (see Hasebe 1981, and references therein). Structurally, the transition is characterized by a change in the hydrogen bond scheme. In the para-electric phase, many of the hydrogen atoms are involved in relatively weak three- and four-centred interactions (i.e., bifurcated and trifurcated H-bonds), whilst in the ferroelectric phase these are replaced with predominantly 'normal' hydrogen bonds, which are much stronger. This strongly exothermic transition is accompanied by large lambda-like discontinuities in the elastic compliances and the heat capacity. The pressure dependence of the transition temperature has been investigated under hydrostatic stress up to 600 MPa (Tsunekawa et al. 1972; Polandov et al. 1972), and under uniaxial stress up to 100 MPa (Jerzak 2002). All three experiments yield $dT_c/dP \approx -4.1 \times 10^{-2}$ K MPa^{-1}. Neither experiment measured the volume change across the transition at high-pressure, which is necessary to evaluate the entropy change from the Clausius-Clapeyron relation. Thermal models of Titan's interior (Grindrod et al. 2008) indicate that the 'geotherm' will intersect the ferro- to para-electric transition in $(NH_4)_2SO_4$ at a depth of ~30 km; hence, this phase transition may have a significant effect on convective flow in the crust. Finally, the possible existence of other higher-pressure polymorphs has consequences for the density profile and heat transport of Titan's deeper mantle. Such phases might also occur in impact craters, being preserved by rapid quenching to ambient temperatures (e.g., 90 K on Titan's surface).

The ambient pressure binary system is shown in Fig. 11; the liquidus field has been contoured to show the values of, respectively, density (left) and viscosity (right), based on polynomial fits to literature data (see figure caption). There have been no high pressure studies of the liquid–solid equilibria in the $(NH_4)_2SO_4$–H_2O system. Also of interest in this system

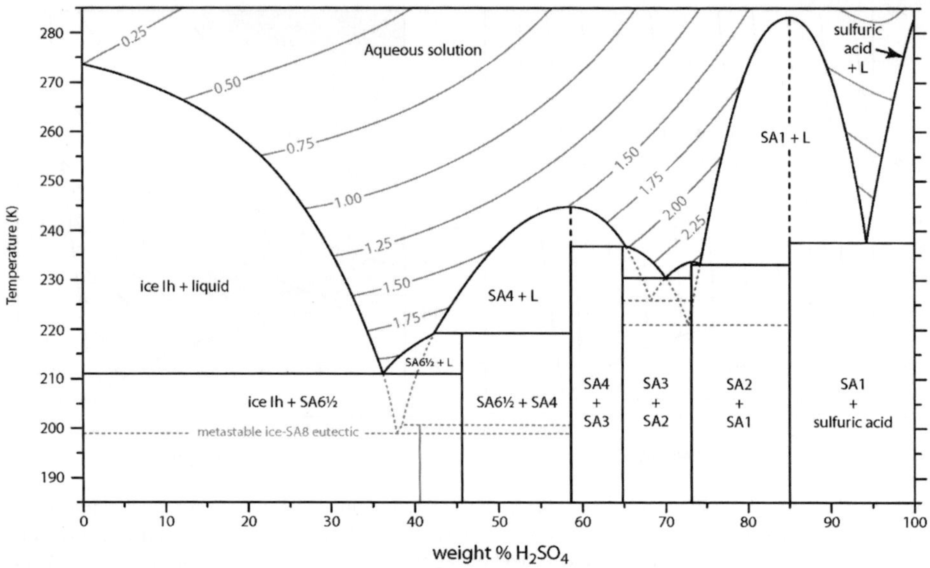

Fig. 12 The binary system H_2SO_4–water at atmospheric pressure, the viscosity of the liquidus field being contoured in units of $\log_{10}$ Pa s (from data in Williams and Long 1995). Redrawn after Beyer et al. (2003), and references therein. *Dashed grey lines* represent metastable phase boundaries. SAx labels indicate the phases $H_2SO_4{\cdot}xH_2O$

is the behaviour of the suggested intermediate compound, ammonium sulfate tetrahydrate—$(NH_4)_2SO_4{\cdot}4H_2O$, or AST—which was identified recently by Xu et al. (1998); see also Hazra et al. (2003). At ambient pressure, AST can apparently not co-exist with the liquid phase, forming by solid state reaction of anhydrous ammonium sulfate with ice below the eutectic temperature. The high-pressure behaviour of this compound remains unknown, and particularly how its presence might affect the phases of ice that may co-exist with a highly concentrated solution.

Questioning which phase of ice will be stable in concentrated aqueous solutions at high pressure is relevant because of the dramatic difference in transport properties between ices II and III/IX. One would expect ice II to be the equilibrium phase below 240 K; however, we have observed crystallization of ice IX in aqueous ammonia at low temperature (Fortes et al. 2003a) and it is possible that slight differences in the structure of the liquid can suppress the nucleation of the stable phase in favour of the metastable phase (see the work regarding the use of nucleating agents in the formation of stable and metastable high-pressure ices by Evans 1967). Metastability of aqueous solutions and compounds is enormous at the laboratory scale; one can speculate if such metastability may play a role in the structure and the dynamics of icy satellites.

7 Aqueous Sulfuric Acid

Europa is known to harbour a liquid ocean beneath a thick icy crust (e.g., Kargel et al. 2000 and references therein). There is good evidence from the NIMS spectrometer, carried to Jupiter aboard the Galileo spacecraft, that large portions of Europa's surface are coated with hydrates of sulfuric acid (Carlson et al. 2005), and it is possible that the crust is underplated with thick deposits of water-rich sulfuric acid hydrates (e.g., Pappalardo and Barr 2003; McKinnon and Zolensky 2003).

The sulfuric acid–water system (Fig. 12) is of interest, not least from a crystal chemical point of view, but also because it plays a role in the formation of aerosols in the terrestrial stratosphere (e.g., Zhang et al. 1993; Bertram et al. 1996), and is likely to be of relevance to understanding acidic groundwaters on Mars. The binary system has been investigated by Pickering (1890), Giran (1913), Hulsmann and Biltz (1934), Gable et al. (1950), Rubin and Giauque (1952), Kunzler and Giauque (1952), Hornung and Giauque (1955), Hornung et al. (1956), and Giauque et al. (1960). High pressure (up to 200 MPa) phase equilibria between ice and $H_2SO_4{\cdot}6\frac{1}{2}H_2O$ were investigated by Hogenboom et al. (2001).

At the water-rich end of the H_2O–H_2SO_4 phase diagram, likely to be applicable to the subsurface ocean of Europa, there are two solid hydrates; a hemitriskaidecahydrate ($H_2SO_4{\cdot}6\frac{1}{2}H_2O$) and an octahydrate ($H_2SO_4{\cdot}8H_2O$). The existence of the $6\frac{1}{2}$-hydrate is well established, whereas there has been some dispute in the literature as to whether or not a higher hydrate (specifically, an octahydrate) even exists. Beyer et al. (2003) demonstrated that there is a higher hydrate, but that it may in fact be a hemi*enneakaideca*hydrate ($H_2SO_4{\cdot}9\frac{1}{2}H_2O$). Beyer et al. (2003) give an account of the dispute regarding the existence of the higher hydrate. Neutron diffraction studies of this hydrate by one of us (ADF) indicate strongly that it is in fact an octahydrate, as originally conceived, although this solid is relatively difficult to produce, due to its metastability with respect to a mixture of the $6\frac{1}{2}$-hydrate and water ice.

The original structural data for the two hydrates is sourced from the single crystal X-ray diffraction studies of Mootz and Merschenz-Quack (1987). The $6\frac{1}{2}$-hydrate is proposed to be monoclinic (space group Cm, $Z = 4$), with all hydrogens ordered. The 8-hydrate is proposed to be proton disordered in space group $I\bar{4}2d$ ($Z = 4$). Hirsch and Ojamäe (2004) carried out a density functional theory (DFT) study of these two sulfuric acid hydrates, proposing new proton ordering schemes, which were subsequently tested—in the case of the $6\frac{1}{2}$-hydrate—using neutron powder diffraction by Fortes et al. (2006a).

No information exists at present concerning the high pressure behaviour of these two water rich hydrates. Fortes et al. (2008b) carried out a study of the high pressure behaviour of perdeuterated sulfuric acid tetrahydrate ($D_2SO_4{\cdot}4D_2O$), which shares some structural similarities with the octahydrate. These authors measured the thermoelastic properties of the tetrahydrate in the range 0–550 MPa, and discovered a new monoclinic polymorph of $D_2SO_4{\cdot}4D_2O$, which is stable above 235 K at 550 MPa. This work places constraints on the slope of the melting curve, and suggests that the tetrahydrate (and also the other water-rich hydrates) will exhibit a rich polymorphism, comparable at least to ammonia dihydrate or even water ice.

8 Summary

As the timing of geological events on the icy satellites of the gas giants has been clarified by higher-resolution mapping, it has become apparent that the evolutionary differences between icy satellites are due to three major considerations:

(1) Formation timescale (i.e., accretion rates), which determines the importance of short-lived radioisotopes (^{60}Fe, ^{26}Al). Bodies taking more than a few million years to form miss out on this source of primordial heating.

(2) Convective or conductive loss of heat from long-lived radiogenic isotopes (^{232}Th, ^{235}U, ^{40}K). Here, the nature of the heat loss mechanism depends upon the phase behaviour, thermoelastic and transport properties of the ice/hydrate layer (see below).
(3) Tidal evolution, giving rise to significant internal heating by dissipation of viscoelastic energy, driven by orbital resonances with neighbouring satellites.

The end product of the evolution of icy satellites is recorded in the geological features on their surfaces, and is a function of some or all of these three factors. As described in subsequent chapters, interpretation of the observed surface morphologies of the icy satellites and understanding of their thermal evolution, requires construction of geophysical models of their interiors. Addressing this problem requires the knowledge of phase behavior described in this chapter.

All of the materials described in this chapter exist in a number of crystal structures over the expected P–T range inside the icy satellites. The density contrasts between the different ice polymorphs are very large ($\Delta V/V$ up to 20%) in comparison to those observed in silicate minerals. In addition, all of these materials are very compressible compared with silicates. As a result, both water ice and ammonia dihydrate, for example, undergo density increases of $\sim$35% in the range $0 < P < 2$ GPa; achieving a similar densification in the silicate minerals in the Earth's mantle requires a pressure of $\sim$100 GPa. The pressure-induced contrasts in other physical properties are also large, varying from factors of a few (for thermal conductivity or thermal expansion) to orders of magnitude (for viscosity). However, temperature induced contrasts are also important, far more so than inside the Earth. In the Earth's interior, temperatures are much greater than the characteristic Debye temperatures (θ_D) of silicate minerals, and so their thermal properties vary by $<$15% across the entirety of the mantle. The temperature regime inside icy moons is at or far below θ_D for the constituent ices and there are consequent large changes in thermal properties, such as the volume thermal expansion (α_V), specific heat capacity (C_P), and thermal conductivity (K). For example, C_P of water ice drops by 60% as the temperature is reduced from 273–100 K. By themselves, the phase boundaries between different polymorphs, with their large changes in physical properties, can influence heat flow in an icy mantle by dividing it into a series of thinner shells defined by each polymorph's thermodynamic stability. Within each shell both conductive and convective heat transport will be affected by the thermoelastic and transport properties of each polymorph. Hence, accurate modeling is strongly dependent on accurate laboratory measurements.

Acknowledgements ADF is supported by an Advanced Fellowship from the UK Science and Technology Facilities Council (STFC), grant number PP/E006515/1. MC is supported by a NASA Postdoctoral Fellowship, administered by Oak Ridge Associated Universities. Part of this work has been conducted at the Jet Propulsion Laboratory, California Institute of Technology. Government sponsorship acknowledged.

References

Y. Alibert, O. Mousis, Astron. Astrophys. **465**, 1051–1060 (2007)
L.J. Allamandola, M.P. Bernstein, S.A. Sandford, R.L. Walker, Space Sci. Rev. **90**, 219–232 (1999)
P. Andersson, R.G. Ross, G. Backstrom, J. Phys. C **13**(4), L73–L76 (1980)
O. Andersson, A. Inaba, Phys. Chem. Chem. Phys. **7**(7), 1441–1449 (2005)
G.P. Arnold, E.D. Finch, S.W. Rabideau, R.G. Wenzel, J. Chem. Phys. **49**(10), 4365–4369 (1968)
G. Baume, W. Borovsky, J. Chim. Phys. **12**, 276–281 (1914)
R. Benjamin, Recherches sur le coefficient d'activité thermodynamique dans les solutions aqueuses de composés organiques. Thèse. Bruxelles (1932)

M.P. Bernstein, S.A. Sandford, L.J. Allamandola, S. Chang, M.A. Scharberg, Astrophys. J. **454**, 327–344 (1995)
J.E. Bertie, M.R. Shehata, J. Chem. Phys. **81**(1), 27–30 (1984)
A.K. Bertram, D.D. Patterson, J.J. Sloan, J. Phys. Chem. **100**(6), 2376–2383 (1996)
K.D. Beyer, A.R. Hansen, M. Poston, J. Phys. Chem. A **107**(12), 2025–2032 (2003)
D. Blake, L. Allamandola, S. Sandford, D. Hudgins, F. Freund, Science **254**, 548–551 (1991)
E.A. Block, Z. Phys. Chem. Stoich. Verwandtschaftslehre **82**, 403–438 (1913)
S. Bobev, K.T. Tait, Am. Mineral. **89**, 1208–1214 (2004)
S.C. Boone, Mixtures of ammonia and water and of methane and water at high pressures. Ph.D Thesis. University of California, Los Angeles (1989)
S. Boone, M.F. Nicol, Proc. Lunar Planet. Sci. Conf. **21**, 603–612 (1991)
H.E.A. Brand, A.D. Fortes, I.G. Wood, K.S. Knight, L. Vočadlo, Phys. Chem. Miner. **36**(1), 29–46 (2009)
H.E.A. Brand, A.D. Fortes, I.G. Wood, L. Vočadlo, Phys. Chem. Miner. **41** (2010). doi:10.1007/s00269-009-0331-1
P.W. Bridgman, Proc. Am. Acad. Arts Sci. **47**, 441–558 (1912)
P.W. Bridgman, J. Chem. Phys. **3**(10), 597–605 (1935)
P.W. Bridgman, J. Chem. Phys. **5**(12), 964–966 (1937)
P.W. Bridgman, Proc. Am. Acad. Arts Sci. **76**, 71–87 (1948a)
P.W. Bridgman, Proc. Am. Acad. Arts Sci. **76**, 89–99 (1948b)
B. Buffet, Ann. Rev. Earth Planet. Sci. **28**, 477–507 (2000)
G.M. Cardoso, Los modernos métodos roentgenográficos aplicados en la determinación de la estuctura cristalina de la epsomita. Trabajos del Museo Nacional de Ciencias naturales., Serie Geológica **37**, 5–133. Madrid (1930)
R.W. Carlson, M.S. Anderson, R. Mehlman, R.E. Johnson, Icarus **177**(2), 461–471 (2005)
J.P. Chan, W.F. Giauque, J. Phys. Chem. **68**(10), 3053–3057 (1964)
B. Chazallon, W.F. Kuhs, J. Chem. Phys. **117**(1), 308–320 (2002)
V.E. Chizhov, J. Appl. Mech. Tech. Phys. **34**(2), 253–262 (1993)
M. Choukroun, O. Grasset, J. Chem. Phys. **127**(12), 124506 (2007)
M. Choukroun, Y. Morizet, O. Grasset, J. Raman Spectrosc. **38**, 440–451 (2007)
M. Choukroun, O. Grasset, G. Tobie, C. Sotin, Lunar Planet. Sci. Conf. (2008) **39**, Abstract 1837
M. Choukroun, O. Grasset, G. Tobie, C. Sotin, Icarus **205**(2), 581–593 (2010). doi:10.1016/j.icarus.2009.08.011
I.L. Clifford, E. Hunter, J. Phys. Chem. **37**(1), 101–118 (1933)
F.H. Conrad, E.F. Hill, E.A. Ballman, Ind. Eng. Chem. **32**(4), 542–543 (1940)
S.K. Croft, J.I. Lunine, J.S. Kargel, Icarus **73**(2), 279–293 (1988)
D.P. Cruickshank, Space Sci. Rev. **116**, 421–439 (2005)
H.C. Cynn, S. Boone, A. Koumvakalis, M. Nicol, D.J. Stevenson, Proc. Lunar Planet. Sci. **19**, 433–441 (1989)
J.B. Dalton III, Astrobiology **3**(4), 771–784 (2003)
J.B. Dalton III, O. Prieto-Ballesteros, J.S. Kargel, C.S. Jamieson, J. Jolivet, R. Quinn &, Icarus **177**(2), 472–490 (2005)
S. Day, E. Asphaug, L. Bruesch, EOS, Transactions of the American Geophysical Union, vol. 84, no 46, Fall Meeting Supplement, Abstract P72B-0507 (2002)
R. De Forcrand, Ann. Chim. Phys. **6**(27), 525–540 (1892)
T. Dong, L. Wang, A. Liu, X. Guo, Q. Ma, G. Li, Q. Sun, Petroleum Sci. **6**(2), 188–193 (2009). doi:10.1007/s12182-009-0030-2
A.J. Dougherty, D.L. Hogenboom, J.S. Kargel, Y.F. Zheng, Lunar Planet. Sci. Conf. XXXVII, abstract 1732 (2006)
W.B. Durham, S.H. Kirby, L.A. Stern, J. Geophys. Res. **98**(B10), 17667–17682 (1993)
W.B. Durham, S.H. Kirby, L.A. Stern, W. Zhang, J. Geophys. Res. **108**(B4), 2182 (2003)
W.B. Durham, O. Prieto-Ballesteros, D.L. Goldsby, J.S. Kargel, Sp. Sci. Rev. (2010, this issue)
Y.A. Dyadin, Ya.E. Aladko, E.G. Larionov, Mendeleev Commun. **1**, 34–35 (1997)
L.D. Elliot, J. Phys. Chem. **28**(8), 887–888 (1924)
H. Engelhardt, B. Kamb, J. Chem. Phys. **75**(12), 5887–5899 (1981)
L.F. Evans, J. Appl. Phys. **38**(12), 4930–4932 (1967)
M. Ewert, Bull. Soc. Chim. Belge **46**(1), 90–103 (1937)
B.J. Falabella, A study of natural gas hydrates. Dissertation, University of Massachusetts, 1975
R. Feistel, W. Wagner, J. Chem. Ref. Data **35**(2), 1021–1047 (2006)
A.D. Fortes, Icarus **146**(2), 444–452 (2000)
A. Fortes, Computational and experimental studies of solids in the ammonia-water system. Ph.D Thesis, University of London (2004)

A.D. Fortes, Chem. Phys. Lett. **431**(4–6), 283–288 (2006)
A.D. Fortes, Icarus **191**(2), 743–748 (2007)
A.D. Fortes, I.G. Wood, K.S. Knight, J.P. Brodholt, M. Alfredsson, G.S. McGrady, L. Vočadlo, J. Chem. Phys. **119**(20), 10806–10813 (2003a)
A.D. Fortes, I.G. Wood, J.P. Brodholt, L. Vočadlo, Icarus **162**(1), 59–73 (2003b)
A.D. Fortes, I.G. Wood, M. Alfredsson, L. Vočadlo, K.S. Knight, J. Appl. Cryst. **38**(4), 612–618 (2005)
A.D. Fortes, I.G. Wood, K.S. Knight, J. Chem. Phys. **125**(14), 144510 (2006a)
A.D. Fortes, I.G. Wood, M. Alfredsson, L. Vočadlo, K.S. Knight, Eur. J. Mineral. **18**(4), 449–462 (2006b)
A.D. Fortes, I.G. Wood, L. Vočadlo, H.E.A. Brand, P.M. Grindrod, K.H. Joy, M.G. Tucker, Lunar Planet. Sci. Conf. **37**, abstract #1029 (2006c)
A.D. Fortes, I.G. Wood, M. Alfredsson, L. Vočadlo, K.S. Knight, W.G. Marshall, M.G. Tucker, F. Fernandez-Alonso, High Press. Res. **27**(2), 201–212 (2007a)
A.D. Fortes, P.M. Grindrod, S.K. Trickett, L. Vočadlo, Icarus **188**(1), 139–153 (2007b)
A.D. Fortes, I.G. Wood, K.S. Knight, Phys. Chem. Miner. **35**, 207–221 (2008a)
A.D. Fortes, I.G. Wood, L. Vočadlo, L. Chapon, K.S. Knight, R.I. Smith, J. Chem. Phys. **128**(5), 054506 (2008b)
A.D. Fortes, I.G. Wood, L. Vočadlo, K.S. Knight, W.G. Marshall, M.G. Tucker, F. Fernandez-Alonso, J. Appl. Cryst. **42**(5), 846–866 (2009a)
A.D. Fortes, E. Suard, M.-H. Lemée-Cailleau, C.J. Pickard, R.J. Needs, J. Am. Chem. Soc. **131**(37), 13508–13515 (2009b)
A.D. Fortes, E. Suard, M.-H. Lemée-Cailleau, C.J. Pickard, R.J. Needs, J. Chem. Phys. **131**(15), 154503 (2009c)
A.D. Fortes, I.G. Wood, K.S. Knight, J. Appl. Cryst. **42**(6), 1054–1061 (2009d)
A.D. Fortes, I.G. Wood, K.S. Knight, J. Appl. Cryst. **43** (2010, in press). doi:10.1107/S0021889810005595
F. Franks, D.J.G. Ives, Q. Rev. **20**, 1–44 (1966)
C.J. Fritzsche, Bulletin Sci. Publ. Acad. Imp. Sci. St.-Pétersb. **2**(13), 193–196 (1837)
H. Fukazawa, S. Ikeda, S. Oguro, T. Fukumura, S. Mae, J. Phys. Chem. B **106**(23), 6021–6024 (2002)
H. Fukazawa, A. Hishikawa, Y. Ishii, B.C. Chakamoukos, J.A. Fernandez-Baca, Astrophys. J. **652**(1), L57–L60 (2006)
C.M. Gable, H.F. Betz, S.H. Maron, J. Am. Chem. Soc. **72**(4), 1445–1448 (1950)
R.E. Gagnon, H. Kiefte, M.J. Clouter, E. Whalley, J. Chem. Phys. **92**(3), 1909–1914 (1990)
D.E. Garrett, *Sodium Sulphate: Handbook of Deposits, Processing, Properties, and Use* (Academic Press, San Diego, 2001)
R.S. Gärtner, F.E. Genceli, D.O. Trambitas, G.J. Witkamp, J. Cryst. Growth **275**(12), 1773–1778 (2005)
D. Gautier, F. Hersant, Space Sci. Rev. **116**, 25–52 (2005)
A. Geller, Z. Krist. **60**, 415–472 (1924)
W.F. Giauque, J.W. Stout, J. Am. Chem. Soc. **58**(7), 1144–1150 (1936)
W.F. Giauque, E.W. Hornung, J.E. Kunzler, T.R. Rubin, J. Am. Chem. Soc. **82**(1), 62–70 (1960)
H. Giran, C. R. Hebd. Séances Acad. Sci. **157**, 221–223 (1913)
L. Godefroy, E. Varenne, C. R. Hebd. Séances Acad. Sci. **138**, 990–992 (1904)
O. Grasset, J. Pargamin, Planet. Space. Sci. **53**, 371–384 (2005)
O. Grasset, C. Sotin, Icarus **123**(1), 101–112 (1996)
O. Grasset, C. Sotin, F. Dechamps, Planet. Space Sci. **48**(7–8), 617–636 (2000)
O. Grasset, C. Sotin, O. Mousis, L. Mevel, Proc. Lunar Planet. Sci. **31** abstract 1386 (2001a)
O. Grasset, L. Mevel, O. Mousis, C. Sotin, Proc. Lunar Planet. Sci. **31**, abstract 1524 (2001b)
P.M. Grindrod, A.D. Fortes, F. Nimmo, D.L. Feltham, J.P. Brodholt, L. Vočadlo, Icarus **197**(1), 137–151 (2008)
E.L. Gromnitskaya, O.V. Stal'gorova, V.V. Brazhkin, A.G. Lyapin, Phys. Rev. B **64**, 094205 (2001)
E.L. Gromnitskaya, O.F. Yagafarov, A.G. Lyapin, V.V. Brazhkin, A.D. Fortes, High Press. Res. (2010). doi:10.1080/08957951003588860
C. Gutt, B. Asmussen, W. Press, M.R. Johnson, Y.P. Handa, J.S. Tse, J. Chem. Phys. **113**(11), 4713–4721 (2000)
A. Hamilton, C. Hall, J. Anal. At. Spectrom. **23**, 840–844 (2008)
K. Hasebe, J. Phys. Soc. Jpn. **50**(4), 1266–1274 (1981)
A. Hazra, S. Paul, U.K. De, S. Bhar, K. Goswami, Progr. Cryst. Growth Charact. Mater. **2003**, 45–61 (2003)
C. Himawan, in Chemical Engineering Transactions, volume 1, Proceedings of the 15th International Symposium on Industrial Crystallisation (2002), pp 951–956
H. Hirai, T. Kondo, M. Hasegawa, T. Yagi, Y. Yamamoto, T. Komai, K. Nagashima, M. Sakashita, H. Fujihisa, K. Aoki, J. Phys. Chem. B **104**(7), 1429–1433 (2000)
H. Hirai, Y. Uchihara, H. Fujihisa, M. Sakashita, E. Katoh, K. Aoki, K. Nagashima, Y. Yamamoto, T. Yagi, J. Chem. Phys. **115**(15), 7066–7070 (2001)

H. Hirai, T. Tanaka, T. Kawamura, Y. Yamamoto, T. Yagi, Phys. Rev. B **68**, 172102 (2003)

H. Hirai, T. Tanaka, T. Kawamura, Y. Yamamoto, T. Yagi, J. Phys. Chem. Solids **65**, 1555–1559 (2004)

T.K. Hirsch, L. Ojamäe, Acta Crystallogr., B Struct. Sci. **60**, 179–183 (2004)

P.V. Hobbs, *Ice Physics* (Clarendon Press, Oxford, 1974)

D.L. Hogenboom, J. Winebrake, G.J. Consolmagno, W. Dalrymple III, Proc. Lunar Planet. Sci. **20**, 420 (1989)

D.L. Hogenboom, J.S. Kargel, T.C. Holden, J. Ganasan, Proc. Lunar Planet. Sci. **25**, 555–556 (1994)

D.L. Hogenboom, J.S. Kargel, J.P. Ganasan, L. Lee, Icarus **115**(2), 258–277 (1995)

D.L. Hogenboom, J.S. Kargel, G.J. Consolmagno, T.C. Holden, L. Lee, M. Buyyounouski, Icarus **128**(1), 171–180 (1997)

D.L. Hogenboom, J.S. Kargel, P.V. Pahalawatta, Lunar Planet. Sci. Conf. XXX, abstract 1793 (1999)

D.L. Hogenboom, J.S. Kargel, M.E. Daly, Lunar Planet. Sci. Conf. **32**, abstract 1739 (2001)

D.L. Hogenboom, J.S. Kargel, M.L. Reiter, Y.N. Khor, Lunar Planet. Sci. Conf. XXXIII, abstract 1638 (2002)

E.W. Hornung, W.F. Giauque, J. Am. Chem. Soc. **77**(11), 2983–2987 (1955)

E.W. Hornung, T.E. Bracket, W.F. Giauque, J. Am. Chem. Soc. **78**(22), 5747–5751 (1956)

R. Howe, R.W. Whitworth, J. Chem. Phys. **90**(8), 4450–4453 (1989)

O. Hulsmann, W. Biltz, Z. Anorg. Allg. Chem. **218**(4), 369–378 (1934)

H. Hussmann, G. Choblet, D.L. Matson, C. Sotin, G. Tobie, T. Van Hoolst, Space Sci. Rev. (2010, this issue)

G.A. Jeffrey, Cryst. Rev. **9**(2–3), 135–176 (2003)

S. Jerzak, Ferroelectrics **275**, 19–27 (2002)

G.P. Johari, J. Chem. Phys. **118**(1), 242–248 (2003)

M.L. Johnson, M. Nicol, J. Geophys. Res. **92**, 6339–6349 (1987)

M.L. Johnson, A. Schwake, M. Nicol, Proc. Lunar Planet. Sci. **15**, 405–406 (1984)

M.L. Johnson, A. Schwake, M. Nicol, in *Ices in the Solar System*, ed. by J. Klinger et al. (ed.) (Reidel, Dordrecht, 1985), pp. 39–47

H.C. Jones, F.H. Getman, Am. Chem. J. **32**, 308–338 (1905)

B. Kamb, Science **150**, 205–207 (1965)

J.S. Kargel, Cryomagmatism in the outer solar system. PhD Thesis, University of Arizona (1990)

J.S. Kargel, Icarus **94**(2), 369–390 (1991)

J.S. Kargel, Icarus **100**(2), 556–574 (1992)

J.S. Kargel, D.L. Hogenboom, Proc. Lunar Planet. Sci. **26**, 725–726 (1995)

J.S. Kargel, J.I. Lunine, Clathrate hydrates in the solar system, in *Solar System Ices*, ed. by B. Schmitt, C. de Bergh, M. Festou (1998)

J.S. Kargel, S.K. Croft, J.I. Lunine, J.S. Lewis, Icarus **89**(1), 93–122 (1991)

J.S. Kargel, J.Z. Kaye, J.W. Head, G.M. Marion, R. Sassen, J.K. Crowley, O. Prieto-Ballesteros, S.A. Grant, D.L. Hogenboom, Icarus **148**, 226–265 (2000)

J. Klinger et al. (ed.), *Ices in the Solar System* (Reidel, Dordrecht, 1985)

H. König, Z. Krist. **105**, 279–246 (1943)

A. Koumvakalis, High pressure study of ammonia monohydrate. Ph.D Thesis, University of California, Los Angeles (1988)

A.I. Krivchikov, B.Ya. Gorodilov, O.A. Korolyuk, V.G. Manzhelii, H. Conrad, W. Press, J. Low Temp. Phys. **139**(5–6), 693–702 (2005)

P.A. Kryukov, V.I. Manikhin, Russ. Chem. Bull. **9**(12), 2077–2078 (1960)

W.F. Kuhs, M.S. Lehmann, J. Phys. Chem. **87**(21), 4312–4313 (1983)

W.F. Kuhs, J.L. Finney, C. Vettier, D.V. Bliss, J. Chem. Phys. **81**(8), 3612–3623 (1984)

T. Kumazaki, Y. Kito, S. Sasaki, T. Kume, H. Shimizu, Chem. Phys. Lett. **388**, 18–22 (2004)

J.E. Kunzler, W.F. Giauque, J. Am. Chem. Soc. **74**(21), 5271–5274 (1952)

A. Kurnosov, L. Dubrovinsky, A. Kuznetsov, V. Dmitriev, Z. Naturforsch. B **61**(12), 1573–1576 (2006)

S.J. La Placa, W.C. Hamilton, B. Kamb, A. Prakash, J. Chem. Phys. **58**, 567–580 (1973)

A.J. Leadbetter, R.C. Ward, J.W. Clark, P.A. Tucker, T. Matsuo, H. Suga, J. Chem. Phys. **82**(1), 424–428 (1985)

A. Léger, F. Selsis, C. Sotin, T. Guillot, D. Despois, D. Mawet, M. Ollivier, A. Labèque, C. Valette, F. Brachet, B. Chazelas, H. Lammer, Icarus **169**(2), 499–504 (2004)

J. Leliwa-Kopystyński, M. Maruyama, T. Nakjima, Icarus **159**(2), 518–528 (2002)

J.S. Lewis, Icarus **15**(2), 174–185 (1971)

J.S. Lewis, Icarus **16**(2), 241–252 (1972)

J.S. Lewis, R.G. Prinn, Astrophys. J. **238**, 357–364 (1980)

D.R. Lide (ed.), *CRC Handbook of Chemistry and Physics*, 86th edn. (Chemical Rubber Company Press, London, 2006)

C.M.B. Line, R.W. Whitworth, J. Chem. Phys. **104**(24), 10008–10013 (1996)

L.D. Livshits, Yu.S. Genshaft, Yu.N. Ryabin, Russ. J. Inorg. Chem. **8**, 676–678 (1963)

C. Lobban, J.L. Finney, W.F. Kuhs, Nature **391**, 268–270 (1998)

C. Lobban, J.L. Finney, W.F. Kuhs, J. Chem. Phys. **112**(16), 7169–7180 (2000)
C. Lobban, J.L. Finney, W.F. Kuhs, J. Chem. Phys. **117**(8), 3928–3934 (2002)
T. Loerting, I. Kohl, C. Salzmann, E. Mayer, A. Hallbrucker, J. Chem. Phys. **116**(8), 3171–3174 (2002)
J.D. Londono, J.L. Finney, W.F. Kuhs, J. Chem. Phys. **97**(1), 547–552 (1992)
R.M.C. Lopes et al., Icarus **186**(2), 395–412 (2007)
P. Loubeyre, R. LeToullec, E. Wolanin, M. Hanfland, D. Hausermann, Nature **397**, 503–506 (1999)
J.S. Loveday, R.J. Nelmes, Phys. Rev. Lett **83**(21), 4329–4332 (1999)
J.S. Loveday, R.J. Nelmes, in *Science and Technology of High Pressure: Proceedings of AIRAPT-17*, ed. by M.H. Manghnani, W.J. Nellis, M.T. Nicol (Universities Press, Hyderabad, 2000), pp. 133–136
J.S. Loveday, R.J. Nelmes, High Press. Res. **24**(1), 45–55 (2004)
J.S. Loveday, R.J. Nelmes, Phys. Chem. Chem. Phys. **10**(7), 937–950 (2008)
J.S. Loveday, M. Guthrie, R.J. Nelmes, Structural changes under pressure in ammonia dihydrate and ammonia hemihydrate. *ISIS Facility Experimental Report* (POLARIS) RB 9859. Rutherford Appleton Laboratory, RAL-TR-1999-050 (1999)
J.S. Loveday, R.J. Nelmes, M. Guthrie, S.A. Belmonte, D.R. Allan, D.D. Klug, J.S. Tse, Y.P. Handa, Nature **410**, 661–663 (2001a)
J.S. Loveday, R.J. Nelmes, M. Guthrie, D.D. Klug, J.S. Tse, Phys. Rev. Lett. **87**, 215501 (2001b)
J.S. Loveday, R.J. Nelmes, D.D. Klug, J.S. Tse, S. Desgreniers, Can. J. Phys. **81**(1–2), 539–544 (2003)
J.S. Loveday, R.J. Nelmes, C.L. Bull, H.E. Maynard-Casely, M. Guthrie, High. Press. Res. **29**(3), 396–404 (2009)
J.I. Lunine, D.J. Stevenson, Astrophys. J. Suppl. Ser. **58**, 493–531 (1985)
J.I. Lunine, D.J. Stevenson, Icarus **70**(1), 61–77 (1987)
J.I. Lunine, S.K. Atreya, J.B. Pollack, in *Origin and Evolution of Planetary and Satellite Atmospheres* (University of Arizona Press, Tucson, 1989)
J.I. Lunine, M. Choukroun, D.J. Stevenson, G. Tobie, The origin and evolution of Titan, in *Titan from Cassini-Huygens*. Springer (accepted)
S. Machida, H. Hirai, T. Kawamura, Y. Yamamoto, T. Yagi, Phys. Earth Planet. Int. **155**, 170–176 (2006)
A.Yu. Manakov, V.I. Voronin, A.V. Kurnosov, A.E. Teplykh, V.Yu. Komarov, Yu.A. Dyadin, J. Incl. Phenom. Macrocycl. Chem. **48**, 11–18 (2004)
V.G. Manzhelii, Yu.A. Freiman, M.L. Klein, A.A. Maradudin, *Physics of Cryocrystals* (AIP, New York, 1996)
C. McCarthy, R.F. Cooper, S.H. Kirby, K.D. Rieck, L.A. Stern, Am. Mineral. **92**(10), 1550–1560 (2007)
T.B. McCord, G.B. Hansen, D.L. Matson, T.V. Johnson, J.K. Crowley, F.P. Fanale, R.W. Carlson, W.D. Smythe, P.D. Martin, C.A. Hibbitts, J.C. Granahan, A. Ocampo, J. Geophys. Res. **104**, 11827–11852 (1999)
T.B. McCord, G.B. Hansen, C.A. Hibbitts, Science **292**, 1523–1525 (2001)
W.B. McKinnon, in *Solar System Ices*, ed. by B. Schmitt, C. de Bergh, M. Festou (Kluwer, Dordrecht, 1998), pp. 525–550
W.B. McKinnon, M.E. Zolensky, Astrobiology **3**(4), 879–897 (2003)
G.A. Miller, D.K. Carpenter, J. Chem. Eng. Data **9**, 371–373 (1964)
K.E. Mironov, Zh. Obschei Khimii **25**(6), 1081–1086 (1955)
O. Mishima, Phys. Rev. Lett. **85**(2), 334–336 (2000)
D. Mootz, A. Merschenz-Quack, Z. Naturforsch., B **42**(10), 1231–1236 (1987)
O. Mousis, J. Pargamin, O. Grasset, C. Sotin, Geophys. Res. Lett. **29**(24), 2192 (2002)
S.S.N. Murthy, J. Phys. Chem. A **103**(4), 7927–7937 (1999)
R.S. Nakamura, Eos Trans. AGU **84**(46), Fall Meeting Suppl., abstract P51B-0449 (2003)
H. Nakayama, D.H. Brouer, Y.P. Handa, D.D. Klug, J.S. Tse, C.I. Ratcliffe, J.A. Ripmeester, Am. Chem. Soc. Div. Fuel Chem. **42**(1), 516–520 (1997)
A.S. Negi, S.C. Anand, *A Textbook of Physical Chemistry* (Wiley, New York, 1985)
R.J. Nelmes, J.S. Loveday, Diffraction studies of the low-pressure phases of ammonia hydrates. ISIS Experimental Report RB9411. CCLRC Rutherford Appleton Laboratory (1998)
R.J. Nelmes, J.S. Loveday, M. Guthrie, Structural changes under pressure in ammonia dihydrate and ammonia hemihydrate. ISIS Experimental Report RB 9859, CCLRC Rutherford Appleton Laboratory (1999)
K. Nishibata, Jpn. J. Appl. Phys. **11**(11), 1701–1708 (1972)
P. Novotný, O. Söhnel, J. Chem. Eng. Data **33**(1), 49–55 (1988)
A.G. Ogienko, A.V. Kurnosov, A.Y. Manakov, E.G. Larionov, A.I. Anchorov, M.A. Sheremov, A.N. Nesterov, J. Phys. Chem. B **110**, 2840–2846 (2006)
I.D.H. Oswald, A. Hamilton, C. Hall, W.G. Marshall, T.J. Prior, C.R. Pulham, J. Am. Chem. Soc. **130**, 17795–17800 (2008)
J.B. Ott, J.R. Goates, B.A. Waite, J. Chem. Thermodyn. **11**, 739–746 (1979)
K. Pachler, M. von Stackelberg, Z. Kristalogr. **119**, 15–29 (1963)
R.T. Pappalardo, A.C. Barr, Bull. Am. Astron. Soc. **35**, 919 (2003)

R.C. Peterson, R. Wang, Geology **34**(11), 957–960 (2006)
R.C. Peterson, W. Nelson, B. Madu, H.F. Shurvell, Am. Mineral. **92**(10), 1756–1759 (2007)
V.F. Petrenko, R.W. Whitworth, *Physics of Ice* (Oxford University Press, Oxford, 1999)
S.U. Pickering, J. Chem. Soc., Trans. **57**, 331–369 (1890)
S.U. Pickering, J. Chem. Soc., Trans. **63**, 141–195 (1893a)
S.U. Pickering, J. Chem. Soc., Trans. **63**, 998–1027 (1893b)
I.N. Polandov, V.P. Mylov, M.E. Levina, Russ. J. Phys. Chem. **46**(2), 281 (1972)
M. Polo, N. Gérard, M. Lallemont, C.R. Hebd. Séances Acad. Sci., Sér. C, Sci. Chim. **272**, 642–645 (1971)
S. Postma, Rec. Trav. Chim. Pays-Bas **39**, 515–536 (1920)
R.G. Prinn, B. Fegley Jr., Astrophys. J. **249**, 308–317 (1981)
N.A. Pushin, A.A. Glagoleva, J. Chem. Soc., Trans. **71**, 2813–2822 (1922)
F. Raulin, Space Sci. Rev. **135**(1–4), 37–48 (2008)
J.A. Ripmeester, C.I. Ratcliffe, J. Phys. Chem. **94**(25), 8773–8776 (1990)
A.P. Rollet, G. Vuillard, C. R. Acad. Sci. Paris **243**, 383–386 (1956)
H.K. Ross, Ind. Eng. Chem. **46**(3), 601–610 (1954)
R.G. Ross, P. Andersson, G. Backstrom, Nature **259**, 553–554 (1976)
R.G. Ross, P. Andersson, G. Bäckström, High Temp. High Press. **9**, 87–96 (1977)
T.R. Rubin, W.F. Giauque, J. Am. Chem. Soc. **74**(3), 800–804 (1952)
F.F. Rupert, J. Am. Chem. Soc. **31**(8), 866–868 (1909)
F.F. Rupert, J. Am. Chem. Soc. **32**(6), 748–749 (1910)
C.G. Salzmann, P.G. Radaelli, A. Hallbrucker, E. Mayer, J.L. Finney, Science **311**, 1758–1761 (2006)
C.G. Salzmann, P.G. Radaelli, E. Mayer, J.L. Finney, Phys. Rev. Lett. **103**(10), 105701 (2009)
J.M. Schicks, J.A. Ripmeester, Angew. Chem., Int. Ed. **43**, 3310–3313 (2004)
B. Schmitt, C. de Bergh, M. Festou (eds.), *Solar System Ices* (Kluwer, Dordrecht, 1998). 826 pp.
G. Schubert, T. Spohn, R.T. Reynolds, in *Satellites*, ed. by J.A. Burns, M. Matthews (University of Arizona Press, Tucson, 1986), pp. 224–292
G.H. Shaw, J. Chem. Phys. **84**(10), 5862–5868 (1986)
H. Shimizu, T. Kumazaki, T. Kume, S. Sasaki, J. Phys. Chem., B **106**, 30–33 (2002)
V.P. Shpakov, J.S. Tse, C.A. Tulk, B. Kvamme, V.R. Belosludov, Chem. Phys. Lett. **282**(2), 107–114 (1998)
M.B. Simakov, in Proc. First European Workshop on Exo-/Astro-Biology. ESA SP-496 (2001), pp. 211–214
G.A. Slack, Phys. Rev. B **22**(6), 3065–3071 (1980)
E.D. Sloan, C. Koh, *Clathrate Hydrates of Natural Gases*, 3rd edn. (Dekker, New York, 2007)
S.J. Smith, B.E. Lang, S. Liu, J. Boerio-Goates, B.F. Woodfield, J. Chem. Thermodyn. **39**(5), 712–716 (2007)
A. Smits, S. Postma, Verslagen Akad. Wet. **12**, 110–118 (1910)
A. Smits, S. Postma, Z. Anorg. Chem. **71**(1), 250–253 (1911)
A. Smits, S. Postma, Proc. K. Ned. Akad. Wet. **17**(1), 182–191 (1914)
F. Sohl, H. Hussmann, B. Schwenker, T. Spohn, J. Geophys. Res. **108**, 1–11 (2003)
F. Sohl, M. Choukroun, J. Kargel, J. Kimura, R. Pappalardo, S. Vance, M. Zolotov, Space Sci. Rev. (2010, this issue)
A. Sprung, Ann. Phys. Chem. **159**(9), 1–35 (1876)
H. Suga, Thermochim. Acta **300**, 117–126 (1997)
A.K. Sum, R.C. Buruss, E.D. Sloan, J. Phys. Chem. B **101**(38), 7371–7377 (1997)
P.J. Tackley, Rev. Geophys. Suppl. **33**, 275–282 (1995)
T. Takamuku, K. Saisho, S. Nozawa, T. Yamaguchi, J. Mol. Liq. **119**, 133–146 (2005)
G. Tammann, Z. Phys. Chem. **46**, 818–826 (1903)
G. Tammann, Z. Anorg. Allg. Chem. **179**(1), 186–192 (1929)
Y. Tanaka, S. Hada, T. Makita, M. Moritoki, Fluid Phase Equilib. **76**, 163–173 (1992)
G. Tobie, O. Grasset, J. Lunine, A. Mocquet, C. Sotin, Icarus **175**, 496–502 (2005)
G. Tobie, J.I. Lunine, C. Sotin, Nature **440**, 61–64 (2006)
G. Tobie, B. Giese, T.A. Hurford, R.M. Lopes, F. Nimmo, F. Postberg, K.D. Retherford, J. Schmidt, J.R. Spencer, T. Tokano, E.P. Turtle, Space Sci. Rev. (2010, this issue)
S. Tsunekawa, Y. Ishibashi, Y. Takagi, J. Phys. Soc. Jpn. **33**, 862 (1972)
C.A. Tulk, H. Kiefte, M.J. Clouter, R.E. Gagnon, J. Phys. Chem. B **101**(32), 6154–6157 (1997)
P.H.G. Van Kasteren, Bull. Inst. Int. Froid **1973–1974**, 81–87 (1973)
J.H. Van't Hoff, W. Meyerhoffer, N. Smith, Untersuchungen über die bildungsverhältnisse der oceanischen salzablagerungen, insbesondere des Stassfurter salzlagen. XXIII. Das auftreten von Kieserit bei 25°. Sitzungsberichte der Königlich Preussischen Akademie der Wissenschaften zu Berlin. Phys.-math. Classe. **1901** (1901) pp. 1034–1044
G. Vuillard, M. Sanchez, Bull. Soc. Chem. France **1961**, 1877–1880 (1961)
W.F. Waite, M.B. Helgerud, A. Nur, J.C. Pinkston, L.A. Stern, S.H. Kirby, W.B. Durham, Ann. New York Acad. Sci. **912**, 1000–1010 (2000)

E. Whalley, in *Ices in the Solar System* (Riedel, Dordrecht, 1985), pp. 9–37
E. Whalley, S.J. Jones, L.W. Gold (eds.), *Physics and Chemistry of Ice* (Royal Society of Canada, Ottawa, 1973)
K.D. Williams, J.P. Devlin, J. Mol. Struct. **416**, 277–286 (1997)
L.R. Williams, F.S. Long, J. Phys. Chem. **99**, 3748–3751 (1995)
J. Xu, D. Imre, R. McGraw, I. Tang, J. Phys. Chem. B **102**(38), 7462–7469 (1998)
Y. Yamamoto, K. Nagashima, T. Kornai, A. Wakisaka, Ann. N. Y. Acad. Sci. **912**, 797–806 (2000)
J. Yarger, J.I. Lunine, M. Burke, J. Geophys. Res. **98**(E7), 13109–13117 (1993)
T.W. Yergovich, G.W. Swift, F. Kurata, J. Chem. Eng. Data **16**(2), 222–226 (1971)
R. Zhang, P.J. Wooldridge, J.P.D. Abbot, M.J. Molina, J. Phys. Chem. **97**, 7351–7358 (1993)

Space Sci Rev (2010) 153: 219–247
DOI 10.1007/s11214-010-9658-7

Spectroscopy of Icy Moon Surface Materials

J.B. Dalton

Received: 28 August 2009 / Accepted: 2 April 2010 / Published online: 9 July 2010

Abstract Remote sensing of icy objects in the outer solar system relies upon availability of appropriate laboratory measurements. Surface deposits of specific substances often provide our most direct route to understanding interior composition, thereby informing theories of endogenic surface modification, exogenic surface processing and processes involving exchange of material with the interiors. Visible to near-infrared reflectance spectra of properly prepared compounds are required to enable retrieval of surface abundances through linear and nonlinear mixture analysis applied to spacecraft observations of icy bodies. This chapter describes the techniques, conditions and approaches necessary to provide reference spectra of use to theoretical models of icy satellite surface compositions, and summarizes the current state of knowledge represented in the published literature.

Keywords Ice · Infrared spectroscopy · Remote sensing · Planetary science

1 Spectral Studies of Icy Satellites

1.1 Historical Background

Most of our knowledge of icy satellite surface composition is derived from ultraviolet, visible, and infrared spectral studies from ground-based telescopes and spacecraft instruments. Water ice was originally detected on Europa and Ganymede (Kuiper 1957) using telescopic instrumentation, with subsequent detections of water ice on other satellites (e.g. Moroz 1965; Pilcher et al. 1972; Fink et al. 1973; Pollack et al. 1978; Clark and McCord 1980; Brown et al. 1988). Additional volatile ices were soon identified, such as SO_2 (Lane et al. 1981), CO_2 (Cruikshank et al. 1993), N_2 (Cruikshank et al. 1984) and O_3 (Johnson and Jesser 1997; Noll et al. 1996, 1997). These are summarized in a number of articles (Sill and Clark 1982; Thomas and Veverka 1985; Clark et al. 1986; Cruikshank and Brown 1986;

J.B. Dalton (✉)
Planetary Ices Group, Jet Propulsion Laboratory, California Institute of Technology, MS 183-301,
4800 Oak Grove Drive, Pasadena, CA 91109-8001, USA
e-mail: dalton@jpl.nasa.gov

Calvin et al. 1995; Cruikshank et al. 1998; Dalton et al. 2010) describing the icy satellites of the gas giants Jupiter, Saturn, Uranus and Neptune, the classical planet Pluto, Charon, and other icy worlds of the outer solar system. Table 1 summarizes ices known to exist as well as other materials expected to occur on each of the outer planet satellites. A number of volatile ices may be stable in these surface environments; specific compounds that have been positively identified from observations are listed under the column labeled "Known Compounds." The "Expected" column includes tentative identifications, materials predicted on the basis of theoretical models or laboratory experiments, and broad classes of materials, such as "Fe-bearing" or "hydroxylated minerals." Average temperatures and temperature ranges for individual satellites may be compared to freezing and sublimation temperatures for various compounds to give some idea of their stability.

1.2 Endogenic and Exogenic Materials

Icy satellite surface materials can be of either endogenic or exogenic origin (Prockter et al. 2010, this issue). In addition to those materials remaining from the original formation of the satellite, new materials may have been created by chemical processing in the interior, and emplaced at the surface by endogenic processes such as cryovolcanism or extensional tectonics (Tobie et al. 2010, this issue). Additional material may be brought by comets, meteorites or micrometeorites. Ultraviolet photolysis or radiolysis driven by magnetospheric charged particle bombardment can chemically modify the surface as well as damage the ice crystal lattice, altering spectral characteristics and complicating the process of determining surface composition from spectral observations (Coustenis et al. 2010, this issue; Burger et al. 2010, this issue). Ice sputtered from the surface will tend to preferentially redeposit in cold regions, creating bright ridges along crater rims and other elevated features. As ice is sputtered away, lag deposits accumulate in basins, crater rims and graben. The dark lag deposits and bright surface ridges create a localized albedo dichotomy of icy and less-icy surface units (Fig. 1). On an active satellite, surface materials may be buried through cryovolcanism or tectonic activity. Radiolysis products such as hydrogen peroxide may be transported to the interior in this way (Chyba 2000) and provide energy sources to potential life forms.

In a convecting ice shell, the ice itself would seem to be the most important compound being exchanged between the surface and interior. Impurities carried by the ice however can be just as important. Endogenically-produced materials, such as hydrated salts, clathrates and aqueous minerals, if detected at the surfaces of icy satellites, can provide direct insight into interior transport phenomena and exchange processes. Surface materials that, after emplacement, have undergone exogenic processing may still retain sufficient character to provide similar insights. While materials produced via entirely exogenic means (such as radiolytically-derived volatile ices or elements implanted into the surface) may not be as important for the study of exchange processes, they may still be of interest if subsequently transported to the interior. Johnson (1998) pointed out that sputtering of multicomponent mixtures preferentially enriches the more refractory species near the surface, resulting in vertical stratification. What is their role in the exchange of material between surface and interior?

1.3 Compounds Identified on Icy Satellite Surfaces

Materials known to exist on the icy satellites are identified in Table 1 and described in more detail in Dalton et al. (2010, this issue). Water ice, as predicted by Lewis (1972), is nearly

Table 1 Some of the major compounds identified on the outer solar system satellites, and others predicted by models or experiments. Sources: Hanel et al. (1979); Orton et al. (1996); Carlson et al. (1996, 2005, 2009); Davies et al. (2000); Schmitt et al. (1998); Nash and Betts (1998); Spencer et al. (1999); Hibbitts et al. (2000); Roush (2001); Clark et al. (2005); Emery et al. (2005); Brown et al. (2006); Waite et al. (2006, 2009); Verbiscer et al. (2006, 2008); Grundy et al. (2006); Cook et al. (2007); Coustenis et al. (2007); Cruikshank et al. (2007); Dalton (2007); Moore et al. (2007); Newman et al. (2007); Hodyss et al. (2009)

Planet	Satellite	Temperature	Known Compounds	Expected Compounds
Jupiter	Io	80–130 K; 150–200 K at hot spots; peaks of 400–1870 K	SO_2, H_2S, H_2O, silicates	S_2O, S_8, S_nO, Mg-rich orthopyroxene
	Europa	86–132 K	H_2O, SO_2, CO_2, H_2O_2, O_2, hydrates of H_2SO_4, $MgSO_4$ and Na_2SO_4	heavily hydrated and/or hydroxylated minerals, tholins
	Ganymede	90–160 K	H_2O, SO_2, CO_2, O_2, O_3	Aliphatic hydrocarbons, heavily hydrated and/or hydroxylated minerals, tholins
	Callisto	80–158 K	H_2O, SO_2, CO_2, O_2, silicates	Aliphatic hydrocarbons, heavily hydrated and/or hydroxylated minerals, tholins
Saturn	Mimas	73	H_2O, CO_2	
	Enceladus	33–85 K; 120–160 K at hot spots; peaks ~195–223 K	H_2O, CO, NH_3, $NH_3{\cdot}H_2O$, H_2O_2, CH_3OH	CH_4, C_2H_2, C_3H_8, C_6H_6, CO, CO_2, HCN, N_2, salt hydrates
	Tethys	86 K	H_2O, CO_2, $NH_3{\cdot}H_2O$	
	Dione	87 K	H_2O, O_3, CO_2	NH_3
	Rhea	53–99 K	H_2O, O_3, CO_2, CH_4	Tholin, amorphous carbon
	Hyperion	60–115 K	H_2O, CO_2	Hydrocarbons, tholin
	Iapetus	40–130 K	H_2O, CO_2, $NH_3{\cdot}H_2O$	XCN, KCN, N-rich tholin, PAHs, amorphous carbon, NH_3
	Phoebe	72–113 K	H_2O, CO_2, HC_3N	Tholin, amorphous carbon, Fe-bearing minerals, aliphatic and aromatic hydrocarbons, nitriles, phyllosilicates, hydrated silicates
	Titan	70–200 K	H_2O, CO, CO_2, CH_4, CH_3D, C_2H_2, C_2H_4, C_2H_6, C_3H_8, HCN, HC_3N, C_4H_2, C_2N_2, C_3H_4, C_6H_6	NH_3, $NH_3{\cdot}H_2O$, tholins, hydrocarbons and other volatiles including alkanes, alkenes and alkynes
Uranus	Miranda	80–86 K	H_2O, NH_3, $NH_3{\cdot}2H_2O$	
	Ariel	60–84 K	H_2O, CO_2	
	Umbriel	64 K	H_2O, CO_2	
	Titania	60–64 K	H_2O, CO_2	
	Oberon	73 K	H_2O, CO_2	
Neptune	Triton	36–39 K	N_2, CH_4, H_2O, CO, CO_2	
	Nereid	60 K	H_2O	
Pluto		38–63 K	N_2, CH_4, CO, H_2O, CO_2	CH_2CHCN, tholin
	Charon	40–53 K	H_2O, $NH_3{\cdot}H_2O$, N_2	NH_3, $NH_3{\cdot}nH_2O$

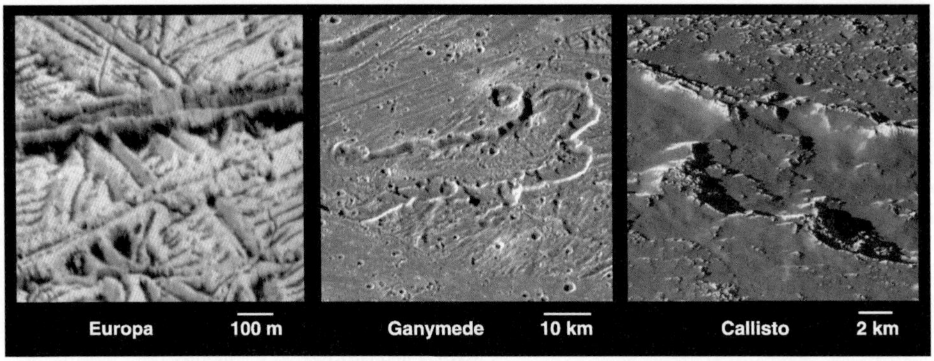

Fig. 1 Surface features on Europa, Ganymede, and Callisto illustrate the influence of composition on geologic processes involving exchange of material between surface and interior. Ridged terrain on Europa (*left*) results from tectonic deformation and is subsequently modified by radiolysis, sputtering and redeposition of volatiles. Schenk et al. (2001) attribute this caldera on Ganymede (*center*) to cryovolcanic flows. Sublimation degradation (Moore et al. 2004) enlarges pits such as this one on Callisto (*right*), providing volatiles that are re-deposited as bright ice limning the ridges

ubiquitous among the large outer solar system satellites, failing to dominate only Io, Titan and Triton. In addition to water ice, several other simple volatile ices occur in the outer solar system, including CO_2, CO, CH_4, H_2S, SO_2 and NH_3. Many were confirmed in comets or their comae well before detection on icy satellites (Chyba and Sagan 1987; Cody et al. 2008; Mumma et al. 2001). CO_2 is found on all three of the icy Galilean satellites. Europa, Ganymede, and Callisto all exhibit evidence for H_2O, CO_2, O_2 and SO_2, but evidence for O_3 embedded in the surface ice has only been found on Ganymede, Rhea and Dione (Lane et al. 1981; Noll et al. 1996, 1997; Hendrix et al. 1999; Hibbitts et al. 2000, 2003; Spencer and Calvin 2002). CO_2 is found on Triton and at varying abundances on many of the Saturnian satellites (Clark et al. 2009; Cruikshank et al. 2010), as well as the Uranian satellites Ariel, Umbriel, Titania, and Oberon (Grundy et al. 2006). In some places, physisorption of CO_2 onto other materials produces unique spectra indicating the influence of the host material on the CO_2 spectrum (Hibbitts and Szanyi 2007; Cruikshank et al. 2010). Mixtures of H_2O and CO_2 ice are common, and laboratory studies (e.g. Bernstein et al. 2005) indicate that spectra of these mixtures are not simply a superposition of the spectra of either taken separately. Otherwise forbidden infrared transitions sometimes occur in mixed ices, some absorption features split into doublets or multiple features, and some expected transitions vanish. This occurs for other ice mixtures as well, such as CH_4 and H_2O (Bernstein et al. 2005, 2006) and as more attention is paid to mixtures of similar compounds in the laboratory, more of this type of behavior is likely to be discovered. This also means that spectral models utilizing spectra of pure materials as endmembers will someday no longer be sufficient to explain spectral observations: eventually, many mixtures will need to be studied empirically in the laboratory to advance our understanding.

1.4 Space Weathering

Irradiation of H_2O-CO_2 ice mixtures is expected to produce a plethora of simple organic compounds (Moore et al. 1991; Delitsky and Lane 1997, 1998) as is irradiation of H_2O-SO_2 or even H_2O-CO_2-SO_2 mixtures. The reaction pathways and equilibrium concentrations for production of materials from radiolysis of H_2O and CO_2 ices are very different in the presence of SO_2 ice (Moore et al. 2007) due to the highly reactive nature of the sulfur products.

By extension, it is clear that relative abundances of radiolysis products on the various satellites depend upon their initial complements of materials. Furthermore, reaction rates are also strongly dependent upon temperature and therefore abundances of reaction products will vary with satellite surface temperatures throughout the outer solar system. Laboratory studies at appropriate temperatures are needed to constrain these rates and produce realistic models of space weathering.

2 Determining Composition of Icy Satellite Surfaces: Interpretation of Observations

2.1 Role of VNIR Lab Spectra

Spectral observations of planetary surfaces are generally analyzed by comparison with spectra of known materials measured under controlled conditions. An exact match suggests that the composition, structure and physical conditions have been replicated; however there is always the possibility that some other material may also provide a match, therefore spectral interpretations are always by definition subject to the question of uniqueness. Most spacecraft spectrometers measure radiance at the sensor, from which the surface reflectance is derived. If an atmosphere is present, then atmospheric transmission and absorption must also be taken into account. In most cases, then, mission observations are of reflected solar ultraviolet, visible, and infrared light.

2.2 Spectral Ranges of Spacecraft Instruments

The Voyager I and II, Galileo, Cassini, and New Horizons missions were all designed with visible wavelength cameras, visible and/or near-infrared imaging spectrometers, and ultraviolet spectrometers (Table 2). Voyager and Galileo also carried photopolarimeters, while Cassini includes a thermal infrared system. Voyager I and II carried Vidicon-based television cameras with filter wheels centered on absorption features of atmospheric species. The Photopolarimeter System (PPS) utilized a polarizer, eight filters, and a photomultiplier tube to observe UV, visible, and IR radiation. An additional broadband radiometer measured .33 to 2 μm (5000 to 30,000 cm^{-1}). Many discoveries were made regarding the surface compositions, textures, and temperatures of the icy bodies, and the atmospheric densities, scattering properties and molecular scale heights for the giant planets. Galileo and Cassini featured sensitive charge-coupled device (CCD)-based instruments that have provided critical information on not only surface compositions, but also distributions of surface materials on the icy Galilean and Saturnian satellites. The New Horizons spacecraft carries an imaging spectrometer system (named Ralph) that is comprised of a six-filter camera system (MVIC, MultiSpectral Visible Imaging Camera) and a near-infrared imaging spectrometer (LEISA, Linear Etalon Imaging Spectral Array) with a resolving power ($\lambda/\Delta\lambda$, see Table 2) that provides spectral resolution up to 3.75 nm. Future missions, such as proposed missions to giant icy moons around Jupiter or Saturn, are likely to carry similar instruments with even higher spectral resolution and coverage.

2.3 Reflectance, Transmittance, Emittance and Absorbance

The measured spectral quantities of reflectance, transmittance, emittance, and absorbance arise from, and thereby carry information regarding, very different physical processes. Each type of infrared spectrum has unique characteristics. Reflectance is defined as the ratio of the

Table 2 Spectral coverage and resolution of some spacecraft mission remote-sensing instruments

Instrument	Spectral Range	Spectral Resolution	Citation
	Voyager		
Infrared Interferometer Spectrometer (IRIS)	180–2500 cm^{-1} (55–4 μm)	4.3 cm^{-1} (7–1300 nm)	Hanel et al. (1980)
	5000–30000 cm^{-1} (2–0.33 μm)	panchromatic	
Ultraviolet Spectrometer (UVS)	50–1700 Å	10 Å	Broadfoot (1977)
Photopolarimeter System (PPS)	2200–7300 Å	8 filters	Lillie et al. (1977)
	Galileo		
Instrument	Spectral Range	Spectral Resolution	Citation
Solid State Imager (SSI)	400–1100 nm	7 filters + clear filter	Belton et al. (1992, 2000)
Near Infrared Mapping Spectrometer (NIMS)	700–5200 nm	12.5 nm below 1 μm; 25 nm above 1 μm	Carlson et al. (1992)
Ultraviolet Spectrometer (UVS)	113–432 nm	0.7 nm below 190 nm 1.3 nm above 190 nm	Hord et al. (1992)
Extreme Ultraviolet Spectrometer (EUVS)	55–125 nm	3.5 nm	Hord et al. (1991)
Photopolarimeter (PPR)	410–945 nm	10 channels: 7–60 nm	Russell et al. (1992)
	Cassini		
Instrument	Spectral Range	Spectral Resolution	Citation
Imaging Science Subsystem (ISS)	200–1100 nm	Two cameras with 42 filters	Porco et al. (2004)
Visible and Infrared Mapping Spectrometer (VIMS)	300–5100 nm	7.3 nm from 350 to 1050 nm; 16.6 nm from 850 to 5100 nm	Brown et al. (2004)
Composite InfraRed Spectrometer (CIRS)	10–1400 cm^{-1} (7–1000 μm)	Selectable 0.5–15.5 cm^{-1}	Flasar et al. (2004)
Ultraviolet Imaging Spectrograph (UVIS)	EUV 563–1182 Å FUV 1115–1912 Å	EUV 2.75, 4.8, 19.4 Å FUV 2.75, 4.8, 24.9 Å	Esposito et al. (2004)
	New Horizons		
Instrument	Spectral Range	Spectral Resolution	Citation
Long Range Reconnaissance Imager (LORRI)	350–850 nm	panchromatic	Cheng et al. (2008)
Visible Imager and Imaging Spectrometer (Ralph)	MVIC 400–975 nm LEISA 1250–2500 nm	MVIC 6 filters LEISA ~5–10 nm ($\lambda/\Delta\lambda = 240$) from 1250 to 2500 nm and ~4 nm ($\lambda/\Delta\lambda = 550$) from 2100–2250 nm	Reuter et al. (2008) Weaver et al. (2008)
Ultraviolet Mapping Spectrometer (Alice)	465–1880 Å	1024 channels at 0.183 nm	Stern (2008)

radiant power reflected from a surface divided by the radiant power incident upon the surface (Sheppard et al. 1985). Transmittance is the ratio of light transmitted through a material to the incident light. Emittance is the proportion of emitted radiation leaving a sample to the radiation emitted by an ideal blackbody of the same temperature. Absorbance is the ratio of the light lost during passage through a material to the incident light. The proportion of incident power lost during passage through a medium cannot strictly be referred to as absorbance if other loss processes such as scattering and reflection are also operative. In thin films or polished samples, this may not be much of an issue, but in planetary regoliths, it is *the* issue.

According to Kirchhoff's law, emittance and absorbance are equivalent: the same vibrational frequencies are responsible for both, and their relative intensities are the same. Thus, in the absence of scattering, the sum of reflectance and absorbance is unity:

$$R = 1 - A \tag{1}$$

However, for particulate surfaces in reality, scattering from grain boundaries, facets, imperfections (scratches, fractures and other asperities) enhances absorption in some frequencies relative to others; multiple scattering (passage of the incident ray through several grains, including their boundaries and imperfections) increases the absorption path length and the number of reflections and refractions. Due to the dependence of multiple scattering on wavelength, grain size, and refractive index, and the extreme complexity of particulate grains in regoliths, the absorbance and transmittance of a polished sample, thin film, or monocrystal in the laboratory can rarely be converted directly into the reflectance of a planetary surface. Kirchhoff's law only holds in the absence of scattering. In reality, this is generally not the case for planetary regoliths, so it is actually the sum of reflected, absorbed, *and* scattered light that sums to unity:

$$R = 1 - A - S \tag{2}$$

In practice, a reflectance spectrum of the anticipated surface materials is usually compared to the observations to interpret surface composition. If an appropriate laboratory spectrum is not available, the observed spectrum can be compared to some other spectrum of candidate surface materials: perhaps in transmittance or absorbance, or taken at an inappropriate temperature (such as room temperature or near absolute zero) or even of a different phase (such as liquid instead of solid). This will at least provide a qualitative indication of whether the candidate material has diagnostic absorption features near those in the observation.

2.4 Linear and Nonlinear Mixtures

Pure materials are rarely found in nature: remote sensing observations are usually of mixtures. Popular methods for approximating remote spectral observations include linear and nonlinear mixture models, also called areal and intimate mixtures, respectively. The simplest is the linear mixture: a surface region composed of discrete patches of pure materials will have an average reflectance spectrum that can be matched by a linear combination of the reflectance spectra of the pure materials (endmembers), with each material weighted by its fractional area (Adams et al. 1993; Clark 1999). The weights correspond to abundances, and in practice, these linear models come surprisingly close to reproducing the observations even when the assumption of discrete areas is not strictly true (Fig. 2). Where particles are mixed but large enough to be optically thick, linear mixing is in many cases quite useful, such as for silicate minerals at mid-infrared wavelengths near their rehstrahlen features (Thompson and Salisbury 1993).

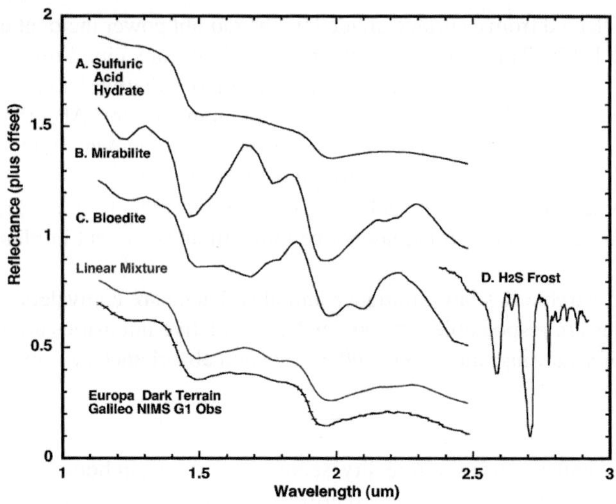

Fig. 2 Linear mixture model and endmember spectra. The spectrum of Europa's dark terrain (*Bottom curve*) was acquired by the Galileo NIMS instrument. The top three curves (A, B, and C) are cryogenic laboratory spectra of materials predicted to occur in Europa's dark terrain. The mixture spectrum (*blue curve*) is a linear combination of these: 70% Sulfuric Acid Hydrate (A, $H_2SO_4{\bullet}8H_2O$), 20% Mirabilite (B, $Na_2SO_4{\bullet}10H_2O$), and 10% Bloedite (C, $Na_2Mg(SO_4)_2{\bullet}4H_2O$). The laboratory spectrum of hydrogen sulfide frost (D, H_2S), despite its high precision and strong diagnostic features, cannot be incorporated into the model because it does not cover the same wavelength range. The model spectrum and endmember spectra A, B, and C have been offset vertically for clarity. (After Fink and Sill 1982; Dalton et al. 2005; and Dalton 2007)

More advanced, nonlinear models have been developed using a combination of physical considerations and theoretical approximations that account, either directly or empirically, for unquantifiable effects while making the equations tractable and computationally realistic (Kubelka and Munk 1931; Chandrasekhar 1960; Hapke and Van Horn 1963; Nicodemus 1965; Emslie and Aronson 1973; Hapke 1981, 1993; Wiscombe 1980; Shkuratov et al. 1999; Shkuratov and Grynko 2005). The most popular formulations are those of Hapke (1981, 1993), Shkuratov et al. (1999) and Shkuratov and Grynko (2005). Both are based on the general treatment of radiative transfer theory as developed by Chandrasekhar (1960) to calculate spectroscopy and energy transfer in stellar atmospheres. For planetary surfaces, the assumption that particles are not in contact with each other is eliminated and new formulations of the equation of radiative transfer, taking into account singly- and multiply-scattered radiation (Hapke 1993) are developed. These physics-based models can be used to model mixtures of grains of varying size and composition (intimate mixtures, also referred to as salt-and-pepper models) and require a number of measured and assumed parameters. Several of the assumed parameters can be adjusted with successive iterations, leading to improved estimates. These include porosity, internal scattering coefficient, and grain sizes. Such models must be applied carefully, since many of the underlying assumptions are strictly true only under certain conditions, such as the requirements that the particle grain size be much greater than the wavelength of light, or the assumption that particles scatter isotropically.

It is critical to note that both types of spectral mixture models need reference spectra which cover the full wavelength range of the modeled observations (Fig. 2), in order to compare relative abundances of the reference compounds and estimate surface abundances.

2.5 Optical Constants

Critical to these nonlinear models are the measured optical constants, or real and imaginary indices of refraction. These can be derived from laboratory measurements of reflectance, transmittance, and/or absorbance through a number of approaches, including classical dispersion theory, Kramers-Kronig analysis and radiative transfer modeling. Each approach is appropriate to different sample properties such as coarse versus fine grain size relative to wavelength; polished, parallel sample surfaces; thin films, thick films, powders, slabs or monocrystals; weakly vs. strongly absorbing materials, and the type of viewing geometry (normal, specular, and diffuse reflection; absorption, transmission; multiple angles) available (Arhenkiel 1971; Warren 1984; Hapke 1993; Lucey 1998; Grundy and Schmitt 1998; Dalton and Pitman 2010).

3 Laboratory Methods for Spectroscopy of Candidate Icy Satellite Surface Materials

Even the seminal discovery of methane on Titan by Kuiper (1944) did not utilize laboratory spectra of methane ice: he detected the 7260 Å methane vapor line, arising from Titan's atmosphere, not surface. By the time Kuiper (1957) and Moroz (1965) confirmed the presence of water ice on Europa and Ganymede, water ice and a few other volatile ices had been characterized sufficiently to enable their detection. Derivation of abundances for most other surface species awaits cryogenic laboratory measurements of these species (Carlson et al. 1999a, 1999b, 2005; Dalton 2007).

3.1 Cryogenic Laboratory Reflectance Spectroscopy

An excellent review of cryogenic laboratory spectroscopic progress and techniques was given by Schmitt et al. (1998), with discussion of nuances involved in transmission versus reflectance measurements. An updated review is provided by De Bergh et al. (2008). Additional considerations for reflectance measurements are described in Clark (1999). The primary aim is to provide spectra that are directly comparable to ground-based and spacecraft observations of icy planetary surfaces. The observations (to date) are passive measurements of reflected sunlight, in the ultraviolet, visible, and near-infrared. The surfaces under investigation may be covered with areas of vapor-deposited, annealed, smooth or crystalline ice (or minerals); regoliths of pulverized ice and/or minerals seem to predominate.

3.2 Sample Preparation Techniques

Sample sizes generally depend upon the characteristics of the spectrometer systems to be used. The area subtended by the optical beam should not contain anything other than the sample, regardless of spectral properties. Typical subtended areas range from a millimeter to one or two centimeters in scale. Maintaining sample consistency beyond those sizes can be challenging. Depth of the sample should be greater than an optical depth of unity for reflectance measurements (i.e. the light should not sample the substrate) but insufficient to produce saturation in transmittance/absorbance measurements. Sample volumes thus are on the order of a few cubic centimeters at most, and sometimes as small as a single atomic monolayer.

3.2.1 *Vapor Deposition of Thin Films*

Thin-film vapor deposition is a very popular and powerful technique for mid-infrared absorbance and transmittance spectroscopy. Gas is introduced to a vacuum chamber at a controlled rate, and deposited onto a cooled mirror or window. Film thickness can be monitored very precisely using a laser and interference fringe counting technique (Adams and Katz 1956; Ferraro et al. 1980; Quirico and Schmitt 1997a, 1997b). A smooth surface is desirable, both for the fringe counting technique and derivation of optical constants. Molecular monolayers can be constructed with care, but films of 1 to 3 microns are more common. Deposition beyond five to ten microns tends to build mound-shaped or irregular profiles, from which neither film thickness nor optical constants can be derived.

3.2.2 *Closed Cell Crystal Growth*

Many of these difficulties can be overcome through the use of a closed cell. Infrared-transparent windows on two sides allow the beam to pass through the sample, while holding the sample thickness at a known value (Quirico and Schmitt 1997a, 1997b; Grundy and Schmitt 1998). Samples may be grown from liquid or vapor phase, and a combination of resistive heaters and temperature sensors embedded in the cell can allow for annealing of the ice as well as temperature control to within ± 0.1 K. Crystal growth occurs under thermodynamic equilibrium, improving reproducibility of results. There are some disadvantages however, compared to ordinary thin-film deposition. The closed cell technique does not lend itself well to forming molecular mixtures of volatiles, as they tend to stratify within the cell. Also, amorphous solids are difficult to form, as they approach thermodynamic equilibrium within the cell. However the improved knowledge of temperature and sample thickness permits accurate derivation of temperature-dependent optical constants for pure volatile ices with high precision.

3.2.3 *Matrix Isolation*

The closed-cell approach also lends itself well to matrix-isolation techniques (Quirico and Schmitt 1997a, 1997b). The objective is to study reactive and unstable species, such as free radicals, by trapping them in a matrix of solid, inert (usually noble) gas, at very low temperature (<20 K). The trapped species, with little thermal energy, and separated from each other by the matrix, are prevented from decomposing and are available for study by infrared spectroscopy and other analytical techniques (Leroi et al. 1961). A major application is the assignment of individual absorption features to specific chemical bonds and vibrational modes. Matrix isolation can also be used to detect and examine products of irradiation chemistry before they decompose; many of these may persist in certain regions of the outer solar system—or at least exist in an equilibrium state between creation and destruction (Mardyukov et al. 2009). This approach is useful for generating infrared spectra of volatile ices (Quirico and Schmitt 1997a; Irvine et al. 1982; Bernstein et al. 2005) and has been used to demonstrate that spectra of species within matrices exhibit frequency shifts from those of pure ices (Leroi et al. 1961; Jödl 1989; Bernstein et al. 2005, 2006). In many cases, the shifts are too subtle to interfere significantly with spacecraft observations, which generally lack the spectral resolution of laboratory instruments; but at times pronounced shifts can complicate attempts to interpret observations (Cruikshank et al. 2010).

3.3 Sample Measurement

Measurements of ice spectra for studies of icy bodies in the outer solar system almost always take place in a purged or evacuated container in order to prevent contamination by condensation of atmospheric water vapor, carbon dioxide or other species on the sample. Design considerations include viewing geometry, influence of scattered and reflected light, and external light sources. Closed chambers require viewports made of materials transparent to the particular wavelength range under study. For both reflectance and transmittance/absorbance configurations, it is critical that the system be aligned such that the first-surface specular reflection does not reach the detector, as this can overwhelm the measured signal from a smooth surface; likewise, care must be taken to ensure that neither incident light nor light scattered from the sample reflects from chamber components into the measured beam.

3.4 Transmittance and Absorbance

For transmittance and absorbance measurements, there are a few popular configurations. Ice samples are usually deposited onto either a transparent window or a mirror. Mirrors and sometimes lenses along the beam may be used to redirect the path, but some signal is lost with each reflection, or to attenuation as the beam traverses a lens. If the sample is on a mirror, the calculated transmittance or absorbance must be adjusted for the light having passed twice through the sample:

$$I_T = I_0 e^{-\kappa x} \tag{3}$$

The transmitted intensity, I_T, is equivalent to the incident intensity I_0 attenuated by an amount determined by the absorption coefficient κ times the distance traveled, x. The absorption coefficient κ is not the same as the imaginary index of refraction k, but they are related by

$$\kappa = \frac{4\pi k}{\lambda} \tag{4}$$

where λ is the wavelength of the light. Usually, the sample thickness is x, but in the mirror configuration, with very small phase angle, x must be doubled; for larger phase angles, the total path x is a trigonometric function of the viewing geometry.

3.5 Reflectance

The preparation of ice and other volatile samples for reflectance spectroscopy measurements draws heavily on techniques developed for reflectance studies of rocks and minerals. In the outer solar system, water and several other ices can be considered minerals; however under terrestrial conditions they behave quite differently and so their preparation relies on methods that control temperature, pressure, humidity and other environmental conditions while still allowing mechanical, physical, and chemical manipulation of the sample. Reflectance measurements are generally conducted at low phase angles from 0 to 30 degrees, but sometimes at higher angles. Accurate determination of photometry and the phase curve of a sample relies on multiple measurements over the full 0–180 degree range of phase angles; closer spacing near zero phase angle provides valuable precision for evaluating the opposition effect.

4 Applications of Thick and Thin Films

A major source of confusion among spectroscopists and non-spectroscopists alike is the distinction between infrared spectra performed on thin films ($\lesssim$10 μm) and those conducted using optically thick samples. In reflectance studies, samples may be several mm to cm thick (typically 2–8 mm but generally $\gtrsim$10–20 μm) specimens including polished slabs, sieved powders, or carefully-grown monocrystals. However, the majority of published cryogenic infrared spectra of volatile ices were measured with the intention of understanding the interstellar medium (ISM). Here irradiation of μm-scale rinds of ice has created organic compounds ranging in complexity from CO, CH_4 and alkanes to larger organic molecules, including polycyclic aromatic hydrocarbons (PAHs) with typical sizes of up to ~20–100 carbon atoms (Léger and Puget 1984; Allamandola et al. 1985, 1988; Bernstein et al. 1997a, 1999, 2001; Charnley et al. 2002; Gerakines et al. 2005). Telescopic infrared observations of the ISM have provided richly detailed spectra that can be used to identify these molecules.

Consider the viewing geometry (Fig. 3, left) of these observations: starlight is transmitted through a cloud of gas, dust, and ice and is received at the sensor. Most of these observations are made in the mid- to far-infrared, where fundamental molecular vibrational modes are found. Though the interstellar medium is cold (20–100 K), the hot stars (thousands of K) provide abundant mid- and far-infrared photons.

In order to interpret these observations, mid- and far-infrared *transmittance* or *absorbance* spectra are needed. Thin films of various ices are deposited from the vapor phase onto a cooled (~20 K) substrate, and the transmittance or absorbance of the sample is measured. The mid- to far-infrared molecular fundamental vibrations of many volatile and organic ice compounds are very strong: after traversing only a few microns of the medium, nearly all of the incident light at these wavelengths may be absorbed. If the sample is too thick (>5–10 μm), the absorption features will *saturate*: the signal within the band will be zero.

Now consider the viewing geometry for an icy planet or satellite (Fig. 3, right). Sunlight is reflected from the top surface of an icy body hundreds to thousands of meters thick.

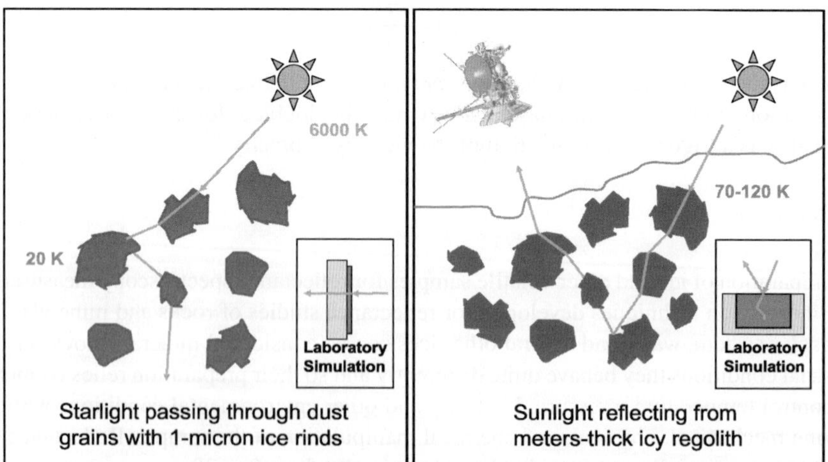

Fig. 3 Typical viewing geometry for observations of the interstellar medium (*left*) and icy regoliths (*right*). Studies of molecular clouds usually measure starlight transmitted through dust grains with optically thin ice rinds having strong mid-infrared absorptions. Observations of icy regoliths measure reflected sunlight dominated by near-infrared single- and multiple-scattering effects

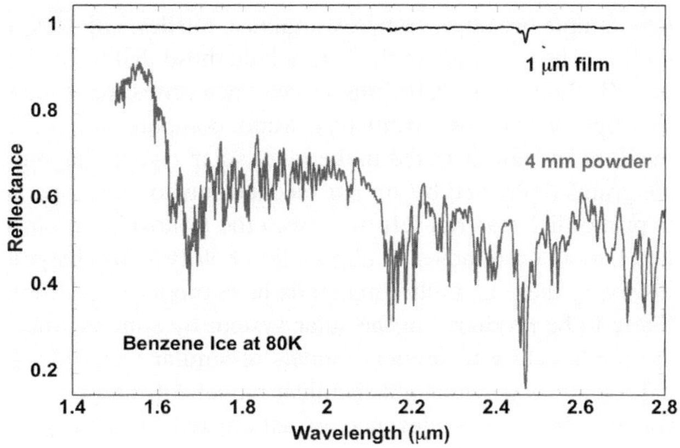

Fig. 4 Reflectance spectrum of benzene ice at 80 Kelvins: Thin film vs. thick sample. No scaling or offsets have been applied to either spectrum. *Top curve*: vapor-deposited thin (1-μm) film (Brunetto et al. 2006). *Bottom curve*: Ground and sieved powder (4-mm thick, mean grain size 50–100 μm; Clark et al. 2009). Only the strongest features appear in the spectrum of the thin film because there is insufficient optical path length. The optically thick sample displays diagnostic absorption features representative of an icy regolith

Very little thermal emission from this 30–100 K body is present in the visible- to mid-infrared spectral range observed by the sensors. The sensors therefore measure *reflectance*—the proportion of sunlight reflected from the surface. In order to interpret these observations, *reflectance* spectra of the candidate surface materials, or their optical constants are required.

Laboratory samples must be sufficiently thick to reproduce the near-infrared combination and overtone vibrational features, which can be orders of magnitude weaker than their corresponding mid-infrared fundamentals (Clark 1999; Clark et al. 2009). Below ∼10–20 microns mean optical path length, many of these features are not apparent at all (Fig. 4). As some wavelengths of light are preferentially refracted and reflected through multiple grains, the mean optical path length increases, enhancing some absorption features by multiple scattering (cf. Buratti and Veverka 1985). Since on a planetary surface, incident light traverses very long path lengths, these features are generally present in observations, so laboratory measurements utilizing similar path lengths are necessary in order to interpret observations of planetary regoliths. This explains the need for optically thick, macroscopic samples to reproduce the diagnostic near-infrared absorptions observed on icy worlds.

5 Review of Published Laboratory Spectra

This section provides an overview of laboratory measurements needed for the direct detection and quantification of icy satellite surface compounds from remote sensing observations. Examination of all of the many high-quality laboratory measurements described here would be beyond the scope of this volume. For over 125 compounds proposed to occur on icy satellites, Table 3 lists published spectra of those compounds under the conditions relevant to icy satellite surfaces; it includes ices, minerals and other chemical compounds, measured below either the freezing point of water, or that of the relevant compound. As explained in Sect. 2.4, in order to derive even the most rudimentary quantitative abundance estimates of icy satellite surface materials from spectral remote sensing observations, laboratory spectra of the appropriate *wavelengths*, *temperatures*, *viewing geometry*, and *sample thickness* are required. Table 3 summarizes the current state of the art. We have tried to be as complete as possible, but in an undertaking of this magnitude some publications have likely been missed, particularly those that are either very old or very recent.

Each material in the table has been proposed in the peer-reviewed literature as a surface compound relevant to an icy object in the outer solar system. The list includes volatile ices, simple organics, complex organics, nitriles, silicates, sulfates, hydrates, clathrates and tholins. The materials in the list include those that have already been observed on icy satellites (Dalton et al. 2010, this issue), meteorites, or in comets (usually by emission spectroscopy or mass spectrometry). Many compounds have been observed in the interstellar medium and are expected in the outer solar system, having either been brought on dust and ice grains or created by similar mechanisms to those operating in the ISM. Another source is primordial material left over from the protosolar, protoJovian and protoSaturnian nebulae. Laboratory studies of ultraviolet photolysis or charged particle irradiation (radiolysis) involving these and other materials have produced additional materials that are considered likely to be produced in the solar system by similar processes. Some materials in this list are predicted by theoretical models of similar experiments or of chemical and geochemical processes (such as geochemical models of interior oceans). Other materials are sought for their biological relevance, or anticipated for their roles in surface processes involving exchange with the interior (such as the role of ammonia hydrates in cryovolcanism). Many other relevant materials are not yet in this table, such as complex hydrocarbons expected at Titan, or isotopic variations.

The first column on the left gives the names of compounds; where possible, their formula is given in the next column. The center of the table depicts the electromagnetic spectrum from ultraviolet to far infrared: wavelengths are given in microns *and* frequencies in inverse centimeters. The VNIR spectral range covered by cameras and imaging spectrometers on Outer Planet Missions including Cassini, Galileo and New Horizons is shaded light brown. The final column contains partial references (first author and year only) to the papers containing cryogenic spectra of these materials. For each published spectrum, a colored bar marks the spectral range of the measurement. The exact beginning and ending wavelengths (in microns) are printed at either end of the bar. The colors of the bars indicate the following temperature ranges:

250–≥273 K light green

200–250 K dark green

150–200 K light blue

50–150 K dark blue (relevant to icy satellites)

0–50 K purple

A *red bar* is used to denote ALL spectra which are in the appropriate wavelength range (.3–5 μm), temperature range (50–150 K), and viewing geometry (reflectance), using sufficient sample optical thicknesses (≥20 μm) to be used in quantitative abundance models of satellite observations.

Many discussions in the literature refer to infrared spectra but do not specify which *type* of infrared spectra: absorbance, transmittance, near infrared, far infrared, etc. In Table 3, a code at the left hand side of each bar denotes the spectral type:

A Absorbance
AC Absorption coefficient κ
Geo Geometric albedo

M_{ext} Molar extinction coefficient
Od Optical depth
T Transmittance
n Real Index of Refraction only
k Imaginary Index of Refraction only
n&k Both Real *and* Imaginary Index of Refraction
R Reflectance

Those published spectra that are in the appropriate *wavelength range* for abundance models of the spacecraft observations, as well as being in either *reflectance* or *complex index of refraction*, that are at *temperatures* appropriate to studies of icy satellite surfaces, and were performed using sufficiently *thick samples* for the VNIR features to manifest, are colored RED. Only those compounds represented by a red bar completely covering the light brown regions can be incorporated into abundance models using the full extent of existing and planned observations.

This is presently possible only for a few compounds: water ice, carbon dioxide ice, ammonia ice, and methane ice. Sulfur dioxide has also been measured over most but not all of the desired range. During the production of this issue, cryogenic reflectance spectra for several alkane ices were published by Clark et al. (2009), more than tripling the number of compounds that can be modeled, at least through linear mixing analysis. Many published efforts are limited to part of the observed spectral range; while these are useful, the full spectral range still needs to be measured in order for complete spectral models to be constructed.

5.1 Summary of State of Knowledge

Most of the published spectra in Table 3 are in the mid-infrared range, performed in transmission or absorption, at extremely low temperatures (0–50 K) or near room temperature (250–300 K). Recent work by Mastrapa et al. (2009) on amorphous ice, and the optical constants of carbon dioxide ice (Warren 1986; Hansen 1997, 2005) permit these volatiles to be quantified on the icy satellites. Absorption coefficients of solid methane (Grundy et al. 2002) enable models of CH_4 ice on Titan and Pluto. Dalton et al. (2005) measured cryogenic reflectance spectra of hydrated salts predicted to be on the surface of Europa, and demonstrated their utility with linear mixture modeling of Europa's surface composition (Dalton 2007; Dalton et al. 2009; Shirley et al. 2010). However, these salt hydrate measurements do not extend above 2.5 μm, while published spectra of hydrogen peroxide ice do not extend below 2.5 μm, so a composite model with all of these materials cannot be constructed. When the best available models are not able to include materials that are known to be present, discrepancies between actual and estimated abundances are inevitable. For most entries in Table 3, a single reflectance spectrum, measured at appropriate wavelengths (0.3 to 5.5 μm), temperature (50–150 K), and sample thickness (>20 μm), would be an improvement to the current state of the art. While optical constants derived from reflectance measurements are necessarily approximate, they are better than no optical constants at all. A properly conducted campaign of laboratory measurements would result in a significant enhancement to the scientific return from planetary missions.

6 Laboratory Measurements Needed for Interpretation of Spacecraft Data

6.1 Critical Properties of Needed Laboratory Measurements

Spectral observations from past, present and planned spacecraft missions require laboratory reference spectra for quantitative abundance estimation. In order to be of use in mixture

Table 3 Published cryogenic spectra of predicted icy satellite surface compounds. Light brown denotes spectral ranges of typical spacecraft instruments (visible and near-infrared cameras and imaging spectrometers). Reflectance spectra (or optical constants) for each material in the list will enable abundance estimates for these materials. This requires measurements at appropriate temperatures, using optically thick samples, over the spectral ranges of the instruments. Adapted from Henning et al. (1999), Dalton (2008), Dalton et al. (2009) and spectral database tables available at http://speclab.cr.usgs.gov, http://mos.seti.org and http://www.astro.uni-jena.de/Laboratory/Database/databases.html

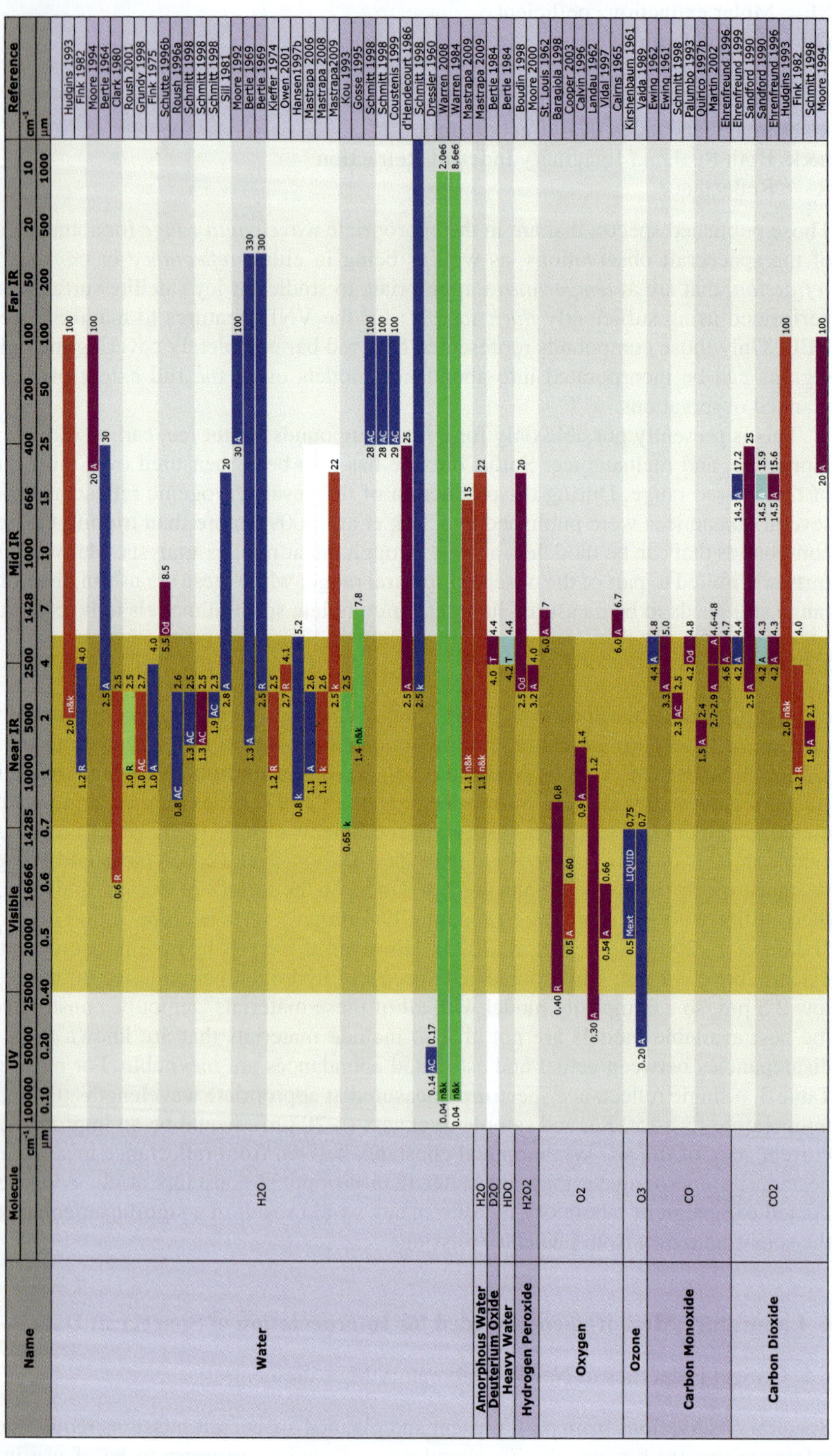

Table 3 *(Continued)*

Spectral axis: UV — Visible — Near IR — Mid IR — Far IR; cm^{-1}: 100000, 50000, 25000, 20000, 16666, 14285, 10000, 5000, 2500, 1428, 1000, 666, 400, 200, 100, 50, 20, 10; μm: 0.10, 0.20, 0.40, 0.5, 0.6, 0.7, 1, 2, 4, 7, 10, 15, 25, 50, 100, 200, 500, 1000

Name	Molecule	Spectral data (μm)	Reference
Carbon Dioxide	CO2	1.3 A – 3.0	Quirico 1997a
		2.5 k –	Schmitt 1998
		1.0 R – 6.0	Smythe 1975
		2.0 AC – 330	Hansen 1997a
		0.17 A – 1.8	Hansen 2005
		1.2 R – 2.5	Kieffer 1974
		0.05 n&k – 2e5	Warren 1986
Carbon Trioxide	CO3	3.2 A – 20	Moll 1966
		3.1-3.3 A; A 4.0-5.6; 9.1 A – 20	Jacox 1971
Formyl ion	HCO	3.9 A – 12.1	Milligan 1964
Formaldehyde	H2CO	20 A – 100	Moore 1994
		2.0 A – 6.25	Bohn 1994
		2.5 A – 20	Gerakines 1996
		2.5 A – 2[illegible]	DelloRusso 1993
		2.0 Od – 10	Schutte 1996a
Formic Acid	H2CO2	5.5 Od – 8.5	Schutte 1996b
		5.0 A – 25	Moore 1996
		5.0 A – 25	Moore 1996
		2.5 A – 10	Grim 1989
		5.6 Od – 7.7	Trakhtenberg 1998
		2.5 A – 25	Khanna 2002
		2.2 A – 15.4	Chapman 1956
Carbonic Acid	H2CO3	5.5 A – 11	Moore 1991
		2.5 A – 22	DelloRusso 1993
		2.8 A – 9	Strazzulla 1996
		2.8-5.0 A; A 5.4-6.4	Hage 1993
		2.5 Od – 20	Brucato 1997
		2.5 A – 40	Gerakines 2000
		2.5 A – 22	DelloRusso 1993
D-Carbonic Acid	D2CO3	2.8-5.0 A – 5.4-6.4	Hage 1993
Carbon Suboxide	C3O2	2.0 A – 25	Gerakines 2001
		4.0 A – 20; 80 A – 1000	Smith 1966
Methanol	CH3OH	3.2-3.8 A; 6.5 A – 8; F 9.1-11.1	Ehrenfreund 1999
		2.3 Od – 25	Palumbo 1999
		2.5 Od – 25	Schutte 1991
		2.0 n&k – 100	Hudgins 1993
		2.5 A – 25	Moore 1996
		20 A – 100	Moore 1994
		2.5 A – 20	Gerakines 1996
		0.8 T – 2.60	Quirico 1997a
		1.4 R – 2.5	Quirico 1997a
		0.6 R – 2.6	Brunetto 2006
		2.5 A – 25	d'Hendecourt 1986
		2.5 A – 25	Dempster 1971
Ethanol	CH3CH2OH	5.0 A – 25	Moore 1996
		2.5 Od – 20	Boudin 1998
Acetone	(CH3)2CO	2.5 A – 22	DelloRusso 1993
Methyl Formate	HCOOCH3	2.5 A – 25	Khanna 2002
Glycolaldehyde	CH2OHCHO	2.5 A – 20	Hudson 2005
		0.22 A – 0.33	Bacher 2001
Ethyl Acetate	CH3CO2CH2CH3	2.5 A – 25	d'Hendecourt 1986
Dimethyl Carbonate	(OCH3)2CO	2.5 A – 22	DelloRusso 1993
Ethylene Glycol	(CH2OH)2	2.5 A – 25	Moore 1996
Methane	CH4	0.12 AC – 0.14	Dressler 1960
		0.7 AC Ice – 5.0	Grundy 2002
		0.7 AC LIQUID – 5.0	Grundy 2002
		2.8 A – 14.3	Comeford 1960
		2.0 n&k – 100	Hudgins 1993
		1.2 R – 4.0	Fink 1982
		1.1 A – 2.7	Quirico 1997a
		2.0 A – 20	Pearl 1991
		2.5 k –	Schmitt 1998
		6.5 A – 14.3	Coustenis 1999
		1.0 R – 6.0	Smythe 1975
		1.2 R – 2.5	Kieffer 1974
		2.5 A – 25	Bohn 1994
		0.6 R – 2.6	Brunetto 2006

Table 3 (*Continued*)

Name	Molecule	Start (µm)	Type	End (µm)	Reference
Methane	CH4	0.8	T	2.6	Quirico 1997a
		0.35	R (LIQUID)	15.5	Clark 2009
		2.5	A	25	d'Hendecourt 1986
Acetylene	C2H2	2.0	A	50	Khanna 1988
		6.5	A	14.3	Coustenis 1999
		2.0	A	16	Bohn 1994
		2.5	Od	20	Boudin 1998
		2.5	A	12.5	Hirabayashi 2002
		2.5	A	12.3	Wieder 1962
		2.8	A	14.3	Comeford 1960
Ethylene	C2H4	1.1	A	2.7	Quirico 1997a
		6.5	A	14.3	Coustenis 1999
		0.35	R	15.5	Clark 2009
		2.0	A	16	Bohn 1994
Ethane	C2H6	2.3	A	14.3	Comeford 1960
		1.1	A	2.7	Quirico 1997a
		2.2; 6.0	A; A	3.5; 20 (Amorphous)	Hudson 2009
		2.2; 6.0; 7.5	A; A; A	3.5; 12.8 (Crystalline)	Hudson 2009
		2.0	A	20	Pearl 1991
		6.5	A	14.3	Coustenis1999
		2.5	Od	20	Boudin 1998
		2-2.5; 6.3	Od; Od	3.2-4; 14.3	Boudin 1998
		0.35	R	15.5	Clark 2009
Propane	C3H8	0.8	T	2.6	Quirico 1997a
		0.35	R	15.5	Clark 2009
Butane	C4H10	0.35	R	15.5	Clark 2009
Pentane	C5H12	0.35	R	15.5	Clark 2009
Hexane	C6H14	0.35	R	15.5	Clark 2009
Heptane	C7H16	0.35	R	15.5	Clark 2009
Octane	C8H18	0.35	R	15.5	Clark 2009
Nonane	C9H20	0.35	R	15.5	Clark 2009
Decane	C10H22	0.35	R	15.5	Clark 2009
Propdiene	C3H4	6.6	k	50	Coustenis 1999
		0.35	R	15.5	Clark 2009
Propene	C3H6	0.35	R	15.5	Clark 2009
Benzene	C6H6	0.35	R	15.5	Clark 2009
		0.6	R	2.6	Brunetto 2006
1,3-Butadiyne	C4H2	2.0	A	25	Khanna 1988
		6.6	k	50	Coustenis 1999
Carbon Trimer	C3	0.38	A	0.42	Weltner 1964
		0.19	A	0.22	Weltner 1964
Cysteinesulfonic Acid	OCS	2.0	n&k	100	Hudgins 1993
		4.8	A	5.0	Palumbo 1995
Nitrogen Oxide	NO	0.8	T	2.6	Quirico 1997a
		1.4	R	2.5	Quirico 1997a
Nitrogen	N2	2.1-2.2	AC; AC	4.0-4.4	Schmitt 1998
		2.1	A	2.2	Tryka 1993
		1.0	A	2.7	Quirico 1997a
		2.1	AC	2.3	Grundy 1993
Ammonia	NH3	0.14	AC	0.20	Dressler 1960
		1.4	AC	200	Sill 1980
		2.0	A	15.0	Staats 1959
		2.0	A	12.0	Bromberg 1977
		2.8	A	20	Sill 1981
		2.5	A		Schmitt 1998
		1.9	k	2.3	Schmitt 1998
		1.2	R	4.0	Fink 1982
		1.0	A	166	Ferraro 1980
		2.6	A	20	Fink 1982
		20	A	100	Moore 1994
		1.0	R	2.5	Brown 2000
		1.2	R	2.5	Kieffer 1974
		0.14	n&k	200	Martonchik 1984
		2.5	A	25	d'Hendecourt 1986
Ammonia hydrate	NH3*H2O	2.8	A	20	Sill 1981
		1.9	AC	2.3	Schmitt 1998
		0.8	R	2.6	Brown 1988
		2.5	A	20	Bertie 1980

Table 3 (*Continued*)

	UV			Visible			Near IR			Mid IR				Far IR				
cm⁻¹	100000	50000	25000	20000	16666	14285	10000	5000	2500	1428	1000	666	400	200	100	50	20	10
μm	0.10	0.20	0.40	0.5	0.6	0.7	1	2	4	7	10	15	25	50	100	200	500	1000

Name	Molecule	Range start (μm)	Type	Range end (μm)	Reference
Hydrazine	N2H4	2.5	Od	20	Boudin 1998
Hydrogen Isocyanide	HNC				
Hydrogen Cyanide	HCN	6.5	k	50	Coustenis 1999
		2.0	A	22	Masterson 1990
		1.0	R	2.5	Brown 2000
		2.0	A	125	DelloRusso 1996
Poly-HCN	poly HCN	2.0	A	2.5	Cruikshank 1991
		0.40	R	3.8	Owen 2001
Acetonitrile	CH3CN	2.0	A	20	Bernstein 1997b
		2.0	A	125	DelloRusso 1996
		2.5	A	25	d'Hendecourt 1986
Ethylamine	CH3CH2CN	7.7	k	50	Coustenis 1999
		2.0	A	125	DelloRusso 1996
Vinyl Cyanide	CH2CHCN	2.0	A	125	DelloRusso 1996
Propinonitrile	C2H5CN	2.0	A	125	DelloRusso 1996
Methylcyanoacetylene	CH3C3N	2.0	A	125	DelloRusso 1996
Cyanoacetylene	HC3N	6.5	k	50	Coustenis 1999
		2.0	A	22	Masterson 1990
		2.0	A	125	DelloRusso 1996
Cyanogen	C2N2	2.0	A	125	DelloRusso 1996
		2.5	A	20	Verderame 1963
Dicyanoacetylene	C4N2	6.6	k	50	Coustenis 1999
		2.0	A	22	Masterson 1990
Ammonium Nitrate	NH4NO3	2.5	A	20	Grim 1989
		2.8	A	15	Keller 1949
Hydroxylamine	NH2OH	2.5	A	20	Grim 1989
Cyanate ion	OCN-	2.8-3	A		Schettino 1970
			A	4.5-4.7	
		7.5-7.9	A	8.1-8.5	Schettino 1970
		14.5	A	18.1	Schettino 1970
	XCN-	4.5	Od	5.0	Grim 1987a
Isocyanic Acid	HNCO	200	A	1000	Cugley 1973
		2.5	A	20	Lowenthal 2002
		2.5	A	25	Khanna 2002
Methyl isocyanate	CH3NCO	2.5	A	25	d'Hendecourt 1986
Formamide	HCONH2	5.5	A	8.3	Grim 1989
		2.5	A	25	Khanna 2002
		2.5	A	50	King 1971
	HCO2NH2	2.5	A	25	Khanna 2002
Urea	H2NCONH2	2.0	A	40	Stewart 1957
		5.5	A	8.3	Grim 1989
Thiourea	H2NCSNH2	2.0	A	40	Stewart 1957
	HCOO- +NH4	2.5	A	10	Grim 1989
Ammonium Hydrosulfide	NH4HS	1.2	R	4.0	Fink 1982
		2.6	A	2[illegible]	Fink 1982
		1.2	R	2.5	Kieffer 1974
Thiocyanic Acid	HNCS	200	A	1000	Cugley 1973
	HSNC	3.8	A	14.7	Wierzejewska 2001
Hydrogen Sulfide	H2S	1.6	R	4.0	Ferraro 1977
		1.5	R	4.0	Fink 1982
		2.6	A	20	Fink 1982
		2.5	A	20	Salama 1990
		1.2	R	2.5	Kieffer 1974
		3.6	A	4.3	Reding 1957
Sulfur Dioxide	SO2	6.0	R	14	Slobodkin 1980
		0.30	R	1.0	Doute 2001
		2.0	R	4.5	Doute 2001
		3.7	A	22	Sandford 1993
		3.3	A	20	Moore 1984
		7.1	A	8.7	Quirico 1996
		4.0	AC	4.2	Schmitt 1998
		1.9	A	2.2	Quirico 1996
		3.6	A	3.8	Quirico 1996
		3.8	A	4.6	Quirico 1996

Table 3 (*Continued*)

Name	Molecule	Spectral coverage (UV / Visible / Near IR / Mid IR / Far IR): start (μm) type – end (μm)	Reference
	cm⁻¹	100000, 50000, 25000, 20000, 16666, 14285, 10000, 5000, 2500, 1428, 1000, 666, 400, 200, 100, 50, 20, 10	cm⁻¹
	μm	0.10, 0.20, 0.40, 0.5, 0.6, 0.7, 1, 2, 4, 7, 10, 15, 25, 50, 100, 200, 500, 1000	μm
Sulfur Dioxide	SO2	2.5 A – 20	Salama 1990
		11.1 A – 50	Khanna 1995
		2.5 k –	Schmitt 1998
		3.2 R – 4.8	McCord 1998
		0.8 T – 2.6	Quirico 1996
		1.4 – 2.5	Quirico 1996
		1.9 A – 2.6	Schmitt 1994
		1.9 AC – 4.0	Schmitt 1994
		2.0 R – 25	Nash 1995
Sulfur Trioxide	SO3	2.5 A – 50	Khanna 1995
		2.5 A – 50	Khanna 1995
		2.5 A – 50	Khanna 1995
Sulfuric Acid	H2SO4	13.3 A – 33	Giguere 1963
		1.0 R – 2.5	Buie 2000
	•8H2O	1.0 R – 2.5	Carlson 1999
	•8H2O	1.0 R – 2.5	Carlson 1999
	•6.5H2O	1.0 R – 2.5	Carlson 1999
Sulfur	S2	0.38 E – 0.5	Grim 1987b
		3.2 A – 3.8	Moore1988
		1.5-2.8 A – A 3.2-3.8	Moore 1988
Magnesium Sulfate	MgSO4	1.0 R – 2.5	McCord 2002
Undecahydrate	MgSO4•11H2O	0.7 R – 2.5	**Dalton 2005**
Epsomite	MgSO4•7H2O	1.0 R – 2.5	Carlson 1999
		0.5 R – 2.5	Buie 2000
		1.0 R – 2.6	McCord 2002
		1.0 R – 2.5	McCord 2002
		0.40 R – 2.5	Crowley 1991
		0.7 R – 2.5	**Dalton 2005**
Hexahydrite	MgSO4•6H2O	1.0 R – 2.5	McCord 1999
		1.0 R – 2.5	Carlson 1999
			McCord 2002
		0.40 R – 2.5	Crowley 1991
		0.7 R – 2.5	**Dalton 2003**
Pentahydrite	MgSO4•5H2O	0.40 R – 2.5	Crowley 1991
Starkeyite	MgSO4•4H2O	1.0 R – 2.7	McCord 2002
		0.40 R – 2.5	Crowley 1991
		0.7 R – 2.5	**Dalton 2005**
Mirabilite	Na2SO4•10H2O	0.40 R – 2.5	Crowley1991
		1.0 R – 2.6	McCord 2002
		0.7 R – 2.5	**Dalton 2005**
Bloedite	Na2Mg(SO4)2•4H2O	0.40 R – 2.5	Crowley 1991
		0.5 R – 2.5	Buie 2000
		0.7 R – 2.5	**Dalton 2003**
Bassanite	2CaSO4•H2O	0.40 R – 2.5	Crowley 1991
Gypsum	CaSO4•2H2O	5.0 n&k – 25	Roush 1996b
		0.40 R – 2.5	Crowley 1991
Syngenite	K2Ca(SO4)2•H2O	0.40 R – 2.5	Crowley 1991
Methyl Clathrate	CH4•H2O	2.5 A – 50	Khare 1989
		1.0 R – 6.0	Smythe 1975
	CO2•H2O	1.0 R – 6.0	Smythe 1975
H2O•graphite		0.30 Geo Albedo – 3.9	Owen 2001
Bisodium Sulfate	Na2SO4	1.0 R – 2.5	McCord 2002
		0.40 R – 2.5	Crowley 1991
Bisodium Carbonate	Na2CO3	1.0 R – 2.5	McCord 2002
		0.7 R – 2.5	**Dalton 2005**
Calcium Carbonate	CaCO3	0.40 R – 2.5	Crowley 1991
		5.0 n&k – 25	Roush 1996b
Dolomite	(Ca,Mg)(CO3)2	2.0 n&k – 100	Henning 1999
Beryl	Be3Al(SiO2)6	3.0 R – 5.25	McCord 1998
Bronzite	(Mg,Fe2+)2(SiO3)2	10 n&k – 100	Henning 1999

Table 3 (*Continued*)

Name	Molecule	Start (µm)	Data	End (µm)	Reference
Amorphic Pyroxene	(Mg,Fe)SiO3	10	n&k	100	Henning 1999
Pyroxene Bronzite	Mg0.8Fe0.2SiO3	10	n&k	100	Henning 1999
Burkeite	Na6CO3(SO4)2	0.40	R	2.5	Crowley 1991
Cordierite	Mg2Al4Si5O18	3.0	R	5.25	McCord 1998
Eugsterite	Na4Ca(SO4)3•2H2O	0.40	R	2.5	Crowley 1991
Gaylussite	Na2Ca(CO3)2•5H2O	0.40	R	2.5	Crowley 1991
Goethite	FeOOH	0.40	R	2.5	Owen 2001
		3.0	R	5.25	McCord 1998
		0.40	R	2.5	Singer 1985
		2.5	R	40	Bell 1995
Hematite	Fe2O3	0.40	R	2.5	Singer 1985
		0.40	R	2.5	Owen 2001
Jarosite	KFe3(SO4)2(OH)6	0.40	R	2.5	Owen 2001
Kaolinite	Al2Si2O(OH)4	0.6	R	4.6	Roush 1987b
		0.40	R	2.5	Singer1985
		0.5	R	2.5	Buie 2000
		1.0	R	2.5	Brown 2000
		5.0	R	25	Roush 1991
Leonite	K2Mg(SO4)2•4H2O	0.40	R	2.5	Crowley 1991
Lepidocrocite	FeOOH	2.5	A	25	Bell 1995
Magnetite	Fe(II)Fe(III)2O4	2.5	A	25	Bell 1995
		0.5	Geo Albedo	5.5	Roush 1990
Palagonite		0.6	R	4.6	Roush 1987b
		0.5	R	5.0	Roush1990
		2.7	R	4.1	Owen 2001
		5.0	R	25	Roush 1991
Montmorillonite		0.6	R	4.6	Roush1987b
		0.5	R	5	Roush 1990
		0.40	R	2.5	Singer1985
		0.5	R	2.5	Buie 2000
		2.7	R	4.1	Owen 2001
		1.0	R	2.5	Brown 2000
		5.0		25	Roush 1991
Nahcolite	NaHCO3	0.40	R	2.5	Crowley 1991
		5.0	A	25	Roush 1996b
Nontronite		0.40	R	2.5	Singer1985
Natron	Na2CO3•10H2O	1.0	R	2.5	Carlson 1999
		0.40	R	2.5	Crowley1991
		1.0	R	2.6	McCord 2002
Olivine	(Mg,Fe)2SiO4	0.40	R	2.5	Roush 1983
		0.6	R	4.6	Roush 1987a
Iron Oxide	FeO	10	n&k	100	Henning 1999
Iron Sulfide	FeS	50	n&k	100	Henning 1999
Plagioclase	(Na,Ca)(Si,Al)4O8	0.6	R	4.6	Roush 1987a
Picromerite	K2Mg(SO4)2•6H2O	0.40	R	2.5	Crowley 1991
Amorphic SiO2 Quartz	SiO2	10	n&k	100	Henning 1999
Pirssonite	Na2Ca(CO3)2•2H2O	0.40	R	2.5	Crowley 1991
Polyhalite	K2Ca2Mg(SO4)4•2H2O	0.40	R	2.5	Crowley 1991
Pyrophyllite	Al2Si4O10(OH)2	0.6	R	4.6	Roush 1987c
		0.5	R	5	Roush 1990
		5.0	R	25	Roush 1991
Ortho-Pyroxene	XY(Si,Al)2O6	5.0	R	25	Roush 1991
		5.0	R	25	Roush 1991
		0.40	R	2.5	Roush 1983
		0.6	R	4.6	Roush 1987a
Clino-Pyroxene	XY(Si,Al)2O6	0.40	R	2.5	Roush 1983
		5.0	R	25	Roush 1991
Serpentine		0.5	Geo Albedo	5.5	Roush 1990
		5.0	R	25	Roush 1991
Tholin	C3H5N2		n&k	25	Imanaka 2004
Kerogen	C200H300SN5O11	0.10	n&k	40	Khare 1989
Trona	Na3CO3HCO3•2H2O	5.0	A	25	Roush 1996b
		0.40	R	2.5	Crowley 1991

Table 4 Critical properties required for laboratory spectra of candidate icy satellite surface compounds. Availability of high-quality laboratory spectra with these properties will permit estimation of surface abundances through modeling of spacecraft observations

Property	Minimum Properties	Functional Enhancements
Wavelength Range	0.3–5.5 μm	0.1–25 (or 1000) μm
Temperature	50–150 K	30–300 K
Unit	Reflectance	Optical Constants ($n + ik$)
Sample Thickness	> 20 μm	Up to several mm

analysis, reference spectra are needed with four specific characteristics appropriate to the observations: these are listed in Table 4.

Section 2 described the wavelength ranges of contemporary spacecraft mission instruments, summarized in Table 2. The infrared imaging spectrometers used for most surface composition studies generally operate in the 1 to 5.5 μm range, but extension can be made to .3 μm to include visible spectrometers and/or camera systems; also, the 5- to 7-μm range is home to many diagnostic mid-infrared fundamentals and may prove useful as instruments become more sensitive. Thermal and far-infrared instruments, with some relevance to surface compositional studies (and much relevance to atmospheres and rings) operate out to at least 25 μm and even 1000 μm (e.g. Cassini CIRS). Ultraviolet spectroscopy of icy satellite surfaces has proven useful as well. So, the wavelength range 0.3–5.5 μm should be considered critical for laboratory work in support of spacecraft observations, but coverage from 0.1 to 25 (or even 1000) μm desirable for eventual compilation of spectral libraries.

Table 1 summarizes temperature ranges of icy satellites. The boiling and melting points of many volatile compounds allow them to persist on some of these. The temperatures of the Jovian and Saturnian satellites, for which we have the most data, range from 50 to 150 K in general, but Pluto/Charon, Triton and Nereid, the Kuiper Belt objects, transNeptunian objects and Centaurs can have temperatures as low as 30 K while Mars is known to reach as high as 300 K; this temperature is also relevant to terrestrial studies. Measurements near STP (standard temperature and pressure: 1 atm and 298 K) are also easier to reproduce and can serve as a valuable check on the veracity of methods. So while the 50–150 K range is of critical interest in quantifying spectral behavior of planetary ices, measurements from 30 to 300 K are desirable. Icy bodies can manifest broad variations in surface temperature, and near-infrared spectra often show noticeable changes over ranges as small as a few Kelvins (cf. Hudgins et al. 1993; Grundy and Schmitt 1998; Hinrichs and Lucy 2002; Dalton et al. 2005). Temperature steps of ten Kelvins would be useful, but may be impractical: ices in vacuum may change over the times required to conduct such detailed measurements. Laboratory measurements separated by more than 50 K are likely to miss subtle and useful temperature-dependent characteristics, so this should be considered a minimum separation while ten or even one Kelvin would be desirable if within the investigators' capabilities.

Reflectance spectra, which are relatively easy to acquire, are needed at a minimum and enable simple linear mixture models to produce first-order estimates of material abundances. If optical constants can be derived as well (whether from a combination of reflectance/transmittance measurements, or by estimation using a combination of radiative transfer theory and subtractive Kramers-Kronig analysis), this will permit more advanced, nonlinear mixture analysis that can also estimate surface properties (such as porosity) and provide improved abundance estimates while assuming a range of grain sizes—including those not measured. So, reflectance spectra at a minimum are needed; and optical constants would be desirable.

Measurements must be on samples sufficiently thick to manifest the near-infrared features of interest to remote sensing of icy satellites, i.e., much greater than the wavelength. For minerals and volatile ices, in the visible to near infrared, this is generally at least 10 to 20 μm but in practice samples of a few mm are typically used to ensure sufficient optical depth. A range of grain sizes should also be examined if possible: for example, a set of measurements using 1–25 μm, 25–100 μm, and 100–250 μm would provide sufficient range to permit optimization of optical constant derivation, and evaluation of relative effects of grain size, volume vs internal scattering, and other parameters (Arhenkiel 1971; Salisbury et al. 1991; Lucey 1998; Dalton and Pitman 2010).

The different compounds in Table 3 vary in likelihood of occurrence and importance to studies of surface composition with attention to exchange processes involving the interiors. For missions to Jupiter and the Galilean satellites, hydrated materials (salts and acids), sulfur frosts and silicate lavas, radiolysis and photolysis products, and simple volatile ices rise to prominence. Simple organic or other hydrocarbons that may be produced by irradiation of these ices are important to studies of all the gas giant satellites, while nitrogen ices, methane ices and more complex organics are anticipated farther from the Sun. Alkanes, alkenes and alkynes, simple short-chain aliphatic hydrocarbons, and volatile organics like C_2H_2 and C_3H_8 ices are important for the Saturnian satellites. Titan is a special case, with a plethora of hydrocarbon species that need to be measured. For the New Horizons mission expected to arrive at Pluto in 2015, CH_4, N_2, CO and ice mixtures require further characterization. Other compounds have varying levels of relevance and importance, but there is always the possibility that any of these materials may be out there. The only way to know for certain is to look for them.

6.2 Summary and Recommendations

Many compounds have been proposed as constituents of icy satellite surfaces in the outer solar system. Laboratory spectroscopy of these materials under carefully controlled conditions will permit researchers to search for them in spectral observations from past, current and planned spacecraft missions. Most importantly, these measurements need to have four characteristics: (1) wavelength coverage from 0.3 to 5.5 μm at a minimum; 0.1 to 25 (and even 1000) μm would provide additional capability likely to be of use in the near future. (2) Temperatures should be relevant to the outer solar system, i.e. 50–150 K (Jovian and Saturnian satellites) or 30–300 K to cover the range from Mars to the Kuiper Belt. (3) Reflectance spectra are required, at a minimum, to permit linear modeling. Optical constants (*both* real and imaginary indices of refraction) if obtained, will enable more advanced modeling with a variety of assumed grain sizes. (4) Samples must be of sufficient thickness for near-infrared absorptions, which are orders of magnitude weaker than their mid-infrared counterparts, to manifest. This means at least 10–20 μm thick samples, though sample sizes on the order of a few mm are generally required to approach optical depth unity. Meeting these four conditions will provide useful reference spectra that have been desperately sought by researchers for decades. Furthermore, measurements of multiple grain size fractions would be particularly useful, as this would provide information on spectral effects of volume and surface scattering, as well as improve derivation of optical constants. Attention to these details has the potential to significantly enhance the scientific return of remote sensing observations of icy bodies, and improve understanding of the processes that produce, destroy and distribute the materials comprising their surfaces and interiors.

Acknowledgements The author wishes to express his gratitude to Barbara Amago, Robert Powers and Marion Hernandez of the Jet Propulsion Laboratory Technical Library for their assistance with this article. Corey Jamieson, Jake Cooper and Saveelah Talib contributed to Table 3. This work was supported by the NASA Outer Planets Research Program and the Jet Propulsion Laboratory Research and Technology Development Program.

References

R.M. Adams, J.J. Katz, J. Opt. Soc. Am. **46**, 895–898 (1956)
J.B. Adams, M.O. Smith, A.R. Gillespie, in *Remote Geochemical Analysis*, ed. by C.M. Pieters, P.J. Englert (Cambridge University Press, Cambridge, 1993)
R.K. Arhenkiel, J. Opt. Soc. Am. **61**, 1651–1655 (1971)
L.J. Allamandola, A.G.G.M. Tielens, R.J. Barker, Astrophys. J. **290**, L25 (1985)
L. Allamandola, S. Sandford, G. Valero, Icarus **76**, 225 (1988)
C. Bacher, G.S. Tyndall, J.J. Orlando, J. Atmos. Chem. **39**(2), 171–189 (2001)
R. Baragiola, D. Bahr, J. Geophys. Res. **103**(E11), 22 (1998) 865
J. Bell, T. Roush, R. Morris, J. Geophys. Res. **100**(E3), 5297 (1995)
M.J. Belton, K.P. Klaasen, M.C. Clary, J.L. Anderson, C. Anger, M.H. Carr, C.R. Chapman, M.E. Davies, R. Greeley, D. Anderson, L.K. Bolef, T.E. Townsend, R. Greenberg, J.W. Head III, G. Neukum, C.B. Pilcher, J. Veverka, P.J. Gierasch, F. Fanale, A.P. Ingersoll, H. Masursky, D. Morrison, J.B. Pollack, Space Sci. Rev. **60**, 413–455 (1992)
M.J. Belton, the Galileo Imaging Team, Adv. Space Res. **26**(10), 1641–1647 (2000)
M. Bernstein, L. Allamandola, S. Sandford, Adv. Space Res. **19**(7), 991 (1997a)
M.P. Bernstein, S.A. Sandford, L.J. Allamandola, Astrophys. J. **476**, 932–942 (1997b)
M.P. Bernstein, S.A. Sandford, L.J. Allamandola, J.S. Gillette, S.J. Clemette, R.N. Zare, Science **283**, 1135–1138 (1999)
M.P. Bernstein, J.P. Dworkin, S.A. Sandford, L.J. Allamandola, Meteorit. Planet. Sci. **36**, 351–358 (2001)
M.P. Bernstein, D.P. Cruikshank, S.A. Sandford, Icarus **179**, 527–534 (2005)
M.P. Bernstein, D.P. Cruikshank, S.A. Sandford, Icarus **181**, 302–308 (2006)
J. Bertie, J.P. Devlin, J. Chem. Phys. **81**, 1559–1572 (1984)
J. Bertie, M. Morrison, J. Chem. Phys. **73**(10), 4832 (1980)
J. Bertie, M.R. Shehata, J. Chem. Phys. **81**, 27–30 (1984)
J. Bertie, E. Whalley, J. Chem. Phys. **40**(6), 1646 (1964)
J. Bertie, H. Labbe, E. Whalley, J. Chem. Phys. **50**(10), 4501 (1969)
R. Bohn, S. Sandford, L. Allamandola, D. Cruikshank, Icarus **111**, 151–173 (1994)
N. Boudin, W. Schutte, J. Greenberg, Astron. Astrophys. **331**, 749 (1998)
A.L. Broadfoot, Space Sci. Rev. **21**(2), 183–205 (1977)
A. Bromberg, S. Kimel, A. Ron, Chem. Phys. Lett. **46**(2), 262 (1977)
M. Brown, W. Calvin, Science **287**, 107 (2000)
R.H. Brown, D.P. Cruikshank, A.L. Tokunaga, R.G. Smith, R.N. Clark, Icarus **74**, 262–271 (1988)
R.H. Brown et al., Space Sci. Rev. **115**, 111–168 (2004)
R.H. Brown, R.N. Clark, B.J. Buratti, D.P. Cruikshank, J.W. Barnes, R.M.E. Mastrapa, J. Bauer, S. Newman, T. Momary, K.H. Baines, G. Bellucci, F. Capaccioni, P. Cerroni, M. Combes, A. Coradini, P. Drossart, V. Formisano, R. Jaumann, Y. Langevin, D.L. Matson, T.B. McCord, R.M. Nelson, P.D. Nicholson, B. Sicardy, C. Sotin, Science **311**, 1425 (2006)
J. Brucato, M. Palumbo, G. Strazzulla, Icarus **125**, 135 (1997)
R. Brunetto, M.A. Barucci, E. Dotto, G. Strazzulla, Astrophys. J. **644**, 646–650 (2006)
M. Buie, W. Grundy, Icarus **148**, 324 (2000)
B. Buratti, J. Veverka, Icarus **64**, 320–328 (1985)
M.H. Burger et al., Space Sci. Rev. (2010, this issue). doi:10.1007/s11214-010-9645-z
B. Cairns, G. Pimentel, J. Chem. Phys. **43**(10), 3432 (1965)
W.M. Calvin, R.N. Clark, R.H. Brown, J.R. Spencer, J. Geophys. Res. **100**, 19041–19048 (1995)
W. Calvin, R. Johnson, J. Spencer, Geophys. Res. Lett. **23**(6), 673 (1996)
R.W. Carlson, P.R. Weissman, W.D. Smythe, J.C. Mahoney, the NIMS team, Space Sci. Rev. **60**, 457–502 (1992)
R.W. Carlson et al., Science **274**, 385–388 (1996)
R.W. Carlson, M.S. Anderson, R.E. Johnson, W.D. Smythe, A.R. Hendrix, C.A. Barth, L.A. Soderblom, G.B. Hansen, T.B. McCord, J.B. Dalton, R.N. Clark, J.H. Shirley, A.C. Ocampo, D.L. Matson, Science **283**, 2062–2064 (1999a)

R.W. Carlson, R.E. Johnson, M.S. Anderson, Science **286**, 5437–5440 (1999b)
R.W. Carlson, M.S. Anderson, R. Mehlman, R.E. Johnson, Icarus **177**, 461–471 (2005)
R.W. Carlson, W.M. Calvin, J.B. Dalton, G.B. Hansen, R.L. Hudson, R.E. Johnson, T.B. McCord, M.H. Moore, in *Europa*, ed. by R.T. Pappalardo, W.M. McKinnon, K. Khurana (2009)
S. Chandrasekhar, *Radiative Transfer* (Dover, New York, 1960)
D. Chapman, J. Chem. Soc. (Lond.) **56**, 225 (1956)
S.B. Charnley, S.D. Rodgers, Y.-J. Kuan, H.-C. Huang, Adv. Space Res. **30**(6), 1419–1431 (2002)
A.F. Cheng et al., Space Sci. Rev. **140**, 189–215 (2008)
C.F. Chyba, Nature **403**, 381–382 (2000)
C. Chyba, C. Sagan, Nature **330**, 350–353 (1987)
R.N. Clark, in *Manual of Remote Sensing*, ed. by A.N. Rencz (Wiley, New York, 1999)
R. Clark, T. McCord, Icarus **41**, 323 (1980)
R.N. Clark, F.P. Fanale, M.J. Gaffey, in *Satellites*, ed. by J. Burns, M.S. Mathews (Univ. Arizona Press, Tucson, 1986)
R.N. Clark et al., Nature **435**, 66–69 (2005)
R.N. Clark, J.M. Curchin, T.M. Hoefen, G.A. Swayze, J. Geophys. Res. **114**(E3), E03001 (2009)
G.D. Cody, H. Ade, C.M. O'D. Alexander, T. Araki, A. Butterworth, H. Fleckenstein, G. Flynn, M.K. Gilles, C. Jacobsen, A.L.D. Kilcoyne, K. Messenger, S.A. Sandford, T. Tyliszczak, A.J. Westphal, S. Wirick, H. Yabuta, Meteorit. Planet. Sci. **43**(1), 353–365 (2008)
J. Comeford, J. Gould, J. Mol. Spectrosc. **5**, 474 (1960)
J.C. Cook, S.J. Desch, T.L. Roush, C.A. Trujillo, T.R. Geballe, Astrophys. J. **663**, 1406–1419 (2007)
A. Coustenis, B. Schmitt, R. Khanna, F. Trotta, Planet. Space Sci. **47**, 1305 (1999)
A. Coustenis et al., Icarus **189**(1), 35–62 (2007)
A. Coustenis et al., Space Sci. Rev. (2010, this issue). doi:10.1007/s11214-009-9615-5
J.K. Crowley, J. Geophys. Res. **96**, 16,231–16,240 (1991)
D.P. Cruikshank, R.H. Brown, in *Planetary Satellites*, ed. by J. Burns, M.S. Mathews (Univ. of Arizona Press, Tucson, 1986)
D.P. Cruikshank, J. Veverka, L.A. Lebofsky, in *Saturn*, ed. by T. Gehrels, M.S. Mathews (Univ. of Arizona Press, Tucson, 1984)
D. Cruikshank, L. Allamandola, W. Hartmann, D. Tholen, R. Brown, C. Matthews, J. Bell, Icarus **94**, 345 (1991)
D.P. Cruikshank et al., Science **261**, 742–745 (1993)
D.P. Cruikshank, R.H. Brown, W.M. Calvin, T.L. Roush, in *Solar System Ices*, ed. by B. Schmitt et al. (Kluwer, Dordrecht, 1998)
D.P. Cruikshank, J.B. Dalton, C.M. Dalle Ore, J. Bauer, K. Stephan, G. Filacchione, A.R. Hendrix, C.J. Hansen, A. Coradini, P. Cerroni, F. Tosi, F. Capaccioni, R. Jaumann, B.J. Buratti, R.N. Clark, R.H. Brown, R.M. Nelson, T.B. McCord, K.H. Baines, P.D. Nicholson, C. Sotin, A.W. Meyer, G. Belluci, M. Combes, J.-P. Bibring, Y. Langevin, B. Sicardy, D.L. Matson, V. Formisano, P. Drossart, V. Mennella, Nature **448**(7149), 54–55 (2007)
D.P. Cruikshank, A.W. Meyer, R.H. Brown, R.N. Clark, R. Ralf Jaumann, K. Stephan, C.A. Hibbitts, S.A. Sandford, R.M.E. Mastrapa, G. Filacchione, C.M. Dalle Ore, P.D. Nicholson, B.J. Buratti, R.M. Nelson, J.B. Dalton, K.H. Baines, D.L. Matson, Icarus **206**(2), 561–572 (2010)
J. Cugley, A. Pullin, Chem. Phys. Lett. **19**(2), 203 (1973)
J.B. Dalton, Geophys. Res. Lett. **34**, L21205 (2007). doi:10.1029/2007GL031497
J.B. Dalton, in *Wkshp. Sci. Solar System Ices*, #9028, Oxnard CA, May 5–8, 2008
J.B. Dalton, K.M. Pitman, J. Geophys. Res. (2010, submitted)
J.B. Dalton, O. Prieto-Ballesteros, C.S. Jamieson, J.S. Kargel, J. Jolivet, R.C. Quinn, Icarus **177**, 472–490 (2005)
J.B. Dalton, J.H. Shirley, L.M. Prockter, L.W. Kamp, in *Proc. Lunar Planet. Sci. XL* (2009), p. 2511
J.B. Dalton, D.P. Cruikshank, K. Stephan, T.B. McCord, A. Coustenis, R.W. Carlson, A. Coradini, Space Sci. Rev. (2010, this issue). doi:10.1007/s11214-010-9665-8
A.G. Davies, R. Lopes-Gautier, W.D. Smythe, R.W. Carlson, Icarus **148**(1), 211–225 (2000)
C. De Bergh, B. Schmitt, L.V. Moroz, E. Quirico, D.P. Cruikshank, in *The Solar System Beyond Neptune*, ed. by M.A. Barucci, H. Boehnhardt, D.P. Cruikshank, A. Morbidelli (University of Arizona Press, Tucson, 2008), pp. 483–506, 592
M.L. Delitsky, A.L. Lane, J. Geophys. Res. **102**, 16,385 (1997)
M.L. Delitsky, A.L. Lane, J. Geophys. Res. **103**, 31,391–31,403 (1998)
N. Dello Russo, R. Khanna, Icarus **123**, 366 (1996)
N. Dello Russo, R. Khanna, M. Moore, J. Geophys. Res. **98**(E3), 5505 (1993)
A.J. Dempster, G. Serbi, J. Chem. Phys. **54**(8), 3600 (1971)
L. d'Hendecourt, L. Allamandola, Astron. Astrophys. Suppl. Ser. **64**, 453 (1986)

S. Doute, B. Schmitt, L. Soderblom, J. Shirley, the Galileo NIMS Team, Icarus **149**, 107 (2001)
K. Dressler, O. Schnepp, J. Chem. Phys. **33**(1), 270 (1960)
P. Ehrenfreund, A. Boogert, P. Gerakines, D. Jansen, W. Schutte, A. Tielens, E. van Dishoeck, Astron. Astrophys. **315**, L34 (1996)
P. Ehrenfreund, O. Kerkhof, W. Schutte, A. Boogert, P. Gerakines, E. Dartois, L. d'Hendecourt, A. Tielens, E. van Dishoeck, D. Whittet, Astron. Astrophys. **350**, 240 (1999)
J.P. Emery, D.M. Burr, D.P. Cruikshank, R.H. Brown, J.B. Dalton, Astron. Astrophys. **435**, 353–362 (2005)
A. Emslie, J. Aronson, Appl. Opt. **12**, 2563–2572 (1973)
L.W. Esposito, C.A. Barth, J.E. Colwell, G.M. Lawrence, W.E. McClintock, A.I.F. Stewart, H.U. Keller, A. Korth, H. Lauche, M.C. Festou, A.L. Lane, C.J. Hansen, J.N. Maki, R.A. West, H. Jahn, R. Reulke, K. Warlich, D.E. Shemansky, Y.L. Yung, Space Sci. Rev. **115**, 294–361 (2004)
G. Ewing, J. Chem. Phys. **37**(10), 2250 (1962)
G. Ewing, G. Pimentel, J. Chem. Phys. **35**(3), 925 (1961)
J. Ferraro, U. Fink, J. Chem. Phys. **67**(2), 409 (1977)
J. Ferraro, G. Sill, U. Fink, Appl. Spectrosc. **34**(5), 525–533 (1980)
U. Fink, H. Larson, Icarus **24**, 411 (1975)
U. Fink, G. Sill, in *Comets*, ed. by L.E. Wilkening (Univ. Arizona Press, Tucson, 1982), pp. 164–202
U. Fink, N.H. Dekkers, H.P. Larson, Astrophys. J. **179**, L155–L159 (1973)
F.M. Flasar et al., Space Sci. Rev. **115**, 169–297 (2004)
P. Gerakines, W. Schutte, P. Ehrenfreund, Astron. Astrophys. **312**, 289 (1996)
P. Gerakines, M. Moore, R. Hudson, Astron. Astrophys. **357**, 793 (2000)
P.A. Gerakines, M.H. Moore, R.L. Hudson, J. Geophys. Res. **106**, 33381–33385 (2001)
P.A. Gerakines, J.J. Bray, A. Davis, C.R. Richey, Astrophys. J. **620**, 1140–1150 (2005)
P. Giguere, R. Savoie, J. Am. Chem. Soc. **85**(1), 287 (1963)
S. Gosse, D. Labrie, P. Chylek, Appl. Opt. **34**(28), 6582 (1995)
R. Grim, J. Greenberg, Astrophys. J. **321**, L91 (1987a)
R. Grim, J. Greenberg, Astron. Astrophys. **181**, 155 (1987b)
R. Grim, J. Greenberg, M. de Groot, F. Baas, W. Schutte, B. Schmitt, Astron. Astrophys. Suppl. Ser. **78**, 161 (1989)
W.M. Grundy, B. Schmitt, J. Geophys. Res. **103**, 25,809–25,822 (1998)
W. Grundy, B. Schmitt, E. Quirico, Icarus **105**, 254 (1993)
W. Grundy, B. Schmitt, E. Quirico, Icarus **155**, 486 (2002)
W.M. Grundy, L.A. Young, J.R. Spencer, R.E. Johnson, E.F. Young, M.W. Buie, Icarus **184**, 543–555 (2006)
W. Hage, A. Hallbrucker, E. Mayer, J. Am. Chem. Soc. **115**, 8,427 (1993)
R. Hanel, B. Conrath, M. Flasar, V. Kunde, P. Lowman, W. Maguire, J. Pearl, J. Pirraglia, R. Samuelson, D. Gautier, P. Gierasch, S. Kumar, C. Ponnamperuma, Science **204**(4396), 972–976 (1979)
R. Hanel et al., Appl. Opt. **19**, 1391 (1980)
G.B. Hansen, J. Geophys. Res. **102**, 21569–21587 (1997)
G.B. Hansen, J. Geophys. Res. E. **110**, E11003 (2005)
G. Hansen, T. McCord, Lunar Planet. Sci. **28**, 1565 (1997)
B.W. Hapke, J. Geophys. Res. **86**, 3039–3054 (1981)
B.W. Hapke, *Reflectance and Emittance Spectroscopy* (Cambridge Univ. Press, Cambridge, 1993)
B. Hapke, H. Van Horn, J. Geophys. Res. **68**, 4545–4570 (1963)
A.R. Hendrix et al., J. Geophys. Res. **104**, 14169–14178 (1999)
Th. Henning, H. Mutschke, Astron. Astrophys. **327**, 743 (1997)
Th. Henning, V.B. Il'in, N.A. Krivova, B. Michel, N.V. Voshchinnikov, Astron. Astrophys. Suppl. **136**, 405–406 (1999)
C.A. Hibbitts, J. Szanyi, Icarus **191**, 371–380 (2007)
C.A. Hibbitts, T.B. McCord, G.B. Hansen, J. Geophys. Res. **105**, 22541–22558 (2000)
C.A. Hibbitts, R.T. Pappalardo, G.B. Hansen, T.B. McCord, J. Geophys. Res. **108**(E5), 5036 (2003)
J.L. Hinrichs, P.G. Lucy, Icarus **155**, 169–180 (2002)
S. Hirabayashi, Y. Hirahara, Chem. Phys. Lett. **361**, 265–270 (2002)
R. Hodyss, C.D. Parkinson, P.V. Johnson, J.V. Stern, J.D. Goguen, Y.L. Yung, I. Kanik, Geophys. Res. Lett. **36**, L17103 (2009). doi:10.1029/2009GL039336
C.W. Hord et al., Science **253**(5027), 1548–1550 (1991)
C.W. Hord, W.E. McClintock, A.I.F. Stewart, C.A. Barth, L.W. Esposito, G.E. Thomas, B.R. Sandel, D.M. Hunten, A.L. Broadfoot, D.E. Shemansky, J.M. Ajello, A.L. Lane, R.A. West, Space Sci. Rev. **60**, 503–530 (1992)
D.M. Hudgins, S.A. Sandford, L.J. Allamandola, A.G.G.M.X. Tielens, Astrophys. J. Suppl. Ser. **86**, 713–869 (1993)
R.L. Hudson, M.H. Moore, A.M. Cook, Adv. Space Res. **36**, 184–189 (2005)

R.L. Hudson, M.H. Moore, L.L. Raines, Icarus **203**, 677–680 (2009)
H. Imanaka, B.N. Khare, J.E. Elsila, E.L.O. Bakes, C.P. McKay, D.P. Cruikshank, S. Sugita, T. Matsui, R.N. Zare, Icarus **168**, 344–366 (2004)
M.J. Irvine, J.G. Mathieson, A.D.E. Pullin, Aust. J. Chem. **35**, 1971–1977 (1982)
M. Jacox, D. Milligan, J. Chem. Phys. **54**(3), 919 (1971)
H.J. Jödl, in *Chemistry and Physics of Matrix-Isolated Species*, ed. by L. Andrews, M. Moskovits (North-Holland, Amsterdam, 1989)
R.E. Johnson, in *Solar System Ices*, ed. by B. Schmitt, C. de Bergh, M. Festou (Kluwer Acad., Dordrecht, 1998)
R.E. Johnson, W.A. Jesser, Astrophys. J. **480**, L79–L82 (1997)
W. Keller, R. Halford, J. Chem. Phys. **17**, 26 (1949)
R. Khanna, M. Ospina, G. Zhao, Icarus **73**, 527 (1988)
R. Khanna, J. Pearl, R. Dahmani, Icarus **115**, 250 (1995)
R. Khanna, M. Lowenthal, H. Ammon, M. Moore, Astrophys. J. Suppl. Ser. **140**, 457 (2002)
B.N. Khare, W. Thompson, B. Murray, C. Chyba, C. Sagan, E. Arakawa, Icarus **79**, 350 (1989)
H.H. Kieffer, W.D. Smythe, Icarus **21**, 506–512 (1974)
S. King, J. Phys. Chem. **75**(3), 405 (1971)
A.D. Kirshenbaum, A.G. Streng, J. Chem. Phys. **35**(4), 1440 (1961)
P. Kubelka, F. Munk, Z. Tch. Fiz. **12**, 593–601 (1931)
L. Kou, D. Labrie, P. Chylek, Appl. Opt. **32**(19), 3531 (1993)
G.P. Kuiper, Astrophys. J. **100**, 378–383 (1944)
G.P. Kuiper, Astron. J. **62**, 295 (1957)
A. Landau, E. Allin, H. Welsh, Spectrochim. Acta **18**, 1 (1962)
A.L. Lane, R.M. Nelson, D.L. Matson, Nature **292**, 38–39 (1981)
A. Léger, J.L. Puget, Astron. Astrophys. **137**, L5 (1984)
G.E. Leroi, G.E. Ewing, G.C. Pimentel, J. Chem. Phys. **35**, 025 (1961)
J.S. Lewis, Icarus **16**, 241–252 (1972)
C.F. Lillie et al., Space Sci. Rev. **21**(2), 159–181 (1977)
M. Lowenthal, R. Khanna, M. Moore, Spectrochim. Acta Part A **58**, 73 (2002)
P.G. Lucey, J. Geophys. Res. **103**, 1703–1713 (1998)
A. Mardyukov, E. Sanchez-Garcia, W. Sander, J. Phys. Chem. A **113**(6), 1086–1095 (2009)
C. Martin, C. Manca, P. Roubin, Surf. Sci. **502**, 280 (2002)
J.V. Martonchik, G.S. Orton, J.F. Appleby, Appl. Opt. **23**, 541–547 (1984)
C.M. Masterson, R.K. Khanna, Icarus **83**, 83–92 (1990)
R.M.E. Mastrapa, R.H. Brown, Icarus **183**, 207–214 (2006)
R.M. Mastrapa, M.P. Bernstein, S.A. Sandford, T.L. Roush, D.P. Cruikshank, C.M. Dalle Ore, Icarus **197**, 307–320 (2008)
R.M. Mastrapa, S.A. Sandford, T.L. Roush, D.P. Cruikshank, C.M. Dalle Ore, Astrophys. J. **701**, 1347–1356 (2009)
T.B. McCord, G.B. Hansen, R.N. Clark, P.D. Martin, C. A Hibbitts, F.P. Fanale, J.C. Granahan, M. Segura, D.L. Matson, T.V. Johnson, R.W. Carlson, W.D. Smythe, G.E. Danielson, the NIMS team, J. Geophys. Res. **103**, 8603–8626 (1998)
T.B. McCord, G.B. Hansen, D.L. Matson, T.V. Johnson, J.K. Crowley, F.P. Fanale, R.W. Carlson, W.D. Smythe, P.D. Martin, C.A. Hibbitts, J.C. Granahan, A. Ocampo, the NIMS team, J. Geophys. Res. **104**, 11,827–11,852 (1999)
T.B. McCord, G. Teeter, G.B. Hansen, M.T. Sieger, T.M. Orlando, J. Geophys. Res. **107**, 4-1–4-6 (2002)
D. Milligan, M. Jacox, J. Chem. Phys. **41**(10), 3032 (1964)
M. Moll, D. Clutter, W. Thompson, J. Chem. Phys. **45**(12), 4469 (1966)
M. Moore, Icarus **59**, 114 (1984)
M. Moore, R. Hudson, Astrophys. J. **401**, 353 (1992)
M. Moore, R. Hudson, Astron. Astrophys. Suppl. Ser. **103**, 45 (1994)
M.H. Moore, R.L. Hudson, Icarus **154**, 282–288 (2000)
M.H. Moore, B. Donn, R.L. Hudson, Icarus **74**, 399–412 (1988)
M.H. Moore, R. Khanna, B. Donn, J. Geophys. Res. **96**, 17,541–17,545 (1991) (ISSN 0148-0227)
M. Moore, R. Ferrante, J. Nuth III, Planet. Space Sci. **44**(9), 927 (1996)
J.M. Moore, B. Bierhaus, C. Chapman, F. Chuang, R.N. Clark, J.B. Dalton, R. Greeley, C. Hibbitts, J. Klemaszewski, P. Schenk, J. Spencer, R. Wagner, in *Jupiter* ed. by F. Bagenal (University of Arizona Press, Tucson, 2004)
M.H. Moore, R.L. Hudson, R.W. Carlson, Icarus **189**, 409–423 (2007)
V.I. Moroz, Astron. Z. **42**, 1287–1295 (1965) (in Russian). Trans Sov. Astron. A. J. **9**, 999–1006

M.J. Mumma, N. Dello Russo, M.A. DiSanti, K. Magee-Sauer, R.E. Novak, S. Brittain, T. Rettig, I.S. McLean, D.C. Reuter, L.-H. Xu, Science **292**(5520), 1334–1339 (2001)
D. Nash, B. Betts, Icarus **117**, 402 (1995)
D.B. Nash, B.B. Betts, in *Solar System Ices*, ed. by B. Schmitt, C. de Bergh, M. Festous (Kluwer Academic, Dordrecht, 1998)
S.F. Newman, B.J. Buratti, R. Jaumann, J.M. Bauer, T.W. Momary, Astrophys. J. **670**, L143–L146 (2007)
F.E. Nicodemus, Appl. Opt. **4**, 767–773 (1965)
K.S. Noll, R.E. Johnson, A.L. Lane, D.L. Domingue, H.A. Weaver, Science **273**, 341–343 (1996)
K.S. Noll, T.L. Roush, D.P. Cruikshank, Y.J. Pendleton, R.E. Johnson, Nature **388**, 45–47 (1997)
G.S. Orton, J.R. Spencer, L.D. Travis, T.Z. Martin, L.K. Tamparri, Science **274**, 389–391 (1996)
T. Owen, D. Cruikshank, C. Dalle Ore, T. Geballe, T. Roush, C. de Bergh, R. Meier, Y. Pendleton, B. Khare, Icarus **149**, 160 (2001)
M. Palumbo, G. Strazzulla, Astron. Astrophys. **269**, 568 (1993)
M. Palumbo, A. Tielens, A. Tokunaga, Astrophys. J. **449**, 674 (1995)
M. Palumbo, A. Castorina, G. Strazzulla, Astron. Astrophys. **342**, 551 (1999)
J. Pearl, M. Ngoh, M. Ospina, R. Khanna, J. Geophys. Res. **96**(E2), 17,477 (1991)
C.B. Pilcher, S.T. Ridgway, T.B. McCord, Science **178**, 1087–1089 (1972)
J.B. Pollack, F.C. Witteborn, F.E. Erickson, D.W. Strecker, J.B. Baldwin, T.E. Bunch, Icarus **36**, 271–303 (1978)
C. Porco et al., Space Sci. Rev. **115**, 363–497 (2004)
L.M. Prockter et al., Space Sci. Rev. (2010, this issue). doi:10.1007/s11214-010-9649-8
E. Quirico, B. Schmitt, Icarus **127**, 354–378 (1997a)
E. Quirico, B. Schmitt, Icarus **128**, 181–188 (1997b)
E. Quirico, B. Schmitt, R. Bini, P. Salvi, Planet. Space Sci. **44**(9), 973 (1996)
F. Reding, D. Horning, J. Chem. Phys. **27**(5), 1024 (1957)
D.C. Reuter et al., Space Sci. Rev. **140**, 129–154 (2008)
T. Roush, LPI **27**, 1107R (1996a)
T. Roush, J. Geophys. Res. **101**(E1), 2215 (1996b)
T. Roush, J. Geophys. Res. **106**(E12), 33,315 (2001)
T. Roush, R. Singer, LPI **14**, 654R (1983)
T. Roush, R. Singer, T. McCord, LPI **18**, 854R (1987a)
T. Roush, R. Singer, T. McCord, LPI **18**, 856R (1987b)
T. Roush, R. Singer, T. McCord, LPI **18**, 858R (1987c)
T.L. Roush, J.B. Pollack, F.C. Witteborn, J.D. Bregman, J.P. Simpson, Icarus **86**, 355–382 (1990)
T.L. Roush, J.B. Pollack, J. Orenberg, Icarus **94**, 191–208 (1991)
E.E. Russell et al., Space Sci. Rev. **60**, 531–563 (1992)
F. Salama, L. Allamandola, F. Witteborn, D. Cruikshank, S. Sandford, J. Bregman, Icarus **83**, 66 (1990)
J.W. Salisbury, L.S. Walter, N. Vergo, D.M. D'Aria, *Infrared (2.1–2.5 μm) Spectra of Minerals* (Johns Hopkins University Press, Baltimore, 1991)
S. Sandford, L. Allamandola, Astrophys. J. **355**, 357 (1990)
S.A. Sandford, L.J. Allamandola, Icarus **106**, 478–488 (1993)
V. Schettino, I. Hisatsune, J. Chem. Phys. **52**(1), 9 (1970)
P.M. Schenk, W.B. McKinnon, D. Gwynn, J.M. Moore, Nature **410**, 57–60 (2001)
B. Schmitt, C. De Bergh, E. Lellouch, J.P. Maillard, A. Barbe, S. Doute, Icarus **111**, 79 (1994)
B. Schmitt, E. Quirico, F. Trotta, W.M. Grundy, in *Solar System Ices*, ed. by B. Schmitt et al. (Kluwer, Dordrecht, 1998)
W. Schutte, A. Tielens, S. Sandford, Astrophys. J. **382**, 523 (1991)
W. Schutte, P. Gerakines, T. Geballe, E. van Dishoeck, J. Greenberg, Astron. Astrophys. **309**, 633 (1996a)
W. Schutte, A. Tielens, D. Whittet, A. Boogert, P. Ehrenfreund, T. de Graauw, T. Prusti, E. van Dishoeck, P. Wesselius, Astron. Astrophys. **315**, L333 (1996b)
N. Sheppard, H.A. Willis, J.C. Rigg, Pure Appl. Chem. **57**(1), 105–120 (1985)
J.H. Shirley, J.B. Dalton, L.M. Prockter, L.W. Kamp, Icarus (2010, in press)
Y. Shkuratov, Y.S. Grynko, Icarus **173**(1), 16–28 (2005). doi:10.1016/j.icarus.2003.12.022
Y. Shkuratov, L. Starukhina, H. Hoffman, G. Arnold, Icarus **137**(2), 235–246 (1999)
G.S. Sill, R.N. Clark, in *Satellites of Jupiter*, ed. by D. Morrison (Univ. of Arizona Press, Tucson, 1982)
G.S. Sill, U. Fink, J.R. Ferraro, J. Opt. Soc. Am. **70**, 724–739 (1980)
G. Sill, U. Fink, J. Ferraro, J. Chem. Phys. **74**(2), 997 (1981)
R.B. Singer, T.L. Roush, J. Geophys. Res. **90**(B14), 12434–12444 (1985)
L. Slobodkin, I. Buyakov, N. Triput, R. Cess, J. Caldwell, T. Owen, Nature **285**, 211 (1980)
W.H. Smith, G. Leroi, J. Chem. Phys. **45**(1), 1767 (1966)
W.D. Smythe, Icarus **24**, 421–427 (1975)

J.R. Spencer, W.M. Calvin, Astron. J. **124**, 3400–3403 (2002)
J.R. Spencer, L.K. Tamppari, T.Z. Martin, L.D. Travis, Science **284**, 1514–1516 (1999)
P. Staats, H. Morgan, J. Chem. Phys. **31**, 553 (1959)
S.A. Stern, Space Sci. Rev. **140**, 155–187 (2008)
J. Stewart, J. Chem. Phys. **26**(2), 248 (1957)
R. St. Louis, B. Crawford Jr., J. Chem. Phys. **37**, 2156 (1962)
G. Strazzulla, J. Brucato, G. Cimino, M. Palumbo, Planet. Space Sci. **44**(11), 1447 (1996)
P. Thomas, J. Veverka, Icarus **64**, 414–424 (1985)
J.L. Thompson, J.W. Salisbury, Rem. Sens. Env. **45**, 1–13 (1993)
G. Tobie et al., Space Sci. Rev. (2010, this issue). doi:10.1007/s11214-010-9641-3
S. Trakhtenberg, R. Naaman, Thin Solid Films **327**, 499 (1998)
K. Tryka, R. Brown, V. Anicich, D. Cruikshank, T. Owen, Science **261**, 751 (1993)
V. Vaida, D.J. Donaldson, S.J. Strickler, S.L. Stephens, J.W. Birks, J. Phys. Chem. **93**, 506 (1989)
A.J. Verbiscer, D.E. Peterson, M.F. Skrutskie, M. Cushing, P. Helfenstein, M.J. Nelson, J.D. Smith, J.C. Wilson, Icarus **182**, 211–223 (2006)
A.J. Verbiscer, M. Skrutskie, J. Wilson, M. Nelson, P. Helfenstein, Bull. Am. Astron. Soc. **40**, 510 (2008) #61.02
F. Verderame, J. Nabgen, E. Nixon, J. Chem. Phys. **39**(9), 2274 (1963)
R.A. Vidal, D. Bahr, R.A. Baragiola, M. Peters, Science **27**, 1839–1842 (1997)
J.H. Waite et al., Science **311**, 1419–1422 (2006)
J.H. Waite, W.S. Lewis, B.A. Magee, J.I. Lunine, W.B. McKinnon, C.R. Glein, O. Mousis, D.T. Young, T. Brockwell, J. Westlake, M.-J. Nguyen, B.D. Teolis, H.B. Niemann, R.L. McNutt Jr, M Perry, W.-H. Ip, Nature **460**, 487–490 (2009)
S.G. Warren, Appl. Opt. **23**, 1206–1225 (1984)
S.G. Warren, Appl. Opt. **25**, 2650–2674 (1986)
S.G. Warren, R.E. Brandt, J. Geophys. Res. **113**(D14220) (2008). doi:10.1029/2007JD009744
H.A. Weaver, W.C. Gibson, M.B. Tapley, L.A. Young, S.A. Stern et al., Space Sci. Rev. **140**, 75–91 (2008)
W. Weltner, P. Walsh, C. Angell, J. Chem. Phys. **40**(5), 1299 (1964)
G. Wieder, D. Dows, J. Chem. Phys. **37**(12), 1962
M. Wierzejewska, Z. Mielke, Chem. Phys. Lett. **349**, 227 (2001)
W.J. Wiscombe, Appl. Opt. **19**, 1505–1509 (1980)

Space Sci Rev (2010) 153: 249–271
DOI 10.1007/s11214-010-9642-2

Radar Signal Propagation and Detection Through Ice

Wlodek Kofman · Roberto Orosei · Elena Pettinelli

Received: 1 September 2009 / Accepted: 1 March 2010 / Published online: 31 March 2010

Abstract In this paper we describe the existing and planned radar measurements of the planetary bodies. The dielectric properties of water ice and other potential surface and subsurface materials are discussed, as well as their dependency on temperature and structure. We then evaluate the performance of subsurface sounding radars using these parameters. Finally we describe some laboratory technique to help interpret the radar data, presenting some results obtained using dielectric spectroscopy methods.

Keywords Dielectric properties · Radar · Laboratory measurements

1 Introduction—What Can Be Learned from Radar Observations?

Radio-echo sounding (RES), or ice-penetrating radar, is a well established geophysical technique, similar to ground-penetrating radar, which has been employed for more than five decades to investigate the structure of ice sheets and glaciers in Antarctica, Greenland, and the Arctic. It is based on the transmission of radar pulses at frequencies in the MF, HF and VHF portions of the electromagnetic spectrum into the surface to detect reflected signals from subsurface structures (see e.g. Bogorodsky et al. 1985).

Ground-penetrating radar is particularly effective on glaciers because ice is the most transparent natural material in this range of frequencies. The ice-covered Lake Vostok in

W. Kofman (✉)
Laboratoire de Planetologie de Grenoble, CNRS/UJF, BP 53, 38041 Grenoble, France
e-mail: wlodek.kofman@obs.ujf-grenoble.fr

R. Orosei
Istituto Nazionale di Astrofisica, Istituto di Fisica dello Spazio Interplanetario,
Via del Fosso del Cavaliere, 100, 00133 Roma, Italy

E. Pettinelli
Dipartimento di Fisica "E. Amaldi", Università Roma Tre, Via della Vasca Navale 84,
00146 Roma, Italy

Antarctica was characterized through more than 4 km of polar ice. Cold ice is more transparent than warm ice, so that it is expected that at the temperatures of the outer Solar System this technique could provide detection of much deeper subsurface structures (Blankenship et al. 2010).

An electromagnetic wave propagating through a medium is attenuated, scattered and dispersed. Its distortion depends on the dielectric characteristics of the medium and its homogeneity at the wavelength scale. The electromagnetic pulse emitted from orbiting ice-penetrating radar would be subject to scattering from the irregular natural surface of an icy moon, and it would be scattered by volume inhomogeneities and attenuated in the dielectric subsurface medium before reaching any subsurface dielectric discontinuity causing a reflection.

The capability of subsurface sounding radar to detect an underground dielectric discontinuity is thus strongly dependent on the medium through which the radar pulse propagates. For icy bodies the physical properties of water ice and the effect of impurities in the ice play a major role.

Radar sounding is the only remote sensing technique allowing the study of the subsurface of a planet from orbit. By detecting dielectric discontinuities associated with compositional and/or structural discontinuities, it is possible to map the stratigraphy, i.e. the distribution characteristics of ice and/or rocks vs. depth, which can be of fundamental importance to better understand the dynamics and the history of the first meters to kilometers of the subsurface (depending on frequency and dielectric properties of the material). On the other hand, the estimation of the propagation time and the signal attenuation, also allows us to retrieve some information on the subsurface petrophysical properties like density and material composition. Moreover, the form of the received signal gives some information about the surface and the volume scattering, from which the size distribution of inhomogeneities can be inferred. All this nonwithstanding, the information obtained with subsurface sounding radars should be compared and integrated with other observations (infrared, visible or in situ analysis) to interpret inferred petrophysical properties in an unambiguous way.

In this chapter we describe the existing and planned radar measurements of the planetary bodies. The dielectric properties of water ice and other potential surface and subsurface materials are discussed, as well as their dependency on temperature and structure. We then evaluate the performance of subsurface sounding radars using these parameters. Finally we describe some laboratory technique to help interpret the radar data, presenting some results obtained using dielectric spectroscopy methods.

2 Overview of Radar Observations of Icy Satellites

Radar observations of the icy satellites were obtained from the Earth using the Goldstone and Arecibo radar (Ostro et al. 1992). The measurements of the reflected signal by the surface and subsurface of Galilean satellites and of Saturn's icy satellites were made in the 3.5–70 cm wavelength range. Recently also observations of the Saturn's satellites were made using Cassini 2.2 cm radar. The Arecibo and Goldstone measurements used essentially transmitted circularly polarized and, rarely, linearly polarized signals. The reception of returning signal was in the same circular (SC) (or same linear SL) and opposite circular (OC) (orthogonal linear OL) polarizations. The radar target's properties are measured by its albedo that is defined by the radar cross section divided by the projected area of the observed surface. The total radar albedo is TP = SL + OL = SC + OC and the polarization ratio is $\mu_c = \mathrm{SC}/\mathrm{OC}$ (OL/SL). For a polished metal sphere of the radius of the observed bodies the single back reflections would be entirely OC (or SL) and the polarization ratios equal zero. In this case the radar albedo would equal the surface's Fresnel coefficient.

The reflections from rough surfaces (or produced by multiple reflections or refractions) produce power on two polarizations. For instance, for randomly oriented dipoles, $SC/OC = 1$ and $OL/SL = 1/3$. The echoes from Earth-like planets are characterised by the quasi specular reflection having $SC/OC \cong 0.3$ due to the much stronger direct reflection compared to the power scattered by the surface and subsurface. The total albedo for Earth-like planets or for asteroids are of the order of $TP \sim 0.1$. Figure 1 (Black et al. 2007) summarizes the measurements of total radar albedo and of polarization ratio obtained during the last three decades for icy satellites. One can see that the results are really different from that obtained for the rocky bodies. The polarization ratio is larger than 1 and the total albedo are between 0.3 and 2.6. The total albedo value larger than 1 means that the back reflection is stronger than the reflection from the perfect metal sphere with the same radius as observed body! What is the process responsible for this anomalous high albedo and polarization ratio? The models proposed in the geometrical optics limit included: buried craters effect on the refracted signal (Eshleman 1986), another refraction scattering from subsurface lenses (Hagfors et al. 1997) and volume multiple scattering due to the scatterers embedded in the ice (Hapke 1990). The models were improved (Baron et al. 2003) and the present agreement is that this "anomalous" behaviour results from the multiple scattering within a random medium whose intrinsic microwave absorption is very low. The regoliths of the planetary bodies are very heterogeneous and that combined with the radio transparency of icy regoliths gives the observed radar behaviour. Measurements of radar properties and any inter-satellite variations address issues of surface characteristics and formation histories.

The unusual radar properties of icy satellite surfaces are thus determined by a combination of the structure of the regolith, providing many dielectric discontinuities by which an incident wave is reflected, and its composition, a transparent material that allows several internal reflections of the same wave without attenuating it significantly.

Subsurface sounding radars operate at much lower frequencies, and electromagnetic propagation at wavelengths of tens of meters is probably unaffected by the small-scale structure of the regolith producing the anomalous reflection observed at cm wavelengths. However, the fact that the surface material is so transparent to electromagnetic waves testifies to a mostly icy composition with a small content of impurities, a very favorable scenario for radar sounding experiments.

3 Interpretation of Radar Measurements

Subsurface radar sounding is now an established technique in planetary exploration, with an history of success (see Phillips et al. 1973; Picardi et al. 2004; Seu et al. 2007a; Ono et al. 2009) and a promising future (e.g. Lebreton et al. 2009). The Consert tomographic radar (Kofman et al. 1998, 2007), the next application of this technique, is on the way to its cometary target on the Rosetta mission. However, radar data have features that depend on instrument characteristics and on the environment in which observations take place, which have to be taken into account for a correct interpretation.

The main parameters determining the performance of an orbiting subsurface sounding radar are frequency and bandwidth of the transmitted pulse. Frequency determines the penetration capability of the radar, while bandwidth controls resolution.

For a wide range of frequencies ranging from MHz to GHz and beyond, dielectric losses (loss tangent) in most natural materials are independent of frequency, so that the number

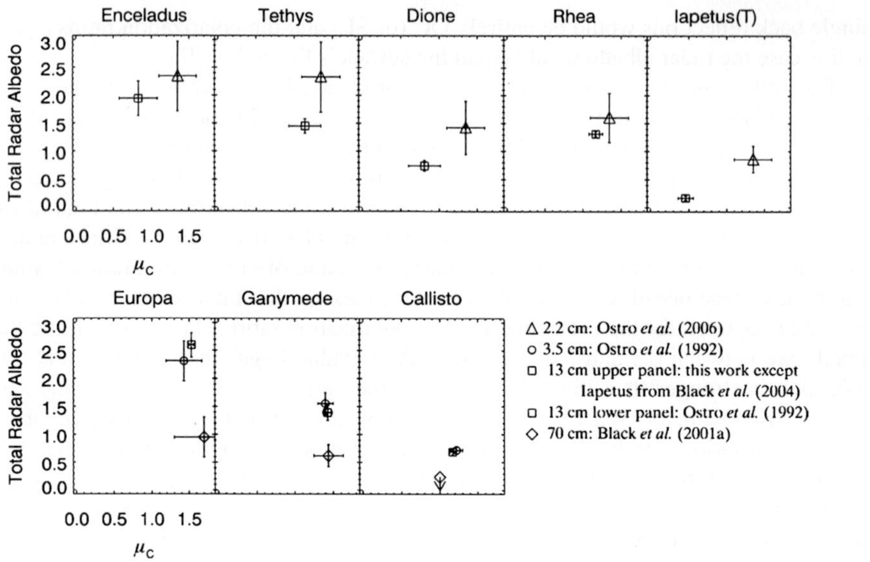

Fig. 1 Plots of total albedo versus circular polarization ratio for each target. Panels with data for Iapetus' trailing (icy) hemisphere and for the icy Galileam satellites are included for comparison. Each point includes data from all available longitudes except Iapetus. The abscissa of each panel is the same as the first panel's. Saturn satellite data at 13 cm wavelength are plotted with the quadrature sum of statistical uncertainties and a 25% systematic uncertainty. At 2.2 cm wavelength the error bars further include assumptions in the ranges of polarization ratios (see text and Ostro et al. 2006). Error bars for the Galilean satellites are the larger of the statistical uncertainties or the range of any variation with longitude. Figure taken from Black et al. (2007)

of wavelengths that an electromagnetic wave can penetrate into the material before being attenuated to a given fraction of its initial amplitude is approximately the same regardless of frequency. This implies that the longer the wavelength, the deeper the radar signal can penetrate before being attenuated below the detection capabilities of the radar. Thus, deep penetration requires that the radar operates at the lowest possible frequency.

In most orbiting radar, the transmitted waveform is a chirp, a long pulse that is linearly modulated in frequency. Chirps are used when the length of the pulse for the desired range resolution is so short that to achieve good signal-to-noise ratio the pulse would require a peak power exceeding the limits imposed by the mission design. The chirp allows a resolution that depends on the bandwidth of the pulse rather than on its duration, but requires processing of the received signal: with a bandwidth B, the approximate time resolution of the output pulse, after processing, is $1/B$. This translates in a vertical resolution of the radar Δr equal to

$$\Delta r = c/(2B\varepsilon^{1/2}) \tag{1}$$

where ε is the relative dielectric permittivity of the material through which the pulse propagates.

Because the bandwidth of the chirp can be at most equal to the highest frequency of the radar pulse (although for practical purposes this limit is never reached), it can be seen that vertical resolution requires the radar to operate at high frequency. Thus, penetration and resolution are conflicting requirements, and no orbiting subsurface sounding radar can achieve one without compromising on the other.

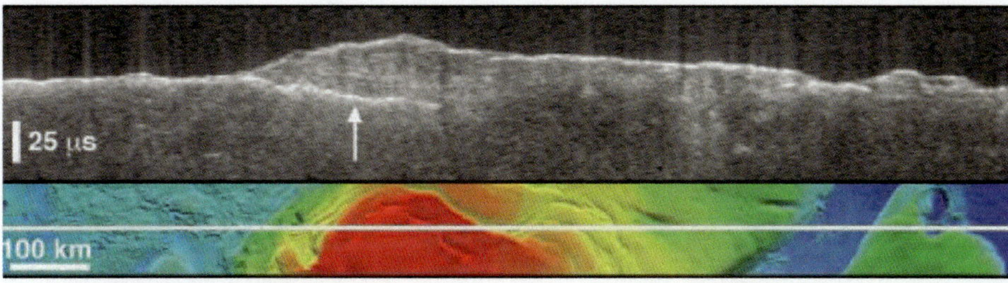

Fig. 2 (*Top*) MARSIS radargram for orbit 2682 above Planum Australe, Mars, showing a bright basal reflector (*arrow*) beneath the ice-rich South Polar Layered Deposits (SPLD). The vertical dimension is round-trip travel time. The apparent curvature of the reflector is an artefact of the time representation of the data. The maximum depth at which the basal reflector could be detected, converting travel time for propagation through water ice, is about 3.5 km. (*Bottom*) Ground track of the spacecraft during data acquisition, shown on a shaded relief topographic map of the south polar region of Mars (adapted from Plaut et al. 2007)

An example of this trade off is the design of the two radars currently operating at Mars, MARSIS (Picardi et al. 2004) and SHARAD (Seu et al. 2007a). They are both synthetic-aperture, orbital sounding radars, carried respectively by ESA's Mars Express and NASA's Mars Reconnaissance Orbiter. MARSIS is capable of transmitting at four different bands between 1.3 MHz and 5.5 MHz, with a 1 MHz bandwidth. SHARAD operates at a central frequency of 20 MHz transmitting a 10 MHz bandwidth. Whereas MARSIS is optimized for deep penetration, having detected echoes down to a depth of 3.7 km over the South Polar Layered Deposits (Plaut et al. 2007), SHARAD is capable of a tenfold-finer vertical resolution, namely 15 m or less, depending on the dielectric constant of the material being sounded (e.g. Seu et al. 2007b).

The difference in the performance of the two radars can be perhaps better appreciated through the examination of Figs. 2 and 3. Both figures show radargrams that are representations of radar echoes acquired continuously during the movement of the spacecraft as grey-scale images, in which the horizontal dimension is distance along the ground track, the vertical dimension is the round trip time of the echo (Fig. 2) or depth (Fig. 3), and the brightness of the pixel is a function of the strength of the echo.

The capability of MARSIS for deep penetration at the expense of vertical resolution can be seen in the radargram in Fig. 2, in which subsurface echoes almost as bright as surface ones are detected down to depths of 3.5 km, but only very faint details of the stratigraphy can be discerned. The minimal attenuation of subsurface reflections—evident through the very weak change of the basal echo brightness with depth—allows one to infer that MARSIS would be able to discern the base of the South Polar Layered Deposits of Mars through at least three or four times their actual thickness.

In Fig. 3 the tenfold-greater vertical resolution of SHARAD, achieved at the expense of penetration, is made evident by the richness of details discernible in the subsurface stratigraphy. The North Polar Layered Deposits of Mars contain much less impurity than the SPLD, and are much more transparent to the radar signal. SHARAD is thus able to see the base of the NPLD and also part of the Basal Unit through almost 1500 m of material, whereas penetration in the SPLD has been found to be approximately half this amount (Seu et al. 2007b).

Subsurface sounding radar data are affected by a number of artifacts that can make it problematic to correctly interpret echoes as either surface or subsurface reflections.

Because of the long (10–100 m) wavelength of radar pulses required in subsurface sounding, it is extremely challenging to produce an antenna that has both an acceptable gain and

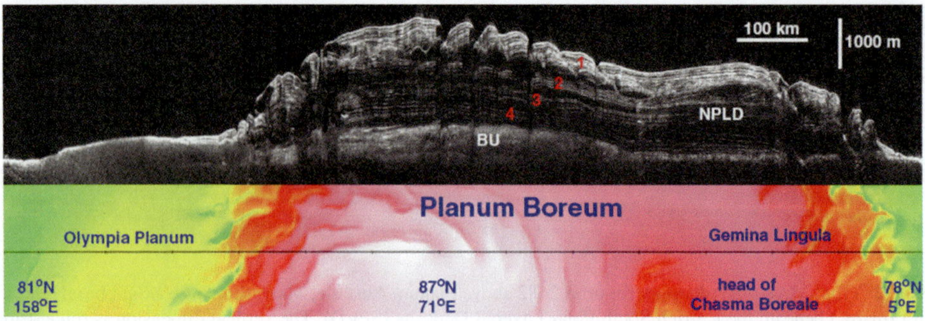

Fig. 3 (*Top*) Radargram from SHARAD orbit 5192 above Planum Boreum, Mars. Range time delay, the usual ordinate in a radargram, has been converted to depth by assigning real permittivities of 1 (corresponding to vacuum) and 3 (water ice) above and below the detected ground surface, respectively. The North Polar Layered Deposit geological unit (NPLD) and the Basal Unit (BU) beneath the NPLD are labelled. Internal radar reflections arise from boundaries between layers differing in their fractions of ice, dust, and sand. The radar reflections in Planum Boreum are clustered into distinct packets of reflectors (numbered). (*Bottom*) Ground track of orbit 5192 shown on a topographic map of the north polar region of Mars (adapted from Phillips et al. 2008)

good directivity. Both MARSIS and SHARAD in fact transmit through dipoles, which have negligible directivity, with the consequence that the radar pulse illuminates the entire surface beneath the spacecraft and not only the near-nadir portion from which subsurface echoes are expected. The electromagnetic wave can then be scattered by any roughness of the surface. If the surface of the body being sounded is not smooth at the wavelength scale, i.e. if the r.m.s. of topographic heights is greater than a fraction of the wavelength, then part of the incident radiation will be scattered in directions different from the specular one. This means that areas of the surface that are not directly beneath the radar can scatter part of the incident radiation back towards it, and thus produce surface echoes that will reach the radar after the echo coming from nadir, which can mask, or be mistaken for, subsurface echoes. This surface backscattering from off-nadir directions is called "clutter". Clutter can produce in a radargram the impression of subsurface structures where there are in fact none. An example is shown in Fig. 4, in which a radargram is compared with a simulated surface echo. What appears in the data to be volume scattering from sediments filling a crater is revealed by simulations to be in fact surface scattering from a crater floor that is flat from the topographic point of view, but must possess a relatively high roughness at meter scale. For data shown in Fig. 4, the effect of roughness below the resolution of the Mars Orbiting Laser Altimeter (MOLA) dataset has been simulated through the use of an analytical formula for the scattering from a random rough surface, but other methods are possible, such as interpolation.

Numerical electromagnetic models of surface scattering, such as those by Nouvel et al. (2004) or Russo et al. (2008), have been routinely used to validate the detection of subsurface interfaces in radargrams. To be effective, however, such models require a knowledge of the topography of the observed surface at scales that are comparable or, better still, much smaller than the wavelength. For example, the MOLA Martian topographic dataset (Smith et al. 2001) has a resolution at the equator of less than 500 m, which is adequate for the MARSIS wavelength range (50–100 m) but requires approximations or interpolations at SHARAD wavelengths (around 15 m).

The presence of a dispersive medium, such as a plasma, between the radar and its observation target can cause a distortion (dispersion) of the transmitted waveform. Such is the

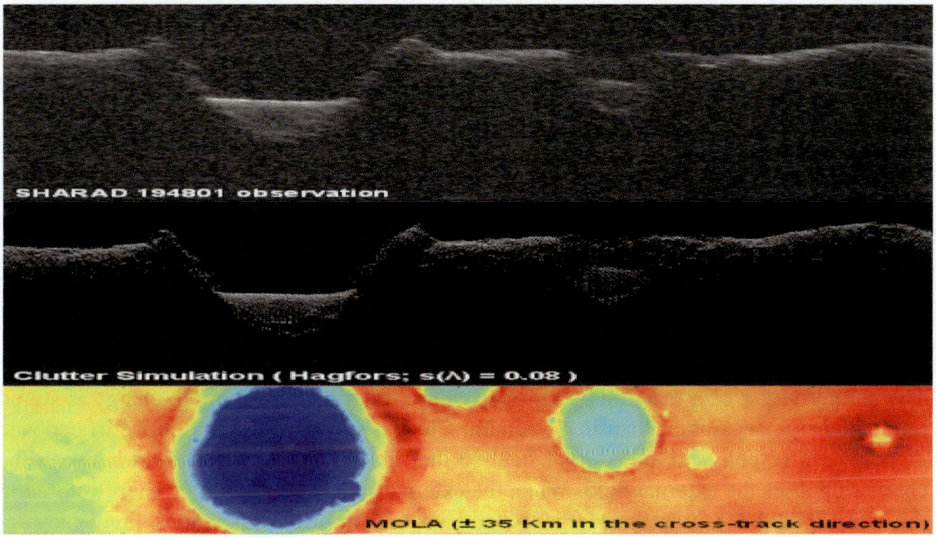

Fig. 4 (*Top*) Radargram from SHARAD orbit 1948 over Meridiani Planum, in which a diffuse echo past the first bright surface echo is visible in flat crater floors, giving the impression that sediments filling the craters are penetrated down to the bedrock. (*Middle*) Simulated radargram obtained through a numerical electromagnetic model of surface scattering making use of the MOLA (Smith et al. 2001) topographic dataset. (*Bottom*) Topographic map centered along the ground track of the MRO spacecraft during observation. It can be recognized that the diffuse echoes seen both in the radargram and in the simulated data originate from the topographically flat yet rough crater floor (adapted from Russo et al. 2008)

case for Mars, which is surrounded by an atmosphere whose upper layers become ionized and excited through solar radiation, producing an ionosphere. As noted above, almost all planetary radar experiments transmit a long chirp to achieve a high signal-to-noise ratio in spite of the usually low power available, with processing on ground required to achieve range resolution. From the mathematical point of view, this processing consists in the correlation of the received signal with the transmitted chirp, and it fails if the phase of the chirp is altered as it propagates to and from the surface. To overcome this problem, several different methods have been proposed, as in Safaeinili et al. (2003), Ilyushin and Kunitsyn (2004), Mouginot et al. (2008), Picardi et al. (2008) and Zhang et al. (2009). This phenomenon should not be encountered by radar sounders exploring the icy moons of the giant planets, however, as most of them (with the notable exception of Titan) are surrounded only by an extremely tenuous exosphere.

In passing, we note that a more serious reason of concern for the performance of a radar sounder for the icy moons of the Solar System is the noise introduced by the giant planets' radio emission at decametric and hectometric wavelengths, especially Jupiter's (see e.g. Zarka et al. 2004). However, the radio noise from Jupiter should not impact the radar performance on the anti-Jovian side of a moon.

A final effect that needs to be taken into account in the geological interpretation of radargrams is the distortion of subsurface interface shapes and slopes caused by the change of the electromagnetic wave propagation velocity in a medium with a relative dielectric permittivity ε greater than that of vacuum. Propagation velocity scales with the inverse of the square root of ε, and is thus lower in a medium with $\varepsilon > 1$, causing a delay in the return from any subsurface reflection with respect to a reflection at the same distance in vacuum. Visually, this results in an apparent "sagging" of subsurface interfaces such as in Fig. 2, in which the basal echo beneath the SPLD (pointed by an arrow) is in reality at the same level of the topography surrounding the SPLD, but appears to be sloping because of the increasing thickness of the SPLD through which the radar signal propagates. When the value of ε can be derived through other means, it is possible to correct this distortion, such as in Fig. 3, where the dielectric constant of ice ($\varepsilon \approx 3$) has been assumed for the NPLD.

4 Radar Signal Propagation and Detection Through Ice

4.1 Dielectric Properties of Ice

The permittivity ε of a material is a property describing how much more energy is stored through charge separation than in a vacuum. Charges of opposite signs move in opposite directions in response to an external field so that the resultant internal field between the charges opposes the external field. When the external field is removed, the energy stored in this internal field decays as the charges revert back to their original positions. Frequency dependence of permittivity occurs because charge separation does not happen instantaneously.

Charges separate with finite velocities, thus if the external field is reversing polarity too quickly the charges cannot move fast enough to keep up. The frequency at which the charges fully separate and are in constant motion is called the relaxation frequency. At frequencies below the relaxation frequency the permittivity plateaus at the low frequency, or static, limit ε_s, and is often call dielectric constant. At high frequencies above the relaxation frequency the permittivity plateaus at the high frequency limit ε_∞.

The response of the ice crystal is the conduction and the polarization. The polarization will develop in time with some delay which duration depends on temperature. The temperature dependency is described by the Arrhenius function of the form of $\exp(E/kT)$ where E is the activation energy (enthalpy) of the formation process, k the Boltzmann's constant and T temperature.

The classical Debye model takes into account this dependency and the dielectric constant is described by the following equations:

$$\varepsilon = \varepsilon_\infty + \Delta\varepsilon \frac{\Delta\varepsilon}{1 + j\omega\tau}, \qquad \omega = 2\pi f \text{ and } \Delta\varepsilon = \varepsilon_s - \varepsilon_\infty \tag{2}$$

where ε_s, ε_∞ are respectively the static and high frequency values of the permittivity and τ is the dielectric relaxation time. The permittivity is a complex function of frequency and usually is described by its real and imaginary part.

$$\varepsilon_R = \varepsilon_\infty + \frac{\Delta\varepsilon}{1 + \omega^2\tau^2}, \qquad \varepsilon_I = \frac{\Delta\varepsilon\omega\tau}{1 + \omega^2\tau^2} \tag{3}$$

The loss tangent $\tan\delta$ is defined by the ratio of these two parts and characterizes the attenuation of the electromagnetic waves in a medium due to ohmic conductivity σ.

$$\tan\delta_d = \frac{\varepsilon_I}{\varepsilon_R} = \frac{\sigma}{\omega\varepsilon_R} \tag{4}$$

The conductivity σ of the medium is directly proportional to the imaginary part of the dielectric constant.

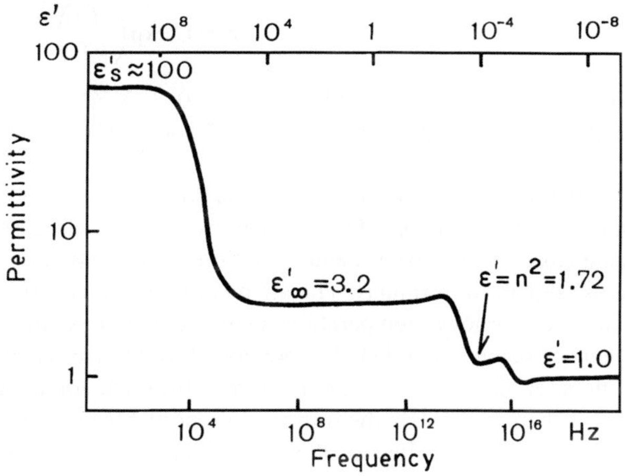

Fig. 5 The real part of the dielectric constant of ice normalised by ε_0 (Fletcher 1970) as function of frequency near −10°C

An approximated expression of the attenuation is given by the formula (5)

$$\alpha = 0.129\sqrt{\varepsilon_R} f[\sqrt{(1+\tan^2\delta} - 1]^{1/2} \approx 0.091\sqrt{\varepsilon_R} f \tan\delta \approx 0.0009\sigma \text{ dB/m} \qquad (5)$$

where σ is in $\mu S\, m^{-1}$. This approximate form of (5) is valid for non-magnetic media with a low loss tangent, such as water ice.

The imaginary part characterizes the attenuation and reaches the maximum at $\omega = \tau^{-1}$.

From (5), one sees that the attenuation coefficient is directly proportional to the frequency. Another way to see this result is that the losses are directly proportional to the conductivity of the medium (5).

The static dielectric constant of a pure ice is very large and anisotropic, and depends on the orientation of the electric field with respect to the crystal's axis (Bogorodsky et al. 1985) changing from about 105 to 90 at 0°C for parallel and perpendicular field. The values increase at −50°. The effect of pressure is small and about 1% per kbar for polycrystalline ice (Chan et al. 1965).

The high-frequency (i.e. at typical radar frequencies) dielectric value is constant over a large frequency band, and is close to 3.2 at 0°C and isotropic within at least 0.5% (Johari and Jones 1978). The temperature dependence of ε_∞ is weak and changes about 1% from 0 to −15°C and another 1% down to −75°C (Johari and Charette 1975; Johari and Jones 1978). Figure 5 shows the behaviour of the real part of the permittivity as function of the frequency.

The temperature dependence of ε_∞ can be modelled as linear (Chyba et al. 1998).

$$\varepsilon_\infty = 3.02 + 6.41 \times 10^{-4} T \qquad (6)$$

and $\Delta\varepsilon$ obeys the Curie-Weiss law:

$$\varepsilon_0 = \varepsilon_\infty + A_C/(T - T_C) \qquad (7)$$

with T_c ranging from 15 K to 46 K (Johari and Jones 1978; Kawada 1978).

Chyba et al. (1998) fitted and interpolated Johari and Charette (1975) measurements of the imaginary part of the permittivity (for temperature range 248–272 K) with $T_c = 35$ K and obtained $A_C = 3.54 \times 10^4$. Finally the relaxation time is obtained using the Arrhenius function with the activation energy of about 0.58 eV (Humbel et al. 1953; von Hippel et al. 1971):

$$\tau = C \exp\left(\frac{E}{kT}\right) \tag{8}$$

where $C = 5.3 \times 10^{-16}$ and $k = 8.61 \times 10^{-5}$ eV K^{-1}. The pressure effect is to increase the relaxation time of about 10% between 0 and 500 bars, independent on temperature (Ruepp 1973).

Other data were used by Thompson and Squyres (1990) to obtain slightly different formulas that were compared in the paper by Chyba et al. (1998). The results are similar and remain close. For the experiment definition and the estimation of radar capabilities it's enough to use the above formulas. The attenuation of the electromagnetic waves in the ice can be calculated for lower temperatures using the previous equations. However, for very low frequencies below a few kHz the previous formula can greatly underestimate the attenuation (Chyba et al. 1998). But for a future subsurface sounding radars working at frequencies higher than a few MHz the formulas are probably correct.

In general the values of dielectric parameters for pure polycrystalline ice are close to those for single-crystal ice (Bogorodsky et al. 1985).

Because of the low losses expected in pure water ice, the major effect on the absorption of radar waves will in fact depend on the nature and concentration of impurities in the ice. This is of course difficult to evaluate due to uncertainties and lack of knowledge of the physical nature of impurities for icy satellites. For Europan ice this problem was studied by Chyba et al. (1998) and Moore (2000). The latter considered three types of water ice, produced by three basic processes occurring on Earth: meteoric ice formed by atmospheric precipitations, sea ice formed by the freezing of water close to the atmospheric interface, and marine ice forming beneath ice shelves directly from ocean water. Moore (2000) concluded that similar processes are likely to occur on Europa as well, and that the most probable form of ice would be marine ice.

Chyba et al. (1998) studied the electromagnetic properties of a medium resulting from the mixing of different impurities (rock, dust) into an ice matrix. They used a mixing formula (Sihvola and Kong 1988, 1989) to calculate the dielectric constant of the mixture and the properties of lunar materials as a model for the impurities within the Europan ice. This approach requires many assumptions and gives only some estimation of the dielectric constant that can be used in the evaluation of the radar capabilities.

In addition to the Chyba et al. (1998) approach, Moore (2000) was interested in soluble impurities in ice like F^-, Cl^-, NH_4^+, SO_4^{2-} and H^+ ions. In Table 1 (Moore 2000) the attenuation for different types of impurities in ice is shown, based on laboratory measurements, ice temperature modelling for Europa and some scaling from Earth ice measurements. These data are valid for electromagnetic frequencies of a few tens of MHz.

These calculations are not taking into account a possible scattering of the electromagnetic waves by the ice/pore interfaces that can exist in the crust. The scattering effect plays the same role as attenuation and depends strongly on the dimension of cavities, (voids) in the medium compared to the wavelength. The Mie or Rayleigh approaches (see e.g. Ishimaru 1978) can be used to calculate the extinction of the radar signal.

The scattering of radio waves by surface and by volume irregularities is an important factor which one has to take into account to evaluate the penetration of the radar wave, and the ratio of any subsurface echo to surface clutter. These two parameters are essential to predict the radar performance. The scattering depends strongly, as we wrote above, on the radar wavelength and surface parameters of the body under investigation. This frequency dependence of attenuation requires that subsurface sounding radars operate at low-frequency (<100 MHz) in order to achieve a deep penetration. The choice of the frequency has an influence on the instrument characteristics, and especially on the size of the antenna. The

Table 1 Radar absorptions for various ice types and temperatures, taken from Moore (2000)

M	Ice type	Impurity content	α	I	II	III	Notes
0	pure ice	nil	0.0045	1.4–2.4	0.5–0.5	0.2–0.3	Glen and Paren (1975)
1	chloride-dominated Europa ice/ocean	3.5 ppt chlorinity ocean	0.016	4–7	2.5–4	1.6–2.8	Marine ice scaled from Earth ocean, $k_{MI} = 7 \times 10^4$
2	rock/ice	1% lunar soil	0.008	5–6	4–4.7	3.6–4.1	Chyba et al. (1998) recalculated
3	rock/ice	10%lunar soil	0.01	8–9	7–8	6–7	Chyba et al. (1998) recalculated
3	sultato-dominated Europa ice/ocean	10 ppt chlorinity ocean	0.037	9–16	6–11	4–7	Kargel (1991); marine ice, $k_{MI} = 7 \times 10^4$
1	chloride-dominated Europa ice/ocean	3.5 ppt chlorinity ocean	0.05	14–24	10–17	7–12	Marine ice scaled from Earth ocean. $k_0 = 3 \times 10^{-3}$
1	rock/ice	50% lunar soil	0.021	30–33	29–32	28–31	Chyba et al. (1998) recalculated
2	depth-dependent Ronne Ice Shelf marine ice	0–400 μM CI linear rise surface to bottom	varies	34–56	24–40	17–28	0–100% Ronne Ice Shelf ion Content
2	sulfate-dominated Europa ice/ocean	10 ppt chlorinity ocean	0.15	36–61	25–44	18–31	Kargel (1991); marine ice, $k_0 = 3 \times 10^{-3}$
2	Ronne Ice Shelf marine ice	400 μM CI (0.025 ppt salinity) ice	0.15	36–61	25–44	18–31	Moore et al. (1994) LF σ from core samples
0	Baltic Sea ice	ice grown in ≈3 ppt sea water	0.85 (at 270 K)	50–85	26–43	16–27	200 MHz radar measurement

Note. Attenuation, α, is in dB/m at 251 K. Columns I, II, and III are computed two-way attenuations, in dB/km, for ice shells with base temperatures of 270, 260, and 250 K, respectively. The range of values for each of these corresponds to surface temperatures of 50 and 100 K. These values are independent of shell thickness since the temperature profile is stretched to the ice thickness. The M column represents the plausibility of the ice type for Europa; 0 is least likely while 3 is more likely, given the present understanding of Europa. The distribution coefficients k_0 and k_{MI} affecting the marine ice models come from laboratory experiments and Ronne Ice Shelf marine ice measurements, respectively exact choice of the frequency results from a trade-off between science requirements and technical limitations.

Due to the lack of knowledge of physical parameters controlling scattering, it is rather difficult to predict its effects with accuracy. For example, Eluszkiewicz (2004) concluded that the presence of a regolith about 1 km thick with an 1% of cavities whose size is comparable to the radar wavelength will make radar wave penetration impossible.

In spite of all these uncertainties, experience has demonstrated that data such as those presented in Table 1 provide the capability to evaluate the radar performance with sufficient accuracy. At the time in which the MARSIS (Picardi et al. 2004) and SHARAD (Seu et al. 2007a) radar sounding experiments were proposed, radar sounding of planetary bodies was deemed problematic if not impossible, in spite of data obtained by the ALSE experiment on board the Apollo 17 spacecraft (Phillips et al. 1973), but results at Mars (e.g. Picardi et al. 2005; Plaut et al. 2007; Seu et al. 2007b; Mouginot et al. 2009) have conclusively demonstrated that this technique is effective in the investigation of planetary bodies from orbit.

4.2 Radar Performance

The data from Table 1 can be used to evaluate penetration depth in the subsurface of icy bodies. As we want to use the radar to study interiors using the reflection from discontinuities and/or diffraction of signals, the reflection and transmission coefficients are also factors determining how much of the incoming signal amplitude is reflected and transmitted. These quantities allow the estimation of the power of the incoming radar signal and therefore the radar capabilities. The following equations summarize the factors affecting radar performance:

$$P_{Rx} = \frac{P_{Tx}(1-|r_{ice}|^2)\tau_p B G_{Tx} G_{Rx} \lambda^2 r_{layer}^2 10^{-\alpha z} L_{sys}}{(4\pi)^2 (2(R+z))^2} \tag{9}$$

$$SNR = \frac{P_{Rx}}{P_{noise}} = \frac{P_{Tx}(1-|r_{ice}|^2)\tau_p G_{Tx} G_{Rx} \lambda^2 r_{layer}^2 10^{-\alpha z} L_{sys}}{(4\pi)^2 (2(R+z))^2 k T_{ref} N_f} \tag{10}$$

where

P_{Tx} = peak RF power during transmission of the pulse
G_{Tx} = antenna gain on transmission
G_{Rx} = antenna gain on reception
λ = operating wavelength
r_{ice} = effective reflectivity of the ice-atmosphere interface
r_{layer} = effective reflectivity of an internal layer
α = 2-way attenuation through ice of thickness z
L_{sys} = system losses including internal transmission losses and losses within the antenna during transmission
R = range from satellite to ice surface
z = depth from the surface to the internal layer
P_{noise} = $k T_{ref} B N_f$ where
k = Boltzmann constant
T_{ref} = antenna reference temperature
B = bandwidth of transmitted pulse

N_f = system noise figure
τ_p = pulse duration

Modern radars transmit a modulated signal that provides a compression gain so that the effective echo power received within the pulse bandwidth is increased by a factor $\tau_p B$.

This equation is given here only to highlight different factors playing a role in radar detection capabilities. The noise level is one of the factors limiting performance. For low frequency radars (1–50 MHz) the galactic noise temperature varies between 10^7 to 10^4 K. In the Jovian environment its radio noise increases the noise temperature of about 10^{10} to 10^6 K but on the anti-Jovian side of a moon the influence is negligible.

The altitude of the spacecraft also plays an important role. For a radar placed on an orbit at 100 km of altitude, at the frequency of 20 MHz (the SHARAD radar frequency), with the noise temperatures of the order of 5×10^4 K, the antenna gain of 10 dB, the peak power of 100 W and a modulated pulse of 200 μs, the signal to noise ratio is:

$$SNR \approx 130 - \alpha Z + 20 \log r_{layer} + 10 \log(1 - |r_{ice}|^2) + L_{sys} \tag{11}$$

The transmission loss between the air/ice interface is about 1 dB, and the reflection loss at the interface within the ice is estimated to be between −10 to −20 dB. L_{sys} is assumed to be −3 dB.

Finally, as one needs about 20 dB SNR in order to adequately process the signal, the acceptable losses inside the ice are found to be of about 85 dB. Coherent processing can reduce the noise power and then increase the SNR. Typical gains for coherent processing are of the order of 10–20 dB. In this analysis surface clutter effects are not taken into account, but they can easily reduce the signal detection capability by a comparable amount, although they can in turn be reduced through Doppler filtering. This is why the acceptable value for attenuation can still be maintained at 85 dB.

From Table 1 it can be seen that radar penetration inside pure ice can exceed 40 km. The most probable losses, however, are between 6 and 16 dB/km, which will clearly limit the penetration to 5 to 15 km.

5 Laboratory Studies Relevant to Interpretation of Radar Observations

As mentioned above, in order to estimate the maximum penetration depth of a subsurface radar signal in the icy moons crust of the Solar System, the electromagnetic properties of the materials forming the solid and/or liquid layering sequence should be known. Several laboratory techniques can be used to measure the complex permittivity of ice or icy soil simulants: some of them operate in the frequency domain (FD) regime, others in the time domain (TD) regime. The FD techniques (dielectric spectroscopy) allows one to perform a straightforward measurement of both real and imaginary part of the permittivity, in a limited range of frequency (in fact, it is not possible to use just one instrument to measure the permittivity from zero-frequency up to the microwave region). Conversely, to extract the same parameters as a function of frequency, the TD techniques need a transformation from time to frequency domain and a dielectric model to fit the experimental data. However, if no dispersion is observed, time domain analysis can be a very fast and reproducible technique to estimate the dielectric permittivity and the DC conductivity of a material.

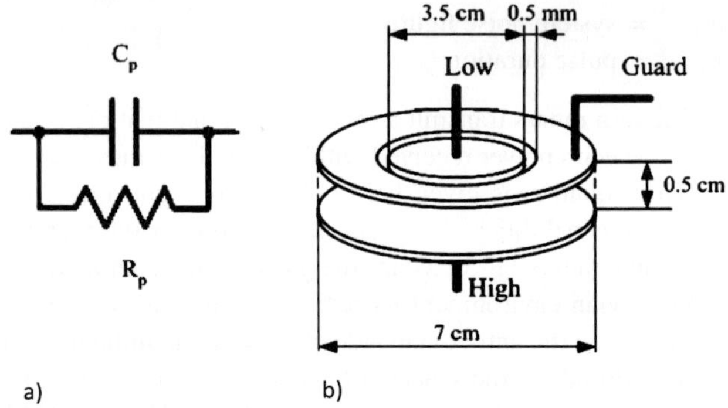

Fig. 6 (**a**) Equivalent circuit; (**b**) Schematic of a capacitance cell to measure electrical properties of granular or liquid materials

5.1 Frequency Domain Measurements

Usually, the laboratory techniques for measuring the electromagnetic properties of granular or liquid materials at hertz-to-megahertz frequencies are based on the equivalent circuit analysis performed through *L-C-R* meters (Pettinelli et al. 2003). Conductivity and permittivity are obtained by measuring the magnitude and phase of the electrical impedance of a capacitive cell (i.e. Parallel Plate Capacitor—PPC) filled with the material being tested.

With reference to the lumped equivalent circuit shown in Fig. 6a, the current in the cell, can be expressed through its complex admittance Y^*:

$$I = V_g \cdot Y^* = V_g \cdot \left(\frac{1}{R_p} + j\omega C_p\right) \tag{12}$$

Using the geometrical features of the cell the measured values of capacitance C_p and parallel resistance R_p can be related to the real and imaginary parts of the complex permittivity of the material enclosed between the electrodes.

In particular, if A is the surface of the guarded electrode (the plate connected to the "low" terminal in Fig. 6) and d is the spacing between the two parallel plates, one obtains:

$$\varepsilon' = \frac{d}{A} \cdot \frac{C_p}{\varepsilon_0} = \frac{C_p}{C_0}; \qquad \varepsilon'' = \frac{d}{A} \cdot \frac{1}{\omega \varepsilon_0 R_p} = \frac{1}{\omega C_0 R_p} \tag{13}$$

Note that, a precise knowledge of the cell geometry is not strictly required to get accurate measurements, as both the real and imaginary terms of permittivity can be calculated using the previously measured value of the capacitance with the cell empty (i.e. C_0). The effect of the fringing field at the edges of the capacitor plates is cancelled by using a guarded electrode (see Fig. 6).

Several types of capacitive cell can be used to estimate the dielectric parameters of a material; the choice depends on the material itself (soil, ice or mixtures) and the pressure and temperature that should be applied to the sample. As an example, Fig. 7a and b show a parallel plate cell and a coaxial cell suitable for granular or icy samples measured at atmospheric pressure.

The maximum frequency achievable with the capacitor cell technique is of the order of few tens of MHz, therefore suitable to support the investigations of orbiting subsurface radars like MARSIS or SHARAD. However, if a higher frequency range should be investigated (from hundreds of MHz to few GHz), the most common technique used to extract the real and imaginary part of the permittivity is the measurement of the scattering parameters

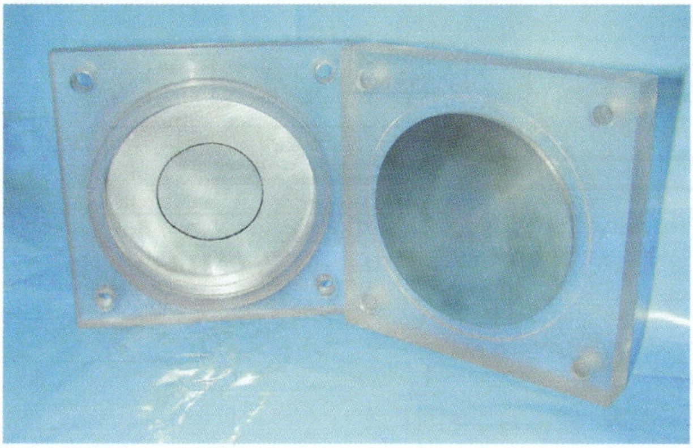

(a) (b)

Fig. 7 (**a**) Parallel plate capacitor cell; (**b**) Coaxial capacitor cell

of a coaxial line, connected to a network analyzer (see for example the review of Stuchly and Stuchly 1980).

5.2 Time Domain Measurements

The use of Time Domain Reflectometry (TDR) technique for permittivity measurements is based on evaluating the velocity of a step-like signal which travels along a transmission line filled with the material under test or embedded in it (for details see Topp and Ferré 2002, and Robinson et al. 2003). The TDR signal has a broad band and the upper frequency, in a sample, can extend up to about 500 MHz.

A basic TDR system consists of a pulse generator, a sampler, an oscilloscope, a coaxial cable, and a transmission line probe (see Fig. 7). The pulse generator applies a fast rise-time voltage step (~200 ps) to a 50-ohm coaxial cable and triggers a sampler. The step pulse travels down the coaxial cable, until it reaches the probe, where part of the signal is reflected back toward the cable tester due to an impedance mismatch, and part passes along the line in the material itself. The signal traveling in the material ultimately reaches the end of the probe, sending a reflection back toward the oscilloscope.

The waveform displayed on the oscilloscope represents the superposition of incoming and reflected waves (in phase or in counter phase), generated at every impedance mismatch in the system. From the propagation time of the wave front along the probe, the pulse velocity can be calculated according to the following equation:

$$v = \frac{2L}{t} \tag{14}$$

where t is the two-way travel time (from t_2 to t_3 in Fig. 8) and L is the probe length. Assuming that the material is homogeneous within the sample, the electromagnetic wave velocity is given by:

$$v = \frac{c}{\mathrm{Re}\sqrt{\varepsilon^* \mu^*}} \tag{15}$$

where c is the electromagnetic wave velocity in the vacuum (3×10^8 m/s), and ε^* and μ^* are the material complex permittivity and complex magnetic permeability, respectively.

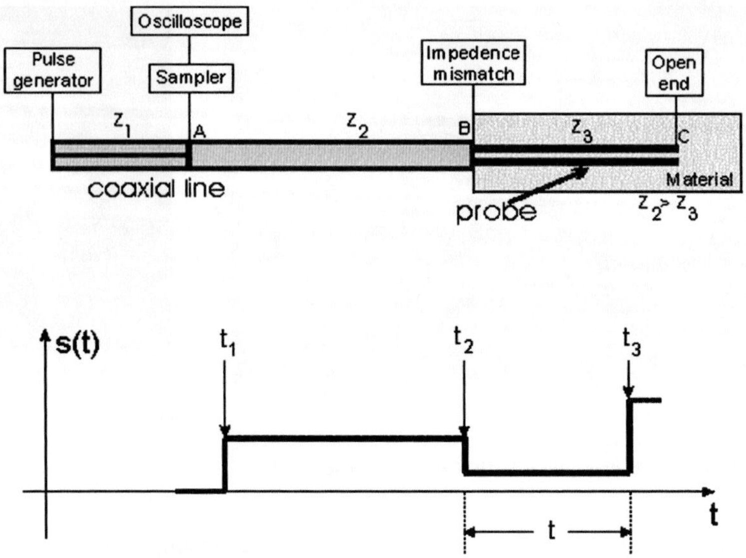

Fig. 8 Schematic of TDR measurement technique

Equation (15) can be expressed in a more explicit form as:

$$v = \frac{c\sqrt{2}}{\sqrt{\mu'\varepsilon' - \mu''\varepsilon'' + |\varepsilon^*||\mu^*|}} \tag{16}$$

If magnetic materials are not present, as in the case of pure ice or icy mixtures, TDR travel-time measurements can be used to estimate the *apparent relative dielectric permittivity*, ε_{ra}, throughout the following

$$\varepsilon_{ra} = \left(\frac{ct}{2L}\right)^2 \tag{17}$$

which takes into account for both real and imaginary parts of the permittivity.

Figure 9a shows a TDR configuration that can be easily used in icy material, where the probe can be directly inserted in the material before freezing, whereas Fig. 9b shows a TDR configuration suitable to measure the dielectric properties of liquids or granular materials.

Furthermore, the analysis in time domain of the TDR wave forms allows one to estimate also the DC conductivity using the equation (Giese and Tiemann 1975):

$$\sigma_{GT} = \sqrt{\frac{\varepsilon_0}{\mu_0}}\frac{1}{L}\left(\frac{Z_0}{Z_c}\right)\left(\frac{2V_0 - V_f}{V_f}\right) \tag{18}$$

where L is the probe length, Z_0 is the probe impedance in air and Z_c is the coaxial cable impedance, V_0 is the input voltage and V_f is the asymptotic voltage value (see Fig. 10). For a detail description of the DC conductivity evaluation with TDR see Robinson et al. (2003).

5.3 Determination of Electromagnetic Parameters via Scatter Function

The TDR wave form carries information on the electromagnetic properties of the sample under test in a wide band of frequency. The frequency response of a TDR device, can be

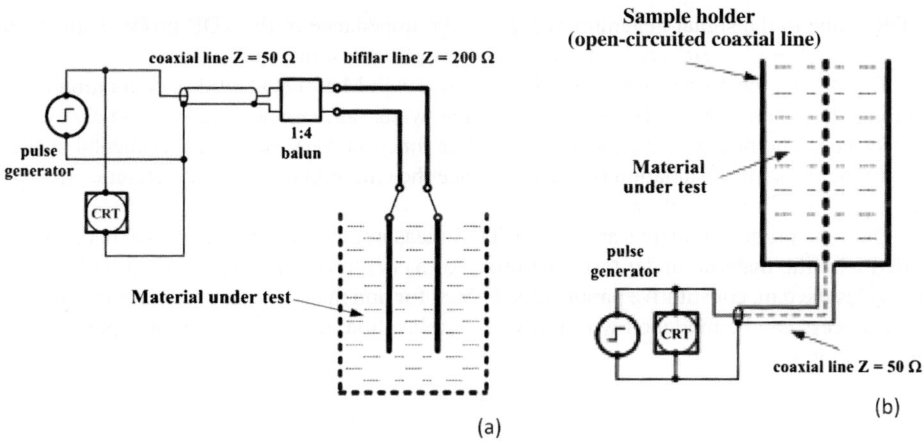

Fig. 9 (**a**) Schematic drawing of a TDR system where the probe consists of two parallel metal rods which are inserted in the material to be tested; (**b**) the same TDR configuration connected to a coaxial probe

Fig. 10 TDR wave form and relevant parameters used to estimate DC conductivity

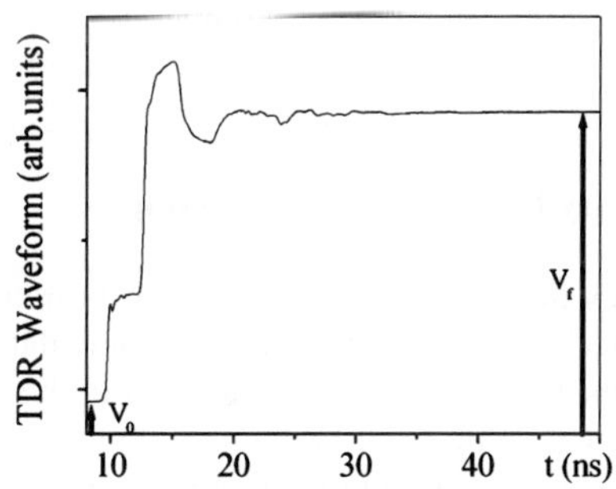

conveniently studied by using its transfer function $S(\nu)$. This function allows one to estimate both the material electromagnetic properties and the transmission line parameters, because it can be directly compared with the theoretical expressions.

The theoretical scatter function $S_T(\nu)$ of an open-ended coaxial probe is given by Clarkson et al. (1977), Ramo et al. (1984)

$$S_T(\nu) = \frac{(\rho^* + \exp(-2\gamma L))}{(1 + \rho^* \exp(-2\gamma L))} \tag{19}$$

where L is the probe length, ρ^* is the reflection coefficient

$$\rho^* = \frac{(1 - z\sqrt{(\varepsilon^*(\nu)/\mu^*(\nu))})}{(1 + z\sqrt{(\varepsilon^*(\nu)/\mu^*(\nu))})} \tag{20}$$

and γ is the propagation function

$$\gamma = \frac{j2\pi f}{c}\sqrt{\varepsilon^*(\nu)\mu^*(\nu)} \tag{21}$$

where c is the velocity of light in a vacuum. The quantity z in (20) represents a normalized impedance given by $z = Z_C/Z_P$, where Z_C is the impedance of the cable connecting the

TDR probe to the signal generator and Z_P is the impedance of the TDR probe in air. $\varepsilon^* = \varepsilon' - j\varepsilon''$, $\mu^* = \mu' - j\mu''$ are the e.m. parameters of the test material.

To estimate the probe parameters (Feng et al. 1999; Mattei et al. 2006), calibration measurements in some reference materials (such as water, air, ethanol, etc.) should always be performed. The fitting of the theoretical scatter function $S_T(\nu)$ to the experimental scatter function, allows the evaluation of z and L, since the e.m. properties of the reference materials are known from literature.

Once the probe parameters are known, TDR measurements can be performed in the probe filled with the material under test. The fitting of $S_T(\nu)$ to the experimental scatter function provides the e.m. constitutive parameters. In fact, for non magnetic materials, if the polarization process can be modelled with Debye-like relaxation, the relative complex permittivity is given by

$$\varepsilon^*(\nu) = \varepsilon_\infty + \frac{(\varepsilon_S - \varepsilon_\infty)}{1 + i\nu/\nu_{\text{rel}}} + i\frac{\sigma_{dc}}{2\pi\nu\varepsilon_0} \tag{22}$$

where ε_S is the static value of permittivity, ε_∞ its high frequency limit, ν_{rel} is the relaxation frequency, σ_{dc} is the static conductivity, and ε_0 is the permittivity in a vacuum. $S_T(\nu)$ can be obtained applying (19) and (20), whereas the minimization procedure allows the evaluations of the model parameters and, consequently, the frequency dependence of test material e.m. properties.

6 Laboratory Results on Icy Soil Samples

As examples of the information that can be retrieved from dielectric measurements with FD or TD techniques, we present here some data acquired on CO_2 ice and some icy mixtures. The measurements were performed in the frequency domain (100 Hz–1 MHz) using a capacitive cell (PPC) and in the time domain using a two-prong transmission line. These two techniques are particularly suitable to perform dielectric measurements on icy samples because they offer several advantages: (i) data acquisition is simple and fast; (ii) sample volume is large (i.e. the bulk properties of a material are measured); and (iii) the experimental techniques are reproducible and the results are consistent.

Figure 11a and b show real and imaginary parts of permittivity measured in the same frequency range on two samples prepared with CO_2 ice (ice powder) and CO_2 ice mixed with volcanic sand. For comparison, in the figures the plots for the same dry sand are shown as well. Table 2 summarises volume fractions and density of the materials under test. The data confirm the non-polar behaviour of CO_2 ice, as the real part of the permittivity is very low (about 1.4). The $\varepsilon' \cong 2$ measured for the mixture CO_2 ice and sand is in general agreement with what is reasonably expected. Indeed the dielectric constant of the mixture should exhibit a sort of average between the $\varepsilon' \cong 1.4$ of CO_2 ice and that of the anhydrous sand ($\varepsilon' \cong 3$).

The plot of the imaginary parts of permittivity shows the general tendency of low temperature mixtures to behave like good insulators because the mobility of ions significantly slowed down by the freezing process.

More extended measurements have been conducted on CO_2 ice and icy mixtures with TDR techniques. Table 3 summarises the information on volume fractions and density of the samples and the apparent permittivity values (with the uncertainties) measured with the TD method (for detail see Pettinelli et al. 2003). Note that there are little differences in the apparent permittivity values of the samples presented in the table, with different materials having approximately the same permittivity.

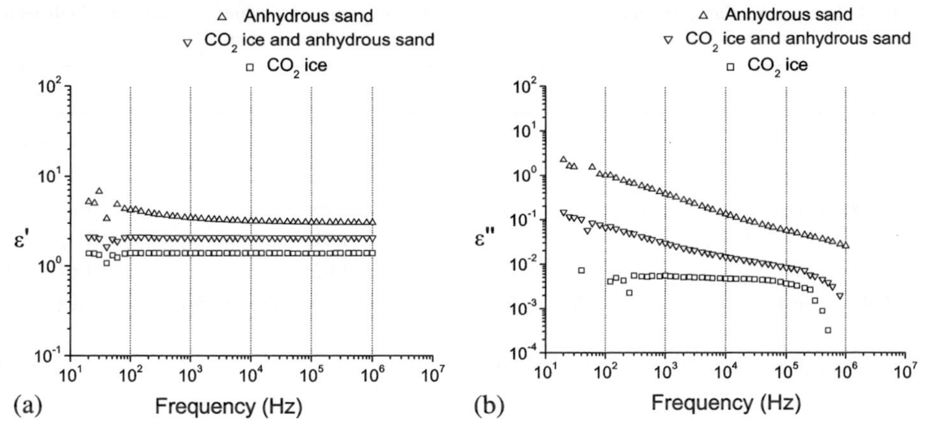

Fig. 11 Real (**a**) and imaginary (**b**) part of permittivity of CO_2 ice and volcanic sand. Comparison with the values measured for anhydrous volcanic sand at room temperature

Table 2 Volume fractions and density of the samples shown in Figs. 11

Material	Volume fractions (%)	Density ($g\,cm^{-3}$)
CO_2 powder (−75°C)	$V_{CO_2} = 70.3\%$ $V_{air} = 29.7\%$	1.05
dry Vulcano sand (24.4°C)	$V_{sand} = 55.9\%$ $V_{air} = 44.1\%$	2.58 (grain density)
CO_2 powder + sand	$V_{sand} = 31.7\%$ $V_{CO_2} = 35.7\%$ $V_{air} = 32.6\%$	–

7 Laboratory Measurements Needed to Support Interpretation of Radar Observations

At the present stage an accurate survey of the literature available, shows a lack of data on dielectric properties of icy materials, gas hydrates (clathrates), and salt doped ice in the frequency range of interest (i.e. tens of MHz or higher) to define the performance properties (the maximum penetration depth) of future subsurface radars capable of investigating the interior structure of the icy moons of the Solar System. The vast majority of dielectric measurements, in fact, have been collected at low frequency (below 1 MHz) (Thompson and Squyres 1990; Lorenz 1998; Chyba et al. 1998; Moore 2000; Pettinelli et al. 2003; Grimm et al. 2008). These data show, in general, that the impurities in the ice have the effect of shifting the relaxation time at higher frequencies, however, never exceeding few tens of kHz. In the MHz range, the ice or the icy mixtures seem to have low values of the real part of the permittivity and small values of the conductivity, behaving like non-polar low-loss materials. Therefore, we should expect a transparent behavior of the icy materials in a broadband of frequencies well above the relaxation frequency.

Nevertheless, the definition of the best radar frequency to use to investigate the internal structure of an icy moon crust should be strongly supported by adequate and extensive

Table 3 Volume fractions, density and apparent permittivity experimental values, measured by TDR technique for various samples of solid CO_2, volcanic sand, and mixtures (Pettinelli et al. 2003)

Material	Volume fraction	Density ($g\,cm^{-3}$)	$\varepsilon_{ra} \pm \Delta\varepsilon_{ra}$
CO_2 ice	–	1.50	2.12 ±0.04
CO_2 snow	–	–	1.48 ±0.04
CO_2 powder (1)	–	0.99	1.67 ±0.06
CO_2 powder (2)	–	1.01	1.63 ±0.06
CO_2 powder (3)	–	1.02	1.67 ±0.06
dry Vulcano sand	$V_{sand} = 57.5\%$ $V_{air} = 42.5\%$	2.58 (grain density)	3.83 ±0.09
CO_2 snow + glass beads (1)	$V_{glass1} = 37.47\%$ $V_{CO_2} = 62.53\%$	–	2.32 ±0.03
CO_2 snow + glass beads (2)	$V_{glass2} = 20.75\%$ $V_{CO_2} = 79.25\%$	–	1.56 ±0.03
CO_2 powder + sand (1)	$V_{sand1} = 38.2\%$ $V_{CO_2} = 23.8\%$ $V_{air} = 38.0\%$	–	2.6 ±0.2
CO_2 powder + sand (2)	$V_{sand2} = 25.5\%$ $V_{CO_2} = 38.6\%$ $V_{air} = 35.9\%$	–	2.28 ±0.07

laboratory investigations on a sufficient number of composite ice samples. In particular, because the temperature, the impurities, and the density of the icy materials are the parameters that affect the most the dielectric behavior of the simulants, these measurements should be performed systematically changing these two parameters. These types of measurements, however, require a certain level of accuracy: in fact, as shown in Table 3, the permittivity values obtained in the MHz range for different icy soil, might be very close to each other, and in order to distinguish between different materials a small uncertainty should be associated to the measurement. This ambiguity in the dielectric properties also affects strongly the possibility to perform some data inversion and extract some information on the dielectric properties of the deeper layers, like an ocean under an icy crust. Moreover, laboratory measurements can be used to validate a dielectric model. As an example, Table 4 compares, the apparent permittivity values obtained using TDR technique with those calculated using different dielectric models (for detail see Pettinelli et al. 2003).

Among the models presented in Table 4 to predict the average dielectric properties of a mixture, the Tinga model is the most flexible and therefore it was applied to simulate the dielectric behavior of almost all measured mixtures. The results for mixtures of glass beads and air or water are in excellent agreement with experimental data, due to the fact that this model is particularly well suited to describe the dielectric properties of a mixture consisting of a uniform known host material with homogeneous, equally-sized spherical inclusions. Note that, also Rayleigh mixing formula is suitable to estimate the real part of the permittivity of a material on the basis of its density, if the permittivity value of the solid phase is known. However, many other dielectric mixing formulas are present in the literature (see Sihvola 2000), and in order to assess the prediction capability of such models for the icy mixtures, an extensive study combining laboratory-measured and theoretical permittivity values should be performed

Table 4 Comparison between TDR experimental apparent permittivity values and calculated values from models

Material	$\varepsilon_{ra} \pm \Delta\varepsilon_{ra}$	Model		
		Rayleigh 1892	Tinga et al. 1973	Topp et al. 1980
dry glass beads	3.19 ± 0.08	–	3.0	3.3
saturated glass beads (r.t.)	30.0 ± 0.2	–	30.6	29.6
drained glass beads (r.t.)	4.09 ± 0.08	–	–	5.3
CO_2 powder (1)	1.67 ± 0.06	1.6	1.6	–
CO_2 powder (2)	1.63 ± 0.06	1.7	1.7	–
CO_2 powder (3)	1.67 ± 0.06	1.7	1.7	–
dry Vulcano sand	3.83 ± 0.09	–	3.9	3.0
CO_2 snow + glass beads (1)	2.32 ± 0.03	–	2.7	–
CO_2 snow + glass beads (2)	1.56 ± 0.03	–	1.5	–
CO_2 powder + sand (1)	2.6 ± 0.2	–	3.6	–
CO_2 powder + sand (2)	2.28 ± 0.07	–	2.8	–

Vulcano sand: $\varepsilon_{grain} \cong 17$ (calculated with Rayleigh mixing formula)
Glass beads: $\varepsilon_{grain} = 7.2$

Finally, laboratory measurements are fundamental to help calculate the total attenuation coefficient (ohmic + volumetric scattering) for an electromagnetic wave propagating through an inhomogeneous medium. Tables 5a and b show the total attenuation and the maximum penetration depth of a 225 MHz radar signal in different icy media with basaltic inclusions (for detail see Pettinelli et al. 2007). Again in this case, both dielectric and magnetic properties of the materials forming the mixtures and the inclusions have been measured using the above described FD techniques.

8 Summary and Recommendations

In this chapter, we discussed the application of subsurface radar sounding from orbit to the study of icy moons. We first summarized known radar properties of icy moon surfaces, and discussed existing subsurface radar observations of other planetary bodies. We then presented the main physical parameters of a planetary crust affecting the performance of a radar sounder, concluding that in icy moons a radar sounder should be able to detect dielectric discontinuities at depths of several kilometers to a few tens of kilometers, depending on the nature of the materials through which the electromagnetic pulse propagates. We finally reviewed methods and results for the measurements of dielectric properties of natural materials affecting radar performance, providing suggestions for laboratory activities in support of future subsurface radar observations of icy moons.

Subsurface radar sounding is the only existing remote sensing technique for the study of planetary subsurfaces. Radar sounders have been successfully flown to the Moon and are currently operating at Mars, while another instrument of this kind is on its way to a comet nucleus. Radar experiments are being considered for future missions to near-Earth asteroids, to Vesta and to the icy moons of giant planets. In spite of this, the body of knowledge associated with this technique is still relatively small compared to other remote sensing methods such as spectroscopy, and notions on the correct use and interpretation of data are not yet widespread in the planetary community.

Table 5 Maximum penetration depth at which an interface can be detected below an icy soil with 10% volume fraction of basaltic inclusions (a) and 20% volume fraction of basaltic inclusions (b)

a (10% volume fraction)

Scenarios	225 MHz Radius of the inclusions (mm)	α_{tot} (dB/m)	L (m)
CO_2/Basalt	200	0.13	43
CO_2&DrySoil/Basalt	200	0.11	40
IcySoil/Basalt	200	0.082	22

b (20% volume fraction)

Scenarios	225 MHz Radius of the inclusions (mm)	α_{tot} (dB/m)	L (m)
CO_2/Basalt	200	0.49	21
CO_2&DrySoil/Basalt	200	0.42	21
IcySoil/Basalt	200	0.084	27

We are convinced that radar sounding has the potential to provide an enormous increase in our knowledge of the geology of planetary bodies, because of its unique capability to explore the third dimension and its yet limited application to planetary exploration. Because of the transparency of ice-rich materials to low-frequency electromagnetic waves, we strongly advocate the use of this technique in the study of icy bodies in the Solar System. We also deem that more work is required to fully utilize radar sounding data for the quantitative determination of subsurface properties, an inverse problem of no small complexity. Laboratory measurements of the dielectric parameters of natural materials occurring on planetary bodies are a key activity towards this end.

References

J.E. Baron, G.L. Tyler, R.A. Simpson, Icarus **164**, 404–417 (2003)

G.J. Black, D.B. Campbell, L.M. Carter, Icarus **191**, 702–711 (2007)

D.D. Blankenship, D.A. Young, W.B. Moore, J.C. Moore, in *EUROPA*, ed. by R.T. Pappalardo, W.B. McKinnon, K. Khurana (The University of Arizona Press, Tucson, 2010), pp. 631–653

V. Bogorodsky, C. Bentley, P. Gudmandsen, *Radioglaciology* (Reidel, Dordrecht, 1985). ISBN 90-277-1893-8

R.K. Chan, D.W. Davidson, E. Whalley, J. Chem. Phys. **43**, 2376–2383 (1965)

C.F. Chyba, S.J. Ostro, B.C. Edwards, Icarus **134**, 292–302 (1998)

T.S. Clarkson, L. Glasser, R.W. Tuxworth, G. Williams, Adv. Mol. Relax. Proc. **10**, 173–202 (1977)

V.R. Eshleman, Science **234**, 587–590 (1986)

J. Eluszkiewicz, Icarus **170**, 234–236 (2004)

N.H. Fletcher, *The Chemical Physics of Ice* (Cambridge University Press, New York, 1970)

W. Feng, C.P. Lin, R.J. Dechamps, V.P. Drnevich, Water Resour. Res. **35**, 2321–2331 (1999)

K. Giese, R. Tiemann, Adv. Mol. Relax. Proc. **7**, 45–59 (1975)

J.W. Glen, J.G. Paren, J. Glaciol. **15**, 15–37 (1975)

R.E. Grimm, D.E. Stillman, S.F. Dec, M.A. Bullock, J. Phys. Chem. B **112**, 15382–15390 (2008)

T. Hagfors, I. Dahlstrom, T. Gold, S.-E. Hamran, R. Hansen, Icarus **130**, 313–322 (1997)

B. Hapke, Icarus **88**, 407–417 (1990)

F. Humbel, F. Jona, P. Scherrer, Helv. Phys. Acta **26**, 17–32 (1953)

Y.A. Ilyushin, V.E. Kunitsyn, J. Commun. Technol. Electron. **49**, 154–165 (2004)
A. Ishimaru, *Wave Propagation and Scattering in Random Media* (Academic Press, New York, 1978)
G.P. Johari, P. Charette, J. Glaciol. **14**, 293–303 (1975)
G.P. Johari, S.J. Jones, J. Glaciol. **21**, 259–276 (1978)
J.S. Kargel, Icarus **94**, 368–390 (1991)
S. Kawada, J. Phys. Soc. Jpn. **44**, 1881–1886 (1978)
W. Kofman et al., Adv. Space. Res. **21**(11), 1589–1598 (1998)
W. Kofman, A. Herique, J.-P. Goutail, T. Hagfors, I.P. Williams, E. Nielsen, J.-P. Barriot, Y. Barbin, C. Elachi, P. Edenhofer, A.-C. Levasseur-Regourd, D. Plettemeier, G. Picardi, R. Seu, V. Svedhem, Space Sci. Rev. **128**(1–4), 413–432 (2007)
J.-P. Lebreton et al., in *Lunar and Planetary Institute Science Conference Abstracts*, vol. 40 (2009), p. 2383
R.D. Lorenz, Icarus **136**(2), 344–348 (1998)
E. Mattei, A. Di Matteo, A. De Santis, G. Vannaroni, E. Pettinelli, Water Resour. Res. **42** (2006). doi: 10.1029/2005WR004728
J.C. Moore, Icarus **147**, 292–300 (2000)
J.C. Moore, A.P. Reid, J. Kipfstuhl, J. Geophys. Res., Oceans **99**(C3), 5171–5180 (1994)
J. Mouginot, W. Kofman, A. Safaeinili, C. Grima, A. Herique, J.J. Plaut, Icarus **201**(2), 454–459 (2009)
J. Mouginot, W. Kofman, A. Safaeinili, A. Herique, Planet. Space Sci. **56**, 917–926 (2008)
J.-F. Nouvel, A. Herique, W. Kofman, A. Safaeinili, Radio Sci. **39**(2004), 1013 (2004)
T. Ono, A. Kumamoto, H. Nakagawa, Y. Yamaguchi, S. Oshigami, A. Yamaji, T. Kobayashi, Y. Kasahara, H. Oya, Science **323**, 909–912 (2009)
S.J. Ostro, D.B. Campbell, R.A. Simpson, R.S. Hudson, J.F. Chandler, K.D. Rosema, I.I. Shapiro, E.M. Standish, R. Winkler, D.K. Yeomans, R. Velez, R.M. Goldstein, J. Geophys. Res. **97**, 18227–18244 (1992)
S.J. Ostro et al., Icarus **183**, 479–490 (2006)
E. Pettinelli, G. Vannaroni, A. Cereti, F. Paolucci, G. Della Monica, M. Storini, F. Bella, J. Geophys. Res., Planets **108**(E4), 8029–8040 (2003)
E. Pettinelli, P. Burghignoli, A. Galli, A.R. Pisani, F. Ticconi, G. Vannaroni, F. Bella, in *IEEE-TGRS 2007*, vol. 48 (2007), pp. 1271–1281
R.J. Phillips et al., NASA Spec. Publ. **330**(22), 1–26 (1973)
R.J. Phillips et al., Science **320**, 1182–1185 (2008)
G. Picardi et al., in *Mars Express: The Scientific Payload*, ed. by A. Wilson, A. Chicarro vol. SP-1240 (ESA, Noordwijk, 2004), pp. 51–69
G. Picardi et al., Science **310**, 1925–1928 (2005)
G. Picardi et al., in *Radar Conference, RADAR'08*, IEEE, 26–30 May 2008 (2008), pp. 1–5
J.J. Plaut et al., Science **316**, 92–95 (2007)
S. Ramo, J.R. Whinnery, T.V. Duzer, *Fields and Waves in Communication Electronics* (Wiley, New York, 1984)
Lord Rayleigh J.W. Strutt, Philos. Mag. **4**, 481–502 (1892)
D.A. Robinson, S.B. Jones, J.M. Wraith, D. Or, R.P. Fredman, Vadose Zone J. **2**, 444–475 (2003)
R. Ruepp, in *Physics and Chemistry of Ice*, ed. by E. Whalley, S.J. Jones, L.W. Gold (R. Soc. Can., Ottawa, 1973), pp. 179–186
F. Russo, M. Cutigni, R. Orosei, C. Taddei, R. Seu, D. Biccari, E. Giacomoni, O. Fuga, E. Flamini, in *Radar Conference, RADAR'08*, IEEE, 26–30 May 2008 (2008), pp. 1–4
A. Safaeinili, W. Kofman, J.-F. Nouvel, A. Herique, R.L. Jordan, Planet. Space Sci. **51**, 505–515 (2003)
D.E. Smith et al., J. Geophys. Res. **106**, 23 689–23 722 (2001)
R. Seu, R.J. Phillips, D. Biccari, R. Orosei, A. Masdea, G. Picardi, A. Safaeinili, B.A. Campbell, J.J. Plaut, L. Marinangeli, S.E. Smrekar, D.C. Nunes, J. Geophys. Res. **112**, E05S05 (2007a)
R. Seu et al., Science **317**, 1715–1718 (2007b)
M.A. Stuchly, S.S. Stuchly, IEEE Trans. Instrum. Measur. **29**(3) (1980)
A.H. Sihvola, Subsurf. Sens. Technol. Appl. **1**, 393–415 (2000)
A.H. Sihvola, J.A. Kong, IEEE Trans. Geosci. Remote Sens. **26**, 420–429 (1988)
A.H. Sihvola, J.A. Kong, IEEE Trans. Geosci. Remote Sens. **27**, 101–102 (1989)
W.R. Thompson, S.W. Squyres, Icarus **86**, 336–354 (1990)
W.R. Tinga, W.A.G. Voss, D.F. Blossey, J. Appl. Phys. **44**, 3897–3902 (1973)
G.C. Topp, P.A. Ferré, in *Methods of Soil Analysis: Part 4, Physical Methods*, ed. by J.H. Dane, G.C. Topp, Soil Sci. Soc. Am. Book Ser, vol. 5 (Springer, Berlin, 2002), pp. 417–421
G.C. Topp, J.L. Davis, A.P. Annan, Water Resour. Res. **16**(3), 574–582 (1980)
P. Zarka, B. Cecconi, W.S. Kurth, J. Geophys. Res. (Space Phys.) **109**, 9 (2004)
Z. Zhang, E. Nielsen, J.J. Plaut, R. Orosei, G. Picardi, Planet. Space Sci. **57**, 393–403 (2009)
A. von Hippel, D.B. Knoll, W.B. Westphal, J. Chem. Phys. **54**, 134–134 (1971)

Space Sci Rev (2010) 153: 273–298
DOI 10.1007/s11214-009-9619-1

Rheological and Thermal Properties of Icy Materials

W.B. Durham · O. Prieto-Ballesteros · D.L. Goldsby · J.S. Kargel

Received: 26 May 2009 / Accepted: 1 December 2009 / Published online: 20 February 2010

Abstract Laboratory measurements of physical properties of planetary ices generate information for dynamical models of tectonically active icy bodies in the outer solar system. We review the methods for measuring both flow properties and thermal properties of icy planetary materials in the laboratory, and describe physical theories that are essential for intelligent extrapolation of data from laboratory to planetary conditions. This review is structured with a separate and independent section for each of the two sets of physical properties, rheological and thermal. The rheological behaviors of planetary ices are as diverse as the icy moons themselves. High-pressure water ice phases show respective viscosities that vary over four orders of magnitude. Ices of CO_2, NH_3, as well as clathrate hydrates of CH_4 and other gases vary in viscosity by nearly ten orders of magnitude. Heat capacity and thermal conductivity of detected/inferred compositions in outer solar system bodies have been revised. Some low-temperature phases of minerals and condensates have a deviant thermal behavior related to paramount water ice. Hydrated salts have low values of thermal conductivity and an inverse dependence of conductivity on temperature, similar to clathrate hydrates or glassy solids. This striking behavior may suit the dynamics of icy satellites.

Keywords Thermal properties · Heat capacity · Thermal conductivity · Cryomagmatism · Thermal gradient · Rheology · Flow strength · Grain size · Grain boundary sliding

W.B. Durham (✉)
Massachusetts Institute of Technology, Cambridge, MA 02127, USA
e-mail: wbdurham@mit.edu

O. Prieto-Ballesteros
Centro de Astrobiología-INTA-CSIC, Ctra. Ajalvir km. 4, 28850 Torrejón de Ardoz, Spain
e-mail: prietobo@inta.es

D.L. Goldsby
Brown University, Providence, RI 02912, USA
e-mail: David_Goldsby@brown.edu

J.S. Kargel
Department of Hydrology, University of Arizona, Tucson, AZ 85721, USA
e-mail: jkargel1054@earthlink.net

1 Introduction to Rheological Properties

The rheology of water ice and other frozen volatiles has primary influence on the evolution and appearance of low-density moons in the outer solar system. The diverse kinds of past and present activity on these moons is driven by internal and external energy sources (radiogenic, gravitational, tidal), and includes planetary-scale convection to smaller-scale relaxation of surface topography. Internally, if ice flows readily, the moon is efficiently cooled; if ice resists flow, in the extreme the moon will overheat, begin to melt internally, and differentiate catastrophically. Warm ice flowing in the subsurface exerts tractions on a cold icy crust, which, if sufficient to overcome the strength of the crust, may cause the crust to fail. On an icy crust, impact craters and other landforms relax under gravitational forces at a rate controlled by the viscosity of ice.

The flow and fracture of water ice under terrestrial conditions has seen extensive study, and a great body of literature exists describing its rheological properties. The "Glen law" for the creep of polycrystalline ice, relating strain rate to the third power of shear stress, owes its origins to one of the first systematic laboratory studies of ice flow (Glen 1952, 1955). The Glen law is still the rule of thumb for glaciologists, although the implications for glaciology of the new rheological descriptions for ice flow described below are being explored. Most of the ice in planetary environments is mostly either much colder or exists at far higher pressures than on Earth, and flow under these conditions may differ significantly from terrestrial ice flow. Laboratory experiments undertaken in support of planetary exploration have begun to provide a catalog of rheological characterizations applicable to models of planetary dynamics and evolution and to the dynamics of terrestrial glaciers and ice sheets. Here we provide a survey of those experiments and measurements as well as a brief introduction to our current understanding of the physics of deformation in planetary settings.

1.1 The Flow Law

Rheology is the study of deformation of materials. The fundamental rheological relationship can be written as

$$\dot{\varepsilon} = A' f(\sigma), \tag{1}$$

where $\dot{\varepsilon}$ and σ are, respectively, strain rate and an appropriate scalar representation of the state of deviatoric stress (differential stress or octahedral shear stress, for example). The factor A' comprises the effect of all relevant environmental conditions on the relationship between $\dot{\varepsilon}$ and σ. Relationship (1) describes flow, and so implicitly applies to large strain or "steady state" deformation that does not vary with strain. The term "viscosity" refers generally to the ratio of $\sigma/\dot{\varepsilon}$. There are important aspects of the strength of materials that have relevance to near-surface geologic and planetary settings: transient deformation at low absolute strains, brittle faulting, and shock deformation during impact. We limit our discussion here to steady-state ductile flow.

Most of the steady-state flow of planetary ices encountered in laboratory experiments and that presumably exists in planetary settings can be described using a power-law form of the flow law:

$$\dot{\varepsilon} = A d^{-p} \sigma^n \exp\left(-\frac{E^* + PV^*}{RT}\right) \tag{2}$$

where we now make explicit the environmental variables: grain size d, hydrostatic pressure (i.e., mean stress) P, and temperature T; and the material-specific parameters: grain size

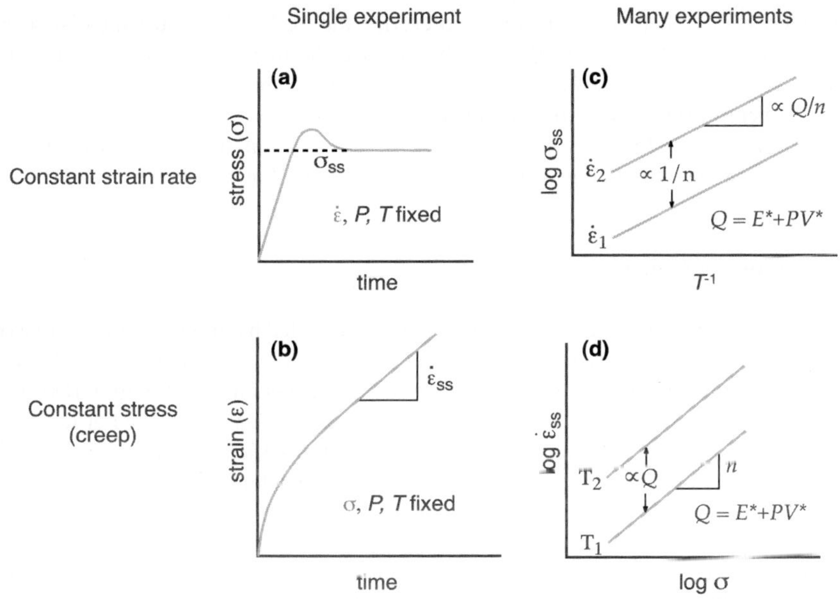

Fig. 1 The two basic laboratory approaches to measuring the parameters in (2): (**a**) Measure the stress required to hold a constant strain rate, or (**b**) measure the strain rate that results from the application of constant stress; (**c**) and (**d**), results from many experiments plotted as a locus of points can be used to solve graphically for material constants

exponent p, stress exponent n, activation energy E^*, and activation volume V^*, along with the pre-exponential constant A comprising all dependencies that have not yet been identified. The functional form of the temperature and pressure dependence will be recognized as that describing a thermally activated process.

1.2 Laboratory Measurement of Rheological Properties

An axisymmetric geometry is one of the practical designs for our mechanical testing instruments because of the necessity in some cases (for example, for relatively large-grained samples) to apply high hydrostatic confinement to samples in order to suppress fracturing and remain in the more planetarily relevant ductile field of deformation.[1] Samples are thus often cylindrical in shape, and deformation is induced by applying a load to the end of the sample with a moving piston, usually in compression. In this condition, the stress term in (2) becomes the differential stress, i.e., the difference between the maximum stress σ_1 applied to the ends of the sample and the gas pressure applied to the outside of the sample. The strain rate is obtained from the rate of shortening of the cylinder.

We approach rheological measurements in either of two ways suggested by the flow law. In a so-called "constant displacement rate" experiment we impose a strain rate with the deforming piston and measure as the dependent variable the stress (load divided by cross-sectional area of the sample) required to maintain that rate (Fig. 1a). Alternatively, in a

[1]The application of hydrostatic pressure during deformation is required to suppress fracturing of coarse-grained polycrystals, e.g., $d > 1$ mm at stresses $\sigma > 1$ MPa for ice, since the fracture strength is inversely proportional to $d^{1/2}$. The need to apply hydrostatic pressure can thus be negated by using fine-grained samples.

constant stress or "creep" experiment, we prescribe the load and measure the displacement rate required to maintain that load (Fig. 1b). Among the many ways of presenting results graphically, there are two styles employed in this paper: the Arrhenius plot of $\log\sigma$ vs. inverse temperature (Fig. 1c), which is convenient for displaying the results of constant strain rate experiments, and log $\dot{\varepsilon}$ vs. log σ (Fig. 1d). Activation energies and stress exponents, respectively, can be identified graphically on such plots as shown in Fig. 1, c and d.

2 Mechanisms of Creep

In broad terms, the steady-state flow of materials is controlled by grain size-insensitive (GSI) dislocation mechanisms (low-temperature plasticity and power-law creep) at high stresses, and grain size-sensitive (GSS) mechanisms involving grain boundary sliding (GBS) at lower stresses (Frost and Ashby 1982). At the highest stresses, typically achieved at low temperatures in the laboratory, the creep rate is limited by the lattice resistance to the glide of dislocations and is characterized by relatively weak dependences of flow stress on strain rate and temperature; in the limit of 'perfect plasticity,' the flow stress is independent of deformation rate. At lower stresses, typically associated with higher temperatures in the laboratory, "power-law creep" (or dislocation creep) occurs via either the glide of dislocations or via the combined glide and climb of dislocations. Dislocation creep is characterized by a power-law dependence of creep rate on stress and an exponential dependence of creep rate on temperature (e.g., see (2)). In cases where glide and climb occur simultaneously as serial mechanical processes, the creep rate is often limited by dislocation climb. At the lowest stresses and at high temperatures, various GSS creep mechanisms become dominant over dislocation creep. The creep rate for these low-stress mechanisms increases strongly with decreasing grain size, so that these mechanisms become accessible at practical strain rates in the laboratory by deforming samples of very fine grain size. Such creep mechanisms can dominate the flow of materials in low-differential-stress planetary environments, for example, in flow of ice in polar ice caps (Nye 2000; Nye et al. 2000), in 'viscous flow features' (Milliken et al. 2003) on Mars, and in the interiors of the icy satellites (Dombard and McKinnon 1996, 2000; Pappalardo et al. 1998; Nimmo and Manga 2002). These same mechanisms likely control the flow of ice under most conditions that occur within terrestrial glaciers and ice sheets (Goldsby and Kohlstedt 2001; Goldsby 2006). Here we focus on the creep mechanisms most relevant for planetary flow of ice, dislocation creep and the various GSS mechanisms.

2.1 Grain Size-Insensitive Creep

Power-law creep or dislocation creep occurs via the motion of lattice dislocations. When a stress of sufficient magnitude is applied to a polycrystalline sample, grains within the sample that are well oriented for slip deform via dislocation glide. These gliding dislocations may encounter other dislocations, or dislocation structures (e.g., tilt walls, dislocation pile-ups, dislocation tangles) which impede their motion. For steady-state deformation to occur, dislocations must 'climb,' allowing for blocked dislocations to glide on their glide planes. Dislocation creep can be either 'glide-limited,' i.e., limited by the glide of lattice dislocations, or 'climb-limited,' i.e., limited by the climb of lattice dislocations, depending on which is the slower process. When climb is rate limiting, the activation energy for creep is equivalent to that for volume diffusion. Typical values of the stress exponent for dislocation creep are $n = 3-5$, and dislocation creep is often independent of grain size. Dislocation creep requires five independent deformation systems (e.g., glide, climb) for homogeneous deformation on the grain scale, or possibly fewer if locally heterogeneous deformation is allowed (see, e.g., Paterson 1969).

2.2 Grain Size-Sensitive Creep

2.2.1 Grain Boundary Sliding

At low stresses, deformation mechanisms involving GBS control the creep rate of materials. However, it must be emphasized that GBS cannot occur in isolation, i.e., it is not an independent creep mechanism. Upon application of a stress to a polycrystalline sample, sliding occurs along grain boundaries, inducing a stress concentration at triple junctions. For small strains, GBS is accommodated by elastic deformation of the material at triple junctions. To achieve larger strains (as during steady-state creep), i.e., to deform the material in a volume-conservative fashion without cavitation or microfracturing, GBS must be accommodated by transfer of material away from the stress concentrations induced by sliding. During steady-state flow, GBS is accommodated by either diffusional flow of material or dislocation flow of material.

2.2.2 Diffusion-Accommodated GBS (Diffusion Creep)

In the so-called diffusion creep regime, GBS and diffusional flow are mutually accommodating creep processes that allow grain-scale stress concentrations induced by GBS to relax (Raj and Ashby 1971), allowing for steady-state deformation. The diffusional flow of material away from and the counterflux of vacancies toward areas of stress concentration occur within grain interiors or along grain boundaries, with the creep rate limited by the faster of these two diffusion steps. First-principles derivations of the diffusion creep equation (Nabarro 1948; Herring 1950; Coble 1963) predict a linear dependence of creep rate on stress for both volume diffusion creep ("Nabarro–Herring creep") and grain boundary diffusion creep ("Coble creep"), with grain size dependences of $p = 2$ and 3, respectively, and activation energies equivalent to those for volume diffusion and grain boundary diffusion, respectively. The diffusion creep equation is given in general form by

$$\dot{\varepsilon} = \frac{42\sigma}{3RT}\left[\frac{D_v}{d^2} + \frac{\pi\delta D_b}{d^3}\right] \tag{3}$$

where δ is the grain boundary width, and D_v and D_b are diffusion coefficients for volume and grain boundary diffusion, respectively, each of the form

$$D_i = D_0 \exp\left(\frac{-Q_i}{RT}\right) \tag{4}$$

where Q is the activation energy and the subscript i indicates v or b. Experiments on a wide range of materials have corroborated these theoretical predictions (Burton 1977); diffusion creep data for ice, however, are lacking (see below).

2.2.3 Dislocation-Accommodated GBS ('Superplastic Flow')

Grain boundary sliding during steady-state creep can also be accommodated by the motion of lattice dislocations. Such behavior is thought to underlie so-called "superplasticity," the phenomenological behavior whereby materials can be deformed to large strains in tension, sometimes to spectacular strains of thousands of percent, without necking and failure. Although considerable controversy exists as to the precise physical mechanism underlying superplasticity, many workers have emphasized the importance of intragranular dislocation slip as well as GBS during superplastic flow (e.g., Ball and Hutchinson 1969; Langdon 1970; Mukherjee 1971; Gifkins 1976; Kaibyshev 1992). Values of flow law parameters n, p and Q ($= E^* + PV^*$) in (2) obtained from experiments on materials deformed in the superplastic flow regime (Region II in the superplasticity parlance) vary somewhat, but values of $n = 2$, $p = 2$, and $Q = Q_b$ are typical. Hence, models of superplastic flow typically yield values of $n = 2$, $p = 2$, and $Q = Q_b$.

The rheological behavior of water ice—Within the past decade, critical advances have been made in understanding the rheological behavior of water ice (ice I and to a lesser extent ice II) relevant to flow in low-differential-stress planetary environments (Goldsby et al. 1993; Goldsby and Kohlstedt 1995, 1997a, 1997b, 2001; Stern et al. 1997; Durham et al. 2001; Kubo et al. 2006). These discoveries were made possible in large part by the development of novel methods for fabricating ice samples of exceedingly fine grain size (<200 μm), which allow for exploration of grain size-sensitive creep mechanisms like diffusion creep and superplastic flow, as well as dislocation creep, at practical laboratory strain rates ($>1 \times 10^{-8}$ s^{-1}).

3 Rheological Behavior of Planetary Ices

3.1 Dislocation Creep of Planetary Ices

3.1.1 Water Ices I and High-Pressure Phases

Flow of ice in the power-law creep regime is classically described by the Glen flow law, which holds that stress exponent $n = 3$ in (2). In spite of the widespread adoption of the Glen law in the terrestrial glaciological community, more recent laboratory experiments on relatively coarse-grained ice samples consistently yield a stress exponent of $n \approx 4$ at differential stresses >1 MPa (e.g., Durham et al. 1997). Higher-stress data in the power-law creep regime are also in excellent agreement with relatively high-stress creep data on fine-grained samples (Goldsby and Kohlstedt 2001; Durham et al. 2001), demonstrating the grain size insensitivity of this mechanism. Although Glen law flow of ice in the power-law creep regime has been most often attributed to a dislocation glide mechanism (e.g., Frost and Ashby 1982), the equivalence of the value of the activation energy for dislocation creep with that for volume diffusion (Ramseier 1967a, 1967b), ~60 kJ/mol, and the good agreement of the value of $n = 4$ in the dislocation creep regime with theoretical models of climb-limited dislocation creep (e.g., Weertman 1968) suggest that dislocation creep of ice may be limited by dislocation climb.

Dislocation creep laws for ice I as well as for high-pressure phases II, III, V, and VI are illustrated in Fig. 2. The broad field associated with ice I is described with a rheology of $n = 4$, $E^* = 60$ kJ/mol and $V^* = -11$ m^3/mol (i.e., ice I becomes *weaker* at higher pressure). This regime applies to a temperature range of roughly 150–250 K at planetary strain rates (see, e.g., the curve labeled 3.5×10^{-14} s^{-1} in Fig. 2). The flow law constants associated with this lower-temperature regime can be expected to apply to most planetary conditions in those cases when the ice is deforming via dislocation creep. An important exception is at the base of an ice shell overlying an internal ocean where the water in that ocean is sufficiently

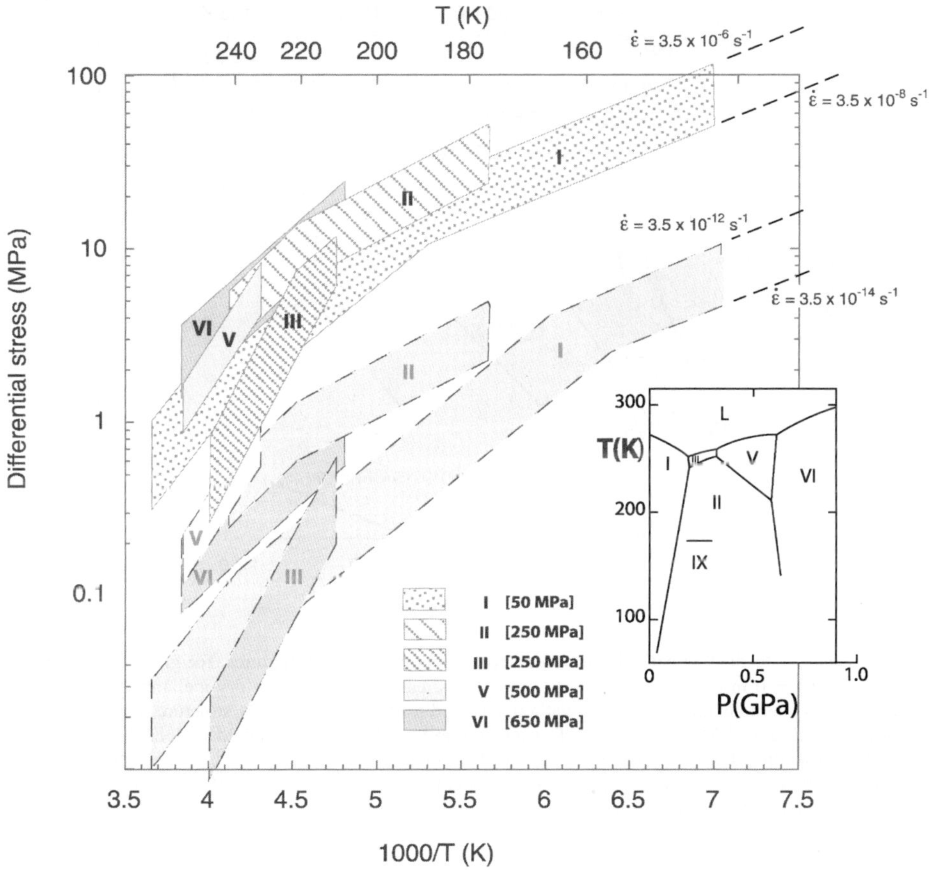

Fig. 2 Arrhenius plot (see Fig. 1(c)) of flow stress for several phases of ice I at fixed strain rate. *Inset* phase diagram shows the equilibrium locations of the several phases in T, P space. *Numbers in brackets in the legend* give the hydrostatic pressure. Fields (labeled by phase) in the Arrhenius plot are bounded by locations of phase transformations (except for the right hand boundaries for ices I, II, and VI) and on the top and bottom by the locus of differential stress values at two strain rates (see labels on the ice I field) two orders of magnitude apart. The top set of fields are directly measured at lab strain rates; the ghosted bottom set of fields are extrapolated to planetary rates using (2). Higher differential stress at a given strain rate means higher viscosity. It can be seen that ices II and VI are the strongest phases at lab rates, but the slightly lower stress sensitivity of ice II means that at lower planetary rates, ice II becomes stronger than ice VI. Ice III is the weakest phase at all strain rates, with ice I being the next weakest

pure that its melting temperature is above 250 K. In that case, the apparent temperature sensitivity of flow will be higher, $E^* \geq 90$ kJ/mol. The flow of ice at warm conditions has obvious terrestrial applications and is for the most part beyond the "planetary" scope of the present paper. Many excellent reviews of the flow of ice in terrestrial settings are available (Alley 1992; Budd and Jacka 1989; Paterson 1994; Petrenko and Whitworth 1999; Schulson and Duval 2009).

The several phases of water ice are rheologically distinct (Fig. 2). It can also be seen in Fig. 2 that relative viscosities of the ice phases in the dislocation creep regime are different at laboratory strain rates (of order 10^{-6} s^{-1}) than they are at planetary rates ($<10^{-12}$ s^{-1}), as extrapolated based on (2). (As discussed below, different creep mechanisms appearing at

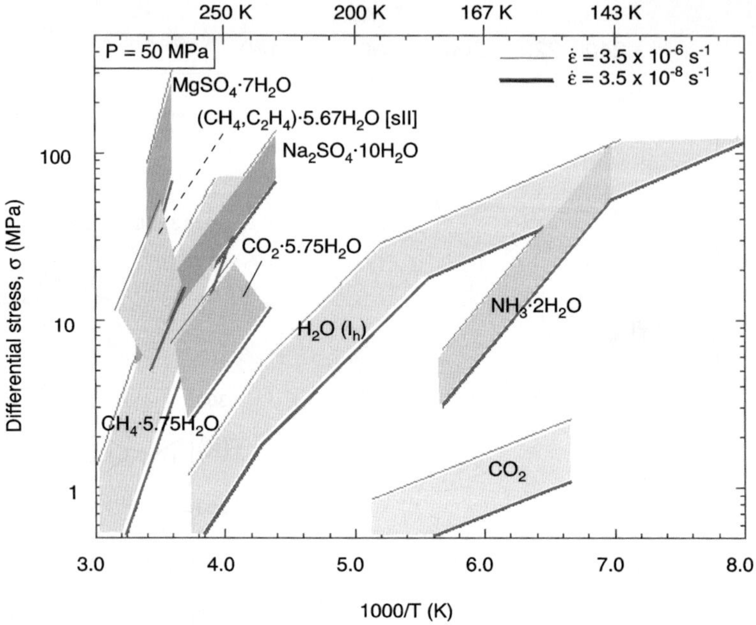

Fig. 3 Steady-state rheological behavior of several planetary ices and mixtures, for grain size-insensitive mechanisms ($p = 0$ in (2)). In approximate order of increasing viscosity are dry ice, ammonia dihydrate, water ice, CO_2 clathrate, methane (CH_4) clathrate, mirabilite ($Na_2SO_4 \cdot 10H_2O$), structure II methane-ethane clathrate, and epsomite ($MgSO_4 \cdot 7H_2O$). Top and bottom boundaries for each field are loci of constant strain rate that are two orders of magnitude apart. Plot is Arrhenius-style (as in Fig. 1(c)), so stress exponent n can be immediately seen as inversely proportional to vertical distance between lines of constant strain rate, and Q/n proportional to the slope of those lines

low stress may change these relationships in the planetary setting.) Each field in Fig. 2 is bounded by fixed strain rates that are two orders of magnitude apart. Perhaps the strongest viscosity contrast is that along the ice I–III boundary, roughly two orders of magnitude at both laboratory and planetary strain rates, and along the ice II–III boundary, where the contrast is roughly 6 orders of magnitude at planetary rates. Given that ice III displays such profound weakness, it is likely that the II–III boundary is an attractor for temperature-depth profiles, meaning that the layer beneath an ice I layer would be comprised of mixed ice II+III. If this is not the case, then the horizontal viscosity layering between I and the much weaker (III) or much stronger (II) phase will likely disrupt large-scale convection.

Figure 2 applies strictly to the behavior of GSI creep mechanisms. GSS creep in ice I has been intensively studied by two research groups, and in ice II by one group. Those results are discussed in the GSS creep section below.

3.1.2 Comparative Rheology of Other Icy Materials

In contrast to ice I, whose rheological behavior and deformation physics has been studied with rigor, other ices of interest in planetology have been the subject of only cursory lab research of their behavior under stress. Most of that research has been conducted since images from the Voyager spacecraft in the late 1970s revealed that icy moons had stunningly varied and active dynamic histories.

Figure 3 is an Arrhenius plot comparing the rheological behaviors of the more common ices that are proven or anticipated building materials of the icy moons of the outer solar system, and of the near surface of Mars. The measurements have been carried out for the most part in the laboratory of one of the authors (WBD), and Fig. 3 is compiled from studies carried out over a wide interval of time (Durham et al. 1983, 1993, 1999, 2003, 2005a, 2005b). In the absence of detailed microscopy or studies of the effect of grain size on creep for most of these materials, we take the behavior to be dislocation creep, and the curves in Fig. 2 are fit to (2) with the parameter $p = 0$, i.e. with GSS processes assumed insignificant. The various curves in Fig. 2 in fact all show $n \geq 3$ (see (2)), so the assumption of dislocation creep is apparently borne out.

The shaded fields in Fig. 3 are defined as in Fig. 2. They apply to laboratory strain rates, and generally reflect directly measured (i.e., not extrapolated) values. For reference, the ice I field from Fig. 2 is reproduced in Fig. 3. Again using a viscosity contrast scale that is two orders of magnitude vertically across any field, it is obvious that the range of viscosities of planetary ices is huge, far larger, say, than the contrasts among water ice phases (Fig. 2). Keeping in mind the composition of Earth's crust and mantle, we should expect that many of these phases are likely to be mixed in planetary environments. The rheological implications of mixing phases of different viscosities can be very complex, and are discussed near the end of this section. A mixture of dry ice (CO_2) and ice I will obviously involve very little deformation of the ice I phase given the >6 order-of-magnitude viscosity contrast between the two phases, except for very ice-rich compositions. Given the extremely low viscosity of dry ice, and knowledge of the age and shape of the Martian South Polar Cap, for example, Nye et al. (2000) concluded that the cap contains very little dry ice in its interior. Clathrate hydrates on the other hand are essentially undeformable compared with ice I, so significant concentrations of clathrates in ice I layers will be significant obstacles to deformation of that layer. Likewise, hydrated sulfate salts such as may exist in the Europan ice shell, technically not ices but appearing in Fig. 3 nonetheless, cannot actively deform with ice I.

3.2 Dislocation-Accommodated GBS in Ice I

With decreasing stress, for ice I samples of sufficiently small grain size (<200 μm), a transition occurs from dislocation creep, with $n = 4$, to a creep mechanism characterized by $n = 1.8$, as shown schematically in Fig. 4. The creep rate in this regime increases systematically with decreasing grain size, yielding a grain size exponent of $p = 1.4$. The value of the activation energy in this creep regime is ~49 kJ/mol, ~0.8 times that of the activation energy for volume diffusion. These values of the flow law parameters are consistent with various models of dislocation-accommodated GBS, constructed in part to explain observations of 'superplastic flow' (e.g., Ball and Hutchinson 1969; Langdon 1970; Mukherjee 1971; Gifkins 1976; Kaibyshev 1992). In addition to these mechanical data, microstructural observations of samples deformed within this creep regime to large strains of up to 0.8 reveal a lack of significant grain flattening and ubiquitous 4-grain junctions (Goldsby and Kohlstedt 1997a, 1997b), both of which have been described as indicators of GBS and grain-switching events during superplastic flow (e.g., Ashby and Verrall 1973).

With further decreasing stress, a transition occurs from a stress exponent of $n = 1.8$ to $n = 2.4$. In addition, the activation energy changes from $Q = 49$ kJ/mol in the $n = 1.8$ regime to $Q = 60$ kJ/mol in the $n = 2.4$ regime. The creep data for polycrystalline samples in this regime are in excellent agreement with data for single crystals of ice oriented for optimal slip on the basal dislocation slip plane (e.g., Wakahama 1967). The creep rate (or flow stress) of ice within this $n = 2.4$ regime is independent of grain size.

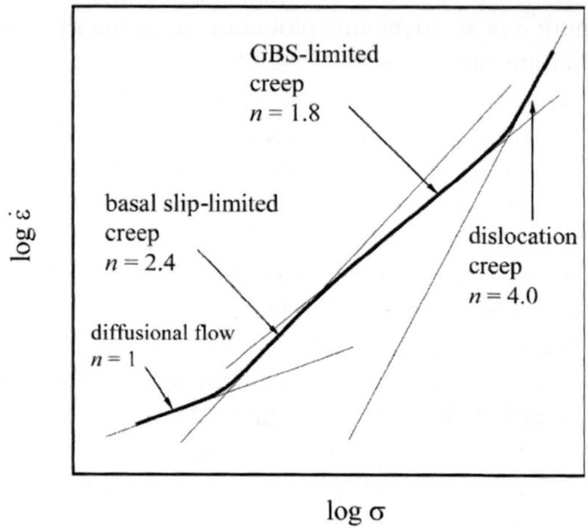

Fig. 4 Schematic diagram showing the various creep regimes for ice I (from Goldsby and Kohlstedt 2001)

3.3 Diffusion-Accommodated GBS in Ice I

Diffusion-accommodated GBS (i.e., diffusion creep) is anticipated to occur at lower stresses than for dislocation-accommodated GBS (e.g., Mukherjee 1971); however, no experimental data for diffusion creep yet exist for ice. There is, however, no reason to doubt its existence (Duval et al. 1983). The question for experimentalists is: can we devise experiments to access the diffusion creep regime of ice in the laboratory?

3.4 A New Rheological Model for Ice I

The rheological behavior described above suggests the following interpretation of experimental (and anticipated, for the case of diffusion creep) data represented schematically in Fig. 4: At sufficiently high stresses, ice samples deform via dislocation creep which is characterized by $n = 4$. Dislocation glide in ice occurs most easily on the basal plane, and with increasing difficulty on the prismatic and pyramidal slip systems. Dislocation creep requires activation of dislocations on both the basal slip system and harder, non-basal slip systems, such that the creep strength at a given strain rate is an appropriate average of easy and hard slip systems.

With decreasing stress, the rate of GBS becomes sufficiently fast, so that compatible deformation can occur via dislocation slip on the basal plane acting in concert with GBS. The effect of GBS is to ease compatibility requirements and effectively 'lose' the hard slip systems required for deformation in the dislocation creep regime, allowing for compatible deformation via only basal slip and GBS. The creep rate then becomes limited by the slower of these two processes. When basal slip is faster than GBS, the creep rate is limited by the slower process, GBS, yielding a stress exponent of $n = 1.8$; when GBS is fast compared to basal slip, a stress exponent of $n = 2.4$ is obtained, and the creep data for polycrystalline samples agree with those for single crystals deformed via basal slip. With further decreasing stress, a transition to diffusion-accommodated GBS, or diffusion creep, is anticipated (Mukherjee 1971), but thus far this creep mechanism has eluded experimentalists (Goldsby and Kohlstedt 2001).

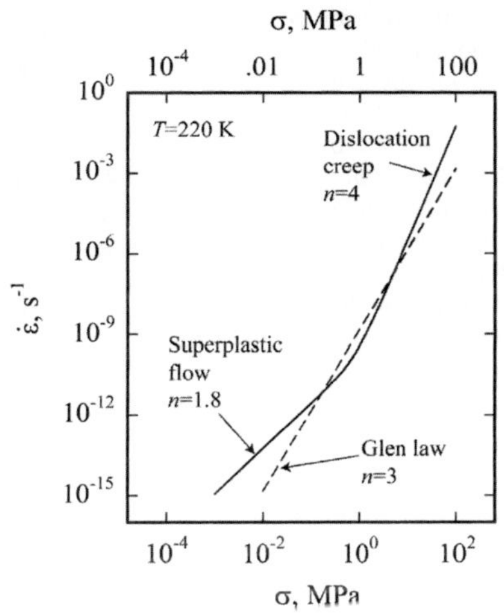

Fig. 5 Plot showing the extrapolation of the composite flow law for creep of ice I (see (5)) to 220 K and a grain size of 1 mm. *Solid line* is the composite flow law, *dashed line* is the Glen flow law with a low-temperature B parameter value (see p. 39 in Paterson 1994). The Glen law, obtained over a limited range of stresses from ~0.1 to <10 MPa, is an average of the flow behaviors in the superplastic flow and dislocation creep regimes

The rheological behavior described above is embodied in the following semi-empirical composite flow law (Goldsby and Kohlstedt 2001):

$$\dot{\varepsilon} = \dot{\varepsilon}_{\text{diff}} + \left(\frac{1}{\dot{\varepsilon}_{\text{basal}}} + \frac{1}{\dot{\varepsilon}_{\text{GBS}}} \right)^{-1} + \dot{\varepsilon}_{\text{disl}}. \tag{5}$$

The subscripts from left to right denote diffusion creep, basal slip-limited GBS creep, GBS-limited basal slip creep, and dislocation creep. Each term is a flow law of the general form of (2).

3.5 GSS Creep in Ice II

GSS creep can generally be expected on theoretical grounds to operate in any crystalline materials if stresses are sufficiently low. Ice II has recently become the second planetary ice phase to exhibit unambiguously such rheological behavior in the laboratory (Kubo et al. 2006). Synthesizing fine-grained high-pressure water ice is problematic because of the high pressures required, but Kubo et al. were able to synthesize fine-grained ice II samples by repeatedly cycling pressure across the ice I–II phase boundary. They measured a flow law for ice II that predicts that at planetary stresses <0.01 MPa, ice II will be softer than ice I for grain sizes below 1–10 mm. It is plausible therefore that GSS creep in ice II defines the rheological behavior of the potentially large volume of ice II that must exist within the large icy moons of the outer solar system.

3.6 Implications for the Glen Law

A simple extrapolation of the composite flow law in (5) to a larger grain size of 1 mm, using experimentally derived flow-law parameters for dislocation creep and GBS-limited creep, and estimated or measured parameters in the diffusion creep equation (given in Goldsby and Kohlstedt 2001), is shown in Fig. 5. The Glen law for ice is also plotted in the figure for comparison. As can be seen in Fig. 5, the Glen law essentially intersects the transition region between GBS-limited flow and dislocation creep of ice. This comparison indicates that the Glen law is not representative of one creep mechanism, but is rather a composite

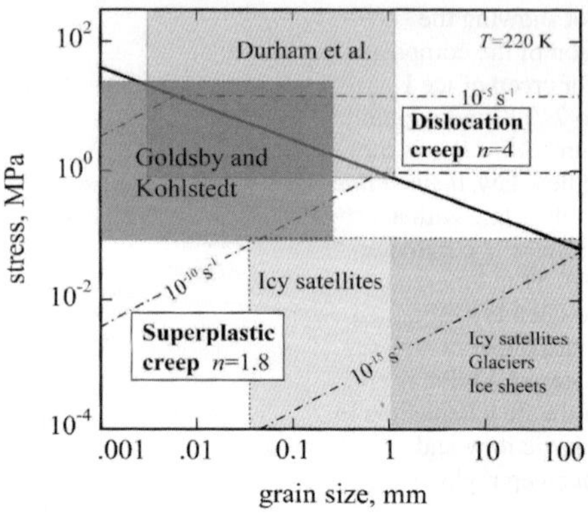

Fig. 6 Deformation map for ice I at 220 K. Conditions of experiments of Goldsby and Kohlstedt (see text for the references) are shown in *red box*; conditions of Durham et al. experiments are shown in *green* (see text for the references). The *solid line* is the boundary between the superplastic flow and dislocation creep regimes

of (i.e., contains contributions from) two creep mechanisms, dislocation creep proper, with $n = 4$, and GBS-limited creep, with $n = 1.8$, leading to an intermediate value of the stress exponent ($n = 3$). This is likely not a coincidence, but rather may be a confirmation of the 'field boundary hypothesis' of de Bresser and co-workers, described below. Hence, the Glen law significantly underestimates the strain rate in natural ice bodies and in planetary environments. This is consistent with the very common observation that ice flow in glaciers and ice sheets, for example, is faster than predicted by the Glen law, requiring rather arbitrary 'enhancement factors' to compensate (Paterson 1994).

3.7 Glaciological and Planetary Implications

A simple extrapolation of laboratory-derived flow laws for superplastic flow and dislocation creep to grain sizes encountered in terrestrial glacier and ice sheets, and to reasonable planetary grain sizes and stresses, strongly suggests that GBS limits the flow of ice over most temperatures at stresses <0.1 MPa. An example of such an extrapolation is illustrated in the deformation mechanism map in Fig. 6. The map is drawn on axes of grain size and stress, for a constant temperature of 220 K. The solid line in the figure delineates the boundary between regions of stress–grain size wherein dislocation creep and GBS-limited creep dominate the strain rate, and lines of constant strain rate are overlain in the figure. As shown on the map, at nearly all conditions of grain size and stress encountered within icy satellites and terrestrial ice bodies, the creep rate is expected to be dominated by GBS-limited flow. It is only at the base of glaciers and ice sheets, for example, where stresses are highest, that the ice nears or encounters the transition to the dislocation creep regime. Similarly, estimates of, for example, convective stresses in icy satellites indicate stresses $\ll 1$ MPa; for reasonable estimates of ice grain sizes based on terrestrial analogs, one also concludes that the flow of planetary ices is limited by GBS-limited flow. Thus, the flow law for superplastic creep has been employed in successfully modeling a host of planetary ice flow phenomena, including crater relaxation on Ganymede (Dombard and McKinnon 2000), flow of the Martian polar ice caps (Nye 2000; Nye et al. 2000), grooved terrain formation on Ganymede (Dombard and McKinnon 1996), convection in Europa's outer icy shell (Pappalardo et al. 1998; Nimmo and Manga 2002) and flow of periglacial features on Mars (Milliken et al. 2003).

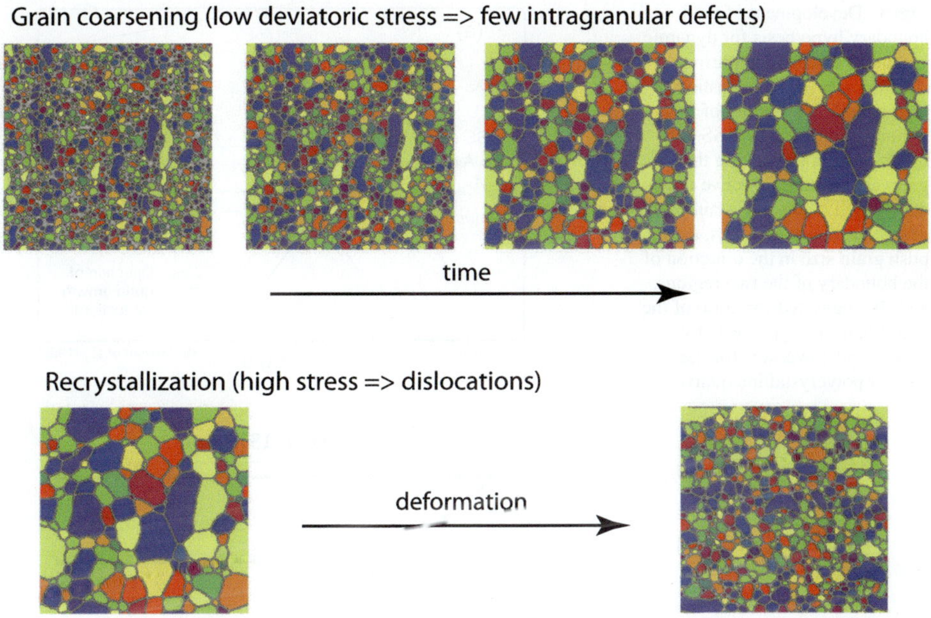

Fig. 7 Computer-generated cartoon of grain growth (*top*) and grain size reduction by recrystallization under stress (*bottom*) in a polycrystal. Grain growth requires only time and is ongoing, so when stress is applied, the two processes act simultaneously, suggesting that grain size may reach a dynamic balance (after Y. Suwa and Y. Saito, Waseda University, Tokyo, http://www.msm.cam.ac.uk/phase-trans/abstracts/grain.movies.html)

3.8 Recent Topics of Interest

3.8.1 Grain Size Distribution and the Dynamic Balance of GSS and GSI Creep

As described above, GSS creep in ice I and ice II, deformation mechanisms of the very utmost planetary relevance, have been observed and measured in the laboratory. However, it is important to recognize that grain size in solid materials is not necessarily a free variable. As depicted in Fig. 7, grains can grow (coarsen) by annealing, the process of lowering internal energy by lowering the amount of grain boundary area per unit volume. Grain size can also decrease, especially during deformation, through processes of dynamic recrystallization. Both processes can be understood from a thermodynamic perspective: in grain coarsening, internal energy is lowered by lowering the density of grain boundary area per unit volume; in dynamic recrystallization, the internal energy of large, dislocation-filled grains is lowered by replacing them with dislocation-free grains. In the latter case, self-organizational instabilities during the recrystallization process lead to an increase in grain boundary area per unit volume (i.e., smaller grains), but the net change in internal energy is still a decrease.

Operation of the GSS mechanism of diffusion creep does not promote the multiplication of line defects (dislocations) so that there is no thermodynamic driving force for reducing the density of these defects by nucleating new grains; grain growth therefore occurs readily in the diffusion creep regime. Dislocation slip in a dislocation-accommodated GBS regime may increase the density of dislocations and provide a driving force for nucleation of new grains, and thus grain growth may be at least somewhat suppressed compared to that in the diffusion creep regime. In contrast, GSI creep can be expected to generate planar and line

Fig. 8 Development of field boundary hypothesis for dynamic grain size distribution during deformation. (**a**) Deformation map including a portion of regions dominated by GSS and by GSI creep, illustrating that grain size reduction, active only during GSI creep, and grain growth, active at all times, act to push grain size in the direction of the boundary of the two regions, and (**b**) suggested variation of the field boundary hypothesis by Austin and coworkers for the case of polycrystalline quartz

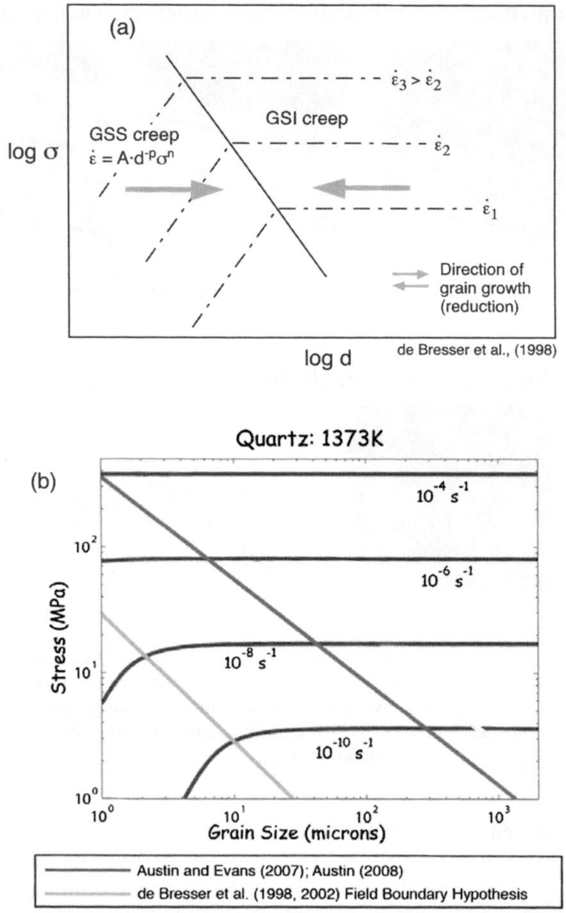

defects in large numbers and therefore will strongly promote grain-size reduction processes, i.e., dynamic recrystallization. The driving force for grain nucleation due to grain boundary migration and bulging is particularly large in the GSI creep regime, where slip on multiple slip systems oriented for easy and hard slip can yield grains of very different internal strain energies; this strain energy mismatch drives boundary migration and recrystallization. De Bresser and colleagues (De Bresser et al. 1998, 2001; Kellermann Slotemaker and de Bresser 2006) noted that the differing effects of GSI and GSS creep on grain size act at cross purposes with respect to their respective dominance of creep rate (Fig. 8a). They hypothesized that during creep grain sizes evolve to a distribution such that GSS and GSI mechanisms contribute approximately equally to the overall creep rate, their so-called "field-boundary" hypothesis. Austin and Evans (2007) have taken this one step further to hypothesize that the dynamic distribution of grain size thus created is also influenced by the rate at which GSI creep generates intracrystalline defects, which in turn is proportional to the rate of work (i.e., power) dissipated in GSI creep. Where paleo-strain rate can be estimated in an exposed rock formation that has undergone natural deformation, measured grain size of the rock compared against that at the theoretical field boundary (Fig. 8b) becomes an effective "paleowattmeter" for the particular tectonic setting.

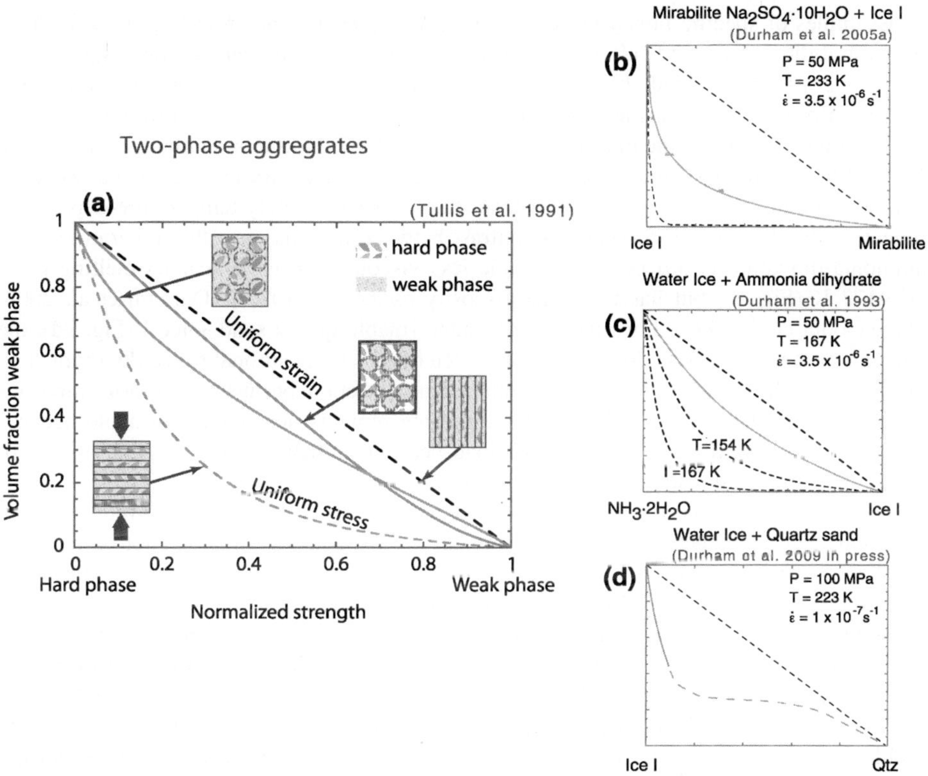

Fig. 9 Rheological behavior of two-phase mixtures. (**a**) General concepts, (**b**)–(**d**) specific icy mixtures of planetary interest. In all plots, normalized strength on the *horizontal axis* varies linearly from 0 to 1, where 0 is the strength of the harder phase and 1 is that of the softer

3.8.2 Rheological Behavior of Multiphase Systems

The rheological properties of systems of more than one phase are complex and diverse. Efforts to date to generalize relationships for the flow of multiphase system have often been inspired but have mostly been unsuccessful (Bloomfield and Covey-Crump 1993; Ji 2004; Treagus 2003; Tullis et al. 1991). Predictably, the best quantitative models occur where phases have the least contrasting rheologies. Figure 9(a), taken from Tullis et al. (1991), illustrates a range of behaviors for simple geometric arrangements of two-phase systems. The "normalized strength" on the horizontal axis of the plot runs from 0, where the strength of the aggregate has the strength of the weaker phase, to 1, where the aggregate strength is that of the harder phase. At one extreme, where the two phases are constrained to strain at the same rate, as with a geometric arrangement of vertical layers with the direction of a uniaxial stress parallel to the layers, the strength of the aggregate is linearly proportional to the volume fractions of the phases. At the other extreme, where both phases experience the same stress, as with horizontal layers lying normal to the uniaxial stress, deformation of the weaker phase will dominate the flow, and the strength of the aggregate for any given choice of volume fractions between 0 and 1 will be much weaker than for the case of uniform strain. Intermediate geometries exhibit intermediate behaviors.

Examples of specific binary systems of planetary ices are shown in Fig. 9, (b to d). In 9(a), mirabilite, a hydrated sulfate salt that is a surrogate for candidate hydrated salts on the Europan crust (Dalton et al. 2010, this issue) is far stronger than ice I, but still deformable. Quasi-uniform mixtures (i.e. not layers or columns) of mirabilite + ice I show dominance of ice I rheology for all but the mirabilite-richest (say, >0.67) mixtures (Durham et al. 2005b). In Fig. 9(c), mixtures of ammonia dihydrate (the elusive planetary ice phase) and water ice I show a composition-strength curve whose shape is very strongly temperature-dependent, reflecting the decrease in rheological contrast between the phases with temperature. Pure ammonia dihydrate is orders of magnitude less viscous than ice I near its melting point at ~173 K (Fig. 3), but has the same viscosity as ice I at 143 K (Durham et al. 2003; Hogenboom et al. 1997). For mixtures of undeformable quartz sand + ice I (Fig. 9d), the strength of the weak phase dominates over much of the compositional range (Durham et al. 1992), although an inflection point is introduced into the curve because at low concentrations of ice, <0.25 (Durham et al. 2009), the aggregate remains nearly undeformable because quartz particles continue to form a rigid, load-bearing framework after a small amount of strain. Note the similar shape in one of the curves in Fig. 9(a).

In real systems of more than one phase, special textures which influence grain–grain interactions can be expected to affect rheological behavior in ways that cannot be modeled employing idealized geometries. In particular, compositional foliation (Jordan 1987) and other forms of phase segregation (Holtzman et al. 2003; Holtzman and Kohlstedt 2007) can cause unexpected rheological influences of one phase or the other, depending on the stress conditions. Finely textured, complex, intimate intergrowths, like alloy structures of metals, are seen in frozen magnesium sulfate–water mixtures (McCarthy et al. 2003, 2004, 2007). Such intergrowths of rheologically disparate minerals as ice and epsomite are expected to yield deformational behavior unlike either of the end-members; however, this behavior has not been fully investigated. One may expect that crustal domains of Europa (and other bodies) that contain large concentrations of such intergrowths (such as frozen cryovolcanic brine flows) would be rheologically decoupled from the ice-dominated crust as a whole. Hence, salt-rich domains may have distinct tectonic structures associated with them, thus helping to explain the distinctive geological nature of the salt-rich zones observed on Europa. The tidal response of ice shells overlying oceans in bodies such as Europa, including heat generation from tidal dissipation, might also be affected by any widespread occurrence of such materials in the icy shell. Such phenomenology has not been investigated to date.

4 Thermal Properties of Some Minerals in Icy Satellites. The Case of Hydrated Salts

4.1 Introduction

Thermal properties of a planetary body's materials determine its activity, geological evolution, and in the case of the icy satellites, even the existence of potential habitable environments. Planetary thermal energy may be produced by accretion and global differentiation (release of gravitational potential energy), impacts (release of kinetic energy), decay of radioactive elements, dissipation of tidal stress, dissipation of frictional energy during faulting and release of stored lattice strains caused by other tectonic motions, electrical resistance, and exogenic chemical reactions during chemical equilibration. Some of these energy sources are linked and flow from one energy pool to another (for instance, radioactive heating can lead to buoyant forces, and those may cause faulting and shear heating). The end result is heating of planetary interiors; this energy then can be dissipated through the body by advective processes such as magmatism and solid-state convection as well as by thermal conduction.

Heat transfer and establishment of the thermal state and geophysical dynamics of a body are governed by melting points, latent heats of phase transitions, thermal conductivity, specific heat and thermal expansivity, and temperature-dependent rheological behavior governing solid-state deformation, tectonics, and advective heat transfer, as well as all relevant length scales (e.g., satellite radius). Partial melting of the original accreted materials (or of reequilibrated/recrystallized primitive materials), and then advective mass transfer (e.g., magmatism) driven by differences in density of the products of partial melting, are primarily responsible for the large-scale structure of planetary interiors. The global thermal state and tectonics of a body are in turn affected by the material layering and will consequently evolve with the configuration of the materials. This endogenous activity may be detected at the surface since it is manifested in recognizable geological features, such as flows, dispersed deposits around a local source, domes, and tectonic patterns of faults. Formation of global aqueous oceans or disconnected liquid reservoirs in icy satellites, which may be maintained for long periods of interest both geologically and biochemically, is either allowed or restricted by the thermal state history of the object. Accordingly, habitability of the aqueous environments in the outer solar system is controlled in some way by the thermal conductivity and other thermodynamic properties of major compounds that accompany the required liquid water. Not only are non-water substances important as sources of biochemicals, but also for control of heat transfer and stability of liquid water habitat.

Classically, studies of icy satellites have considered only pure compositions (usually water ice, sometimes different phases) in calculations, neglecting the influence of other constituents on the thermal state. The predominance of exotic cryogenic solid materials in the icy satellites, compared with the familiar silicates of terrestrial planets, represents a challenge in modeling endogenous planetary processes. As new data about surface and interior chemistry and mineralogy of icy satellites and geophysical constraints on their interior structures are obtained from space missions, as geological observations are made about key processes, and theoretical models of key processes are produced, laboratory experimentalists generate new data relevant to these materials and processes. Through this interactive science community-based effort of observation, interpretation, and theoretical application, we gain an improved understanding of the outer solar system icy bodies and derive insights relevant to planetary science in general. The significance of major non-ice constituents in icy satellites is therefore largely due to their contrasting physical-chemical properties, especially different thermal conductivities, relative to water ice.

Ices in the outer solar system are diverse, and their distributions depend on the location, especially distance from the Sun. Water ice is the most abundant and widespread volatile and may appear in different phases depending on the stability conditions of pressure and temperature and abundances of other constituents with which it may either react chemically or form hydrates. Molecular ices, salt hydrates and clathrate hydrates are also relevant to the solid compositions of the icy satellites. Hydrated salts are usually formed in terrestrial planets by evaporation of water from concentrated solutions, and by crystallization at low temperatures in icy satellites. Signatures of hydrates associated with some geological features have been detected by near-infrared spectroscopy on the surface of Europa and Ganymede. They have been interpreted as hydrated salts or hydrates of sulfuric acid (McCord et al. 1998; Carlson et al. 1999; Dalton et al. 2005). As we will see below, they may play an important role in the activity of the Jovian satellites (Kargel et al. 2000; Prieto-Ballesteros and Kargel 2005).

Aqueous cryomagmatism is interesting not just for geology and geophysics, but also for astrobiology. Chemical systems with water, which are present in the icy satellites, have lower melting points than pure water (see Fortes and Choukroun 2010, this issue). This means that aqueous liquids may occur in the interior, even if the temperature is extremely low. Ammonia and methanol are very effective melting point depressants; for example, the peritectic of the ammonia–water system is at 176.1 K, and a eutectic occurs at 175.4 K (Hogenboom et al. 1997). Methanol–ammonia–water and some other frozen aqueous systems melt at even lower temperatures (Kargel 1992). Because ammonia hydrate is a predicted condensable volatile in the Solar Nebula, expected to be the second most widespread volatile after water ice (Lewis 1972), and because of its low melting temperature and expected high abundance in some nebula condensation scenarios, ammonia–water has commonly been assumed to be a system that leads to formation of the aqueous reservoir in the interior of icy satellites, such as Enceladus (Croft et al. 1988; Kargel et al. 1991, 1992; Kargel and Pozio 1996; Roberts and Nimmo 2008) or Titan (Mitri et al. 2008; Choukroun et al. 2009).

Other expected freezing-point depressants include salts, which may be present in the satellites of Jupiter, Europa or Ganymede (Kargel 1991; Kargel et al. 2000; Prieto-Ballesteros and Kargel 2005). The expected presence of salts in icy satellites is due to a set of hydrolysis reactions similar to those caused by aqueous weathering of the Earth's surface; however, differences in modeled compositions, conditions, and reaction pathways in icy satellites lead to different modeled relative and absolute abundances of various salts, and in some cases some different types of salts, in icy satellites compared to Earth and other mainly silicate bodies (Kargel et al. 2000; Zolotov and Kargel 2008). Chlorides have lower eutectic points than sulfates in water systems, but they have not been detected yet in any icy satellite, with the exception of small amounts of sodium chloride discerned in particulate ejecta from Enceladus' plumes (Schneider et al. 2009; Postberg et al. 2009). Sodium chloride also is believed to be present on Io, a non-icy outer planet satellite (Lellouch et al. 2003; Carlson et al. 2006); in that case, volcanic exhalative processes appear to be responsible for the salt. Pure or mixed anhydrous molecular ices (CH_4, CO_2, CO, N_2) have extremely low melting points. However, methane and ethane may exist in liquid phase at the surface of Titan, stabilized by the pressure of the atmosphere. They form ponds or even seas (Lunine et al. 1983; Lorenz et al. 2003; Stofan et al. 2007), which are part of the dynamical exogenous cycle of this satellite.

4.2 Thermal Behavior of Hydrates at Low Temperatures

Heat capacity and thermal conductivity are properties of chemical compounds that may be measured experimentally, but it is common to find some discrepancies among authors because of the use of differing methodologies and configurations of sample materials.

Vibrations on the crystal lattice, crystal disorder, and magnetic and electric properties determine the heat capacity and thermal conductivity of solids. Each parameter affects to different degrees each type of mineral or condensate. For instance, electric and magnetic contributions are important in metal-bearing minerals, while crystallinity strongly affects the silicate frameworks or lattices of hydrated salts. To explain the theory of conduction of covalent and ionic minerals is beyond the scope of this chapter, but it is important to note that heat transport is due to a phonon movement, which is analogous to photon movement in radiation energy. Some scattering is produced when phonons interact, due to the non-harmonic nature of thermal vibrations. The mean free path between collisions will determine the value of thermal conductivity, and this value will generally be higher when the path is longer. Crystalline materials usually have higher thermal conductivity at low and medium temperatures (which are the temperatures of interest for studies of icy satellites) than at high. This is because at low temperatures there are longer mean free paths at lower amplitudes and

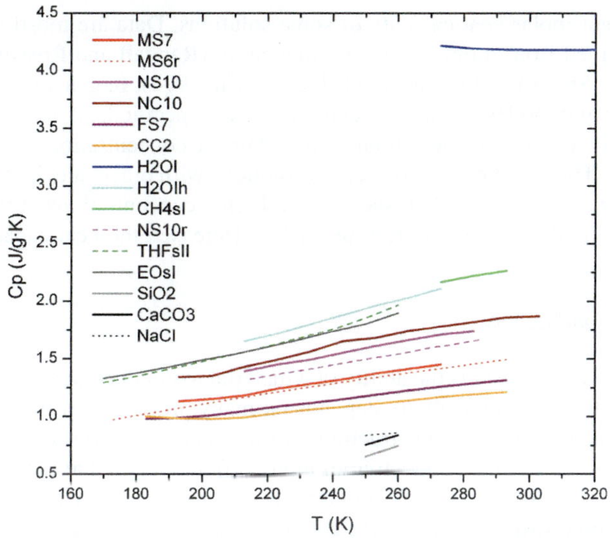

Fig. 10 Heat capacity (C_p) of some compounds of planetological interest as function of temperature. Liquid water, H_2O-I, SiO_2, $CaCO_3$, and NaCl data are from Lide (2008); THF-sII (tetrahydrofuran structure II clathrate) and EO-sI (ethylene oxide structure I) are from White and MacLean (1985), CH_4-sI is from Waite et al. (2007); NS10-r is from Brodale and Giauque (1958); MS6-r is from Cox et al. (1955); MS7 ($MgSO_4 \cdot 7H_2O$, epsomite), NS10 ($NaSO_4 \cdot 10H_2O$, mirabilite), NC10 ($Na_2CO_3 \cdot 10H_2O$, natron) are from Prieto-Ballesteros and Kargel (2005); FS7 ($FeSO_4 \cdot 7H_2O$, melanterite) and CC2 ($CaCO_3 \cdot H_2O$, gypsum) are new data. Data for the last two hydrated salts have been obtained following the same methodology than those published in Prieto-Ballesteros and Kargel (2005), by MDSC

more harmonicity of the thermal vibrations. Phonon scattering is also less when cations are close to the atomic weight of anions. Another observed fact is that it is common that crystals with complex structures have lower thermal conductivities than more simple structures.

4.2.1 Heat Capacity (C_p)

Heat capacity is defined as the amount of heat required to increase by one degree the temperature of one mole of a substance; alternatively, it may be defined on a per-unit-mass basis. Here we use units of joules/gram-Kelvin. Heat capacity depends on the mass, chemical composition, the thermodynamical state and the process that is involved in the heat transfer. Heat capacity in aqueous systems depends on the composition and concentration of solutes.

Heat capacity at atmospheric pressure as a function of temperature for several species of planetological interest, including hydrated salts, water ice I and clathrate hydrates may be seen in Fig. 10. Laboratory experiments show that hydrated salts have specific heat values 15–65% lower than that of water ice, with the lowest values found for anhydrous salts (Fig. 10). Comparison of data obtained at room temperature for several salt hydration states demonstrates that the specific heat increases as the number of water molecules increases (Prieto-Ballesteros and Kargel 2005). Clathrate hydrates also have values of specific heat similar to that of water ice I. It has been observed that specific heat values for clathrates are slightly dependent on guest molecule and occupancy of the cages (Andersson and Ross 1983; Handa and Tse 1996; Tombari et al. 2006).

Considering that internal aqueous oceans may exist in some icy satellites, the specific heat of aqueous solutions has also been explored. There are many references in the literature about the apparent molar heat capacity of some solutions. Data are mostly for halides and are usually obtained from room to higher temperature (Randall and Rossini 1929; Randall and Taylor 1941; Simard and Fortier 1981; Picker et al. 1974). In general, salty aqueous solutions with low and moderate salt concentrations have specific heat values that are similar to or slightly less than that of pure liquid water. Higher concentrations of solutes increase the specific heat. The effect of sulfuric acid in solutions with other salts is remarkable. Bhattacharyya and Bhattacharyya (1979) showed that higher concentrations of the acid (25%) in an aqueous system also containing ferrous sulfate decreased the heat capacity of the solution.

4.2.2 Thermal Conductivity

Thermal conductivity measures the ability of a material to conduct heat. In the same manner as specific heat, thermal conductivity depends on many properties of the material, particularly its chemical composition and structure. For instance, pure crystalline substances may exhibit very different thermal conductivities along different crystal axes.

The thermal conductivity of a rock is determined by the conductivities of its mineral constituents, by the pore structure, and by the conductivities of the fluids that saturate the pores. A similar situation exists for assemblages of low-temperature solid phases in icy satellites. Micro- and macro-structure of the assemblage is critical in determining the conductivity, so porosity, lattice anisotropy, or schistosity of samples must be controlled in laboratory experiments and taken into account when modeling.

In considering the thermal conductivity materials are in general grouped as being either conducting or insulating. Values of thermal conductivity in the range 10 to 400 W/m·K are typical for conductors, while insulators are below 2 W/m·K, more or less the value of water ice at around 273 K. Most terrestrial rocks and minerals have conductivities between 1 and 10 W/m·K. For our purposes, we consider materials that are more conductive than water ice to be conducting and materials less conductive than water ice to be insulating. An extensive review of thermal conductivity of ices in the solar system is given in Ross and Kargel (1998). These authors compiled data for the most relevant ices of the solar system (water ice, CO_2, CH_4, N_2, amorphous ice, clathrate hydrate), showing the different conductivities of the polymorphs of ice with pressure, as well as the inverse dependence of conductivity on temperature of amorphous solids and crystalline clathrate hydrates. Thermal conductivity of water ice I at atmospheric pressure is very well known (Slack 1980), but there are few data for other ices. Only limited data exist for thermal conductivity at high pressure. The best-studied ice at high pressure is water ice. High pressure polymorphs of ices II, III, IV, V, VI, and IX have lower thermal conductivities than I, while VII and VIII have higher values. All the water phases have the dependence with temperature typical of crystalline solids (Ross et al. 1978; Dharma-Wardana et al. 1983).

Because of possible economic importance of some gas clathrates, their properties are being analyzed in detail. All clathrates studied to date have similar thermal conductivities (about 0.5–0.7 W/m·K) for temperatures from 50 K to near the melting point (Ross et al. 1981; Andersson and Ross 1983; White and MacLean 1985; Handa and Cook 1987; Waite et al. 2007; Rosenbaum et al. 2007). It is worth nothing that the thermal conductivity of clathrate hydrates exhibits a slightly inverse dependence with temperature. This behavior is typical of glassy solids, but it has been alternatively explained by a different hypothesis: by the low frequency rattling of the guest molecule in the cage, resulting in strong scattering of the thermal phonons (Tse and White 1988; Ross and Kargel 1998), and by tunneling states arising from proton disorder (Phillips 1972; Ahmad and Phillips 1987). In addition to this, the thermal conductivity of clathrate hydrates is apparently independent of pressure.

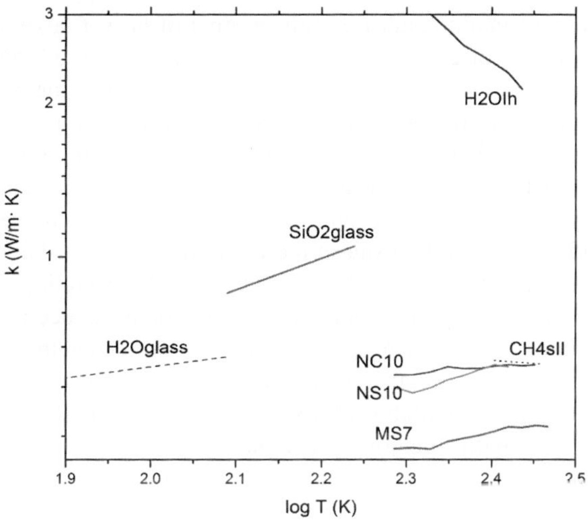

Fig. 11 Dependence of thermal conductivity (k) on temperature for some minerals and condensates of planetological interest. Material abbreviations same as Fig. 10. Data for H_2O-I are from Slack (1980), glassy-H_2O from Andersson and Suga (1994), glassy SiO_2, crystalline SiO_2 and NaCl are from Lide (2008), THF-sI is from Tse and White (1988), CH_4-sI (structure I methane clathrate) is from Waite et al. (2007); MS7, NS10 and NC10 are from Prieto-Ballesteros and Kargel (2005)

It is noticeable that measurements of thermal conductivity of some hydrated salts show very similar behavior (Fig. 11). Thermal conductivity data obtained at atmospheric pressure from some magnesium- and sodium-hydrated sulfates and carbonates have values a third or fourth lower than that of water ice (Prieto-Ballesteros and Kargel 2005). In addition, at low to medium temperatures the thermal conductivities of hydrated sulfates and carbonates also decrease when temperature increases, in the same manner as for clathrate hydrates. The molecular structures of clathrate hydrates and hydrated sulfates may be described in a simple way as being formed by guest molecules surrounded by a cortex of water molecules. If the similar thermal behavior is due to the water molecules' configuration, then arguments previously mentioned for clathrate hydrates may be applied to hydrated salts. In sharp contrast with the hydrates, anhydrous salts, notably halite and anhydrite, are more conductive than water ice.

Some authors have explored the thermal properties of porous materials (H_2O, CO_2, mineral powders and some of their mixtures), which are typical of comets (Komle et al. 1996; Seiferlin et al. 1995, 1996; Steiner and Kömle 1991). Measurements of these mixtures were made at vacuum conditions and, as expected, gave very low values. As an example, porous fine-grained ice measured from 80 to 240 K have values of thermal conductivity in the range 0.005–0.15 W/m·K, while mixtures of ice and silicate minerals have values from 0.1 to 4 W/m·K (Seiferlin et al. 1996; Prialnik et al. 2004).

4.2.3 Planetological Thermal Effects of the Paradoxical Hydrates

Traditionally, ice I has been used to model processes occurring in the icy satellites. However, the other compounds that have been detected by spacecraft sensors must also be taken into account. In some cases, these materials are associated with geological features such as fractures or cryomagmatic structures. This is the case for the hydrated salts on the surface of Europa.

The analysis of the thermal properties of these hydrates indicates that they may favor geological activity more than pure water ice; either the hydrates are the product of this activity, or they spur this activity, or both. Assuming that there exist significant amounts of hydrated salts on Europa, the low thermal conductivity of the salt-rich zones may result in better heat retention than if the mineralogy was dominated by water ice. Steeper thermal gradients would result for a given input of heat if large amounts of hydrates are present in the interior of Europa (Prieto-Ballesteros and Kargel 2005). Some effects of this gradient are a thinner floating crust and an ocean closer to the surface, a shallower depth to partial melting, or a shallower onset of convective conditions.

On the other hand, aqueous solutions, including pure liquid water, may store heat due to their high specific heat and latent heats, while hydrates may act as thermal insulators. If hydrated salts are present in an icy body, the combination of heat capacity and thermal conductivity would result in the liquid reservoirs sustained for long times. Insulation tends to prevent heat loss and to stabilize subsurface liquids (brine plutons or an ocean), which may be more potent than effects due to solute-induced freezing-point depression, but both insulation and freezing point depression work together to make melting more likely than in the absence of salts. The high strength and low ductility of salt hydrates compared to ice also helps to inhibit solid-state convection, and thereby also increase chances of melting compared with situations where the salts are absent. The inverse dependence of conductivity on temperature for salt hydrates and clathrate hydrates means that they are better insulators at lower temperatures (close to the surface of the satellite), but become more ice-like in their conductivity as the water melting point is approached. This temperature dependence is opposite that of crystalline water ice and most other crystalline materials.

Considering the presence of a briny ocean, the impact of salt hydrates on thermal gradient is probably even greater in the suboceanic region, where a thick subcrust of pure salt hydrates may occur (Kargel et al. 2000; Prieto-Ballesteros and Kargel 2005). (Note, however, the possibility of other scenarios, where the salts would not be so abundant; see Zolotov and Kargel 2008.) Salts would form from aqueous solution and precipitate to the ocean bottom due to their higher density. If hydrates are not mixed with other conducting minerals, the thermal gradient may be even steeper here than in the floating salty ice crust. However, this effect may be self-limiting, because high temperatures would cause dehydration of salts and an increase in thermal conductivity to values more like those of silicate rock and ordinary ice I. This effect has been explored also for Mars (Kargel et al. 2007).

If salts are distributed inhomogeneously in Europa's crust, large lateral gradients in subsurface thermal structure and geologic activity would be likely. The next mission to the Jupiter system and Europa should try to confirm the association of local thermal anomalies with the material type and geological structure. Intruded brine dikes and plutons and extruded brine lava flows would cool much more slowly than equivalent water bodies. This may have significance for the future thermal prospecting of geologic activity on Europa and other icy bodies, since thermal emissions due to shallow brine bodies may be extended over greater periods of time before the bodies freeze solid and cool to ambient conditions. The main objective of a future mission to Europa and the rest of the Jovian system may be to search for potentially habitable environments, characterizing any aqueous reservoir in the satellites in a first step. It will be essential to know the thermal properties of the materials in order to understand the thermal state of the satellite. Although there are already some existing thermal properties data for planetary ices, most were obtained at terrestrial conditions. Hence, much work on thermal properties under specific planetary conditions (from the interior to the surface) still needs to be done in order to prepare for future exploration of the icy bodies of the solar system.

References

N. Ahmad, W.A. Phillips, Solid State Commun. **63**(2), 167–171 (1987)

R.B. Alley, J. Glaciol. **38**, 245–256 (1992)
P. Andersson, R.G. Ross, J. Phys. C. Solid State Phys. **16**, 1423–1432 (1983)
O. Andersson, H. Suga, Solid State Comm. **91**(12), 985–988 (1994)
M.F. Ashby, R.A. Verrall, Acta Metall. **21**, 149–163 (1973)
N.J. Austin, B. Evans, Geology **35**(4), 343–346 (2007)
A. Ball, M.M. Hutchinson, Metal Sci. J. **3**, 1–7 (1969)
S. Bhattacharyya, S.N. Bhattacharyya, J. Chem. Eng. Data **24**(2), 93–96 (1979)
J.P. Bloomfield, S.J. Covey-Crump, J. Struct. Geol. **15**(8), 1007–1019 (1993)
G.E. Brodale, W.F. Giauque, J. Am. Chem. Soc. **80**, 2042–2024 (1958)
W.F. Budd, T.H. Jacka, Cold Reg. Sci. Technol. **16**, 107–144 (1989)
B. Burton, *Diffusional Creep of Polycrystalline Materials* (Trans. Tech. Publications, Germany, 1977), 119 pp.
R.W. Carlson, R.E. Johnson, M.S. Anderson, Science **286**, 97–99 (1999)
R.W. Carlson, J.S. Kargel, S. Doute, L.A. Soderblom, J.B. Dalton, in *Io after Galileo*, ed. by R.M.C. Lopes, J.R. Spencer (Springer, Berlin, 2006)
M. Choukroun, O. Grasset, G. Tobie, C. Sotin, Icarus (2009). doi:10.1016/j.icarus.2009.08.011
R.L. Coble, J. Appl. Phys. **34**, 1679–1682 (1963)
W.P. Cox, E.W. Hornung, W.F. Giauque, J. Am. Chem. Soc. **77**(15), 3935–3938 (1955)
S.K. Croft, J.I. Lunine, J. Kargel, Icarus **73**, 279–293 (1988)
J.B. Dalton et al., Space Sci. Rev. (2010, this issue)
J.B. Dalton, O. Prieto-Ballesteros, J.S. Kargel et al., Icarus **177**, 472–490 (2005)
J.H.P. De Bresser, C.J. Peach, J.P.J. Reijs, C.J. Spiers, Geophys. Res. Lett. **25**, 3457–3460 (1998)
J.H.P. De Bresser, J.H. Ter Heege, C.J. Spiers, J. Earth Sci. **90**(1), 28–45 (2001)
M.W.C. Dharma-Wardana, F. Perrot, G.C. Aers, Phys. Rev. A **28**, 344–349 (1983)
A.J. Dombard, W.B. McKinnon, Lunar Planet. Sci. Conf. **27**, 317–318 (1996)
A.J. Dombard, W.B. McKinnon, Geophys. Res. Lett. **27**, 3663–3666 (2000)
W.B. Durham, H.C. Heard, S.H. Kirby, J. Geophys. Res. **88**, B377–B392 (1983)
W.B. Durham, S.H. Kirby, L.A. Stern, J. Geophys. Res. **97**, 20883–20897 (1992)
W.B. Durham, S.H. Kirby, L.A. Stern, J. Geophys. Res. **98**, 17667–17682 (1993)
W.B. Durham, S.H. Kirby, L.A. Stern, J. Geophys. Res. (Planets) **102**, 16293–16302 (1997)
W.B. Durham, L.A. Stern, S.H. Kirby, Geophys. Res. Lett. **26**, 3493–3496 (1999)
W.B. Durham, S.H. Kirby, L.A. Stern, J. Geophys. Res. **106**, 11,031–11,042 (2001)
W.B. Durham, S.H. Kirby, L.A. Stern, W. Zhang, J. Geophys. Res. **108**(B4), 2182 (2003). doi:2110.1029/2002JB001872
W.B. Durham, L.A. Stern, T. Kubo, S.H. Kirby, J. Geophys. Res. (Planets) **110**(E12), E12010 (2005a). doi:12010.11029/12005JE002475
W.B. Durham, L.A. Stern, S.H. Kirby, S. Circone, in *Proceedings of the 5th International Conference on Gas Hydrates* (paper #2030), Trondheim, Norway (2005b), pp. 607–614
W.B. Durham, A.V. Pathare, L.A. Stern, H.J. Lenferink, Geophys. Res. Lett. (2009). doi:10.1029/2009GL040392
P. Duval, M.F. Ashby, I. Anderman, J. Phys. Chem. **87**(21), 4066–4074 (1983)
A.D. Fortes, M. Choukroun, Space Sci. Rev. (2010, this issue)
H.J. Frost, M.F. Ashby, *Deformation Mechanism Maps* (Pergamon, New York, 1982) 167 pp.
R.C. Gifkins, Metall. Trans. **7A**, 1225–1232 (1976)
J.W. Glen, J. Glaciol. **2**, 111–114 (1952)
J.W. Glen, Proc. R. Soc. Lon., Ser. A **228**, 519–538 (1955)
D.L. Goldsby, in *Glaciology and Earth's Changing Environment*, ed. by P. Knight (Blackwell, Oxford, 2006), p. 527
D.L. Goldsby, D.L. Kohlstedt, in *Proceedings of the 35th U.S. Rock Mechanics Symposium*, ed. by J.K. Daemen, R.A. Schultz (Balkema, United States 1995), pp. 199–206
D.L. Goldsby, D.L. Kohlstedt, Scr. Mater. **37**, 1399–1406 (1997a)
D.L. Goldsby, D.L. Kohlstedt, in *Proceedings of the 28th Annual Lunar and Planetary Science Conference* (Lunar and Planetary Institute, Houston, 1997b), pp. 429–430
D.L. Goldsby, D.L. Kohlstedt, J. Geophys. Res. **106**, 11,017–11,030 (2001)
D.L. Goldsby, D.L. Kohlstedt, W.B. Durham, in *Abstracts of the 24th Lunar and Planetary Science Conference*, Houston, TX, 15–19 March 1993 (1993), p. 543
Y.P. Handa, J.G. Cook, J. Phys. Chem. **91**(25), 6327–6328 (1987)
Y.P. Handa, J.S. Tse, J. Phys. Chem. **90**, 5917–5921 (1996)
C. Herring, J. Appl. Phys. **21**, 437–444 (1950)
D.L. Hogenboom, J.S. Kargel, G.J. Consolmagno, T.C. Holden, L. Lee, M. Buyyounouski, Icarus **128**, 171–180 (1997)

B.K. Holtzman et al., Science **301**(5637), 1227–1230 (2003)
B.K. Holtzman, D.L. Kohlstedt, J. Petrol. **48**(12), 2379–2406 (2007)
S.C. Ji, J. Geophys. Res. (Solid Earth) **109**(B10), B10207 (2004)
P.G. Jordan, Tectonophysics **135**(1–3), 185–197 (1987)
O.A. Kaibyshev, *Superplasticity of Alloys, Intermetallides, and Ceramics* (Springer, New York, 1992), 317 pp.
J.S. Kargel, Icarus **94**, 368–390 (1991)
J.S. Kargel, S.K. Croft, J.I. Lunine, J.S. Lewis, Icarus **89**, 93–112 (1991)
J.S. Kargel, Icarus **100**, 556–574 (1992)
J.S. Kargel, S. Pozio, Icarus **119**, 385–404 (1996)
J.S. Kargel, J.Z. Kaye, J.W. Head, G.M. Marion, R. Sassen, J.K. Crowley, O. Prieto-Ballesteros, S.A. Grant, D.L. Hogenboom, Icarus **148**, 226–265 (2000)
J.S. Kargel, R. Furfaro, O. Prieto-Ballesteros, J.A.P. Rodriguez, D.R. Montgomery, A.R. Gillespie, G.M. Marion, S.E. Wood, Geology **35**(11), 975–978 (2007) (2007)
A. Kellermann Slotemaker, J.H.P. de Bresser, Tectonophysics **427**(1–4), 73–93 (2006)
N.I. Komle, G. Kargl, K. Thiel, K. Seiferlin, Planet. Space Sci. **44**(7), 675–689 (1996)
T. Kubo, W.B. Durham, L.A. Stern, S.H. Kirby, Science **311**(5765), 1267–1269 (2006)
T.G. Langdon, Phil. Mag. **22**, 689–700 (1970)
E. Lellouch, G. Paubert, J.I. Moses, N.M. Schneider, D.F. Strobel, Nature **421**, 45–47 (2003)
J.S. Lewis, Icarus **16**(2), 241–252 (1972)
D.R. Lide (Editor in chief). *Crc Handbook of Chemistry and Physics 2008–2009*. Editorial: Taylor & Francis 89th edition (2008)
R.D. Lorenz, E. Kraal, E. Asphaug, R. Thomson, Eos **84**, 125–132 (2003)
J.I. Lunine, D.J. Stevenson, Y.L. Yung, Science **222**(4629), 1229–1230 (1983)
C.M. McCarthy, S.H. Kirby, W.B. Durham, L.A. Stern, Eos Trans. AGU **84**(46), Fall Meet. Suppl., Abstract T42A-0275 (2003)
C.M. McCarthy, S.H. Kirby, W.B. Durham, L.A. Stern, Eos Trans. AGU **85**(47), Fall Meet. Suppl., Abstract P31A-0955 (2004)
C. McCarthy, R.F. Cooper, S.H. Kirby, K.D. Rieck, L.A. Stern, Am. Mineral. **92**(10), 1550–1560 (2007)
T.C. McCord et al., Science **280**, 1242–1245 (1998)
R.E. Milliken, J.F. Mustard, D.L. Goldsby, J. Geophys. Res. **108**, 5057 (2003) 10.1029/2002JE002005
G. Mitri et al., Icarus **196**(1), 216–224 (2008)
A.K. Mukherjee, Mater. Sci. Eng. **8**, 83–89 (1971)
F.R.N. Nabarro, in *Report of a Conference on Strength of Solids (Bristol)* (The Physical Society, London, 1948), pp. 75–90
F. Nimmo, M. Manga, Geophys. Res. Lett. **29**(23) (2002)
J.F. Nye, J. Glaciol. **46**, 438–444 (2000)
J.F. Nye, W.B. Durham, P.M. Schenk, J.M. Moore, Icarus **144**(2), 449–455 (2000)
R.T. Pappalardo, J.W. Head, R. Greeley, R.J. Sullivan, C. Pilcher, G. Schubert, W.B. Moore, M.H. Carr, J.M. Moore, M.J.S. Belton, D.L. Goldsby, Nature **391**, 365–368 (1998)
M.S. Paterson, in *Physics of Strength and Plasticity*, ed. by A.S. Argon (MIT, Cambridge, 1969), pp. 377–392
W.S.B. Paterson, *The Physics of Glaciers*, 3rd edn. (Pergamon, Oxford, 1994), 250 pp.
V.F. Petrenko, R.W. Whitworth, *Physics of Ice* (Oxford University Press, New York, 1999), 373 pp.
W.A. Phillips, Bull. Am. Phys. Soc. **17**(1), 115 (1972)
P. Picker, E. Tremblay, C. Jolicoeur, J. Solut. Chem. **3**, 377 (1974)
D. Prialnik, J. Benkhoff, M. Podolak, Comets **II**, 359–387 (2004)
F. Postberg, S. Kempf, J. Schmidt, N. Brilliantov, A. Beinsen, B. Abel, U. Buck, R. Srama, Nature **459**, 1098–1101 (2009)
O. Prieto-Ballesteros, J.S. Kargel, Icarus **173**, 212–221 (2005)
R. Raj, M.F. Ashby, Metall. Trans. **2**, 1113–1127 (1971)
R.O. Ramseier, Self-diffusion in ice monocrystals. US Army Cold Regions Research and Engineering Laboratory Hanover, NH, Research Report, 232 (1967a)
R.O. Ramseier, J. Appl. Phys. **38**, 2553–2556 (1967b)
M. Randall, F.D. Rossini, J. Am. Chem. Soc. **51**, 323–345 (1929)
M. Randall, M.D. Taylor, J. Phys. Chem. **45**(6), 959–967 (1941)
J.H. Roberts, F. Nimmo, Icarus **194**, 675–689 (2008)
E.J. Rosenbaum, N.J.K. English Johnson, D.W. Shaw, R.P. Warzinski, J. Phys. Chem. B **111**(46), 13194–13205 (2007)
R.G. Ross, J.S. Kargel, in *Solar System Ices*, ed. by C. de Bergh, M. Festou, B. Schmitt (Kluwer, Dordrecht, 1998), pp. 33–62
R.G. Ross, P. Andersson, G. Backstrom, J. Chem. Phys. **68**(9), 3967–3972 (1978)

R.G. Ross, P. Andersson, G. Backstrom, Nature **290**(5804), 322–323 (1981)
K. Seiferlin, T. Spohn, J. Benkhoff, Lab. Planetol. **15**(10), 35–38 (1995)
K. Seiferlin, N. I Komle, G. Kargl, Planet. Space Sci. **44**(7), 691–704 (1996)
N.M. Schneider, M.H. Burger, E.L. Schaller, M.E. Brown, R.E. Johnson, J.S. Kargel, M. Dougherty, N. Achilleos, Nature **459**, 1102–1104 (2009)
E.M. Schulson, P. Duval, *Creep and Fracture of Ice* (Cambridge University Press, New York, 2009)
M.A. Simard, J.L. Fortier, Canadian J. of Chemistry-Revue Canadienne de Chimie **59**(22), 3208–3211 (1981)
G.A. Slack, Phys. Rev. B **22**, 3065–3071 (1980)
G. Steiner, N.I. Kömle, Planet. Space Sci. **39**(3), 507–513 (1991)
L.A. Stern, W.B. Durham, S.H. Kirby, J. Geophys. Res. **102**, 5313–5325 (1997)
E.R. Stofan et al., Nature **445**, 61–64 (2007)
E. Tombari, S. Presto, G. Salvetti, G.P. Johari, J. Chem. Phys. **124**(15), 154507 (2006)
S.H. Treagus, Tectonophysics **372**(3–4), 121–133 (2003)
J.S. Tse, M.A. White, J. Phys. Chem. **92**, 5006–5011 (1988)
T.E. Tullis, F.G. Horowitz, J. Tullis, J. Geophys. Res. **96**, 8081–8096 (1991)
G. Wakahama, in *Proceedings of the International Conference on Low Temperature Science*, vol. 1 (Inst. of Low Temp. Sci., Hokkaido Univ., Sapporo, 1967), pp. 292–311
W.F. Waite, L.A. Stern, S.H. Kirby, W.J. Winters, D.H. Mason, Geophys. J. Int. **169**, 767–774 (2007)
J. Weertman, Trans. ASM **61**, 681–694 (1968)
M.A. White, M.T. MacLean, J. Phys. Chem. **89**, 1380–1383 (1985)
M.Yu. Zolotov, J.S. Kargel, in *Europa*, ed. by R. Pappalardo. Space Science (University of Arizona Press, Tucson, 2008, in press)

Science 299, 1193–1195 (2003)
... [illegible] ... (1991)

Space Sci Rev (2010) 153: 299–315
DOI 10.1007/s11214-009-9625-3

Radiolysis and Photolysis of Icy Satellite Surfaces: Experiments and Theory

T. Cassidy · P. Coll · F. Raulin · R.W. Carlson · R.E. Johnson · M.J. Loeffler · K.P. Hand · R.A. Baragiola

Received: 14 May 2009 / Accepted: 16 December 2009 / Published online: 23 February 2010

Abstract The transport and exchange of material between bodies in the outer solar system is often facilitated by their exposure to ionizing radiation. With this in mind we review the effects of energetic ions, electrons and UV photons on materials present in the outer solar system. We consider radiolysis, photolysis, and sputtering of low temperature solids. Radiolysis and photolysis are the chemistry that follows the bond breaking and ionization produced by incident radiation, producing, e.g., O_2 and H_2 from irradiated H_2O ice. Sputtering is the ejection of molecules by incident radiation. Both processes are particularly effective on ices in the outer solar system. Materials reviewed include H_2O ice, sulfur-containing compounds

T. Cassidy (✉) · R.E. Johnson
Engineering Physics Program, University of Virginia, Charlottesville, VA 22904, USA
e-mail: tac2z@virginia.edu

Present address:
T. Cassidy
Jet Propulsion Laboratory, California Institute of Technology, Pasadena, CA 91109, USA

P. Coll
Laboratoire Interuniversitaire des Systèmes Atmosphériques, Université Paris 7 - Denis Diderot, Créteil, France

F. Raulin
Laboratoire Interuniversitaire des Systèmes Atmosphériques, UMR 7583, Université Paris 7 et Paris 12, CNRS, Créteil, France

R.W. Carlson · K.P. Hand
Jet Propulsion Laboratory, California Institute of Technology, Pasadena, CA 91109, USA

R.E. Johnson
Physics Dept., New York University, New York, NY 10003, USA

M.J. Loeffler
Astrochemistry Laboratory, NASA Goddard Spaceflight Center, Greenbelt, MD, USA

R.A. Baragiola
Laboratory for Atomic and Surface Physics, University of Virginia, Charlottesville, VA 22904, USA

(such as SO_2 and S_8), carbon-containing compounds (such as CH_4), nitrogen-containing compounds (such as NH_3 and N_2), and mixtures of those compounds. We also review the effects of ionizing radiation on a mixture of N_2 and CH_4 gases, as appropriate to Titan's upper atmosphere, where radiolysis and photolysis produce complex organic compounds (tholins).

Keywords Radiolysis · Photolysis · Sputtering · Tholins · Ices

1 Introduction

This chapter describes the effects of ultraviolet and ionizing radiation on materials in the outer solar system, from the erosion and modification of icy surfaces to the production of aerosols in Titan's atmosphere. Although there have been a number of reviews of radiation effects, we emphasize the signatures in reflectance spectra that indicate radiation-induced modifications. We also describe the importance of radiation in producing gas phase species which can populate the neutral and plasma tori that are so important to inter-satellite exchange processes. The exchange processes mentioned here are described in detail in Krupp et al. (2010, this issue) and Burger et al. (2010, this issue).

When the ionizing radiation consists of energetic ions, electrons and very energetic photons, the processes induced are often referred to as radiolysis, whereas photolysis typically refers to the effect of UV photons. Radiolysis and photolysis may be broadly defined as the chemistry that follows the breakup of molecular bonds. The incident radiation ionizes and fragments the molecules; the resulting radicals can then diffuse and react to form new species in non-equilibrium processes. Such reactions are often followed by the release of energy and sputtering, the ejection of molecules from the surface. Although the studies of radiolysis and photolysis have been of interest since the discovery of radiation sources, particularly as it relates to biology (health physics), the study of the radiation effects in *ices* increased following the experiments of Brown, Lanzerotti and coworkers on radiation processes relevant to the icy bodies in the outer solar system (e.g., Lanzerotti et al. 1978).

The effects seen in laboratory irradiation of low-temperature condensed-gas solids (H_2O, SO_2, CO_2, etc.) were often observed on solar system bodies, particularly the icy surfaces of outer planet moons. Reflectance spectra of some water ice surfaces exhibited the presence of common radiolytic products such as trapped H_2O_2, O_2 and O_3, and tenuous radiation-induced O_2 atmospheres are common. Sulfur and carbon containing surfaces showed the expected radiation-induced darkening.

Sputtered molecules are often ionized and "picked up" by the local magnetic fields. Since the plasma so formed can diffuse throughout the magnetosphere, material from one body can be implanted in a neighboring body. Sulfur from Io populates the Jovian magnetosphere and may contribute to the sulfur signature on Europa's trailing hemisphere (Lane et al. 1981; Carlson et al. 2009). The description of such processes has led to considerable experimental and modeling efforts both in the planetary science and the radiation effects communities. Using earlier reviews, this effort is summarized here with an emphasis on the implications for exchange processes.

2 Water Ice Radiolysis

Following the discovery of Jovian decametric radiation and the Pioneer mission Jupiter flybys it became clear that the ice-covered moons of Jupiter were imbedded in a relatively intense, by space standards, flux of ions and electrons. This inspired a series of

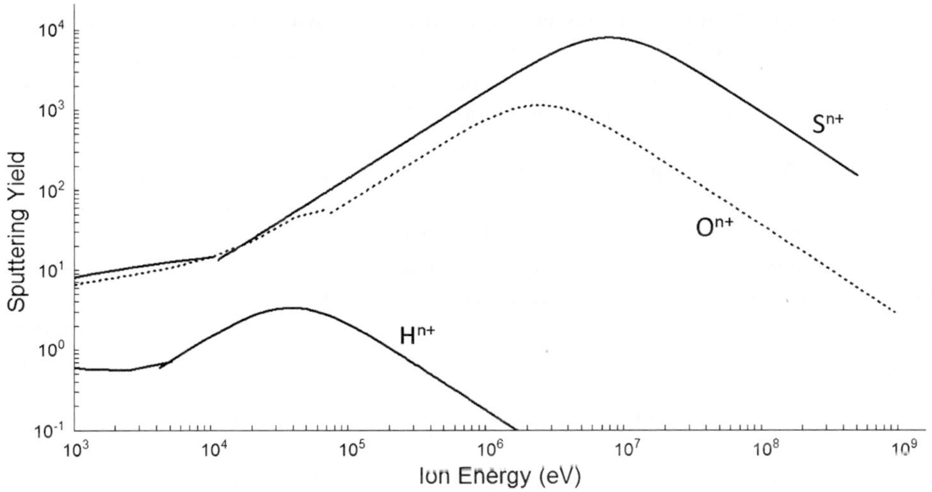

Fig. 1 Water ice sputtering yields (number of equivalent H_2O molecules ejected per incident ion of a given energy) for various ions for $T \lesssim 100$ K. Adapted from Johnson et al. (2009). Below ∼80 K H_2O dominates the ejecta; with increasing temperature O_2 and H_2 become a large fraction of the ejecta

experiments on energetic plasma ion irradiation of low temperature water ice. *Sputtering*, the ejection of molecules from the surface due to incident radiation, was found to be orders of magnitude larger than predicted (Brown et al. 1978). The incorrect predictions were based on a theory that modeled collisions between molecules ("knock-on collision cascades") whereas fast ions traveling through a condensed-gas solid, such as H_2O ice, primarily deposit energy as a track of electronic excitations and ionizations. Although there have been a number of reviews of the radiation effects in ice (e.g., Johnson 2010; Baragiola 2003) we describe the principal results as they apply to material transport and as a guide to understanding the subsequent descriptions of more complex ices.

One of the principal measurements relevant to the exchange of materials between outer Solar system bodies is the sputtering yield, the number of molecules ejected per incident ion. Sputtering yields for ions found in the outer planet magnetospheres (H^+, O^+ and S^+) are given in Fig. 1, which shows semi-empirical fits to a large body of experimental data. Note that the sputtering yield can be as large as $\sim 10^4$ H_2O (for ∼100 MeV S^+).

The principal sputter products from condensed gases are whole molecules and not radicals, atoms or ions, as is the case for the more extensively studied metallic and ionic solids. At low temperatures water molecules are the dominant sputter product of water ice, but the incident radiation also chemically alters the ice to produce, among other products, molecular hydrogen and oxygen which can dominate the sputtering yield at high temperatures ($T \gtrsim 120$ K). It is now generally agreed that the radiolysis of ice by energetic ions, electrons and solar UV photons produces the oxygen, ozone and peroxide seen at many icy satellites.

Any H_2 formed in ice diffuses and escapes much more efficiently than O_2 at the relevant temperatures in the outer solar system, and, in turn, escapes from the icy moons because of their relatively weak gravitational fields. Therefore the irradiation can preferentially populate the magnetosphere with hydrogen, as is the case at Europa (Lagg et al. 2003; Mauk et al. 2003), leaving behind an oxygen-rich satellite surface (e.g., Johnson et al. 2009).

Extensive laboratory studies of the irradiation of water ice have been carried out using incident electrons, UV photons and a large range of ion energies and types (see Johnson 2010).

In these experiments the ice samples and ejecta are monitored as a function of different radiation doses, incident radiation types, energies and sample temperatures. The sputtering yield is typically monitored via the change in sample thickness (measured by, e.g., microbalance gravimetry) while the identity of the sputter products is determined through mass spectrometry.

Though differing quantitatively, depending on the radiation type and energy, the results are qualitatively the same. Incident ions and electrons, which produce multiple ionizations, promptly eject H_2O so that the sputtering of water molecules is independent of dose (time-integrated number of ions or electrons incident per unit area), but the production of gas phase O_2 starts at zero and grows with fluence, eventually reaching a steady state. The other major product of H_2O irradiation is H_2, which, though ejected promptly, eventually is lost from ice stoichiometrically in a 2 to 1 ratio with O_2. The observed O_2 fluence dependence may result from a two-step process: precursor species are created by the incident radiation from which subsequent radiation ejects O_2. The yield of O_2 produced from vapor deposited ice samples depends on the temperature and radiation dose, but not the dose *rate* (Reimann et al. 1984; Teolis et al. 2005). The temperature dependence reveals an activation energy; the yields increase with temperature according to the formula: $Y = Y_o + Y_a \exp(-E_a/kT)$, where the activation energy E_a is on the order of 0.05 eV. Y_o is the portion of the sputtering yield that is independent of temperature (primarily due to the ejection of H_2O) and the temperature dependent component Y_a is primarily associated with loss of O_2 and H_2.

Although laboratory data on the radiolysis and photolysis of ice is now extensive, the chemical pathways are not yet agreed upon. Whereas the principal dissociation process in ice, as in liquid water, is dissociation to form H and OH, the dominant gas phase species ejected from irradiated ice are those with the lowest binding energies (H_2, O_2 and H_2O). These ejecta are the end result of reactions among the many other species produced; e.g., H_2O_2, HO_2 and O have all been suggested as precursor species for the formation of O_2 (Laffon et al. 2006). For strongly ionizing particles, such as magnetospheric heavy ions, O_2 itself can be considered a precursor species: radiation-damaged ice traps the radiolytically produced O_2, which is only released as the ice itself is sputtered away (Teolis et al. 2005, 2009).

Since hydrogen is preferentially lost at the relevant temperatures, oxygen-rich molecules are the principal trace species trapped in the irradiated surface. In addition to the observation of gas-phase oxygen, solid-state O_2 was inferred by the signature of interacting pairs of O_2 molecules in the reflectance spectra of Ganymede (Calvin et al. 1995), Europa and Callisto (Spencer and Calvin 2002). The two signature absorptions for dimer O_2, at 577.2 nm and 627.5 nm, respectively (Vidal et al. 1997; Cooper et al. 2003b), suggested that the O_2 is locally dense though the band shapes and relative band depths are distorted from solid or liquid O_2. The presence of O_2 in the solid state at these temperatures is surprising, given O_2's volatility. Although O_2 can diffuse out of unirradiated H_2O ice in laboratory experiments at relevant planetary temperatures (<150 K), radiolytically-produced O_2 can be trapped in radiation-damaged ice at up to 155 K (e.g., Teolis et al. 2009). The trapping appears to occur in radiation-induced defects that coalesce to form voids in ice (Johnson and Jesser 1997; Loeffler et al. 2006).

Related to the oxygen observations, the Hartley band of O_3 was tentatively identified in the ultraviolet (UV) reflectance spectra of Ganymede and in the icy satellites of Saturn (Noll et al. 1997), although the band shape suggests additional absorbing species may be present. The O_3 is likely formed in the O_2 voids described above (Johnson and Jesser 1997); however, O_3 is not observed on pure ice irradiated in vacuum except when irradiation and H_2O vapor deposition occur simultaneously (Teolis et al. 2006). The O_3-like feature is superimposed on a very broad UV absorption shortward of 0.4 μm seen on most of the icy satellites. The broad UV absorption has been attributed to another radiolytically-produced oxidant, H_2O_2 (Carlson et al. 1999a), but may also include absorption by trapped HO_2 and OH (Johnson and Quickenden 1997). Identification of specific radiolytic products in the UV is difficult because the bands are very broad and overlap, but H_2O_2 has been detected via infrared absorption on Europa (e.g. Carlson et al. 2009). H_2O_2 is produced by the principal dissociation product OH in the absence of H (e.g., Laffon et al. 2006).

Although the identifications of the radiolysis and photolysis products (OH, H_2O_2, O_2, O_3) are uncertain, the reflectance spectra taken as a whole give a fairly clear picture. The irradiated surfaces of the icy solar system bodies contain trapped molecules that would be expected from radiolysis or photolysis of ice. Consistent with this, oxygen-rich molecules in ice containing trace amounts of sulfur (SO_2 and a sulfate) and carbon (CO_2 and a carbonate) have been identified as well a dark feature consistent with the degradation of organics ice (Hendrix and Johnson 2008) as discussed in later sections of this chapter.

What is most important for the exchange of material between bodies are the radiation-induced chemical processes that eject molecules from the surface. Unfortunately, laboratory data on ejecta energy are sparse to non-existent, so that new results are urgently needed. Reimann et al. (1984) measured the energy spectra of sputtered D_2O for several incident ion energies and Haring et al. (1984) measured the energy spectra of sputtered H_2 and O_2 for several incident ions (Ar^+, He^+, H^+). The H_2O energy spectrum is clearly non-thermal and usually fit, for convenience, to a standard sputter distribution $2UE/(E+U)^3$, where E is the sputter product energy, and U is an "effective binding energy" of 0.055 eV (much less than the actual H_2O binding energy of about 0.45 eV). The spectra of the H_2 and O_2 are primarily thermal, consistent with their diffusion from ice, but also have small high energy tails consistent with non-thermal contributions. When considering escape energies from a regolith, the ejecta energy spectra are also modified by collisions with neighboring grains (Cassidy and Johnson 2005).

3 Sulfur Compound Radiolysis

Water ice is the most abundant compound in the near-surfaces of outer solar system objects, it is the "rock" of the outer solar system, but other compounds are responsible for much of the outer solar system's character, from the multicolored surface of Io to the dark regions of icy moons that come in a bewildering variety patterns (e.g. Iapetus). The rest of this chapter is devoted to these other compounds, beginning with radiolytic chemistry of sulfur. Sulfur is a common element in the solar system, with a cometary molar abundance relative to H of $\sim 1\%$ (Anders and Grevesse 1989). This is comparable to the relative abundances of nitrogen and silicon ($\sim 4\%$ and 2.5%, respectively), so it is not surprising that sulfur compounds are found in the surfaces and atmospheres of planets, satellites, and comets.

The phase and chemical composition of these sulfur compounds are diverse and related to major planetary features and processes, in part because many sulfur compounds are highly volatile. For example, Venus's pervasive clouds are composed of sulfuric acid droplets produced from the photolysis of gaseous sulfur dioxide and carbonyl sulfide, while gaseous native sulfur may be present in Venus's hot lower atmosphere. Volcanoes on Jupiter's moon Io emit enormous amounts of elemental sulfur and sulfur oxides. Condensed sulfur and sulfur dioxide ice are present on both Io and its neighboring satellite Europa, while the latter also exhibits sufate compounds, derived perhaps from a subsurface ocean. The outer planets' atmospheres contain hydrogen sulfide gas and their clouds are thought to contain hydrosulphides and possibly sulphides that may provide the unknown pigments that color these atmospheres (Coustenis et al. 2010, this issue).

Depending on the particular environment—gaseous or condensed, oxidizing or reducing—different sulfur compounds are found or expected. While numerous radicals and reactive molecules are produced during photon or energetic particle excitation, as in the case of water ice, the most stable and abundant end products will be closed-shell molecules. In the following paragraphs we briefly describe some relevant sulfur compounds and their photolytic and radiolytic chemistry. Summaries of sulfur properties relevant for Io and Europa are given in Carlson et al. (2007, 2009).

The most stable form of elemental or native sulfur is the octahedral molecule S_8, which is shaped like a puckered or crown-like ring. Sulfur rings (cyclo-S) exist with five to twenty or more sulfur atoms. Small sulfur molecules, S_2, S_3 and S_4 form chains (catena-S), while molecules with five or more atoms can form chains or rings, with rings generally being more stable. Condensed mixtures of sulfur rings and chains, termed polymeric sulfur, are produced by thermal, photolytic, or radiolytic processes and can exhibit a variety of colors ranging from yellow to red, orange, and black, and even green and purple; Moore (1984) irradiated white, cryogenic, SO_2 with energetic protons and found that irradiation darkened the sample and produced different colors for different sample temperatures. The ultraviolet and visible spectrum can be used to indicate the amount of polymerization and the transient presence of tetrasulfur, S_4. High-temperature gaseous sulfur contains small rings, chains, and S_2, S_3, and S_4.

As discussed above, radiolysis tends to oxidize icy satellite surfaces through the production and loss of H_2. Under these conditions, the stable sulfur-containing molecules are expected to be sufates, sulfur dioxide, and elemental sulfur (as opposed to reducing compounds such as H_2S). In the presence of water, sufate is likely present as sulfuric acid (H_2SO_4+ H_2O or dissociated into its ions $H_3O^+ + SO_4^{2-}$). The clouds and atmosphere of Venus, for example, contain sulfuric acid, SO_2, and probable polymeric sulfur.

Irradiation of S_8 grains in H_2O ice produces sulfuric acid (Carlson et al. 2002), similar to the irradiation of sulfur grains in H_2O liquid (DellaGuardia and Johnston 1980; Donaldson and Johnston 1971). It is observed through the presence of its components: the hydronium cation (H_3O^+) and the sufate anion (SO_4^{2-}). Continued irradiation destroys the previously made sufate anions to produce SO_2 (Hochanadel et al. 1955), elemental S, and, based on results from irradiated Li_2SO_4 (Sasaki et al. 1978), possibly H_2S. Proton irradiation of SO_2 in H_2O ice also yields sulfuric acid (Moore et al. 2007). When annealed, various hydrates of sulfuric acid ($H_2SO_4 \cdot nH_2O$) are formed. In radiolytic equilibrium, when there is as much sulfuric acid destroyed as that being produced, the proportions of S_n, SO_2, and H_2SO_4 will become constants that depend on the initial sulfur to water ratio (Carlson et al. 1999, 2002, 2009; Moore et al. 2007). This is called the radiolytic sulfur cycle. Relevant rates are summarized in Johnson et al. (2004). Similar results are obtained by the photolysis of SO_2 in H_2O ice (Schriver-Mazzuoli et al. 2003).

Relevant exchange processes occur at Europa where sulfur is likely delivered to the water ice surface by two exchange processes: plasma implantation of sulfur ions (ultimately derived from neighboring Io) and sufate salts from geological processes that deliver briny water from the subsurface ocean. The intense bombardment of electrons and ions transforms these compounds into the S_n, SO_2, and H_2SO_4 of the radiolytic sulfur cycle, hiding their original identity and complicating interpretation of the surface reflectance spectra (see Chap. X). Implantation of sulfur ions in H_2O ice does not make SO_2 directly, but instead produces H_2SO_4 (Strazzulla et al. 2007). Upon continued irradiation, the H_2SO_4 is decomposed, producing SO_2 (Moore et al. 2007)

If the initial material is from sufate salts, the presence of the metal cation increases the number of chemical pathways so that metal oxides and hydroxides are possible. If, for example, one started with $MgSO_4{\cdot}7H_2O$ (epsomite), then one expects $MgSO_4{\cdot}nH_2O$, MgO, $Mg(OH)_2$, $MgS{\cdot}nH_2O$, $H_2SO_4{\cdot}nH_2O$, S_μ, and SO_2 (Johnson 2001). The product yields are not well known and more experimental work is needed in this area.

In reducing environments such as the atmospheres of the giant planets, hydrogen sulfide (H_2S) is the dominant form. At about 200 K, ammonia (NH_3) and H_2S condense to form the molecular solid ammonium hydrosulfide (NH_4SH). Ultraviolet photolysis of NH_4SH, which exists as a cloud layer in the giant planet atmospheres, produces ammonium sulfide and ammonium polysulfide, $(NH_4)_2S$ and $(NH_4)_2S_n$, respectively. H_2S can be photo-dissociated by ultraviolet radiation with wavelengths $\lambda < 0.314$ µm (Huebner et al. 1992), so photolysis of Jupiter-like mixtures of gaseous H_2S, NH_4, and H_2 will likely produce similar compounds and perhaps native sulfur and sulfanes (sulfur chains with terminal hydrogen atoms, HS_nH). These compounds could account for some of the red, yellow and brown hues of Jupiter's clouds (Taylor et al. 2004).

4 Carbon Compound Radiolysis

Like sulfur, carbon is an important and abundant compound on the surfaces and in the interiors of many outer solar system bodies. The radiolysis of carbon compounds, however, has added significance in the context of astrobiology. Exchange of radiolytically processed carbon compounds with liquid water in the sub-surface of icy worlds could be critical to their habitability (Hand 2009).

The photolysis and radiolysis of carbon-containing compounds and their mixtures with H_2O have many similarities with sulfur. As with sulfur compounds, irradiation of hydrocarbons tends to darken them. In addition, carbon may undergo a radiolytic cycle analogous to the radiolytic sulfur cycle, with carbonic acid and CO_2 replacing sulfuric acid and SO_2. Gomis and Strazzulla (2005) demonstrated this by irradiating thin H_2O films on two different complex, refractory organics with a variety of ions and found a buildup of about a monolayer of CO_2 (see also Mennella et al. 2004). Evidence for radiolytic CO_2 production is seen throughout the outer solar system. Hibbitts et al. (2000) suggested that the CO_2 abundance in Callisto's dark terrains follows the bulls-eye pattern of magnetospheric plasma impact, and Grundy et al. (2006) found the same for several of Uranus' moons. The CO_2 is presumed to be in the solid state, which is surprising given its volatility (Sandford and Allamandola 1990). Radiation-induced defects may trap the CO_2, as discussed for O_2 above.

Irradiation of pure hydrocarbons produces dark, refractory material (e.g., Strazzulla et al. 1995). H_2 is a primary radiation product that easily escapes both solids and gravitational wells, leaving behind polymerized carbon compounds. The simplest example of this is the irradiation of CH_4 (Lanzerotti et al. 1987): cleavage of the H from CH_4 leaves behind the radicals H and CH_3. CH_3 can combine with an adjacent CH_3 to form C_2H_6, and the H with another H to produce H_2. The escape of H_2 and further bond breaking leads to larger polymers (Bennett et al. 2006). The redness of some outer solar system objects might result from the photolysis and radiolysis of small hydrocarbons in icy surfaces (Trilling and Brown 2000). Though the production of refractory organic material in irradiated ices in the laboratory is well documented (see e.g. Khare et al. 1989; Bernstein et al. 1995; Moore and Hudson 1998), the production of dark, refractory organics on the surfaces of icy worlds may, at a minimum, require an initial ice composition of simple organics at relatively high concentration (>10% by number abundance). For instance, radiolytically and photolytically processed ices containing only CO_2 and H_2O do not form hydrocarbons as efficiently (Hand et al. 2007).

Experiments at astrophysical temperatures (~5–20 K) have also shown that formaldehyde (H_2CO) is a minor product during ion and UV irradiation of CO_2-rich ices (Gerakines et al. 2000; Brucato et al. 1997; Moore et al. 1991). The predominant carbon-bearing products under such conditions were CO and carbonic acid, H_2CO_3. Similarly, Hand et al. (2007) found carbonic acid and carbon monoxide to be the primary products of electron radiolysis of H_2O and CO_2 ices at temperatures comparable to those found on the icy Jovian and Saturnian moons. Starting with a ~30%, by number abundance, of CO_2 in water ice they found that ~97% of the evolved carbon remained as CO_2 or went into CO or H_2CO_3. No definitive signature of H_2CO, CH_3OH, or even CH_4 was found, though growth of features at 2044 cm^{-1} and 1878 cm^{-1}, may indicate production of the CO_3 radical (Moore et al. 1991).

Figure 2 shows the spectra of CO_2 and H_2O ice at 90 K before and after irradiation by 10 keV electrons with cumulative dose of $\sim 2 \times 10^{21}$ eV cm^{-2} (the figure was adapted from Hand et al. 2007). The strong lines that emerged during irradiation are indicated at 2570, 2141, 1714, 1485, 1380, and 1298 cm^{-1}. The 2340 cm^{-1} and 2278 cm^{-1} bands show persistence of CO_2 in the irradiated ice and the feature at 2141 cm^{-1} indicates production of CO. The formation of H_2CO_3 is indicated by the combination of the remaining bands (Brucato et al. 1997; Moore and Hudson 1998; Gerakines et al. 2000). Hand et al. concluded that upper limits for H_2CO, CH_3OH, and CH_4 production are ~0.5% by number of the remaining, or observed, CO_2. For bodies such as Europa, where CO_2 has been observed but where no organics have yet been observed, this has important consequences for geological exchange processes that could bring irradiated surface material to the subsurface ocean, and thus to a possible ecosystem (Chyba 2000). Chyba and Phillips (2002) argued that an ecosystem in a sub-surface ocean on Europa could be powered solely by radiolytically produced H_2O_2 and H_2CO delivered from the surface ice. The astrobiological aspects of exchange processes are considered in greater detail in Raulin et al. (2010, this issue).

5 Nitrogen Compound Radiolysis

NH_3 ice is present, though not common, on the outer planet moons (see summary in Moore et al. 2007). It is more common on trans-Neptunian objects. The reason for this seems to be its volatility and susceptibility to decomposition by radiation. The sputtering yield (number of molecules ejected per incident ion) of pure NH_3 ice is an order of magnitude larger than it is for water ice (Lanzerotti et al. 1984).

The main sputter products are NH_3, N_2 and H_2 and, as with the sputtering of H_2O, the relative amounts depend on the incident ion used, its energy, and the ice temperature. If ions transfer energy via mainly elastic collisions (knock-on sputtering, as opposed to electronic sputtering), then the principal sputter product is NH_3. This is the case for the ~keV Ar^+ ions used by Haring et al. (1984). However, if the ions have a significant electronic component, like the ~MeV ions used by Lanzerotti et al. (1984), then the H_2 and N_2 are the dominant ejecta (Haring et al. 1984). Thus the high sputtering yield observed is likely explained by the rapid decomposition of NH_3 into more volatile species. Infrared transmission studies on the decomposition of NH_3 via ion irradiation (Strazzulla and Palumbo 1998) are consistent with these findings. UV photolysis of NH_3 also produces N_2H_4 and H_2, as identified by IR spectroscopy (Gerakines et al. 1996). An absorption at 1505 cm^{-1} has been identified as either NH_2 (Gerakines et al. 1996) or NH_4^+ (Moore et al. 2007).

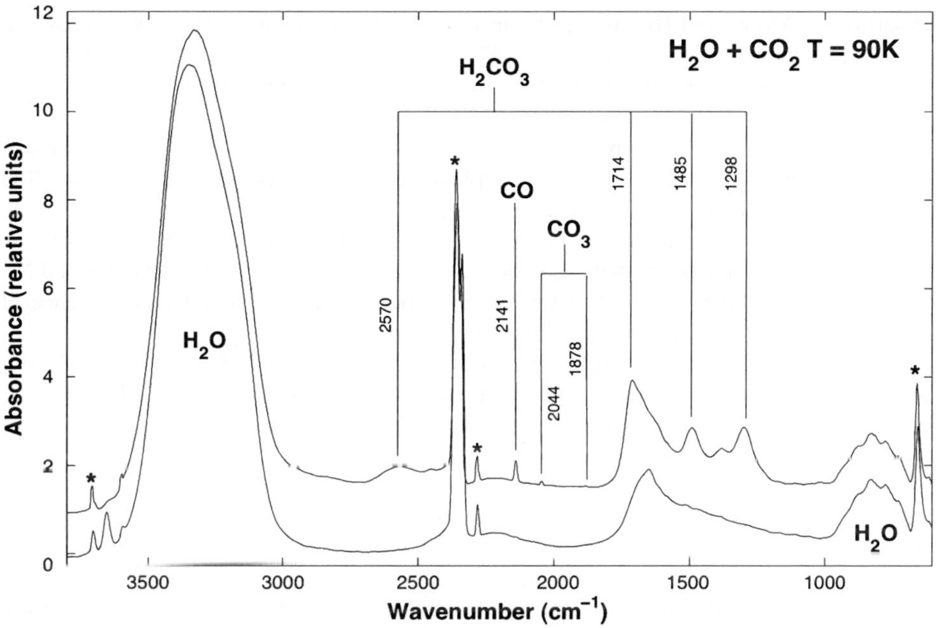

Fig. 2 Electron radiolysis of $CO_2 + H_2O$ ices at 90 K. The film was deposited at 50 K and then warmed to 90 K. The bottom spectrum shows the ice before irradiation while the upper spectrum shows the ice after irradiation. The peaks marked with *asterisks* are CO_2 bands. This figure was reproduced from Hand et al. (2007)

Observed NH_3 is usually identified as a hydrate, a stoichiometric mixture of H_2O and NH_3. The effect of ion irradiation of ammonia-water mixtures has been studied using infrared spectroscopy to monitor the loss of the ammonia absorption feature at either 4523 cm^{-1} ($\upsilon_1 + \upsilon_2$) or 1115 cm^{-1} (υ_2) (Strazzulla and Palumbo 1998; Loeffler et al. 2006; Moore et al. 2007) and using mass spectroscopy to monitor the composition of the species sputtered during irradiation and thermal desorption (Loeffler et al. 2006). Infrared spectroscopy showed that ammonia radiolytically decomposes faster at 80 K than at 10 K (Strazzulla and Palumbo 1998; Moore et al. 2007); specifically Strazzulla and Palumbo (1998) showed that 30 keV He^+ ions decreased the ammonia-water ratio from 0.6 to 0.3 at 10 K and from 0.4 to 0.15 at 80 K. Moore et al. (2007) found that ammonia decomposes faster in a matrix of water ice than when it is pure. Loeffler et al. (2006) irradiated a 2:1 H_2O NH_3 mixture with 100 keV protons and found a peak sputtering yield $\sim$110 molecules/ion (assuming mass = 17 amu) at a fluence of $\sim 3 \times 10^{15}$ ions/cm^2, with the yield first increasing and then decreasing with ion dose. This fluence dependence is related to the formation of new species formed in the ice and the destruction of NH_3. Other species detected via infrared spectroscopy, which are estimated to be minor compared to the H_2 and N_2, are NO (20 K; 1871 cm^{-1}) and N_2O (120 K; 2225 cm^{-1}). The sputtered flux ejected from the ammonia-water sample, judging from mass spectrometry, is mainly N_2 and H_2.

N_2 is another volatile solid component of the outer solar system. Neptune's moon Triton, for example, has abundant N_2 ice (Quirico et al. 1999). Gerakines et al. (1996) performed UV photolysis of solid N_2 at 10 K with photons from a hydrogen discharge lamp. After irradiation, they detected a shoulder on top of a stronger CO infrared absorption feature (from background contamination) and tentatively identified it as the N_3 radical. In more recent experiments, Moore and Hudson (2003) irradiated solid N_2 ice with 0.8 MeV protons and UV photons. However, the N_3 radical (1657 cm^{-1}) was only observed after ion irradiation and not photolysis. Strazzulla et al. (2003) irradiated N_2:H_2O mixtures with 60 keV Ar^{++} at 20 K and found the formation of NO, NO_2, and N_2O.

Strazzulla et al. (2003) implanted water ice with 15 keV N^+ to see if any new nitrogen species (N_xO_y) would form, just as implantation of S^+ in water ice can produce S-O compounds (see earlier section). After a fluence of 4.2×10^{16} ions/cm^2, they detected no new species, other than hydrogen peroxide, which could be expected for *any* projectile. In recent unpublished experiments Loeffler and coworkers implanted 100 keV N^+ and N_2^+ into water ice at 80 K and looked for nitrogen species. After irradiating to a fluence of 2×10^{16} ions/cm^2, they also detected only hydrogen peroxide.

Cottin et al. (2003) performed UV photolysis on pure NH_3 and NH_3 in mixtures of water and nitrogen. Interestingly they found that the destruction of pure NH_3 and NH_3 mixed in water was similar, while the destruction of NH_3 mixed in N_2 actually increased by a factor of seven compared to pure NH_3. The latter result is intriguing, because the N_2 was transparent to the UV lamp and thus was thought to play no direct role in the radiation chemistry.

Since the transfer of materials between solar system bodies often requires that they be driven into the gas-phase, NH_3 radiolysis and photolysis is of interest as NH_3 might enhance cryovolcanism by suppressing the freezing point of water. The identification of such a process is made difficult by the efficient destruction of NH_3 by plasma and UV photons. Further, unlike sulfur and carbon compounds, but much like H_2O, radiolytic destruction of NH_3 does not produce refractory byproducts. NH_3 has been theorized to play a role in the outgassing of Enceladus (see Sohl et al. 2010, this issue). Although initially difficult to detect (Emery et al. 2005; Verbiscer et al. 2006) its presence has been confirmed (Smith et al. 2008; Waite et al. 2009).

6 Tholins: Photolysis and Radiolysis in Titan's Upper Atmosphere

So far we have discussed the effects of ionizing radiation on ices, as appropriate for the nearly airless moons of the outer solar system, but at Titan it is *gases* that absorb the incident radiation. While ionizing radiation tends to produce gas from solids, it tends to produce solids from the gases of Titan's atmosphere. The transformation from simple molecules (CH_4, NH_3, N_2, CO, CO_2, H_2O...) to more complex compounds appears to be a key step in the context of planetary atmosphere evolution. Such transformations may have happened in the reducing atmosphere of the early Earth (Chyba and Sagan 1991; Kobayashi et al. 1998).

Titan's atmosphere consists of mainly N_2 (95% to 98.5% depending on the altitude), CH_4 (5% to 1.5%) and H_2. As with the solid-state radiolysis described above, the destruction of carbon-hydrogen bonds tends to result in more complex molecules such as HCN, C_2H_2, C_2H_4, C_2H_6. As radiolysis continues, bigger molecules, known as aerosols or, more specifically, "tholins",[1] are formed. These aerosols form a planet-wide haze that gives Titan's atmosphere its characteristic orange colour. This section describes what *in situ* and laboratory studies have revealed about these aerosols.

Sagan and Thompson (1984) made the first estimates of energy deposition in the atmosphere of Titan as a function of altitude. They showed that the short wavelength

[1] The name "tholins" proposed by Carl Sagan refers to any *complex organic solid produced from cosmically abundant molecules*.

(<150 nm) solar ultraviolet light is mainly absorbed by N_2 and CH_4 at high altitudes, producing free radicals that initiate chemistry. Wilson and Atreya (2004) correlated the escape of hydrogen produced by photolysis of CH_4 in Titan's upper atmosphere with the formation and precipitation of aerosols. Early work estimated that the aerosols were formed at altitudes of several hundred km, but Cassini's Ion and Neutral Mass Spectrometer INMS team found that aerosol formation begins at much higher altitudes (~1000 km). The observed organic mix strongly implies a series of chemical reactions and physical processes that lead from simple molecules (CH_4 and N_2) to larger, more complex molecules (80 to 350 daltons) to negatively charged massive molecules (~8000 daltons), which are identified as aerosols (Waite et al. 2007).

Besides photolysis, other forms of radiation also contribute to the radiation chemistry of Titan. Titan orbits most of the time within Saturn's magnetosphere, and the magnetospheric plasma can easily dissociate molecules. The magnetosphere is also a source of O atoms which are ultimately derived from Enceladus' outgassing of H_2O (Hartle et al. 2006). The addition of O could affect tholin formation chemistry (e.g., Sittler et al. 2009). When Titan is outside Saturn's magnetosphere, the solar wind and galactic cosmic rays bombard the atmosphere. These penetrating particles ionise species in the lower atmosphere, producing the primary ions N_2^+ and N^+, and electrons. The nitrogen ions are rapidly converted, by their interaction with CH_4 molecules, into secondary ions and ion clusters. A simplified scheme of these main processes is shown in Fig. 3.

Another external source of energy possibly relevant to Titan is the plasma created during the influx of cometary material. This would have been important in Titan's early history, especially for the contribution of oxygen-bearing molecules to the atmosphere, but in the present day impacts are too infrequent to provide a significant supply of oxygen (Lorenz 1997).

The most accurate information about the properties of aerosols came from direct observation, first in the 1980s flybys of Titan by the Voyager mission, and most recently by the Cassini-Huygens mission (especially the Huygens probe, which descended into Titan's atmosphere in January 2005). Light propagation measurements suggest that the particles are a fractal arrangement of spherical particles (with an average diameter of 0.1 μm) (Tomasko et al. 2005). Huygens probe studies confirmed that the aerosols are solid and have an organic composition. Furthermore, the surface reflectance spectra of Titan (Tomasko et al. 2005) and the surface permittivity measurements (Fulchignoni et al. 2005) suggest that the aerosols coat the landing site surface. Despite these observations, our knowledge of the composition and structure of Titan's haze particles is limited. This is why it is necessary to use laboratory simulations of Titan's chemistry to study the processes that occur in this environment.

7 Laboratory Work

Analogues of Titan's tholin aerosols can, in principle, be synthesized in the laboratory to study their physical and chemical properties (Sagan and Khare 1979). A variety of experimental conditions have been used to synthesize tholins (e.g., Coll et al. 1999; Imanaka et al. 2004 and references therein). Differences between experiments include the radiation source (UV photons or plasma produced from electrical discharge), gas composition, temperature, and pressure. Figure 4 shows a laboratory-synthesized tholin aerosol.

Tholins are usually generated from N_2/CH_4 gas mixtures, the two major components of Titan's atmosphere. In experiments, the ratio of CH_4 to N_2 is usually varied from 2 to 50%. These variations do not affect the hydrogen content of the tholins, but seriously affect their

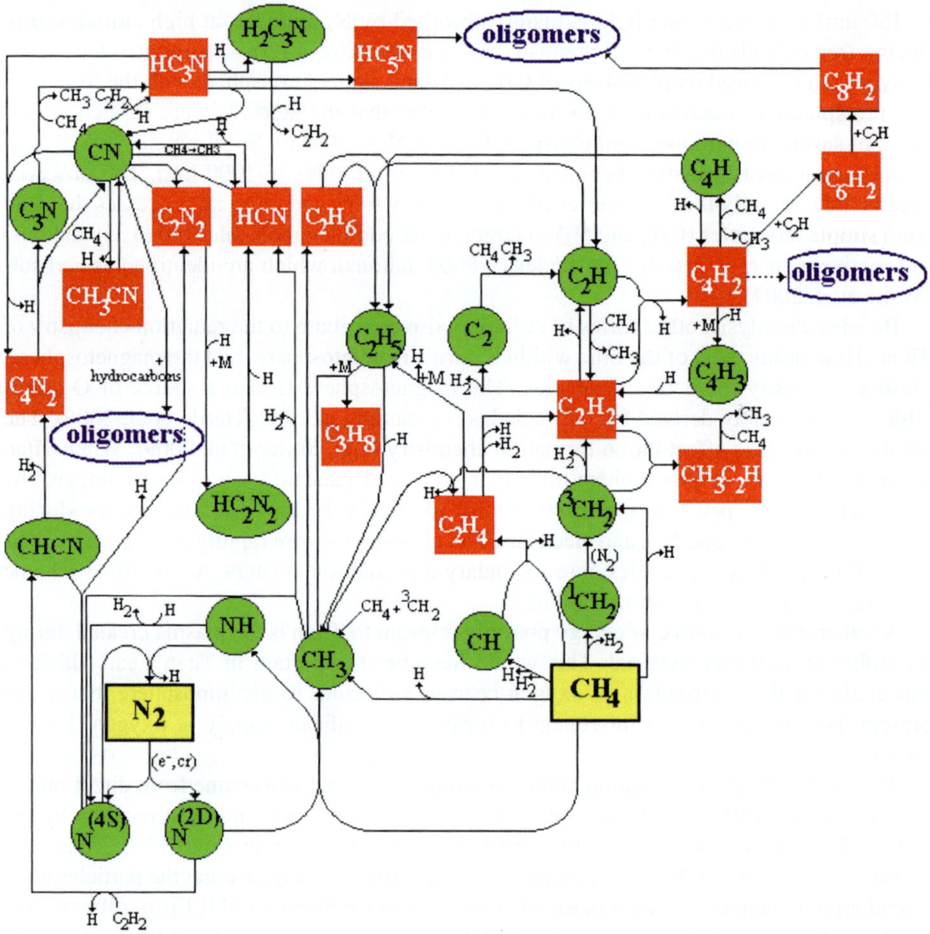

Fig. 3 Illustration of the chemical pathways from N_2 and CH_4 (*yellow squares*) in Titan's atmosphere, leading to secondary gaseous products (*red squares*) solid oligomers or aerosols (*blue ellipses*)

nitrogen content. This observation emphasizes the need for experiments using conditions that are clearly representative of tholin formation (Tobie et al. 2006).

IR spectroscopy has frequently been used to monitor the chemical structure of tholins. Despite differences in the experimental conditions, IR spectroscopy consistently revealed the same structural groups: -CH-, $-CH_2-$, $-CH_3$, aromatic -CH-, C=C and C≡N groups, though differences in terms of aromaticity and N content were observed (e.g., Bernard et al. 2006).

A complementary way to study the structure is pyrolysis coupled with gas chromatograph mass spectrometry (GC-MS) to identify the fragments produced by thermal degradation. Compounds such as aliphatic hydrocarbons (propane, heptane, propene, butene, pentene...), nitriles (hydrogen cyanide, acetonitrile, propanenitrile, isobutanenitrile, crotonitrile...) and aromatics (benzene, pyrrole, toluene, imidazole, xylene, pyridine, pyrazine, benzonitrile, triazine) have been identified using this technique (Ehrenfreund et al. 1995; Khare et al. 1984; Coll et al. 1999). The main evolved products are unsaturated hydrocarbons, nitriles and

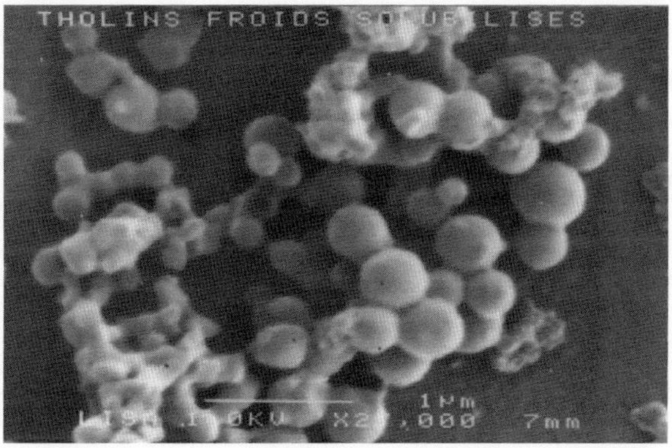

Fig. 4 Electron microscope view of a tholin aerosol analogue synthesized in a laboratory (Coll et al. 1999)

monoaromatic species including N-containing aromatics. However, it is important to note that a significant refractory fraction remains after pyrolysis, and thus remains undetected by the GC-MS. Further, some of the species detected could be formed *during* pyrolysis.

Trainer et al. (2004) provided additional information on the molecular composition of tholins using aerosol mass spectrometry, which flash vaporizes the aerosols after measuring their size. They identified PAHs (polycyclic aromatic hydrocarbons, with a maximum of 4 rings) and alkylated PAHs when high CH_4/N_2 ratios were used in the initial gas mixtures, and, when CH_4/N_2 was less than 0.5%, aliphatic structures, demonstrating that the aerosol-forming chemical pathways (aromatic versus aliphatic) are sensitive to the CH_4/N_2 ratio in the initial mixture.

Overall, tholin analogue experiments have produced two families of tholins. First, the family produced with *photochemistry* is characterized by low N content, organic polymers, aliphatic unsaturated chains and low or no aromatic content. Second, the family of tholins produced by electrical-discharge *plasma* is characterized by higher N content, organic oligomers, aromatic structures including N-heterocycles and aliphatic chains.

Aerosols in the lower atmosphere might interact with liquid water in episodic surface pools produced by cometary impact or by cryovolcanism. Somogyi et al. (2006) exposed tholins produced by AC plasma to hot liquid water with dissolved ammonia. They concluded that laboratory tholins are reactive enough with the water/ammonia mixture to generate oxidized solid material. If the aerosols of Titan's environment are from the same chemical family as Somogyi's tholins, they can be considered a potential starting material for several synthetic processes that can provide organic compounds of pre-biotic interest (Fig. 5).

Other simulations investigated acidic hydrolysis of plasma-generated tholins (Khare et al. 1986; Nguyen et al. 2007), hydrolysis at neutral pH (Raulin et al. 2007) and at low temperature (Neish et al. 2009). They produced amino-acids, such as glycine, β-alanine and aspartic acid, and urea. Such water-tholin interactions might not be possible, however: Tran et al. (2003) showed that photolytically-produced tholins are not soluble in most solvents, and McKay (1996) and Coll et al. (1999) concluded that plasma-produced tholins have very low solubility in liquid hydrocarbons (which have been detected on Titan's surface).

8 Conclusion

Radiation-induced processing of materials is an ubiquitous feature at outer solar system bodies. The satellites of the outer solar system are exposed to intense ionizing radiation over

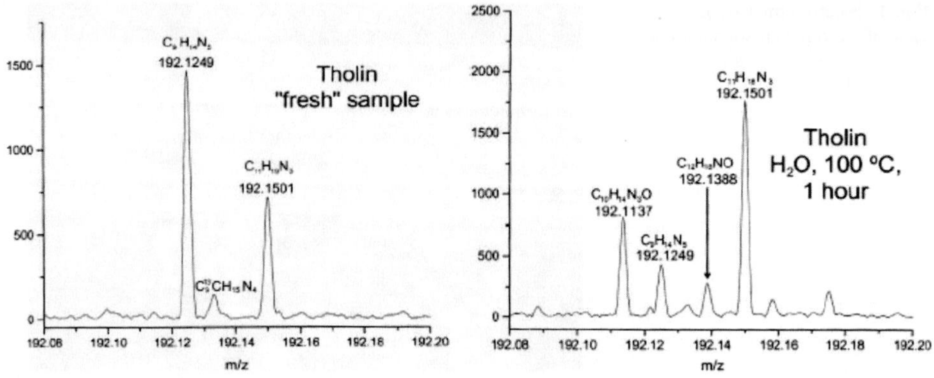

Fig. 5 Mass spectrum of a tholin (*left*). Mass spectrum after one hour of reaction with water (*right*). This investigation shows the integration of O atoms into the tholin ($C_{12}H_{14}N_3O$ on the left and $C_{12}H_{18}NO$ on the right). Hydrolysis reactions occur with rate constants of 2–10 hour^{-1} at room temperature. Adapted from Somogyi et al. (2006)

geologically long periods of time, producing volatiles from the solids and solid particulates from gases. Ions, electrons, and, to a lesser extent, UV photons, erode and chemically alter satellite surfaces. Although the affected layer is small ($<$ a few meters, the gardened layer), it is this layer that is sensed remotely. Therefore, inferring bulk properties from surface properties requires observers to understand the effects of radiolysis and photolysis. For instance, NH_3 reaching Enceladus' surface will be quickly converted to N_2 and H_2 by magnetospheric plasma and solar UV photons. Similarly, a definitive interpretation of the source of the sulfur on Europa has not been possible because material from the two proposed sources, the neighbouring moon Io or the sub-surface ocean, is radiolytically processed in a relatively short time period into materials that are spectroscopically similar. Radiolysis does more than just complicate planetary science, though. It is responsible for some of the exchange processes described in this issue. Materials driven into the gas phase can be ionized and transported by plasma diffusion to neighbouring bodies.

As discussed above, the effects of radiation of materials of interest in the outer solar system are primarily determined in the laboratory and then applied to the observations. However, there have been counter examples, as in the HST detection of solid-state oxygen trapped in irradiated ice before it was observed in the laboratory. Although analytic and computer models have been useful for extrapolating and interpreting laboratory data, there is a dearth of quantitative models. Even a problem as relatively simple as the sputtering of pure H_2O has not been accurately simulated. Therefore, for the foreseeable future, our understanding of the space weathering of outer solar system surfaces will rely heavily on extrapolations from experimental data.

References

E. Anders, N. Grevesse, Geochim. Cosmoch. Acta. **53**, 197–214 (1989)

R.A. Baragiola, Planet. Space Sci. **51**, 953–961 (2003)

C.J. Bennett, C.S. Jamieson, Y. Osamura, R.I. Kaiser, Astrophys. J. **653**, 792–811 (2006)

J.-M. Bernard, E. Quirico, O. Brissaud, G. Montagnac, B. Reynard, P. Mc Millan, P. Coll,, M.J. Nguyen, F. Raulin, B. Schmitt, Icarus **185**, 301–307 (2006)

M.P. Bernstein, S.A. Sandford, L.J. Allamandola, S. Chang, M.A. Scharberg, Astrophys. J. **454**, 327 (1995)

W.L. Brown, L.J. Lanzerotti, J.M. Poate, W.M. Augustyniak, Phys. Rev. Lett. **40**, 1027–1030 (1978)

J.R. Brucato, A.C. Castorina, M.E. Palumbo, M.A. Satorre, G. Strazzulla, Planet. Space Sci. **45**, 835–840 (1997)
Burger et al., Space Sci. Rev. (2010, this issue)
W.M. Calvin, R.N. Clark, R.H. Brown, J.R. Spencer, J. Geophys. Res. **100**, 19041–19048 (1995)
R.W. Carlson, R.E. Johnson, M.S. Anderson, Science **286**, 97–99 (1999)
R.W. Carlson, M.S. Anderson, R.E. Johnson, M.B. Schulman, A.H. Yavrouian, Icarus **157**, 456–463 (2002)
R.W. Carlson, J.S. Kargel, S. Doute, L.A. Soderblom, B. Dalton, in *Io after Galileo*, ed. by R.M.C. Lopes, J.R. Spencer (Springer-Praxis, Chichester, 2007), pp. 193–229
R.W. Carlson, W.M. Calvin, J.B. Dalton, G.B. Hansen, R.L. Hudson, R.E. Johnson, T.B. McCord, M.H. Moore, in *Europa*, ed. by R.T. Pappalardo, W.B. McKinnon, K.K. Khurana (Univ. Arizona Press, Tucson, 2009)
T.A. Cassidy, R.E. Johnson, Icarus **176**, 499–507 (2005)
C.F. Chyba, Nature **403**, 381–382 (2000)
C.F. Chyba, C.B. Phillips, Orig. Life Evol. Biosph. **32**, 47–67 (2002)
C. Chyba, C. Sagan, OLEB **21**, 3–17 (1991)
P. Coll, D. Coscia, N. Smith, M.-C. Gazeau, S.I. Ramirez, G. Cernogora, G. Israel, F. Raulin, Planet. Space Sci. **47**, 1331–1340 (1999)
P.D. Cooper, R.E. Johnson, T.I. Quickenden, Planet. Space Sci. **51**, 183–192 (2003b)
H. Cottin, M.H. Moore, Y. Benilan, Astrophys. J. **590**, 874 (2003)
A. Coustenis et al., Space Sci. Rev. (2010, this issue). doi:10.1007/s11214-009-9615-5
R.A. DellaGuardia, F.J. Johnston, Radiat. Res. **84**, 259–264 (1980)
G.W. Donaldson, F.J. Johnston, J. Phys. Chem. **75**, 756–763 (1971)
P. Ehrenfreund, J.J. Boon, J. Commandeur, C. Sagan, W.R. Thompson, B. Khare, Adv. Space Res. **15**, 335–342 (1995)
J.P. Emery, D.M. Burr, D.P. Cruikshank, R.H. Brown, J.B. Dalton, Astron. Astrophys. **435**, 353–362 (2005)
M. Fulchignoni (HASI team), Nature **438**, 785–791 (2005)
P.A. Gerakines, W.A. Schutte, P. Ehrenfreund, Astron. Astrophys. **312**, 289 (1996)
P.A. Gerakines, M.H. Moore, R.L. Hudson, Astron. Astrophys. **357**, 793–800 (2000)
O. Gomis, G. Strazzulla, Icarus **177**, 570–576 (2005)
W.M. Grundy, L.A. Young, J.R. Spencer, R.E. Johnson, E.F. Young, M.W. Buie, Icarus **184**, 543–555 (2006)
K.P. Hand, in *Astrobiology: From Simple Molecules to Primitive Life*, ed. by V.A. Basiuk, R. Navarro-Gonzalez (2009)
K.P. Hand, R.W. Carlson, C.F. Chyba, Astrobiology **7**, 1006–1022 (2007)
R.A. Haring, R. Pedrys, D.J. Oostra, A. Haring, A.E. De Vries, Nucl. Instrum. Methods Phys. Res. B **5**, 476–482 (1984)
R.E. Hartle, E.C. Sittler, F.M. Neubauer, R.E. Johnson, H.T. Smith, F. Crary, D.J. McComas, D.T. Young, A.J. Coates, D. Simpson, S. Bolton, D. Reisenfeld, K. Szego, J.J. Berthelier, A. Rymer, J. Vilppola, J.T. Steinberg, N. Andre, Planet. Space Sci. **54**, 1211–1224 (2006)
A.R. Hendrix, R.E. Johnson, Astrophys. J. **687**, 706–713 (2008)
C.A. Hibbitts, T.B. McCord, G.B. Hansen, J. Geophys. Res. **105**, 22541–22557 (2000)
C.J. Hochanadel, J.A. Ghormley, T.J. Sworski, J. Am. Chem. Soc. **77**, 3215 (1955)
W.F. Huebner, J.J. Keady, S.P. Lyon, Astrophys. Space Sci. **195**, 1–294 (1992)
H. Imanaka, B.N. Khare, J.E. Elsila, E.L.O. Bakes, C.P. McKay, D.P. Cruikshank, S. Sugita, T. Matsui, R.N. Zare, Icarus **168**, 344–366 (2004)
R.E. Johnson, in *Chemical Dynamics in Extreme Environments*, ed. by R.A. Dressler (World Scientific, Singapore, 2001), pp. 390–419
R.E. Johnson, in *Physics and Chemistry at Low Temperatures*, ed. by L. Khriachtchev (World Scientific, Singapore, 2010)
R.E. Johnson, W.A. Jesser, Astrophys. J. **480**, L79 (1997)
R.E. Johnson, T.I. Quickenden, J. Geophys. Res. **102**, 10985–10996 (1997)
R.E. Johnson, R.W. Carlson, J.F. Cooper, C. Paranicas, M.H. Moore, M. Wong, in *Jupiter*, ed. by F. Bagenal, W. McKinnon (Cambridge University Press, Cambridge, 2004)
R.E. Johnson, M.H. Burger, T.A. Cassidy, F. Leblanc, M. Marconi, W.H. Smyth, in *Europa*, ed. by R.T. Pappalardo, W.B. McKinnon, K.K. Khurana (Univ. of Arizona Press, Tucson, 2009)
B.N. Khare, C. Sagan, E.T. Arakawa, F. Suits, T.A. Callcott, M.W. Williams, Icarus **60**, 127–137 (1984)
B.N. Khare, C. Sagan, H. Ogino, B. Nagy, C. Er, K. Schram, E.T. Arakawa, Icarus **68**, 176–184 (1986)
B.N. Khare, W.R. Thompson, B.G.J.P.T. Murray, C.F. Chyba, C. Sagan, E.T. Arakawa, Icarus **79**, 350–361 (1989)
K. Kobayashi, T. Kaneko, T. Saito, T. Oshima, OLEB **28**, 155–165 (1998)
Krupp et al., Space Sci. Rev. (2010, this issue)
C. Laffon, S. Lacombe, F. Bournel, Ph. Parent, J. Chem. Phys. **125**, 204714 (2006)

A. Lagg, N. Krupp, J. Woch, D.J. Williams, Geophys. Res. Lett. **30**, 110000-1 (2003)
A.L. Lane, R.M. Nelson, D.L. Matson, Nature **292**, 38 (1981)
L.J. Lanzerotti, W.L. Brown, J.M. Poate, W.M. Augustyniak, Geophys. Res. Lett. **5**, 155–158 (1978)
L.J. Lanzerotti, W.L. Brown, K.J. Marcantonio, R.E. Johnson, Nature **312**, 139 (1984)
L.J. Lanzerotti, W.L. Brown, C.G. Maclennan, A.F. Cheng, S.M. Krimigis, R.E. Johnson, J. Geophys. Res. **92**, 14949–14956 (1987)
M.J. Loeffler, U. Raut, R.A. Baragiola, Astrophys. J. **649**, L133 (2006)
R.D. Lorenz, Planet. Space Sci. **45**, 1009–1019 (1997)
B.H. Mauk, D.G. Mitchell, S.M. Krimigis, E.C. Roelof, C.P. Paranicas, Nature **421**, 920–922 (2003)
C.P. McKay, Planet. Space Sci. **44**, 741–747 (1996)
V. Mennella, M.E. Palumbo, G.A. Baratta, Astrophys. J. **615**, 1073–1080 (2004)
M.H. Moore, Icarus **59**, 114–128 (1984)
M.H. Moore, R.L. Hudson, Icarus **135**, 518–527 (1998)
M.H. Moore, R.L. Hudson, Icarus **161**, 486 (2003)
M.H. Moore, R. Khanna, B. Donn, J. Geophys. Res. **96**, 17541 (1991)
M.H. Moore, R.F. Ferrante, R.L. Hudson, J.N. Stone, Icarus **190**, 260–273 (2007)
C.D. Neish, A. Somogyi, J.I. Lunine, M.A. Smith, Icarus **201**, 412–421 (2009)
M.J. Nguyen, F. Raulin, P. Coll, S. Derenne, C. Szopa, G. Cernogora, G. Israël, J.M. Bernard, Planet. Space Sci. **55**, 2010–2014 (2007)
K.S. Noll, T.L. Roush, D.P. Cruikshank, R.E. Johnson, Y.J. Pendleton, Nature **388**, 45–47 (1997)
E. Quirico, S. Doute, B. Schmitt, C. de Bergh, D.P. Cruikshank, T.C. Owen, T.R. Geballe, T.L. Roush, Icarus **139**, 159–178 (1999)
F. Raulin, M.J. Nguyen, P. Coll, in *Instruments, Methods, and Missions for Astrobiology X*, ed. by R.B. Hoover, G.V. Levin, A.Y. Rozanov, P.C.W. Davies. Proc. SPIE, vol. 6694 (2007). doi:10.1117/12.732883
F. Raulin, et al., Space Sci. Rev. (2010, this issue). doi:10.1007/s11214-009-9610-x
C.T. Reimann, J.W. Boring, R.E. Johnson, J.W. Garrett, K.R. Farmer, W.L. Brown, Surf. Sci. **147**, 227–240 (1984)
C. Sagan, B.N. Khare, Nature **277**, 102–107 (1979)
C. Sagan, W.R. Thompson, Icarus **59**, 133–161 (1984)
S.A. Sandford, L.J. Allamandola, Astrophys. J. **355**, 357–372 (1990)
T. Sasaki, R.S. Williams, J.S. Wong, D.A. Shirley, J. Chem. Phys. **68**, 2718–2724 (1978)
L. Schriver-Mazzuoli, A. Schriver, H. Chaabouni, Can. J. Phys. **81**, 301–309 (2003)
E.C. Sittler, A. Ali, J.F. Cooper, R.E. Hartle, R.E. Johnson, A.J. Coates, D.T. Young, Planet. Space Sci. **57**, 1547–1557 (2009)
H.T. Smith, M. Shappirio, R.E. Johnson, D. Reisenfeld, E.C. Sittler, F.J. Crary, D.J. McComas, D.T. Young, J. Geophys. Res. (Space Phys.) **113**, 11206 (2008)
Sohl et al., Space Sci. Rev. (2010, this issue)
A. Somogyi, C.-H. Oh, M.A. Smith, J.I. Lunine, J. Am. Soc. Mass Spectrom. **16**, 850–859 (2006)
J.R. Spencer, W.M. Calvin, Astron. J. **124**, 3400–3403 (2002)
G. Strazzulla, M.E. Palumbo, Planet. Space Sci. **46**, 1339 (1998)
G. Strazzulla, A.C. Castorina, M.E. Palumbo, Planet. Space Sci. **43**, 1247–1251 (1995)
G. Strazzulla, G. Leto, O. Gomis, M.A. Satorre, Icarus **164**, 163 (2003)
G. Strazzulla, G.A. Baratta, O. Gomis, Icarus **192**, 623–628 (2007)
F.W. Taylor et al., in *Jupiter: The Planet, Satellites and Magnetosphere*, ed. by F. Bagenal, T.E. Dowling, W. McKinnon (Cambridge Univ. Press, Cambridge, 2004). Chap. 4
B.D. Teolis, R.A. Vidal, J. Shi, R.A. Baragiola, Phys. Rev. B **72**, 245422-9 (2005)
B.D. Teolis, M.J. Loeffler, U. Raut, M. Fama, R.A. Baragiola, Astrophys. J. **644**, L141–L144 (2006)
B.D. Teolis, J. Shi, R.A. Baragiola, J. Chem. Phys. **130**, 134704-1-9 (2009)
G. Tobie, J.I. Lunine, C. Sotin, Nature **440**, 61–64 (2006)
M. Tomasko (DISR Huygens team), Nature **438**, 765–778 (2005)
M.G. Trainer, A.A. Pavlov, J.L. Jimenez, C.P. McKay, D.R. Wornsnop, O.B. Toon, M.A. Tolbert, Geophys. Res. Lett. **31**, L17S08 (2004)
B.N. Tran, J.C. Joseph, J.P. Ferris, P.D. Persans, J.J. Chera, Icarus **165**, 379–390 (2003)
D.E. Trilling, R.H. Brown, Icarus **148**, 301–306 (2000)
A.J. Verbiscer, D.E. Peterson, M.F. Skrutskie, M. Cushing, P. Helfenstein, M.J. Nelson, J.D. Smith, J.C. Wilson, Icarus **182**, 211–223 (2006)
R.A. Vidal, D.A. Bahr, R.A. Baragioa, M. Peters, Science **276**, 1839–1842 (1997)

J.H. Waite, D.T. Young, T.E. Cravens, A.J. Coates, F.J. Crary, B. Magee, J. Westlake, Science **316**, 870–875 (2007)

J.H. Waite Jr., W.S. Lewis, B.A. Magee, J.I. Lunine, W.B. McKinnon, C.R. Glein, O. Mousis, D.T. Young, T. Brockwell, J. Westlake, M.-J. Nguyen, B.D. Teolis, H.B. Niemann, R.L. McNutt, M. Perry, W.-H. Ip, Nature **460**, 487–490 (2009)

E.H. Wilson, S.K. Atreya, J. Geophys. Res. **109**, E06002 (2004). doi:10.1029/2003JE002181

Space Sci Rev (2010) 153: 317–348
DOI 10.1007/s11214-010-9636-0

Implications of Rotation, Orbital States, Energy Sources, and Heat Transport for Internal Processes in Icy Satellites

Hauke Hussmann · Gaël Choblet · Valéry Lainey · Dennis L. Matson · Christophe Sotin · Gabriel Tobie · Tim Van Hoolst

Received: 13 May 2009 / Accepted: 8 February 2010 / Published online: 17 March 2010

Abstract Internal processes in icy satellites, e.g. the exchange of material from the subsurface to the surface or processes leading to volcanism and resurfacing events, are a consequence of the amount of energy available in the satellites' interiors. The latter is mainly determined shortly after accretion by the amount of radioactive isotopes incorporated in the silicates during the accretion process. However, for satellites—as opposed to single objects—important contributions to the energy budget on long time-scales can come from the interaction with other satellites (forcing of eccentricities of satellites in resonance) and consequently from the tidal interaction with the primary planet. Tidal evolution involves both changes of the rotation state—usually leading to the 1:1 spin orbit coupling—and long-term variations of the satellite orbits. Both processes are dissipative and thus connected with heat production in the interior. The way heat is transported from the interior to the surface (convection, conduction, (cryo-) volcanism) is a second main aspect that determines how internal processes in satellites work. In this chapter we will discuss the physics of heat production and heat transport as well as the rotational and orbital states of satellites. The

H. Hussmann (✉)
DLR Institute of Planetary Research, Rutherfordstr. 2, 12489 Berlin, Germany
e-mail: hauke.hussmann@dlr.de

G. Choblet · G. Tobie
Laboratoire de Planétologie et Géodynamique, Université de Nantes, 2, rue de la houssinière, BP 92208, 44322 Nantes, France

V. Lainey
Institut de Mécanique Céleste et de Calcul des Éphémérides (IMCCE), Observatoire de Paris 77, Avenue Denfert-Rochereau, 75014 Paris, France

D.L. Matson · C. Sotin
Jet Propulsion Laboratory/California Institute of Technology, 4800 Oak Grove Drive, Pasadena, CA 91109, USA

T. Van Hoolst
Royal Observatory of Belgium, Ringlaan 3, 1180 Brussels, Belgium

relevance of the different heat sources for the moons in the outer solar system are compared and discussed.

Keywords Satellites · Energy sources · Rotation · Tides · Orbital dynamics · Heat transfer

1 Introduction

The first evidence that dynamical processes play a significant role for the evolution and present states of satellites in the outer solar system was obtained in the 1970s when ground-based observations revealed an anomalous output of thermal emission and unusual spectral reflectance at the surface of Io, the innermost of the Galilean Satellites of Jupiter. Additionally, the existence of Io's ion and neutral gas clouds and plasma torus were detected at that time (Nash et al. 1986, and references therein). Dynamical variations—"outbursts"—in Io's infrared spectra (Witteborn et al. 1979) were first recognized shortly before the *Voyager 1* encounter in 1979. The *Voyager 1* flyby revealed Io's intense geologic activity and vigorous volcanism (Smith et al. 1979), confirming the presence of enormous internal energy sources that had been theoretically predicted by Peale et al. (1979) shortly before the encounter. Io is an excellent example showing that internal energy sources can be closely connected to a satellite's orbital state, here its orbital eccentricity and its tidal interaction with the primary planet. Additionally, the importance of mean motion resonances forcing the eccentricities of satellites was recognized at that time as an essential part of the energy available within moons evolving in satellite systems (Yoder 1979). Tidal friction has been recognized as the main heat source in Io's interior giving rise to the thermal and volcanic activity observed by the *Galileo* mission (Geissler 2003; McEwen et al. 2004), by recent ground-based campaigns (Marchis et al. 2005) and by the *New Horizons* fly-by of the Jupiter system in 2007 on its way to Pluto (Spencer et al. 2007).

With the observation of anomalously high surface temperatures and large eruptions of water-vapor and dust from the icy surface of the Saturnian satellite Enceladus, the *Cassini* spacecraft detected strong internal activity of a relatively small icy moon ($R = 252$ km) in the outer solar system (Porco et al. 2006; Spencer et al. 2006). Enceladus' activity presumably implies liquid water and high temperatures within the moon's ice shell. However, the processes generating such a locally confined activity are still debated. Especially, with the presence of liquid water tidal dissipation can be considered as the major heat source.

During the flyby of *Voyager 2* at Neptune's moon Triton the imaging data showed plumes rising in the satellites' tenuous atmosphere (Soderblom et al. 1990). However, the energy involved in this kind of activity is considerably smaller as compared to the eruptions on Io and Enceladus. It is not clear whether Triton's activity is related to internal processes or merely a consequence of solar insolation (Brown et al. 1990).

So far Io, Enceladus and Triton are the only satellites of the solar system showing present-day activity on their surfaces.[1] However, there is ample evidence of past activity on icy moons which is related to intense heating and exchange of energy in their interiors. Io, Europa, Ganymede, Enceladus, Titan, Dione, Miranda, Ariel, and Triton show signs of substantial resurfacing on global scales. Iapetus, Titania, Tethys and Rhea display regional terrain characterized by tectonism and/or other geological processes that occurred early in their histories. In contrast Callisto, Mimas, Oberon and Umbriel did not pass through active periods

[1] Titan's surface is currently modified by interaction with its dense atmosphere. Since we focus on processes related to the interior of the moons this kind of resurfacing is not discussed here.

in their histories, as indicated by their heavily cratered surfaces. These examples show that activity in icy satellites is not a simple consequence of their mass, size, composition, or semi-major axis. On the basis of the satellites' bulk properties the state of activity—be it early or late in their histories—is unpredictable. It is rather a consequence of the satellites properties in combination with external influence, e.g., impacts or gravitational interaction with other bodies. Early activity may be linked to non-synchronous rotation during early epochs before the 1:1 spin-orbit coupling, which is the current state occupied by all large satellites, was established. During that early epoch large impacts may have significantly contributed to the energy budget of the moons. Later, orbital resonances and tidal heating in the synchronous state in elliptical orbits have played an important role for some of the satellites.

Earlier reviews on orbital and rotational states are given by Peale (1986) and on the thermal evolution and convection by Schubert et al. (1986) and McKinnon (1998).

In this chapter we discuss the significance of energy sources that are available for the satellites and the internal processes that can be triggered during their evolution. This requires a description of their rotational and dynamical states, as well as their composition and state of differentiation. The connection from energy sources to dynamical processes is provided by heat transport mechanisms finally releasing energy at the surface. The available energy sources and their relevance for internal processes and satellite evolution are compared for the individual objects.

2 Rotation

The large and medium-sized (radius > 200 km) satellites of the Solar System all have rotation periods that are almost exactly equal to their revolution periods. Non-synchronous rotation might occur for some satellites. The typical time-scale of 10^4–10^5 years for non-synchronous rotation, however, significantly exceeds the periods of revolution, of the order of days or tens of days. The state of synchronous rotation is stable due to the asymmetric mass distribution with the longest axis pointing towards the central planet at pericenter (Goldreich and Peale 1966, see also Sect. 3.2.2). However, stability does not imply that the rotation rate is constant. Both internal and external forces may cause the rotation rate to vary about its equilibrium value. In particular, due to the eccentricity of all the satellite orbits, the long axis is not pointing toward the central planet but approximately in the direction to the empty focus of the orbit (Murray and Dermott 1999) so that the central planet exerts a time-variable gravitational torque on the satellite. As a result, the satellite periodically varies its rotation rate about the mean rate and the direction of the long axis librates with respect to its orientation at constant mean rotation rate (forced libration). As is well known for the Earth, the rotation rate of a celestial body can also vary as a result of angular momentum exchange between the atmosphere and the solid body. These variations may have been observed for Titan (Stiles et al 2008, 2010).

Observations of librations provide information on the moments of inertia of the satellite since the gravitational torque exerted by the central planet is proportional to the difference between moments of inertia and the rotational reaction depends on the moment of inertia about the rotation axis. Therefore, together with measurements of the gravitational field, rotation can be used to determine the degree of differentiation of a satellite. Moreover, a subsurface ocean strongly affects the rotation variations and leads to a transfer of angular momentum and energy between the solid and liquid layers. Rotation observations could prove the existence of an internal ocean and could even constrain the thickness of the overlying ice shell. Besides providing constraints on the interior structure, librations can also increase the tidal heating rate of a satellite (e.g., Wisdom 2004).

2.1 Static Tides and Shape

Rotation flattens a synchronously rotating satellite at its poles and creates an equatorial bulge, whereas the static tides stretch the satellite in the direction to the central planet and cause a contraction in the direction perpendicular to it. As a result, the longest axis is expected to be along the planet-satellite line at pericenter and the short axis is parallel to the rotation axis. For a satellite in hydrostatic equilibrium, rotation and static tides are the only causes of deviations from spherical symmetry and the satellite's shape can be calculated by applying Clairaut theory (see, e.g. Jeffreys 1952; Hubbard 1984; Moritz 1990). In view of the study of variations in the rotation speed of satellites, the difference between the long and short axes of the equator is dynamically important. For a given spherical reference model of the interior structure with density profile $\rho(r_0)$, where r_0 is the radial coordinate from the center of the model, the equatorial flattening β of the model due to the static tides raised by the central planet can be determined from Clairaut's second-order differential equation

$$\frac{d^2\beta}{dr_0^2} + \frac{6}{r_0}\frac{\rho}{\overline{\rho}}\frac{d\beta}{dr_0} - \frac{6}{r_0^2}\left(1 - \frac{\rho}{\overline{\rho}}\right)\beta = 0, \tag{1}$$

and the associated boundary condition

$$\frac{d\beta}{dr_0}(R) = \frac{1}{R}\left[\frac{15}{2}q - 2\beta(R)\right] \tag{2}$$

(e.g. Hubbard 1984; Van Hoolst et al. 2008). Here, R is the mean radius, $q = (n^2R^3)/(Gm_s)$ the ratio of the centrifugal acceleration to the gravitational acceleration, and m_s the total mass of the satellite; $\overline{\rho}$ is the mean density in a sphere of radius r_0 and n the orbital mean motion. The flattening $\beta(r_0) = [a(r_0) - b(r_0)]/a(r_0)$, where $a(r_0) > b(r_0)$ are the lengths of the two principal axes in the equatorial plane of the ellipsoidal surface with mean radius r_0.

Instead of the difference in length of the long and short equatorial axes, the equatorial moment of inertia difference $B - A$ is dynamically more relevant and can be calculated by integration over the radial coordinate:

$$B - A = \frac{8\pi}{15}\int_0^R \rho(r_0')\frac{d(\beta r_0'^5)}{dr_0'}dr_0'. \tag{3}$$

Data on the geometric shape and on the gravitationally determined moment of inertia difference are available for a selection of satellites. The shape of many but not all large and medium-sized satellites seems to be consistent to first order with that of a relaxed satellite in hydrostatic equilibrium (e.g. Thomas et al. 2007; Van Hoolst et al. 2008, 2009). A necessary but not sufficient condition for hydrostatic equilibrium is that the ratio J_2/C_{22} of the degree two coefficients of the gravitational field is equal to 10/3. At least two different close flybys—one equatorial and one polar—are required to test this hypothesis. So far, Io is (apart from the Moon) the only satellite for which polar and equatorial flybys could be used. The hydrostatic assumption was found to be consistent with the gravity data analysis from these flybys (Schubert et al. 2004). Data from one polar and three equatorial passes was also obtained at Ganymede. The independent determination of J_2 and C_{22} was, however, impossible. Detailed analysis suggests the presence of mass anomalies inside Ganymede that may be related to the different rock content of Ganymede's dark and bright terrain (Schubert et al. 2004; Palguta et al. 2009).

The shape of Saturn's satellite Iapetus is remarkable. It corresponds to a body in hydrostatic equilibrium that rotates with a period of about 16 h (Thomas et al. 2007), much faster than at its current period of 79.33 days. This implies that its shape has been conserved from a much faster spin-state early in its evolution (for details see Schubert et al. 2010, this issue).

2.2 Libration

2.2.1 *Librations of a Rigid Body*

Since satellite orbits are eccentric (eccentricity between about 0.001 and 0.03 for the large and medium-sized satellites), the orbital speed is variable and satellites do not always show exactly the same face to their central planet. As a consequence, the central planet exerts a time-variable gravitational torque on the satellite that tends to modify the satellite's rotation rate. Longitudinal librations are the corresponding changes in the rotation angle. If a satellite can be assumed to respond as a rigid body to an external gravitational force, conservation of angular momentum $H = C\dot{\theta}$, where C is the largest (polar) principal moment of inertia and θ the angle of rotation between the long axis of the satellite and the major axis of the orbit, can be expressed as (e.g. Goldreich and Peale 1966)

$$C\ddot{\theta} + \frac{3}{2}n^2(B - A)\left(\frac{a}{r}\right)^3 \sin 2(\theta - f) = 0, \tag{4}$$

where the second term is equal and opposite to the gravitational torque of the central planet on the satellite and zero obliquity is assumed. Here, a is the semi-major axis, f the true anomaly, and r the distance between the centers of mass of the central planet and the satellite. The rigid rotational response is a good approximation for an entirely solid satellite without internal ocean since the effect of elastic deformation on the moment of inertia of the satellite is small (Van Hoolst et al. 2008).

It is convenient to introduce the small libration angle $\gamma = \theta - M$ and expand the equations as a series in mean anomaly M and eccentricity e by using well-known expansions for $(a/r)^3$, $\cos f$, and $\sin f$ (Cayley 1861). After linearization of (4) in eccentricity and libration angle, we then have

$$C\ddot{\gamma} + 3n^2(B - A)\gamma = 6en^2(B - A)\sin M. \tag{5}$$

The right-hand side causes the satellite to librate with a period equal to the orbital period of the satellite and amplitude (e.g., Murray and Dermott 1999)

$$A_\gamma = \frac{2\omega_f^2 e}{n^2 - \omega_f^2}. \tag{6}$$

The natural free frequency ω_f follows directly from the left-hand side of (5) (e.g., Comstock and Bills 2003):

$$\omega_f = n\sqrt{\frac{3(B - A)}{C}}. \tag{7}$$

It is orders of magnitude smaller than the mean motion n since $B - A$ is much smaller than C. Therefore, the amplitude of the forced libration can be approximated by (Eckhardt 1981)

$$A_\gamma = 6\frac{B - A}{C}e. \tag{8}$$

Dissipation processes due to tides and libration damp the long-periodic free libration, so that it can be assumed to have a (almost) zero amplitude. Comstock and Bills (2003) provide a survey of libration amplitudes of satellites calculated by assuming that the relative moment of inertia difference $(B - A)/C$ corresponds to that of a homogeneous body in hydrostatic equilibrium. Both assumptions could lead to errors of the order of some ten percent for the larger and rapidly rotating satellites (e.g., Van Hoolst et al. 2008 in application to Europa) or even larger for bodies that have a strong non-hydrostatic contribution to their flattening. Typical maximum displacements of the long axis with respect to its equilibrium position at the surface range from a few tens to a few hundreds of meters. The largest amplitude is expected for Mimas (about 1.4 km) and for the non-synchronous rotator Hyperion (about 1.3 km) (Comstock and Bills 2003).

For small deviations from a spherically symmetric shape the relative contribution of the forced libration to tidal heating of a synchronous satellite is of the order of $3(B - A)/C$ (Wisdom 2004). This ratio is smaller than 0.01 for the large satellites but of the order of 0.1 for more aspherical satellites like Mimas, Enceladus, and Tethys (Thomas et al. 2007). A resonance between the orbital frequency and the free libration frequency could further enhance the heating (Wisdom 2004).

2.2.2 *Effect of a Subsurface Ocean*

When a subsurface ocean exists, the rotation variations of the ice shell can be different from those of the interior. It is tempting to assume that the shell is dynamically decoupled from the interior by the ocean, but pressure and gravitational forces between the shell, ocean, and solid interior are likely to be efficient in coupling the rotational motion of the shell to that of the solid interior (Van Hoolst et al. 2008, 2009). Internal gravitational and pressure coupling arises when the principal axes of the shell are not aligned with those of the interior, for example as a result of the different rotational response of the shell and interior to the gravitational torque of the central planet. As a result of internal coupling, energy and angular momentum are exchanged between the internal layers.

Friction on the boundaries of an internal ocean can be neglected because of the small ocean viscosity, provided that variations in ocean-floor topography are small. Furthermore, electromagnetic coupling between the ocean and the shell and between the ocean and the interior is not considered to be important, at least not when the satellite is lacking a self-generated magnetic field. By assuming that the ocean dynamics can be expressed essentially as a small rigid rotation and by neglecting topography on the ellipsoidal boundaries between the interior, the ocean, and the shell, Van Hoolst et al. (2009) showed that the existence of a subsurface ocean increases the amplitude of the forced librations. For example, for a highly differentiated model of Titan with a 224 km thick ammonia–water ocean beneath an ice shell of 69 km thickness (Sohl et al. 2003), the amplitude of libration equals 1.5×10^{-4} rad, whereas in the rigid assumption, the amplitude is only 1.7×10^{-5} rad (Van Hoolst et al. 2009). These values are calculated by assuming a hydrostatic flattening profile. In that case the libration amplitude increases about linearly with decreasing shell thickness (Van Hoolst et al. 2008), offering the possibility to determine the thickness of the ice shell from libration observations. For thin shells, for example a few km for Europa, the forced libration can become resonantly enhanced due to a resonance with a natural mode of the satellite. This mode is essentially a free libration of the ice shell and its eigenfrequency increases with decreasing moment of inertia of the shell C_s (Van Hoolst et al. 2008). Because the librations are larger when a satellite has a subsurface ocean, tidal heating is expected to increase with respect to heating for a completely solid body. Moreover, the relative rotation of a viscous ocean with respect to the solid boundaries could be an additional heat source.

2.3 Length-of-Day Variations of Titan

Cassini radar observations show that Titan's spin is slightly faster than synchronous leading to a shift of 0.12° per year in apparent longitude (Stiles et al. 2010). It is thought that long-periodic exchanges of angular momentum between Titan's surface and atmosphere cause the orientation of Titan to vary with an amplitude of about 0.005 rad. Periodic length-of-day (LOD) variations with main period of 14.74 yrs were predicted by Tokano and Neubauer (2005) on the basis of their general circulation model of the atmosphere. For a solid Titan, their predicted amplitude is smaller than the observed value. If Titan has an ocean and the rotation of the shell could be considered decoupled from the interior, the atmosphere would only force rotation variations of the shell and the amplitude of the surface rotation variations would be the observed value (Tokano and Neubauer 2005). However, misalignment of the long axis of Titan with the direction to Saturn due to the atmosphere results in a strong gravitational restoring torque of Saturn exerted on Titan, which limits the amplitude of the orientation variations to a value of about 10^{-4} rad, or about 50 times smaller than observations indicate (Van Hoolst et al. 2009). Although this suggests that Titan cannot be entirely solid, the combined effect of the gravitational torque due to Saturn and internal coupling when a subsurface ocean exists in Titan also strongly reduces the rotation variations with respect to that for a decoupled shell. By assuming hydrostatic flattening and taking into account the atmospheric torque predicted by Tokano and Neubauer (2005), the rotation variations for Titan with an internal ocean can be shown to be about as small as expected for an entirely solid Titan. Internal coupling contributes to reducing the amplitude of the rotation variations by its tendency to lock the rotation of the shell to that of the interior (Karatekin et al. 2008). Besides the much smaller predicted than observed LOD variations, Saturn's torque is also predicted to cause the rotation to be slower than synchronous in contrast to the Cassini observations. These large differences with the observations in amplitude and in phase suggest that non-hydrostatic effects in Titan are important. In particular, if non-hydrostatic effects would strongly reduce the equatorial flattening of the ice shell above an internal ocean, the amplitude and phase of the calculated rotation variations were similar to the observed values. Alternatively, the calculated LOD variations would increase if the atmospheric torque was larger than predicted or if fast viscous relaxation of the ice shell would reduce the gravitational coupling (Van Hoolst and Karatekin 2008). However, it remains to be studied whether an almost two order of magnitude increase is possible and if these effects can also explain the phase difference of the predicted rotation variations. On the other hand, Noyelles (2008) suggests that a resonantly excited long period wobble could explain the apparent faster rotation even for an entirely solid Titan, which would nevertheless require fine-tuning of Titan's moment of inertia.

3 Orbital Dynamics

3.1 Resonances Among Satellites

Many satellites in the Solar system are locked in resonance. An orbital resonant state occurs when a linear combination of orbital frequencies is zero on average. Such configurations can be a consequence of expansion or regression of the orbits under tidal interactions. However, a primordial origin as suggested by e.g., Peale and Lee (2002) for the Galilean satellites would be an alternative scenario for the formation of resonances.

When satellites are in a resonant state they exchange their orbital energy and angular momentum. The computation of orbital variations induced by tidal effects is then rendered more complex by the mutual gravitational interaction of the moons in resonance. For the main satellites of the giant planets, several resonances are observed. With L being the mean longitude, Ω the longitude of the node and ϖ the longitude of the periapsis, the following resonances and their respective arguments are observed in the Saturnian system:

(i) Mimas (subscript M) and Tethys (T)

$$4L_{\mathrm{T}} - 2L_{\mathrm{M}} - \Omega_{\mathrm{T}} - \Omega_{\mathrm{M}} = \phi; \tag{9}$$

conjunctions librate about ϕ, the midpoint of the node (amplitude 43.6° and period 71.8 years) (Murray and Dermott 1999);

(ii) Enceladus (subscript E) and Dione (D)

$$L_{\mathrm{E}} - 2L_{\mathrm{D}} + \varpi_{\mathrm{E}} = 0, \tag{10}$$

with a libration amplitude of 0.297° and a period of 11.1 years (Murray and Dermott 1999);

(iii) Titan (subscript T) and Hyperion (H)

$$3L_{\mathrm{T}} - 4L_{\mathrm{H}} + \varpi_{\mathrm{H}} = \pi, \tag{11}$$

with libration amplitude of 36° over a 640 days period.

The three inner Galilean satellites (Io (I), Europa (E), and Ganymede (G)) are locked in the Laplace resonance (who first demonstrated the stability of this configuration) and satisfies the longitude relation

$$L_{\mathrm{I}} - 3L_{\mathrm{E}} + 2L_{\mathrm{G}} = \pi, \tag{12}$$

with a small amplitude of 0.175° over a 2059.6 days period (Lainey et al. 2006). No orbital resonances are known among the main Uranian satellites.

The main aspect of resonances for satellite evolution is the forcing of eccentricities (e-type resonance) and inclinations (i-type resonance). For satellites, locked in an e-type resonance the forced eccentricities are maintained, although dissipation within the satellites tends to reduce them. This can keep tidal heating ongoing on geological timescales. The forced eccentricities and inclinations, associated with the resonances above are given in Table 1.

3.2 Orbital Variation Due to Tides

Tidal effects arise because of the differences in gravitational attraction due to an external mass experienced on different parts of a body. This gives rise to tidal bulges that are directed toward the tide-raising body, if no dissipation occurs. Because most satellites are in a 1:1 spin orbit resonance, it is convenient to separately consider the tidal effects in the planet and in the satellite.

3.2.1 Tides Raised in the Planet

If no dissipation occurs in the planet, the tidal bulge would be perfectly aligned with the planet–satellite line. Because of tidal dissipation inside the primary, a phase lag arises and leads to an exchange of angular momentum between the spin of the planet and the orbit of

Table 1 Forced eccentricities and inclinations of satellites in resonance (e.g., Murray and Dermott 1999). The eccentricity of Europa is forced due to two resonances. The perturbations by Enceladus and Hyperion are too small to significantly force the eccentricities of Dione and Titan, respectively. Titan has a large free eccentricity of 0.0292. Also in the cases of Ganymede and Dione the free eccentricities of 0.0015 and 0.0022, respectively (Peale 1986), significantly exceed the forced eccentricities

Satellite	Ratio of mean motions	e (forced)	i (forced)
Io	2:1 Io–Europa	0.0041	–
Europa	2:1 Io–Europa and 2:1 Europa–Ganymede	0.0101	–
Ganymede	2:1 Europa–Ganymede	0.0006	–
Enceladus	2:1 Enceladus–Dione	0.0044	–
Dione	2:1 Enceladus–Dione	$\approx 10^{-5}$	–
Mimas	4:2 Mimas–Tethys	–	1.53°
Tethys	4:2 Mimas–Tethys	–	1.09°
Titan	4:3 Titan–Hyperion	–	–
Hyperion	4:3 Titan–Hyperion	0.1042	–

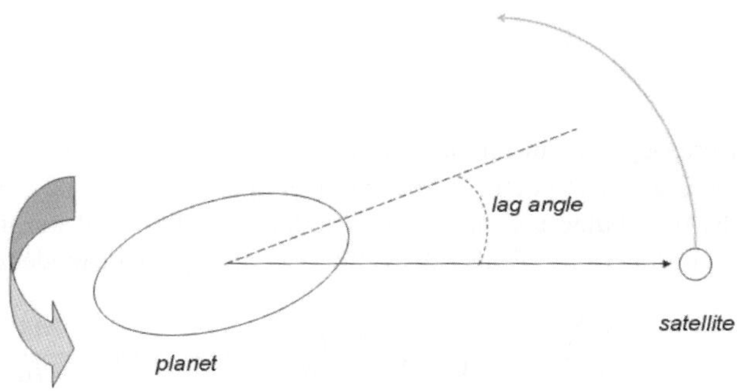

Fig. 1 Tidal bulges configuration induced by a tide raising satellite on its primary, assuming a faster rate for the planet spin than the satellite mean motion

the satellite. In the following we assume prograde motion of the satellite (Triton's retrograde orbit is the only exception). If the satellite revolves slower than the planet rotates—which is the case for all large and medium-sized satellites—, the tidal bulge of the planet will be ahead of the sub-planet line (see Fig. 1). The torque exerted by the satellite on the planet's tidal bulge will cause the planet to rotate slower, while conservation of angular momentum will increase the satellite's semi-major axis. On the other hand, if the satellite revolves faster than the planet rotates—which is the case for Phobos and also for several small inner satellites of the giant planets—, the satellite will cause the planet rotation rate to increase while the satellite's orbit will regress.

Under the assumption of low eccentricity, low inclination and outside of resonance we have (Goldreich and Soter 1966)

$$\frac{da}{dt} = +3\left(\frac{k_2}{Q}\right)_p \frac{m_s}{m_p}\left(\frac{R_p}{a}\right)^5 na, \quad \text{and} \tag{13}$$

$$\frac{de}{dt} = +\frac{57}{8}\left(\frac{k_2}{Q}\right)_p \frac{m_s}{m_p}\left(\frac{R_p}{a}\right)^5 ne. \tag{14}$$

Here t is the time, m the mass, k_2 the second-degree potential Love number, Q the quality factor, R the radius, and p and s denote parameters referring to the planet and the satellite, respectively.

3.2.2 Tides Raised in the Satellite

Synchronous Spin-Orbit Resonance The rotation rate of most of the natural satellites has slowly evolved to an equilibrium state as a result of tidal dissipation in the satellites' interiors. In equilibrium, the satellites rotate synchronously with their orbital motion, meaning that they have on average (see above) equal rotation and revolution periods.

Satellite tides are caused by the gravitational field of a planet, which periodically deforms the natural satellites orbiting the planet. Since friction is associated with tidal deformation, tidal energy is converted to heat and the tides lag in phase with respect to the driving force (as for a damped forced oscillator). As a result, the tidal bulge of the satellite is misaligned with respect to the line joining the mass centers of the satellite and the planet. For a zero-obliquity satellite in a circular equatorial orbit, the tidal torque Γ on the satellite as a result of the misaligned tidal bulge can be written as (e.g. Goldreich 1963; MacDonald 1964; for the more general case of eccentric and inclined orbits, see Ferraz-Mello et al. 2008)

$$\Gamma = \frac{3}{2} k_2 \frac{G m_p^2}{R_s} \left(\frac{R_s}{a} \right)^6 \sin 2\epsilon. \tag{15}$$

Here, R_s is the mean radius of the satellite, a the semi-major axis of the satellite's orbit, k_2 the degree two potential Love number of the satellite, and ϵ the geometric phase lag between the tidal bulge and the planet-satellite line. The tidal torque changes the rotation rate Ω_s according to e.g. Peale (1999) (the second approximate identity is for a homogeneous body)

$$\frac{d\Omega_s}{dt} = -\frac{3}{2} \left(\frac{k_2}{Q} \right)_s \frac{G m_p^2 R_s^5}{C_s a^6} \mathrm{sign}(\Omega_s - n) \approx -\frac{45}{76} \frac{\rho_s R_s^2}{\mu_s Q_s} n^4 \mathrm{sign}(\Omega_s - n), \tag{16}$$

and the rotation angular velocity decreases in a despinning time scale τ given by

$$\tau \approx \frac{76}{45} \frac{\Omega_s \mu_s Q_s}{\rho_s R_s^2} \frac{1}{n^4} \mathrm{sign}(\Omega_s - n). \tag{17}$$

Here, Q_s is the quality factor, C_s the polar moment of inertia, ρ_s the mean density, and μ_s the mean bulk modulus of the satellite. By taking typical values as $Q_s = 100$ and $\mu_s = 10^{10}$ Pa, this equation shows that the larger satellites despin in a short period compared to the age of the Solar System (e.g. Peale 1977). Values for τ are given in Table 8. A notable exception is the distant Iapetus, whose despinning time-scale is of the order of the age of the solar system.

Because of the strong distance dependence of the tidal torque, satellites despin more slowly with increasing distance from their central planet. Therefore, distant small satellites have almost retained their primordial spin up to the present time.

The final equilibrium rotation state for a spherically symmetric satellite is the synchronous rotation for a circular orbit but has a rotation rate slightly faster than synchronous for an eccentric orbit (we neglect obliquity here) due to the larger accelerating effect on rotation near pericenter than the decelerating effect near apocenter (see, e.g., Murray and Dermott 1999). Since despinning is the fastest tidally driven evolution, satellites still have an elliptical orbit upon reaching an equilibrium rotation state. The synchronous equilibrium state for an eccentric orbit is stable if the torque of the central planet on a permanent asymmetry of the satellite is larger than the tidal torque at synchronous rotation averaged over an orbital period (Goldreich and Peale 1966), which can be expressed for the small eccentricities of the satellites as

$$C_{22} > 3e^2 \left(\frac{k_2}{Q}\right)_s \frac{m_p}{m_s} \left(\frac{R_s}{a}\right)^3 \approx \frac{9}{19} \frac{\rho_s R_s^2}{\mu_s Q_s} n^2 e^2. \tag{18}$$

Here a constant tidal lag is assumed. Other tidal lag models can lead to a different dependency on e (e.g., Greenberg and Weidenschilling 1984). All large and medium-sized (radius > 200 km) satellites are synchronous rotators. Equation (18) shows that relaxed satellites in hydrostatic equilibrium have a sufficiently large C_{22}, typically of the order of 10^{-3} to 10^{-5}, for the synchronous rotation state to be stable. However, since the direction of the long axis of a satellite differs slightly from the direction to the central planet at pericenter due to the balance between the tidal torque and the torque on the asymmetry for synchronous rotation, the satellite's figure will tend to change slowly by creep to adapt to the tidal potential. To maintain the equilibrium orientation, the satellite will rotate slightly asynchronously (Greenberg and Weidenschilling 1984; Ferraz-Mello et al. 2008). Evidence for non-synchronous rotation of Europa, which has been suggested by Greenberg and Weidenschilling (1984) and by Ojakangas and Stevenson (1989b) has been obtained based on analysis of surface features (Hoppa et al. 2001).

Orbital Variations As the main satellites of the giant planets are locked in a 1:1 spin orbit resonance, no exchange of angular momentum is expected from tidal effects. Nevertheless, if a satellite is on an eccentric orbit, it will still experience tidal dissipation as, first, the distance to the primary changes with time, and second, the revolution rate will be higher at periapsis than at apoapsis. These two effects give rise to radial and librational tides, and hence dissipation. As the energy source is the orbital energy (equal to $-Gm_p m_s/2a$), tidal dissipation decreases the satellite semi-major axis. Moreover, conservation of angular momentum $m_s na^2\sqrt{1-e^2}$ will tend to decrease the eccentricity. After some computations and based on the assumption of small eccentricity and absence of resonances, we obtain for the evolution of a and e (Goldreich and Soter 1966)

$$\frac{da}{dt} = -21 \left(\frac{k_2}{Q}\right)_s \frac{m_p}{m_s} \left(\frac{R_s}{a}\right)^5 nae^2, \quad \text{and} \tag{19}$$

$$\frac{de}{dt} = -\frac{21}{2} \left(\frac{k_2}{Q}\right)_s \frac{m_p}{m_s} \left(\frac{R_s}{a}\right)^5 ne. \tag{20}$$

As all major satellites of the giant planets evolve beyond the synchronous orbit (defined by the distance to the primary for which orbital period equals the rotation period of the planet), competition arises between tidal dissipation in the planet and in the satellite. The induced expansion or regression of the orbit will be the result of the two opposing effects.

These changes in mean distance to the planet and orbital eccentricity are important for the evolution of resonances and for the tidal heat production on long time-scales. The amount of energy dissipated in the satellite interiors will be discussed in Sect. 4.4.1.

3.3 Observed Tidal Accelerations

Since the satellite semi-major axis and mean motion are related by Kepler's third law $n^2a^3 = G(m_s + m_p)$, tidal dissipation can, in principle, be observed from monitoring of the satellite orbits as a mean motion acceleration or deceleration.

Table 2 Selection of tidal secular accelerations for the Galilean satellites in the form $\dot{n}/n$ (units are in 10^{-10} yr^{-1})

Reference	Io	Europa	Ganymede
Lieske (1987)	-0.074 ± 0.087	-0.082 ± 0.097	-0.098 ± 0.153
Goldstein and Jacobs (1995)	$+4.54 \pm 0.95$	$+5.6 \pm 5.7$	$+2.8 \pm 2.0$
Vasundhara (1996)	$+2.27 \pm 0.70$	-0.67 ± 0.80	$+1.06 \pm 1.00$
Aksnes and Franklin (2001)	$+3.6 \pm 1.0$	–	–

3.3.1 The Galilean Satellite System

The first estimates of the secular tidal accelerations among the Galilean system have to be credited to De Sitter (1928). He derived a significant tidal acceleration for all three inner Galilean satellites (as they are locked in the Laplace resonance). Decades passed since new estimations were computed by different authors (see Table 2). Using an accurate orbital model first developed by Sampson (1921) and improved by Lieske (1977), Lieske (1987) derived secular accelerations from the use of old eclipses. He found no accelerations within the errors, which could suggest that the tides raised in Jupiter by the Galilean moons counterbalance the tides raised by Jupiter on Io. Using Sampson-Lieske's orbital theory with different observational sets, subsequent authors found discordant results (i.e., Vasundhara et al. 1996 and Aksnes and Franklin 2001).

Comparing Sampson-Lieske theory to numerical simulation, Avdyushev (2004) and Lainey et al. (2004) showed that the internal precision of Sampson-Lieske theory was of the order of a few hundred of kilometers. Such lack of precision compared to the tidal acceleration effect was unexpected and attributed to the lack of long period terms in the model. This probably explains the disagreement in the quantification of orbital accelerations from astrometry in the past. Using a full numerical model to compute the orbital evolution of the Galilean system, Lainey et al. (2009) considerably improved the internal accuracy on the satellite positions at the level of few hundred of meters, only. Moreover, tidal accelerations were not derived from a quadratic fit on astrometric residuals, but obtained directly as a consequence of modeling the lag of the tidal potential in the dynamics. This allowed the authors to obtain independently the tidal dissipation in both Io and Jupiter. Considering a set of astrometric observations covering the period (1891–2007) (and neglecting over such a short time span the influence of tidal dissipation in Europa and Ganymede on the Galilean orbits), they found $(k_2/Q)_{\text{Io}} = 0.015 \pm 0.003$ and $(k_2/Q)_{\text{J}} = (1.102 \pm 0.203) \times 10^{-5}$ for Io and Jupiter, respectively.

Since the three innermost Galilean moons are locked in the Laplace resonance, strong correlations arise when dissipation is considered in Europa or Ganymede. Hence, this method may not provide strong constraints for tidal dissipation in Europa, unless tidal dissipation in Io is fixed. Using very precise gravity field determinations and altimetry an orbiter around Europa may provide better options to derive Europa's rate of dissipation by detecting the lag angle of the tidal response. Alternatively, a radio beacon at the surface might be able to transmit the tidal signal with sufficient accuracy. Both methods are technically challenging, mainly because of the harsh radiation environment at Europa and the resulting relatively short duration of measurements.

3.3.2 The Saturnian System

Mimas is the only Saturnian satellite that has been suspected to accelerate when comparing its expected position from orbital model to astrometric observations. An anomalous acceleration of about $2^\circ\ \mathrm{cy}^{-2}$ was found successively by Kozai (1957) and Dourneau (1987). A few years after Dourneau's result, Vienne et al. (1992) using a more accurate orbital theory of the Saturnian satellites proved that the anomalous acceleration of Mimas was most probably a long-period term that was missing in the former orbital models. No other accelerations have been reported recently among the Saturnian system.

4 Energy Sources

Many different forms of energy play a role in the evolution of satellites. However, their significance may vary considerably for individual objects and with time. In the following, the energy sources and their corresponding internal heat production, the main driver for internal processes, are described.

4.1 Nuclear Energy

Nuclear energy can be released by the decay of unstable isotopes in the rock component of planets and satellites. The energy available due to the mass defect is—apart from the small amount of neutrino energy—eventually converted into heat within the rocky material. Whereas radioactive isotopes of uranium, thorium, and potassium (Table 3) with half-lives of the order of 10^9 years are important throughout the evolution, three isotopes—^{26}Al, ^{60}Fe, and ^{53}Mn—having half-lives of the order of million years, can be relevant during the satellites' accretion and possibly during the very early stages of evolution. With the given mass m_{sil} of the silicate component of a specific satellite the total time dependent heat production rate $\dot{E}_{\mathrm{rad}}$ due to long-lived radioactive isotopes in watts is calculated from

$$\dot{E}_{\mathrm{rad}} = m_{\mathrm{sil}} \sum_{i=1}^{4} c_i C_i H_i \exp\left[\lambda_i (t - t_{\mathrm{pr}})\right], \tag{21}$$

where i stands for the four long-lived isotopes. C is the abundance of uranium, thorium, potassium typical for a specific class of meteorites and has to be assumed according to Tables 4 or 5. From the half-lives (Table 3) the decay constants can be calculated by $\lambda_i = \ln 0.5/\tau_i$. t is the time variable, starting with $t = 0$ at the time of the satellites' formation and $t_{\mathrm{pr}} = 4.6 \times 10^9$ yr is the present time. In Fig. 2 the total heat production rate per unit mass, $\dot{E}_{\mathrm{rad}}/m_{\mathrm{sil}}$ is shown as a function of time for selected types of meteorites. Variation due to the different elemental abundances can reach a factor of about 1.7 with the ordinary LL chondrites (high) and the carbonaceous CI chondrites (low) being the most extreme cases. Representative present-day heat production rates per unit mass are about $3.45\ \mathrm{pW\,kg^{-1}}$ (CI), $4.55\ \mathrm{pW\,kg^{-1}}$ (EH), $5.20\ \mathrm{pW\,kg^{-1}}$ (CO), and $5.75\ \mathrm{pW\,kg^{-1}}$ (LL). The obtained present-day total heat production rates of the satellites, accounting for the silicate mass fraction, are summarized in Table 6.

Table 3 Half-lives τ, present (for long-lived isotopes) and initial (for short-lived isotopes) composition c, and heat release H for long-lived isotopes and—separated by a horizontal line—short-lived isotopes (Audi et al. 1997; Lodders and Fegley 1998; Spohn 1997; Castillo-Rogez et al. 2007)

i	Isotope	τ, years	c, kg kg^{-1}	H, W kg^{-1}
1	^{238}U	4.468×10^9	0.992745	9.48×10^{-5}
2	^{235}U	0.7038×10^9	7.2×10^{-3}	5.69×10^{-4}
3	^{232}Th	14.05×10^9	1.0	2.69×10^{-5}
4	^{40}K	1.277×10^9	1.17×10^{-4}	2.92×10^{-5}
5	^{26}Al	0.740×10^6	$^{26}\mathrm{Al}/^{27}\mathrm{Al} = 5 \times 10^{-5}$	0.146
6	^{60}Fe	1.5×10^6	$^{60}\mathrm{Fe}/^{56}\mathrm{Fe} = 0.1\text{–}1 \times 10^{-6}$	0.068–0.074
7	^{53}Mn	3.74×10^6	$^{53}\mathrm{Mn}/^{55}\mathrm{Mn} = 1\text{–}4 \times 10^{-5}$	0.027

Table 4 Elemental abundance of radionuclides and densities of carbonaceous chondrites (Lodders and Fegley 1998)

i	Element	CI	CM	CV	CO	CK	CR	CH
1,2	U, ppb	8	12	17	18	15	13	
3	Th, ppb	29	41	58	80	58	42	
4	K, ppm	550	370	360	360	290	315	200
5	Al, wt.%	0.865	1.13	1.68	1.40	1.47	1.15	1.05
6	Fe, wt.%	18.2	21.3	23.5	25.0	23.0	23.8	38.0
7	Mn, ppm	1940	1650	1520	1620	1440	1660	1020
	ρ, kg m^{-3}	2230	2710	3420	3630	3400	3270	4200

Table 5 Elemental abundance of radionuclides and densities of ordinary and enstatite chondrites (Lodders and Fegley 1998)

i	Element	H	L	LL	R	Acap.	K	EH	EL
1,2	U, ppb	13	15	15	<25			9.2	7.0
3	Th, ppb	38	42	47	<50			30	38
4	K, ppm	780	920	880	780	475	710	840	700
5	Al, wt.%	1.06	1.16	1.18	1.06	1.2	1.3	0.82	1.00
6	Fe, wt.%	27.2	21.75	19.8	24.4	23.5	24.7	30.5	24.8
7	Mn, ppm	2340	2590	2600	2960	3000	2400	2120	1580
	ρ, kg m^{-3}	3800	3600	3550	3600	3700	3700	3670	3580

4.2 Potential Energy

The potential energy U of a radially symmetric mass distribution $M(r)$ is obtained by integrating $-(GM(r)/r)dm$ over all mass elements dm or, equivalently, over the volume V if the density distribution is given:

$$U = -\int_V \frac{GM(r)}{r} \rho(r) dV. \tag{22}$$

Fig. 2 Specific heat production rates due to long-lived isotopes for different classes of meteorites (Carbonaceous CI and CO-chondrites, ordinary LL-chondrites and enstatite EH-chondrites) as a function of time. Present-day values are obtained after 4.6 Ga

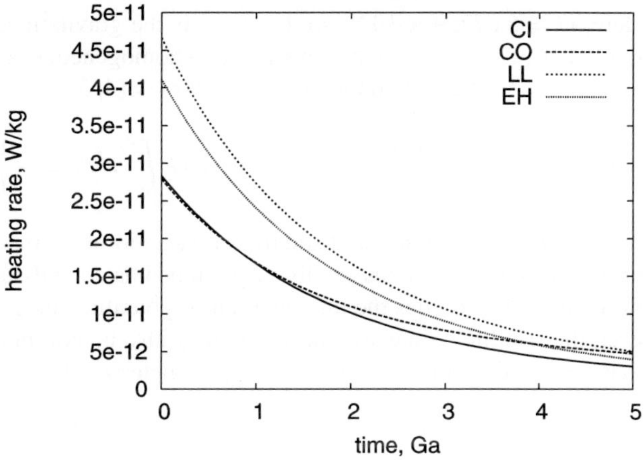

Table 6 Rock-mass fraction M_{rock} (including metal) and heat production rates $\dot{E}_{\mathrm{rad}}$ due to long-lived isotopes for the largest satellites. Io's density of about 3529 kg m^{-3} is assumed as a typical value for the density of rock. The ranges correspond to the heat production rates assuming a CI and LL-chondritic composition, respectively. Other types of chondrites range between these two. In case of Io, the CI-chondrite would be inconsistent with the satellite's mean density. Numbers are given for the initial (t_0) and present-day (t_{pr}) heating rates, respectively

Satellite	M_{rock}, wt.%	$\dot{E}_{\mathrm{rad}}(t_0)$, W	$\dot{E}_{\mathrm{rad}}(t_0)/V$, W m^{-3}	$\dot{E}_{\mathrm{rad}}(t_{\mathrm{pr}})$, W	$\dot{E}_{\mathrm{rad}}(t_{\mathrm{pr}})/V$, W m^{-3}
Io	100	$(2.50–4.15) \times 10^{12}$	$\sim 1.31 \times 10^{-7}$	$(3.08–5.14) \times 10^{11}$	$\sim 1.62 \times 10^{-8}$
Europa	92	$(1.24–2.05) \times 10^{12}$	$\sim 1.03 \times 10^{-7}$	$(1.52–2.54) \times 10^{11}$	$\sim 1.27 \times 10^{-8}$
Ganymede	67	$(2.78–4.62) \times 10^{12}$	$\sim 4.84 \times 10^{-8}$	$(3.43–5.71) \times 10^{11}$	$\sim 5.98 \times 10^{-9}$
Callisto	64	$(1.93–3.20) \times 10^{12}$	$\sim 4.38 \times 10^{-8}$	$(2.38–3.96) \times 10^{11}$	$\sim 5.41 \times 10^{-9}$
Mimas	18	$(1.92–3.18) \times 10^{8}$	$\sim 7.81 \times 10^{-9}$	$(2.36–3.93) \times 10^{7}$	$\sim 9.65 \times 10^{-10}$
Enceladus	53	$(1.63–2.71) \times 10^{9}$	$\sim 3.24 \times 10^{-8}$	$(2.01–3.35) \times 10^{8}$	$\sim 4.00 \times 10^{-9}$
Tethys	<0.1	$(1.04–1.73) \times 10^{7}$	$\sim 2.18 \times 10^{-11}$	$(1.28–2.14) \times 10^{6}$	$\sim 2.70 \times 10^{-12}$
Dione	45	$(1.39–2.30) \times 10^{10}$	$\sim 2.48 \times 10^{-8}$	$(1.71–2.85) \times 10^{9}$	$\sim 3.07 \times 10^{-9}$
Rhea	27	$(1.75–2.90) \times 10^{10}$	$\sim 1.24 \times 10^{-8}$	$(2.15–3.59) \times 10^{9}$	$\sim 1.53 \times 10^{-9}$
Titan	65	$(2.45–4.07) \times 10^{12}$	$\sim 4.56 \times 10^{-8}$	$(3.02–5.03) \times 10^{11}$	$\sim 5.63 \times 10^{-9}$
Iapetus	12	$(0.61–1.01) \times 10^{10}$	$\sim 4.85 \times 10^{-9}$	$(0.75–1.25) \times 10^{9}$	$\sim 5.99 \times 10^{-10}$
Miranda	23	$(4.25–7.06) \times 10^{8}$	$\sim 1.03 \times 10^{-8}$	$(5.24–8.73) \times 10^{7}$	$\sim 1.27 \times 10^{-9}$
Ariel	56	$(2.12–3.52) \times 10^{10}$	$\sim 3.47 \times 10^{-8}$	$(2.61–4.35) \times 10^{9}$	$\sim 4.28 \times 10^{-9}$
Umbriel	40	$(1.31–2.18) \times 10^{10}$	$\sim 2.08 \times 10^{-8}$	$(1.61–2.69) \times 10^{9}$	$\sim 2.57 \times 10^{-9}$
Titania	58	$(5.73–9.52) \times 10^{10}$	$\sim 3.71 \times 10^{-8}$	$(0.71–1.18) \times 10^{10}$	$\sim 4.58 \times 10^{-9}$
Oberon	54	$(4.55–7.56) \times 10^{10}$	$\sim 3.27 \times 10^{-8}$	$(5.61–9.35) \times 10^{9}$	$\sim 4.04 \times 10^{-9}$
Triton	72	$(4.31–7.16) \times 10^{11}$	$\sim 5.54 \times 10^{-8}$	$(5.32–8.86) \times 10^{10}$	$\sim 6.84 \times 10^{-9}$
Charon	60	$(2.72–4.52) \times 10^{10}$	$\sim 3.90 \times 10^{-8}$	$(3.35–5.59) \times 10^{9}$	$\sim 4.82 \times 10^{-9}$

Here, $G = 6.67259 \times 10^{-11}$ m^3 kg^{-1} s^{-2} is the gravitational constant, r the distance from the center of mass, and ρ the density. For a homogeneous ($\rho = $ const.) sphere with radius R, and by using spherical coordinates and $M(r) = (4/3)\pi r^3 \rho$ we obtain

$$U_{\rm h} = -4\pi\rho \int_0^R \frac{GM}{r} r^2 dr = -\frac{16}{3}\pi^2\rho^2 G \int_0^R r^4 dr = -\frac{16}{15}\pi^2\rho^2 G R^5 = -\frac{3GM^2}{5R}. \quad (23)$$

Accretion The formation of satellites by accretion of smaller bodies releases gravitational energy. If a homogeneous satellite is assumed as an end-state of the accretional processes, Equation (23) provides the total amount of potential energy available in the satellite. However, the satellite's state after accretion may also be non-homogeneous. The energy per unit mass released during accretion is thus parameterized by

$$E_{\rm ac} = \chi GM/R, \quad (24)$$

where χ is a factor of order unity accounting for different accretion processes (Schubert et al. 1986). In case of a homogeneous end-state $\chi = 3/5$. In that case all the energy released due to later differentiation will be available in the satellite's interior. In general, however, part of this energy will heat the satellite's interior while the other part is radiated into space. The latter fraction is not well-constrained and the amount of heat released during accretion is uncertain. In contrast to later stages of evolution the near-surface temperature is high compared to the temperature in the deep interior during and shortly after accretion (Schubert et al. 1986).

Differentiation Energy of differentiation will be released by core formation and/or the separation of other distinct layers which differ in density (e.g., HP-ice, liquid water, ice-I). For the two-layer case (core radius and density R_1 and ρ_1, respectively; mantle radius $R_2 = R$ and density ρ_2, respectively) the total potential energy $U_{\rm d}$ is that of the core (23)

$$U_1 = -\frac{16}{15}\pi^2\rho_1^2 G R_1^5, \quad (25)$$

plus that of the spherical mantle shell. The latter is—analogous to the integration above (22)—given by

$$U_2 = -\frac{16}{3}\pi^2 G \left(\frac{\rho_1\rho_2 R_1^3}{2}(R_2^2 - R_1^2) + \frac{\rho_2^2}{5}(R_2^5 - R_1^5) - \frac{\rho_2^2 R_1^3}{2}(R_2^2 - R_1^2) \right). \quad (26)$$

The mass distribution is in this case given by $M(r) = 4/3\pi\rho_1 r^3$ for $0 \le r \le R_1$, and $M(r) = 4/3\pi(\rho_1 R_1^3 + \rho_2(r^3 - R_1^3))$ for $R_1 \le r \le R_2$. The potential energy of a two-layer satellite is then given by $U_{\rm d} = U_1 + U_2$. The energy released by transition from the homogeneous to the completely differentiated two-layer state is thus given by

$$U_{\rm pot} = U_{\rm h} - U_{\rm d} = -\frac{16}{15}\pi^2\rho^2 G R_2^5 - (U_1 + U_2), \quad (27)$$

where $\rho = 3M/(4\pi R_2^3)$ is the mean density of the satellite with total mass M.

In Table 7 the potential energies for homogeneous and two-layer differentiated satellites are compared. The differentiated states assume densities of 1000 and 3500 kg m^{-3} for the outer ice-layer and rock-core, respectively. Exceptions are Io (3275 and 5150 kg m^{-3}), the large icy satellites—Ganymede, Callisto and Titan—for which we assume an ice density of

Table 7 Potential energy U_h and U_d for the homogeneous (23) and the two-layer differentiated (27) case. The heating rate due to differentiation $\dot{E}_{diff}$ is estimated assuming a time scale of 100 Myr

Satellite	U_h, J	U_d, J	$\dot{E}_{diff}$, W	$\dot{E}_{diff}/V$, W m^{-3}
Io	1.75×10^{29}	1.81×10^{29}	1.70×10^{12}	6.72×10^{-8}
Europa	5.91×10^{28}	6.35×10^{28}	1.41×10^{12}	8.85×10^{-8}
Ganymede	3.34×10^{29}	4.66×10^{29}	4.20×10^{13}	5.50×10^{-7}
Callisto	1.92×10^{29}	2.05×10^{29}	3.89×10^{12}	6.64×10^{-8}
Mimas	2.92×10^{23}	3.04×10^{23}	3.97×10^{6}	1.22×10^{-10}
Enceladus	1.92×10^{24}	2.11×10^{24}	5.92×10^{7}	8.82×10^{-10}
Tethys	2.89×10^{25}	2.91×10^{25}	5.43×10^{7}	8.56×10^{-11}
Dione	8.62×10^{25}	9.66×10^{25}	3.27×10^{9}	4.40×10^{-9}
Rhea	2.80×10^{26}	2.89×10^{26}	3.00×10^{9}	1.61×10^{-9}
Titan	2.81×10^{29}	3.06×10^{29}	7.85×10^{12}	1.10×10^{-7}
Iapetus	1.78×10^{26}	1.88×10^{26}	3.15×10^{9}	1.89×10^{-9}
Miranda	7.40×10^{23}	8.04×10^{23}	2.05×10^{7}	3.73×10^{-10}
Ariel	1.26×10^{26}	1.45×10^{26}	6.01×10^{9}	7.40×10^{-9}
Umbriel	9.37×10^{25}	1.06×10^{26}	3.97×10^{9}	4.75×10^{-9}
Titania	6.32×10^{26}	7.35×10^{26}	3.24×10^{10}	1.58×10^{-8}
Oberon	4.76×10^{26}	4.50×10^{26}	2.33×10^{10}	1.26×10^{-8}
Triton	1.36×10^{28}	1.53×10^{28}	5.65×10^{11}	5.45×10^{-8}
Charon	1.74×10^{26}	2.02×10^{26}	9.11×10^{9}	9.82×10^{-9}

1300 kg m^{-3}, Rhea (1000 and 1465 kg m^{-3}) and Tethys (quasi homogeneous assuming 917 and 1000 kg m^{-3}). In cases where there are constraints available for the moment of inertia values—Io, Europa, Ganymede, Callisto, Rhea—these are matched by adjusting the core sizes accordingly. This procedure fails for Ganymede which at present is highly differentiated and cannot be described by a two-layer model. For Ganymede and also for Europa and other satellites, which presumably are further differentiated into three or more layers this estimate may thus represent an earlier stage of evolution.

4.3 Kinetic Energy

Kinetic energy plays a major role in many processes inside icy moons. In the early stages of evolution the kinetic energy of impactors is converted into heat. The resulting (near-) surface temperature is much higher as compared to the internal temperature in this phase. This changes when the radiogenic heating from the deep interior becomes the dominant heat source after accretion has ceased.

Flows of material—be it solid ice and rock or liquid water—transport heat by convection from the interior to the surface. In the case of electrically conducting material these processes can generate magnetic fields. The intrinsic field of Ganymede, generated in the satellite's liquid iron core, is an important example of such motion on macroscopic scales in satellite interiors.

4.3.1 Rotation

A satellite's energy of rotation is given by

$$E_{\text{rot}} = \frac{1}{2} C_s \Omega^2 = \frac{1}{5} m_s R_s^2 \Omega^2, \tag{28}$$

where C_s is the polar moment of inertia and Ω the rotation rate. At the right hand side the homogeneous case $C_s = 2/5 m_s R_s^2$ has been assumed. In that approximation the rate of energy change is given by

$$\dot{E}_{\text{rot}} = \frac{2}{5} m_s R_s^2 \Omega \dot{\Omega}. \tag{29}$$

After accretion, the rotation rates of the satellites are decreasing due to tidal dissipation in the satellites' interiors. Assuming an initial rotation period of, e.g. 10 h, the timescales for despinning can be calculated from (16). The results for the despinning timescales and the corresponding heating rates are shown in Table 8. The initial rotation period is unknown and the dissipative response (k_2/Q) may not be constant. However, the values show that despinning to the synchronous state can significantly contribute to the energy budget of satellites early in their histories. The volumetric heating rates are high especially for Io, Europa, Ganymede, Tethys, Ariel and Triton. An extreme case is Iapetus where the heating rates are very small and the despinning timescale is of the order of the age of the solar system, presumably indicating that Iapetus' k_2/Q (time averaged) was lower (for details, see Schubert et al. 2010, this issue).

4.4 Energy of Deformation

As discussed above, satellites are distorted mainly due to rotation and due to external forces, i.e. tides. Whereas the former changes only on very long-timescales and leads to a quasi-static deformation, the frequency of the forcing for satellite tides is of the order of days or weeks. This can cause dynamical deformations of the satellite that are—depending on the planetary material and the frequency of the forcing—partly elastic (energy conserved) and partly inelastic (dissipation of energy). Internal friction involved with the latter can significantly contribute to the energy balance of satellites, that are on eccentric orbits, close to their primary planets. In the following we will discuss the relevance of inelastic tidal deformation of the solid part (mainly ice and rock) of a satellite for internal processes. Tyler (2008) has suggested that substantial dissipation within an ocean on Europa can occur due to the resonant excitation of Rossby waves. However, in the following we will focus on solid-body tides, only.

4.4.1 Inelastic Deformation: Tidal Dissipation Processes in Icy Moons

Most of the icy moons are subjected to a periodic variation of gravitational forces along their orbit about their central planet, owing to the orbital eccentricity. The materials that compose the moon's interior deform according to these periodic force variations. For a layered satellite, the response of each internal layer is determined by the thermo-mechanical properties of the dominating material (e.g., liquid iron/iron sulfide for the core, olivine for the silicate mantle, liquid water for the ocean, high-pressure phase layer and hexagonal ice-I for the icy shells). Owing to the anelastic properties of the materials in their interiors, part of the energy involved in the tidal deformation is dissipated into heat. As detailed below,

Table 8 Despinning time-scales τ (17), despinning heating rates $\dot{E}_{\rm rot}$ assuming an initial period of 10 h, volumetric despinning heating rate $\dot{E}_{\rm rot}/V$, present-day tidal heating rate $\dot{E}_{\rm sync}$ in synchronous rotation and present-day volumetric tidal heating rate $\dot{E}_{\rm sync}/V$ of the largest satellites. Homogeneous density distributions and for $\dot{E}_{\rm sync}$ Maxwell rheologies are assumed. For the rocky moons (only Io and Europa) and for the icy moons elastic rigidities of $\mu = 65$ GPa and $\mu = 3.5$ GPa, respectively, are assumed. For the viscosity we use 10^{16} Pa s and 10^{15} Pa s for rock and ice, respectively. Such low rock-viscosities are applicable only for 'Io-like' partially molten rock which is assumed here also for Europa. Furthermore, the entire volume of the satellite is assumed to dissipate. The obtained dissipation rates should therefore be regarded as possible maximum values

Satellite	τ, yr	$\dot{E}_{\rm rot}$, W	$\dot{E}_{\rm rot}/V$, W m^{-3}	$\dot{E}_{\rm sync}$, W	$\dot{E}_{\rm sync}/V$, W m^{-3}
Io	1400	3.9×10^{16}	1.53×10^{-3}	2.7×10^{13}	1.1×10^{-6}
Europa	42,000	5.4×10^{14}	3.36×10^{-5}	2.5×10^{12}	1.6×10^{-7}
Ganymede	21,000	9.2×10^{15}	1.20×10^{-4}	2.8×10^{11}	3.6×10^{-9}
Callisto	830,000	1.5×10^{14}	2.48×10^{-6}	1.6×10^{11}	2.7×10^{-9}
Mimas	1100	2.1×10^{11}	6.34×10^{-6}	1.5×10^{9}	4.7×10^{-8}
Enceladus	2750	4.5×10^{11}	6.63×10^{-6}	2.1×10^{8}	3.1×10^{-9}
Tethys	4200	7.8×10^{12}	1.22×10^{-5}	1.8×10^{6}	2.8×10^{-12}
Dione	12,000	5.5×10^{12}	7.41×10^{-6}	6.4×10^{8}	8.6×10^{-10}
Rhea	61,000	4.2×10^{12}	2.25×10^{-6}	1.1×10^{8}	5.7×10^{-11}
Titan	591,000	2.9×10^{14}	4.07×10^{-6}	4.5×10^{12}	6.3×10^{-8}
Iapetus	7.8×10^{9}	2.4×10^{7}	1.44×10^{-11}	5.3×10^{5}	3.2×10^{-13}
Miranda	4900	1.3×10^{11}	2.40×10^{-6}	4.8×10^{6}	8.8×10^{-11}
Ariel	7100	1.2×10^{13}	1.48×10^{-5}	4.1×10^{8}	5.1×10^{-10}
Umbriel	65,000	1.2×10^{12}	1.41×10^{-6}	4.5×10^{8}	5.4×10^{-10}
Titania	599,000	7.1×10^{11}	3.44×10^{-7}	2.2×10^{7}	1.1×10^{-11}
Oberon	3.9×10^{6}	8.5×10^{10}	4.62×10^{-8}	4.4×10^{6}	2.4×10^{-12}
Triton	34,300	2.2×10^{14}	2.12×10^{-5}	1.3×10^{7}	2.4×10^{-12}
Charon	284,000	4.0×10^{11}	4.32×10^{-7}	5.0×10^{7}	5.4×10^{-11}

several kinds of dissipation processes can occur within a satellite. The total energy that can be generated by these dissipation processes depend on both the orbital configuration and the mechanical properties of each internal layer. In some circumstances, the tide-generated energy can exceed all the other sources of energy and induce a thermal runaway.

Viscous friction in convective layers Each layer that composes the satellite's interior deforms differently depending on its viscoelastic properties. The amplitude of deformation in liquid layers (liquid metallic core or water ocean) is large as there is little resistance to shear motions except at boundary layers. In contrast, the deformation of cold rigid lithosphere is small as they possess a high resistance to shear motions. In the liquid case, the tidal motion is large, but as the resistance is extremely low, little viscous dissipation occurs. In the rigid case, the resistance is so large that viscous motions are prohibited, and the deformation is only determined by the elastic properties of the layer. Anelastic deformation can contribute to the total motion, but only if the solid material has sufficient time to deform as a viscous body. This can occur for forcing periods close and beyond the Maxwell time, which is the ratio between the viscosity of the materials and the rigidity.

The global tidal heating rate in synchronous rotation on a slightly eccentric orbit $\dot{E}_{\text{sync}}$ can be calculated by (Segatz et al. 1988)

$$\dot{E}_{\text{sync}} = -\frac{21 R_s^5 n^5 e^2}{2G} \text{Im}(k_2), \tag{30}$$

where besides previously defined quantities, $\text{Im}(k_2)$ is the imaginary part of the second degree potential Love number. This formulation is independent of specific rheology. However, to calculate $\text{Im}(k_2)$ assumptions as to the rheological model (e.g., Maxwell or alternative models) and on the density distribution of the satellite are required. Assuming a Maxwell body the rigidity μ_c is a complex number with real and imaginary part given by

$$\text{Re}(\mu_c) = \eta^2 n^2 \mu / (\mu^2 + \eta^2 n^2) \quad \text{and} \quad \text{Im}(\mu_c) = \eta n \mu^2 / (\mu^2 + \eta^2 n^2), \tag{31}$$

where μ and η are the elastic rigidity, and the viscosity of the body, respectively. In the homogeneous case the complex degree-two Love number is given by

$$k_2 = 1.5 / \left(1 + \frac{19}{2} \frac{\mu_c}{\rho g R_s}\right), \tag{32}$$

where g is the gravitational acceleration at the surface. The factor $3/2$ is the Love number in the fluid limit. Tidal heating rates for the homogeneous case can thus be obtained from (30) by inserting (32). Results for the major satellites are summarized in Table 8. Because of the approximations (homogeneous satellite) these are only very rough estimates of the tidal heating rates. The real values can be considerably lower especially for larger viscosities (as is the case for many icy satellites, e.g. for Mimas) or if the dissipating volume is only a small fraction of the satellite. Additionally, the presence of liquid layers would enhance the deformation of the overlying shells significantly. Nevertheless, the estimates are useful to generally compare possible maximum tidal heating rates.

On microscopic scales, the viscous mechanisms are mainly related to the motions of defects and their interaction with the crystal lattice, resulting in friction. During the first quarter of a tidal cycle, strain energy is progressively stored as compressive (or tensile) stress increases. During the second quarter, strain energy is restored as stress decreases. The same process operates during the other half period when stress is tensile (or compressive). The back-and-forth motions of crystal defects (or atomic and molecular diffusion) induced by alternating compression and tension convert part of the strain energy into heat. The efficiency of such dissipative processes depends on the population of existing defects (or dislocations) and therefore is very sensitive to the density of existing dislocations within the crystal lattice (e.g. Cole et al. 1998). From a macroscopic point of view, the anelastic behavior of a material sample results in a time delay of the material response relative to tidal forcing, in a smaller effective shear modulus and in the production of heat. As the mobility of crystal defects is strongly temperature-dependent, the amount of dissipated energy at a given frequency strongly varies within a convective layer, which may be characterized by strong lateral and radial temperature contrast. Different models are proposed in the literature to describe the temperature dependence of dissipation processes. The Maxwell model is classically used to describe the viscoelastic response of both ice and silicate mantles (e.g. Ojakangas and Stevenson 1989a; Sohl et al. 1995; Hussmann et al. 2002; Moore 2003; Tobie et al. 2005a). It is adapted to describe material relaxation properties when the forcing period is close to their Maxwell time, but it fails to quantify dissipation processes over a wide range of frequencies and temperature (Sotin et al. 2009). Other models, based on laboratory measurements (e.g., Tatibouet et al. 1986; Cole 1995 on cyclic loading) or geophysical measurements (e.g., Reeh et al. 2003 on tidal bending of polar glaciers) are probably more appropriate for describing the viscoelastic response over a wide range of temperature (see also Durham et al. 2010, this issue). Whatever the rheological models, significant viscous dissipation is expected to occur within convective mantles for temperature ranging between 200–270 K for icy mantles and 1000–1600 K for silicate mantles.

Even if viscous dissipation in a water ocean might be negligible, the presence of an ocean strongly increases the deformation amplitude of the above ice shell, and hence the amount of dissipated energy. In a convective ice shell on Europa, owing to the decoupling effect of the ocean, the tidal heating rate can reach values between 10^{-6} and 10^{-5} W m^{-3}, which is more than one hundred times the volumetric radiogenic heating rate in the silicate mantle (e.g. Tobie et al. 2003). The dissipation rate could vary by one order of magnitude between cold and hot plumes, and the localization of maximum heating rate (in hot or cold plumes) is very sensitive to the rheological parameters (Tobie et al. 2003; Mitri and Showman 2005). On Enceladus, the viscous dissipation rate at the south pole can reach values of the order of 10^{-7} and 10^{-6} W m^{-3} if an internal ocean is present (Tobie et al. 2008). On Titan, solid tidal dissipation is expected to be much smaller (Sohl et al. 1995), but may still reach values comparable to radiogenic heating (of the order of 10^{-8} W m^{-3}) (Tobie et al. 2005b). On Io, the dissipation rate in the silicate mantle must reach values of at least 10^{-5} W m^{-3} in order to explain the observed surface heat flow (Segatz et al. 1988). On Europa, the dissipation rate in the silicate mantle is highly uncertain, but could reach values of roughly 10^{-7} W m^{-3} if the mantle is as hot and dissipative as Io's mantle (see also Moore and Hussmann 2009).

Shear friction along faults Localized deformation along faults within the cold icy lithosphere can also provide a significant source of dissipation. This process may be important for satellites that exhibit large-scale faulting and are subjected to large tidal deformations (e.g. on Europa, on Enceladus, and possibly on Triton). By applying a shear deformation model initially developed for Earth applications (Yuen et al. 1978), Nimmo and Gaidos (2002) proposed that shear heating owing to tidally-induced strike-slip motion along fractures could locally raise the temperature, possibly up to the melting point at shallow depths. Dissipation in this model is also related to viscous deformation in the warmer part of the layer, but in this case, the rate of deformation is locally controlled by the displacements along the faults, which is very high near the faults. This model initially adapted to Europa has been more recently applied to Enceladus (Nimmo et al. 2007) and Triton (Prockter et al. 2005). If faults are subjected to large shear displacements (of the order of 0.1–0.5 m per day), the temperature could locally rise up to the melting point of water ice at the base of the faults. Such large shear velocities can occur only if the lithosphere is subjected to very large tidal deformations, implying the existence of a decoupling liquid layer a few tens of kilometers below the surface. In their model, Nimmo and Gaidos (2002) prescribed the displacement rate, independently of the amount of work that might be needed to create the shear motions. Preblich et al. (2007) and Smith-Konter and Pappalardo (2008) showed that tidal displacements along tectonic faults are smaller than initially anticipated by Nimmo and Gaidos (2002). This mainly follows from the fact that the shear velocity rates prescribed by Nimmo and Gaidos (2002) are valid for two sliding rigid blocks above a totally fluid layer. It does not take into account the resistance due to friction between the two rigid blocks and the resistance due to the viscous ice layer below, which would tend to reduce the shear motion rate. Despite these limitations, shear heating along tidally-forced faults remains a viable mechanism, but is probably overestimated in existing models. Further modeling efforts are required to better quantify this process.

4.5 Latent Heat and Chemical Energy

Phase Transitions Depending on internal temperature and pressure conditions the state of aggregation of the planetary material may change during satellite evolution. For icy satellites the crystallization of water and the associated release of latent heat is the most relevant process. To melt the mass m of ice, the latent heat of $\Delta H = Lm$ must be provided. $L =$ 334 kJ/kg is the latent heat of fusion of water. Latent heat released by crystallization of liquid H_2O can significantly contribute to the internal energy budget. If we for instance assume an H_2O-layer thickness of Europa of 150 km, the latent heat released by solidification of the entire layer would amount to 1.4×10^{27} J. Assuming a timescale of 100 Myr this would imply a heating rate of 4.4×10^{11} W, comparable to the present radiogenic heating rate. With a larger amount of water the corresponding values for Ganymede are 1.80×10^{28} J and 5.69×10^{12} W. For likely situations in icy satellites the latent heat energy would be comparable to the product of heat capacity and the temperature change.

For large satellites which may contain liquid water layers, melting and freezing of water can be relevant for both the ocean floor, where high-pressure ices form and the top of the ocean where ice-I forms. Other phase transitions, e.g., the transition from one phase of high-pressure ice to another (depending on temperature) may release additional energy .

Exogenic Reactions Energy can be provided by chemical reactions. An important example for a possible reaction in icy satellites is the process of serpentinization (see also Sohl et al. 2010, this issue). Provided that liquid water is available in icy moons and that a rock-ocean interface exists, for instance on Europa, serpentine may be formed from olivine by rock-water interactions, e.g.

$$\text{forsterite} + \text{water} \leftrightarrow \text{brucite} + \text{serpentine}, \tag{33}$$

$$2Mg_2SiO_4 + 3H_2O \leftrightarrow Mg(OH)_2 + Mg_3Si_2O_5(OH)_4. \tag{34}$$

The serpentine-producing reaction is exogenic and provides an average enthalpy of $\Delta H =$ -83 kJ per mole serpentine. The efficiency of the reaction will depend on temperature and pressure conditions and mainly on the percolation depth of the water into the rock. The latter can be estimated by applying a model of micro-cracking to the rock-layer of the icy satellites at the specific temperature and pressure conditions. Vance et al. (2007) have found that the heat flux due to serpentinization is of the order of 1% of the heat flux due to radiogenic heating. However, it may be significant especially for small icy moons where the percolation depth is greater (as compared to large icy moons) because of the smaller pressures. Cracks can penetrate deeper and the rock-volume available for the above reaction is correspondingly larger. For a completely differentiated Enceladus, for instance, the cracking-depth could reach down to the center of the moon. This could imply around 6% of heating due to serpentinization as compared to radiogenic heating, i.e. a quite significant contribution (Vance et al. 2007), if the core was initially ultramafic.

5 Heat Transfer in Icy Moons

Heat within a satellite interior is mainly transported by thermal diffusion and thermal convection. In the present section, we present the basic concepts of thermal diffusion and thermal convection and discuss the conditions for which each of these two processes is predominant in the context of icy moons. In addition to thermal diffusion and convection, vapor and/or cryomagma emission can also significantly contribute to the heat transfer at shallow depths, in some particular circumstances. These specific processes are discussed in more detail in other contributions of this volume and will not be further detailed here. Finally, coupling between tidal heating and heat transfer is also a key process in icy moons.

5.1 Conductive Transfer Within Icy Moons

Thermal conduction is the dominant mechanism for heat transport where the viscosity is large enough to inhibit the initiation of convective motions. Under these conditions, the efficiency of heat transfer is mainly controlled by the thermal conductivity and heat capacity of the materials in the satellites' interiors.

Such stable conditions are reached in the smallest satellites, which remain cold during their entire history. Similar conditions are also reached within the inner part of mid-sized and large satellites just after the accretion: owing to the deposition of accretional energy, the outer part of a satellite is warmer than the inner part just after the accretion whereas the center has a low temperature (e.g. Schubert et al. 1986). As this kind of thermal structure is stable against thermal convection, heat transfer is dominated by conduction alone during the early stage of the satellite evolution. Finally, the upper part of the outer icy layer remains cold during most of the evolution, even on the largest satellites, and internal heat is mainly transferred through thermal diffusion toward the surface.

5.1.1 Thermal Diffusion in a Cold Ice-Rock Core

The conductive transfer of energy in a cold core consisting of a mixture of rock and ices can be described by the following time-dependent heat conduction equation for radial symmetry, where all materials parameters may be functions of temperature and pressure (e.g., Hillier and Squyres 1991):

$$\rho c_p \frac{\partial T}{\partial t} = \frac{k}{r}\frac{\partial^2 (rT)}{\partial r^2} + \frac{\partial k}{\partial r}\frac{\partial T}{\partial r} + \rho H(t,r), \tag{35}$$

where t is time, k is thermal conductivity, ρ is density, c_p is heat capacity, and $H(t,r)$ is the rate of heat production per unit mass.

For any initial temperature profile and any surface temperature, the above equation can be easily solved provided that conductivity, heat capacity, density and heating profiles are prescribed. For a mixture of silicate and ices, thermal conductivity and heat capacity can be estimated from the volume and mass fraction of each phase:

$$k = f_s k_s + (1 - f_s) k_i, \tag{36}$$

$$c_p = x_s c_{p,s} + (1 - x_s) c_{p,i}, \tag{37}$$

where f_s and x_s are the silicate volume and mass fraction of the mixture, respectively. Whereas (37) is exact, it should be noted that (36) is valid only for rock and ice units (layers) aligned in parallel to the direction of heat flow. Heat sources includes radioactive decay of isotopes in the silicate phase, tidal friction in the whole mixture, and possibly phase changes.

During the early stage of evolution, the excess energy deposited in the outer region during accretion is diffused toward the surface, resulting in a progressive cooling of the outer layer. In the inner portion, as thermal diffusion is not very efficient, the heat produced by radioactive decay cannot be entirely evacuated and the interior progressively warms up. At some point, as discussed in the next section, thermal convection can initiate or the melting point of ice can be reached. In this latter condition, a rapid differentiation of the interior can occur. For the smallest satellites, a temperature warm enough to initiate convection or to melt the interior is never reached, except if a significant amount of short-lived isotopes was initially present (e.g., Castillo-Rogez et al. 2007).

For the largest satellites, a segregation of rock and ice is very likely to occur, thus leading to the formation of a discrete rock core. A full segregation would result from a relatively hot start of the satellite involving melting of the ice component (e.g., Kirk and Stevenson 1987 in the case of Ganymede). Another mechanism may be a gradual but incomplete solid-state differentiation of the ice-rock mixture (e.g., Nagel et al. 2004 for Callisto). Just after differentiation, the temperature in the core is also too low to permit convective motions of rock. Heat is transferred by conduction only and the evolution of the temperature profile can be determined by (35) by setting a rock mass fraction of 1. As the rocky core is relatively cold during this diffusive stage, heating by tidal friction is negligible.

Although only the time integration of (35) can provide the exact solution of the conductive evolution of the core, an approximate solution can also be derived by assuming an initially isothermal core. Assuming that there is no conductive heat loss, the increase in core temperature as a function of time can be estimated from the rate of radiogenic heating $H_{0,i}e^{-\lambda_i t}$ for each unstable isotope (e.g., Kirk and Stevenson 1987; Grasset et al. 2000):

$$\Delta T_c(t) = \frac{1}{c_{p,c}} \sum_i H_{0,i} \frac{1 - e^{-\lambda_i t}}{\lambda_i}. \tag{38}$$

In fact, part of the energy is lost by thermal diffusion through a boundary region of width $\sim (\kappa_c t)^{1/2}$, where $\kappa_c = k_c/(\rho_c c_{p,c})$ is the thermal diffusivity of the core. The diffusive heat flux across this boundary is:

$$F_c = \frac{2k_c \Delta T_c}{\sqrt{\pi \kappa_c t}}. \tag{39}$$

As this conductive heat flow remains small relative to the heat production, the core warms and subsolidus convection becomes eventually possible. In the case of Titan and Ganymede, this diffusive stage lasts about 1.5 to 2 billion years (Kirk and Stevenson 1987; Tobie et al. 2006). The duration of the diffusive stage in the rocky core is mainly determined by the initial content in radiogenic elements and the degree of hydration of the silicates, which strongly influence the rock viscosity and hence the onset of convection (Tobie et al. 2006, Fig. 3).

5.1.2 Thermal Diffusion Through an Icy Shell

Even for satellites with very warm interiors (e.g. Enceladus, Europa, Titan etc.), a significant part of the internal heat is transported by thermal conduction through the last kilometers below the surface. The evolution of the temperature profile in this upper layer can be calculated from (35) by imposing a constant temperature at the surface and the heat flux coming from the inner part at the base. The only source of heating is provided by tidal friction. As tidal energy is only efficiently converted into heat for ice having a temperature larger than 200 K, this source of energy is only located in the warmer part of the conductive layer, i.e. at its base (e.g. Ojakangas and Stevenson 1989a; McKinnon 1999). As this source of energy remains small, it has no major impact on the

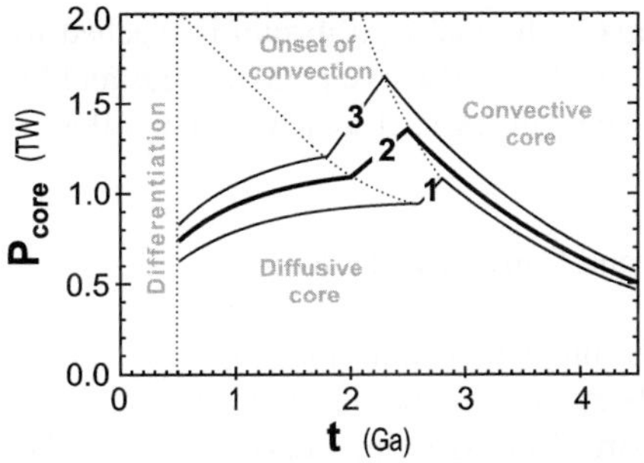

Fig. 3 Estimation of global heat power provided by the rocky core of Titan. Three different mineral compositions, reflecting different hydration states, have been used to estimate the possible evolution of the silicate core over a reasonable range: 1, $\rho_{sil} = 3000\ \mathrm{kg\,m^{-3}}$, $H_0 = 3 \times 10^{-11}\ \mathrm{W\,kg^{-1}}$, $T_i = 1400$ K; 2, $\rho_{sil} = 3300\ \mathrm{kg\,m^{-3}}$, $H_0 = 4 \times 10^{-11}\ \mathrm{W\,kg^{-1}}$, $T_i = 1500$ K; 3, $\rho_{sil} = 4000\ \mathrm{kg\,m^{-3}}$, $H_0 = 5 \times 10^{-11}\ \mathrm{W\,kg^{-1}}$, $T_i = 1600$ K, where ρ_{sil} is the silicate density, H_0 is the initial radiogenic heating rate and T_i is the interior temperature required to trigger thermal convection in the silicate core. Evolution of the silicate core is characterized by a diffusive stage following the differentiation and a convective stage once the interior is sufficiently hot. Depending on the composition of the silicate core, we estimate an onset time of convection of between 1.8 and 2.8 Gyr (from Tobie et al. 2006)

thermal structure of the conductive layer. Therefore, at equilibrium ($\partial T/\partial t = 0$), the conductive heat flow across a layer of thickness d having a surface temperature of T_{surf} and a bottom temperature of T_{bot} is described by a simple Fourier law:

$$\phi_{cond} = k\frac{T_{surf} - T_{bot}}{d}. \tag{40}$$

However, owing to a significant temperature dependence ($k(T) = 488.12/T(\mathrm{K}) + 0.4685$), the thermal conductivity of water ice is expected to vary from 2–2.5 $\mathrm{W\,m^{-1}\,K^{-1}}$ near the melting point to more than 5 $\mathrm{W\,m^{-1}\,K^{-1}}$ at a temperature of 100 K. Due to this effect, heat is more efficiently transferred at shallow depths where temperature increases more slowly with depth whereas temperature gradients are larger at deeper levels. As a consequence, for an equivalent heat flow and for a bottom temperature of 220–240 K (Tobie et al. 2003), the thickness between the surface and the bottom is 1.5 to 2 times larger if the temperature dependence of thermal conductivity is considered.

Other factors may also influence the thermal conductivity of an icy layer. The presence of other compounds with a lower conductivity, such as gas clathrates, ammonia hydrates, salts (e.g., Ross and Kargel 1998) can significantly reduce the effective thermal conductivity of the layer. On Titan, for instance, the presence of methane clathrate in large quantities is expected to strongly influence the cooling rate of the whole satellite (Tobie et al. 2006). On Europa also, various gas clathrates, if present, are expected to modify the thermal structure of the shell (e.g., Prieto-Ballesteros et al. 2005). The thermal conductivity is also expected to be reduced if the upper part of the icy layer includes a significant porous fraction (e.g., Eluszkiewicz 1990; Leliwa-Kopystyński and Kossacki 2000). As ice resists compaction at low temperature, a significant porosity can be preserved during a very long period of time and hence influence the thermal evolution of the satellite, in particular on small and cold bodies (Eluszkiewicz 1990).

Finally, the presence of significant tidal dissipation at the base of conductive layer (e.g. Ojakangas and Stevenson 1989a; Tobie et al. 2003) can reduce the heat flow across a conductive layer just above a liquid water ocean and hence limits the crystallization rate of the liquid layer. On Europa, a thermal equilibrium is reached for a tidally-heated conductive layer having a thickness of 25–30 km (e.g. Ojakangas and Stevenson 1989a). However, as it is discussed below, when an icy shell reaches such a thickness, thermal convection is expected to occur.

5.2 Onset of Convection and Convective Transfer

Convective motion can be initiated within the solid-state interior of a satellite due to the potentially unstable top-heavy configuration induced by the conjunction of a temperature field hotter in the inner part of the satellite and of a positive thermal expansivity α. In his seminal paper (see Strutt 1916), Lord Rayleigh demonstrated in the simple framework of a plane fluid layer heated from below, that the occurrence of such convective instabilities is dictated by the value of the non-dimensional parameter Ra,

$$\mathrm{Ra} = \frac{\alpha \rho g \Delta T d^3}{\kappa \eta}, \tag{41}$$

measuring the relative contribution of buoyancy forces ($\alpha \rho g \Delta T$, with g gravity and, $\Delta T = T_{\mathrm{bot}} - T_{\mathrm{surf}}$, the temperature difference across the layer) to the factor inhibiting fluid motion, i.e., viscosity η.

5.2.1 Basic Equations of Thermal Convection

The equations describing thermal convection in an incompressible fluid correspond to the three conservation equations for mass, momentum and energy,

$$\nabla \cdot \mathbf{V} = \mathbf{0}, \tag{42}$$

$$\rho \left(\frac{\partial V}{\partial t} + \mathbf{V} \cdot \nabla \mathbf{V} \right) = -\nabla \mathbf{p} + \nabla \cdot \tau + \mathbf{f}, \tag{43}$$

$$\rho c_p \left(\frac{\partial T}{\partial t} + \mathbf{V} \cdot \nabla T \right) = k \nabla^2 T + \rho H. \tag{44}$$

with the notations introduced earlier. The velocity ($\mathbf{V}$) and pressure (p) fields describe the flow induced by the buoyancy force ($\mathbf{f} = \rho(T)\mathbf{g}$) given that a viscous rheology is introduced that relates the strain rate tensor $((\nabla \mathbf{V}) +^t (\nabla \mathbf{V}))$ to the deviatoric stress tensor τ. The above formulation results from the hypothesis that thermodynamical coefficients (such as α, κ, $\ldots$) vary only by very moderate amounts so that they can be neglected as well as variations of ρ, except in the buoyancy term that drives convection (*Boussinesq approximation*). As indicated above, this assumption is only partly valid for satellites' interiors, especially for icy layers.

In addition, conservation of momentum (or *Navier-Stokes equation* (43)) is often simplified as, for icy or rocky materials, the value of the diffusivity of momentum (or kinematic viscosity η/ρ) is extremely large when compared to the thermal diffusivity κ (*infinite Prandtl number approximation*). As a consequence, inertial terms in the left hand side of (43) are neglected and the time evolution of the system resides entirely in the conservation of energy (44).

In the context of satellites, as indicated above, the "fluid" layer models either a sphere (corrresponding to a global undifferentiated interior or solely to the silicate core) or a spherical shell (corrresponding, for example, to an icy layer). Solid state creep of icy or rocky materials enables the description of long term motion with viscous fluid dynamics. A major characteristic is that the effective viscosity associated with these thermally activated processes varies strongly with temperature. Typically, the temperature differences ΔT associated with internal layers of icy satellites lead to several orders of magnitude variations for the viscosity of these materials, ices or rocks.

As a consequence, one specificity of thermal convection within icy satellites is that these solid state dynamics, if they operate, often occur in the so called *stagnant lid regime* obtained asymptotically for very large viscosity contrasts. In this case, as mentioned in the previous section, convective flow develops only in the warmer part of the layer, beneath a cold conductive lid. Since most of the viscosity variations occur within this lid, heat transfer in the convective sublayer below can be described as isoviscous convection (e.g., Solomatov 1995; Grasset and Parmentier 1998). Equation (44) is a generalized version of (35) when convection occurs: it indicates the dual nature of convective transfer where both conduction ($k\nabla^2 T$), as described in Sect. 5.1 of this paper, and advection (transfer associated to flow, $\rho c_p \mathbf{V}\cdot\nabla T$) coexist. Usually, advection is dominant in the bulk of the layer and diffusion becomes larger in regions near the boundaries (*thermal boundary layers*).

5.2.2 Conditions for Initiating Convection

The evaluation of the Rayleigh number Ra (according to (41)) should be sufficient to decide whether convective instabilities initiate or not in a (given layer of a) given satellite, if the simple isoviscous fluid, plane layer approximation is valid—the critical value of Ra is about 1000; even when the more realistic geometries of a sphere or a spherical shell are introduced, the linear stability analysis provides similar orders of magnitudes (Chandrasekhar 1961). If all icy satellites are considered as undifferentiated mixtures of ices and rocks, variations in values of κ and α are probably not significant when evaluating the order of magnitude of the Rayleigh number for a given satellite. Neither are the variations in mean density among satellites (so that g can also be approximated by a linear function of the satellite's radius R_s) or the global temperature difference potentially driving convection. Reasonable values for these parameters lead to $\mathrm{Ra} \simeq (R_s^4/\eta)$. However, unfortunately, due to a very large range of possible values for the reference viscosity η (induced by the nature of the creeping material and its temperature, mostly), the plausible range for Ra generally forbids a definitive conclusion.

The criterion for the onset of convection has been studied in the context of fluids with strongly temperature-dependent viscosity (e.g. Stengel et al. 1982): additional complexities arise when considering that the rheology of ices and rocks may also depend on stress, grain size and pressure. At large grain sizes and/or high temperatures, for example, materials flow by strongly stress-dependent dislocation creep. The onset of convection for such a viscosity, $\eta \sim \tau^{n-1}$ (with n, stress exponent), is then a highly non-linear problem addressed by only few studies (e.g. Solomatov and Barr 2006).

The critical value of Ra is often considered as a mean to investigate whether convection instabilities will develop in the outer icy shell of large Galilean satellites and eventually what is the depth of plausible internal oceans. The existence of such deep oceans is supported by electromagnetic induction signatures for Europa (Kivelson et al. 2000) and Callisto (Zimmer et al. 2000). The magnetic field data returned by several flybys of Ganymede is also interpreted in terms of a layer of liquid water at depth (Kivelson et al. 2002). While the evaluation of the critical thickness for convection to occur in the outer icy crust is addressed by many authors assuming a Newtonian rheology for the ice I shell (generally leading to values of a few tens of kilometers), the consideration of a more complex rheology demonstrates the important role of the grain size distribution and initial thermal state (Barr and Pappalardo 2005).

While the above problem of the stability of an icy layer can be interesting in terms of the present-day internal structure of satellites, a different approach is to consider the following transient problem: given that the interior is potentially heated uniformly (this would be the case for a silicate core or an ice/rock mixture, see (38)) and cannot lose enough heat by conduction alone (39), what is the critical time for convection to occur? In this case, the first instabilities take place where the largest temperature gradient is present, for example, beneath the boundary region near the surface of the core, mentioned in the previous section. In the isoviscous framework, it has been shown that the onset time for this transient problem can again be related to a critical value of the local Rayleigh number defined using the depth $\sim (\kappa t)^{1/2}$ of the cold front rather than the whole thickness of the layer Howard (1964). In the stagnant lid regime, only the fraction of this cold boundary layer where viscosity η_c is sufficiently low to participate to convective motion should be considered (e.g. Kirk and Stevenson 1987). The local Rayleigh number can then be approximated by

$$\mathrm{Ra}_\delta(t) = \frac{\alpha \rho g \Delta T_c(t)(\kappa t)^{3/2}}{\kappa \eta_c(t)}. \tag{45}$$

5.2.3 *Efficiency of Heat Transfer*

Once convection is initiated, a measure of the efficiency of heat transfer is provided by the Nusselt number Nu. This second dimensionless parameter is defined as the ratio between the surface heat flux ϕ induced by the convective transfer and the conductive heat flux (ϕ_{cond}, (40)) that would be obtained for the same layer if advective motions were not present. Scaling relationships of the form

$$\mathrm{Nu} \sim \mathrm{Ra}^{\beta} \tag{46}$$

are obtained for various convective set-ups: for isoviscous fluids, values of $\beta \simeq -1/3$ are generally reported. Since real configurations may imply both basal and internal heating, a potentially important effect is the precise distribution of mixed heat sources (Sotin and Labrosse 1999; Moore 2008; Choblet and Parmentier 2009). Departures from this simple framework are thus plausible.

A parameterized evolution of the satellite's interior can then be considered, following the method first proposed for terrestrial planets by Schubert et al. (1979): assuming steady-state heat transfer, the global heat balance for the internal spherical shell between radii R_{bot} and R_{top} results from the integration of (44),

$$\rho c_p \frac{\partial T}{\partial t} = \rho H + 3\frac{\phi_{\mathrm{top}} R_{\mathrm{top}}^2 - \phi_{\mathrm{bot}} R_{\mathrm{bot}}^2}{R_{\mathrm{top}}^3 - R_{\mathrm{bot}}^3}, \tag{47}$$

the values of fluxes ϕ being associated to the convective vigor of the layer through (46).

The value of the exponent β is naturally modified in the context of the stagnant lid regime that inhibits the efficiency of heat transfer when such an insulating viscous layer forms on top of the convective region (appropriate values of β are thus smaller, cf. for example Solomatov 1995). Applications to the outer shell of large icy satellites at present-day are specifically proposed by Deschamps and Sotin (2001) indicating the role of convective heat transfer on the possible radial structures of these objects (preservation or complete crystallization of a residual internal ocean). Thermal evolutions of the Galilean satellites

(Spohn and Schubert 2003), Titan (Tobie et al. 2006) or smaller mid-sized icy satellites of Saturn (Multhaup and Spohn 2007) are based on a similar conceptual framework. These studies indicate that specific parameters attributable to each of these bodies may induce different evolutionary paths. One such factor is the amount of ammonia in the water-ice system that largely affects the phase diagram (e.g. Sotin et al. 1998). Methane clathrates possibly controlling the coupled evolution of the interior and atmospheric processes are also likely candidates.

The general cases discussed here are the basis for more complex models in which different aspects of satellite evolution must be combined. The history of tidal heating induced by the orbital evolution of some of the satellites, implies that in some cases coupled methods (internal dynamics, shape relaxation, orbital evolution) are required to properly describe their evolution.

6 Concluding Remarks

In this chapter we have discussed the main energy sources which drive processes in the satellites' interiors. Furthermore, we have described the rotational and orbital states, which are closely linked to the internal energy budget by tidal friction. Depending on the size, different rock mass fraction, internal composition, and interaction with other satellites and the primary planet through tides and resonances, the relevance of these energy sources may strongly vary for different objects. Furthermore, there can be strong variations with time. E.g., whereas tidal friction plays a major role as a heat source on Io and Enceladus, such processes can at present be neglected for most of the satellites. Tidal despinning, however, may have been an important process involving dissipation of heat for all major satellites early in their histories. The conversion of different energy contributions into heat and the transport of the latter triggers processes, which are very diverse for the individual satellites. Such processes and their relevance for the evolution of satellites are discussed in other contributions to this volume. Examples are the interaction of thermal convection and despinning of Iapetus, the melting induced by tidal heating in plumes in convecting satellites, the chemical cycles on Titan, serpentinization and ocean chemistry on Europa, or localized heating in the ice shell of Enceladus.

To combine the effects relevant for the evolution of specific satellites is a challenging task. *Cassini* and *Galileo* data as well as ground-based astrometry of satellite orbits and laboratory studies on ices are major tools to improve our current models on satellite evolution. Furthermore, such tools are essential for the preparation of the future exploration of the outer planets and their moons.

Acknowledgements We would like to thank W.B. McKinnon and an anonymous reviewer for the very helpful and constructive comments.

Part of this work was carried out at the Jet Propulsion Laboratory, California Institute of Technology, under contract with NASA. Part of this work has been supported by the Helmholtz Association through the alliance 'Planetary Evolution and Life'.

References

K. Aksnes, F.A. Franklin, Astron. J. **122**, 2734–2739 (2001)
G. Audi, O. Bersillon, J. Blachot, A.H. Wapstra, Nucl. Phys. A **624**, 1–124 (1997)
V.A. Avdyushev, Sol. Syst. Res. **38**, 238–240 (2004)
A.C. Barr, R.T. Pappalardo, J. Geophys. Res. **110**, E12 (2005). CiteID E12005

R.H. Brown, T.V. Johnson, R.L. Kirk, L.A. Soderblom, Science **250**, 431–435 (1990)
J.C. Castillo-Rogez, D.L. Matson, C. Sotin, T.V. Johnson, J.I. Lunine, P.C. Thomas, Icarus **190**, 179–202 (2007)
A. Cayley, Mem. R. Astron. Soc. **29**, 191–306 (1861)
S. Chandrasekhar, *Hydrodynamic and Hydromagnetic Stability*. International Series of Monographs on Physics (Clarendon, Oxford, 1961)
G. Choblet, E.M. Parmentier, Phys. Earth Planet. Inter. **173**, 290–296 (2009)
D.M. Cole, Philos. Mag. A **72**, 231–248 (1995)
D.M. Cole, R.A. Johnson, G.D. Durell, J. Geophys. Res. **103**(C10), 21751–21758 (1998)
R.L. Comstock, B.G. Bills, J. Geophys. Res. **108**(E9), 5100 (2003). doi:10.1029/2003JE00210
W. De Sitter, Leiden Ann. **16**(2), 1–92 (1928)
F. Deschamps, C. Sotin, J. Geophys. Res. **106**, 5107–5121 (2001)
G. Dourneau, Thesis, Bordeaux, 1987
W.B. Durham, O. Prieto-Ballesteros, D.L. Goldsby, J.S. Kargel, Space Sci. Rev. (2010, this issue). doi:10.1007/s11214-009-9619-1
D.H. Eckhardt, Moon Planets **25**, 3–49 (1981)
J. Eluszkiewicz, Icarus **84**, 215–225 (1990)
S. Ferraz-Mello, A. Rodriguez, H. Hussmann, Celest. Mech. Dyn. Astron. **101**, 171–201 (2008)
P.E. Geissler, Annu. Rev. Earth Planet. Sci. **31**, 175–211 (2003)
P. Goldreich, Mon. Not. R. Astron. Soc. **126**, 257–268 (1963)
P. Goldreich, S.J. Peale, Astron. J. **71**, 425–437 (1966)
P. Goldreich, S. Soter, Icarus **5**, 375–389 (1966)
O. Grasset, E.M. Parmentier, J. Geophys. Res. **103**, 18171–18181 (1998)
O. Grasset, C. Sotin, F. Deschamps, Planet. Space Sci. **48**, 617–636 (2000)
R. Greenberg, S.J. Weidenschilling, Icarus **58**, 186–196 (1984)
J. Hillier, S.W. Squyres, J. Geophys. Res. **96**, 15665–15674 (1991)
G.V. Hoppa, B.R. Tufts, R. Greenberg, T.A. Hurford, D.P. O'Brien, P.E. Geissler, Icarus **153**, 208–213 (2001)
L.N. Howard, in *Proceedings of the Eleventh International Congress of Applied Mechanics*, vol. 153, ed. by H. Gortler (Springer, New York, 1964), pp. 1109–1115
W.B. Hubbard, *Planetary Interiors* (Van Nostrand Reinhold, New York, 1984)
H. Hussmann, T. Spohn, K. Wieczerkowski, Icarus **156**, 143–151 (2002)
H. Jeffreys, *The Earth. Its Origin, History and Physical Constitution* (Cambridge University Press, Cambridge, 1952)
Ö. Karatekin, T. Van Hoolst, T. Tokano, Geophys. Res. Lett. **35**, L16202 (2008). doi:10.1029/2008GL034744
R.L. Kirk, D.J. Stevenson, Icarus **69**, 91–134 (1987)
M.G. Kivelson, K.K. Khurana, C.T. Russell, M. Volwerk, R.J. Walker, C. Zimmer, Science **289**, 1340–1343 (2000)
M.G. Kivelson, K.K. Khurana, M. Volwerk, Icarus **157**, 507–522 (2002)
Y. Kozai, Ann. Tokyo Astron. Obs. **73**, 1 (1957)
V. Lainey, L. Duriez, A. Vienne, Astron. Astrophys. **420**, 1171–1183 (2004)
V. Lainey, L. Duriez, A. Vienne, Astron. Astrophys. **456**, 783–788 (2006)
V. Lainey, J.E. Arlot, O. Karatekin, T. Van Hoolst, Nature **459**, 957–959 (2009)
J. Leliwa-Kopystyński, K.J. Kossacki, Planet. Space Sci. **48**, 727–745 (2000)
J.H. Lieske, Astron. Astrophys. **176**, 146–158 (1987)
J.H. Lieske, Astron. Astrophys. **56**, 333–352 (1977)
K. Lodders, B. Fegley Jr., *The Planetary Scientists Companion* (Oxford Univ. Press, New York, 1998)
G.J.F. MacDonald, Rev. Geophys. Space Phys. **2**, 467–541 (1964)
F. Marchis et al., Icarus **176**, 96–122 (2005)
A.S. McEwen, L.P. Keszthelyi, R. Lopes, P.M. Schenk, J.R. Spencer, in *Jupiter: The Planet, Satellites, and Magnetosphere*, ed. by F. Bagenal, T.E. Dowling, W.B. McKinnon (Cambridge Univ. Press, Cambridge, 2004), pp. 307–328
W.B. McKinnon, in *Solar System Ices*, ed. by B. Schmitt, C. de Bergh, M. Festou (Kluwer, Dordrecht, 1998), p. 525
W.B. McKinnon, Geophys. Res. Lett. **26**, 951–954 (1999)
G. Mitri, A.P. Showman, Icarus **195**, 757–764 (2005)
W.B. Moore, J. Geophys. Res. **108**, E8 (2003). doi:10.1029/2002JE001943
W.B. Moore, J. Geophys. Res. **113**, B11407 (2008). doi:10.1029/2006JB004778
W.B. Moore, H. Hussmann, in *Europa*, ed. by R.T. Pappalardo, W.B. McKinnon, K.K. Khurana (University of Arizona Press, Tucson, 2009), pp. 369–380
H. Moritz, *The Figure of the Earth* (Herbert Wichmann, Karlsruhe, 1990)

K. Multhaup, T. Spohn, Icarus **186**, 420–435 (2007)
C.D. Murray, S.F. Dermott, *Solar System Dynamics* (Cambridge University Press, Cambridge, 1999)
K. Nagel, D. Breuer, T. Spohn, Icarus **169**, 402–412 (2004)
D.B. Nash, M.H. Carr, J. Gradie, D.M. Hunten, C.F. Yoder, in *Satellites*, ed. by J.A. Burns, M.S. Matthews (Univ. of Arizona Press, Tucson, 1986), pp. 629–688,
F. Nimmo, E. Gaidos, J. Geophys. Res. **107**, E4 (2002). doi:10.1029/2000JE001476
F. Nimmo, J.R. Spencer, R.T. Pappalardo, M.E. Mullen, Nature **447**, 289–291 (2007)
B. Noyelles, Celest. Mech. Dyn. Astron. **101**, 13–30 (2008)
G.W. Ojakangas, D.J. Stevenson, Icarus **81**, 220–241 (1989a)
G.W. Ojakangas, D.J. Stevenson, Icarus **81**, 242–270 (1989b)
J. Palguta, G. Schubert, K. Zhang, J.D. Anderson, Icarus **201**, 615–625 (2009)
S.J. Peale, in *Planetary Satellites*, ed. by J.A. Burns (Univ. of Arizona Press, Tucson, 1977), pp. 87–111
S.J. Peale, in *Satellites*, ed. by J.A. Burns, M.S. Matthews (Univ. of Arizona Press, Tucson, 1986), pp. 159–223
S.J. Peale, Annu. Rev. Astron. Astrophys. **37**, 533–602 (1999)
S.J. Peale, M.H. Lee, Science **298**, 593–597 (2002)
S.J. Peale, P. Cassen, R.T. Reynolds, Science **203**, 892–894 (1979)
C.C. Porco et al., Science **311**, 1393–1401 (2006)
B. Preblich, R. Greenberg, J. Riley, D. O'Brien, Planet. Space Sci. **55**, 1225–1245 (2007)
O. Prieto-Ballesteros, J.S. Kargel, M. Fernández-Sampedro, F. Selsis, E.S. Martínez, D.L. Hogenboom, Icarus **177**, 491–505 (2005)
L. Prockter, F. Nimmo, R.T. Pappalardo, Geophys. Res. Lett. **32**, 14 (2005). CiteID L14202
N. Reeh, E. Lintz Christensen, C. Mayer, O. Olesen, Ann. Glaciol. **37**, 83–89 (2003)
R.G. Ross, J.S. Kargel, in *Solar System Ices*, ed. by B. Schmitt, C. de Bergh, M. Festou (Kluwer, Dordrecht, 1998), p. 33
R.A. Sampson, Mem. R. Astron. Soc. **63**, 1–270 (1921)
G. Schubert, P. Cassen, R.E. Young, Icarus **38**, 192–211 (1979)
G. Schubert, T. Spohn, R.T. Reynolds, in *Satellites*, ed. by J.A. Burns, M.S. Matthews (Univ. of Arizona Press, Tucson, 1986), pp. 224–292
G. Schubert, J.D. Anderson, T. Spohn, W.B. McKinnon, in *Jupiter: The Planet, Satellites, and Magnetosphere*, ed. by F. Bagenal, T. Dowling, W.B. McKinnon (Cambridge Univ. Press, Cambridge, 2004), pp. 281–306
G. Schubert, H. Hussmann, V. Lainey, D. Matson, W.B. McKinnon, F. Sohl, C. Sotin, G. Tobie, D. Turrini, T. Van Hoolst, Space Sci. Rev. (2010, this issue)
M. Segatz, T. Spohn, M.N. Ross, G. Schubert, Icarus **75**, 187–206 (1988)
B.A. Smith et al., Science **204**, 951–957 (1979)
B. Smith-Konter, R.T. Pappalardo, Icarus **198**, 435–451 (2008)
L.A. Soderblom, T.L. Becker, S.W. Kieffer, R.H. Brown, C.J. Hansen, T.V. Johnson, Science **250**, 410–415 (1990)
F. Sohl, W.D. Sears, R.D. Lorenz, Icarus **115**, 278–294 (1995)
F. Sohl, H. Hussmann, B. Schwentker, T. Spohn, R.D. Lorenz, J. Geophys. Res. **108**, 5130 (2003)
F. Sohl, M. Choukroun, J. Kargel, J. Kimura, R.T. Pappalardo, S. Vance, M. Zolotov, Space Sci. Rev. (2010, this issue)
V.S. Solomatov, Phys. Fluids **7**, 266–274 (1995)
V.S. Solomatov, A.C. Barr, Phys. Earth Planet. Inter. **155**, 140–145 (2006)
C. Sotin, S. Labrosse, Phys. Earth Planet. Inter. **112**, 171–190 (1999)
C. Sotin, O. Grasset, S. Beauchesne, in *Solar System Ices*, ed. by B. Schmitt, C. de Bergh, M. Festou (Kluwer, Dordrecht, 1998), pp. 79–96
C. Sotin, G. Tobie, J. Wahr, W.B. McKinnon, in *Europa*, ed. by R.T. Pappalardo, W.B. McKinnon, K.K. Khurana (University of Arizona Press, Tucson, 2009), pp. 85–118
J.R. Spencer, J.C. Pearl, M. Segura, F.M. Flasar, A. Mamoutkine, P. Romani, B.J. Buratti, A.R. Hendrix, L.J. Spilker, R.M.C. Lopes, Science **311**, 1401–1405 (2006)
J.R. Spencer et al., Science **318**, 240 (2007)
T. Spohn, in *Bergmann, Schaefer: Lehrbuch der Experimentalphysik, Band 7: Erde und Planeten*, ed. by W. Raith (1997), pp. 427–525
T. Spohn, G. Schubert, Icarus **161**, 456–467 (2003)
K.C. Stengel, D.C. Oliver, J.R. Booker, J. Fluid Mech. **120**, 411–431 (1982)
B.W. Stiles, et al., Astron. J. **135**(5), 1669–1680 (2008). doi:10.1088/0004-6256/135/5/1669
B.W. Stiles et al. (the Cassini Radar Team), Astron. J. **139**, 311 (2010)
J.W. Strutt, Philos. Mag. **32**, 529–546 (1916)
J. Tatibouet, J. Perez, R. Vassoille, J. Phys. **47**, 51–60 (1986)

P.C. Thomas et al., Icarus **190**, 573–584 (2007)
G. Tobie, G. Choblet, C. Sotin, J. Geophys. Res. **108**, E11 (2003). doi:10.1029/2003JE002099
G. Tobie, A. Mocquet, C. Sotin, Icarus **177**, 534–549 (2005a)
G. Tobie, O. Grasset, J.I. Lunine, A. Mocquet, C. Sotin, Icarus **175**, 496–502 (2005b)
G. Tobie, J.I. Lunine, C. Sotin, Nature **440**, 61–64 (2006)
G. Tobie, O. Čadek, C. Sotin, Icarus **196**, 642–652 (2008)
T. Tokano, F.M. Neubauer, Geophys. Res. Lett. **32**, L24203 (2005)
R. Tyler, Nature **456**, 770–772 (2008)
T. Van Hoolst, N. Rambaux, Ö. Karatekin, V. Dehant, A. Rivoldini, Icarus **195**, 386–399 (2008). doi:10.1016/j.icarus.2007.12.011
T. Van Hoolst, Ö. Karatekin, in *European Planetary Science Congress 2008*. Extended abstract, 2008
T. Van Hoolst, N. Rambaux, Ö. Karatekin, Icarus **200**, 256–264 (2009). doi:10.1016/j.icarus.2008.11.009
S. Vance, J. Harnmeijer, J. Kimura, H. Hussmann, B. DeMartin, J.M. Brown, Astrobiology **7**, 987–1005 (2007)
R. Vasundhara, J.-E. Arlot, P. Descamps, in *Dynamics, Ephemerides, and Astrometry of the Solar System: Proceedings of the 172nd Symposium of the International Astronomical Union*, Held in Paris, France, 38 July 1995, ed. by S. Ferraz-Mello, B. Morando, J.-E. Arlot (1996), p. 145
A. Vienne, J.M. Sarlat, L. Duriez, in *Chaos, Resonance, and Collective Dynamical Phenomena in the Solar System: Proceedings of the 152nd Symposium of the International Astronomical Union*, ed. by S. Ferraz-Mello, Angra dos Reis, Brazil, 15–19 July 1991 (Kluwer Academic, Dordrecht, 1992), p. 219
J. Wisdom, Astron. J. **128**, 484–491 (2004)
F.E. Witteborn, J.D. Bregman, J.B. Pollack, Science **203**, 643–646 (1979)
C.F. Yoder, Nature **279**, 767–770 (1979)
D.A. Yuen, L. Fleitout, G. Schubert, C. Froidevaux, Int. J. Geophys. **54**, 93–119 (1978)
C. Zimmer, K.K. Khurana, M.G. Kivelson, Icarus **147**, 329–347 (2000)

Space Sci Rev (2010) 153: 349–374
DOI 10.1007/s11214-010-9645-z

Effects of the External Environment on Icy Satellites

Matthew H. Burger · Roland Wagner · Ralf Jaumann · Timothy A. Cassidy

Received: 15 May 2009 / Accepted: 1 March 2010 / Published online: 3 April 2010

Abstract In order to understand the evolution of planetary satellite surfaces and atmospheres it is important to understand their external environments. In this paper we look at the interactions between plasma in planetary magnetospheres and satellite atmospheres responsible for the production and loss of atmospheric mass. We focus on the processes which take place in the tenuous atmospheres of the Galilean satellites and in the Enceladus water plume. We also review the impact histories of satellites in the outer solar system, and compare the record of impacts in the inner and outer solar system.

Keywords Planetary science · Atmospheres · Magnetospheres · Outer solar system · Satellites · Solar system evolution · Surfaces · Impact histories

1 Introduction

When considering the on-going evolution of satellites in planetary systems, it is important to consider the satellites within the context of the entire system: objects are not isolated from their surrounding environments. In this chapter we discuss two of the ways in which satellites interact with their surroundings: the chemistry of satellite exospheres and their interactions with magneospheric plasma, and the record of external impactors left on the satellite surfaces. Due to space constraints we have left out a number of relevant topics, such as the transfer of dust to and from the satellites, and the signatures of mass loading in the plasma and magnetic field.

M.H. Burger (✉)
University of Maryland and NASA/GSFC, Greenbelt, MD, USA
e-mail: Matthew.H.Burger@nasa.gov

R. Wagner · R. Jaumann
DLR, Berlin, Germany

T.A. Cassidy
University of Virginia, Charlottesville, VA, USA

Section 2 describes the physical processes important for understanding the sources, sinks, and observable emissions in the exospheres of the Galilean satellites and the neutral tori at Jupiter and Saturn. We emphasize the chemistry between exospheric neutrals and magnetospheric plasma. Observations relevant to this discussion are presented in previous chapters (Krupp et al. 2010, this issue; Dalton et al. 2010, this issue; Coustenis et al. 2010, this issue). Physical processes between the magnetosphere and satellite surfaces are discussed in Cassidy et al. (2010, this issue), although we consider the effects of sputtering on atmospheric production and escape. We also refer readers to recent reviews in the books *Jupiter* (Bagenal et al. 2004), *Saturn from Cassini-Huygens* (Dougherty et al. 2009), and *Europa* (Pappalardo et al. 2009) which discuss other aspects of the magnetospheric interactions with icy satellites. Our goal has been to complement, rather than supplant, the discussions presented in those works.

Section 3 presents a review of the impact histories of the outer planet satellites and should be seen as distinct from the discussion in Sect. 2. Section 3 concentrates on (a) projectile fluxes, (b) crater size-frequency distributions, and (c) surface ages of the icy satellites of Jupiter and Saturn. Images returned by spacecraft from these two planets were used for measurements of crater distributions. A large image data set is available, obtained by the SSI camera aboard the Galileo spacecraft (in orbit around Jupiter from 1995 to 2003), and by the ISS cameras aboard Cassini (in orbit around Saturn since July 2004). Crater distributions on these satellites are compared to size distributions of candidate impactor families responsible for cratering these satellites. Also, cratering chronology models derived from impact rates of potential impactors and implications for the geological histories of these satellites are discussed.

2 Plasma Interaction with Tenuous Satellite Atmospheres

The characteristics of tenuous atmospheres in the outer solar system are described in Chap. I.4. Here were describe the external processes responsible for shaping these atmospheres; i.e., we describe the interactions between satellite atmospheres and the planetary magnetospheres. We focus on processes taking place within the volcanic, SO_2 dominated atmosphere of Io, the O_2 atmospheres of Europa and Ganymede, and the Enceladus water plume.

Charged particles in the magnetosphere are classified into two components. The "plasma" refers to the approximately thermalized component below ~10 keV. "Energetic particles" are the high energy ions and electrons above 10 keV which have been observed to 100's of MeV (Johnson et al. 2004). The interactions between the plasma, satellite surfaces, and gas produced at the satellites which make up atmospheres, exospheres, and extended neutral tori are summarized in Fig. 1. Magnetospheric charged particles (both the thermal plasma and energetic ions and electrons) are swept into the satellites by the rapidly rotating planetary magnetic fields. Energetic ion implantation on the surface can alter the surface constituents chemically and eject neutral atoms and molecules which form the atmosphere and neutral clouds. The plasma ionizes gas in the atmosphere producing fresh ions which re-supply the plasma, dissociates molecules altering the atmospheric composition, and ejects neutrals which contribute to atmospheric escape. The neutral gas and surfaces interact through the equilibrium of adsorption and desorption of gas forming surface frosts in a diurnal cycle of condensation of the atmosphere followed by evaporation.

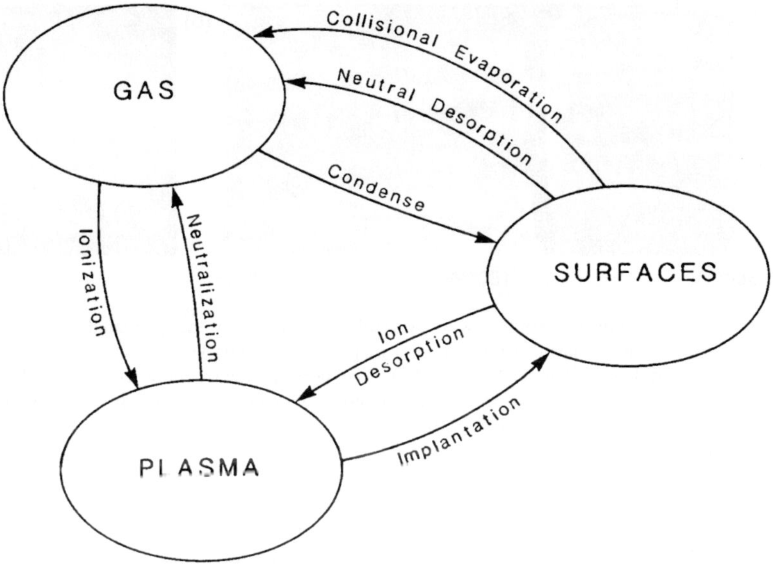

Fig. 1 Summary of interactions between the magnetospheric plasma, exospheric gas, and satellite surfaces. "Plasma" refers to both the thermal and energetic components. The plasma-gas interactions occur in both the satellite atmospheres and exospheres, and in the plasma and neutral tori encircling Jupiter and Saturn (from Johnson 1990)

2.1 Ion Bombardment of Satellite Surfaces

Following the first observations by the Pioneer missions of the Jovian radiation environment, Brown et al. (1978) exposed water ice to high energy ion flux and discovered that the ice was eroded by sputtering orders of magnitude more quickly than expected. Sputtering is the process by which plasma and energetic ions eject material from a surface or atmosphere though a series of knock-on collisions or chemical reactions (Johnson 1990; Johnson et al. 2004). Baragiola et al. (2003) consider sputtering a synonym for impact desorption when applied to the icy surfaces in the outer solar system. Chemical reactions may alter the surface producing gases which become trapped or diffuse outward and escape. The surface of Europa probably suffers the worst erosion of any icy body in the solar system. The erosion rate is estimated to be on the order of tens of meters per billion years, a rate that is roughly globally uniform (Paranicas et al. 2002). Sputtering is also important in Saturn's much less energetic magnetosphere, where E-ring grains at Enceladus' orbit are quickly eroded: a 1 μm grain has a lifetime of ~50 years (Johnson et al. 2008).

Sputtering of water ice by energetic ions ejects H_2O, as well as the decomposition products O_2 and H_2 (Brown et al. 1982). Other species, including H, O, HO_2, and H_2O_2 can also be produced (Baragiola 2003). The production of decomposition products depends on the ice temperature: while the yield of H_2O is constant for ice temperatures below 140 K, the production of H_2 and O_2 increases with temperature (Baragiola et al. 2003; Baragiola 2003).

The energy distribution of escaping material helps to determine the composition and evolution of satellite exospheres. Species ejected at low energies remain bound to the satellites; those produced at higher energies escape to form neutral tori and supply magnetospheric plasma. Unfortunately there are many uncertainties in understanding these distributions due

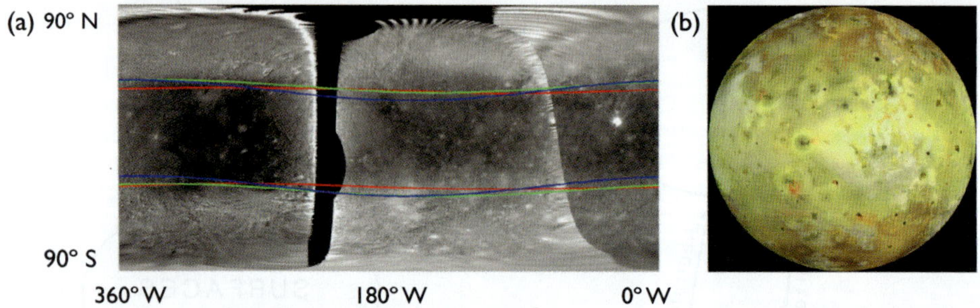

Fig. 2 **a** A map of Ganymede's surface with overlaid lines that show the estimated boundary between open and closed field lines. Plasma from Jupiter's magnetosphere hits Ganymede's surface in the open field regions (adapted from Khurana et al. 2007). The correspondence between open field lines and bright terrain suggests that plasma bombardment brightens Ganymede's surface. **b** Io's dark poles (from http://photojournal.jpl.nasa.gov/catalog/PIA02308)

to the difficulty of laboratory measurements and effects of surface composition, temperature, impactor energy and composition. We present a generalized form for the sputtering energy distribution based on laboratory measurements (e.g., Johnson 1990):

$$F(E,\theta) \sim \frac{E^{\beta}}{(E+U)^{\alpha}} \cos\theta \tag{2.1}$$

where E is the energy of the ejected neutral, θ is the angle of the ejected neutral relative to the normal, U is the surface binding energy, and α and β are experimentally measured parameters. Typical values based on laboratory results that have been used are $\alpha = 3$, $\beta = 1$, and $U = 0.055$ eV for the ejection of H_2O; $\alpha = 2$, $\beta = 0$, and $U = 0.015$ eV for the ejection of O_2 (Shematovich et al. 2005; Johnson et al. 2009). The energy distribution of sodium desorbed from ice by 200 eV electrons can be fit using $\alpha = 2.7$, $\beta = 1$, and $U = 0.052$ eV (Johnson et al. 2002). Molecular hydrogen produced by ion impact is not sputtered directly, but diffuses upward through the ice and escapes thermally with a Maxwellian energy distribution at the local surface temperature (Johnson et al. 2009). Because of the low mass of hydrogen, H_2 emitted from the surface is not gravitationally bound, and is not an important constituent of satellite exospheres.

The incident plasma also chemically alters outer planet moon surfaces. Radiolysis of water ice is responsible for the O_2 detected around Europa, Ganymede, and Saturn's rings. The poles of Io and Ganymede demonstrate the influence of radiolysis on surface properties: both moons have poles exposed to magnetospheric plasma and relatively shielded equatorial regions. Ganymede's poles are exposed to plasma flux because its polar magnetic field lines are open to the Jovian magnetosphere, whereas the lower latitudes are protected by closed field lines. The remarkable correlation between the open field lines and bright terrain (Fig. 2) strongly suggests that the plasma has brightened the terrain, though the exact mechanism responsible is uncertain. Johnson (1985) suggested that brightening was due to enhanced scattering: implanted impurities such as sulfur ions derived from Io's atmosphere scatter light, as do radiation defects. Khurana et al. (2007), however, argued that the brightening is caused by sputtered H_2O, which returns to the surface where it might form a bright frost.

Io's low-mid latitudes are partially shielded by the presence of an SO_2 vapor-pressure controlled atmosphere which is much less dense at the relatively cold poles. Based on laboratory work (Moore 1984), Wong and Johnson (1996) suggested that Io's poles have been darkened by plasma flux. Moore (1984) found that proton-irradiated SO_2 ice, which started out white in color, became yellow and dark. Radiation-induced darkening, rather than brightening, is the norm. Hydrocarbons, common in outer solar system surfaces, are darkened by magnetospheric radiation and UV photons (see Chap. II.4). Radiolysis and photolysis of hydrocarbons tend to produce H_2, which escapes to leave dark, reddish surfaces with large C:H ratios (e.g. Baines et al. 1998).

Another comparison between Galilean satellites illustrates water ice amorphization. Hansen and McCord (2004) used an IR absorption peak to identify amorphous water ice; i.e., ice without crystal structure. Amorphous water ice is in a thermodynamically unfavorable state, but can be created by flash-freezing liquid water at sufficiently low temperatures or by ion bombardment. Hansen and McCord found a range of amorphous ice abundances: at Europa, the most heavily irradiated Galilean icy satellite, water ice was found to be amorphous to depths of $\sim$1 mm, limited by the depth to which energetic ions can penetrate. They found only crystalline water ice on Callisto, where thermal annealing is thought to dominate ion amorphization because Callisto is warmer than Europa (Orton et al. 1996) and receives orders of magnitude less ion flux. Ganymede had a mixture of amorphous and crystalline ice, with the amorphous ice found preferentially at its cold and highly-irradiated poles.

The Jovian magnetosphere also alters surface composition through the implantation of sulfur ions. The best evidence for this exchange process is at Europa, where various sulfur-containing compounds have been identified. Some of these sulfur compounds are likely due to exchange between Europa's surface and subsurface, but a global pattern in surface spectra shows Io's influence. The presence of various sulfur compounds, such as SO_2, H_2SO_4 and S_n, is explained by the presence of a "radiolytic sulfur cycle" in which radiation both destroys and creates each compound, eventually producing equilibrium ratios (Carlson et al. 2009). Such a cycle could also apply to sulfate salts from Europa's and Ganymede's subsurfaces. A similar cycle has been suggested for CO_2, H_2O and carbonic acid, with CO_2 and H_2O dominating. Evidence for this is seen on the Uranian satellites where CO_2 abundance correlates with plasma bombardment, consistent with radiolysis acting on a mixture of H_2O and carbonaceous material (Grundy et al. 2006) (Fig. 3).

2.2 Exospheric Processes

Reactions in tenuous atmospheres or exospheres are dominated by the plasma interaction with neutrals and have components due to electron impacts and ion-neutral charge transfer reactions. In tenuous atmospheres, neutral-neutral collisions are negligible and when studying their behavior each neutral species can be discussed independently. Cross sections for these processes are energy dependent, and therefore the reaction rates depend on the speed distribution of the charged particles relative to the neutrals, as well as the densities of each reactant.

The effect of electron bombardment of the atmosphere is highly dependent on the temperature of the electrons. Possible outcomes of electron impact are excitation, dissociation, and ionization:

$$A + e \rightarrow A^* + e \tag{2.2}$$

$$AB + e \rightarrow A + B + e \tag{2.3}$$

$$A + e \rightarrow A^+ + 2e \tag{2.4}$$

The electron impact cross sections are strong functions of temperature, so that the relative importance of the reactions defined in (2.2)–(2.4) depends on the energies of the electrons

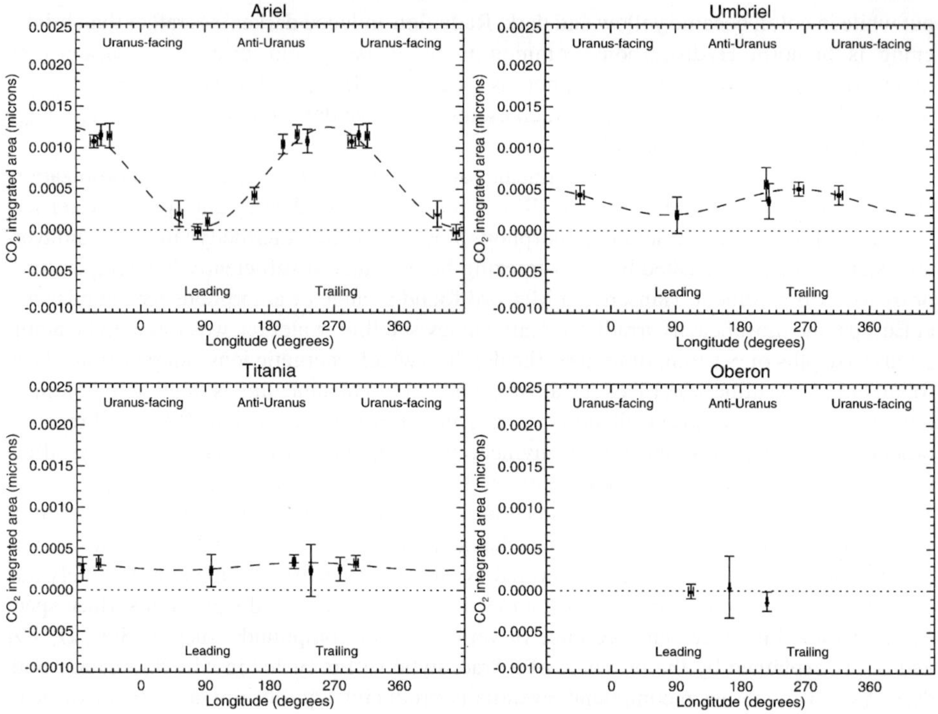

Fig. 3 CO_2 abundance vs. longitude for Uranian satellites. CO_2 is more abundant on the trailing hemispheres, where plasma bombardment is strongest (from Grundy et al. 2006)

flowing through the atmosphere. Lower energy electrons are more likely to put the neutrals into excited states which may decay radiatively, and higher energy electrons will induce ionizations.

The electrons control the emissions seen in tenuous atmospheres, and can cause ionization and atmospheric loss if the temperatures are high enough. Figure 4 shows the excitation cross sections for four emission lines of atomic oxygen, and the cross section for ionization. The excitation energies from the ground state for the transitions shown are 2.0 eV (for the transition $^3P \rightarrow ^1D$, which emits at 6300 Å), 4.2 eV ($^3P \rightarrow ^5S$, 2792 Å), 9.1 eV ($^3P \rightarrow ^5S$, 1356 Å), and 9.5 eV ($^3P \rightarrow ^3S$, 1304 Å), and the ionization energy from the ground state is 13.6 eV. Below ~10 eV, the primary effect of electron impact is the production of an excited state which spontaneously emits a 6300 Å photon; i.e., a visible wavelength emission line is produced. At higher energies, oxygen is ionized with some emission, primarily in the FUV at 1304 Å.

The 1304 Å emission line results from a resonant transition, meaning that it can efficiently scatter sunlight. Resonantly scattered emissions from oxygen have been observed in Europa's extended corona (Hansen et al. 2005) and in the neutral cloud filling Saturn's magnetosphere (Esposito et al. 2005; Melin et al. 2009). Emission from the resonantly scattered sodium D lines at 5890 and 5896 Å have been observed in the exospheres and neutral clouds of Io and Europa (Brown 1974; Brown and Hill 1996). When the gas is optically thin, the intensity of the emission is linearly proportional to the column abundance along the line of

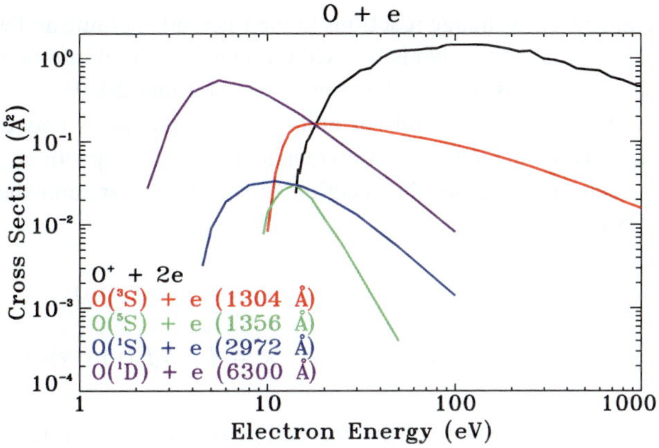

Fig. 4 Cross sections for excitation of the ^3S, ^5S, ^1D states and ionization of atomic oxygen from the ground state. References: Majeed and Strickland (1997), Johnson et al. (2005)

sight:

$$I = N \times g \times 10^{-6} \tag{2.5}$$

where I is the intensity in units of Rayleighs (1 Rayleigh equals 10^6 photons $cm^{-2}\,s^{-1}$ emitted isotropically into 4π steradians), N is the column abundance in units of cm^{-2}, and g is the "g-value," the product of the number of solar photons available for scattering and the scattering probability per atom. Because of deep Fraunhofer absorptions due to atomic species in the solar atmosphere, the g-values are strong functions of the radial velocity of the scattering atoms relative to the sun. The g-values for many species relevant to the exospheres of the Jovian satellites, including Na, O, and S, have been computed by Killen et al. (2009).

Ion-neutral processes refer to the reactions which occur during collisions between plasma ions and neutral species. Charge exchange is a subset of this latter category in which an electron is transferred but the parent species remain unchanged. In general,

$$A^+ + B \rightarrow C + D^+ \tag{2.6}$$

$$A^+ + B \rightarrow A + B^+ \tag{2.7}$$

where (2.6) is the generalized ion-neutral reaction in which chemical changes in the reactants can occur, and (2.7) is a charge-exchange reaction.

In the magnetospheres of Jupiter and Saturn, ions corotating with the magnetic fields are usually traveling much faster than neutrals. During charge transfer reactions, the neutralized ions retain their high speeds, continuing on ballistic orbits no longer influenced by electric and magnetic fields in the system. The distribution of neutrals at Jupiter and Saturn are determined in part from fast, often escaping, neutrals formed through ion-neutral reactions. At Jupiter, a cloud of sodium >1000 Jupiter radii across has been observed (Mendillo et al. 1990). Wilson et al. (2002) showed that this cloud is formed from the combined effects of atmospheric sputtering and dissociative recombination of molecular sodium ions which produces fast sodium neutrals. Closer to Io, a jet of fast sodium atoms atmosphere has been observed that results from plasma-neutral interactions in the near-surface atmosphere (Burger et al. 1999). At Saturn, clouds of OH and O have been detected in the inner Saturnian magnetosphere (Shemansky et al. 1993; Esposito et al. 2005). A mechanism has been proposed for producing this cloud through the redistribution of neutral water and water dissociation products created through ion-neutral and charge exchange reactions in the Enceladus plume and water torus (Johnson et al. 2006). An alternative mechanism based on neutral-neutral interactions between water molecules in the water torus has also been proposed (Farmer 2009).

The rate per neutral (in units s^{-1}) for reactions between charged and neutral species depends on the ion or electron density (n_i or n_e, respectively) multiplied by a rate coefficient, κ, determined from the reaction cross section, σ, and the velocity distribution function, $f(\mathrm{v})$:

$$\nu_{e-n} = n_e \kappa(f_e(\mathbf{v})) \tag{2.8}$$

$$\kappa = \int_0^\infty \mathrm{v} f_e(\mathrm{v}) \sigma(\mathrm{v}) d\mathrm{v} \tag{2.9}$$

where v is the speed of the charged particles measured relative to the neutrals.

The velocity distribution should include the thermal and bulk motions of the charged and neutral particles. For electron impact ionization, the bulk velocities can be ignored since the electron thermal speeds are much larger than either the electron or neutral flow velocities. Therefore, for a thermalized (Maxwellian) electron population, the rate coefficient is a function of the electron temperature:

$$\kappa(T_e) = 4\pi \left(\frac{m_e}{2\pi k T_e} \right)^{3/2} \int_0^\infty \mathrm{v}^3 e^{-m_e \mathrm{v}^2 / 2kT_e} \sigma(\mathrm{v}) d\mathrm{v} \tag{2.10}$$

For reference, Table 1 includes reaction rate coefficients for reactions important for ionizing neutral gas in the Jovian and Saturnian magnetospheres. Not all reactions which may be important are included due to the lack of laboratory measurements. For example, cross sections have not been measured for the electron impact dissociation of OH to produce O and H, or H_2O to produce O and H_2. Unfortunately, it is difficult to observe neutral products in the ground state, and therefore determine the cross sections and branching ratios (Itikawa and Mason 2005). We also plot the rate coefficients as functions of electron temperature in Fig. 5 for some of these reactions. In reactions with neutral molecules, the two most important reactions are dissociation and ionization, with the former tending to dominate at lower energies and the latter becoming more important with increasing electron temperature. Dissociative ionization, which results in both neutral and ionized products, requires more energy and therefore has lower branching ratios than pure dissociation or ionization. However, these reactions are important for changing the chemistry of the exospheres, and products in excited states often produce emissions which are essential for understanding atmospheric composition. For example, dissociative ionization of O_2 preferentially leaves an O atom in state which decays radiatively, producing a 1356 Å to 1304 Å line ratio indicative of O_2 dissociation rather than electron impact excitation. This led Hall et al. (1995, 1998) to conclude the atmospheres of Europa and Ganymede are dominated by molecular oxygen, despite having observed atomic emission lines.

Ion-neutral reaction rates and rate coefficients are also determined by (2.8) and (2.9). Because ions are more massive than electrons, the relative bulk motion between the ions and neutrals is significant. If the cross section varies slowly, or linearly, with velocity, than the velocity distribution function (2.9) can be approximated by the Dirac delta function, $f(\mathrm{v}) = \delta(\mathrm{v} - \mathrm{v}_{flow})$ such that the rate coefficient reduces to:

$$\nu = n_i \mathrm{v}_{flow} \sigma(\mathrm{v}_{flow}) \tag{2.11}$$

Table 1 Electron impact ionization and dissociation reaction rate coefficients. The references provided give the reaction cross sections. Maxwellian electron distributions have been used for calculating rate coefficients

Reaction	κ(1 eV) (cm^3 s^{-1})	κ(10 eV) (cm^3 s^{-1})	κ(100 eV) (cm^3 s^{-1})	Ref.
$O + e \rightarrow O^+ + 2e$	5.1×10^{-15}	6.6×10^{-9}	7.8×10^{-8}	(Johnson et al. 2005)
$Na + e \rightarrow Na^+ + 2e$	3.2×10^{-10}	7.8×10^{-8}	1.3×10^{-7}	(Johnston and Burrow 1995)
$S + e \rightarrow S^+ + 2e$	1.2×10^{-12}	4.7×10^{-8}	2.3×10^{-7}	(Tawara 2003)
$H + e \rightarrow H^+ + 2e$	7.8×10^{-15}	5.0×10^{-9}	2.8×10^{-8}	(Tawara 2003)
$N + e \rightarrow N^+ + 2e$	8.4×10^{-15}	8.9×10^{-9}	8.3×10^{-8}	(Tawara 2003)
$H_2O + e \rightarrow OH + H + e$	1.9×10^{-12}	1.4×10^{-8}	9.6×10^{-8}	(Itikawa and Mason 2005)
$H_2O + e \rightarrow H_2O^+ + 2e$	1.2×10^{-14}	8.0×10^{-9}	7.1×10^{-8}	(Lindsay and Mangan 2003)
$H_2O + e \rightarrow OH^+ + H + 2e$	3.4×10^{-17}	1.5×10^{-9}	2.3×10^{-8}	(Lindsay and Mangan 2003)
$H_2O + e \rightarrow O^+ + H_2 + 2e$	–	7.8×10^{-11}	3.8×10^{-9}	(Lindsay and Mangan 2003)
$H_2O + e \rightarrow H^+ + OH + 2e$	–	6.0×10^{-10}	1.9×10^{-8}	(Lindsay and Mangan 2003)
$O_2 + e \rightarrow O + O + e$	2.4×10^{-13}	6.9×10^{-9}	1.4×10^{-8}	(Cosby 1993)
$O_2 + e \rightarrow O_2^+ + 2e$	1.0×10^{-14}	7.1×10^{-9}	9.0×10^{-8}	(Straub et al. 1996)
$O_2 + e \rightarrow O^+ + O + 2e$	–	1.6×10^{-9}	5.1×10^{-8}	(Straub et al. 1996)
$OH + e \rightarrow OH^+ + 2e$	1.3×10^{-15}	8.8×10^{-9}	9.4×10^{-9}	(Joshipura et al. 2001)
$SO_2 + e \rightarrow SO_2^+ + 2e$	4.0×10^{-14}	1.3×10^{-8}	1.1×10^{-7}	(Lindsay and Mangan 2003)
$SO_2 + e \rightarrow SO^+ + O + 2e$	1.6×10^{-15}	8.7×10^{-9}	8.8×10^{-8}	(Lindsay and Mangan 2003)
$SO_2 + e \rightarrow S^+ + O_2 + 2e$, $O_2^+ + S + 2e$, $SO_2^{++} + 3e$	1.8×10^{-18}	2.4×10^{-9}	4.4×10^{-8}	(Lindsay and Mangan 2003)
$SO_2 + e \rightarrow O^+ + SO + 2e$	4.8×10^{-19}	7.1×10^{-10}	3.7×10^{-8}	(Lindsay and Mangan 2003)
$H_2 + e \rightarrow H + H + e$	6.0×10^{-15}	3.3×10^{-10}	4.8×10^{-11}	(De La Haye 2005)
$H_2 + e \rightarrow H_2^+ + 2e$	1.0×10^{-15}	5.5×10^{-9}	4.3×10^{-8}	(Lindsay and Mangan 2003)
$H_2 + e \rightarrow H^+ + H + 2e$	7.3×10^{-19}	1.5×10^{-10}	3.4×10^{-9}	(Lindsay and Mangan 2003)
$N_2 + e \rightarrow N + N + e$	3.8×10^{-14}	9.5×10^{-9}	4.3×10^{-8}	(Cosby 1993)
$N_2 + e \rightarrow N_2^+ + 2e$	1.1×10^{-15}	8.5×10^{-9}	9.9×10^{-8}	(Lindsay and Mangan 2003)
$N_2 + e \rightarrow N^+ + N + 2e$, $N_2^{++} + 3e$	–	8.5×10^{-10}	3.2×10^{-8}	(Lindsay and Mangan 2003)

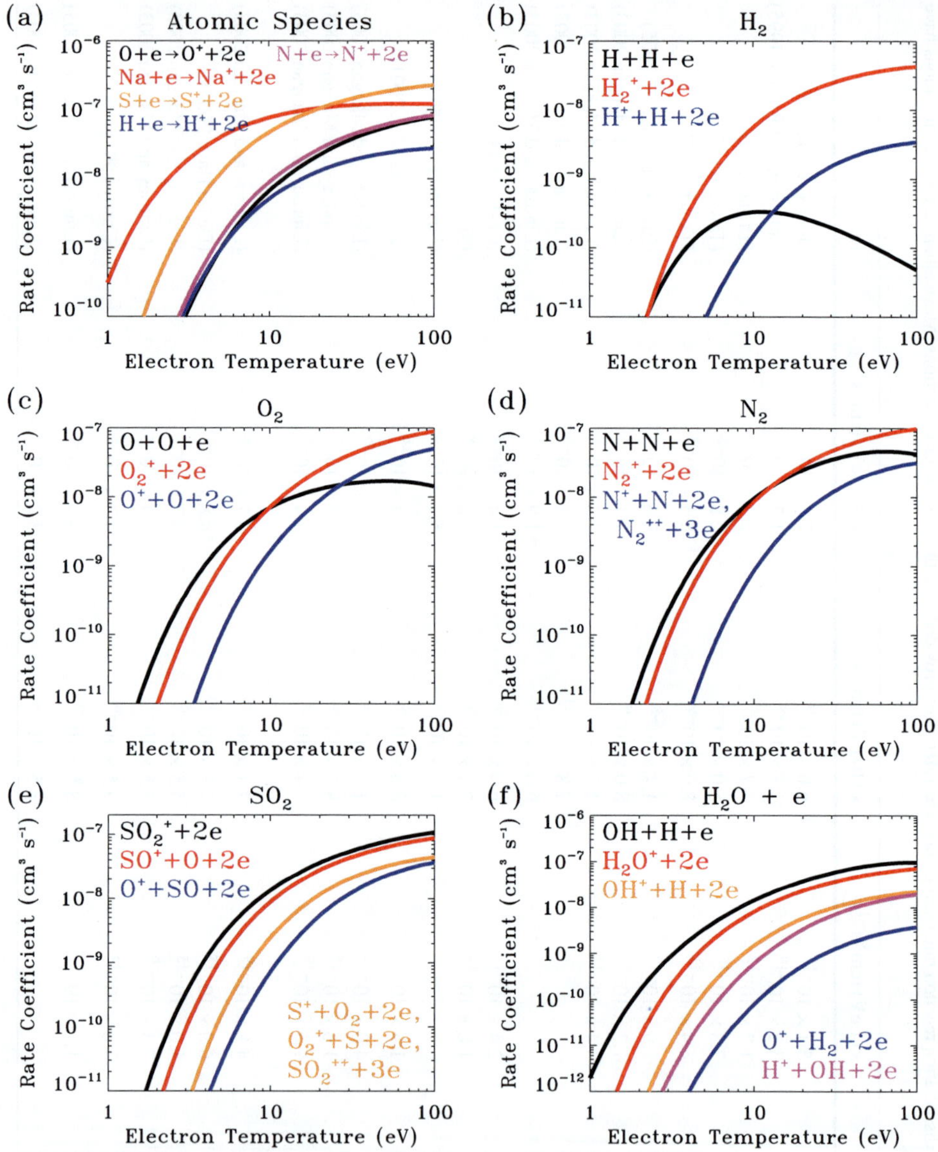

Fig. 5 **a** Electron impact ionization rate coefficients versus electron temperature for atomic O, H, S, Na, and N. **b** Electron impact ionization and dissociation rate coefficients versus electron temperature for molecular hydrogen. **c** Electron impact ionization and dissociation rate coefficients versus electron temperature for molecular oxygen. **d** Electron impact ionization and dissociation rate coefficients versus electron temperature for molecular nitrogen. **e** Electron impact ionization and dissociation rate coefficients versus electron temperature for sulfur dioxide. **f** Electron impact ionization and dissociation rate coefficients versus electron temperature for water

where v_{flow} is the bulk flow velocity of the ions relative to the neutral gas. If the plasma is corotating with the magnetic field and the neutrals are in the satellite exospheres or bound neutral tori, then the relative velocities are well approximated by the difference between

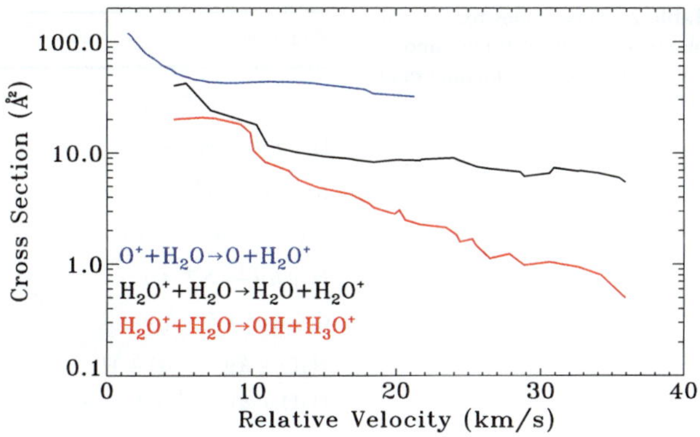

Fig. 6 Charge exchange cross sections versus relative velocity for species important in the Saturnian magnetosphere. References: Lishawa et al. (1990), Dressler et al. (1996)

the magnetic field corotational velocity and the Keplerian orbital velocity: $v_{flow} = |v_{corot} - v_{orb}|$, where $v_{corot} = \Omega_B \rho$ and $v_{orb} = \sqrt{GM/r}$. Figure 6 shows ion-neutral cross sections for reactions between ions important in the Saturnian plasma and H_2O as a function of flow velocity. Charge exchange reactions important in the Io plasma torus and Io's exosphere are discussed in McGrath and Johnson (1989).

If the thermal temperature is large such that the reaction cross section varies non-linearly over the possible ion-neutral speeds, then both the thermal (random) motions of the particles and the bulk motion of the plasma relative to the neutral gas must be considered. This is especially true if plasma is slowed as it flows through satellite exospheres or the Enceladus plume. Johnson et al. (2006) showed that the presence of H_3O^+ in the Saturnian plasma implied reactions between neutral and ionized water molecules at low relative velocities because the cross section for H_3O^+ production is large for speeds below ~10 km/s (Lishawa et al. 1990) (Fig. 6). For high ion temperatures where the thermal velocity is comparable to the flow speed of the plasma through the Enceladus plume, a significant amount of H_3O^+ can be produced without slowing the plasma to a near standstill because the spread in random velocities is such that a significant fraction of the ions are slow, despite a large mean velocity.

Although atmospheric loss from the satellites in the outer solar system is primarily driven by interactions with the magnetospheric plasma, photodissociation and ionization contribute to the loss rates in the neutral tori, especially at Saturn where the plasma is cooler and less dense than at Jupiter. Table 2 lists the photoionization and dissociation lifetimes for important atmospheric species at Jupiter and Saturn. These rates place an upper limit on the residence time for each species, although, in general photo-processes do not significantly contribute to atmospheric loss or plasma mass loading.

2.3 Neutral Lifetimes in the Jovian and Saturnian Magnetosphere

The ionization rates at Jupiter and Saturn vary dramatically with location in the magnetosphere due to radial, latitudinal, and longitudinal variations in the plasma distributions. A full review of the plasma distributions are beyond the scope of this chapter. For a review of the Io plasma torus and plasma distribution in the Jovian magnetosphere, see Thomas et al. (2004). Observations of the Io plasma torus obtained during the approach of Cassini to Jupiter are discussed in a series of papers by Steffl and coworkers (Steffl et al. 2004a, 2004b, 2006, 2008; Delamere et al. 2005). Recent Cassini observations of the Saturnian

Table 2 Time scales for photo-processes at Jupiter and Saturn. Reference: Huebner et al. (1992)

Reaction	τ at Jupiter	τ at Saturn
$O + h\nu \rightarrow O^+ + e$	4.0 yr	14 yr
$H + h\nu \rightarrow H^+ + e$	12 yr	40 yr
$S + h\nu \rightarrow S^+ + e$	290 days	2.7 yr
$Na + h\nu \rightarrow Na^+ + e$	53 days	180 days
$N + h\nu \rightarrow N^+ + e$	4.6 yr	16 yr
$C + h\nu \rightarrow C^+ + e$	2.1 yr	7.1 yr
$H_2O + h\nu \rightarrow OH + H$	30 days	103 days
$H_2O + h\nu \rightarrow H_2O^+ + e$	2.6 yr	8.8 yr
$H_2 + h\nu \rightarrow H + H$	10 yr	35 yr
$H_2 + h\nu \rightarrow H_2^+ + e$	16 yr	54 yr
$O_2 + h\nu \rightarrow O + O$	74 days	251 days
$O_2 + h\nu \rightarrow O_2^+ + e$	1.8 yr	6.3 yr
$O_2 + h\nu \rightarrow O^+ + O + e$	7.8 yr	26.4
$OH + h\nu \rightarrow O + H$	16 days	54 days
$OH + h\nu \rightarrow OH^+ + e$	3.5 yr	12 yr
$SO_2 + h\nu \rightarrow SO + O$	2.0 days	6.7 days
$SO_2 + h\nu \rightarrow S^+ + O_2 + e$	6.2 days	20.9 days
$SO + h\nu \rightarrow S + O$	12 hours	41 hours
$SO + h\nu \rightarrow SO^+ + e$	1.0 yr	3.4 yr
$N_2 + h\nu \rightarrow N_2^+ + e$	2.4 yr	8.3 yr

plasma have been discussed by Sittler et al. (2006, 2008), Wilson et al. (2008), and Persoon et al. (2009).

To estimate the ionization rates in the magnetosphere we use representative plasma distributions based on Bagenal (1994) at Jupiter and Sittler et al. (2006, 2008) at Saturn. Figures 7 and 8 show plasma densities and temperatures in the plasma centrifugal plane at Jupiter and Saturn, respectively, as a function of L-shell, where L is the distance from the planet where a magnetic field line crosses the magnetic equator. Plasma trapped on the rapidly rotating magnetic field lines is subjected to high centrifugal forces. An equilibrium is reached where the field aligned component of the centrifugal forces is zero (Hill et al. 1974; Thomas et al. 2004) and there is no net force along field lines. At Saturn, because the axis of the magnetic dipole is very closely aligned with the rotation axis, the centrifugal plane is coincident with the satellite orbital plane. The satellites, neutral clouds, and the E-ring do not see periodic variations in plasma abundance due to Saturn's rotation. At Jupiter, however, the magnetic dipole axis is tilted $\sim 10°$ relative to the rotation axis. The tilt of the centrifugal plane is $2/3$ of the tilt of the magnetic equatorial plane relative to Jupiter's rotational plane, which is also the satellite orbital plane (Hill et al. 1974). Because of this tilt, the Galilean satellites pass through the densest portion of the Io plasma torus twice each 9.9 hour rotation of Jupiter's magnetic field. The dipole is also offset $0.12\ R_J$ from the center of Jupiter causing a small variation in the L value of the magnetic field line passing through the satellites which also affects the plasma conditions. Outside $\sim 10\ R_J$, non-dipole components of the magnetic field become important and the plasma seen at Ganymede and Callisto is more variable than at the satellites in the inner magnetosphere.

Fig. 7 Jupiter ion and electron temperatures (*top*) and densities (*bottom*) in the centrifugal equator as a function of L shell (from Bagenal 1994)

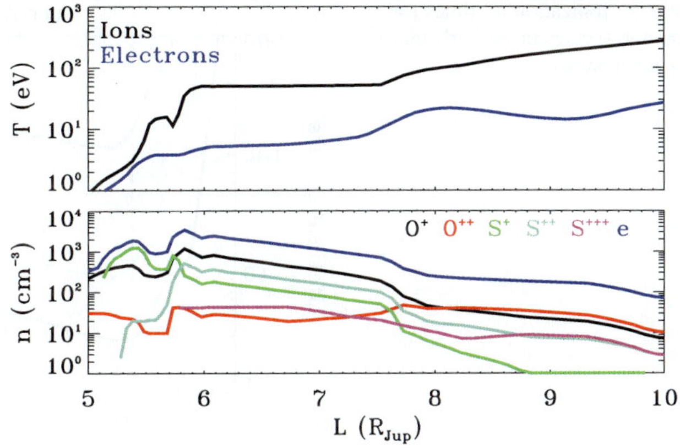

Fig. 8 Equatorial water group (H_2O^+, OH^+, O^+, H_3O^+), proton, and electron temperatures (*top*) and densities (*bottom*) in the inner Saturnian magnetosphere from Sittler et al. (2008)

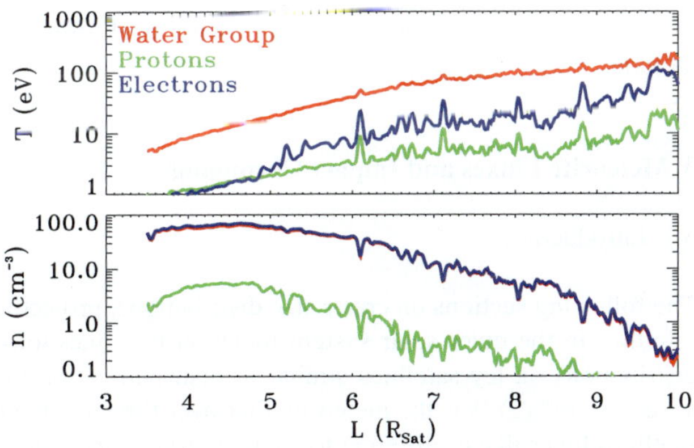

Figures 9 and 10 plot the lifetimes for important neutral species at Jupiter and Saturn versus distance from the planet in the centrifugal plane. The lifetime of a species is the inverse of the reaction rate per neutral (ν) defined in (2.8) and (2.11). The electron impact lifetimes shown only include the cool component of the plasma due to uncertainties in the warm component. However, as Burger et al. (2007) have shown, a small hot electron component can make a large difference in the electron impact ionization and dissociation rates at Saturn. Near Enceladus, where the electron temperature is $\lesssim 1$ eV, a hot electron population with $T_e > 10$ eV making up only a few percent of the total electron density can dominate the electron impact loss rate, and therefore be significant in determining the ion production rates.

The lifetimes for Jupiter shown in Fig. 9 should be considered approximate lower limits on the lifetimes (or upper limits on the ionization rates). As the magnetic field sweeps past them, Io and Europa move in and out of the centrifugal plane, reaching vertical heights of 1 R_J and 1.6 R_J from the equator, respectively. Bagenal (1994) found that the plasma scale height near Io is $\sim$1 R_J and increases outward of Io's L shell. There are, however, longitudinal and local time variations in the ion temperature (and therefore the scale height) and density (Schneider and Trauger 1995; Steffl et al. 2006, 2008). These variations all affect the local plasma conditions, such that estimating the Io plasma torus densities and temperatures during a neutral atmosphere observation requires either simultaneous observations of the

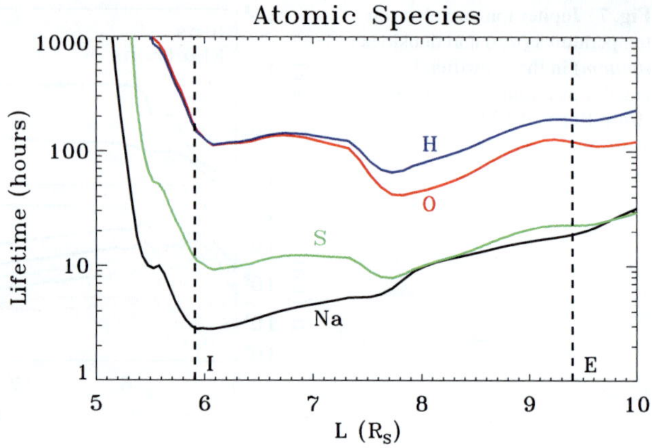

Fig. 9 Ionization lifetimes for atomic species in Jupiter's inner magnetosphere

plasma or a model of plasma variations with local time, magnetic longitude, and magnetic latitude.

3 Meteorite Fluxes and Impact Chronology

3.1 Introduction

The following sections on crater size distributions, projectile fluxes and surface ages on icy satellites in the outer solar system focus on two questions: (1) Are crater size-frequency distributions on icy satellites similar to distributions on the terrestrial bodies in the inner solar system? (2) Was the meteorite flux over time and the bombardment history of the icy satellites lunar-like or different to the terrestrial planets in the inner solar system?

First, we briefly review the current status of knowledge about crater distributions on icy satellites and about possible projectile families which were responsible for the craters on these bodies. Second, we compare existing impact chronology models and infer how differences in these models influence the interpretation of the geologic history of each satellite. Third, we discuss the strengths and weaknesses in our understanding of the cratering chronologies and suggest requirements for further orbiting and landing missions.

The datasets used for determining crater counts on the icy satellites were obtained by spacecraft sent to study the outer solar system. During the six flybys of the Voyager 1 and 2 spacecraft at the four large planets in the outer solar system between March 1979 and August 1989, the cratered landscapes of the icy satellites were first observed at medium resolution with an average of 1 km/pxl or less, and with a maximum of 500 m/pxl for at least four satellites (Ganymede, Rhea, Miranda, Triton). First investigations of the crater distributions on these bodies were presented in a series of Science papers (Smith et al. 1979a, 1979b, 1981, 1986, 1989).

In this section we will concentrate on the Jovian and Saturnian satellites because these have been observed at much higher resolution and areal coverage since Voyager times by the cameras aboard the Galileo and Cassini Orbiter. Imaging data of the Galilean satellites Europa, Ganymede and Callisto (the craterless surface of Io is not treated here) were obtained by the Solid-State Imaging (SSI) System aboard the Galileo spacecraft between December 1995 and September 2003 (Belton et al. 1992). Despite technical problems with the High Gain Antenna (HGA), which could not be fully deployed causing

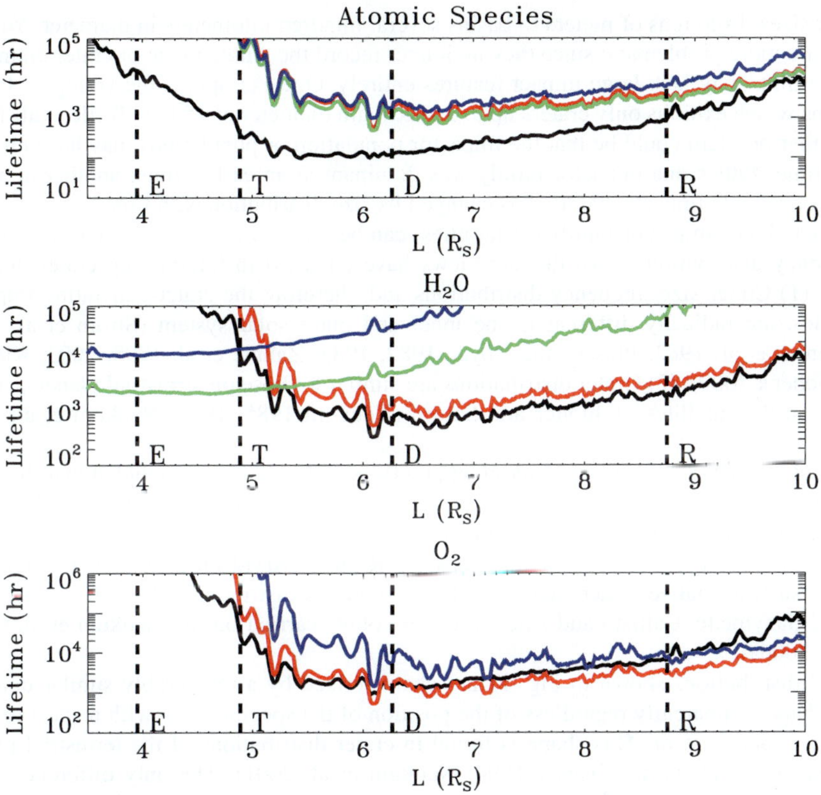

Fig. 10 Neutral species lifetimes at Saturn versus L shell. **a** Atomic species. Electron impact ionization of O (*red*), N (*green*), H (*blue*) and Na (*black*). **b** Water: $H_2O + e \rightarrow OH + H + e$ (*black*), $H_2O + e \rightarrow H_2O^+ + 2e$ (*red*), $H_2O^+ + H_2O \rightarrow H_2O + H_2O^+$ (*blue*), $O^+ + H_2O \rightarrow O + H_2O^+$ (*green*). **c** O_2: $O_2 + e \rightarrow O + O + e$ (*black*), $O_2 + e \rightarrow O_2^+ + 2e$ (*red*), $O_2 + e \rightarrow O^+ + O + 2e$ (*blue*)

a strong reduction in data compared to original imaging plans, SSI was able to greatly expand the image data base of the three icy Galilean satellites (Belton et al. 1992; Carr et al. 1995).

Global coverage of the Galilean satellites by SSI in combination with Voyager at resolutions of 0.5–1 km/pxl are more or less complete, except for Callisto. Higher resolutions <50 m/pxl, however, are available only from a small number of target areas on these satellites.

The Cassini Obiter has been in orbit around Saturn since July 1, 2004 and is currently in its first extensional phase, the Cassini Extended Mission or Cassini Equinox Mission. So far, the two cameras aboard Cassini, a Narrow Angle Camera (NAC) and a Wide Angle Camera (WAC) (Porco et al. 2004), have returned images of all major icy satellites, including some of the smaller ones, more or less globally at resolutions up to 150 m/pxl. Also, higher resolutions <10 m/pxl are available from selected target areas.

3.2 Crater Size-Frequency Distributions on Icy Satellites and Impactor Families

Production crater size distributions, which reflect the crater-generating impactor size distribution, can be measured in the oldest, densely cratered geologic units over a wide range of crater sizes, from tens of meters to basins several hundred kilometers in diameter. Younger units are more problematic since they no longer record the whole range of crater diameters. Young units may lack large impact features entirely. One example is the young surface of Europa which exhibits only craters up to a maximum diameter of 45 km (Turtle et al. 1999). A further problem could be that the impactor population or populations may have changed over time: Either one impactor family was dominant at an earlier time, another at a later time, or a single impactor family has changed its size distribution with time.

A candidate impactor family (or families) can be inferred from the shape of crater size-frequency distributions. Two distinct views have emerged in interpreting crater distributions: (1) Crater size-frequency distributions and, therefore the crater-generating impactor families, are radically different in the inner and outer solar system (Strom et al. 1981; Woronow et al. 1982; Plescia and Boyce 1982, 1983; Zahnle et al. 1998, 2003; Kirchoff and Schenk 2009). (2) Crater distributions are similar to those on terrestrial planets, at least those on the satellites of Jupiter and Saturn (Neukum 1985, 1997; Neukum et al. 1998, 2005, 2006a).

Neukum (1985, 1997) and Neukum et al. (1998, 2005, 2006a) concluded that the crater size distribution on the terrestrial planets and on the icy satellites are similar. Such similarities, or dissimilarities, between crater distributions on different target objects are enhanced in relative crater size-frequency diagrams, or R-plots (Arvidson et al. 1979). Figure 11 shows such a relative crater diagram which compares distributions from the lunar highlands, Ganymede, Callisto and Rhea in crater counts carried out by Neukum et al. (1998, 2005, 2006a).

All distributions shown in Fig. 11 are characterized by a remarkably similar upturned "U"-shape—apparently regardless of the position of the specific body with respect to inner or outer solar system. This shape is found in crater distributions of the terrestrial planets in general (Neukum and Ivanov 1994; Neukum et al. 2001). The only difference is the location of the maximum in crater frequency, which varies in diameter from one body to another. This maximum is approximately located at 60–70 km on the Moon, at 40 km on Ganymede, at 20–30 km on Callisto, and at 10–20 km on Rhea (note that because of erosion which affected craters smaller than 60 km in the distribution from the lunar highlands, this upturned "U" appears more narrow than those of Ganymede, Callisto, and Rhea). Because of these similarities, Neukum (1997) and Neukum et al. (1998, 2005, 2006a) demonstrated that crater distributions on the icy satellites could be fitted by the lunar production function, a polynomial of eleventh degree (see, e.g., Neukum and Ivanov 1994; Neukum et al. 2001).

Different impact conditions on the Moon and on a specific icy satellite must be taken into account for such a fit (Neukum 1997; Neukum et al. 1998, 2005, 2006a). These differences are reflected in the location of the R-plot maxima, as mentioned above and shown in Fig. 11. A production function polynomial for a specific satellite is empirically derived by shifting the lunar polynomial in log-D, approximately by a factor obtained from the ratio between the crater diameters at the R-plot maximum of the Moon and of the satellite. This procedure is based on the relationship between impact condition and the final crater diameter, known as crater scaling (e.g., Chapman and McKinnon 1986, and references therein): A projectile of a given diameter and mass creates craters of different diameters on different objects, reflecting differences in impact conditions, such as impact velocity, projectile, or target properties (density, strength), etc.

A similar complex shape with an upturned "U" in relative diagrams has been shown to be characteristic for Main Belt Asteroids (Durda et al. 1998; Ivanov et al. 2001). Neukum et al. (1998, 2005, 2006a) therefore concluded that Main Belt Asteroids were the dominant impactor family both on terrestrial planets and on the icy satellites of Jupiter and Saturn and created the majority of the craters on these bodies.

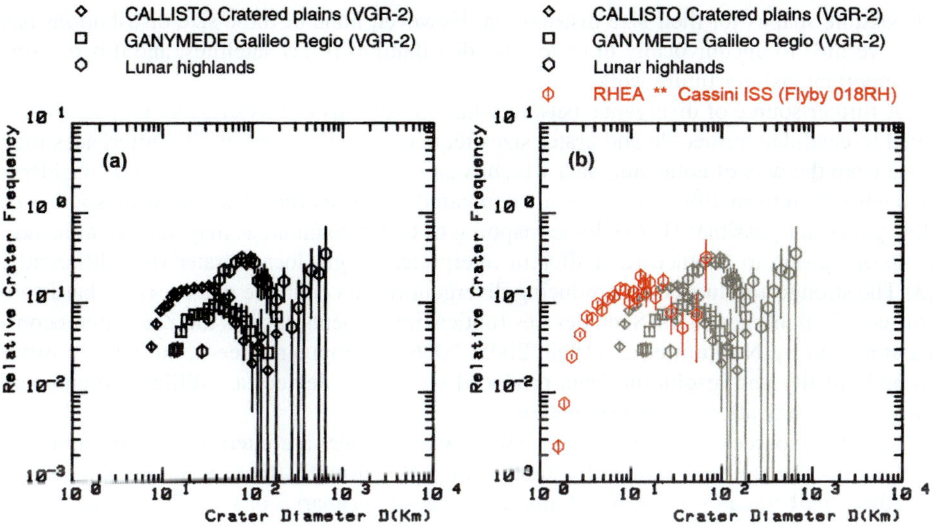

Fig. 11 R-plots showing crater size-frequency distributions from the Moon and from selected icy satellites. **a** Comparison of a lunar highland distribution with distributions from the dark, densely cratered plains on the Jovian satellites Ganymede and Callisto (Neukum et al. 1998). **b** Same distribution as in **a**, shown in comparison with a distribution measured in densely cratered plains of the Saturnian satellite Rhea (*red points*) (Neukum et al. 1998, 2005, 2006a). Further explanation is given in the text

The complex shape of the asteroidal size distribution is caused by the collisional history of these bodies (Durda et al. 1998). In cumulative size-frequency diagrams, this is inferred from a wavy structure at larger projectile sizes (>1 km) and a more or less steep distribution at smaller projectile sizes which in turn is also reflected in the crater size-frequency distribution. In R-plots, this wavy structure is translated into convex and concave parts of the distribution, such as the upturned "U" shown in Fig. 11.

Other groups interpreted the shape of crater size-frequency distributions on the icy satellites as being unlike those on the terrestrial planets and assumed a radically different impactor family in the outer solar system with a specific bombardment history, as will be discussed below in Sect. 3.3 (Strom et al. 1981; Shoemaker and Wolfe 1982; Woronow et al. 1982; Zahnle et al. 1998, 2003). In their view, short-period comets (also termed ecliptic comets), such as the Jupiter Family Comets (JFCs), are responsible for the craters on the icy satellites. Main Belt Asteroids, as well as long-period and Halley-type comets, are thought to be negligible, at least in their contribution to current impact rates. However, this may have been different at earlier times (Zahnle et al. 1998, 2003).

Using new Galileo SSI and Cassini ISS images, several groups of investigators favoring comets as preferential impactors on icy satellites could also show that the shape of the distributions is complex rather than a simple power law as assumed by, e.g., Shoemaker and Wolfe (1982) (Zahnle et al. 1998, 2003; Schenk et al. 2004; Kirchoff and Schenk 2009). Regardless of an asteroidal or cometary source of preferential impactors, one can draw the conclusion that craters on the icy satellites most likely were created by members of a collisionally evolved projectile family (Wagner et al. 2006).

Collisional evolution and a complex shape in the size distribution of potential impactor families from the outer solar system has been discussed. Tancredi et al. (2006) presented possible size distributions of Jupiter Family Comets. Benavidez and Campo Bagatin (2009) discussed collisional evolution of Trans-Neptunian Objects (TNOs), coming up with a series of possible shapes for their size distribution. However, none of these size distributions has been verified to unequivocally fit crater size distributions of icy satellites, and this remains an important task for future work.

A further source of divergence between different groups of investigators may not be related to candidate projectile and crater size-frequency distributions at all: Differences may arise from the way of collecting data which is an issue that should be taken into consideration when data from different groups are compared to one another. The following sources of divergence are possible: (1) Geologic mapping of crater count areas may vary significantly from one group to another. (2) Different interpreters might locate crater rims differently. (3) The strongest influence on producing divergent results could be due to which hardware is used for counting craters. Such devices (optical instruments such as the Zeiss Stereocomparator used by Neukum et al. (1998, 2005, 2006a), computer screens, etc.) might differ in both quality and resolution. Data produced with these devices are difficult to compare directly and may affect the interpretation.

A further source of divergence lies within which craters are interpreted as primaries or secondaries, hence were created by another impact. Currently, there is an ongoing debate whether small craters <1–2 km in diameter are mostly primaries or secondaries and, for the latter case, are recommended not to be used for age dating (e.g., Bierhaus et al. 2004, 2005; McEwen et al. 2005).

An alternative view of other groups of investigators is that the contribution of secondary craters indistinguishable from primaries can cause an uncertainty in age only up to factor of 2 (e.g., Hartmann 2005, Hartmann et al. 2008; Werner 2006a; Werner et al. 2006b; Neukum et al. 2006b). According to these authors, it is permissible to use small craters for crater counts and age dating. However, care must be taken in excluding potential secondaries such as small craters in clusters or rays, or oddly-shaped craters, from crater counts.

A final point mentioned briefly here is the *issue of saturation*: a surface exposed long enough to an incoming flux of projectiles reaches a state of equilibrium between existing craters and newly created ones. For distributions with slopes originally steeper than -2, the slope of the equilibrium distribution has a slope of -2 for craters smaller than a critical diameter while the distribution for craters larger than this diameter still show a production distribution (Neukum and Dietzel 1971; Neukum and Ivanov 1994; Neukum et al. 2001; Chapman and McKinnon 1986). The critical diameter shifts to larger diameters with time.

Galileo and Cassini camera data were used to show that crater size-frequency distributions even in the most densely cratered regions on the satellites of Jupiter and Saturn are still production distributions and reflect the underlying impactor size distributions. Exceptions are found in some locations (or geologic units) and also at small subkilometer craters, e.g. on Ganymede and Callisto, or for craters <10 km in diameter in the oldest terrain on Iapetus (Neukum et al. 1998, 2005, 2006a; Wagner et al. 2006; Wagner 2007).

3.3 Chronology Models for Bodies in the Outer Solar System

For want of surface materials and radiometric data from icy satellites in the outer solar system, the only option to derive surface ages is by using impact cratering chronology models, derived from estimating impact rates in each satellite system and on each individual body. There are two choices of how to derive such a model (Schenk et al. 2004): One approach is to use current impact rates of potential impactors and to extrapolate these rates back in time. A second approach is to compare the bombardment history of the Moon and the terrestrial planets with possible bombardment histories of bodies in the outer solar system and to assume similar early impact and cratering rates and time dependencies. The lunar-like case is well documented for the Moon by radiometric data from rock samples collected on the lunar surface (e.g., Neukum and Ivanov 1994; Neukum et al. 2001).

A lunar-like bombardment history is characterized by an exponentially declining impact and cratering rate during the first 800–900 Ma after planetary formation, a possible spike (or several spikes) in impact rate such as the so-called lunar cataclysm in this time period notwithstanding (see, e.g., Solomon and Head 2007), and a more or less constant rate since about 3 Ga (Neukum and Ivanov 1994, and references therein). For both approaches, an essential indicator for candidate impactors is the shape of a given crater size-frequency distribution which reflects the projectile size distribution, as discussed in the previous section.

Historically, lunar-like chronology models were the first ones to be published for outer solar system bodies. A lunar-like model, based on the assumption of comets as preferential impactors, was derived by Shoemaker and Wolfe (1982) for the Jovian system, and by Plescia and Boyce (1982, 1983, 1985) for the Saturnian system. Contrarily, Neukum (1985, 1997) and Neukum et al. (1998, 2005, 2006a) assumed a lunar-like bombardment history for the Jovian and Saturnian satellites primarily by asteroids from the Main Belt. Thus, a period of heavy bombardment in the first 800–900 Ma, followed by a constant impact and cratering rate until the present, was thought to be common both in the inner and outer solar system.

Mathematically, a lunar-like model (e.g., Neukum and Ivanov 1994) can be formulated as the sum of an exponential and a linear term. For a cumulative frequency N_{cum}, crater diameter D, a reference crater diameter D_{ref} and time t (in Ga), the model has the form:

$$N_{cum}(D \geq D_{ref}) = p_1\left(e^{p_2 t} - 1\right) + p_3 t \tag{3.1}$$

Generally D_{ref} is 1 km (Neukum and Ivanov 1994; Neukum et al. 2001), but the expression can be scaled to any crater diameter by using the slope of the production function (either as a polynomial or a simple power law) to calculate the ratio in crater frequency at a chosen reference crater diameter (e.g. 10 km) and at a diameter of 1 km.

The three coefficients p_1, p_2 and p_3 are specific for each body. To define these coefficients for one satellite, a fixed point in time is needed to correlate a measured crater frequency with time. For Ganymede, Neukum (1997) and Neukum et al. (1998) estimated the age of Gilgamesh, the stratigraphically youngest basin, to be comparable to the age of the Orientale basin on the Moon, on the order of 3.8 Ga. The oldest surface on any of the Saturnian satellites is found on Iapetus and is estimated to be on the order of 4.4 Ga (Neukum et al. 2006a; Castillo-Rogez et al. 2007). Lunar-like model chronologies for every satellite can then be derived comparing inter-satellite impact and cratering rates within each satellite system (e,g., Horedt and Neukum 1984b). For the Zahnle et al. (1998, 2003) models, the uncertainties of each model are also included in the diagrams. The uncertainty of the Neukum et al. (1998) model is on the order of a factor of 2–3, similar to lunar-like cratering chronologies in the inner solar system (e.g., Neukum and Ivanov 1994; Neukum et al. 2001).

A graph of the model function derived for the Jovian satellite Callisto (Neukum 1997; Neukum et al. 1998) compared to the models by Zahnle et al. (1998, 2003) is shown in Fig. 12.

Since the first publication of the Shoemaker and Wolfe chronology model in 1982, based primarily on cometary impacts on the Galilean satellites of Jupiter, the observational database of such objects has been greatly extended. Also, the Edgeworth-Kuiper Belt was discovered as the source of the so-called ecliptic or short-period comets (Zahnle et al. 1998;

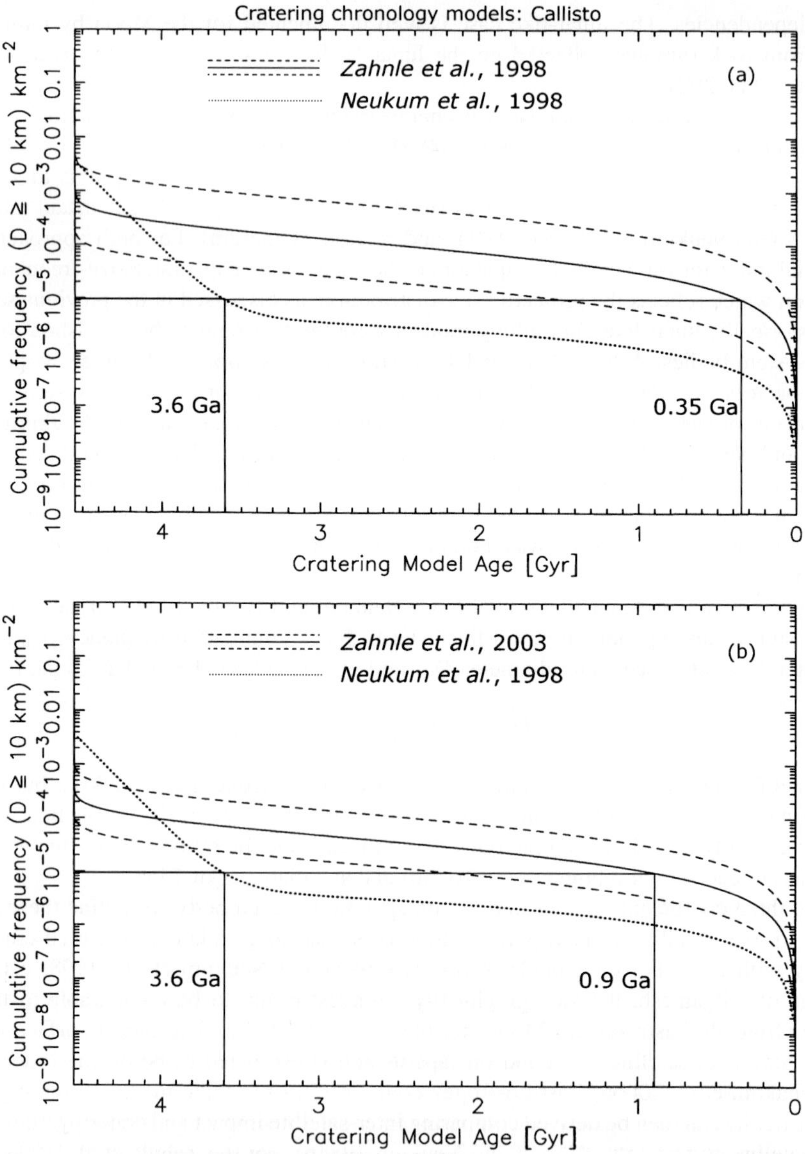

Fig. 12 Comparison of the chronology models by Neukum et al. (1998) (*dotted lines*), and by Zahnle et al. (1998, 2003) (*solid lines*). The *dashed lines* in (**a**) and (**b**) represent lower and upper uncertainties of the Zahnle model, derived from uncertainties in impact and cratering rate of a factor of 5 (1998) and of a factor of 3 (2003). The measurement of a distribution from Callistos smooth dark cratered plains with a cumulative frequency of 9×10^{-6} for $D \geq 10$ km (Wagner 2007), was selected to demonstrate the highly divergent cratering model ages between Neukum et al. (1998) and Zahnle et al. (1998, 2003). Further explanation is given in the text

Schenk et al. 2004, and references therein). Ever since, several groups of investigators have come up with chronology models by extrapolating current impact rates of observable cometary objects back in time (e.g., Shoemaker 1996; Levison and Duncan 1997; Levison et al. 2000; Zahnle et al. 1998, 2003).

The chronology models by Zahnle et al. (1998, 2003), derived for all major icy satellites in the outer solar system, are based on the assumption that the majority of potential impactors are heliocentric ecliptic or short-period comets (including the Jupiter-family comets (JFCs)) from the Kuiper Belt, while Halley-type short-period and long-period comets do not contribute significantly to current impact rates. Also, the contribution of asteroids from the Main Belt to impact rates on these satellites appears to be negligible at present time. However, small primary craters at least on the icy Galilean satellites of Jupiter could have been created by members of specific asteroid families, such as the Trojans (Zahnle et al. 1998, 2003; Kuzmitcheva and Ivanov 2002) or members of the Hilda group (Brunini et al. 2003; Di Sisto et al. 2005).

Zahnle et al. (1998, 2003) pointed out that the highest uncertainties in the derivation of impact and cratering rates were caused by uncertainties in defining numbers, masses (and/or diameters), and the size distribution of these candidate impactors. Another source of uncertainty is introduced by crater scaling used to calculate cratering rates from correlated impact rates (e.g., Zahnle et al. 1998). For the Jovian satellites, Zahnle et al. (2003) discussed uncertainties in cratering rates on the order of a factor 3 (a factor of 5 in their earlier model in 1998). For the satellites of Saturn, Uranus and Neptune uncertainties in cratering rate are a factor of 4, 6 and 10 respectively (Zahnle et al. 2003).

For the Saturnian satellites, there is no unique solution for the chronology model given by Zahnle et al. (2003). The authors distinguish between two cases: Case A is based on a mass distribution of comets at Saturn similar to the one at Jupiter. Case B assumes a mass distribution of comets responsible for creating numerous small craters on Neptune's satellite Triton. Both cases yield different cratering model ages for the same satellite or the same surface area on a specific satellite, as discussed by Zahnle et al. (2003) and shown by examples given therein.

For a measured cumulative frequency N_{cum} equal to, or larger than, a reference diameter $D_{ref} = 10$ km, a cratering rate C specific for each satellite, and time t' (in Ga), the Zahnle et al. (1998, 2003) chronology model is expressed as:

$$N_{cum}(D \geq 10 \text{ km}) = \frac{dC}{dt}t' \tag{3.2}$$

The early cratering rate is uncertain because it is not clear if the present-day constant cratering rate expressed in this equation can be linearly extrapolated back into earliest times. Furthermore, it is not possible to create the high density of impact craters on many of these satellites within 4.56 Ga with a constant cratering rate alone. Therefore, Zahnle et al. (1998) (and references therein) discussed possible secular variations in impact rate over time and extrapolated the cratering rates back by using a collisionless Kuiper Belt dissipating proportional to $1/t$. Taking this into consideration, and using t' from (3.2) and $t_0 = 4.56$ Ga, the model age t of a given surface area then is

$$t = t_0(1 - e^{t'/t_0}) \tag{3.3}$$

Cratering model ages from measured crater frequencies using the chronology given by Zahnle et al. (1998, 2003) can then be obtained by the combination of (3.2) and (3.3).

Heliocentric impactors such as ecliptic comets should show a pronounced asymmetry in crater frequency between the apex and antapex point of orbital motion on synchronously rotating satellites (e.g., Passey and Shoemaker 1982; Shoemaker and Wolfe 1982; Horedt and

Neukum 1984a). On Ganymede, Schenk and Sobieszczyk (1999) found such an asymmetry only in the younger bright terrain, although a factor of about 10 less than predicted. No such asymmetry was found in the dark cratered plains on Callisto (Wagner 2007). Therefore, if heliocentric impactors are abundant, then such a lack of asymmetry, or a weak asymmetry, can only be explained if (a) there was a significant amount of planetocentric impactors which impacted at equal rates over all geographic longitudes (Horedt and Neukum 1984a), or if (b) the body rotated nonsynchronously, at least for some time in the past (Schenk et al. 2004).

After the two Voyager encounters with Saturn and its satellites in 1980 and 1981, Smith et al. (1981, 1982) concluded that these satellites were bombarded by two different impactor populations: population I consists primarily of heliocentric projectiles responsible for creating craters larger than $\sim$20 km on old surfaces such as Rhea, while population II represents members of a planetocentric impactor family and is responsible for smaller craters ($\lesssim$20 km) in resurfaced regions (e.g. on Dione or Enceladus).

The existence of two projectile (or crater) populations on the moons of Saturn was put into doubt by, for example, Lissauer et al. (1988) and Neukum et al. (2005). Instead, crater size-frequency measurements carried out on Cassini ISS data appear to show only a single population, as reported by Neukum et al. (2005, 2006a).

The chronology models by Zahnle et al. (1998, 2003) and Neukum et al. (1998) are shown in comparison in Fig. 12. For simplicity, the average cratering rates given in the Zahnle et al. models were used, neglecting any possible apex-antapex asymmetries. For the Zahnle et al. model, uncertainties are included as dashed lines. No uncertainties are shown for the Neukum et al. (1998) model. Because of the exponential decay in cratering rate postulated in this model for ages prior to 3 to 3.3 Ga, uncertainties in measured crater frequencies translate into model age uncertainties on the order of $\sim$30 Ma. For ages younger than $\sim$3 Ga, the uncertainty in model age can amount up to $\sim$0.5 Ga because of the constant cratering rate and, as a consequence, of the small increase in crater frequency with time.

Both models yield divergent results for the same geologic unit and measured crater frequency, as shown in Fig. 12. The example chosen is from dark smooth cratered plains on Callisto, counted at high resolution (10–20 m/pxl), for an average Neukum et al. (1998) model age of 3.6 Ga (Wagner 2007). There is a factor of 10 difference in age between Neukum et al. (1998) and Zahnle et al. (1998), and still a factor of 3.5 between Neukum et al. (1998) and Zahnle et al. (2003). For ages around 4 Ga, both models are comparable.

3.4 Summary

Despite a now greatly extended crater size-frequency database obtained by the cameras aboard Galileo and Cassini which imaged the major Jovian and Saturnian satellites (and data from the latter mission are still accumulating), different groups of investigators still arrive at divergent conclusions about the preferential impactor source and, consequently, about the most likely bombardment history. These divergences may have several reasons: (1) Different groups use different hardware with different quality and resolution for crater counts. (2) Tabulating or binning crater size-frequency data may be different between individual groups. (3) Different groups, when counting the same area on the same target body, do not agree in terms of geologic units: geologic boundaries may have been mapped differently, or even may have been neglected. (4) The identification of primaries and secondary craters, and the exclusion of secondary craters from crater counts, may be different between different groups of investigators.

A further source of divergence is within a somewhat tacit assumption among members in every group: each group has "their" favourite model which may primarily influence the interpretation of the data. Therefore, we conclude this section by a brief summary of the current status of crater size-frequency distributions from the icy Jovian and Saturnian satellites and possible bombardment histories these bodies. As we will see, there is no final answer yet because all data can be interpreted in multiple ways.

The crater size-frequency distributions of the icy satellites of Jupiter and Saturn have been shown to have a complex shape with a variety of slopes in either cumulative, differential, or relative graphical representation, rather than a simple power law (Neukum et al. 1998, 2006a; Zahnle et al. 1998, 2003). Neukum et al. (1998, 2006a) use the lunar highland distribution, an 11th degree polynomial describing the lunar production function (Neukum and Ivanov 1994; Neukum et al. 2001), to fit their crater distribution data by shifting this distribution in log-D (where D is the crater diameter) to account empirically for different crater scaling on the Moon and on each icy satellite. The authors could demonstrate that this fit worked remarkably well for the icy satellites, at least for the most densely cratered surfaces on these bodies.

The most straightforward explanation for this strong similarity is that the majority of the craters in the most densely cratered units on the icy moons of Jupiter and Saturn were created from Main-belt asteroids, as was pointed out in detail by Neukum (1985, 1997) and Neukum et al. (1998, 2006a). The complex, wave-like appearance of the crater distributions is also found in the size distribution of Main-belt asteroids and is caused by the collisional evolution of these bodies over time (Durda et al. 1998; Ivanov et al. 2001).

Alternatively, the majority of craters on the icy satellites could also have been created by cometary bodies if these bodies have undergone collisional evolution producing a wave-like shape of their size distribution similar to the one of asteroids. Collisional evolution has been discussed for Edgeworth-Kuiper Belt objects (KBOs) (e.g., Davis and Farinella 1997) or Trans-Neptunian objects (e.g., Benavidez and Campo Bagatin 2009). Observations of the possible size distributions, for example of Jupiter-family comets, infer more or less power-law distributions (e.g., Tancredi et al. 2006). While collisional evolution of potential impactors external to the asteroid Main Belt cannot be ruled out, it has not been verified up to now if their size distribution could fit crater size distribution observed on icy satellites.

Finally, several questions still remain unanswered: Does the complex shape of crater distributions measured on the icy satellites of Jupiter and Saturn reflect (a) a preferential asteroidal impactor source, (b) a preferential collisionally evolved cometary (or general outer solar system) impactor source, or (c) even a mixture of both types of impactors, with varying contributions at different times respectively? Determining the answer to this question requires additional work in the field of collisional evolution modelling and observations of potential cometary impactor populations. Also for the part of crater size-frequency measurements, a future mission to Uranus similar to Galileo and Cassini could provide new insights in crater distributions on outer solar system bodies and inter-satellite comparisons between the icy moons of the Jovian planets.

Furthermore and last, a future lander mission to one of the icy satellites (possibly Europa) could provide a radiometric age for at least one data point to calibrate a cratered surface with an absolute age. Such a landing site should be located in a lightly cratered region, with an admixture of secondaries to be mostly ruled out, and with the surface materials to be dated produced by either impact melt or cryovolcanism. Europa's surface offers a number of promising localities for such a task.

References

R.E. Arvidson, J. Boyce, C. Chapman, M. Cintala, M. Fulchignoni, H. Moore, G. Neukum, P. Schultz, L. Soderblom, R. Strom, A. Woronow, R. Young, Icarus **37**, 467–474 (1979)

K.H. Baines, P.A. Yanamandra-Fisher, L.A. Lebofsky, T.W. Momary, W. Golisch, C. Kaminski, W.J. Wild, Icarus **132**, 266–284 (1998)
F. Bagenal, J. Geophys. Res. **99**, 11043–11062 (1994)
F. Bagenal, T.E. Dowling, W.B. McKinnon, *Jupiter: The Planet, Satellites, and Magnetosphere* (Cambridge University Press, Cambridge, 2004)
R.A. Baragiola, Planet. Space Sci. **51**, 953–961 (2003)
R.A. Baragiola, R.A. Vidal, W. Svendsen, J. Schou, M. Shi, D.A. Bahr, C.L. Atteberry, Nucl. Instrum. Methods Phys. Res. B **209**, 294–203 (2003)
M.J.S. Belton et al., Space Sci. Rev. **60**, 413–455 (1992)
P.G. Benavidez, A. Campo Bagatin, Planet. Space Sci. **57**, 201–215 (2009)
E.G. Bierhaus, C.R. Chapman, W.J. Merline, in *Lunar Planet. Sci. Conf. XXXIV*, Abstr. No. 1963 [CD-Rom] (2004)
E.G. Bierhaus, C.R. Chapman, W.J. Merline, Nature **437**, 1125–1127 (2005)
R.A. Brown, in *Exploration of the Planetary System*, ed. by A. Woszczyck, C. Iqaniszewska (Reidel, Boston, 1974)
M.E. Brown, R.E. Hill, Nature **380**, 229–231 (1996)
W.L. Brown, L.J. Lanzerotti, J.M. Poate, W.M. Augustyniak, Phys. Rev. Lett. **40**, 1027–1030 (1978)
W.L. Brown, W.M. Augustyniak, E. Simmons, K.J. Marcantonio, L.J. Lanzerotti, R.E. Johnson, C.T. Reimann, G. Foti, V. Pirronello, Nucl. Instrum. Methods Phys. Res. A **198**, 1–8 (1982)
A. Brunini, R.P. Di Sisto, R.B. Orellana, Icarus **165**, 371–378 (2003)
M.H. Burger, N.M. Schneider, J.K. Wilson, Geophys. Res. Lett. **26**, 3333–3336 (1999)
M.H. Burger, E.C. Sittler, R.E. Johnson, H.T. Smith, O.J. Tucker, V.I. Shematovich, J. Geophys. Res. **112**, A06219 (2007)
R.W. Carlson, W.M. Calvin, J.B. Dalton, G.B. Hansen, R.L. Hudson, R.E. Johnson, T.B. McCord, M.H. Moore, in *Europa*, ed. by R. Papparlardo et al. (University of Arizona Press, Tuscon, 2009)
M.H. Carr et al., J. Geophys. Res. **100**(E9), 18935–18956 (1995)
T. Cassidy et al., Space Sci. Rev. (2010, this issue). doi:10.1007/s11214-009-9625-3
J.C. Castillo-Rogez, D.L. Matson, C. Sotin, T.V. Johnson, J.I. Lunine, P.C. Thomas, Icarus **190**, 179–202 (2007)
C.R. Chapman, W.B. McKinnon, in *Satellites*, ed. by J.A. Burns, M.S. Matthews (Univ. of Arizona Press, Tucson, 1986), pp. 492–580
P.C. Cosby, J. Chem. Phys. **98**, 9560–9569 (1993)
A. Coustenis et al., Space Sci. Rev. (2010, this issue). doi:10.1007/s11214-009-9615-5
Dalton et al., Space Sci. Rev. (2010, this issue)
D.R. Davis, P. Farinella, Icarus **125**, 50–60 (1997)
V. De La Haye, Ph.D. Thesis, University of Michigan (2005)
P.A. Delamere, F. Bagenal, A. Steffl, J. Geophys. Res. **110**, A12223 (2005)
R.P. Di Sisto, A. Brunini, L.D. Dirani, R.B. Orellana, Icarus **174**, 81–89 (2005)
M.K. Dougherty, L.W. Esposito, S.M. Krimigis, *Saturn from Cassini-Huygens* (Springer, Berlin, 2009)
R.A. Dressler, M.J. Bastian, D.J. Levandier, E. Murad, Int. J. Mass Spectrosc. Ion Proc. **159**, 245–256 (1996)
D. Durda, R. Greenberg, R. Jedicke, Icarus **135**, 431–440 (1998)
L.W. Esposito et al., Science **307**, 1251–1255 (2005)
A.J. Farmer, Icarus **202**, 280–286 (2009)
W.M. Grundy, L.A. Young, J.R. Spencer, R.E. Johnson, E. F Young, M.W. Buie, Icarus **184**, 543–555 (2006)
D.T. Hall, D.F. Strobel, P.D. Feldman, M.A. McGrath, H.A. Weaver, Nature **373**, 677–679 (1995)
D.T. Hall, P.D. Feldman, M.A. McGrath, D.E. Strobel, Astrophys. J. **499**, 475–481 (1998)
G.B. Hansen, T.B. McCord, J. Geophys. Res. **109**, E01012 (2004). doi:10.1029/2003JE002149
C.J. Hansen, D.E. Shemansky, A.E. Hendrix, Icarus **176**, 305–315 (2005)
W.K. Hartmann, Icarus **174**, 294–320 (2005)
W.K. Hartmann, G. Neukum, S. Werner, Geophys. Res. Lett. **35**, 2205–2210 (2008)
T.W. Hill, A.J. Dessler, F.C. Michel, Geophys. Res. Lett. **1**, 3–6 (1974)
G.P. Horedt, G. Neukum, Icarus **60**, 710–717 (1984a)
G.P. Horedt, G. Neukum, J. Geophys. Res. **89**(B12), 10405–10410 (1984b)
W.F. Huebner, J.J. Keady, S.P. Lyon, Astrophys. Space Sci **195**, 1–294 (1992)
Y. Itikawa, N. Mason, J. Phys. Chem. Ref. Data **34**, 1–22 (2005)
B.A. Ivanov, G. Neukem, R. Wagner, in *Collisional Processes in the Solar System*, ed. by M.Y. Marov, H. Rickman (Kluwer Acad., Dordrecht, 2001), pp. 1–34
R.E. Johnson, Icarus **62**, 344–347 (1985)
R.E. Johnson, *Energetic Charge-Particle Interactions with Atmospheres and Surfaces* (Springer, Berlin, 1990)
R.E. Johnson, F. Leblanc, B.V. Yakshinskiy, T.E. Madey, Icarus **156**, 136–142 (2002)

R.E. Johnson, R.W. Carlson, J.F. Cooper, C. Paranicas, M.H. Moore, M.C. Wong, in *Jupiter: The Planet, Satellites, and Magnetosphere*, ed. by F. Bagenal, T. Dowling, W. McKinnon (Cambridge University Press, Cambridge, 2004)
P.V. Johnson, J.W. McConkey, S.S. Tayal, I. Kanik, Can. J. Phys. **83**, 589–616 (2005)
R.E. Johnson, H.T. Smith, O.J. Tucker, M. Liu, M.H. Burger, E.C. Sittler, R.L. Tokar, Astrophys. J. **644**, L137–L139 (2006)
R.E. Johnson, M. Fama, M. Liu, R.A. Baragiola, E.C. Sittler, H.T. Smith, Planet. Space Sci. **56**, 1238–1243 (2008)
R.E. Johnson, M.H. Burger, T.A. Cassidy, F. Leblanc, M. Marconi, W.H. Smith, in *Europa*, ed. by R. Pappalardo et al. (University of Arizona Press, Tuscon, 2009)
A.R. Johnston, P.D. Burrow, Phys. Rev. A **51**, R1735–R1737 (1995)
K.N. Joshipura, M. Vinodkumar, U.M. Patel, J. Phys. B **34**, 509–519 (2001)
K.K. Khurana, R.T. Pappalardo, N. Murphy, T. Denk, Icarus **191**, 193–202 (2007)
R. Killen, D. Shemansky, N. Mouawad, Astrophys. J. Suppl. Ser. **181**, 351–359 (2009)
M.R. Kirchoff, P.M. Schenk, in *LPSC XL*, Abstr. No. 2067 [CD-Rom] (2009)
Krupp et al., Space Sci. Rev. (2010, this issue)
M.Y. Kuzmitcheva, B.A. Ivanov, in *Proceedings of Asteroids, Comets, Meteors (ACM 2002)*, Technical University Berlin, Germany (ACM, New York, 2002), pp. 851–852
H.F. Levison, M.J. Duncan, Icarus **127**, 13–32 (1997)
M.F. Levison, M.J. Duncan, K. Zahnle, M. Holman, L. Dones, Icarus **143**, 415–420 (2000)
B. Lindsay, M. Mangan, in *Interactions of Photons and Electrons with Atoms, Molecules, and Ions, Vol. C: Interactions of Photons and Electrons with Molecules*, ed. by Y. Itikawa (Springer, Berlin, 2003)
C.R. Lishawa, R.A. Dressler, J.A. Gardner, R.H. Salter, E. Murad, J. Chem. Phys. **93**, 3196–3206 (1990)
J.J. Lissauer, S.W. Squyres, W.K. Hartmann, J. Geophys. Res. **93**, 13776–13804 (1988)
T. Majeed, D.J. Strickland, J. Chem. Phys. Ref. Data **26**, 335–349 (1997)
A.S. McEwen, B.S. Preblich, E.P. Turtle, N.A. Artemieva, M.P. Golombek, M. Hurst, R.L. Kirk, D.M. Burr, P.R. Christensen, Icarus **176**, 351–381 (2005)
M.A. McGrath, R.E. Johnson, J. Geophys. Res. **94**, 2677–2683 (1989)
H. Melin, D.E. Shemansky, X. Liu, Planet. Space Sci. **57**, 1743–1753 (2009)
M. Mendillo, J. Baumgardner, B. Flynn, W.J. Hughes, Nature **348**, 312–314 (1990)
M.H. Moore, Icarus **59**, 114–128 (1984)
G. Neukum, Adv. Space Res. **5**(8), 107–116 (1985)
G. Neukum, in *The Three Galileos: The Man, The Spacecraft, The Telescope*, ed. by C. Barbieri, J.H. Rahe, T.V. Johnson (Kluwer Acad., Dordrecht, 1997), pp. 201–212
G. Neukum, H. Dietzel, Earth Planet. Sci. Lett. **12**, 59–66 (1971)
G. Neukum, B.A. Ivanov, in *Hazards due to Comets and Asteroids*, ed. by T. Gehrels (Univ. of Arizona Press, Tucson, 1994), pp. 359–416
G. Neukum, R. Wagner, U. Wolf, B.A. Ivanov, J.W. Head, R.T. Pappalardo, J.E. Klemaszewski, R. Greeley, in *Lunar Planet. Sci. Conf. XXIX*. Abstr. No. 1742 [CD-Rom] (1998)
G. Neukum, B.A. Ivanov, W.K. Hartmann, in *Chronology and Evolution of Mars*, ed. by W.K. Hartmann, J. Geiss, R. Kallenbach (Kluwer Acad., Dordrecht, 2001), pp. 53–86
G. Neukum, R.J. Wagner, T. Denk, C.C. Porco, the Cassini ISS Team, in *Lunar Planet. Sci. Conf. XXXVI*. Abstr. No. 2034 [CD-Rom] (2005)
G. Neukum, R. Wagner, T. Denk, C.C. Porco, Cratering chronologies and ages of the major Saturnian satellites. Geophys. Res. Abstracts, 8. Abstr. No. EGU06-A-09252 [CD-Rom] (2006a)
G. Neukum, S.C. Werner, B.A. Ivanov, LPI Contrib. **3120**, 38–39 (2006b)
G.S. Orton, J.R. Spencer, L.D. Travis, T.Z. Martin, L.K. Tamppari, Science **274**, 389–391 (1996)
R.T. Pappalardo, W.B. McKinnon, K. Khurana, *Europa* (University of Arizona Press, Tucson, 2009)
C. Paranicas, J.M. Ratliff, B.H. Mauk, C. Cohen, R.E. Johnson, Geophys. Res. Lett. **29**, 1074 (2002)
Q.R. Passey, E.M. Shoemaker, in *Satellites of Jupiter*, ed. by D. Morrison (Univ. of Arizona Press, Tucson, 1982), pp. 379–434
A.M. Persoon et al., J. Geophys. Res. **114**, A04211 (2009)
J.B. Plescia, J.M. Boyce, Nature **295**, 285–290 (1982)
J.B. Plescia, J.M. Boyce, Nature **201**, 666–670 (1983)
J.B. Plescia, J.M. Boyce, J. Geophys. Res. **90**, 2029–2037 (1985)
C.C. Porco et al., Space Sci. Rev. **113**, 363–497 (2004)
P.M. Schenk, S. Sobieszczyk, Bull. Am. Astron. Soc. **31**, 1182 (1999)
P.M. Schenk, C.R. Chapman, K. Zahnle, J.M. Moore, in *Jupiter: The Planet, Satellites, and Magnetosphere*, ed. by F. Bagenal, T. Dowling, W. McKinnon (Cambridge University Press, Cambridge, 2004)
D.E. Shemansky, P. Matheson, D.T. Hall, H.-Y. Hu, T.M. Tripp, Nature **363**, 329–331 (1993)
V.I. Shematovich, R.E. Johnson, J.F. Cooper, M.C. Wong, Icarus **173**, 480–498 (2005)

M.M. Schneider, J.T. Trauger, Astrophys. J. **450**, 450–462 (1995)
E.M. Shoemaker, in *Europa Ocean Conference*, ed. by D. Matson, D. Nash, San Juan Capistrano (1996), pp. 65–66
E.M. Shoemaker, R.F. Wolfe, in *Satellites of Jupiter*, ed. by D. Morrison (Univ. of Arizona Press, Tucson, 1982), pp. 277–339
E.C. Sittler et al., Planet. Space Sci. **54**, 1197–1210 (2006)
E.C. Sittler et al., Planet. Space Sci. **56**, 3–18 (2008)
B.A. Smith et al., Science **204**, 951–957 (1979a)
B.A. Smith et al., Science **206**, 927–950 (1979b)
B.A. Smith et al., Science **212**, 163–191 (1981)
B.A. Smith et al., Science **215**, 504–537 (1982)
B.A. Smith et al., Science **233**, 43–64 (1986)
B.A. Smith, L.A. Soderblom, D. Banfield, C. Barnet, R.F. Beebe, A.T. Basilevskij, K. Bollinger, J.M. Boyce, G.A. Briggs, A. Brahic, Science **246**, 1422–1449 (1989)
S.C. Solomon, J.W. Head, in *Lunar Planet. Sci. Conf. XXXVIII*. Abstr. No. 1636 [CD-Rom] (2007)
H.C. Straub, P. Renault, B.G. Lindsay, K.A. Smith, R.F. Stebbings, Phys. Rev. A **54**, 2146–2153 (1996)
A.J. Steffl, A.I. Stewart, F. Bagenal, Icarus **172**, 78–90 (2004a)
A.J. Steffl, F. Bagenal, A.I. Stewart, Icarus **172**, 91–103 (2004b)
A.J. Steffl, P.A. Delamere, F. Bagenal, Icarus **180**, 124–140 (2006)
A.J. Steffl, P.A. Delamere, F. Bagenal, Icarus **194**, 153–165 (2008)
R.G. Strom, A. Woronow, M. Gurnis, J. Geophys. Res. **86**, 8659–8674 (1981)
G. Tancredi, J.A. Fernandez, H. Rickman, J. Licandro, Icarus **182**, 527–549 (2006)
H. Tawara, in *Interactions of Photons and Electrons with Atoms, Molecules, and Ions, Vol. A: Interactions of Photons and Electrons with Atoms*, ed. by Y. Itikawa (Springer, Berlin, 2003), pp. 256–282
N. Thomas, F. Bagenal, T.W. Hill, J.K. Wilson, in *Jupiter: The Planet, Satellites, and Magnetosphere*, ed. by F. Bagenal, T. Dowling, W. McKinnon (Cambridge University Press, Cambridge, 2004)
E.P. Turtle, C.B. Phillips, G.C. Collins, A.S. McEwen, J.M. Moore, R.T. Pappalardo, P.M. Schenk, the Galileo SSI Team, in *Lunar Planet. Sci. Conf. XXX*. Abstr. No. 1882 [CD-Rom] (1999)
R. Wagner, PhD. Dissertation, Free University Berlin, Germany, 2007, http://www.diss.fu-berlin.de/2007/806, in German
R. Wagner, G. Neukum, U. Wolf, in *1st Int. Conf. on Impact Cratering, in the Solar System* (ESTEC, Noordwijk, 2006). Abstr. No. 367460 [CD-Rom]
S.C. Werner, PhD. Dissertation, Freie Universitát Berlin, Germany, 2006a. http://www.diss.fu-berlin.de/2006/33/indexe.html
S.C. Werner, B.A. Ivanov, G. Neukum, in *Lunar Planet. Sci. Conf. XXXVII*. Abstr. No. 1595 [CD-Rom] (2006)
J.K. Wilson, M. Mendillo, J. Baumgardner, N.M. Schneider, J.T. Trauger, B. Flynn, Icarus **157**, 476–489 (2002)
R.J. Wilson, R.L. Tokar, M.G. Henderson, T.W. Hill, M.F. Thomsen, D.H. Pontius, J. Geophys. Res. **113**, A12218 (2008)
M.C. Wong, R.E. Johnson, J. Geophys. Res. **101**, 23255–23260 (1996)
A. Woronow, R.G. Strom, M. Gurnis, in *Satellites of Jupiter*, ed. by D. Morrison (Univ. of Arizona Press, Tucson, 1982), pp. 237–276
K. Zahnle, L. Dones, H.F. Levison, Icarus **136**, 202–222 (1998)
K. Zahnle, P. Schenk, H. Levison, L. Dones, Icarus **163**, 263–289 (2003)

Space Sci Rev (2010) 153: 375–410
DOI 10.1007/s11214-010-9641-3

Surface, Subsurface and Atmosphere Exchanges on the Satellites of the Outer Solar System

G. Tobie · B. Giese · T.A. Hurford · R.M. Lopes · F. Nimmo · F. Postberg · K.D. Retherford · J. Schmidt · J.R. Spencer · T. Tokano · E.P. Turtle

Received: 15 May 2009 / Accepted: 17 February 2010 / Published online: 7 April 2010

Abstract The surface morphology of icy moons is affected by several processes implicating exchanges between their subsurfaces and atmospheres (if any). The possible exchange of material between the subsurface and the surface is mainly determined by the mechanical properties of the lithosphere, which isolates the deep, warm and ductile ice material

G. Tobie (✉)
Laboratoire de Planétologie et Géodynamique de Nantes, Université Nantes Atlantique, 44322 Nantes Cedex 03, France
e-mail: gabriel.tobie@univ-nantes.fr

G. Tobie
CNRS, UMR-6112, 2, rue de la Houssinière, 44322 Nantes Cedex 03, France

B. Giese
German Aerospace Center, Institute of Planetary Research, Rutherfordstrasse 2, 12489 Berlin, Germany

T.A. Hurford
Planetary Systems Laboratory, NASA Goddard Spaceflight Center, Greenbelt, MD 20771, USA

R.M. Lopes
Jet Propulsion Laboratory, California Institute of Technology, 4800 Oak Grove Drive, Pasadena, CA 91109, USA

F. Nimmo
Department of Earth and Planetary Sciences, University of California Santa Cruz, Santa Cruz, CA 95064, USA

F. Postberg
Institute for Earth Sciences, Heidelberg University, Im Neuenheimer Feld 236, 69120 Heidelberg, Germany

F. Postberg
Max-Planck-Institut für Kernphysik (MPI for Nuclear Physics), Saupfercheckweg 1, 69117 Heidelberg, Germany

K.D. Retherford
Southwest Research Institute, 6220 Culebra Rd., San Antonio, TX 78228, USA

from the cold surface conditions. Exchanges through this layer occur only if it is sufficiently thin and/or if it is fractured owing to tectonic stresses, melt intrusion or impact cratering. If such conditions are met, cryomagma can be released, erupting fresh volatile-rich materials onto the surface. For a very few icy moons (Titan, Triton, Enceladus), the emission of gas associated with cryovolcanic activity is sufficiently large to generate an atmosphere, either long-lived or transient. For those moons, atmosphere-driven processes such as cryovolcanic plume deposition, phase transitions of condensable materials and wind interactions continuously re-shape their surfaces, and are able to transport cryovolcanically generated materials on a global scale. In this chapter, we discuss the physics of these different exchange processes and how they affect the evolution of the satellites' surfaces.

Keywords Subsurface · Surface · Atmosphere · Icy moons

1 Introduction

The Galileo mission (1995–2003) and the Cassini-Huygens mission (2004–current) in orbit around Jupiter and Saturn, respectively, revealed that these two giant planets host several very active moons. One of the most spectacular discoveries is certainly the observation of active jets of water vapour and icy particles above Enceladus' south pole, which demonstrates how active icy moons can be (e.g. Porco et al. 2006; Spencer et al. 2006). A wide variety of surface features have been identified on these icy moons, which clearly indicates that their surfaces are continuously shaped by interactions with their interiors, and with their atmospheres, if any (see Prockter et al. 2010, this issue).

Numerous cryovolcanic and tectonic structures have been identified on the surface of Europa (e.g. Pappalardo et al. 1999; Fagents 2003), Titan (Sotin et al. 2005; Lopes et al. 2007; Radebaugh et al. 2007), Ganymede (e.g. Schenk and Moore 1995), Triton (e.g. Smith et al. 1989) and other moons. On Europa, the discovery of hydrated salts on its surface (McCord et al. 1998) supports the idea that exchanges between the internal ocean and the icy crust have occurred throughout Europa's history and are probably still occurring. On Titan, the measurements performed by the Huygens mass spectrometer (Niemann et al. 2005) provide circumstantial evidence that the atmospheric methane is supplied by internal outgassing processes associated with cryovolcanic activity, maybe still operating (Tobie et al. 2006). On Enceladus, there is now no doubt that exchanges between the interior and the surface, anticipated since the Voyager era (Squyres et al. 1983), are occurring, and that sampling

J. Schmidt
Nichtlineare Dynamik, Universität Potsdam, Am Neuen Palais 10, 14469 Potsdam, Germany

J.R. Spencer
Department of Space Studies, Southwest Research Institute, 1050 Walnut Street, Suite 400, Boulder, CO 80302, USA

T. Tokano
Institut für Geophysik und Meteorologie, Universität zu Köln, Albertus-Magnus-Platz, 50923 Köln, Germany

E.P. Turtle
Applied Physics Laboratory, Johns Hopkins University, 11100 Johns Hopkins Rd., Laurel, MD 20723, USA

of the water vapour plumes provides a unique opportunity to probe the composition of its subsurface (Waite et al. 2006).

Titan and Triton are the only moons to have atmospheres of mass sufficient to affect surface processes. On Triton, even if the atmosphere is relatively tenuous (<20 μbar, Elliot et al. 2000), seasonal sublimation-condensation of the nitrogen frost affects the surface morphology and activity (e.g. Spencer 1990). On Titan, evidence of aeolian and fluvial features is widespread and shows that interactions between Titan's atmosphere and surface are continuously occurring through the methane precipitation cycle and wind (Tomasko et al. 2005; Lorenz et al. 2006; Lunine et al. 2008). Interestingly, the angular momentum associated with atmospheric global circulation is so large on Titan that it may affect the rotation of the whole satellite (Tokano and Neubauer 2005; Lorenz et al. 2008a; see also Hussmann et al. 2010, this issue).

Even though there is more and more evidence in favor of endogenic activity on icy moons, the internal processes leading to such activity still remain poorly constrained. Interactions between the warm interior and the cold surface of the icy moons are controlled by the upper rigid layer, termed the lithosphere, which naturally forms at shallow depths owing to the decrease of temperature (and resulting increase of viscosity). Exchanges are possible only if this rigid upper layer is sufficiently thin and/or fractured. Various processes, including impact cratering, tectonic stress concentration, or plume-induced melting, can be invoked to induce lithosphere fracture. A good knowledge of the lithospheric structure is required to understand the efficiency of these different processes and their impact on interior-surface exchanges.

In this study, after describing the mechanical properties of icy satellite lithospheres in Sect. 2, we discuss the mechanisms responsible for exchange between the interiors and the surfaces of the icy moons (Sect. 3). Finally, for volatile-rich icy moons, internal outgassing and interior-atmosphere coupling are discussed in Sect. 4.

2 Lithospheric Structure and Faulting

2.1 What is the Lithosphere?

The lithosphere of a planet/satellite is the near-surface part that behaves elastically or in a brittle fashion under applied stress. In contrast, material in the deeper, warmer asthenosphere deforms by viscous creep. The specific behaviour of the material depends on both the thermal state and the stress applied. Thus a lithosphere is not a fundamental attribute of a planetary body, like a crust, but its thickness is in part determined by the applied stress. Lithospheric thickness in turn can be used to obtain information on the stresses and especially, the thermal state of the body, which is important for constraining thermal evolution models.

How can the thickness of the lithosphere be determined? The Maxwell time is the time a substance will flow under an applied load before the creep strain equals the elastic strain (Melosh 1989). It is given by $\tau_M = \eta/G$, where G is the modulus of rigidity and η is the effective viscosity that depends on the strain rate (or stress) of the process and the temperature. Thus, on time scales short relative to the Maxwell time the material behaves elastically (or in a brittle fashion if the yield strength is exceeded), and on longer time scales viscous creep occurs. As the viscosity (and with it τ_M) is a function of depth, this in principle would allow determination of the thickness of the lithosphere at given time scale and viscosity structure, but the viscosity structure is usually unknown. A different, more suitable way to infer the thickness of the lithosphere is therefore to make use of relationships between its thickness and specific surface deformation features.

2.2 Estimation of Lithospheric Thickness from Surface Features

2.2.1 Brittle Surface Deformations

Simple grabens: Simple grabens form when a narrow block drops down under extension. They are bounded by normal faults that converge at the brittle-ductile transition (McKinnon and Melosh 1980; Golombek and Banerdt 1986). Given the graben width w and fault dip δ (the canonical value is 60°), the brittle lithospheric thickness (or brittle layer thickness) d_b is

$$d_b = \frac{w}{2}\tan(\delta). \tag{1}$$

Pinch and swell topography: Extension or compression of a brittle layer atop a ductile substrate can lead to instability (Fletcher and Hallet 1983; Herrick and Stevenson 1990) with characteristic pinches and swells in the surface topography. The formation of the instability is in particular favored by high surface thermal gradients. Predicted instability wavelengths are 3–4 times (Fletcher and Hallet 1983) and 9–10 times (Dombard and McKinnon 2006) the brittle layer thickness for extension and compression, respectively, which allows determination of d_b from measured wavelengths of topography.

2.2.2 Elastic Surface Deformations

Deformation by periodic loading: Loading of an elastic lithosphere by a periodic load $h = h_0 \sin(2\pi x/\lambda)$ results in deflections with maximum amplitude w_0 given by (Turcotte and Schubert 2002, p. 122):

$$w_0 = \frac{h_0}{1 + (\lambda_0/\lambda)^4}, \tag{2}$$

where λ_0 is a characteristic wavelength given by

$$\lambda_0 = \sqrt{2}\pi\alpha, \tag{3}$$

and α is the flexural parameter given by

$$\alpha = \left(\frac{ET_e^3}{3(1-\nu^2)\rho g}\right)^{1/4}. \tag{4}$$

Here, E is the Young modulus, T_e is the elastic thickness of the lithosphere, ν is the Poisson ratio of the material, g is the gravitational acceleration, and ρ is the density of the surface material. In (2), it is assumed that the load's density is equal to ρ, and that the space vacated by the deflected lithosphere is not filled in by the surface material.

At given wavelength λ these equations can be used to estimate T_e if the degree of isostatic compensation of the topography is known. For example, mostly isostatically supported topography, which could be detected by gravity field measurements, implies that elastic forces are negligible and thus deflections are large, in turn implying that λ_0/λ in (2) is smaller than $\sim 1/3$.

Deformation by horizontal loads: Horizontal stress can lead to buckling of the elastic lithosphere. If the stress reaches the critical value $\sigma_c = \alpha^2 \rho g / T_e$ an instability occurs at which the lithosphere forms folds with wavelength λ_0 (Turcotte and Schubert 2002, p. 124). However, for the large icy moons the required stresses are likely too high (e.g., for Ganymede: $\sigma_c \sim 50$ MPa taking $E = 10$ GPa and $T_e = 0.5$ km) for buckling to occur, but under low gravity conditions and for weaker lithospheres such stresses may be available (e.g., for Enceladus: $\sigma_c \sim 4.5$ MPa at $E = 1$ GPa and $T_e = 0.5$ km). The wavelength of instability (3) increases with the elastic thickness of the lithosphere, which in turn is larger at smaller strain rates and thermal gradients.

Deformation by disc-shaped loads: A more sophisticated load model was studied by Brotchie and Silvester (1969).

They calculated the deflection w of an elastic, thin, spherical lithosphere by a disc-shaped load of radius R_d and height h. For radial distances (from the load) $r > R_d$ they obtained:

$$w = h \cdot [s \cdot ker(x) \cdot ber'(x) - s \cdot kei(x) \cdot bei'(s)], \tag{5}$$

where $x = \frac{r}{\beta}$, $s = \frac{R_d}{\beta}$ and $\beta = \alpha/\sqrt{2}$. *ker*, *kei*, *ber* and *bei* are Bessel-Kelvin functions of zero order. If the load's radius and height are known, measured deflection profiles allow α, and thus the elastic thickness T_e to be determined.

Deformation by line loads: The deflection w of a cracked elastic lithosphere under a line load, the distance x_b from the line load to the maximum amplitude of the forebulge, and the distance x_σ where the bending moment or the stress in the lithosphere is a maximum are given by (Turcotte and Schubert 2002)

$$w = w_0 \exp(-x/\alpha) \cos(x/\alpha), \tag{6}$$

$$w_0 = V_0/(\alpha \rho g),$$

$$x_b = \frac{3}{4}\pi\alpha, \tag{7}$$

$$x_\sigma = \frac{1}{4}\pi\alpha. \tag{8}$$

Here V_0 is the line load. Fitting flexural profiles to such model profiles allows determination of w_0 (or V_0) and the flexural parameter α, which yields T_e. Alternatively, α and T_e can easily be obtained if x_b or x_σ can be measured.

The effect of more complicated loads can be determined by Fourier-transforming the load and applying (2) to each wavenumber independently. For some simplified load models (see below) useful analytic solutions still exist.

2.3 Relationship Between Brittle and Elastic Lithospheric Thickness and Temperature Profile

For brittle surface deformations, given the brittle lithospheric thickness (or brittle-ductile transition depth) and the strain rate of the process, the thermal gradient at the time of feature formation can be derived from Yield Strength Envelopes (YSEs) (Golombek and Banerdt 1986) (Fig. 1a).

For elastic surface deformation, the relationship between T_e as determined by loading models and the internal thermal structure goes along similar lines. The point here is to use an

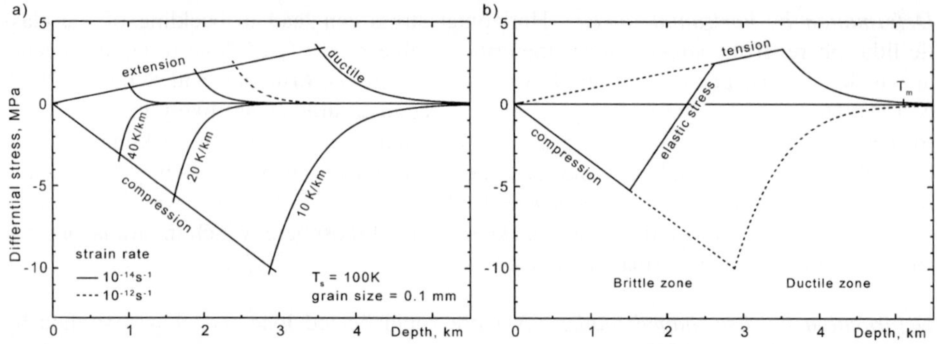

Fig. 1 **a** YSEs adapted for Ganymede. The stresses controlling ductile creep involve dislocation creep, grain boundary sliding, easy slip, and diffusion creep (Goldsby and Kohlstedt 2001). The stresses in the brittle zone (extension/compression) are governed by frictional sliding (Beeman et al. 1988). If the brittle-ductile transition depth and strain rate are known the thermal gradient (or heat flux) can be obtained. The dependence on strain rate is weak: e.g. a simple graben with $d_b = 1.9$ km, which has formed at a strain rate of 10^{-14} s^{-1} yields a thermal gradient of 20 K/km (this figure). If the same simple graben would have formed at a strain rate two orders of magnitude higher the resulting thermal gradient would only be 30% higher. The *dashed line* demonstrates the increase of brittle layer thickness (here in extension) with strain rate at fixed thermal structure (20 K/km in this case). **b** Schematic YSE of an up-warped lithosphere. There is only a small core supporting elastic deformation. Beyond this region brittle failure and ductile creep occur. T_m denotes the mechanical thickness of the lithosphere which marks the depth from where contributions to the total bending moment of the lithosphere are negligible (9)

appropriate stress profile of the lithosphere because purely elastic lithospheres, as considered above, do not exist. Bending stresses induced by warping of an elastic plate and exceeding the yield strength of the material will be relieved by brittle faulting or ductile creep. A stress profile taking into account these effects for a flexed lithosphere was suggested by McNutt (1984) (Fig. 1b). Based on a balance of bending moments, she obtained the equation

$$T_e = \left\{ \frac{12(1-\nu^2)}{E K_{max}} \int_0^{T_m} \sigma(z)(z - z_n) dz \right\}^{1/3}, \qquad (9)$$

where $\sigma(z)$ is the differential stress at depth z, K_{max} is the maximum curvature of the elastic core, T_m is the mechanical thickness of the lithosphere and z_n is a depth fixed by the condition that the stress profile integrates to zero. As $\sigma(z)$ depends on the temperature $T(z)$ in the ductile zone (9) relates T_e to the near-surface thermal structure of the body.

2.4 Lithospheric Thickness and Internal Heat Flux of Icy Moons

The landforms observed on the icy moons (see Prockter et al. 2010, this issue) are specific records of the internal evolution processes and lithospheric structure at the time of formation. Morphologic studies of these landforms in conjunction with modeling thus allow an assessment to be made of the moons' history.

2.4.1 Ganymede

Ganymede exhibits a subdued surface with relief limited to ~2 km (Prockter et al. 2010, this issue). This is suggestive of a generally thin lithosphere during its evolution. Based on Voyager data, Golombek and Banerdt (1986) considered furrows on Ganymede (Prockter et al. 2010, this issue), as simple grabens or the result of an extensional necking instability. Using the relationships given in Sect. 2.2.1 these authors derived brittle layer thicknesses of 5–10 km and, using YSEs, thermal gradients in the range of 1.5–6 K/km (adopting a surface temperature $T_s = 100$ K). Based on higher resolution Galileo data, Nimmo and Pappalardo (2004) related the topography of furrows (Prockter et al. 2010, this issue, Fig. 2) to flexural uplift. Applying (6) and (9) they derived elastic thicknesses of ~ 0.5 km, a brittle-ductile transition depth of 2–3 km, and heat fluxes in the range of 60–80 $\mathrm{mW\,m^{-2}}$ (13–17 K/km in a conductive ice lithosphere at $T_s = 120$ K), which is significantly higher than obtained with Voyager-based analyses (note that a lower surface temperature of 100 K as adopted by Golombek and Banerdt (1986) would increase these gradients further by some K/km).

Likewise, considering the younger grooves on Ganymede (Prockter et al. 2010, this issue) as imaged by Voyager as simple grabens or the result of an extensional necking instability, Golombek and Banerdt (1986) obtained brittle-ductile transition depths of 2–5 km and thermal gradients in the range of 4.5–20 K/km. Subsequent Galileo analysis, however, has strengthened the view that grooved terrain formed by an extensional necking instability (Collins et al. 1998; Dombard and McKinnon 2001). At measured groove spacings of ~ 3–17 km (Prockter et al. 2010, this issue) this implies brittle-ductile transition depths of ~ 1–5 km (Sect. 2.2.1) and thermal gradients ranging from ~ 7 K/km to 40 K/km (Fig. 1a) at the time the grooved lanes were formed. Such high heat fluxes are consistent with those most favored for the development of the extensional necking instability under conditions specific to Ganymede (Dombard and McKinnon 2001).

Flexural uplift as suggested for furrows has also been observed at younger rift zones. For an elevated dark terrain flank (Prockter et al. 2010, this issue, Fig. 3) Nimmo and Pappalardo (2004) derived an elastic lithospheric thickness of 0.18 km associated with a very high heat flux of 154 $\mathrm{mW\,m^{-2}}$ (corresponding to 33 K/km at $T_s = 120$ K). Somewhat lower heat fluxes of 60 $\mathrm{mW\,m^{-2}}$ ($T_e = 0.9$ km) were obtained for an elevated bright terrain flank by Nimmo et al. (2002). The brittle-ductile transition depth in both of these cases is 2–3 km consistent with the results from grooved terrain formation above.

2.4.2 Europa

Europa's surface relief is limited to ~ 1 km (Prockter et al. 2010, this issue). This suggests that, compared with Ganymede, its lithosphere has been thinner in most places during its evolution. Lithospheric thickness estimates for Europa mainly come from flexural studies, notably of double ridges. Double ridges are ubiquitous on Europa and thus can provide important clues to the global lithospheric structure. However, these studies have to be assessed with caution because there is no consensus about how double ridges formed. While Head et al. (1999) relate the formation of double ridges to up-warping of the surface (bottom-loading model), Greenberg et al. (1998) consider double ridges as piles of debris squeezed to the surface by tidal pumping (top-loading model). Conclusive evidence of which model actually applies is lacking as yet. Hurford et al. (2005) identified depressions and forebulges in photoclinometry-derived profiles across double ridges and bands in support of the top loading model. Measuring the distance between the load and the forebulge (7), they obtained an average elastic thickness of the lithosphere of ~ 200 m. Billings and Kattenhorn (2005) measured the distance between the load and the location of maximum stress in the lithosphere (8). They obtained elastic lithospheric thicknesses of ~ 200–1000 m at one site, and ~ 500–2000 m at two other sites, which is significantly larger than found by Hurford et al. (2005). Lithospheric thicknesses of > 1 km are also suggested by stereo-derived topography of a rather big (350 m elevation) double ridge (Prockter et al. 2010, this issue, Fig. 8).

As there are no features indicative of flexure this observation can be used to place a lower bound on the lithospheric thickness. If the individual ridges are (top) loads on it, the simple line load model (6) yields thicknesses in excess of ~ 1 km (at $E = 1$ GPa) in order to withstand (deflection amplitudes < 50 m) these loads.

A thin lithosphere was inferred from a ~ 100 m high dome-like feature (Williams and Greeley 1998). The dome has flexed the lithosphere downwards from which the authors derived elastic lithospheric thicknesses of 0.1–0.5 km. Rather thick lithospheres have been inferred form top-loading induced flexure related to a plateau-like feature located near Europa's crater Cilix (Prockter et al. 2010, this issue, Fig. 8) (Nimmo et al. 2003) and to the emplacement of subsurface material onto Europa's plains (Figueredo et al. 2002). The derived elastic thicknesses are 6^{+5}_{-2} km and 4^{+2}_{-2} km, respectively.

Folds on Europa (Prockter and Pappalardo 2000) provide another means to constrain its elastic thickness. The folds may have formed by buckling of the lithosphere under horizontal stress (Sect. 2.2.2). Using (3), (4) and a wavelength of ~ 25 km (Prockter and Pappalardo 2000) the elastic thickness is ~ 1.5 km. This result is consistent with the elastic thickness estimate of ≥ 1 km obtained for the big double ridge. Rather than assuming an elastic lithosphere, Dombard and McKinnon (2006) considered the lithosphere to be pervasively fractured showing plastic behaviour. Applying the relationship between wavelength and brittle layer thickness at unstable compression of the lithosphere (Sect. 2.2.1) they obtained brittle layer thicknesses in the range of ~ 2.5–2.8 km. Although the assumptions applied to the mechanical state of the lithosphere are different in these two approaches, the results are consistent in that the depths to the ductile ice layer are similar. The instability analysis of Dombard and McKinnon (2006) further indicated high heat fluxes near 100 $\mathrm{mW\,m^{-2}}$ at the time of fold formation.

2.4.3 Enceladus

Global and local topographic data of Enceladus (mean radius $R_m = 252$ km) (Thomas et al. 2007; Giese et al. 2008a) show a surface with relief limited to ~ 2 km. This is much less than observed on Tethys and Iapetus (see below), and much more than observed on Europa ($R_m = 1562$ km) and Ganymede ($R_m = 2632$ km) if divided by the bodies' mean radii. Different from Europa and Ganymede, Enceladus exhibits large-scale (comparable to R_m) depressions that are likely to be supported isostatically (Schenk and McKinnon 2009). Elastically-supported topography has been identified at a rift zone showing an elevated flanking ridge (Fig. 2a). From this observation Giese et al. (2008a) derived an elastic lithospheric thickness of 300 m for an elastic modulus of 1 GPa and assuming the material to be clean water ice. The brittle-ductile transition depth was found at ~ 2 km, and associated heat fluxes range from 200 to 270 $\mathrm{mW\,m^{-2}}$. Near surface porosity however, can reduce these values to ~ 50 $\mathrm{mW\,m^{-2}}$. An independent and consistent estimate of the heat fluxes in that area was obtained by Bland et al. (2007). These authors found evidence for an extensional necking instability (Sect. 2.2.1) and derived heat fluxes of 110–220 $\mathrm{mW\,m^{-2}}$. Further information on the thickness of the lithosphere comes from a feature with simple graben morphology west of the rift zone (Giese et al. 2008a). This suggests a brittle-ductile transition depth of 2–3 km, consistent with the above results.

2.4.4 Tethys

Tethys ($R_m = 531$ km) shows relief of up to 8 km (Giese et al. 2007), which is substantially more than observed on Enceladus. Although larger in size, its density is close to that

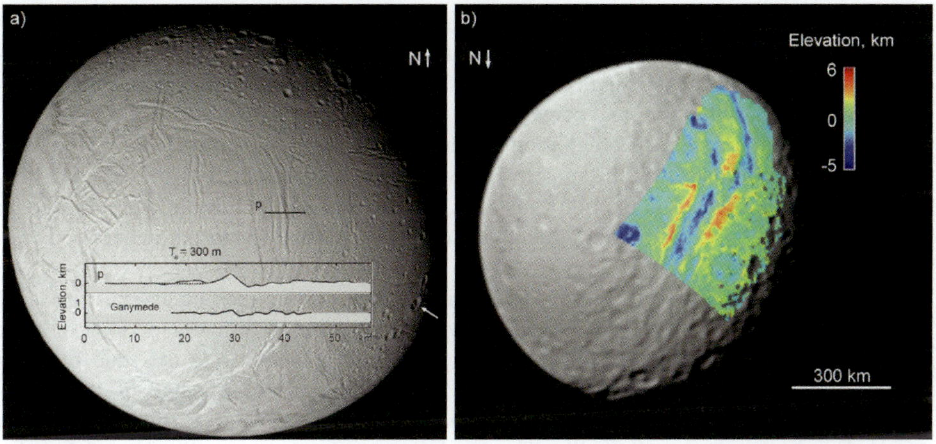

Fig. 2 **a** Cassini frame N1516171253 (930 m/pxl) showing a rift zone on Enceladus with flank-uplift of ~ 1 km (profile) above the local surroundings. The *dotted line* in the upper profile is obtained from fitting the elevation data to the model of a flexed elastic plate (6). The lower profile shows a similar half graben structure on Ganymede, for comparison. Profiles are taken from Giese et al. (2008a). The *white arrow* points to a 20 km central peak crater (22°S, 182°E) with relief of about 1.5 km. **b** Cassini frame N1507664141 (7.6 km/pxl) showing Tethys' ~ 100 km wide rift zone Ithaca Chasma and its topography. Depths reach 2–3 km, and the flanks of the rift are upraised by up to 6 km above the surroundings (Giese et al. 2007)

of pure water ice (984 kg m^{-3}, Thomas et al. 2007), which implies no internal radiogenic heating. The absence of radiogenic heating may have prevented viscous relaxation of topography during most of Tethys' history and might explain the observed large relief. Tethys exhibits a prominent rift zone first seen by Voyager and later, at substantially higher resolution, by Cassini (Fig. 2b). The rift shows elevated flexed flanks from which an elastic lithospheric thickness of 5–7 km, a brittle-ductile transition depth of ~ 12 km, and heat fluxes of 18–30 mW m^{-2} were derived (Giese et al. 2007). The time of formation of Ithaca Chasma was determined to be 4 Gyr before present (BP). Tethys' 400 km impact crater Odysseus is ~ 100 Ma younger and shows as much relief as does Ithaca Chasma (Moore et al. 2004).

2.4.5 *Iapetus*

Among the icy satellites studied so far, Iapetus ($R_m = 735$ km) has the largest relief of more than 20 km (Porco et al. 2005a; Giese et al. 2008b). Its density is close to that of pure water ice (1083 kg m^{-3}, Thomas et al. 2007) and the related lack of radiogenic heating has largely prevented relaxation of topography as suggested above for Tethys. In order to obtain a bound on the thickness of Iapetus' lithosphere the disc-shaped load model (Sect. 2.2.2) has been applied to the rim load associated with an 800 km impact basin (Giese et al. 2008b). This yielded a lower bound on the elastic thickness in the range of 50–100 km at an early stage (4.4–4.5 Gyr BP) in Iapetus' history.

2.5 Lithospheric Structure, Stress and Faulting

The occurrence of faulting and deformation on an icy moon depends on the thermomechanical structure of its lithosphere and on the source of stresses, which both vary during satellite evolution. Accumulation of stress in the satellite lithosphere can result from (in order of decreasing magnitude): internal differentiation; ocean crystallization; thermal expansion/contraction; change of rotation state (despinning, reorientation and non-synchronous rotation); tides and libration; and convection.

Stresses due to internal differentiation and despinning are dominant only during the early stage of the evolution. Only the most primitive bodies have retained some record of this period on their surface. It is likely for instance that the formation of the equatorial ridge on Iapetus is related to its despinning, though the exact mechanism is still not understood (e.g. Castillo-Rogez et al. 2007).

Change of volume due to thermal expansion/contraction or melting/crystallization of an internal ocean is likely to occur on many satellites, but at different periods of their evolution. On the smallest satellites (except Enceladus), thermal stresses due to rapid change of the thermal profile are dominant during the very early stage of the evolution, but become negligible after a few hundred million years (e.g. Hillier and Squyres 1991). For the largest satellites, like Titan and Ganymede, the crystallization of their internal ocean occurs during the major part of their evolution, and accelerates when convection is initiated in the outer ice layer (e.g. Tobie et al. 2006; Mitri and Showman 2008; Bland et al. 2009). This results in a progressive contraction of their interior owing to the formation of a high-pressure ice layer at the base of the ocean. Global expansion due to melting of the high-pressure ice shells can also occur during brief periods of time during the differentiation or at the onset of convection within the silicate core (e.g. Tobie et al. 2006). By contrast, on smaller satellites, the crystallization of the internal ocean results in a global expansion as no high-pressure ice layer forms (e.g. Nimmo 2004; Manga and Wang 2007). On Europa, for instance, periodic change in the ice shell thickness as a function of the orbital evolution is likely to produce alternate global compressive and tensive stresses (Hussmann and Spohn 2004; Mitri and Showman 2005).

During an expansion episode, the stress can exceed several tens of MPa and in this condition a crack can propagate down to several kilometers depth, and reach the base of the lithosphere for a satellite having a thin lithosphere (e.g. Mitri and Showman 2005; Manga and Wang 2007) (e.g. Europa, Enceladus, and many satellites during their early stage). The propagation of tensile cracks through the entire lithosphere may then enable some exchange between the warmer part below the lithosphere and the surface by promoting ice diapirism, intrusion of melt, or volcanism.

Changes in the solid surface position with respect to the tidal bulge can also result in stresses. Non-synchronous rotation (NSR) involves longitudinal shifts in position, while reorientation of the rotational axis, called "true polar wander", also includes latitudinal shifts. Both mechanisms result in predictable global stress patterns (e.g. Melosh 1980; Leith and McKinnon 1996). Non-synchronous rotation is likely in satellites with eccentric orbits (Greenberg and Weidenschilling 1984), while true polar wander can arise if lateral shell thickness variations (Ojakangas and Stevenson 1989) or convective mass anomalies (Nimmo and Pappalardo 2006) occur. Detecting NSR at Europa by analysis of fracture orientations has proven difficult (e.g. Geissler et al. 1998), though crater distributions suggest it has happened elsewhere (Zahnle et al. 2001). Fracture patterns have also been used to argue for true polar wander at Europa (Schenk et al. 2008). Such a reorientation of the rotational axis results from perturbation of the moment of inertia of the body owing to large-scale internal or surface loads, and it is also thought to have a play role in the evolution of Enceladus (Nimmo and Pappalardo 2006).

Icy moons possessing a subsurface ocean are also subjected to significant diurnal tidal stresses. In the absence of an internal ocean, the tidal deformation of the ice shell remains very small as it is coupled to the rocky core and the resulting stress is negligible (Moore and Schubert 2000; Tobie et al. 2005). Tidal stresses are expected to play a significant role on Europa and Enceladus at present time, and also on Triton and Titan in the past, when their eccentricity was potentially higher. As they orbit their giant planet, the moons are constantly being reshaped as the size and location of the tidal bulge changes because of their orbital eccentricity, producing a time dependent diurnal tidal stress field. At any one location on the surface the orientation of the principal stress rotates as the magnitudes of the stresses oscillate daily. On Europa, Hoppa et al. (1999) proposed that cycloidal cracks form as a result of tensile failure in response to this diurnally varying tidal stress field. Cycloid segments initiate as tension fractures and grow perpendicular to the rotating direction of maximum tensile stress, resulting in curved segments. Cracking begins when the tensile strength of Europan ice is overcome and ceases when the tensile stress drops below the crack propagation strength, which is less than the tensile strength. After cessation of growth of one cycloid arc, the next arc starts growing when the tensile strength of the ice is again exceeded, but since the stresses rotate during the period of no crack growth, the crack propagates at an angle to the previous segment, forming a sharp cusp. Once a cycloidal crack has formed, subsequent ridge building occurs, forming the cycloidal ridges observed on Europa.

One prominent rift on Enceladus' south pole consists of arcuate segments, resembling the shape of cycloidal cracks on Europa, which suggests that a similar mechanism may have caused the cracking on Enceladus as on Europa. The cycloidal shape of this tiger stripe rift suggests that its formation may have been similarly controlled by diurnal tidal stresses (Hurford et al. 2007). However, the similarity between tiger stripes and Europan cycloidal ridges is in planform only; otherwise they are quite distinct from one another morphologically. While the initial cracking may have been controlled by similar tidal stresses, subsequent modification of the surface at the tiger stripes on Enceladus may have differed from the ridge-forming processes on Europa.

Even though tidal stresses play a key role in the faulting dynamics on Europa and Enceladus, the amplitudes of the associated stresses are small (e.g. Harada and Kurita 2006). Other source of stresses, such those mentioned above, must be taken into account in order to explain the fracturing of the lithosphere. Once pre-fractured, tidal stress is likely to be the main driver of the propagation and persistence of cracks in the lithosphere, by periodically reactivating them. Although the amplitude of stress is relatively small, the energy associated with tidal motions is quite large and contributes significantly to the dynamics of the lithosphere.

Thermal convection also affects the dynamics of the lithosphere, not directly through fracturing (as the associated stress remains relatively small; Tobie et al. 2003), but mainly by locally modifying the thermal structure of the lithosphere. Thermal plumes lead to thermal erosion of the overlying lithosphere. Tidal dissipation within hot plumes can also promote internal melting (Sotin et al. 2002), thus leading to local volume changes and accumulation of stress in the lithosphere above. In order to understand brittle deformation mechanisms in the lithosphere, all these different sources of stress must be taken into account simultaneously, which requires development of powerful numerical tools, including a viscoplastic description of the lithosphere.

2.6 Diurnal Tidal Stress and Exchange Processes Associated with Faulting

As mentioned above, the diurnal fluctuations of tidal stress can maintain tectonic activity of the icy lithosphere and in some circumstances can trigger eruptive processes. This effect is particularly spectacular on Enceladus, as evidenced by active plumes, and is likely to have occurred on Europa and Triton, though not yet observed.

2.6.1 *Enceladus*

The state of stress at each point on the surface of Enceladus cyclically changes as its tidal shape changes throughout the orbit. In one model, over Enceladus' orbital period, stresses resolved across the tiger stripes alternate from compressive to tensile, perhaps allowing the faults to open at some point in the orbit, and expose a subsurface volatile reservoir, creating an eruption (Hurford et al. 2007). Then, within a few hours, the stress becomes compressive, forcing cracks to close, ending any possibility of an eruption until the next cycle. To test the hypothesis that observed eruption plumes come from these cracks, Hurford et al. (2007) analyzed the stresses along each tiger stripe rift to check whether any of the features were under tension during three Cassini observations in 2005 (Porco et al. 2006). Hurford et al. (2007) found a correlation between tension along tiger strip rifts and plume activity, especially during the November 2005 observations in which individual jets can be seen. However, as further discussed in Sect. 4.2.2, this model fails to explain the observations obtained during following flybys.

2.6.2 *Europa and Triton*

It was hypothesized that similar venting might be observed on Europa and the Galileo spacecraft imaging team included a sequence of high phase observations, specifically designed to search for plumes. However, at the time and place of those observations, the area imaged was under tidal compression, the least likely condition for venting, and no plumes were detected (Hoppa et al. 1999). Given the experience with Enceladus, future missions to Europa can address whether cracks on Europa are conducive to venting by predicting when young cracks are likely to be in tension and planning observations accordingly.

Triton, the largest moon of Neptune, was the first currently active icy satellite discovered in the Solar System, when Voyager 2 imaged eruptions from its surface in 1989. This strange moon of Neptune is believed to be a captured satellite based on its retrograde orbit, which slowly spirals in towards Neptune. Following its capture, tidal dissipation worked to circularize its orbit, depositing large amounts of heat inside Triton (e.g. Goldreich et al. 1989). This heat melted Triton's interior, enhanced tidal stress on its surface, and thereby permitted the formation of cracks and ridges, which resemble the ones imaged on Europa by Galileo (Prockter et al. 2005). While the current wisdom holds that Triton's plumes result from seasonal solar heating of frozen nitrogen on its surface (Moore and Spencer 1990), it may be that internal heat can melt nitrogen under the icy crust and allow this warm liquid to erupt through fissures onto Triton's surface (Brown and Kirk 1994).

3 Interior-Surface Coupling: Cryomagmatism and Excavation Processes

3.1 Cryomagmatic Processes on Icy Moons

Evidence of past and recent cryovolcanic activity is widespread and extremely varied on the icy moons (see Prockter et al. 2010, this issue). By analogy with silicate volcanism on terrestrial planets, cryovolcanism on icy moons results from the melting of icy materials, primarily water ice which is the main icy component on these bodies. However, in contrast to volcanism on rocky planets, cryovolcanism is subject to two major complications: (i) the melting point of water ice increases with decreasing pressure, so that decompression melting, which is a key process in silicate volcanism, cannot occur; (ii) even more problematic,

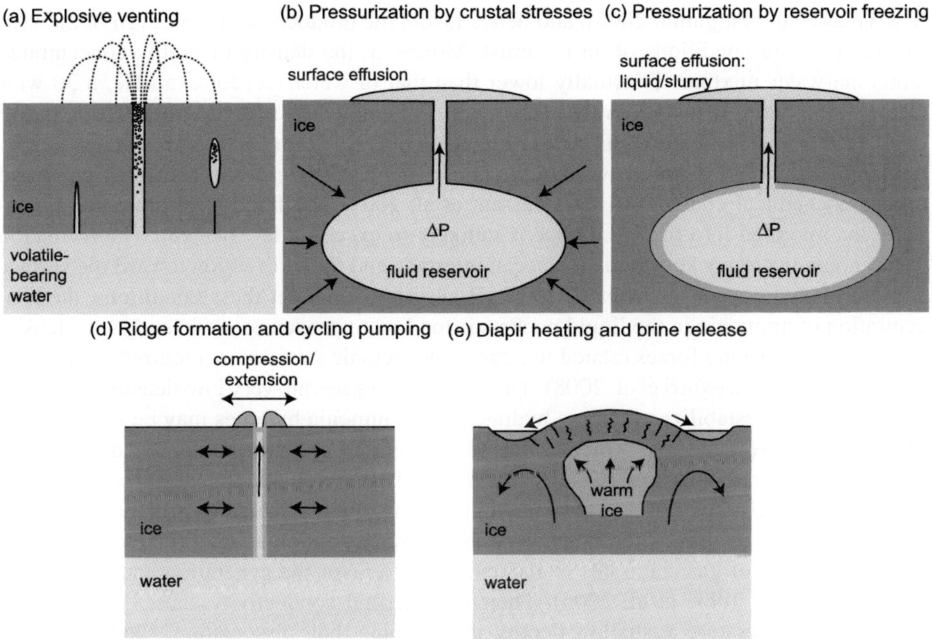

Fig. 3 Synthesis of different mechanisms proposed for delivering fluid to Europa's surface: **a** exsolution of volatiles and explosive venting; **b** pressurization of fluid reservoir in ice lithosphere by lithospheric stresses; **c** pressurization by partial freezing of liquid reservoir; **d** tidal pumping of slush mixtures, and **e** ascent of warm ice diapirs leading to melting and release of near-surface brines (from Fagents 2003)

the liquid water produced upon melting of ice is denser than the solid phase. Therefore, in the pure H_2O system, it is very difficult, if not impossible, to achieve extrusion of a liquid onto the surface of an icy shell.

As decompression melting cannot occur, melting must result from a local increase of temperature, which requires a localized source of heat. Tidal heating concentrated in hot upwelling plumes (Sotin et al. 2002; Tobie et al. 2008) or along faults (Nimmo and Gaidos 2002; Nimmo et al. 2007) has been proposed as a mechanism to concentrate heat production and hence increase the temperature locally. In this situation, the production of a localized water reservoir at shallow depth can be envisaged, in particular on Europa and Enceladus, as well as on Ganymede (Showman et al. 2004), where higher tidal dissipation is hypothesized to have occurred in the past. Liquid cryolavas may also directly originate from a long-lived extensive liquid water layer a few kilometers beneath the surface, as has been proposed for Europa (e.g. Fagents 2003). In this situation, internal melting of the ice shell is not required. However, whatever the nature of the liquid reservoirs, the ascent mechanism remains problematic; for pure water cryolavas, ascent can occur only by means of pressurization of the reservoirs. Such overpressurization of reservoirs may result from crustal stresses related to tidal fluctuations (e.g. Greenberg et al. 1998), topographic variations (e.g. Showman et al. 2004) or global crystallization of the ocean (e.g. Mitri and Showman 2008; Manga and Wang 2007) or from reservoir refreezing (e.g. Fagents 2003) (Fig. 3).

Another solution is to consider the presence of additional components that favor melting and eruption of the resulting liquid. Owing to its very strong anti-freezing effect, ammonia has long been proposed as an additional compound (Lewis 1971; Kargel 1991). Indeed, the presence of a few percent ammonia strongly decreases the melting point, by up to ~ 100 K (e.g. Grasset and Pargamin 2005), and hence favors the production of melt even in the very low temperature conditions of an icy crust. Moreover, the density of highly concentrated ammonia-water mixtures is actually lower than that of water ice; for example, a 30 wt% NH_3 mixture has a density less than 920 $kg\,m^{-3}$ at temperatures higher than 210 K (Croft et al. 1988). Ascent of ammonia-rich cryomagma directly from a subsurface ocean is very unlikely as it would require a very high concentration of NH_3 in the entire ocean. Based on cosmochemical constraints (e.g. Hersant et al. 2008), the fraction of ammonia potentially incorporated into the icy moons is unlikely to exceed a few percent. The formation of ammonia-water pockets at the ice/ocean interface and their advection toward the surface is more likely to occur (Mitri et al. 2008). However, even under these conditions, the concentration of ammonia in the liquid pockets is too low to produce a sufficiently low density and additional driving forces related to large-scale tectonic stresses are required to erupt the liquid to the surface (Mitri et al. 2008). The only way to generate very low-density ammonia-rich melts is to destabilize ammonia hydrate. Such ammonia hydrates may be incorporated into the primitive crust just after satellite formation and could be destabilized by a local increase of internal heat flow (Choukroun et al. 2010). On Titan, for instance, such cryovolcanic events could be related to thermal upwelling plumes, which locally melt ammonia hydrate in the crust, thus generating a very concentrated ammonia-water mixture.

The presence of gas clathrates is also expected to favor the generation and eruption of cryomagma (e.g. Tobie et al. 2006). Their low thermal conductivity causes a significant increase of temperature at shallow depths, thus helping their dissociation. Clathrate dissociation in the icy crust produces a large quantity of gas within the cryogenic liquids which can favor the rise of liquids when gas exsolution occurs (Crawford and Stevenson 1988; Fortes et al. 2007; Choukroun et al. 2010). For instance, at depths of 2 to 3 km, the solubility law at a temperature of 250 K suggests that an aqueous solution in equilibrium with methane clathrate can contain 0.25 to 0.4% of dissolved methane. During the ascent of this solution a significant part of the dissolved gas initially in equilibrium with the cryomagma source would be released. Thus, for material with an initial source depth of 2 to 3 km raised to a depth of 1 km, 45% to 60% of the initially dissolved gas is exsolved. This corresponds to a gas volume fraction of about 10 to 20%, which is expected to have a major effect on the buoyancy of the cryomagmas.

Another possibility is that the icy crust is significantly denser than water ice due to the presence of other heavy components, such as silicate particles, salts (e.g. Pappalardo and Barr 2004), clathrates of CO_2 or SO_2 (e.g. Prieto-Ballesteros et al. 2005), ammonium sulfate (Fortes et al. 2007) etc. In this case, the rise of water-rich cryomagma would be promoted, but the long-term gravitational stability of such an icy crust would then be problematic.

Io is very different in terms of cryovolcanic processes. Volcanism on Io is predominately silicate. However, the possible presence of SO_2 in Io's crust in the form of subterranean "aquifers" or reservoirs, as discussed by McCauley et al. (1979) and Kieffer et al. (2000), may lead to the generation of SO_2 magma. Although temperature measurements from Galileo clearly showed that many hot spots have temperatures far too high for sulfur, the possibility that some sulfur flows occur on the surface remains open (see Prockter et al. 2010, this issue). Furthermore, such flows are likely to be a form of secondary rather than primary volcanism (e.g. Greeley et al. 1984): as silicate magma (primary) nears the surface, it would provide enough heat to melt sulfur-rich country rock, producing "secondary" sulfur flows (as opposed to "primary" flows that originate from molten magmas at depth).

The best candidate regions for SO_2 volcanism are Balder patera and Tohil patera. A nearly pure SO_2 region topographically confined within Balder patera (see Fig. 2 in Prockter et al. 2010, this issue) may have been emplaced by liquid flow of SO_2 rising from the subsurface (Lopes et al. 2001). Tohil patera shows a region of enhanced SO_2 signature and flow-like margins and has also been interpreted as a candidate site for SO_2 cryovolcanism (Williams et al. 2004). SO_2 volcanism on Io, like sulfur volcanism, can also be considered secondary, as the primary magma in Io's interior is silicate. It is also likely rare, as only a few candidate regions have been identified. Most surface changes that show variations in SO_2 are likely to be due to redistribution and/or recrystallization of explosively emplaced SO_2 snow produced by freezing of volcanic gases (Carlson et al. 1997; Douté et al. 2004)

3.2 Thermal Emission, Heat Flow, and Cryovolcanic Transport

Extraction of cryomagma and vapour provide an effective way to transport energy from shallow levels to the surface. Flow and cooling of erupted cryomagma on the surface result in intense thermal emission. In the outer solar system, Io and Enceladus are the only two satellites with detectable thermal emission, as they are radiating measurable power in excess of that provided by the Sun. In both cases the ultimate heat source is presumed to be mainly tidal dissipation. A direct constraint on internal heating is important for several reasons: it provides constraints on the interior structure and rheology; it constrains the orbital and thermal history of the body (e.g. is the eccentricity in equilibrium?); and it can be used to investigate the mass flux from the interior to the surface. From the point of view of subsurface-surface-atmosphere exchanges, it is this last characteristic that is most important. The rest of this section is devoted to investigating how surface thermal emission can be used to measure these exchanges by using the example of Enceladus.

As presented in Prockter et al. (2010, this issue) and Coustenis et al. (2010, this issue), the CIRS instrument on Cassini has observed thermal emission from a series of sub-parallel fractures (dubbed "tiger stripes") at Enceladus' south pole, which are associated with jets of vapour and ice particles (Spencer et al. 2006; Porco et al. 2006). A variety of models have been proposed to explain this activity. While most of the proposed models try to explain the vapour and ice crystals emanating from the tiger stripes, only few of them address the magnitude and pattern of thermal anomalies observed by Cassini. The first class of models investigates whether the eruption implies boiling of liquid water, i.e. a source temperature of ~ 273 K (e.g. Porco et al. 2006) or only sublimation of water ice and dissociation of clathrate, i.e. a source temperature of less than 200 K (e.g. Kieffer et al. 2006). These models are further discussed in Sect. 4.2.2. The second class of models tries to address what heat sources, internal temperatures and heat transport mechanisms are consistent with the observed thermal anomaly (Nimmo et al. 2007; Barr 2008; Tobie et al. 2008; Abramov and Spencer 2009). The shear heating model of Nimmo et al. (2007) proposes that tidally driven fault motion causes local heating and sublimation of solid ice. This steady-state model can reproduce the observed vapour fluxes and velocities, but makes no prediction about vapour composition or ice grain behaviour. The other models focus on the anomalous thermal signature of Enceladus's south polar terrains, rather than the venting. Abramov and Spencer (2009) investigated the thermal implications of a conductive shell with a warm embedded crack (such as an intrusion or vertical water lens). Tobie et al. (2008) investigated the conditions under which tidal dissipation could be increased at the south pole and proposed a mechanism to concentrate energy and melting near the south pole. Barr (2008) and Barr and McKinnon (2007) suggested that the south polar heat flux might be due to upwelling and lateral transport of warm ice, similar to mid-ocean ridge spreading on Earth.

In some models the observed surface heat flow is being conducted from relatively shallow, warm material while in others it is being advected to the near-surface by vapour transport. In the latter case, the heat transported beyond the surface by advection (latent heat of vapour in the plumes) is about an order of magnitude smaller than the heat being radiated from the surface (~ 0.4 GW vs. 5.8 ± 1.9 GW). This in turn implies that exchange between the near-surface and interior is much more rapid than exchange between the interior and the atmosphere.

Most of the models proposed for eruption assume, either implicitly or explicitly, that release of material is triggered by the opening of fractures by extensional tidal stresses (Hurford et al. 2007; Halevy and Stewart 2008). The mass flux should thus be time-dependent, since the stresses depend on the position of Enceladus relative to Saturn at the time of interest. Cassini magnetometer data have been interpreted as evidence for strongly time-variable jet activity (Saur et al. 2008). A comparison of the occultation-derived column densities in 2005 and 2007 revealed higher densities at the latter time (Hansen et al. 2008). The tiger stripes, however, were predicted to be under tension during the 2005 observations (open fractures) and under compression in 2007 (closed fractures). This mismatch between observations and predictions may be alleviated by consideration of potential latitudinal librations of Enceladus (Hurford et al. 2009).

Despite the strong variability of the plume mass flux, the plume heat output and spatial variability do not appear to have varied over an 18 month period (Abramov and Spencer 2009). This lack of variability is in agreement with the shear heating model (Nimmo et al. 2007) which explicitly assumed a steady-state situation.

The tidal dissipation model proposed by Tobie et al. (2008) can match the total power emitted in the south polar region but cannot explain the very high heat flow observed at the tiger stripes. In contrast, the mobile lid model of Barr (2008) can match the regional heat flux at the south pole, but does not provide any explanation for the heat source. In addition, Abramov and Spencer (2009) argue that the temperature distribution derived by Barr (2008) results in a spectrum that does not fit that observed by CIRS. Conversely, a simple conductive model of a vertical fracture with a constant temperature in the range 130–220 K and narrow (< 200 m) width produces spectral energy distributions that closely resemble the CIRS spectra (Abramov and Spencer 2009). These temperatures are consistent with the sublimation and shear heating models. Fractures with temperatures of 273 K do not fit the observed radiance, suggesting that liquid water is not present in the near-subsurface ($\lesssim 1$ km). The requirement for narrow, relatively warm regions implies that advection of heat along fractures, rather than distributed heat transport from depth by conduction, is the dominant mechanism. Similar conclusions have been reached in a theoretical study of heat and vapour transport by Ingersoll and Pankine (2010). These authors also consider that multiple parallel fractures are probably required to explain the tiger stripe venting behaviour; the effect of multiple fractures has not yet been considered in thermal analyses.

Although the shear heating model did not make predictions of the resulting spectral energy distribution, it did predict the spatial variation in heating along the individual tiger stripes (Nimmo et al. 2007). Taking the time-averaged tidal shear stress as a proxy for heating rate, Fig. 4 compares the predicted heat flow distribution along the Damascus tiger stripe with that observed. Two predictions are shown: the first uses the stresses calculated in Nimmo et al. (2007), while the second makes an additional correction for the fact that shear stresses (and velocities) will be larger where the tiger stripes are more widely separated. In general the match between the observations and the predictions is quite good.

There is currently not a satisfactory model that explains both the venting and thermal characteristics of the tiger stripes. Tiger-stripe thermal emission appears constant (over the timescale of the Cassini observations), and the mass flux variability does not agree with predictions based on tidal stresses (Hurford et al. 2007; Halevy and Stewart 2008). The spectral energy distribution of individual tiger stripes matches that generated by a warm

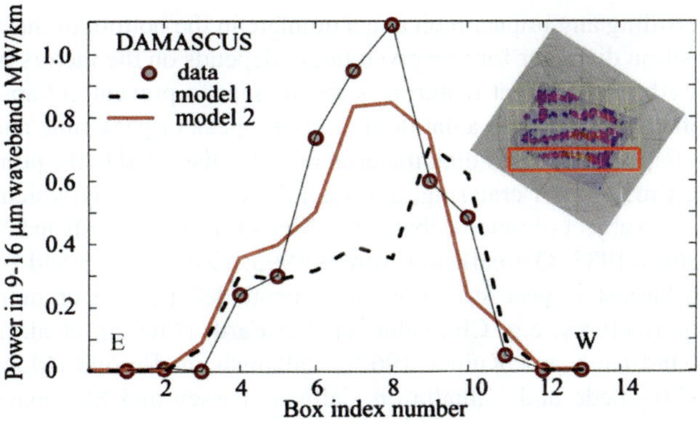

Fig. 4 *Dots* are observed power in 9.1–16.7 μm waveband by Cassini CIRS as a function of distance along Damascus tiger stripe, measured during March 2008 Cassini flyby (Spencer et al. 2008). *Thick lines* are models where power is based on time-averaged shear stress calculated by the shear heating model of Nimmo et al. (2007). The total power over the entire imaged domain is constrained to be equal to the measured power (2.3 GW). Model 1 is the original shear heating model assuming a constant brightness temperature; Model 2 corrects for the varying separation between tiger stripes, and the resulting variation in brightness temperature. The *inset* shows the area imaged by CIRS, where the yellow boxes denote individual power measurements and the colours denote relative brightness temperatures (Nimmo and Spencer 2008)

vertical fracture (temperature 130–220 K) in a conductive medium (Abramov and Spencer 2009), but does not match predictions based on a simple mobile lid model (Barr 2008). The energy distribution cannot be reproduced when higher wall temperature are considered, suggesting that either near-surface liquid water is not present or the energy distribution does not result from conductive heating of the surface by warm subsurface fractures. Further modeling efforts are required to reconcile the models for eruptive processes, heat transfer and thermal emission.

3.3 Exchange Processes due to Impact and Excavation

Over the age of the Solar System, impact cratering contributes significantly to exchange processes on the satellites by excavating fresh underground material to the surface, redistributing material globally, and, in extreme cases, by penetrating into their interiors.

Upon impact, a shock wave is generated in the target (as well as the projectile), in the wake of which substantial amounts of target material and the majority of the projectile are vaporized and melted (e.g. Roddy et al. 1977; Melosh 1989, and references therein; Pierazzo et al. 1997; Turtle and Pierazzo 2001). As it expands outward through the target, the shock wave also imparts velocity to the target material (e.g. Maxwell and Seifert 1975; Maxwell 1977). Initially this velocity is radial, compressing material outward and downward into the surrounding target. Interaction with the free surface results in a rarefaction wave, which adds an upward component to the velocity thereby ejecting some material on ballistic trajectories out of the crater. The combination of these motions opens up a transient crater.

The final morphology of impact craters depends on several factors, including the energy of the impact and the size (i.e., the surface gravity), rheology, and near-surface structure (primarily that of the lithosphere) of the target (e.g., Melosh 1989; and for icy satellites in particular: Schenk 1989, 1991, 2002). For smaller impacts, the final simple crater is only somewhat modified from the transient cavity: melt and breccia that are spread along the interior of the transient crater cavity during excavation subsequently slide back down the crater walls, overriding any impact melt to accumulate in the bottom of the crater. For larger craters (the transition diameter for any given target depends on the factors listed above), the post-shock strength of the target material is insufficient to prevent collapse under gravity, resulting in a complex crater with a flat floor, a central peak or peak ring, and a terraced rim. To allow such collapse the surrounding material must be disrupted to the point that it behaves as a fluid on the timescale of cratering, although the mechanism by which this weakening occurs has been the subject of much debate (e.g. Melosh 1982; Melosh and McKinnon 1978; Melosh and Gaffney 1983; O'Keefe and Ahrens 1993, 1999; Melosh and Ivanov 1999).

Some of the largest impact structures are surrounded by one or more external concentric scarps or fractures, e.g. Chicxulub on the Earth (Morgan et al. 1997), Orientale on the Moon (Hartmann and Kuiper 1962), Callanish on Europa (Moore et al. 2001), Gilgamesh on Ganymede and Valhalla on Callisto (Passey and Shoemaker 1982). There are two different classes of such multi-ring structures, and, as is the case for other aspects of impact craters, their morphologies reflect target properties. Melosh and McKinnon (1978) (also McKinnon and Melosh 1980) demonstrated the formation of external rings as a result of cratering into a rigid layer overlying a more fluid layer, which moves inward during crater collapse inducing radial extensional stresses in the upper layer. Numerical simulations suggest that Europa's ringed impact structures Tyre and Callanish may be examples of impacts into an ice layer overlying liquid water (Turtle 1998; Moore et al. 1998).

One mechanism by which impact cratering exchanges material is the excavation flow. However, the original depth of material that is ejected by the cratering process are actually less than 30% of the depth of the transient crater or roughly 10% of the final diameter for simple craters and 5% for complex craters (e.g. Melosh 1989; Maxwell and Seifert 1975). So, for all but the largest impact craters, the material exchanged in this way is limited to the relatively near sub-surface (i.e. a few to at most several kilometers), although ejected material can be distributed globally. For example, for Europa, which many lines of evidence suggest has a relatively shallow ocean, the largest impact structure, Tyre, is estimated to have a transient crater diameter of 22–30 km (Moore et al. 2001). So the deepest material excavated by Tyre would come from less than 3 km depth, far shallower than the base of the ice, which is constrained by impact crater morphologies to be at least a few tens of kilometers (Turtle and Pierazzo 2001; Turtle and Ivanov 2002; Schenk 2002).

Material is also exchanged locally within and beneath impact craters. As mentioned above, there is substantial disruption of the surrounding material, especially for complex craters which undergo extensive collapse. For icy satellites where impact melt is composed of water and is therefore likely to be denser than the ice bedrock, extensive fracturing in the target may allow the melt to drain deeper into the crust.

Central peaks consist of deeper target bedrock that has been fractured and uplifted stratigraphically (Grieve et al. 1981; Pieters 1982). The largest impact craters have the most uplift and thus the highest potential for surface-subsurface exchange of material, which, again, is especially intriguing in the cases of satellites that have relatively shallow internal oceans. Modeling impact cratering in an ice shell over liquid water, Turtle and Ivanov (2002) showed that, even in cases where the initial stages of the cratering process do not penetrate the ice shell, liquid water from a relatively shallow ocean (depth comparable to the diameter of the transient crater diameter) can reach the surface as a result of central uplift during the later modification stage (Fig. 5). In such cases exchange would be possible between an interior ocean and the surface; however, even for Europa's relatively thin ice shell, impacts large enough to allow water to reach the surface are rare. Oceans within other icy satellites are thought to be significantly deeper, with the possible exception of Enceladus; however, earlier in their histories their icy crusts, too, might have been breached.

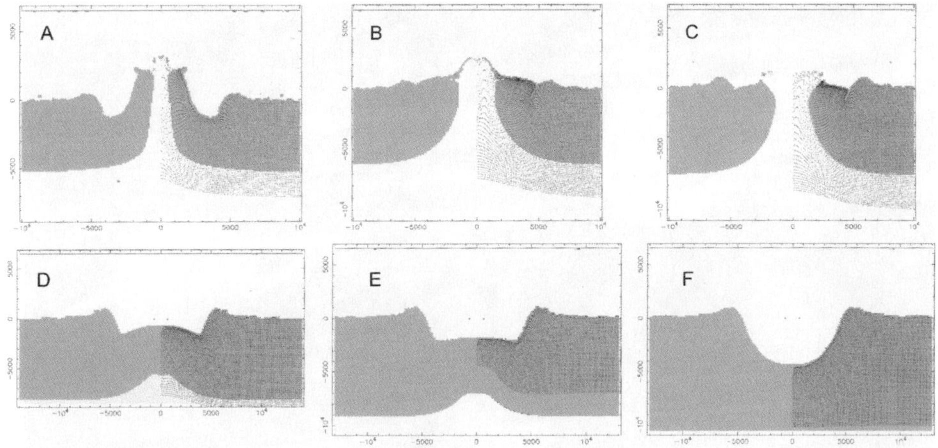

Fig. 5 Cross-sections of the final stages of crater collapse from numerical simulations of impacts into ice overlying water (after Turtle and Ivanov 2002). A through F illustrate ice (*grey*) layers 5, 6, 7, 8, 9, and 11 km thick, respectively, and demonstrate penetration of water (*white*) to the surface during crater collapse in ice ≤ 7 km thick. (*Black points* on the right side of each figure are the locations of tracer particles in the simulations)

4 Atmosphere-Surface Coupling

4.1 Processes Relevant for Titan

4.1.1 Angular Momentum Exchanges Between the Atmosphere and Surface and Its Influence on the Rotation Rate

One obvious way of atmosphere-surface coupling is the surface exchange of angular momentum between the atmosphere and solid Titan by surface friction or mountain torque. Since the total angular momentum in the solid body-atmosphere system is conserved in the absence of external torques this angular momentum exchange gives rise to a change in the planetary rotation rate or length-of-day (LOD). The basic pacemaker of the angular momentum exchange on Earth is the seasonal cycle of the atmospheric circulation resulting in a seasonal LOD variation, which amounts to ∼ 1 millisecond (Munk and MacDonald 1960). A similar effect can be expected on Titan given the presence of a dense atmosphere and seasonal cycle (Fig. 6). Tokano and Neubauer (2005) showed by general circulation modelling that the winds near the surface undergo seasonal variation and even complete reversal as a result of heating of the near-surface air from the surface. A variable surface wind field causes a temporal variation in the globally integrated surface friction torque, which modifies the body's angular momentum. If eastward surface winds prevail globally, Titan's rotation accelerates at the expense of the atmospheric rotation, and vice versa. The exchanged atmospheric angular momentum calculated with the global zonal wind field predicted by a GCM (general circulation model) amounts to 3×10^{25} kg m^2 s^{-1}, which is comparable to the terrestrial value (Tokano and Neubauer 2005).

The calculation of Titan's response to this angular momentum exchange requires knowledge of Titan's interior and, particularly, of its effective moment of inertia. Assuming a spherical shape and an entirely solid interior, the resulting LOD variation amounts to 30 s over one Titan season (∼ 7.5 yr). If one instead assumes that a liquid water-ammonia ocean

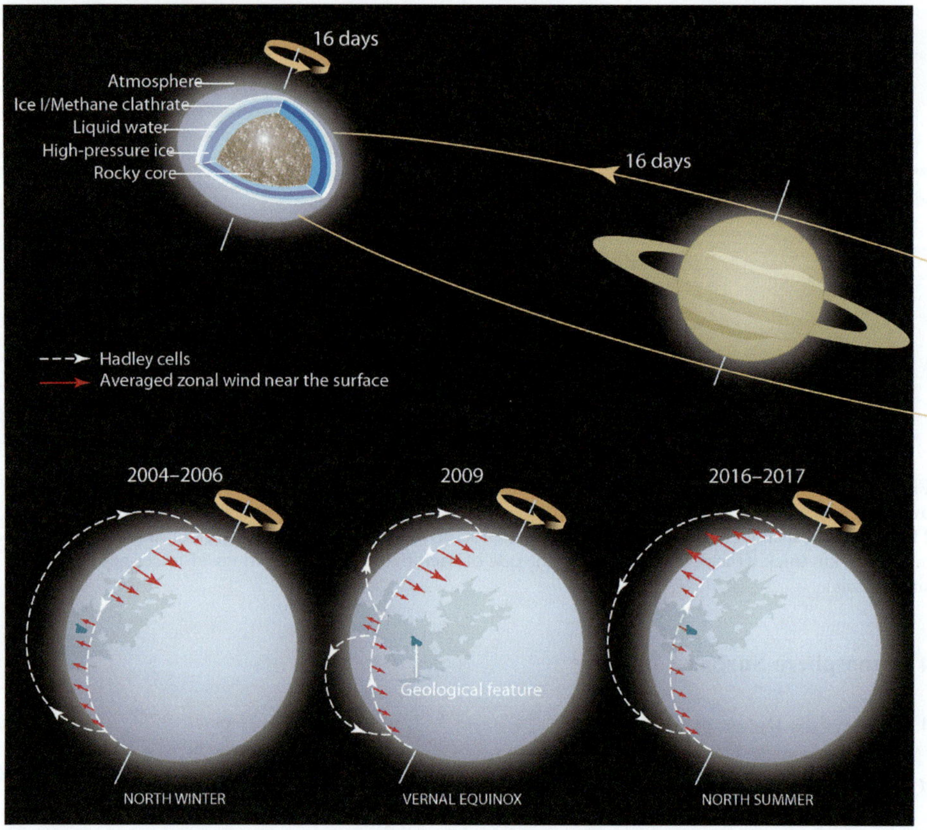

Fig. 6 (*Top*) Titan orbits around Saturn within about 16 terrestrial days, as shown. Titan and the saturnian system travel around the Sun in about 30 years, and it is the orbit around the Sun that controls the seasonal cycle on Titan due to the obliquity of the Titan-Saturn system. Lorenz et al. (2008a) suggest that an ocean may exist below the solid surface of Titan (see cut-open section), but other explanations of the observed results cannot yet be ruled out. (*Bottom*) During the nominal Cassini mission, it was winter in Titan's northern hemisphere (*left*). Seasonal variations of wind directions near the surface would explain why the observed spin rate is apparently nonsynchronous, as proposed by Tokano and Neubauer (2005), whose model predicts variations in the direction of the winds that affect the spin rate. These predictions may be checked if the Cassini mission is extended to cover the vernal equinox (*middle*) and the northern summer (*right*) (Sotin and Tobie 2008)

exists underneath the outer icy shell, this would mechanically decouple the outer shell from the mantle/core and amplify the LOD variation by an order of magnitude because of the smaller moment of inertia of the shell (Tokano and Neubauer 2005). However, if all solid layers of Titan have an ellipsoidal shape as a result of Titan's rotation and the tidal bulge exerted by Saturn, gravitational coupling arises between the shell and the interior across the ocean (Karatekin et al. 2008) as well as between the shell and Saturn's gravitational field (Van Hoolst et al. 2009), which both drastically reduce the LOD variation.

The Cassini radar imaged selected regions during multiple fly-bys. A systematic eastward shift of surface landmarks was inferred, which indicates that Titan's rotation was slightly faster than synchronous (Lorenz et al. 2008a; Stiles et al. 2008, 2010). Note that the spin rate initially reported in Stiles et al. (2008) was incorrect owing to a coding error in software used to make fits to overlapping radar images. The corrected value still indicates that Titan's rotation is not in a perfectly synchronous rotation state, but the deviation from the synchronous rate value is three times smaller than the previously reported value. Assuming that an angular momentum exchange by variable winds is responsible for the apparent deviation of the rotation from synchronous rotation and using the theoretical prediction of Tokano and Neubauer (2005), the erroneous initially reported drift rate of 0.36 deg/year was interpreted as the result of atmospheric forcing of an ice shell decoupled from the interior by a liquid internal ocean. The newly reported value of 0.12 deg/year is still compatible with the interpretation by atmospheric forcing and does not rule out the existence of a decoupling layer at depth. The new value is also more consistent with the theoretical calculations of Karatekin et al. (2008) and Van Hoolst et al. (2009) which predicted that internal and external gravitational coupling should reduce the atmosphere-induced drift rate of a decoupled ice shell. Future observations are required to confirm the observed shift and its possible time variation, and further modelling efforts would be required to interpret the observed shift in term of possible internal structure.

4.1.2 Influence of Winds on Surface Morphology

Wind can also affect the surface landscape by transport of particles on the surface. A wide portion of Titan's tropics is covered by ridges with a height of ~ 150 m, most of which are aligned parallel to the latitudinal belts (Lorenz et al. 2006). These ridges were interpreted as longitudinal "dunes" with eastward orientation. Besides the availability of suitable particles on the surface, a restricted range of wind speed and direction is crucial for the dune formation. The low gravity and dense atmosphere of Titan are generally conducive to the formation of dunes, as lower surface wind velocities are required to lift particles from the surface. Longitudinal (linear) dunes are generally thought to require a moderately variable wind field that varies around a mean or alternates between two widely spaced directions. However, there is no consensus so far as to exactly what type of wind is responsible for the transport that shapes longitudinal dunes on Titan (e.g. Rubin and Hesp 2009). Lorenz et al. (2006) proposed that the wind in the tropics is predominantly eastward, but the wind direction fluctuates diurnally due to wind induced by Saturn's tide. Tokano (2008), on the other hand, showed by general circulation modelling that near the equatorial surface the meridional wind (in north-south direction) is stronger than the zonal wind (in west-east direction) (see wind roses on Fig. 7), in agreement with the probe tracking of Huygens immediately before landing (Karkoschka et al. 2007). This wind is explained as the surface branch of the cross-equatorial Hadley circulation that reverses after each equinox. Consequently, two wind directions separated from each other by an obtuse angle prevail during a Titan year, and would transport material towards the ridges from both sides. The eastward orientation of the longitudinal dunes (Lorenz et al. 2006; Radebaugh et al. 2008) apparently contradicts wind directions predicted by GCMs. As a matter of fact, not a single GCM predicts eastward winds right at the equatorial surface, and it is currently unknown what mechanism controls the zonal wind at the surface (Tokano 2009).

4.1.3 Influence of Precipitation (Methane Hydrological Cycle) on Surface Landscape

Among the various possible erosion mechanisms, methane precipitation has attracted the most attention in the case of Titan. The first suggestion of possible rainfall erosion came from the speculation that the large bright area, now referred to as Xanadu, seen in Hubble

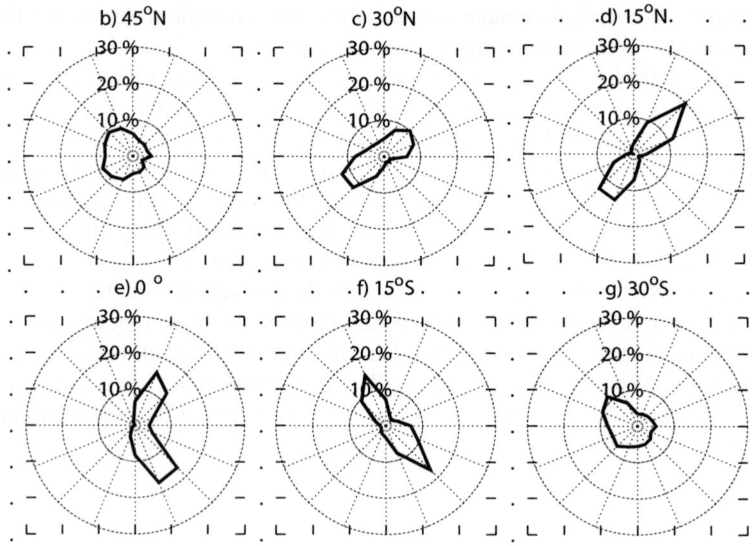

Fig. 7 Wind rose at 6 different latitudes on Titan. The diagram shows the probability of the direction the wind comes from over one Titan year. Longitudinal dunes require the presence of two major wind directions, which exist only at low latitudes. Adopted from Tokano (2008)

Space Telescope images might represent a large hill whose summit was washed out by methane rainfall that exposed the bright bedrock (Smith et al. 1996).

The Cassini-Huygens mission showed more convincing geomorphologic features suggestive of rainfall erosion. Fluvial features, although apparently dried out, have been observed at many places on Titan (Jaumann et al. 2008; Lorenz et al. 2008b) including the Huygens landing site (Tomasko et al. 2005; Soderblom et al. 2007). A broad consensus has since emerged that such dendritic features are most likely generated by surface run-off of liquids. Modelling of hydraulic sediment transport indicates that sediment could be more easily transported by surficial flows of liquid methane on Titan than by water on Earth (Burr et al. 2006). Precipitation rates required to carve the observed dendritic networks have been estimated as 0.5–60 mm hour^{-1} (Perron et al. 2006; Jaumann et al. 2008). Such precipitation rates are predicted by mesoscale convective cloud models (Hueso and Sánchez-Lavega 2006; Barth and Rafkin 2007). Rapidly evolving clouds detected by ground-based observations (Griffith et al. 2000; Brown et al. 2002) as well as by the Cassini camera (Porco et al. 2005b) resemble those simulated by these models and are thus suggestive of moist convection and occasional, violent precipitation.

Although falling intense precipitation per se has never been directly observed on Titan, there is some circumstantial evidence for precipitation. The inferred south polar lakes appear to have darkened after the development of large convective clouds, possibly as a result of ponding by precipitation (Turtle et al. 2009). Lunine and Atreya (2008) suggested that polar precipitation may fill the polar lakes/seas, which in turn may evaporate in another season. Polar clouds downwind of the northern seas ("lake-effect clouds") have been explained primarily by enhanced heat exchange between the lakes and atmosphere rather than methane evaporation (Brown et al. 2009).

More critical is the issue of precipitation at low latitudes (including the Huygens site) as a cause of valley formation. The presence of extended dunes in the equatorial region would argue against frequent or strong precipitation (Lorenz et al. 2006). Rain is expected to reach the surface if the ethane humidity in the troposphere is high enough to keep compositional equilibrium of the ternary mixture (methane-nitrogen-ethane) between the raindrops and surrounding air (Graves et al. 2008). However, no convective cloud has ever been reported in this area. Only stratiform clouds with at most drizzle are reported (Tokano et al. 2006; Ádámkovics et al. 2007). Such drizzle would be too weak to have an erosive effect. Global models of Titan's methane hydrology do not provide a unanimous picture of tropical precipitation. The predictions for the tropics range from perennially large precipitation (Tokano et al. 2001) via perennially little precipitation (Rannou et al. 2006) to precipitation confined to equinoxes (Mitchell et al. 2006). The humidity and temperature lapse rate at the Huygens site is not sufficient to trigger moist convection (Tokano et al. 2006; Hueso and Sánchez-Lavega 2006; Barth and Rafkin 2007). It is doubtful that the humidity and temperature could appreciably change with season, so the fluvial features near the Huygens site may not result from recent seasonal precipitation, but from precipitation in a wetter climate in the past or mechanisms not related to precipitation (Griffith et al. 2008).

4.2 Eruption Processes, Plume-Surface Interactions and Deposits

Io, Enceladus and Triton are the only bodies on which eruptions have been observed. Plumes on Io, Enceladus, and Triton provide insights into their sub-surface volatiles and processes. These plume activities are responsible for the generation of transient and tenuous atmospheres around these moons and for deposits that give Io and Triton their distinctive appearance, and subtly affect features on Enceladus. Europa and Titan are two other moons on which eruptive processes are likely to have occurred in the very recent past, but direct evidence is still lacking. The presence of methane and argon on Titan suggests that outgassing and hence eruptions have occurred during much of its history. On Europa, several surface features suggest that effusive processes have recently occurred.

4.2.1 Io

Ionian volcanism is the ultimate source of its atmosphere and surface frosts. As described below, the sulfur-rich volcanism on Io, which involves sub-surface reservoir of liquid sulfur as well as surface deposits of solid sulfur, share many characteristics with the water-rich (cryo)volcanism on Enceladus and on other icy moons. Direct production through volcanic plumes is likely the primary source for Io's nightside atmosphere. Indirect production through subsequent sublimation of materials deposited around plumes, discussed further below, is likely the primary source for Io's dayside atmosphere. Determining the relative roles of volcanism, sublimation, and sputtering has been the topic of intense research since Io's SO_2 atmosphere was first detected (Pearl et al. 1979), and many questions still remain despite several recent advances.

Direct volcanic atmosphere sources on Io include (1) eruption from sub-surface aquifers, (2) vaporization of surface frosts by traversing lava flows, and (3) surface lava outgassing. The process of erupting gas and dust through conduits from sub-surface reservoirs, described by Kieffer (1982), is supersonic, and may erode the surrounding surfaces to form craters, though imaging of such vents is not yet available. Kieffer et al. (2000) and Milazzo et al. (2001) describe the Promethean plume, shown to move several tens of kilometers between Voyager and Galileo imaging, as in large part explained by lava flows traversing SO_2 frost deposits with active venting occurring near the flow front. Outgassing directly from lava at the surface in lakes and flows (i.e., gases exsolved from silicates) is another source of plumes such as Tvashtar (Geissler and Goldstein 2007).

Giant plumes extend a few hundred kilometers above Io's surface, and dramatically affect the local atmosphere. The gas is ejected supersonically at 1 km/s. Zhang et al. (2004) describe the relation between volcanic source temperature, conduit size, and gravity needed for a canopy shock to form when a sublimation component is present. Models used to describe different Io plume sizes and types include simple lateral hydrostatic adjustment for the general description of localized atmospheres (Ingersoll 1989), stochastic-ballistic motions for the free-molecular regime (e.g. Strom and Schneider 1982; Lellouch 1996), rarefied gas dynamics for the transitional regime (Austin and Goldstein 2000; Zhang et al. 2004), and computational fluid dynamics for the continuum regime (Moreno et al. 1991). Spencer et al. (2007) report images of the large Tvashtar plume obtained with the New Horizons spacecraft that show remarkable time-variable structures from the upper part of the plume but no central column of upgoing particles. These images and a movie suggest that the observed particles may largely condense from gas within the plume instead of being entrained. Such "snowflakes" would be consistent with earlier measurements of 500–1000 amu particles with the Galileo Plasma Spectrometer (Frank and Paterson 2002).

The mass flux from Io's plumes and lava flows resurfaces the satellite at a rate of 0.55–1.06 cm/year based on heat flow estimates (McEwen et al. 2004). No impact craters have been detected on Io (i.e., over the 80% of it observed at high resolution). This fact requires resurfacing to bury 20 km sized craters formed on timescales of one every ~ 3 Myr (Zahnle et al. 2003), which is a minimum rate of 0.05 cm/year. Daily sublimation and re-adsorption of SO_2 covers the surface in frost, which is found everywhere except hot volcanic centers (Carlson et al. 1997; Douté et al. 2001). These SO_2 frosts are enhanced in an equatorial band and are distributed with different depths, mixing ratios, and grain sizes (McEwen et al. 1988; Laver and de Pater 2008). The correlation of SO_2 rich surface regions with active plumes makes it difficult to observationally distinguish between direct volcanic and indirect sublimation sources for Io's atmosphere using spatial associations.

Observations of Io's auroral emissions when passing into and out of eclipse by the New Horizons mission constrain the volcanic and sublimation sources while avoiding this difficulty. Together with models of Io's plasma interaction (Saur and Strobel 2004) and estimates of changes to the sublimation atmosphere in response to surface temperature changes in shadow (Sinton and Kaminski 1988; Spencer et al. 2000), these observations suggest that 1–3% of Io's dayside atmosphere is volcanic, and that the sublimation source is largely halted on the nightside (Retherford et al. 2007). More information on Io's surface temperature, thermal inertia, and the sticking coefficients for SO_2 frosts interacting with supersonic plume gas are needed to better understand the partial atmospheric collapse of Io's atmosphere in shadow and the relative role of the plume atmospheres.

Giant plumes deposit rings of material several hundred kilometers around volcanos. These rings and other shaped deposits display an interesting canvas of coloration that evolves with time. Molecular sulfur gas (S_2) and other species expected from plume photochemistry models (Moses et al. 2002) such as S_3, S_4, and S_2O are deposited on the surface. On the surface, S_2 may anneal to form the S_4 molecule thought to explain the red color of the largest rings (e.g., seen at Pele). Silicate dust particles likely explain the darkest features. In the solid state, sulfur allotropes such as S_8, itself formed from S_3 and S_4, can form long chains of polymeric sulfur, S_∞. Surface processes such as photolysis and radiolysis can modify these sulfur constituents mixed with SO_2 and other frosts such as SO_3 and possibly Cl_2SO_2 and H_2S, resulting in the colors of white, yellow, green, red, and black observed near plumes (cf. reviews by McEwen et al. 2004; Carlson et al. 2007; Geissler and Goldstein 2007). Evidence for water on the surface of Io (e.g. Salama et al. 1990; Carlson et al. 2007) is scant. The sublimation and adsorption/condensation cycle for the primary constituents tends to cover deposits of trace plume species, and converts fine grained ices to coarse grained ices. Shocks formed when the supersonic plume gas impacts the surface and/or the near-surface sublimation atmosphere can at times remove ring material faster than it deposits it, depending on the surface temperature and related sticking coefficient (Zhang et al. 2004).

4.2.2 Enceladus

The Enceladus plume emerges from discrete sources (Spitale and Porco 2007) located on a system of cracks in the ice crust (Spencer et al. 2006). Although a certain variability of the plume activity is expected from the response of the cracks to tidal stresses (Hurford et al. 2007), the plume appears surprisingly steady over the time it has been observed. Similar gas densities are derived from stellar plume occultations recorded by the Cassini Ultraviolet Imaging Spectrograph (UVIS) with an increase by a factor of 1.7 from 2005 (Hansen et al. 2006) (1.5×10^{16} molecules per cm^2 for a line of sight passing at 15 km altitude over the south pole) to 2007 (Hansen et al. 2008). Moreover, unchanged individual dust jets are observed in images taken over more than two years (Spitale and Porco 2007), which further supports the idea that the plume sources remain active over long periods.

Such a continuous plume activity requires a steady mechanism for the production of gas and grains. Explosive and self-limiting processes appear implausible (Brilliantov et al. 2008). This excludes geyser-like processes, with a build up of pressure, suddenly released in an eruption, and it rules out the so-called cold faithful model (Porco et al. 2006). Also the decomposition of clathrates (Kieffer et al. 2006) seems difficult to reconcile with the steady production of gas and grains, although this idea is attractive since it offers an explanation for the observed abundance of roughly 10% of volatile gases (CO_2, N_2, CO) in the plume gas (Waite et al. 2006).

Alternative scenarios for the production of gas are direct evaporation either from liquid water (Schmidt et al. 2008; Postberg et al. 2009) or warm ice (Nimmo et al. 2007) at depth under the ice crust. In these models the gas flows through cracks in the ice to surface. At the site of evaporation the gas is in near thermal equilibrium with the water and/or ice. When the vapour is accelerated it expands and cools, in accordance with the laws of thermodynamics, and thus becomes super-saturated, and ice grains may condense from the vapour.

When a new crack is opened in the ice crust both models (sublimation from liquid or warm ice) lead to a nearly steady state after a transition period, in which the gas slowly heats the ice in the vicinity of the crack. Initially, the vapour will condense entirely at depth on the cold crack walls and by latent heat of condensation the ice will gradually warm up. Eventually, the process approaches a steady state when the heat flow in the ice is in equilibrium with the radiative loss at the surface and the heat supplied by advection by the gas from depth. This transition to the steady state may take hundreds of years for a crack of several kilometers depth (Ingersoll and Pankine 2010). During this period narrow cracks might even be sealed by the condensing vapour before the gas flow reaches vacuum. Systems of nearby cracks may act together heating the ice in their vicinity. In steady state condensation at the walls will be mostly limited to the region close to the surface, since there the temperature gradient, and thus the heat flow, is largest. In this way a conduit should naturally develop its narrowest point near the surface.

Averaging over more complex channel profiles, Schmidt et al. (2008) obtained a differential particle size distribution for the condensed grains which has a slope corresponding to an exponent of -4 for grains in the size range of one micron but globally falling off steeper than a power law. Also, the distribution exhibits a local minimum for submicron sized grains. This distribution is consistent with the narrow E ring size distribution derived from pre Cassini data (Showalter 1991; Nicholson et al. 1996) and Cassini CDA data (Kempf et al. 2008).

From the structure of the dust plume seen in images it was early concluded that most of the grains must fall back to the surface, ejected at a mean speed of 120 m/s (Porco et al. 2006), which is smaller than Enceladus' escape speed of 240 m/s. The plume model of Schmidt et al. (2008) gives a mean grain speed of 100 m/s, which is practically the same number. The velocity distribution was recently constrained (Hedman et al. 2009) from a detailed analysis and modeling of spectral slopes of the plume obtained from data taken by the Cassini Visual and Infrared Mapping Spectrometer (VIMS). The authors find clear evidence that large particles (around 3 micron radius) are ejected at low velocities, practically all falling back to the surface and populating the lower parts of the plume, while smaller particles are systematically faster.

Altogether, the low speed of the ejected grains appears surprising, since large gas speeds on the order of 500 m/s were inferred from Cassini UVIS data (Hansen et al. 2006; Tian et al. 2007). If grains condense in the gas, one would expect them to have the same velocity. If the grains are formed by some other mechanism, then the gas must be sufficiently dense to accelerate them. Also in this case the grain will rapidly reach gas speed, again in contradiction to the observations.

There are in principle two ways to understand the slow velocity of the dust compared to the gas. One possibility (Schmidt et al. 2008) is that frequent collisions of grains with the vent walls repeatedly decelerate the grains relative to the gas. Such collisions are in practice unavoidable in a realistic channel, i.e. one that is not perfectly straight, since the streamlines of the gas and the trajectories will differ. Owing to their smaller speeds, large grains populate the plume at lower altitudes. This model reproduces the particle densities measured by CDA (Schmidt et al. 2008), the brightness of the plume seen in images, and the gas production rate inferred by INMS and UVIS. Another possibility to obtain in principle slow grains and fast gas is that the gas and dust decouple only in the uppermost, funnel shaped, part of a channel. In this region the gas is further accelerated by the pressure drop from the channel to vacuum, while it simultaneously dilutes by large factors. If this mechanism works quantitatively, simulating the observed particle speed-size distribution (Hedman et al. 2009) remains to be verified.

The composition of ejected material addresses the question of whether liquid water or warm ice is Enceladus' primary plume source (Zolotov 2007). If the jets originate from sublimating ice, trapped gases would be the major non-water compound, whereas minerals leached from the large Enceladian rock core should be present in a possible subsurface ocean. In the latter case Na^+ and Cl^- ions, are expected to be the most abundant non-water species, followed by bicarbonate (HCO_3^-) and K^+ (Zolotov 2007). During the slow downward freezing expected after formation of icy planetary bodies, alkali salts always stay in the liquid phase and the ice crust remains practically salt free. On one hand the abundant detection of CO_2, CH_4, N_2 and/or CO in the plume vapour (Waite et al. 2006) suggest the presence of clathrate hydrates and gave rise to the proposition that the decomposition of such ices is the actual plume driver (Kieffer et al. 2006). On the other hand the in situ measurements of Cassini CDA show alkali metals in 93% of mass spectra from E ring ice grains (Postberg et al. 2009). The E ring can be considered as a storage ring for previously ejected plume grains. Whereas most of these grains show only traces of sodium and potassium ($Na/H_2O \approx 10^{-7}$), about 6% exhibit a salt content increased by several orders of magnitude ($Na/H_2O \approx 10^{-2}$–10^{-3}). Particles with intermediate concentrations are rare ($<3\%$).

The Na-rich CDA spectra imply 0.05–0.2 M/kg NaCl and 0.02–0.1 M/kg $NaHCO_3$ yielding an alkaline pH of 8.5–9 in a liquid with the ice grains composition. This is in compelling similarity with the model calculations (Zolotov 2007). However, the inferred Na/K ratio of 100–300 is about 10 times higher than expected from the solar abundance. The presence of salts in concentration as found in Na-rich E ring grains is difficult to reconcile with ice sublimation (Nimmo et al. 2007) or clathrate decomposition as the main plume producing process (Kieffer et al. 2006). The observed composition strongly favors an origin from water that is or has been in contact with the rocky core.

The non-water molecules found in the plume vapour (Waite et al. 2006) suggest clathrate decomposition as a parallel process. Although probably not relevant as a plume driver, gas release from clathrates might take place at ice/water interfaces of the source or in ice layers above such a reservoir. However, a "Soda Ocean" rich in bicarbonates could also work as a source of CO_2. Hydrothermal processes (Matson et al. 2007) might also be considered for the production of N_2, CH_4, and other organic compounds. Postberg et al. (2009) introduce a scenario in which Na-rich and Na-poor grain populations are both produced from liquid plume sources (Fig. 8): Na-rich ice grains are formed by freezing of aerosols created by ascending gas bubbles (e.g. from CO_2) or other turbulent processes in the liquid. The Na-poor grains (which represent the main E ring population) are suggested to stem from nucleation of the supersaturated vapour (Fig. 8B) as suggested by Schmidt et al. (2008). Trace amounts of NaCl molecules (present in low concentration in the liquid Zolotov 2007) can traverse from liquid solution into the gas phase. Due to its high energy of solution, a significant Na escape via the plume gases is not expected and in agreement with the non detection by Cassini instruments and Earth bound spectroscopy (Schneider et al. 2009).

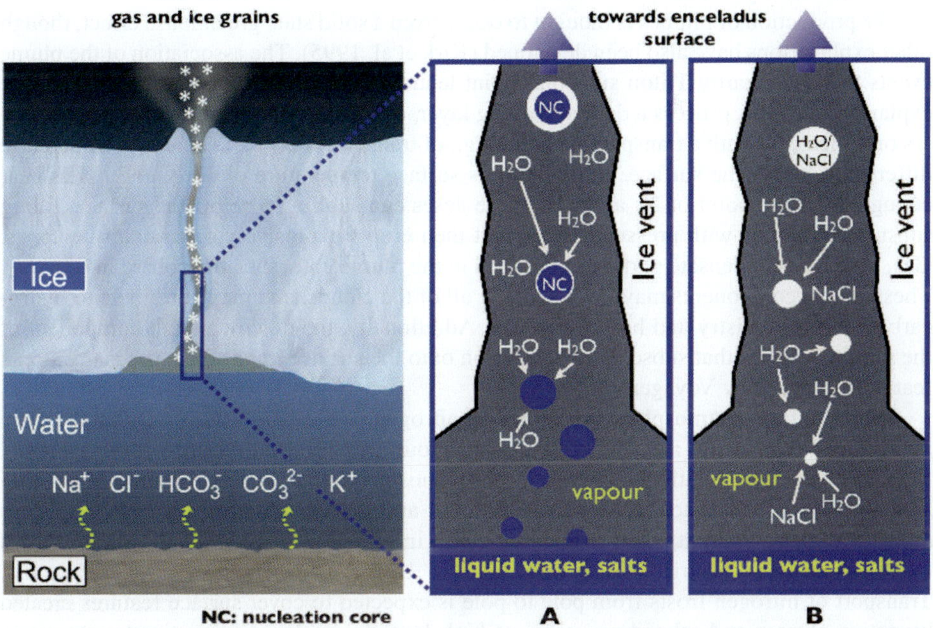

Fig. 8 Plume generation from a liquid source according to Postberg et al. (2009). **A** Production of Na-rich grains: Aerosol-like droplets (*blue*) are generated by dispersion above the salty liquid, preserving the ocean's composition. Such sub-micron sized droplets would condense some additional water from the supersaturated gas that drags them to the surface. Therefore, the salt content of Na-rich grains can be considered as a lower limit for an Enceladus ocean. **B** Production of Na-poor grains: The formation of ice grains by condensation of water vapour ascending from a liquid reservoir with subsequent growth and transport in the vent (Schmidt et al. 2008). The particles initially have the same low ratio of NaCl/H_2O as the gas

One might consider the possibility of parallel venting mechanisms from a liquid and from warm ice. The first would produce the Na-rich grains, the latter the majority of non-water plume gases and Na-poor grains. However, the formation of Na-rich particles requires a liquid reservoir, so evaporation above such a liquid is probably the most plausible driver for Enceladus' plumes in general.

From heat flow arguments it can be shown that the liquid/gas interface below Enceladus' surface must be orders of magnitude larger than the vent cross sections, otherwise implausibly large temperature gradients would be necessary to balance the loss due to latent heat (Postberg et al. 2009). Therefore large vapour chambers which narrow to the vent channels are required above a liquid plume source.

4.2.3 Triton

Two plume features on Triton, Hili and Mahilani, were clearly identified in Voyager 2 images (Smith et al. 1989). These plumes have a central vertical column of height ~ 8 km (and radii of 20 m to 2 km), above which a trailing cloud extends over 100 km downwind toward the west (Soderblom et al. 1990). The images also show evidence for two or more plumes, and a few persistent cloud features viewed near the terminator were speculated to be associated with plume sources. The composition of the plumes is thought to be nitrogen and possibly methane, with dark features being entrained N_2 ices or hydrocarbon (soot) particles. These hydrocarbon particles are likely produced by methane photochemistry (photolysis and radiolysis), but primordial/exogenic sources can not be ruled out.

The production of plumes is thought to occur from a solid state greenhouse effect, though other explanations have also been developed (Kirk et al. 1995). The association of the plume events with the nearby Triton sub-solar point lends credence to the solid state greenhouse explanation. In this process a dark absorbing layer, (e.g., organic compounds or deeper layers of N_2 ice), beneath a transparent layer, (e.g., of translucent N_2 ice), creates a temperature differential below the surface. Since Triton's surface temperature of 37 K to 39 K is near enough the triple point of N_2 and CH_4 these gases can sublimate below the surface, filling subsurface regions with pressurized gas that then erupts explosively or gradually escapes. Gas, dust, and condensate particles entrained in the plume enter the atmosphere in this way. These plume components may form some or all of the clouds, and may contribute to hydrocarbon photochemistry and haze formation. Additionally, the downwind tail component of the plumes implies that subsequent deposition onto the surface might form the many streak features observed by Voyager.

Triton's nitrogen atmosphere is buffered by nitrogen surface frosts, meaning that the main atmospheric constituent is sublimated and in vapour pressure equilibrium with the surface. Voyager 2 radio occultations detected a thick ionosphere and signs of a thermal inversion near the surface consistent with this substantial atmosphere (Tyler et al. 1989). Triton's retrograde orbit, obliquity, and precession result in an interesting seasonal cycle in which the sub-solar latitude varies widely from $\pm 10°$ to $\pm 52°$ during a Triton year (~ 165 years). Transport of nitrogen frosts from pole to pole is expected to cover surface features created by active plumes and related processes at high latitudes with a few centimetres to a few meters of material per year. However, Voyager 2 observed a bright southern hemisphere in summer, and the darker northern hemisphere and equatorial region appeared to lack similar frost deposits (Yelle et al. 1995). Detections of methane and carbon monoxide on the surface imply their presence in the atmosphere owing to their volatility (Brown et al. 1995), with methane confirmed by Voyager UV occultation observations (Broadfoot et al. 1989; Strobel and Summers 1995). Geological features called guttae and aureoles may best be explained as formed by volatile deposits early in Triton's history (McKinnon et al. 1995) and shaped by sublimation and wind erosion, though the composition of the material is not determined (Croft et al. 1995). The age of Triton's surface is constrained by impact cratering records to be 6–50 Myr (Schenk and Zahnle 2007). Observed craters did not show rays, which were likely obscured by a veneer of volatiles (Smith et al. 1989).

4.2.4 Titan

The most convincing evidence for past eruption and internal outgassing over Titan's history is provided by the GCMS onboard the Huygens probe that detected a significant amount of ^{40}Ar in Titan's atmosphere (Niemann et al. 2005). Radiogenic ^{40}Ar is the decay daughter of ^{40}K, which is contained in the rocky part of Titan's interior. Its detection in the atmosphere indicates that significant outgassing from the interior has occurred over Titan's history. Moreover, carbon and nitrogen isotopic ratios measured by the GCMS indicate that nitrogen and methane, the two major compounds in Titan's atmosphere, have followed two distinct evolutionary paths. The carbon isotopic ratio ^{13}C/^{12}C in hydrocarbon molecules, which is relatively close to the terrestrial value (Niemann et al. 2005) reflects a bulk carbon inventory that did not participate in the massive escape phase that followed the formation of an ancient post-accretional atmosphere. In contrast, the isotopic ratio ^{15}N/^{14}N shows strong enrichment relative to the terrestrial value and hence suggests a large escape. This constraint together with the low abundance in ^{36}Ar, which implies that the nitrogen was captured as NH_3 and converted just after accretion (Niemann et al. 2005), indicate that nitrogen has been present in Titan's atmosphere since its formation, whereas today's methane has been injected much later. The exact timing of volatile outgassing is difficult to constrain from only these measurements, but they clearly indicate that internal outgassing has occurred at some period during Titan's history.

Owing to the presence of a very thick atmosphere on Titan, eruption styles are likely to be very different from those of Enceladus. On Titan, explosive eruption are expected to occur only if a large reservoir of clathrate, volatile-rich sediments or crustal methanifer, is suddenly disrupted (Fortes et al. 2007; Choukroun et al. 2010). Even if explosive activity is predicted to be the dominant eruptive style on Titan (Fortes et al. 2007), the absence of direct observation make difficult to conclude if such a process is a common feature on Titan.

5 Concluding Remarks

In this chapter, we have discussed the different physical processes that control the exchanges between the interior, surface and atmosphere on a variety of icy moons. Even though the icy moons are very different in aspect, they are all affected by similar physical processes. Despite recent progresses in the understanding of exchange processes from spacecraft observations (Galileo and Cassini-Huygens), numerical modellings and laboratory experiments, many questions remains unanswered. For instance, even though cryovolcanic activity has been speculated to occur on icy moons for several decades, the exact mechanism that leads to the eruption of liquids on the surface of an icy moon is still not understood. Key measurements are still lacking to constrain the different external and internal processes affecting the satellite surface. In particular the subsurface structure remains unknown. Ice-penetrating radar on future spacecrafts would be of great values to better understand the exchange with the underground. Moreover, flyby observations prevent homogeneous global mapping and long-term monitoring of atmosphere and surface features. Only observations from an orbiter would permit to accurately map the surface and detect subtle surface changes. Future explorations of Europa and Ganymede by the joint NASA/ESA Europa and Jupiter System Mission should provide these missing measurements. In situ measurements such as the analysis of plume particles or of surface materials, or the deployment of geophysical instruments on the surface can also be envisaged in the future, in particular in the Saturnian system where landing on Titan is relatively easy and where Enceladus plumes can be very easily sampled. In parallel, modelling and experimental efforts are required to interpret all the data that are continuously gathered by the Cassini spacecraft as well as to prepare the analysis and interpretation of the new dataset that the future exploration missions will collect.

Acknowledgements We thank L. Soderblom for helpful suggestions and corrections. Supports from INSU's Programme National de Planétologie, NASA's Planetary Geology and Geophysics and Outer Planets Research programmes are gratefully acknowledged.

References

O. Abramov, J.R. Spencer, Icarus **199**, 189–196 (2009). doi:10.1016/j.icarus.2008.07.016

M. Ádámkovics, M.H. Wong, C. Laver, I. de Pater, Science **318**, 962 (2007). doi:10.1126/science.1146244

J.V. Austin, D.B. Goldstein, Icarus **148**, 370–383 (2000). doi:10.1006/icar.2000.6466

A.C. Barr, J. Geophys. Res. **113**, 7009 (2008). doi:10.1029/2008JE003114

A.C. Barr, W.B. McKinnon, Geophys. Res. Lett. **34**, 9202 (2007). doi:10.1029/2006GL028799

E.L. Barth, S.C.R. Rafkin, Geophys. Res. Lett. **34**, 3203 (2007). doi:10.1029/2006GL028652

M. Beeman, W.B. Durham, S.H. Kirby, J. Geophys. Res. **93**, 7625–7633 (1988). doi:10.1029/JB093iB07p07625

S.E. Billings, S.A. Kattenhorn, Icarus **177**, 397–412 (2005). doi:10.1016/j.icarus.2005.03.013

M.T. Bland, R.A. Beyer, A.P. Showman, Icarus **192**, 92–105 (2007). doi:10.1016/j.icarus.2007.06.011

M.T. Bland, A.P. Showman, G. Tobie, Icarus **200**, 207–221 (2009). doi:10.1016/j.icarus.2008.11.016

N.V. Brilliantov, J. Schmidt, F. Spahn, Planet. Space Sci. **56**, 1596–1606 (2008)

A.L. Broadfoot, S.K. Atreya, J.L. Bertaux, J.E. Blamont, A.J. Dessler, T.M. Donahue, W.T. Forrester, D.T. Hall, F. Herbert, J.B. Holberg, D.M. Hunten, V.A. Krasnopolsky, S. Linick, J.I. Lunine, J.C. Mcconnell, H.W. Moos, B.R. Sandel, N.M. Schneider, D.E. Shemansky, G.R. Smith, D.F. Strobel, R.V. Yelle, Science **246**, 1459–1466 (1989). doi:10.1126/science.246.4936.1459

J.F. Brotchie, R. Silvester, J. Geophys. Res. **74**, 5240–5252 (1969). doi:10.1029/JB074i022p05240

R.H. Brown, R.L. Kirk, J. Geophys. Res. **99**, 1965–1981 (1994). doi:10.1029/93JE02618

R.H. Brown, D.P. Cruikshank, J. Veverka, P. Helfenstein, J. Eluszkiewicz, in *Neptune and Triton*, ed. by D.P. Cruikshank, M.S. Matthews, A.M. Schumann (1995), pp. 991–1030

M.E. Brown, A.H. Bouchez, C.A. Griffith, Nature **420**, 795–797 (2002)

M.E. Brown, E.L. Schaller, H.G. Roe, C. Chen, J. Roberts, R.H. Brown, K.H. Baines, R.N. Clark, Geophys. Res. Lett. **36**, 1103 (2009). doi:10.1029/2008GL035964

D.M. Burr, J.P. Emery, R.D. Lorenz, G.C. Collins, P.A. Carling, Icarus **181**, 235–242 (2006). doi:10.1016/j.icarus.2005.11.012

R.W. Carlson, W.D. Smythe, R.M.C. Lopes-Gautier, A.G. Davies, L.W. Kamp, J.A. Mosher, L.A. Soderblom, F.E. Leader, R. Mehlman, R.N. Clark, F.P. Fanale, Geophys. Res. Lett. **24**, 2479 (1997). doi:10.1029/97GL02609

R.W. Carlson, J.S. Kargel, S. Douté, L.A. Soderblom, J.B. Dalton, *Io's Surface Composition*. Springer Praxis Books/Geophysical Sciences (2007), p. 193

J.C. Castillo-Rogez, D.L. Matson, C. Sotin, T.V. Johnson, J.I. Lunine, P.C. Thomas, Icarus **190**, 179–202 (2007). doi:10.1016/j.icarus.2007.02.018

M. Choukroun, O. Grasset, G. Tobie, C. Sotin, Icarus **205**, 581–593 (2010)

G.C. Collins, J.W. Head, R.T. Pappalardo, Icarus **135**, 345–359 (1998). doi:10.1006/icar.1998.5978

A. Coustenis, T. Tokano, M.H. Burger, T.A. Cassidy, R.M. Lopes, R.D. Lorenz, K.D. Retherford, G. Schubert, Space Sci. Rev. (2010, this issue). doi:10.1007/s11214-009-9615-5

G.D. Crawford, D.J. Stevenson, Icarus **73**, 66–79 (1988). doi:10.1016/0019-1035(88)90085-1

S.K. Croft, J.I. Lunine, J. Kargel, Icarus **73**, 279–293 (1988). doi:10.1016/0019-1035(88)90098-X

S.K. Croft, J.S. Kargel, R.L. Kirk, J.M. Moore, P.M. Schenk, R.G. Strom, in *Neptune and Triton*, ed. by D.P. Cruikshank, M.S. Matthews, A.M. Schumann (1995), pp. 879–947

A.J. Dombard, W.B. McKinnon, Icarus **154**, 321–336 (2001). doi:10.1006/icar.2001.6728

A.J. Dombard, W.B. McKinnon, J. Struct. Geol. **28**, 2259–2269 (2006). doi:10.1016/j.jsg.2005.12.003

S. Douté, B. Schmitt, R. Lopes-Gautier, R. Carlson, L. Soderblom, J. Shirley, T.G.N.T. The Galileo NIMS Team, Icarus **149**, 107–132 (2001). doi:10.1006/icar.2000.6513
S. Douté, R. Lopes, L.W. Kamp, R. Carlson, B. Schmitt, Icarus **169**, 175–196 (2004). doi:10.1016/j.icarus.2004.02.001
J.L. Elliot, M.J. Person, S.W. McDonald, M.W. Buie, E.W. Dunham, R.L. Millis, R.A. Nye, C.B. Olkin, L.H. Wasserman, L.A. Young, W.B. Hubbard, R. Hill, H.J. Reitsema, J.M. Pasachoff, T.H. McConnochie, B.A. Babcock, R.C. Stone, P. Francis, Icarus **148**, 347–369 (2000). doi:10.1006/icar.2000.6508
S.A. Fagents, J. Geophys. Res. **108**, 5139 (2003). doi:10.1029/2003JE002128
P.H. Figueredo, F.C. Chuang, J. Rathbun, R.L. Kirk, R. Greeley, J. Geophys. Res. **107**, 5026 (2002). doi:10.1029/2001JE001591
R.C. Fletcher, B. Hallet, J. Geophys. Res. **88**, 7457–7466 (1983). doi:10.1029/JB088iB09p07457
A.D. Fortes, P.M. Grindrod, S.K. Trickett, L. Vočadlo, Icarus **188**, 139–153 (2007). doi:10.1016/j.icarus.2006.11.002
L.A. Frank, W.R. Paterson, J. Geophys. Res. **107**, 1220 (2002). doi:10.1029/2002JA009240
P.E. Geissler, D.B. Goldstein, Plumes and their deposits, 2007
P.E. Geissler, R. Greenberg, G. Hoppa, A. McEwen, R. Tufts, C. Phillips, B. Clark, M. Ockert-Bell, P. Helfenstein, J. Burns, J. Veverka, R. Sullivan, R. Greeley, R.T. Pappalardo, J.W. Head, M.J.S. Belton, T. Denk, Icarus **135**, 107–126 (1998)
B. Giese, R. Wagner, G. Neukum, P. Helfenstein, P.C. Thomas, Geophys. Res. Lett. **34**, 21203 (2007). doi:10.1029/2007GL031467
B. Giese, R. Wagner, H. Hussmann, G. Neukum, J. Perry, P. Helfenstein, P.C. Thomas, Geophys. Res. Lett. **35**, 24204 (2008a). doi:10.1029/2008GL036149
B. Giese, T. Denk, G. Neukum, T. Roatsch, P. Helfenstein, P.C. Thomas, E.P. Turtle, A. McEwen, C.C. Porco, Icarus **193**, 359–371 (2008b). doi:10.1016/j.icarus.2007.06.005
P. Goldreich, N. Murray, P.Y. Longaretti, D. Banfield, Science **245**, 500–504 (1989). doi:10.1126/science.245.4917.500
D.L. Goldsby, D.L. Kohlstedt, J. Geophys. Res. **106**, 11017–11030 (2001)
M.P. Golombek, W.B. Banerdt, Icarus **68**, 252–265 (1986). doi:10.1016/0019-1035(86)90022-9
O. Grasset, J. Pargamin, Planet. Space Sci. **53**, 371–384 (2005). doi:10.1016/j.pss.2004.09.062
S.D.B. Graves, C.P. McKay, C.A. Griffith, F. Ferri, M. Fulchignoni, Planet. Space Sci. **56**, 346–357 (2008). doi:10.1016/j.pss.2007.11.001
R. Greeley, E. Theilig, P. Christensen, Icarus **60**, 189–199 (1984). doi:10.1016/0019-1035(84)90147-7
R. Greenberg, S.J. Weidenschilling, Icarus **58**, 186–196 (1984). doi:10.1016/0019-1035(84)90038-1
R. Greenberg, P. Geissler, G. Hoppa, B.R. Tufts, D.D. Durda, R. Pappalardo, J.W. Head, R. Greeley, R. Sullivan, M.H. Carr, Icarus **135**, 64–78 (1998)
R.A.F. Grieve, P.B. Robertson, M.R. Dence, in *Multi-Ring Basins: Formation and Evolution*, ed. by R.B. Merill, P.H. Schultz (1981), pp. 37–57
C.A. Griffith, J.L. Hall, T.R. Geballe, Science **290**, 509–513 (2000). doi:10.1126/science.290.5491.509
C.A. Griffith, C.P. McKay, F. Ferri, Astrophys. J. **687**, 41–44 (2008). doi:10.1086/593117
I. Halevy, S.T. Stewart, Geophys. Res. Lett. **35**, 12203 (2008). doi:10.1029/2008GL034349
C.J. Hansen, L. Esposito, A.I.F. Stewart, J. Colwell, A. Hendrix, W. Pryor, D. Shemansky, R. West, Science **311**, 1422–1425 (2006)
C.J. Hansen, L.W. Esposito, A.I.F. Stewart, B. Meinke, B. Wallis, J.E. Colwell, A.R. Hendrix, K. Larsen, W. Pryor, F. Tian, Nature **456**, 477–479 (2008). doi:10.1038/nature07542
Y. Harada, K. Kurita, Planet. Space Sci. **54**, 170–180 (2006). doi:10.1016/j.pss.2005.12.001
W.K. Hartmann, G.P. Kuiper, Commun. Lunar Planet. Lab. **1**, 131–137 (1962).
J.W. Head, R.T. Pappalardo, R. Sullivan, J. Geophys. Res. **104**, 24223–24236 (1999). doi:10.1029/1998JE001011
M.M. Hedman, P.D. Nicholson, M.R. Showalter, R.H. Brown, B.J. Buratti, R.N. Clark, Astrophys. J. **693**, 1749–1762 (2009)
D.L. Herrick, D.J. Stevenson, Icarus **85**, 191–204 (1990). doi:10.1016/0019-1035(90)90110-U
F. Hersant, D. Gautier, G. Tobie, J.I. Lunine, Planet. Space Sci. **56**, 1103–1111 (2008). doi:10.1016/j.pss.2008.02.007
J. Hillier, S.W. Squyres, J. Geophys. Res. **96**, 15665 (1991). doi:10.1029/91JE01401
G.V. Hoppa, B.R. Tufts, R. Greenberg, P.E. Geissler, Science **285**, 1899–1902 (1999). doi:10.1126/science.285.5435.1899
R. Hueso, A. Sánchez-Lavega, Nature **442**, 428–431 (2006). doi:10.1038/nature04933
T.A. Hurford, R.A. Beyer, B. Schmidt, B. Preblich, A.R. Sarid, R. Greenberg, Icarus **177**, 380–396 (2005). doi:10.1016/j.icarus.2005.06.019
T.A. Hurford, P. Helfenstein, G.V. Hoppa, R. Greenberg, B.G. Bills, Nature **447**, 292–294 (2007)

T.A. Hurford, B.G. Bills, P. Helfenstein, R. Greenberg, G.V. Hoppa, D.P. Hamilton, Icarus **203**, 541–552 (2009). doi:10.1016/j.icarus.2009.04.025
H. Hussmann, T. Spohn, Icarus **171**, 391–410 (2004). doi:10.1016/j.icarus.2004.05.020
H. Hussmann et al., Space Sci. Rev. (2010, this issue)
A.P. Ingersoll, Icarus **81**, 298–313 (1989). doi:10.1016/0019-1035(89)90055-9
A.P. Ingersoll, A.A. Pankine, Icarus **206**, 594–607 (2010). doi:10.1016/j.icarus.2009.09.015
R. Jaumann, R.H. Brown, K. Stephan, J.W. Barnes, L.A. Soderblom, C. Sotin, S. Le Mouélic, R.N. Clark, J. Soderblom, B.J. Buratti, R. Wagner, T.B. McCord, S. Rodriguez, K.H. Baines, D.P. Cruikshank, P.D. Nicholson, C.A. Griffith, M. Langhans, R.D. Lorenz, Icarus **197**, 526–538 (2008). doi:10.1016/j.icarus.2008.06.002
Ö. Karatekin, T. Van Hoolst, T. Tokano, Geophys. Res. Lett. **35**, 16202 (2008). doi:10.1029/2008GL034744
J.S. Kargel, Icarus **94**, 368–390 (1991). doi:10.1016/0019-1035(91)90235-L
E. Karkoschka, M.G. Tomasko, L.R. Doose, C. See, E.A. McFarlane, S.E. Schröder, B. Rizk, Planet. Space Sci. **55**, 1896–1935 (2007). doi:10.1016/j.pss.2007.04.019
S. Kempf, U. Beckmann, G. Moragas-Klostermeyer, F. Postberg, R. Srama, T. Economou, J. Schmidt, F. Spahn, E. Grün, Icarus **193**, 420–437 (2008). doi:10.1016/j.icarus.2007.06.027
S.W. Kieffer, in *Satellites of Jupiter*, ed. by D. Morrison (1982), pp. 647–723
S.W. Kieffer, R. Lopes-Gautier, A. McEwen, W. Smythe, L. Keszthelyi, R. Carlson, Science **288**, 1204–1208 (2000). doi:10.1126/science.288.5469.1204
S.W. Kieffer, X. Lu, C.M. Bethke, J.R. Spencer, S. Marshak, A. Navrotsky, Science **314**, 1764 (2006). doi:10.1126/science.1133519
R.L. Kirk, L.A. Soderblom, R.H. Brown, S.W. Kieffer, J.S. Kargel, in *Neptune and Triton*, ed. by D.P. Cruikshank, M.S. Matthews, A.M. Schumann (1995), pp. 949–989
C. Laver, I. de Pater, Icarus **195**, 752–757 (2008). doi:10.1016/j.icarus.2007.12.023
A.C. Leith, W.B. McKinnon, Icarus **120**, 387–398 (1996). doi:10.1006/icar.1996.0058
E. Lellouch, Icarus **124**, 1–21 (1996). doi:10.1006/icar.1996.0186
J.S. Lewis, Icarus **15**, 174–185 (1971). doi:10.1016/0019-1035(71)90072-8
R.M.C. Lopes, L.W. Kamp, S. Douté, W.D. Smythe, R.W. Carlson, A.S. McEwen, P.E. Geissler, S.W. Kieffer, F.E. Leader, A.G. Davies, E. Barbinis, R. Mehlman, M. Segura, J. Shirley, L.A. Soderblom, J. Geophys. Res. **106**, 33053–33078 (2001). doi:10.1029/2000JE001463
R.M.C. Lopes, K.L. Mitchell, E.R. Stofan, J.I. Lunine, R. Lorenz, F. Paganelli, R.L. Kirk, C.A. Wood, S.D. Wall, L.E. Robshaw, A.D. Fortes, C.D. Neish, J. Radebaugh, E. Reffet, S.J. Ostro, C. Elachi, M.D. Allison, Y. Anderson, R. Boehmer, G. Boubin, P. Callahan, P. Encrenaz, E. Flamini, G. Francescetti, Y. Gim, G. Hamilton, S. Hensley, M.A. Janssen, W.T.K. Johnson, K. Kelleher, D.O. Muhleman, G. Ori, R. Orosei, G. Picardi, F. Posa, L.E. Roth, R. Seu, S. Shaffer, L.A. Soderblom, B. Stiles, S. Vetrella, R.D. West, L. Wye, H.A. Zebker, Icarus **186**, 395–412 (2007). doi:10.1016/j.icarus.2006.09.006
R.D. Lorenz, S. Wall, J. Radebaugh, G. Boubin, E. Reffet, M. Janssen, E. Stofan, R. Lopes, R. Kirk, C. Elachi, J. Lunine, K. Mitchell, F. Paganelli, L. Soderblom, C. Wood, L. Wye, H. Zebker, Y. Anderson, S. Ostro, M. Allison, R. Boehmer, P. Callahan, P. Encrenaz, G.G. Ori, G. Francescetti, Y. Gim, G. Hamilton, S. Hensley, W. Johnson, K. Kelleher, D. Muhleman, G. Picardi, F. Posa, L. Roth, R. Seu, S. Shaffer, B. Stiles, S. Vetrella, E. Flamini, R. West, Science **312**, 724–727 (2006). doi:10.1126/science.1123257
R.D. Lorenz, B.W. Stiles, R.L. Kirk, M.D. Allison, P. Persi del Marmo, L. Iess, J.I. Lunine, S.J. Ostro, S. Hensley, Science **319**, 1649 (2008a). doi:10.1126/science.1151639
R.D. Lorenz, R.M. Lopes, F. Paganelli, J.I. Lunine, R.L. Kirk, K.L. Mitchell, L.A. Soderblom, E.R. Stofan, G. Ori, M. Myers, H. Miyamoto, J. Radebaugh, B. Stiles, S.D. Wall, C.A. Wood, Planet. Space Sci. **56**, 1132–1144 (2008b)
J.I. Lunine, S.K. Atreya, Nat. Geosci. **1**(3), 159–164 (2008)
J.I. Lunine, C. Elachi, S.D. Wall, M.A. Janssen, M.D. Allison, Y. Anderson, R. Boehmer, P. Callahan, P. Encrenaz, E. Flamini, G. Franceschetti, Y. Gim, G. Hamilton, S. Hensley, W.T.K. Johnson, K. Kelleher, R.L. Kirk, R.M. Lopes, R. Lorenz, D.O. Muhleman, R. Orosei, S.J. Ostro, F. Paganelli, P. Paillou, G. Picardi, F. Posa, J. Radebaugh, L.E. Roth, R. Seu, S. Shaffer, L.A. Soderblom, B. Stiles, E.R. Stofan, S. Vetrella, R. West, C.A. Wood, L. Wye, H. Zebker, G. Alberti, E. Karkoschka, B. Rizk, E. McFarlane, C. See, B. Kazeminejad, Icarus **195**, 415–433 (2008). doi:10.1016/j.icarus.2007.12.022
M. Manga, C.Y. Wang, Geophys. Res. Lett. **34**, 7202 (2007). doi:10.1029/2007GL029297
D.L. Matson, J.C. Castillo, J. Lunine, T.V. Johnson, Icarus **187**, 569–573 (2007). doi:10.1016/j.icarus.2006.10.016
D.E. Maxwell, in *Impact and Explosion Cratering: Planetary and Terrestrial Implications*, ed. by D.J. Roddy, R.O. Pepin, R.B. Merrill (1977), pp. 1003–1008
D.E. Maxwell, K. Seifert, Modeling of cratering, close-in displacements, and ejecta, 1975
J.F. McCauley, L.A. Soderblom, B.A. Smith, Nature **280**, 736–738 (1979). doi:10.1038/280736a0

T.B. McCord, G.B. Hansen, F.P. Fanale, R.W. Carlson, D.L. Matson, T.V. Johnson, W.D. Smythe, J.K. Crowley, P.D. Martin, A. Ocampo, C.A. Hibbitts, J.C. Granahan, Science **280**, 1242 (1998). doi:10.1126/science.280.5367.1242
A.S. McEwen, L.A. Soderblom, T.V. Johnson, D.L. Matson, Icarus **75**, 450–478 (1988). doi:10.1016/0019-1035(88)90157-1
A.S. McEwen, L.P. Keszthelyi, R. Lopes, P.M. Schenk, J.R. Spencer, in *The Lithosphere and Surface of Io*, ed. by F. Bagenal, T.E. Dowling, W.B. McKinnon (2004), pp. 307–328
W.B. McKinnon, H.J. Melosh, Icarus **44**, 454–471 (1980). doi:10.1016/0019-1035(80)90037-8
W.B. McKinnon, J.I. Lunine, D. Banfield, in *Neptune and Triton*, ed. by D.P. Cruikshank, M.S. Matthews, A.M. Schumann (1995), pp. 807–877
M.K. McNutt, J. Geophys. Res. **89**, 11180–11194 (1984). doi:10.1029/JB089iB13p11180
H.J. Melosh, Icarus **44**, 745–751 (1980). doi:10.1016/0019-1035(80)90141-4
H.J. Melosh, J. Geophys. Res. **87**, 371–380 (1982). doi:10.1029/JB087iB01p00371
H.J. Melosh, Impact cratering: A geologic process, 1989
H.J. Melosh, E.S. Gaffney, J. Geophys. Res. **88**, 830 (1983). doi:10.1029/JB088iS02p0A830
H.J. Melosh, B.A. Ivanov, Annu. Rev. Earth Planet. Sci. **27**, 385–415 (1999). doi:10.1146/annurev.earth.27.1.385
H.J. Melosh, W.B. McKinnon, Geophys. Res. Lett. **5**, 985–988 (1978). doi:10.1029/GL005i011p00985
M.P. Milazzo, L.P. Keszthelyi, A.S. McEwen, J. Geophys. Res. **106**, 33121–33128 (2001). doi:10.1029/2000JE001410
J.L. Mitchell, R.T. Pierrehumbert, D.M.W. Frierson, R. Caballero, Proc. Natl. Acad. Sci. **103**, 18421–18426 (2006). doi:10.1073/pnas.0605074103
G. Mitri, A.P. Showman, Icarus **177**, 447–460 (2005). doi:10.1016/j.icarus.2005.03.019
G. Mitri, A.P. Showman, Icarus **193**, 387–396 (2008). doi:10.1016/j.icarus.2007.07.016
G. Mitri, A.P. Showman, J.I. Lunine, R.M.C. Lopes, Icarus **196**, 216–224 (2008). doi:10.1016/j.icarus.2008.02.024
W.B. Moore, G. Schubert, Icarus **147**, 317–319 (2000)
J.M. Moore, J.R. Spencer, Geophys. Res. Lett. **17**, 1757–1760 (1990). doi:10.1029/GL017i010p01757
J.M. Moore, E. Asphaug, R.J. Sullivan, J.E. Klemaszewski, K.C. Bender, R. Greeley, P.E. Geissler, A.S. McEwen, E.P. Turtle, C.B. Phillips, B.R. Tufts, J.W. Head, R.T. Pappalardo, K.B. Jones, C.R. Chapman, M.J.S. Belton, R.L. Kirk, D. Morrison, Icarus **135**, 127–145 (1998). doi:10.1006/icar.1998.5973
J.M. Moore, E. Asphaug, M.J.S. Belton, B. Bierhaus, H.H. Breneman, S.M. Brooks, C.R. Chapman, F.C. Chuang, G.C. Collins, B. Giese, R. Greeley, J.W. Head, S. Kadel, K.P. Klaasen, J.E. Klemaszewski, K.P. Magee, J. Moreau, D. Morrison, G. Neukum, R.T. Pappalardo, C.B. Phillips, P.M. Schenk, D.A. Senske, R.J. Sullivan, E.P. Turtle, K.K. Williams, Icarus **151**, 93–111 (2001). doi:10.1006/icar.2000.6558
J.M. Moore, P.M. Schenk, L.S. Bruesch, E. Asphaug, W.B. McKinnon, Icarus **171**, 421–443 (2004). doi:10.1016/j.icarus.2004.05.009
M.A. Moreno, G. Schubert, M.G. Kivelson, D.A. Paige, J. Baumgardner, Icarus **93**, 63–81 (1991). doi:10.1016/0019-1035(91)90164-O
J.V. Morgan, M.R. Warner, J. Brittan, R. Buffl er, A. Camargo, G.L. Christeson, P. Denton, A. Hildebrand, R. Hobbs, H. Macintyre, G. Mackenzie, P. Maguire, L. Marín, Y. Nakamura, M. Pilkington, V.L. Sharpton, D. Snyder, G. Suarez, A. Trejo, Nature **390**, 472–476 (1997). doi:10.1038/37291
J.I. Moses, M.Y. Zolotov, B. Fegley, Icarus **156**, 76–106 (2002). doi:10.1006/icar.2001.6758
W.H. Munk, G.J.F. MacDonald, The rotation of the earth; a geophysical discussion, 1960
P.D. Nicholson, M.R. Showalter, L. Dones, R.G. French, S.M. Larson, J.J. Lissauer, C.A. McGhee, B. Sicardy, P. Seitzer, G.E. Danielson, Science **272**, 509–516 (1996)
H.B. Niemann, S.K. Atreya, S.J. Bauer, G.R. Carignan, J.E. Demick, R.L. Frost, D. Gautier, J.A. Haberman, D.N. Harpold, D.M. Hunten, G. Israel, J.I. Lunine, W.T. Kasprzak, T.C. Owen, M. Paulkovich, F. Raulin, E. Raaen, S.H. Way, Nature **438**, 779–784 (2005). doi:10.1038/nature04122
F. Nimmo, J. Geophys. Res. **109**, 12001 (2004). doi:10.1029/2004JE002347
F. Nimmo, E. Gaidos, J. Geophys. Res. **107**, 5021 (2002). doi:10.1029/2000JE001476
F. Nimmo, R.T. Pappalardo, Geophys. Res. Lett. **31**, 19701 (2004). doi:10.1029/2004GL020763
F. Nimmo, R.T. Pappalardo, Nature **441**, 614–616 (2006). doi:10.1038/nature04821
F. Nimmo, J. Spencer, Comparison of March 2008 Composite Infrared Spectrometer (CIRS) tiger stripe observations with shear heating predictions. AGU Fall Meeting Abstracts, 6 (2008)
F. Nimmo, R.T. Pappalardo, B. Giese, Geophys. Res. Lett. **29**(7), 070000-1 (2002). doi:10.1029/2001GL013976
F. Nimmo, B. Giese, R.T. Pappalardo, Geophys. Res. Lett. **30**(5), 050000-1 (2003). doi:10.1029/2002GL016660
F. Nimmo, J.R. Spencer, R.T. Pappalardo, M.E. Mullen, Nature **447**, 289–291 (2007)
G.W. Ojakangas, D.J. Stevenson, Icarus **81**, 220–241 (1989). doi:10.1016/0019-1035(89)90052-3

J.D. O'Keefe, T.J. Ahrens, J. Geophys. Res. **98**, 17011–17028 (1993). doi:10.1029/93JE01330
J.D. O'Keefe, T.J. Ahrens, J. Geophys. Res. **104**, 27091 (1999). doi:10.1029/1998JE000596
R.T. Pappalardo, A.C. Barr, Geophys. Res. Lett. **31**, 1701 (2004). doi:10.1029/2003GL019202
R.T. Pappalardo, M.J.S. Belton, H.H. Breneman, M.H. Carr, C.R. Chapman, G.C. Collins, T. Denk, S. Fagents, P.E. Geissler, B. Giese, R. Greeley, R. Greenberg, J.W. Head, P. Helfenstein, G. Hoppa, S.D. Kadel, K.P. Klaasen, J.E. Klemaszewski, K. Magee, A.S. McEven, J.M. Moore, W.B. Moore, G. Neukum, C.B. Phillips, L. Prokter, G. Schubert, D.A. Senske, R.J. Sullivan, B.R. Tufts, E.P. Turtle, R. Wagner, K.K. Williams, J. Geophys. Res. **104**, 24015–24056 (1999)
Q.R. Passey, E.M. Shoemaker, in *Satellites of Jupiter* (1982), pp. 379–434
J. Pearl, R. Hanel, V. Kunde, W. Maguire, K. Fox, S. Gupta, C. Ponnamperuma, F. Raulin, Nature **280**, 755–758 (1979). doi:10.1038/280755a0
J.T. Perron, M.P. Lamb, C.D. Koven, I.Y. Fung, E. Yager, M. Ádámkovics, J. Geophys. Res. **111**, 11001 (2006). doi:10.1029/2005JE002602
E. Pierazzo, A.M. Vickery, H.J. Melosh, Icarus **127**, 408–423 (1997). doi:10.1006/icar.1997.5713
C.M. Pieters, Science **215**, 59–61 (1982). doi:10.1126/science.215.4528.59
C.C. Porco, E. Baker, J. Barbara, K. Beurle, A. Brahic, J.A. Burns, S. Charnoz, N. Cooper, D.D. Dawson, A.D. Del Genio, T. Denk, L. Dones, U. Dyudina, M.W. Evans, B. Giese, K. Grazier, P. Helfenstein, A.P. Ingersoll, R.A. Jacobson, T.V. Johnson, A. McEwen, C.D. Murray, G. Neukum, W.M. Owen, J. Perry, T. Roatsch, J. Spitale, S. Squyres, P.C. Thomas, M. Tiscareno, E. Turtle, A.R. Vasavada, J. Veverka, R. Wagner, R. West, Science **307**, 1237–1242 (2005a). doi:10.1126/science.1107981
C.C. Porco, E. Baker, J. Barbara, K. Beurle, A. Brahic, J.A. Burns, S. Charnoz, N. Cooper, D.D. Dawson, A.D. Del Genio, T. Denk, L. Dones, U. Dyudina, M.W. Evans, S. Fussner, B. Giese, K. Grazier, P. Helfenstein, A.P. Ingersoll, R.A. Jacobson, T.V. Johnson, A. McEwen, C.D. Murray, G. Neukum, W.M. Owen, J. Perry, T. Roatsch, J. Spitale, S. Squyres, P. Thomas, M. Tiscareno, E.P. Turtle, A.R. Vasavada, J. Veverka, R. Wagner, R. West, Nature **434**, 159–168 (2005b). doi:10.1038/nature03436
C.C. Porco, P. Helfenstein, P.C. Thomas, A.P. Ingersoll, J. Wisdom, R. West, G. Neukum, T. Denk, R. Wagner, T. Roatsch, S. Kieffer, E. Turtle, A. McEwen, T.V. Johnson, J. Rathbun, J. Veverka, D. Wilson, J. Perry, J. Spitale, A. Brahic, J.A. Burns, A.D. DelGenio, L. Dones, C.D. Murray, S. Squyres, Science **311**, 1393–1401 (2006)
F. Postberg, S. Kempf, J. Schmidt, N. Brilliantov, A. Beinsen, B. Abel, U. Buck, R. Srama, Nature **459**, 1098–1101 (2009)
O. Prieto-Ballesteros, J.S. Kargel, M. Fernández-Sampedro, F. Selsis, E.S. Martínez, D.L. Hogenboom, Icarus **177**, 491–505 (2005). doi:10.1016/j.icarus.2005.02.021
L.M. Prockter, R.T. Pappalardo, Science **289**, 941–944 (2000). doi:10.1126/science.289.5481.941
L.M. Prockter, F. Nimmo, R.T. Pappalardo, Geophys. Res. Lett. **32**, 14202 (2005). doi:10.1029/2005GL022832
L.M. Prockter, R.M. Lopes, B. Giese, R. Jaumann, R.D. Lorenz, R.T. Pappalardo, G.W. Patterson, P.C. Thomas, E.P. Turtle, R. Wagner, Space Sci. Rev. (2010, this issue)
J. Radebaugh, R.D. Lorenz, R.L. Kirk, J.I. Lunine, E.R. Stofan, R.M.C. Lopes, S.D. Wall, the Cassini Radar Team, Icarus **192**, 77–91 (2007). doi:10.1016/j.icarus.2007.06.020
J. Radebaugh, R.D. Lorenz, J.I. Lunine, S.D. Wall, G. Boubin, E. Reffet, R.L. Kirk, R.M. Lopes, E.R. Stofan, L. Soderblom, M. Allison, M. Janssen, P. Paillou, P. Callahan, C. Spencer, The Cassini Radar Team, Icarus **194**, 690–703 (2008)
P. Rannou, F. Montmessin, F. Hourdin, S. Lebonnois, Science **311**, 201–205 (2006). doi:10.1126/science.1118424
K.D. Retherford, J.R. Spencer, S.A. Stern, J. Saur, D.F. Strobel, A.J. Steffl, G.R. Gladstone, H.A. Weaver, A.F. Cheng, J.W. Parker, D.C. Slater, M.H. Versteeg, M.W. Davis, F. Bagenal, H.B. Throop, R.M.C. Lopes, D.C. Reuter, A. Lunsford, S.J. Conard, L.A. Young, J.M. Moore, Science **318**, 237 (2007). doi:10.1126/science.1147594
D.J. Roddy, R.O. Pepin, R.B. Merrill (eds.), in *Proceedings of the Symposium on Planetary Cratering Mechanics, Impact and Explosion Cratering: Planetary and Terrestrial Implications*, Flagstaff, AZ, 13–17 September 1976 (1977)
D.M. Rubin, P.A. Hesp, Nat. Geosci. **2**, 653–658 (2009)
F. Salama, L.J. Allamandola, F.C. Witteborn, D.P. Cruikshank, S.A. Sandford, J.D. Bregman, Icarus **83**, 66–82 (1990). doi:10.1016/0019-1035(90)90006-U
J. Saur, D.F. Strobel, Icarus **171**, 411–420 (2004). doi:10.1016/j.icarus.2004.05.010
J. Saur, N. Schilling, F.M. Neubauer, D.F. Strobel, S. Simon, M.K. Dougherty, C.T. Russell, R.T. Pappalardo, Geophys. Res. Lett. **35**, 20105 (2008). doi:10.1029/2008GL035811
P.M. Schenk, J. Geophys. Res. **94**, 3813–3832 (1989). doi:10.1029/JB094iB04p03813
P.M. Schenk, J. Geophys. Res. **96**, 15635 (1991). doi:10.1029/91JE00932
P.M. Schenk, Nature **417**, 419–421 (2002)

P.M. Schenk, W.B. McKinnon, Geophys. Res. Lett. **36**, L16202 (2009). doi:10.1029/2009GL039916
P.M. Schenk, J.M. Moore, J. Geophys. Res. **100**, 19009–19022 (1995). doi:10.1029/95JE01854
P.M. Schenk, K. Zahnle, Icarus **192**, 135–149 (2007). doi:10.1016/j.icarus.2007.07.004
P. Schenk, I. Matsuyama, F. Nimmo, Nature **453**, 368–371 (2008). doi:10.1038/nature06911
J. Schmidt, N. Brilliantov, F. Spahn, S. Kempf, Nature **451**, 685–688 (2008)
N. Schneider, M. Burger, E. Schaller, B.M. Brilliantov, R. Johnson, J. Kargel, M. Dougherty, N. Achilleos, Nature **459**, 1102–1104 (2009)
M.R. Showalter, Nature **351**, 709–713 (1991)
A.P. Showman, I. Mosqueira, J.W. Head, Icarus **172**, 625–640 (2004). doi:10.1016/j.icarus.2004.07.011
W.M. Sinton, C. Kaminski, Icarus **75**, 207–232 (1988). doi:10.1016/0019-1035(88)90002-4
B.A. Smith, L.A. Soderblom, D. Banfield, C. Barnet, R.F. Beebe, A.T. Bazilevskii, K. Bollinger, J.M. Boyce, G.A. Briggs, A. Brahic, Science **246**, 1422–1449 (1989). doi:10.1126/science.246.4936.1422
P.H. Smith, M.T. Lemmon, R.D. Lorenz, L.A. Sromovsky, J.J. Caldwell, M.D. Allison, Icarus **119**, 336–349 (1996). doi:10.1006/icar.1996.0023
L.A. Soderblom, T.L. Becker, S.W. Kieffer, R.H. Brown, C.J. Hansen, T.V. Johnson, Science **250**, 410–415 (1990). doi:10.1126/science.250.4979.410
L.A. Soderblom, R.L. Kirk, J.I. Lunine, J.A. Anderson, K.H. Baines, J.W. Barnes, J.M. Barrett, R.H. Brown, B.J. Buratti, R.N. Clark, D.P. Cruikshank, C. Elachi, M.A. Janssen, R. Jaumann, E. Karkoschka, S.L. Mouélic, R.M. Lopes, R.D. Lorenz, T.B. McCord, P.D. Nicholson, J. Radebaugh, B. Rizk, C. Sotin, E.R. Stofan, T.L. Sucharski, M.G. Tomasko, S.D. Wall, Planet. Space Sci. **55**, 2025–2036 (2007). doi:10.1016/j.pss.2007.04.014
C. Sotin, G. Tobie, Science **320**(5883), 1588 (2008)
C. Sotin, J.W. Head, G. Tobie, Geophys. Res. Lett. **29**, 1233–1010292001013844 (2002)
C. Sotin, R. Jaumann, B.J. Buratti, R.H. Brown, R.N. Clark, L.A. Soderblom, K.H. Baines, G. Bellucci, J.P. Bibring, F. Capaccioni, P. Cerroni, M. Combes, A. Coradini, D.P. Cruikshank, P. Drossart, V. Formisano, Y. Langevin, D.L. Matson, T.B. McCord, R.M. Nelson, P.D. Nicholson, B. Sicardy, S. Lemouelic, S. Rodriguez, K. Stephan, C.K. Scholz, Nature **435**, 786–789 (2005). doi:10.1038/nature03596
J.R. Spencer, Geophys. Res. Lett. **17**, 1769–1772 (1990). doi:10.1029/GL017i010p01769
J.R. Spencer, K.L. Jessup, M.A. McGrath, G.E. Ballester, R. Yelle, Science **288**, 1208–1210 (2000). doi:10.1126/science.288.5469.1208
J.R. Spencer, J.C. Pearl, M. Segura, F.M. Flasar, A. Mamoutkine, P. Romani, B.J. Buratti, A.R. Hendrix, L.J. Spilker, R.M.C. Lopes, Science **311**, 1401–1405 (2006)
J.R. Spencer, S.A. Stern, A.F. Cheng, H.A. Weaver, D.C. Reuter, K. Retherford, A. Lunsford, J.M. Moore, O. Abramov, R.M.C. Lopes, J.E. Perry, L. Kamp, M. Showalter, K.L. Jessup, F. Marchis, P.M. Schenk, C. Dumas, Science **318**, 240 (2007). doi:10.1126/science.1147621
J.R. Spencer, J.C. Pearl, C.J. Howett, M.E. Segura, C. Team, High spatial resolution observations of thermal emission from Enceladus' active south pole. AGU Fall Meeting Abstracts, 1372 (2008)
J.N. Spitale, C.C. Porco, Nature **449**, 695–697 (2007)
S.W. Squyres, R.T. Reynolds, P.M. Cassen, Icarus **53**, 319–331 (1983). doi:10.1016/0019-1035(83)90152-5
B.W. Stiles, R.L. Kirk, R.D. Lorenz, S. Hensley, E. Lee, S.J. Ostro, M.D. Allison, P.S. Callahan, Y. Gim, L. Iess, P. Perci del Marmo, G. Hamilton, W.T.K. Johnson, R.D. West, The Cassini RADAR Team, Astron. J. **135**, 1669–1680 (2008). doi:10.1088/0004-6256/135/5/1669
B.W. Stiles, R.L. Kirk, R.D. Lorenz, S. Hensley, E. Lee, S.J. Ostro, M.D. Allison, P.S. Callahan, Y. Gim, L. Iess, P. Perci del Marmo, G. Hamilton, W.T.K. Johnson, R.D. West, The Cassini RADAR Team, Astron. J **139**, 311 (2010). doi:10.1088/0004-6256/139/1/311
D.F. Strobel, M.E. Summers, in *Neptune and Triton*, ed. by D.P. Cruikshank, M.S. Matthews, A.M. Schumann (1995), pp. 1107–1148
R.G. Strom, N.M. Schneider, in *Satellites of Jupiter*, ed. by D. Morrison (1982), pp. 598–633
P.C. Thomas, J.A. Burns, P. Helfenstein, S. Squyres, J. Veverka, C. Porco, E.P. Turtle, A. McEwen, T. Denk, B. Giese, T. Roatsch, T.V. Johnson, R.A. Jacobson, Icarus **190**, 573–584 (2007). doi:10.1016/j.icarus.2007.03.012
F. Tian, A.I.F. Stewart, B.T. Owen, K.M. Larsen, L.W. Esposito, Monte Carlo simulations of the water vapour plumes on Enceladus. Icarus (2007). doi:10.1016/j.icarus.2006.11.010
G. Tobie, G. Choblet, C. Sotin, J. Geophys. Res. **108**, 5124 (2003). doi:10.1029/2003JE002099
G. Tobie, A. Mocquet, C. Sotin, Icarus **177**, 534–549 (2005). doi:10.1016/j.icarus.2005.04.006
G. Tobie, J.I. Lunine, C. Sotin, Nature **440**, 61–64 (2006)
G. Tobie, O. Čadek, C. Sotin, Icarus **196**, 642–652 (2008). doi:10.1016/j.icarus.2008.03.008
T. Tokano, Icarus **194**, 243–262 (2008). doi:10.1016/j.icarus.2007.10.007
T. Tokano, R. Soc. Lond. Philos. Trans., Ser. A **367**, 633–648 (2009). doi:10.1098/rsta.2008.0163
T. Tokano, F.M. Neubauer, Geophys. Res. Lett. **32**, 24203 (2005). doi:10.1029/2005GL024456
T. Tokano, F.M. Neubauer, M. Laube, C.P. McKay, Icarus **153**, 130–147 (2001). doi:10.1006/icar.2001.6659

T. Tokano, C.P. McKay, F.M. Neubauer, S.K. Atreya, F. Ferri, M. Fulchignoni, H.B. Niemann, Nature **442**, 432–435 (2006). doi:10.1038/nature04948

M.G. Tomasko, B. Archinal, T. Becker, B. Bézard, M. Bushroe, M. Combes, D. Cook, A. Coustenis, C. de Bergh, L.E. Dafoe, L. Doose, S. Douté, A. Eibl, S. Engel, F. Gliem, B. Grieger, K. Holso, E. Howington-Kraus, E. Karkoschka, H.U. Keller, R. Kirk, R. Kramm, M. Küppers, P. Lanagan, E. Lellouch, M. Lemmon, J. Lunine, E. McFarlane, J. Moores, G.M. Prout, B. Rizk, M. Rosiek, P. Rueffer, S.E. Schröder, B. Schmitt, C. See, P. Smith, L. Soderblom, N. Thomas, R. West, Nature **438**, 765–778 (2005). doi:10.1038/nature04126

D.L. Turcotte, G. Schubert, *Geodynamics*, 2nd edn. (2002). doi:10.2277/0521661862

E.P. Turtle, PhD thesis, The University of Arizona, 1998

E.P. Turtle, B.A. Ivanov, in *Lunar and Planetary Institute Science Conference Abstracts*. Lunar and Planetary Inst. Technical Report, vol. 33 (2002), p. 1431

E.P. Turtle, E. Pierazzo, Science **294**, 1326–1328 (2001). doi:10.1126/science.1062492

E.P. Turtle, J.E. Perry, A.S. McEwen, A.D. DelGenio, J. Barbara, R.A. West, D.D. Dawson, C.C. Porco, Geophys. Res. Lett. **36**, 2204 (2009). doi:10.1029/2008GL036186

G.L. Tyler, D.N. Sweetnam, J.D. Anderson, S.E. Borutzki, J.K. Campbell, E.R. Kursinski, G.S. Levy, G.F. Lindal, J.R. Lyons, G.E. Wood, Science **246**, 1466–1473 (1989). doi:10.1126/science.246.4936.1466

T. Van Hoolst, N. Rambaux, Ö. Karatekin, R.M. Baland, Icarus **200**, 256–264 (2009). doi:10.1016/j.icarus.2008.11.009

J.H. Waite, M.R. Combi, W.H. Ip, T.E. Cravens, R.L. McNutt, W. Kasprzak, R. Yelle, J. Luhmann, H. Niemann, D. Gell, B. Magee, G. Fletcher, J. Lunine, W.L. Tseng, Science **311**, 1419–1422 (2006)

K.K. Williams, R. Greeley, Geophys. Res. Lett. **25**, 4273–4276 (1998). doi:10.1029/1998GL900144

D.A. Williams, P.M. Schenk, J.M. Moore, L.P. Keszthelyi, E.P. Turtle, W.L. Jaeger, J. Radebaugh, M.P. Milazzo, R.M.C. Lopes, R. Greeley, Icarus **169**, 80–97 (2004). doi:10.1016/j.icarus.2003.08.024

R.V. Yelle, J.I. Lunine, J.B. Pollack, R.H. Brown, in *Neptune and Triton*, ed. by D.P. Cruikshank, M.S. Matthews, A.M. Schumann (1995), pp. 1031–1105. Provided by the SAO/NASA Astrophysics Data System. http://adsabs.harvard.edu/abs/1995netr.conf.1031Y

K. Zahnle, P. Schenk, S. Sobieszczyk, L. Dones, H.F. Levison, Icarus **153**, 111–129 (2001). doi:10.1006/icar.2001.6668

K. Zahnle, P. Schenk, H. Levison, L. Dones, Icarus **163**, 263–289 (2003). doi:10.1016/S0019-1035(03)00048-4

J. Zhang, D.B. Goldstein, P.L. Varghese, L. Trafton, C. Moore, K. Miki, Icarus **172**, 479–502 (2004). doi:10.1016/j.icarus.2004.06.016

M. Zolotov, Geophys. Res. Lett. **34**, 23203 (2007)

Space Sci Rev (2010) 153: 411–429
DOI 10.1007/s11214-009-9611-9

From Gas to Satellitesimals: Disk Formation and Evolution

A. Coradini · G. Magni · D. Turrini

Received: 8 June 2009 / Accepted: 19 November 2009 / Published online: 16 February 2010

Abstract The subject of satellite formation is strictly linked to the one of planetary formation. Giant planets strongly shape the evolution of the circum-planetary disks during their formation and thus, indirectly, influence the initial conditions for the processes governing satellite formation. In order to fully understand the present features of the satellite systems of the giant planets, we need to take into account their formation environments and histories and the role of the different physical parameters. In particular, the pressure and temperature profiles in the circum-planetary nebulae shaped their chemical gradients by allowing for the condensation of ices and noble gases. These chemical gradients, in turn, set the composition of the satellitesimals, which represent the building blocks of the present regular satellites.

Keywords Solar system: formation · Planets and satellites: formation · Accretion disks · Planets and satellites: Jupiter · Planets and satellites: Saturn

1 Introduction

The satellite systems of Jupiter, Saturn and Uranus have been considered as miniature planetary systems since a long time (Greenberg 1975). While they exhibit very diverse features, their study greatly helped to improve our understanding of the behavior of complex systems formed around a central body in terms of dynamical and physical properties. The regular satellites, which experience negligible solar perturbations, display dynamical properties similar to those of the planets. Moreover, the regular satellites are thought to have formed

A. Coradini (✉) · D. Turrini
INAF-IFSI, Via del Fosso del Cavaliere 100, 00133 Rome, Italy
e-mail: angioletta.coradini@ifsi-roma.inaf.it

D. Turrini
e-mail: diego.turrini@ifsi-roma.inaf.it

G. Magni
INAF-IASF, Via del Fosso del Cavaliere 100, 00133 Rome, Italy
e-mail: gianfranco.magni@iasf-roma.inaf.it

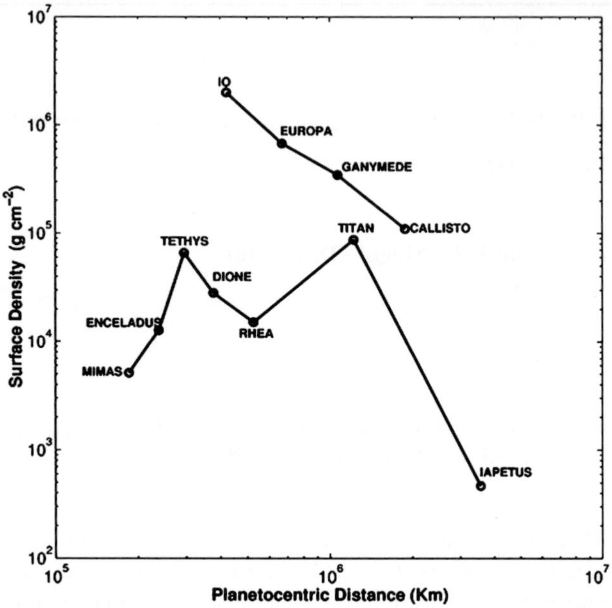

Fig. 1 Surface density of the circum-planetary disks of Jupiter and Saturn estimated from the reconstructed masses of the satellite systems. The reconstructed mass is obtained by adding to each satellite the needed amount of volatile elements to achieve solar composition. Surface density is expressed in $g\,cm^{-2}$ while radial distance from the central planet is expressed in km. Figure adapted from Pollack and Consolmagno (1984). The proto-satellite disk surface density profile was constructed by spreading the mass of each satellite, augmented to match a solar proportion of elements, over an annulus centered on the current satellite orbit. Similarly dense disks have been used in previous models of Galilean satellite formation

around their respective planets through a process similar to the one hypothesized for the accretion of the planets around the Sun. Any theory of the origins and evolution of the Solar System must then account for the presence and properties of satellite systems. When studying the origins of the satellites, we cannot avoid putting them in the context of the formation scenarios of their central planets. In fact, it is difficult to imagine satellite formation as totally independent on the environment generated and strongly affected by the central body. However, one has also to take into account that the formation times of the satellites are very short compared to those of their central bodies, and it is therefore possible for several generations of satellites to exist, each formed under different environmental conditions.

2 From Gas to Satellitesimals

If we examine the different satellite systems present in our Solar System we can find both regions characterized by negative outward density gradients, and others, which seem to have been dominated by stochastic collisional processes. The system of Jupiter exhibits a noticeable regularity. If we plot the so-called reconstructed mass of the satellites (i.e. the mass obtained by adding to each satellite the needed amount of volatile elements, in order to estimate the mass distribution in their formation regions) as a function of the distance from Jupiter (see Fig. 1), the resulting density distribution is very regular. The Galilean satellites are a key example of a satellite system which is co-formed with its parent planet, the forming satellites accumulating within a circum-planetary accretion disk during the final stages of the planetary formation process. The same does not apply to Saturn's system, which shows a large degree of variability (Greenberg 1975). A large fraction of the mass in the Saturnian system resides in the largest satellite, Titan, which comprises more than 96% of the mass in orbit around the planet. Six other moons, prevailingly of icy composition, constitute roughly 4% of the total mass of the satellite system while the remaining 54 small moons, together with the ring system, only comprise about 0.02%. The surface density computed through the reconstructed mass of the icy satellites (see Fig. 1) shows alternating trends, both positive and negative, likely a product of stochastic phenomena at the time of the formation of the system (see e.g. Lissauer 1995, for a review). As a consequence, it is not straightforward to infer the characteristics of the original disk from the present features of the system like in the case of Jupiter. Moreover, the Saturnian system got subject to an intense bombardment whose presence can be deduced from the overall structure of the icy bodies and from their surface crater distribution. Even more complex is the situation in the satellite system of Uranus, where regular satellites lying on the central plane of Uranus exist, thus orbiting the planet with an inclination of 98° with respect to the ecliptic plane. Thus, Uranus and its satellites form a pretty standard system while considered as an isolated dynamical system; yet, once considered as part of the Solar System, they represent an anomalous property which must be accounted for by any theory of planetary formation and evolution. Following Greenberg (1975), one can make two hypotheses, depending on the time scale of the change of the planetary obliquity. The first possibility is that, if the tilting time scale was long (i.e. the change was gradual), the satellites could have undergone an adiabatic variation of their orbital angular momentum and could have survived the process. The second hypothesis assumes that the change was abrupt (e.g. due to a giant impact of a planetary embryo whose mass was $\sim$7% of that of Uranus, Safronov 1969). If the tilting took place before the formation of the satellite system, it is plausible to assume that the circum-planetary disk disaggregated and reformed along the new equatorial plane and the satellites then formed in the new disk. If instead, the tilting took place after the formation of a satellite system, the original system was probably destroyed, and new bodies formed from the debris of the original ones. In this case the link between the satellite system and the formation of Uranus can be weak.

Neptune represents another, different case: Goldreich et al. (1989) conjectured that Triton was captured from a heliocentric orbit as the result of a collision with one of Neptune's primordial regular satellites. The immediate post-capture orbit was highly eccentric, with its pericentric distance oscillating periodically. Dissipation due to tides raised by Neptune on Triton caused its orbit to evolve to its present state in 10^9 years, with Triton being almost entirely molten for most of this time. While its orbit was evolving, Triton cannibalized most of the regular satellites of Neptune and also perturbed Nereid, thus accounting for the satellite's highly eccentric and inclined orbit. The only regular satellites of Neptune that survived were the ones which formed well within 5 Neptune's radii (R_N in the following) and they moved on inclined orbits as the result of chaotic perturbations forced by Triton. Goldreich et al. (1989) calculated that if Triton entered an elliptical orbit with a pericenter at $7R_N$ and semimajor axis of $10^3 R_N$, then Triton could evolve, through tidal dissipation, to its present situation in less than one billion years. One consequence of the model proposed by Goldreich et al. (1989) was that Neptune would be devoid of satellites between $5R_N$ and Triton's current orbit. When Voyager arrived three months later, this was found to be precisely the case. The capture scenario described by Goldreich et al. (1989) was later extended by Banfield and Murray (1992) to take into account the existence and the features of the inner satellites discovered by Voyager. Recently, an alternative scenario was suggested by Agnor and Hamilton (2006) to account for the existence of the outermost irregular satellites. While the implications of the capture of Triton for the Saturnian system will be discussed in detail by Mosqueira et al. (2010, this issue) we emphasize that, also in the case of Neptune, the history of the present regular satellites is too complex to trace back the formation process.

Despite the differences previously described, Canup and Ward (2006) noted that the ratios of the overall masses of the regular satellite systems to those of their parent planets are about 10^{-4} for the systems of Jupiter, Saturn and Uranus. This suggests that a common process took place in all the satellite systems and in particular that the mass fraction of the satellite systems is regulated by the balance in the disk between the supply of inflowing material and the satellite losses due to orbital decay driven by the interaction with the surrounding nebula. As a consequence, several satellite generations could have formed and got lost toward the central body, until the mass inflow onto the disk ended. The present satellite systems are likely the surviving members of the last satellite generation. However, in their work, the authors did not relate the mass accretion rate and the disk features to the formation time scale and characteristics of the central body, so that it is almost impossible to date the formation time of the satellites.

Irregular satellites, instead, are believed to have formed in the Solar Nebula and to have been gravitationally trapped by the giant planets at a later time. Irregular satellites can supply information on the conditions of the surrounding space, as well as on the characteristics of the planets at the time when their trapping efficiencies were higher. The details and implications of the capture of the irregular satellites will be discussed by Mosqueira et al. (2010, this issue). A complete and exhaustive model of satellite formation shall take into account the conditions existing in the Solar Nebula, since they represent the boundary conditions to be considered. However, connecting a model of the Solar Nebula to a model of giant planet formation that includes the formation of regular satellites is quite a difficult problem. Due to the strong link existing between the formation of the disk and that of the central body, we cannot avoid mentioning the current theories on the formation of giant planets.

In the case of Jupiter and Saturn, the study of the formation conditions of the central bodies can be more directly linked to those of their surrounding disks, whose thermodynamic conditions, in turn, represent the boundary conditions for satellite formation and thus determine the initial chemistry of the satellites. Jupiter and Saturn are mostly gaseous and likely accreted a large part of their masses from the proto-Solar Nebula when the gas was still abundant. The state of the art on this subject has been reviewed by several authors (see e.g. Lunine et al. 2004; Lissauer and Stevenson 2007).

The formation of giant planets is of paramount importance not only in order to understand our Solar System, but also to shed light on the possible nature and evolution of extraterrestrial planets. Moreover, following Guillot et al. (2004), we can state that Jupiter, owing to its large mass and rapid formation, played a crucial role in shaping the Solar System as we know it today. Jupiter mostly contains hydrogen and helium (more than 87% by mass) and as such it bears a close resemblance to the Sun. However, the Sun has only 2% of its mass in heavy elements, whereas Jupiter has between 3–13%. Spectral observations of Saturn from the far-infrared spectrometer aboard the Cassini spacecraft revealed that the C/H ratio in the planet is in fact about twice higher than previously derived from ground-based observations and in agreement with the C/H value derived from Voyager IRIS (for a discussion see Hersant et al. 2008). The exact amount of these heavy elements in the planet and their distribution are key issues to understand how the Solar System formed (ibid.), and the only way to correctly evaluate them is to know what the mass of the planetary core is.

In fact, the idea that the atmospheres of the giant planets were similar in composition to the primordial nebula from which they formed has been largely disproved thanks to their study both from space and from Earth (Encrenaz 1994). Encrenaz (1994) noticed that significant differences were measured in the abundances of helium, deuterium and carbon of the four giant planets, both between themselves and in respect to the Solar Nebula. In particular, the variations in the C/H and D/H ratios gave support to the "nucleation" formation scenario, where the four giant planets first accreted a nucleus of about 10 Earth masses, big enough to gravitationally capture the surrounding gaseous nebula (ibid.). One of the problems still open in this scenario is the mass of the central core that should be considered.

Goldreich et al. (2004) revised the problem of initial formation of planets and suggested that the region populated by the outer planets could have been characterized by the presence of many large bodies, about the size of Uranus and Neptune, which could have formed earlier than the inner planets provided that the outer disk was richer in mass than the inner one. This means that the original disk, from which the planets evolved, contained the mass of a "minimum mass nebula" in the inner regions, but a much larger amount in the outer ones. Therefore, the core of Jupiter and Saturn could have formed at this time, much before the time when the nebular gas dissipated.

Another important consideration on the possibility to quickly form the central core of giant planets has been given by Kornet et al. (2007), who showed that the frequently adopted approximation that the surface density of the planetesimals is a constant fraction of the gas surface density cannot be accepted without care. In fact, these authors showed that the surface density of planetesimals can be locally enhanced by a large factor with respect to the gas, thus speeding up the subsequent formation of the planets. This is due to the interaction between the gas and the solid particles in the initial phases of the disk evolution, as well as to the mutual interactions between the growing planetesimals. Therefore, a detailed analysis of the initial conditions present in the Solar Nebula could help solving the problem of the formation of the cores of the giant planets.

3 Gas Capture and the Formation of Disks and Planets

The formation of Jupiter and Saturn can be modeled through the gas accretion onto a pre-existing solid core. This approach, which has been adopted by several authors, can be followed in all its different phases, provided that a detailed description of the gas accretion process and the correct evolution of the central body thermodynamics are considered. In fact, the central body evolution is not irrelevant for the thermal evolution of the disk surrounding it, and, in turn, the hydrodynamic evolution of the disk is responsible for the mass accretion rate onto the planet, which at this stage is the main energy source of the growing object.

For the rest of this section we will consider mainly the cases of Jupiter and Saturn, for which several authors make the following working hypotheses. First, a scenario in which the regular satellites form within circum-planetary accretion disks produced during the final stages of gas accretion (e.g. Lubow and Ogilvie 2001; D'Angelo et al. 2002; Magni and Coradini 2004) is assumed. Second, for a given inflow rate of gas and solids, the circum-planetary gaseous disk is assumed to be in a quasi-steady state due to the balance of the inflow supply and the disk's internal viscous evolution, assuming that the disk viscous spreading time is short compared to the time scale over which the inflow changes. Up to now, the existing hydrodynamic models generally used spherically symmetric proto-planets, consisting of a growing solid core and a solar composition gaseous envelope, surrounded by the accretion region.

The accretion of giant planets, however, can hardly be assumed to be spherically symmetric, since it takes place in a strongly asymmetric region of several AU, dominated by the gravity of the Sun and crossed by density waves. On the contrary, three-dimensional models of the evolution of giant solar and extrasolar planets did not include satisfactory physical description of the interfaces between the growing planet and the nebula (e.g. Artymowicz 1998; Rauer and Hatzes 2007 and references therein). Lissauer et al. (2009) consider the gas accretion in a turbulent disk, whose evolution is computed in a three-dimensional space. This allows these authors to improve their modeling of Jupiter's formation, taking into account the time evolution of the gas flux throughout the Hill sphere. Their modeling, however, is based on ad hoc assumptions on the gas capture criterion. Their results suggest accretion times for Jupiter's gaseous envelope from the circum-Jovian disk ranging between several 10^4 years to a few 10^5 years, the accretion time scale strongly depending on the chosen physical parameterization and the disk turbulence. Moreover, while the influence of the gas accretion onto the planet is correctly taken into account, the feedback effects of planetary formation on the disk are not explicitly considered (ibid.).

The physical nature and the development of the core instability need to be discussed more deeply. First we have to introduce the concept of *critical mass*, i.e. the mass of the core beyond which the core starts collecting the surrounding gas. Theoretical works have modeled the process of gas accretion invoking hydrodynamic instability (Magni and Coradini 2004) or quasi-hydrostatic, irradiation-driven compression (Lissauer et al. 2009). At present, the observational constrains are too limited to discriminate between the two mechanisms. Analytical arguments (Stevenson 1982; Wuchterl 1993) and numerical calculations (Mizuno 1980) suggest that the onset of core instability occurs when the mass of the gaseous atmosphere around the core becomes comparable to the mass of the core itself. In a recent paper, Rafikov (2006) discussed the subject, determining the value of the critical mass in different conditions. This value depends on the physical characteristic of the atmosphere of the accreting object and mostly on the gas opacity. Rafikov (2006) systematically studied quasi-static atmospheres of accreting proto-planetary cores for different opacity behaviors and realistic planetesimal accretion rates in various parts of the proto-planetary nebula. In the region of the giant planets, he considered values of the critical mass as high as 20–60 Earth masses (for opacity values in the range 0.1–1 $cm^2\ g^{-1}$) if planetesimal accretion is fast enough for proto-planetary cores to form prior to the dissipation of the nebular gas. If the mass of the forming planet is much less than the critical value, the envelope/core mass ratio is very low. As a consequence, the self-gravitation of the gas is negligible and the maximum equilibrium radius of the envelope is much larger than the Hill radius. Therefore, there is a limit to the amount of gas that can be collected by the planet from the surrounding nebula.

When the mass of the forming planet is close to the critical value, also the envelope radius is comparable to the Hill radius (about a quarter of the Hill radius following Lissauer et al. 2009). For proto-planetary masses greater than the critical value, a further increase produces a growth of the Hill radius greater than the corresponding value of the envelope radius, and more and more gas can be collected from the nebula. At the same time, the heating due to the in-falling gas produces an expansion of the envelope up to the boundaries of the Hill lobe. The gas accretion is thus driven by the thermodynamic and hydrodynamic conditions near the growing planet and, in turn, depends on the ratio of the energy carried by the in-falling gas and the energy associated with pressure effects. In addition, angular momentum is redistributed in the in-falling gas under the action of the gravitational fields of the Sun and of the growing planet. Moreover, fast radial hydrodynamical instabilities can grow until they stop the accretion, producing a rebound of a significant fraction of the gas envelope out of the sphere of influence of the growing planet, as shown by Wuchterl (1991). Such an effect, however, strongly depends on the assumed boundary conditions, particularly on the density and temperature of the gas in the Solar Nebula (Wuchterl 1993) and on the adopted geometry. As a result of these combined effects, the accretion time scale is the maximum between the radiative time scale and the time scale needed by the external regions of the envelope to reach a hydrostatic quasi-equilibrium state. However, dissipative processes due to turbulence are present in the Solar Nebula and in particular around the growing planet. Here, angular momentum is dissipated inside the accretion disk formed by the in-falling gas producing a substantial flux of matter towards the core of the proto-planet, strongly shortening the accretion time scale.

Thus, the final gaseous accretion of the central planet occurs via formation and replenishment of a disk-shaped envelope embedded in the Hill lobe, which suggests again that the rapid collapse of a spherical gaseous envelope is not a valid description of the process. A 3D hydrodynamic code must be used to trace the accretion. In the next section we shall describe our three-dimensional model of planetary accretion in some detail to supply a template of the relevant processes and their treatment.

4 Evolution of the Circum-Planetary Disk: A Template Model

A few years ago Stevenson (2001) discussed the formation of the Galilean moons, which was thought to take place with a mechanism similar to that of planetary formation in a disk of gas and dust swirling around the proto-Jupiter. As Stevenson pointed out, this mechanism works only if the collapse time of the growing planet is shorter than the accretion time on the planet of the forming satellites, and these constraints are not respected in the Jovian system (ibid.). Stevenson noticed that Callisto's composition could be an indication that the satellites were formed at the end of Jupiter's gas accretion, when the gas density in the disk was tenuous. Several authors have attempted to quantitatively approach this problem: here we report some of the most important results obtained up to now.

In what follows we will describe the structure and the results of different models of the evolution of circum-planetary disks, in order to show what are the critical parameters affecting the satellite formation process. To do that, we have to set some preliminary definitions. We will call *growing planet* the central body in nearly hydrostatic equilibrium, containing a planetary core composed by a mixture of rocky and icy materials plus a spherical gaseous envelope. Surrounding the growing planet is a deeply anisotropic region where gas accretion is supersonic and turbulence is present: this region will be referred to as the *accretion disk* (Shakura and Sunyaev 1973). Outside the accretion disk is the *Hill lobe*, the large spheroid region where gas accretion takes place and through which material captured from the Solar Nebula evolves inward. These three regions represent a useful schematization of the history of the gas which enters the sphere of influence of the growing planet. The accretion on the planet can be treated assuming an annular region, centered on the Sun, as the planet-feeding zone. The radial extension of the feeding zone should be large enough to include the region where the gravitational and dynamical effects (i.e. the density waves induced by the planet on the gas of the Solar Nebula) due to the growing planet are significant.

Alibert et al. (2005) modeled the Jovian disk following the time evolution of its surface density in steady conditions. They also tried to take into account the local vertical disk structure by solving the barometric equation. The disk is fed by the gas captured from the proto-Solar Nebula. The thermal effects of radiation emitted by the growing planet on the disk structure are neglected, as well as the interactions between the large atmosphere surrounding the proto-Jupiter and the disk. Once the proto-Solar Nebula is dissipated, the disk is not fed anymore while it keeps losing mass toward Jupiter. The associated angular momentum transfer causes an outward expansion of the disk. The effects of this expansion are not explained in the paper (ibid.). The two phases considered are totally decoupled and it is not clear how they are related to the overall evolution of the central planet. The disk is turbulent and the strength of the turbulence is modeled through the parameter α, which is treated as a free parameter. The external radius of the sub-disk is treated as well as a free parameter. Owing to these assumptions, the authors show that the satellites can survive both Type I and II migration effects (Ward 1997). Even when giving a good phenomenological description of the involved processes, this approach is however non-predictive, since the parameters describing the disk are poorly constrained. Moreover, the authors reduce the effect of Type I migration by modifying it through a suitable factor f that can vary between 1–0.001. The authors modify the values of α until the temperature of the disk allows the formation of an object with the same rock/ice ratio assumed for Callisto, while the low value of f permits its survival to Type I migration. This method can supply some insight in the thermodynamical conditions of the satellite disk, but does not permit to relate them to the overall evolution of Jupiter, nor to the mass present in the disk, which in turn is a function of f.

This evolutionary model has been used in order to compute, in a following paper (Alibert and Mousis 2006) the detailed thermodynamic conditions of the disk, as well as the associated condensation sequence. In conclusion, this model could be a useful first order evaluation of the thermodynamic conditions of the disk. However, the large number of unknown parameters present in the work does not allow for using it in a predictive way.

Another approach is the one suggested by Canup and Ward (2002), based on the idea that the disk formed at the end of the gas accretion and therefore was gas-poor. The main argument of these authors is that a massive disk would be characterized by too high temperatures to allow for ice condensation in the region of Ganymede and Callisto. Such a consideration holds true unless the disk has very low viscosity ($\alpha \sim 10^{-6}$), with a correspondingly long disk viscous lifetime of $\sim 10^6$ yr. In massive disks, the lifetimes of Galilean-sized satellites against Type I orbital decay due to disk torques and aerodynamic gas drag could be extremely short (10^2 and 10^3 yr respectively). Due to Type II orbital decay, also larger satellites can be lost on short time scales if the satellites do not succeed in opening gaps in the disk. This implies that only satellites that are large enough to open a gap can survive. The model by Canup and Ward (2002) is the one that better describes the present configuration of the Galilean satellites; yet, it is not clear when such a "starved disk" formed across the formation process of Jupiter. Moreover, the model does not satisfactory explain the weak density gradient presently measured in Jupiter's system.

Following Ruskol (2006), who reports the results obtained by the Russian school on this subject, it is very difficult to decide whether large mass or small mass disks should be favored. Makalin et al. (1999) explored the evolution of a large mass disk, similar to the one described in Coradini and Magni (1984). In a later work, Makalkin and Ruskol (2003) investigated the evolution of a low mass disk. In their modeling, the authors used an α model to describe the disk, assuming a value of $\alpha \sim 10^{-3}$. In both cases the authors were able to reconcile the disk's thermodynamic conditions with the chemical composition of the Galilean satellites. This ambiguity is substantially due to the fact that the assumptions on the disk derive by the present observations on the satellites. Therefore, as in any inversion problem, there are several parameters that can be adjusted and tuned. In particular, the mass accretion rate and the strength of the viscosity, as defined through the parameter α, are strongly model dependent. A more robust approach could be to relate the characteristics of the disk to the evolution of Jupiter, as we describe here, following the approach used by Magni and Coradini (2004).

Let us then consider a forming proto-planet, collecting gas and dust from the proto-Solar Nebula. The reference system is rigidly rotating with an angular velocity equal to the Keplerian orbital velocity at the location of the planet. The numerical integration of the system of partial differential equations is performed by using a first order explicit numerical scheme, the so-called *donor cell* method described by Boss (1980). An advantage of this method is that it ensures rigorous conservation of the global mass and momentum through the numerical grid. Moreover, in those regions of severe mass depletion, the method prevents negative densities by taking sufficiently small time steps. Note that tidal and Coriolis forces are present, since the grid is rotating with the growing planet. With this scheme we can properly account for the mass and angular momentum transfer from the Solar Nebula to the planet.

However, due to the discrete structure of the hydrodynamic equations, angular momentum is no longer exactly conserved as it would be in an inertial, non-rotating reference frame. The shorter is the time scale of the spurious diffusion, the more the angular momentum conservation is violated. The time scale of the spurious diffusion can be increased by increasing the mesh grid-points number, at the cost of increasing the computational time: how to achieve a balance between these two competing needs has to be evaluated depending on the aims of the simulations. The grid covers a shell around the present orbit of the planet which has more elements in the volume surrounding the proto-planet itself (i.e. the accretion zone) than elsewhere around the orbit. The angular vertical thickness of the feeding zone is 10°, i.e. large enough to contain more than one pressure height-scale. Due to the reasons previously presented, the radial extension of the feeding zone has been assumed ranging from 2.7 AU to 13.7 AU.

The hydrodynamic evolution is driven by the ordinary mass, momentum, energy conservation and radiative diffusion equations. The external forces are the pressure forces, plus non-inertial forces, i.e. Coriolis and centrifugal, and gravity forces due to the Sun and to the growing planet. In addition, the numerical model also includes a first order treatment of the self-gravitation of the gas, which may be non-negligible inside the Hill lobe and near the growing planet due to the strong density enhancement. The self-gravity term is computed starting from the radially averaged density around the growing planet. The equations are expressed in spherical coordinates and in the Eulerian approach.

The effects of turbulent viscosity on the gas are directly put into the hydrodynamic equations of momenta adding the ordinary nonlinear terms (Landau and Lifschitz 1971) like pseudo-acceleration terms. The main parameter involved in the turbulent terms is the so-called α parameter (Shakura and Sunyaev 1973) that tunes the strength of the turbulence itself. Turbulent dissipation causes a general outward transfer of angular momentum in the nebula and an inward flux of matter. In the region around the growing planet, the geometry of the forces causes respectively outward and inward fluxes of angular momentum and mass. So, turbulent viscosity causes an enhancement of the mass accretion rate of the growing planet. Since around the growing planet the adopted grid geometry contrasts with the natural planet-centered geometry of the fluxes, we considered averaged values of the radial fluxes. The value of α has been assumed to be 0.05 in the case of Jupiter, and 0.01 in the case of Saturn. This has been done since the examination of the structure of the Jovian disk exhibits a much higher degree of turbulence than the Saturnian one. The accretion time scale for Saturn has been found to be from two to four times that of Jupiter. In this calculation, the accretion time scales for Jupiter and Saturn vary between a few 10^3 and about 10^5 years, depending on the value of the α parameter.

In Fig. 2 we plot the density distribution on the mid-plane for the accretion regions around Jupiter and Saturn respectively. The two strongly anisotropic patterns show the regions where the accreting gas is concentrated before entering the envelope of the growing planet. The central region is nearly isotropic due to the combined effects of the non-inertial forces, the pressure and the viscosity, which cause the rotation to change from retrograde to prograde. Therefore, in the region close to the planet a Keplerian, gravitationally bound prograde disk can be identified. The region from which the proto-planet collects the gas, i.e. the feeding zone, is shown in Fig. 3. The feeding zones of Jupiter and Saturn do not overlap between themselves, thus allowing us to treat the formation of the two planets independently. The thermal evolution of the disks has been evaluated integrating the total energy equation and including the radiative time-dependent diffusive terms.

Fig. 2 Density distributions in the accretion regions of the growing planets during the final stages of their formation. For each planet, the *upper plot* represents the mid-plane of the Solar Nebula, while the *lower plot* shows the vertical section along the Sun-planet direction. All distances are expressed in astronomical units, while density is expressed in $g\,cm^{-3}$ and in logarithmic scale

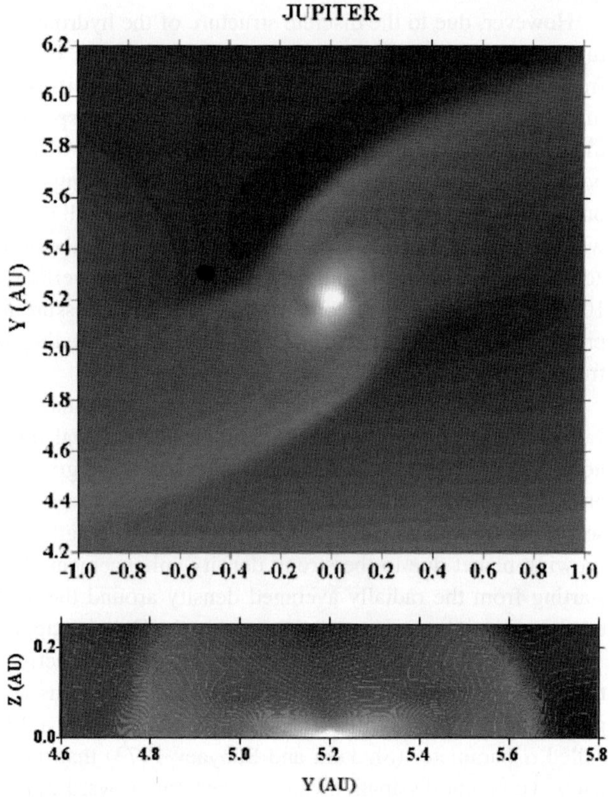

5 The Disk Evolution and Chemistry

As already stated, the gas is captured by the planet either directly or through the formation of an accretion disk. The actual mechanism depends on the distribution of the angular momentum of the gas. The structure of the growing planet, and, in particular, its mass, radius and luminosity, depends on the rate of mass accretion. The distribution of the gas swirling around the planet depends, in turn, on the planet's thermodynamic characteristics. A fraction of the energy collected by the planet through mass capture is in fact released back to the disk as thermal energy. The region surrounding the growing planet is heated up by irradiation, the pressure in the region increases, and a pressure barrier reduces the mass flow (Magni and Coradini 2004). This is a negative feedback effect that tends to increase the planetary growth time scale and affects the overall disk thermodynamics.

The energy production rate per unit mass is assumed to be constant throughout the planet and the energy acquired by the planet is virialized. The energy flux rate can be computed simply as the ratio of the total energy flux per unit mass that enters the region occupied by the planet. The opacity coefficient has been computed by combining the results of different authors: we refer the readers to Magni and Coradini (2004) for details on the opacity treatment. The equation of state, linking the pressure to the other thermodynamic variables, includes both partial dissociation of H_2, ionization of H and He, as well as electronic degeneracy, which might become important in the inner part of growing planet. The dimension, the structure and the global parameters of the circum-planetary accretion disk are evaluated

Fig. 2 (*Continued*)

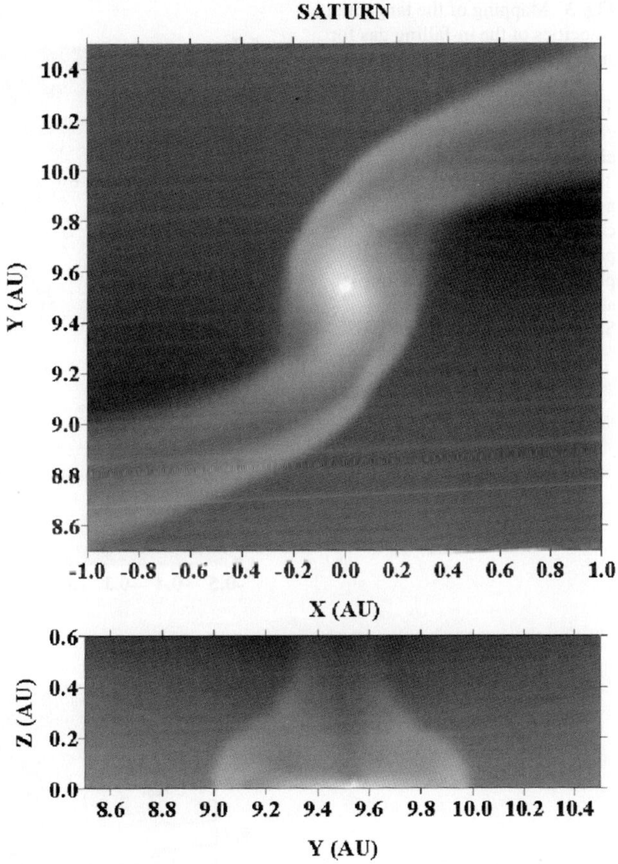

basing on the angular momentum distribution (referred to the growing planet) of the cells inside the Hill lobe.

In order to supply realistic information on the accretion time scales, the mass distribution and the thermal evolution of the region where satellite formation takes place, the authors considered a forming planet embedded in a *minimum mass* Solar Nebula (i.e. about 0.02 solar masses). The growth of the planet is followed from the initial, critical mass (i.e. about 15 Earth masses) up to the final planetary mass. The mass of the feeding zone has been artificially depleted after the planet reached its final mass in order to simulate the natural process of depletion due to gap formation and/or the process of dissipation of the nebula driven by the solar wind. We obtained a sequence of quasi-stationary models, which are characterized by decreasing masses and optical depths, of the formation region of the satellites. Based on these results, we studied the saturation pressures of the main chemical volatile components, like H_2O, CO, CO_2, CH_4, NH_3, Ar, Ne. In Fig. 5 we show the temperature distribution in the disks around Jupiter and Saturn at different phases of the accretion process of their respective planets. As we previously said, these two disks are characterized by different values of the average α parameter: in the case of Jupiter ($\alpha = 0.05$), the disk is more turbulent than in the case of Saturn ($\alpha = 0.01$).

The final phases of accretion here described are slow: the disk gradually cools down, allowing for the condensation of different materials. To follow these phases in a quantitative

Fig. 3 Mapping of the tangential velocities of the in-falling gas for Jupiter and Saturn at the same evolutionary stage as Fig. 2. Tangential velocities are expressed in units of the local circular Keplerian velocity, while distances are expressed in astronomical units. The iso-level curves refer to the regions of prograde motion around the planet. The Jovian Hill radius is about 0.35 AU, while the Saturnian one is about 0.43 AU

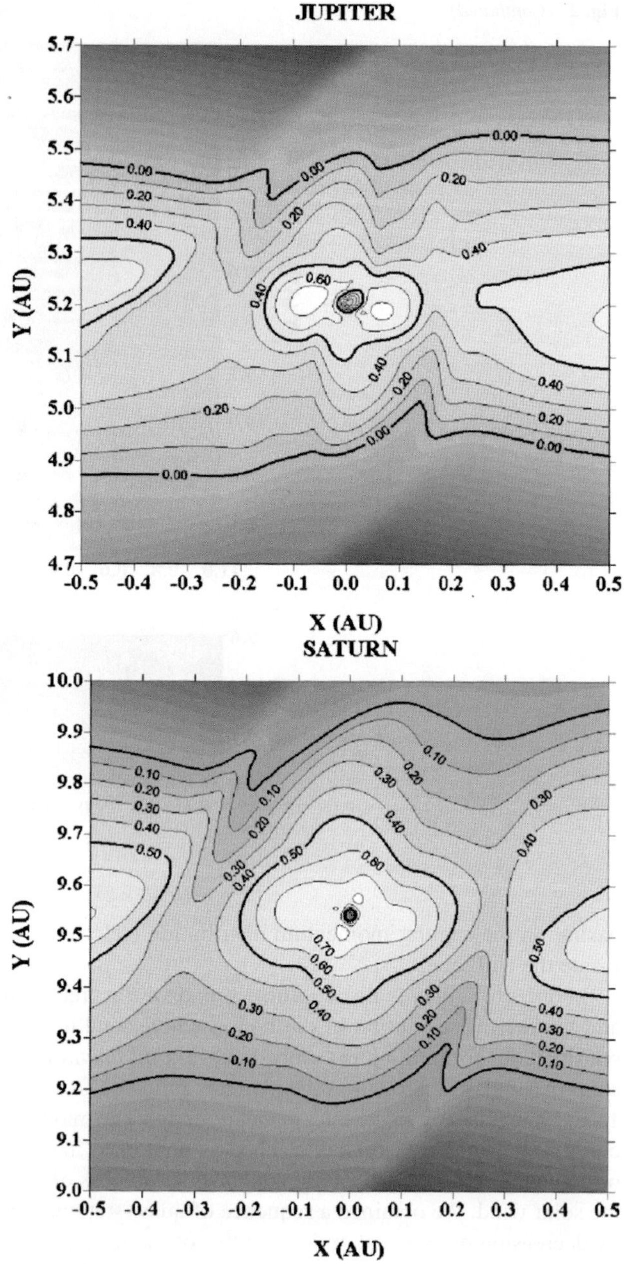

way, we computed the evolution of the temperature in the disk. These calculations take into account the average variation of the temperature, since different parts of the disk can be characterized locally by different values of the thermodynamic quantities. The structure of the disk is in fact extremely complex: the external region is strongly perturbed by the interactions with the gas of the Solar Nebula, while the internal regions are dominated by the interaction with the proto-planet.

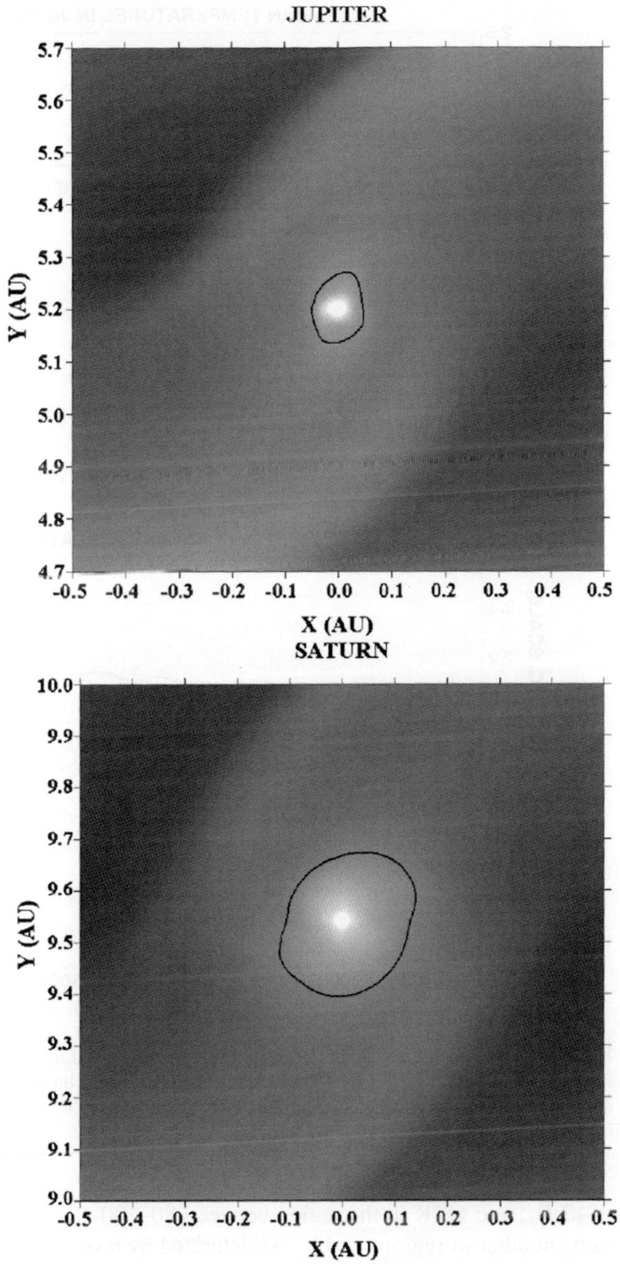

Fig. 4 Surface density distribution for Jupiter and Saturn at the same evolutionary stage as Fig. 2. The *black* iso-curve refers to a density value of 1/30 of the central density. The surface density is expressed in logarithmic scale in cgs units, while distances are expressed in astronomical units

The temperature of the Jovian disk keeps almost constant for a large fraction of its history in all the regions in which we have divided the disk for sake of simplicity. These regions correspond respectively to 0–40 R_J, 40–60 R_J, 60–80 R_J, 80–100 R_J and 100–150 R_J. Obviously, the inner region (0–40 R_J, i.e. the one which is relevant for satellite formation) is the hottest one. Toward the end of the accretion process, when the mass inflow reduces, the disk cools down. At this stage, the Jovian disk reaches about 125°K in the region between

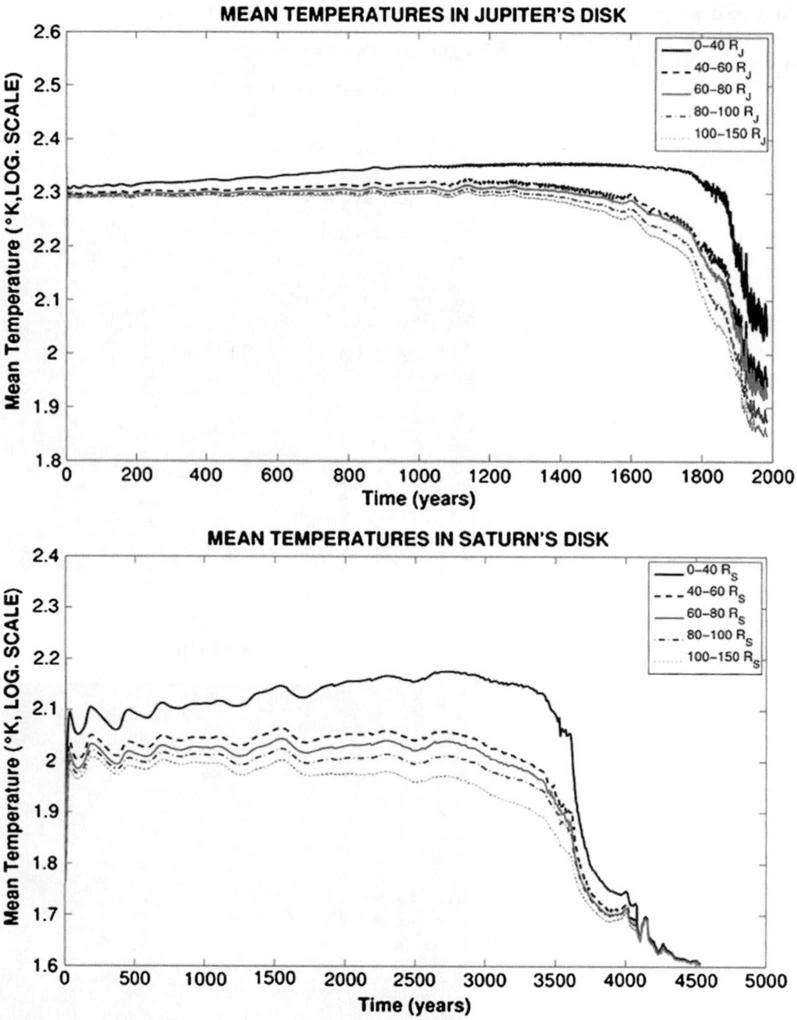

Fig. 5 Mean temperatures in the Jovian and Saturnian disks during the formation of the respective planets. Each disk has been divided into five annular shells centered on the forming planets and the temperatures are averaged over each shell. Temperatures are expressed in °K and in logarithmic scale, while the time is expressed in years

0–40 R_J, and 68°K in the region between 80–100 R_J. The Saturnian disk is divided in the same number of regions and is characterized by a similar behavior, but the disk is generally colder. At the end of the accretion phase, the temperature becomes quite uniform in the entire disk, stabilizing at about 39°K.

The gas pressure in the disks can be computed also as a function of the accretion time: the results are shown in Fig. 6. As the temperature, the global gas pressure and the partial pressures of the different gases decrease as the mass inflow slows down. The mean values of the ratios between the disk pressure and the saturation pressure of the different gases present in the disk are shown in the same Figure. When the ratio is greater than unity, the component is in gaseous form; when it is lower, the thermodynamic conditions in the disk allow

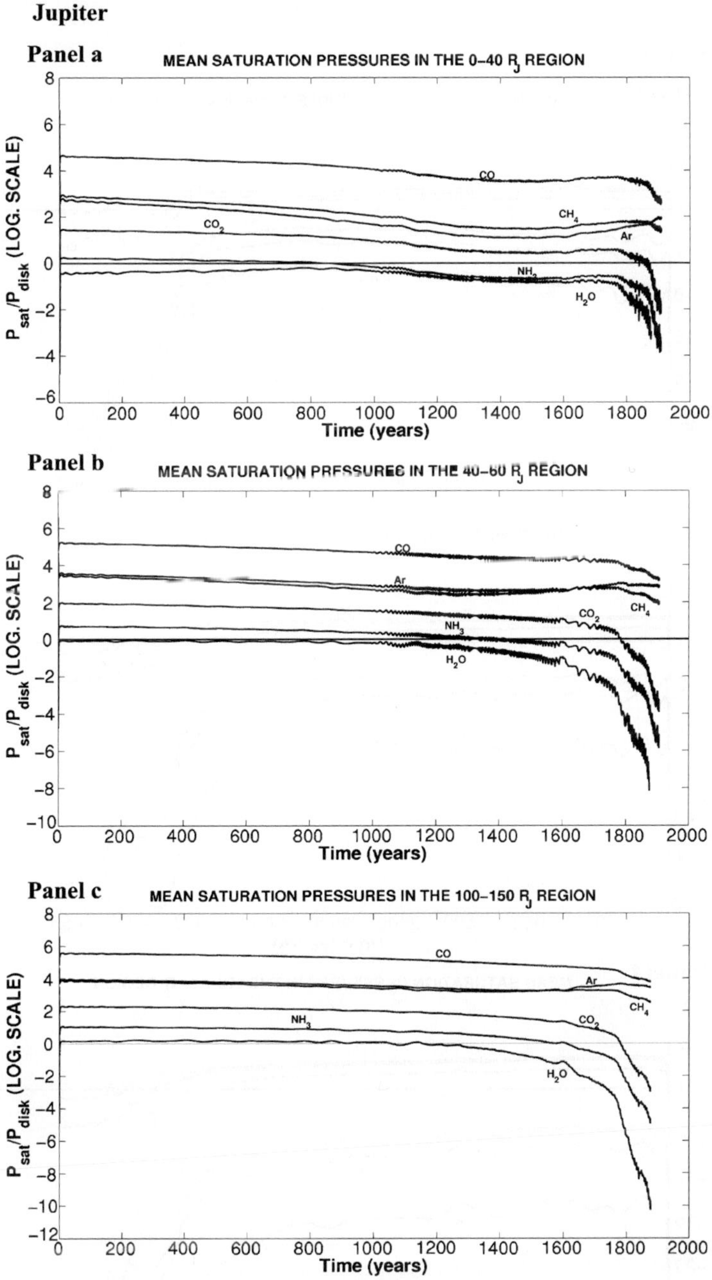

Fig. 6 Mean vapor saturation pressures in the Jovian (panels **a**–**c**) and the Saturnian (panels **d**–**f**) disks as a function of time. Each panel shows the partial saturation pressures of different chemical species (CO, Ar, CH_4, CO_2, NH_3, H_2O) in selected annular shells as those described in Fig. 5. The saturation pressures are expressed in units of the local pressure and in logarithmic scale while the time is expressed in years. As mentioned in the text, these figures do not represent an exact evaluation of the disk chemical evolution, since we do not compute which are the species that are present in the disk at each time step. Our calculations are indication whether, provided certain species are present in the disk, the thermodynamic conditions could allow their condensation

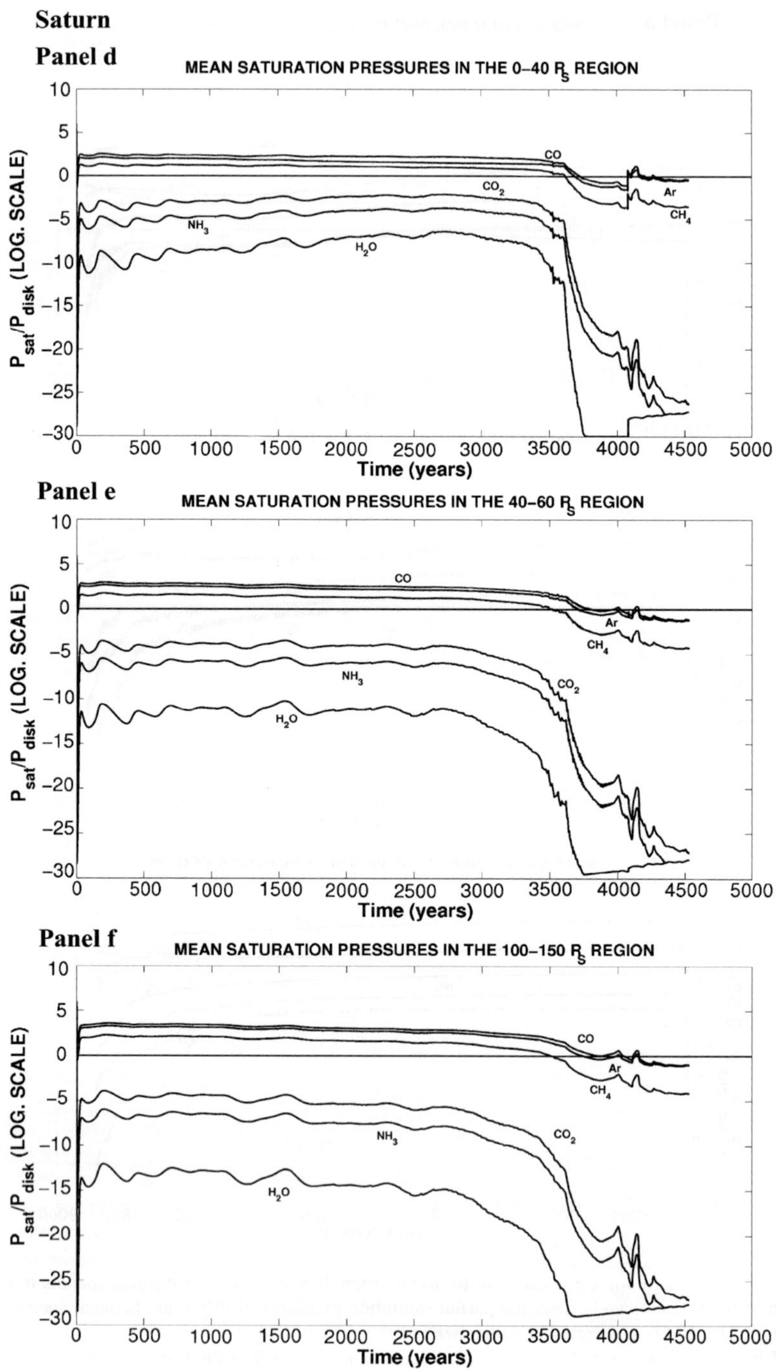

Fig. 6 (*Continued*)

for a phase transition from gas to solid. It is obvious that this is not a complete chemical evolution model, since we do not compute exactly which are the species that are present in the disk at each time step. Our calculations are an indication that, provided certain species are present in the disk, the thermodynamic conditions could allow their condensation. We plan to improve such an evaluation in the near future by developing a more complete model of the gas chemistry, which would use as its input the results of the hydrodynamical model.

From a qualitative point of view, therefore, in Jupiter's case (Fig. 6, panels a–c) only H_2O condensation is possible in the formation region of the satellites (0–40 R_J) and only in the final phases of the accretion. When the accretion stops, other species start to condense (e.g. NH_3 and CO_2 ices) both in the inner and outer regions. The emerging picture is very interesting, since it hints that a certain amount of ammonia should have been incorporated in the satellites, possibly in the form of ice or in the more complex molecules of clathrates, provided that the relative abundances of N and H could allow for their formation. The presence of ammonia is important since it significantly lower the freezing point of the material in which it has been incorporated, therefore influencing the thermal history of the satellite. Our analysis implies that the Jovian disk was dominated since the beginning of its history by the presence of H_2O grains. Silicate particles were also present in the Jovian disk. Icy particles contributed, with the silicate grains, to the formation of *satellitesimals* in the disk: the combination in different proportions of these two main components played the most important role in the determination of the average density of the satellites.

In the case of Saturn the lower temperatures allow for the condensation of CO_2, NH_3 and CH_4 and grains (see Fig. 6, panels d–f). At this stage, in fact, most of the Saturnian disk is as cold as the surrounding Solar Nebula from which the gas is collected. Only the internal regions of the disk are hotter, having temperatures which cause the sublimation of different ices, and thus permit a complex chemistry to evolve. Highly volatile gases can be incorporated in the mixture since the very beginning of the life of the disk. Even Ar can be incorporated in the satellitesimals in the external part of the disk, in agreement with the puzzling features of the atmosphere of Titan, where no primordial noble gases other than Ar were detected by the Gas Chromatograph Mass Spectrometer (GCMS) aboard the Huygens probe during its descent to Titan's surface on January 14, 2005. Other gases were perceptually below the detection level of GCMS (Kr and Xe). The CO ice is not condensed in the Saturnian sub-nebula. As a consequence, the satellitesimals formed in the disk could be depleted of this ice as suggested by Mousis et al. (2008). This would explain the deficiency of CO observed in Titan's atmosphere (Alibert and Mousis 2007). H_2O, NH_3, CO_2, and CH_4 would instead remain incorporated in solids. The complex chemistry present in the Saturnian disk might thus be responsible for the internal evolution and differentiation of the satellites.

On the surfaces of the icy satellites, high volatility ices could be lost during the further evolution of these bodies: the icy moons, in fact, were intensively bombarded and underwent strong interactions with both solar and planetary radiations, as well as by high energy electrons and protons trapped in the radiation belts of Saturn. Only Titan has preserved a large part of the original chemistry.

Water ice is abundant on Saturnian satellites, as shown by the near-infrared spectra returned by VIMS (see chapter by Dalton et al. 2010, this issue). The ice phase can be used to study the crystalline order of the surface ice on all the satellites. It could also be a way to measure the lattice order of ice that, in turn, depends on its condensation temperature and rate, its temperature history, and its radiation environment. Inner satellites are characterized by a larger content of crystalline ice, with respect to the amorphous one (which is almost absent), as measured on the basis of the strength of the absorption bands at 2 and 2.05 μm.

The role of carbon compounds was also important for the overall chemistry together with the water ice particles. Satellitesimals formed in the Jovian and the Saturnian disks would probably maintain records of the original composition. A similar analysis performed on the CO_2 ice seen by VIMS shows that a large amount of "organic contaminants" was present in the Saturnian system. The radial distribution of the "contaminants" across the Saturnian system can be measured through the visible spectral slopes (ibid.). The satellites have red (positive) slopes in this range, probably caused by the presence of a still unidentified UV organic absorber. Therefore, we can argue that the original chemistry is not totally cancelled by the overall evolution of the satellites.

At this point of the evolution of the disks the model developed by Canup and Ward (2002) becomes important. In fact, the end of the accretion of the planet is accompanied by the formation of a tenuous disk like the one described by these authors. For this reason, the previously presented results are not in disagreement with their modeling: on the contrary, they confirm their assumptions in a rigorous way. The further evolution of the disk could imply displacements of the satellitesimals as well as capture and bombardment from external bodies. These subjects will be discussed in the second part of this chapter (see Mosqueira et al. 2010, this issue).

References

C.B. Agnor, D.P. Hamilton, Nature **441**, 192–194 (2006)

Y. Alibert, O. Mousis, Lunar Planet. Sci. **XXXVII**, 1141–1142 (2006)

Y. Alibert, O. Mousis, Astron. Astrophys. **465**, 1051–1060 (2007)

Y. Alibert, O. Mousis, C. Mordasini, W. Benz, Bull. Am. Astron. Soc. **37**, 673 (2005)

P. Artymowicz, in *Brown Dwarfs and Extrasolar Planets*, ed. by R. Rebolo, E.L. Martin, M.R. Zapatero Osorio. ASP Conference Series, vol. 134 (Astronomical Society of the Pacific, San Francisco, 1998), pp. 152–161

D. Banfield, N. Murray, Icarus **99**, 390–401 (1992)

A.P. Boss, Astrophys. J. **236**, 619–627 (1980)

R.M. Canup, W.R. Ward, Astron. J. **124**, 3404–3423 (2002)

R.M. Canup, W.R. Ward, Nature **441**, 834–839 (2006)

A. Coradini, G. Magni, Icarus **59**, 376–391 (1984)

G. D'Angelo, T. Henning, W. Kley, Astron. Astrophys. **385**, 647–670 (2002)

Dalton et al., Space Sci. Rev. (2010, this issue)

T. Encrenaz, Earth Moon Planets **67**, 77–87 (1994)

R. Greenberg, Icarus **24**, 325–332 (1975)

P. Goldreich, N. Murray, Y. Longaretti, D. Banfield, Science **245**, 500–504 (1989)

P. Goldreich, Y. Lithwick, R. Sari, Astrophys. J. **614**, 1024–1037 (2004)

T. Guillot, D.J. Stevenson, W.B. Hubbard, D. Saumon, in *Jupiter. The Planet, Satellites and Magnetosphere*, ed. by F. Bagenal, T.E. Dowling, W.B. McKinnon (Cambridge University Press, Cambridge, 2004), pp. 35–57

F. Hersant, D. Gautier, G. Tobie, I.J. Lunine, Planet. Space Sci. **56**, 1103–1111 (2008)

K. Kornet, S. Wolf, M. Rozyczka, Planet. Space Sci. **55**, 536–546 (2007)

L. Landau, E. Lifschitz, *Mechanique des Fluides* (Mir Editions, Moscow, 1971)

J.J. Lissauer, Icarus **114**, 217–236 (1995)

S.H. Lubow, G.I. Ogilvie, Astrophys. J. **560**, 997–1009 (2001)

J.I. Lunine, A. Coradini, D. Gautier, T.C. Owen, G. Wuchterl, in *Jupiter. The Planet, Satellites and Magnetosphere*, ed. by F. Bagenal, T.E. Dowling, W.B. McKinnon (Cambridge University Press, Cambridge, 2004), pp. 19–34

J.J. Lissauer, D.J. Stevenson, in *Protostars & Planets V*, ed. by B. Reipurth, D. Jewitt, K. Keil (University of Arizona Press, Tucson, 2007), pp. 591–606

J.J. Lissauer, O. Hubickyj, G. D'Angelo, P. Bodenheimer, Icarus **199**, 338–350 (2009)

G. Magni, A. Coradini, Planet. Space Sci. **52**(5–6), 343–360 (2004)

A.B. Makalin, V.A. Dorofeeva, E.L. Ruskol, Sol. Syst. Res. **33**, 578 (1999)

A.B. Makalkin, E.L. Ruskol, Astron. Vestn. **57**, 545–554 (2003)

H. Mizuno, Prog. Theor. Phys. **64**, 544–557 (1980)
I. Mosqueira, P. Estrada, D. Turrini, Space Sci. Rev. (2010, this issue). doi:10.1007/s11214-009-9614-6
O. Mousis, J.I. Lunine, C. Thomas, M. Pasek, U. Marbœuf, Y. Alibert, V. Ballenegger, D. Cordier, Y. Ellinger, F. Pauzat, S. Picaud, Astrophys. J. **691**, 1780–1786 (2008)
J.B. Pollack, G. Consolmagno, in *Saturn*, ed. by T. Gehrels, M.S. Matthews (University of Arizona Press, Tucson, 1984), p. 811
R.R. Rafikov, Astrophys. J. **648**, 666–682 (2006)
H. Rauer, A. Hatzes, Planet. Space Sci. **55**, 535 (2007)
E.L. Ruskol, Astron. Vestn. **40**(6), 499–504 (2006)
V.S. Safronov, *Evolution of the Protoplanetary Cloud and Formation of the Earth and Planets* (Nauka, Moscow, 1969), translation into English by the Israel Program for Scientific Translations, 1972, NASA TTF-677
N.I. Shakura, R.A. Sunyaev, Astron. Astrophys. **24**, 337–355 (1973)
D.J. Stevenson, Planet. Space Sci. **30**, 755–764 (1982)
D.J. Stevenson, Science **294**, 71–72 (2001)
W.R. Ward, Astrophys. J. **482**, L211–L214 (1997)
G. Wuchterl, Icarus **91**, 5364 (1991)
G. Wuchterl, Icarus **106**, 323–338 (1993)

Space Sci Rev (2010) 153: 431–446
DOI 10.1007/s11214-009-9614-6

Planetesimals and Satellitesimals: Formation of the Satellite Systems

Ignacio Mosqueira · Paul Estrada · Diego Turrini

Received: 29 May 2009 / Accepted: 19 November 2009 / Published online: 20 February 2010

Abstract The origin of the regular satellites ties directly to planetary formation in that the satellites form in gas and dust disks around the giant planets and may be viewed as mini-solar systems, involving a number of closely related underlying physical processes. The regular satellites of Jupiter and Saturn share a number of remarkable similarities that taken together make a compelling case for a deep-seated order and structure governing their origin. Furthermore, the similarities in the mass ratio of the largest satellites to their primaries, the specific angular momenta, and the bulk compositions of the two satellite systems are significant and in need of explanation. Yet, the differences are also striking. We advance a common framework for the origin of the regular satellites of Jupiter and Saturn and discuss the accretion of satellites in gaseous, circumplanetary disks. Following giant planet formation, planetesimals in the planet's feeding zone undergo a brief period of intense collisional grinding. Mass delivery to the circumplanetary disk via ablation of planetesimal fragments has implications for a host of satellite observations, tying the history of planetesimals to that of satellitesimals and ultimately that of the satellites themselves. By contrast, irregular satellites are objects captured during the final stages of planetary formation or the early evolution of the Solar System; their distinct origin is reflected in their physical properties, which has implications for the subsequent evolution of the satellites systems.

Keywords Planet formation · Satellite formation · Origin of satellites · Giant planets

The satellites of Jupiter and Saturn have received renewed interest as a consequence of the Cassini mission to Saturn, which has resulted in a growing body of literature concerning both irregular (see Jewitt and Haghighipour 2007 for a review) and regular satellites (see Estrada et al. 2009 for a review; and Canup and Ward 2009 for a different take). We make no

I. Mosqueira (✉) · P. Estrada
SETI Institute, 515 N Whisman Rd, Mountain View, CA 94043, US
e-mail: imosqueira@seti.org

D. Turrini
INAF-IFSI, Via del Fosso del Cavaliere 100, 00133, Rome, Italy

claim to completeness—the aim of this submission is not to provide a comprehensive review of the field, which has already been attempted in those publications, but to describe key physical processes and observational constraints. While there is unambiguous observational evidence that the regular and irregular satellites are distinct populations, there exist at least indirect ties between them when it comes to the processes involved in the capture of objects from heliocentric to circumplanetary orbits. There is also a hint of a direct connection in terms of the locations of the innermost irregular satellites and the size of the disk formed by gas inflow through a gap ($\sim R_H/5$; Mosqueira and Estrada 2003a; Estrada et al. 2009; Ayliffe and Bate 2009); irregular satellites captured into orbits with smaller semi-major axes may be removed by gas drag from the outer disk and possibly contribute mass to the regular satellites, though this link remains to be quantitatively explored.

It is also worth noting that the processes shaping the formation of satellite systems are strongly coupled to those governing planetary formation and the evolution of the early Solar System. In fact, planetary formation sets the initial conditions for satellite formation which, therefore, cannot be treated in isolation. The forming satellite systems can exchange material with the outer Solar System, receiving solids mainly in the form of planetesimals captured or ablated due to the interaction with the circumplanetary gas disks, and even scattering satellitesimals out into heliocentric orbits. Moreover, the regular satellites, once formed, are still subject to exchanges with the outer Solar System, as dramatically illustrated by the disruptive capture of Triton in Neptune's satellite system. Below we discuss the formation and properties of both regular and irregular satellites in the framework of such exchange processes.

1 Formation of the Regular Satellites of Giant Planets

In the core accretion model of planet formation (Bodenheimer and Pollack 1986) a core must first form by accretion of planetesimals. In this mode of planet formation most of the mass of solids is taken to reside in planetesimals of size $\sim$10 km (Wetherill and Stewart 1993). Planetesimals may dissolve in the envelopes of the forming giant planets, thus enhancing the planet's metallicity (Pollack et al. 1986). Most of this high-Z mass delivery takes place before the cross-over time, when the mass of the gaseous envelope grows larger than that of the core. This stage is then followed by a dilution during runaway accretion, depending on the amount of gas accreted during this brief phase of the planet's growth (Pollack et al. 1996). As the giant planet reaches its final mass, planetesimals in its feeding zone undergo collisional grinding. In the Jupiter–Saturn region, the collisional timescale for kilometer-sized objects is similar to the ejection timescale $\sim 10^4$ yrs, so that a fraction of the mass of solids will be fragmented into objects smaller than 1 km (Stern and Weissman 2001; Charnoz and Morbidelli 2003). The collisional cascade facilitates planetesimal delivery to the circumplanetary disk (in $\sim 10^4$ yrs) because smaller planetesimals are easier to capture.

1.1 Satellite Migration and Growth

Mosqueira and Estrada (2003a) start with a disk populated by satellite embryos ($\sim 10^3$ km) and satellitesimals. These authors adopt a bimodal size distribution or two-groups method, as has been done in a number of publications that treat protoplanetary growth by the binary accretion of planetesimals (e.g., Pollack et al. 1996; Goldreich et al. 2004). There are close parallels, but some differences as well. For both planets and satellites, the migration of the embryos is dominated by the tidal interaction with the gas disk, whereas planetesimals and satellitesimals migrate due to aerodynamic gas drag. But in the satellite case the transition from one type of migration to the other overlaps and leaves no size unaffected, i.e., all object sizes from kilometers to full-sized satellites undergo migration. Another difference involves the capture cross-section of migrating planetesimals versus satellitesimals. Satellite embryos have a larger ratio of physical radius to Hill radius ($\sim$0.1 at the location of Ganymede) than do protoplanets ($\sim 10^{-3}$ at the location of Jupiter). As a result, the impact probability of passing objects is larger (cf. Fig. 9 from Kary et al. 1993). Also, the resulting gravitational focusing factor ($F_g \sim 10$) for satellites is never as large as it is for planetary embryos during their runaway phase, before they grow large enough to "heat" up their own food and their growth switches to oligarchic accretion.[1]

1.1.1 Timescale for Disk Formation and Cooling

While the gas accretion rate is high the planet's envelope takes up a fraction of the planet's Roche lobe. Thus, there is agreement in the literature that the formation of icy satellites close in to the planet must await for gas accretion onto the planet to wane (Coradini et al. 1989; Makalkin et al. 1999; Ayliffe and Bate 2009). The timescale for Jupiter and Saturn to clear a gap in between ($\tau \sim 10^4$ years or less in a compact initial configuration) sets the timescale for the end of gas accretion.[2] The timescale for envelope collapse is given by the Kelvin–Helmholtz time of $\sim 10^4$ years (Hubickyj et al. 2005), following which a circumplanetary disk will form. We stress that even after envelope collapse continued inflow onto the disk may keep it too hot for the concurrent accretion of close-in, icy satellites (see Estrada et al. 2009 and Ayliffe and Bate 2009 for further details). Once the circumplanetary disk forms and gas inflow dies down (as the giant planets open a combined gap), the timescale for the disk to cool sufficiently for ice to condense depends on the disk opacity. The Rosseland mean thermal opacity for the gas is $K_{gas} \sim 10^{-4}$ cm^2 g^{-1}, while that of micron-sized dust grains can be of order $\sim$1 cm^2 g^{-1} (Lunine and Stevenson 1982; Chiang et al. 2001). As particles grow, the opacity of the nebula decreases. A sustained high opacity may keep the subnebula weakly turbulent and hotter. Conversely, rapid coagulation allows the subnebula to become optically thin, cooling quickly and allowing turbulence to subside. Indeed, while the incremental growth of grains and dust may take place irrespective of the level of nebula turbulence, growth past the decoupling size likely requires very low levels of turbulence ($\alpha < 10^{-6}$) (Youdin and Shu 2002; Cuzzi and Weidenschilling 2006), which allows for a cool subnebula where condensation of volatiles takes place (Mosqueira and Estrada 2003a). Following envelope collapse, ablation of meter-to-kilometer planetesimal disk crossers (see Estrada et al. 2009 and references therein) and other mechanisms enhanced the concentration of solids in the satellite disk from a starting condition with condensable content enhanced by a factor of 3–4 from cosmic mixtures, as suggested by the high-Z content of giant planet envelopes (Atreya et al. 1999). Note that the circumplanetary disk and the giant planet envelope are both enriched in solids as a result of a common mechanism, i.e., planetesimal ablation (see Fig. 1 upper left panel; Mosqueira et al. 2009); however, the quantitative degree of enrichment may differ

[1] At this stage, satellitesimals should be interpreted as those objects which determine embryo growth, containing a significant fraction of the mass of solids in the disk (though other sizes may also contribute to the mass distribution).

[2] Provided the protoplanetary disk at the location of Jupiter and Saturn is weakly turbulent (Bryden et al. 1999). In that case the gas inflow through the gap ceases.

substantially between the two.[3] Given a density of $\sim$1 g cm^{-3}, a planetesimal of size <1 km encounters a gas column equal to its mass when crossing the circumplanetary gas disk. Such a population of objects is likely to result from the fragmentation cascade following giant planet formation (e.g., Charnoz and Morbidelli 2003). Thus, we consider a solids-enhanced, quiescent disk with optical depth $\tau \gtrsim 1$ and gas surface density $\Sigma_g \sim 10^4$ g/cm^2. The disk cooling time is then given by $\tau_{KH} \approx \Sigma_g C T_c/(2\sigma_{SB} T_d^4) \sim 10^2 (250K/T_c)^3 \tau$ yrs, where σ_{SB} is the Stephan–Boltzmann constant, C is the specific heat, and T_c and T_d are the midplane and photospheric temperatures.

1.1.2 Disk Scales

Gas flowing into the protoplanet's Hill sphere forms a circumplanetary disk. One can obtain an estimate of the expected disk size by assuming that in-falling gas elements conserve specific angular momentum. Assuming that *prior* to gap-opening the giant planet accretes gas with semi-major axis originating within $\sim R_H$ from its location, centrifugal balance yields a characteristic disk size of $\sim R_H/50$ (Cassen and Pettibone 1976; Safronov et al. 1986; Mosqueira and Estrada 2003a). For Jupiter and Saturn these radii are located close to the positions of Ganymede and Titan, respectively. On the other hand, *after* gap-opening accretion continues through the planetary Lagrange points. In this case, the estimated characteristic disk size formed by the inflow is significantly larger, of the order of $\sim R_H/3$ (Mosqueira and Estrada 2003a, but note that this paper settles on an outer disk size of $\sim R_H/5$, based on the locations of irregular satellites). In the absence of a gap, 3D effects allow low angular momentum gas to be accreted directly onto the planet, or in a compact disk of characteristic size $<0.1 R_H$ (D'Angelo et al. 2003), whereas in the presence of a well-formed gap a 2D treatment applies and the resulting disk size will be $>0.1 R_H$. D'Angelo et al. (2003) conclude that, for nominal disk parameters, two-dimensional computations are reliable for secondary to primary mass ratio $\mu \gtrsim 10^{-4}$, such that the Hill radius of the protoplanet is larger than the disk scale-height. Estrada et al. (2009) and Ayliffe and Bate (2009) show that the specific angular momentum of the inflow through the gap is about a factor of 3–4 larger than that of the Galilean satellites 1.1×10^{17} cm^2/s, which means that the gas will achieve centrifugal balance at a radial location of $\sim 100 R_J$. This disk is compact compared to the Hill radius, but it is extended in terms of the locations of the regular satellites, and possibly linked to the location of the irregular satellites.

1.1.3 Model Parameters

Mosqueira and Estrada (2003a) divide the circumplanetary disk into inner and outer regions. For Jupiter, they compute the solids-enhanced minimum mass (SEMM) gas densities in the inner and outer disks based on the solid mass required to form Io, Europa (both re-constituted for missing volatiles) and Ganymede in the inner disk and Callisto in the outer disk. Inside of the centrifugal radius the surface gas density Σ_g is 10^4–10^5 g cm^{-2} (corresponding to pressures $\sim$0.1 bar), which yields an optical depth due to absorption by hydrogen molecules (assuming dust coagulation) $\tau_v \sim \Sigma K_{gas} \sim 1$. Outside the centrifugal radius $r_c \sim 30 R_J$, the gas surface densities are in the range 10^2–10^3 g cm^{-2}, which results in a low optical depth in the range $\tau_v \sim 0.01$–0.1. These authors apply the same procedure

[3] The runaway gas accretion phase is expected to result in a dilution of the solid content of the planet's envelope, as most of the mass of solids is in planetesimals that are not coupled to the gas (see Estrada et al. 2009 for a detailed discussion).

to the circumplanetary disk of Saturn by employing the masses of Titan and Iapetus to set the inner and outer disk masses, respectively, and choose a simple model where the gas density follows a simple $1/a$ dependence inside of $r_a \sim 20R_J$ and outside of $r_b \sim 26R_J$. The transition region has a width of (at least) $\sim 2H_c$, where H_c is the subnebula scale-height at the centrifugal radius. This choice ensures that the gradient in gas density is not so steep as to lead to a Rayleigh–Taylor instability (e.g. Lin and Papaloizou 1993). The use of a two-component subnebula hinges on the assumption of decaying turbulence as Roche-lobe gas inflow ebbs. But it should be noted that detailed simulations of the formation of such a disk remain to be carried out.[4] We tie the disk's temperature profile to the planetary luminosity at the tail end of giant planet formation (Hubickyj et al. 2005). For now, we use a heuristic temperature profile of the form $T = 3600R_J/a$ for Jupiter (e.g., Lunine and Stevenson 1982) and $T = 2\,000R_S/a$ for Saturn. The outer disks of Jupiter and Saturn have roughly constant temperatures in the range of 70–130 K for Jupiter and 40–90 K for Saturn, depending on solar nebula parameters.

Inner Disk In analogy to planetary accretion, we begin by calculating characteristic object sizes in such a disk. The characteristic size over which the self-gravity of the solids overcomes the shear is $l_1 = r_h$, where $l_1 = (m_1/\pi\Sigma_s)^{1/2}$, $\Sigma_s \sim 10^3$ g/cm^2 is the surface density of solids, and $r_h = a(m_1/3M_P)^{1/3}$ is the Hill radius of a satellitesimal of mass m_1. The second characteristic mass can be obtained from the condition $l_2 = r_H$, where $l_2 = m_2/(4\pi a\Sigma_s)$ is the distance over which the mass m_2 in a disk annulus with semimajor axis a can force close encounters with passing satellitesimals, and $r_H = a(m_2/3M_P)^{1/3}$ is the Hill radius of an embryo of mass m_2. For object densities of $\rho_s \sim 1$ g/cm^3, these characteristic masses correspond to sizes of $r_1 \sim$ 1–10 km and $r_2 \sim 1\,000$ km. We now consider a gaseous disk with surface density $\Sigma_g \sim 10^4$ g/cm^2 and aspect ratio $H/a \sim 0.1$ (or a temperature of ~250 K at Ganymede [$15R_J$], and ~100 K at Titan [$20R_s$]), and obtain a Toomre parameter $Q \sim 100$, which shows that self-gravity is unimportant. The particle decoupling size is $\Omega t_s = 16\rho_s r_p v_k/(3C_D\Sigma_g c\Omega) \sim 1 \rightarrow r_p \sim 10$ m, where t_s is the stopping time, v_k is the Keplerian velocity, $C_D \sim 1$ is the drag coefficient[5] (see e.g. Adachi et al. 1976), and c is the sound speed. The gas-drag time for a satellitesimal of size r_1 located at $15R_J$ is $\tau_{drag} = 1/2(v_k/c)^2 t_s \sim 100$ years. The gas-drag time for an embryo of size r_2 is $\sim 10^5$ years. Such an embryo will also migrate due to the tidal interaction with the gas disk in a similar timescale $\tau_I \approx C_I\Omega a^2c^2/(\Sigma_g G^2 m_2) \sim 10^5$ years, where G is the gravitational constant and $C_I \lesssim 1$ in a 3D disk (Tanaka et al. 2002; D'Angelo et al. 2003; Bate et al. 2003). Thus, the crossover size between the regimes of gas-drag and tidal-torque migration is ~500–1 000 km. The Type I migration timescale for Ganymede is $\sim 10^4$ years, which is comparable to its formation time given by the gas-drag time of satellitesimals of size ~100 km.

The question arises whether the satellite growth times are faster than the migration times. Ignoring gravitational focusing, the Safronov binary accretion timescale for embryos is $\tau_{acc} \sim \rho_s r_2/(\Sigma_s\Omega) \sim 10^3$ years, which is faster than either the Type I or gas-drag migration for such objects in the SEMM disk. But note that if all the satellitesimals resided in objects of size $r_1 \sim 1$ km, then the time to clean the disk by gas-drag migration would be ~100

[4]Here we rely mostly on the overall gas surface density of the subnebula, which we justify on observational grounds.

[5]There is an issue whether this drag coefficient should include a correction in cases where the satellite can gravitationally perturb the gas flow (e.g., Canup and Ward 2002). Ostriker (1999) argues that this enhanced drag is already incorporated into the migration rate due to differential tidal torques.

years. The mass growth timescale due to dust and rubble sweep up as a result of differential drift is $\tau_d = 4\rho_s r_2/(3\rho_p \Delta v_p) \sim \rho_s r_2 a/(\Sigma_g \Omega)(a/H) \sim 10^3$ years, where $\Delta v_p \sim \eta v_k$ and $\eta = (v_k - v_g)/v_k$ is the fractional difference between the local Keplerian velocity v_k and the gas velocity v_g due to gas pressure support. This estimate assumes that the density of the dust and rubble layer settles to a state such that $\rho_p \approx \rho_g$, where ρ_g is the nebular gas density, which is consistent with a quiescent disk stirred only by weak, local production of turbulence at the shear layer. Thus, satellitesimals grow as they drift: we interpret Hyperion (whose radius is 135 ± 4 km, Thomas et al. 2007) as a satellitesimal that was captured into resonance by gas-drag inward migration (see Sect. 1.2.1). For objects of this ($\sim$100 km) size, satellitesimal migration, and satellite migration and growth all take place on a similar timescale. This is because satellite growth is regulated by the rate at which the embryo's feeding zone is replenished by satellitesimals or embryos. Also, the Type I migration timescale of satellites is comparable to the gas-drag migration timescales of such satellitesimals (cf. Titan–Hyperion).

Outer Disk Further out the characteristic sizes are smaller despite the increase in semi-major axis. The reason is that the surface density is taken to drop outside the centrifugal radius. Typical sizes in the outer disk for $\Sigma_s \sim 10$ g/cm^2 are $r_1 \sim 1$ km and $r_2 \sim 100$ km. The Safronov binary accretion time to form an embryo of size r_2 in the outer disk is $\sim 10^5$–10^6 years. The time for gas drag to clear the outer disk of such embryos is also 10^5–10^6 years, which Mosqueira and Estrada (2003a) tie to Callisto's formation time.

1.1.4 Satellite Survival

So far we have implicitly assumed that the relatively massive gas component of the circumplanetary disk is static and unaffected by the evolution of the solids. However, the SEMM model of Mosqueira and Estrada (2003b) relies on gap-opening for satellite survival, which means that at some point during their migration and growth satellites become sufficiently massive to open a gap in the gas. There are uncertainties involved in estimating the value for the gap-opening masses, such as the role of 3D effects and stratification, but they are unlikely to greatly affect the inviscid gap-opening criterion of Rafikov (2002) (see also Li et al. 2009), which is based on damping of 2-D acoustic waves by wave steepening (Goodman and Rafikov 2001). For Jupiter and Saturn Rafikov's criterion yields gap-opening masses at $15R_J$ and $20R_S$ in disks with $h \simeq 0.1$ equal to the masses of Ganymede and Titan for disk surface densities of $\sim 2 \times 10^4$ g/cm^2 and 1×10^4 g/cm^2, respectively. It should be noted that the gap-opening mass depends on the semi-major axis a. Also, we assume a surface density profile $\Sigma_g \propto 1/a$, and an extended giant planet following envelope collapse with radius $R_P \sim 2R_J$ (e.g., Hubickyj et al. 2005), where R_J is Jupiter's radius. As a satellite migrates inwards, it finds it easier to open a gap for a given surface density, but the surface density is taken to increase closer to the planet. While the timescale for accretion decreases closer-in, the satellite's feeding zone becomes smaller, and its inward migration speeds-up. Also, both the migration rate and the gap-opening condition are sensitive to the aspect ratio H/a.

1.2 Observational Constraints

1.2.1 Capture of Hyperion into a 4:3 Resonance with Titan

The origin of Hyperion in the 4:3 mean-motion resonance with Titan presents a significant challenge. A tidal origin of resonance capture as may apply to Galilean satellites (Showman and Malhotra 1997) does not apply to the case of Titan and Hyperion. Given Titan's size and distance from Saturn, significant expansion of its orbit would require Saturn's dissipation parameter Q to be much lower than the lower limit set by the proximity of Mimas (Lee and Peale 2000). Work by Weidenschilling and Davis (1985) showed that the combination of gas drag and perturbations due to mean-motion resonances (MMR's) with a planet can have important consequences for the orbital evolution of small planetesimals. Kary et al. (1993) obtained a critical value for the drag parameter at which resonance trapping by a protoplanet breaks down. Malhotra (1993) points out that resonance trapping is vulnerable to planetesimal interactions. This issue also arises in the satellite case, as the capture of Hyperion in a 4:3 MMR with Titan illustrates.

1.2.2 Callisto's Internal State

A challenging result is that the Galileo mission moment of inertia data are consistent with a fully differentiated Ganymede, but only a partially differentiated Callisto (Anderson et al. 2001). While it is possible that non-hydrostatic effects in Callisto's core could be large enough to allow for complete differentiation of Callisto (see e.g. McKinnon 1997; Stevenson et al. 2003), its internal state may be a result of its accretion history. In this view, Callisto forms by accreting the volatile-rich condensables present in the extended, low density outer disk; its long formation timescale is tied to the disk clearing time, which is the time it takes for gas drag to clear the circumplanetary disk of solids (Mosqueira et al. 2001; Mosqueira and Estrada 2003a; Alibert et al. 2005). That is, accretion in the sparse, extended outer disk takes a sufficiently long time that the heat of accretion can escape, and the satellite may not differentiate fully. Callisto could then be said to be the result of slowly assembling thousands of volatile-rich, "cold" embryos. Thus, the issue of Callisto's state of differentiation is linked to the observation of the empty space outside Callisto and inside the irregulars. A related question can be posed for Saturn: namely, why is Iapetus stranded far from the planet? In each case, it is natural to expect that gas-drag clearing of swaths of the circumplanetary disk is involved in the answer.[6] Iapetus' separation from Titan provides strong indirect support for a two-component subnebula: a dense inner region roughly out to the centrifugal radius near Titan, and a lower surface density tail extending out to as far as irregular satellite Phoebe. Similarly, Callisto derives its mass from a more extended region than does Ganymede, which can explain its partially differentiated state without resorting to fine-tuning poorly known parameters; that is, in a two-component subnebula it is natural to expect both that (a) some satellites form in the outer disk and others in the inner disk, and (b) that satellites forming in the outer disk take significantly longer to accrete than those forming in the inner disk.

The issue arises whether typical impactors would bury heat in Callisto so deeply that it would lead to full differentiation of this satellite (McKinnon 2006). Mosqueira and Estrada (2003a) estimate that outer disk satellitesimals could grow as large as several hundred kilometers as they migrate inwards due to gas drag. These sizes are computed by assuming that all collisions among satellitesimals are accretionary. However, as satellitesimals approach a satellite's feeding zone they are excited to velocity dispersions well in excess of their escape speeds, which is likely to result in non-accretionary, disruptive collisions (as Hyperion's irregular shape may attest to). Thus, it is reasonable to expect that Hyperion-like or smaller satellitesimals are typical impactors during the late-stage accretion of both Callisto and Titan.

[6]Note that the capture of Hyperion into resonance also involves gas-drag inward migration.

1.2.3 Compositional Constraints

Regular satellites may provide a probe of the compositional and thermal state in the subnebula at the tail end of giant planet formation. Early models propose the formation of satellites in the circumplanetary nebulae of Jupiter and Saturn (Pollack and Reynolds 1974; Pollack et al. 1976). These models envision a condensation sequence analogous to that of the nebula. Still, the physical conditions in the circumplanetary disks are thought to be sufficiently different from those of the nebula that a sharp contrast has been drawn between objects forming in the outer nebula and giant planet subnebulae (Prinn and Fegley 1981). Yet, such ideas retain a phenomenological character, as the nature of the interaction between the two environments remains poorly understood. The compositional gradient of the Galilean satellites may provide a link to the environment in which they form (see Estrada et al. 2009 and references therein). The similarities in the bulk properties of the regular satellites of Jupiter and Saturn strongly favour a unified framework for their origin; yet, the inner, icy satellites of Saturn exhibit no clear compositional trend (and may have been collisionally disrupted).[7] In fact, recent Cassini results make a strong case that the medium-sized Saturnian satellites have densities (Jacobson et al. 2006) that preclude a solar composition ice/rock ratio (e.g., Johnson and Lunine 2005 and references therein; Mosqueira and Estrada 2005). On the other hand, a persistently hot subnebula can prevent the condensation of volatiles close to Jupiter (Pollack and Reynolds 1974; Lunine and Stevenson 1982). Thus, the thermal conditions under which coagulation and growth take place may hold a key to the formation of the regular satellites.

Compositional Constraints of the Outer Regular Satellites Mosqueira et al. (2009) focus on the large, outer regular satellites of each satellite system: Ganymede and Callisto in the case of Jupiter, and Titan and Iapetus for Saturn. For objects the size of Iapetus or larger, the porosity is likely to be small not only because the internal pressure is large enough to close pore spaces, but also because the presence of short-lived radioactive nuclides heats the interior, causing ice to flow. For such large satellites, densities can be interpreted in terms of rock/ice fractions. Iapetus' low density, and correspondingly low rock/ice fraction, presents a puzzle when compared to the other three satellites, each of which is roughly 50% ice and rock. In turn, the rock/ice fractions for Ganymede, Titan and Callisto are comparable to that of (captured) Saturnian irregular satellite Phoebe (or even smaller if Phoebe were porous).

Tying the properties of solar nebula planetesimals to subnebula satellitesimals involves three distinct aspects: first, characterising the properties of the first generation of planetesimals in terms of sizes and degree of heterogeneity; second, quantifying the collisional evolution of the planetesimal swarm following giant planet formation; and third, delivering planetesimal fragments to the circumplanetary disk. Planetesimal break-up in tandem with delivery via ablation of planetesimal fragments crossing the subdisk provides a framework for understanding the mass budget and compositions of regular satellites (compared to that of solar nebula planetesimals). In particular, ablation can result in fractionation, and account for the observed density of Iapetus provided this satellite formed in situ (Mosqueira and Estrada 2005). For this to work planetesimals of size $\sim$10 km need to be partially differentiated, which indicates that the first generation of planetesimals in the Jupiter-Saturn region (and possibly beyond) incorporated significant quantities of ^{26}Al.

[7] Indeed, a number of factors, such as the stochastic compositional gradient of the inner Saturnian satellites, the majestic icy rings, and Titan's isolation and eccentricity, argue in favour of collisional processes (see Mosqueira and Estrada 2005 for an intriguing, though intricate, possibility).

Thermal ablation occurs because friction of the fast moving heliocentric interloper as it crosses the circumplanetary gas disk heats up the body. At low gas densities impactors lose mass and energy through ablation. For large kilometer-sized bolides mechanical destruction rather than thermal ablation may dictate the fate of the object (see Mosqueira et al. 2009 and references therein). Ablation of planetesimal fragments <1 km may deliver the bulk of the solids needed to form the satellites. For objects in that size range, the heat transfer coefficient $C_{\rm H} \sim 0.1$ may be reasonably obtained from observations of terrestrial meteorites (Bronshten 1983). The rate at which energy is transferred to the planetesimal is given by $E \sim c_{\rm H} \rho v^3$, where ρ is the gas density, v is the speed of the planetesimal through the gas, and some planetesimal flattening (which increases its cross section) takes place. One can then obtain an estimate of the surface temperature T_s by balancing this heating and the radiative cooling. Ablation thus results in delivery of material to the circumplanetary disk. Gas drag may also result in capture of material. To quantify these processes we first need to characterise the properties of the circumplanetary disk.

In the SEMM model one expects a factor of ~ 10 enhancement in solids over cosmic mixtures, resulting in a gas surface density of $\sim 10^4$ g cm^{-2}, which is consistent with and quantitatively constrained by the gap-opening condition for Ganymede and Titan in a inviscid disk with aspect ratio ~ 0.1 (Rafikov 2002; Mosqueira and Estrada 2003b), and the Type I migration of full-sized satellites in such a disk (Mosqueira and Estrada 2003b; Bate et al. 2003). Following envelope collapse, planetesimal ablation of meter-to-kilometer disk crossers (Mosqueira and Estrada 2005) and other mechanisms enhanced the concentration of solids in the satellite disk. If we use a gas surface density of 10^4–10^5 g cm^{-2} for the Jovian and Saturnian subnebulae, then a planetesimal of density ~ 1 g cm^{-2} will encounter a gas column equal to its mass if its radius is in the range 0.1–1 km. Following giant planet formation collisional fragmentation may replenish the 0.01–1 km size range, even if most of the mass was originally in objects >1 km. This means that a significant fraction of the mass in planetesimals crossing the gas disk may be ablated. Such a planetesimal may deposit a significant fraction of its mass in the gas disk.

In Fig. 1 we show (see Mosqueira et al. 2009 for further details) the mass delivery by ablation and gas-drag capture of planetesimal fragments crossing the SEMM circumplanetary disk. These results are obtained assuming a differentiated population of icy and rocky interloper fragments with a rock fraction of 30% by mass, and placing ~ 10 Earth masses in between Jupiter and Saturn. The bottom left panel shows that it is possible to deliver enough mass to account for the regular satellites of Jupiter and Saturn (only the Saturn results are shown). Furthermore, the ablation and the capture cross-section of meter-sized objects can lead to ice/rock fractionation and account for the composition of Iapetus, as well as those of Titan, Ganymede and Callisto (top right panel). This is the only explanation currently available for Iapetus' icy composition, and may also indicate that regular satellites overall may be depleted in rock with respect to solar composition mixtures, which might explain why their densities are lower than that of similar size Kuiper belt objects.

We stress that while full ice/rock separation is unlikely, only partial separation is needed to account for Iapetus' icy composition. Such a population can result from the fragmentation of a partially differentiated, ^{26}Al-heated first generation of ~ 10 km planetesimals (Mosqueira et al. 2009). Indeed, it is likely that most of the mass in the first generation of planetesimals resided in objects 10–100 km in the first 10^5–10^6 years, so that these objects may have incorporated significant amounts of ^{26}Al. If so, ~ 10–100 km planetesimals can differentiate (Prialnik and Podolak 1995) (Phoebe itself may be at least partially differentiated, see Johnson et al. 2009). A study by Merk and Prialnik (2006) that includes

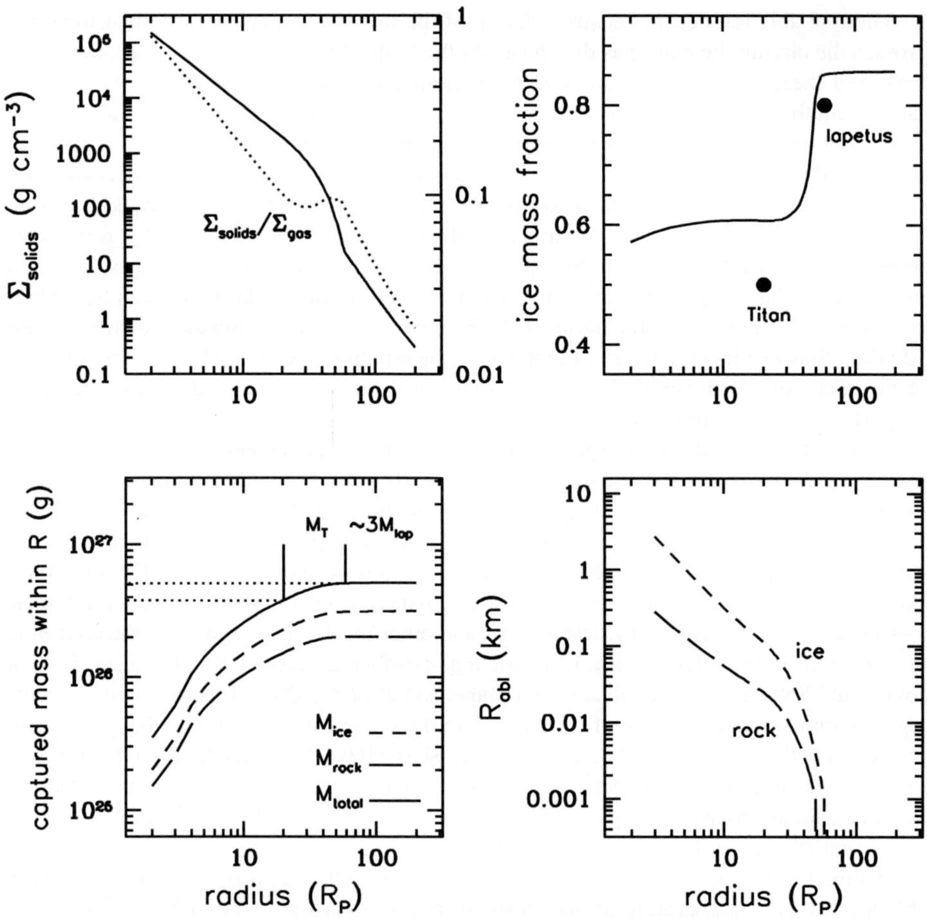

Fig. 1 *Left upper panel*: the solid-to-gas ratio (*dotted line*) and surface density of solids (*solid line*) as a function of radial distance from Saturn. *Right upper panel*: the ice to rock ratio as a function of radial distance. *The empty dots* correspond to Titan and Iapetus. The ice mass ratio of solar nebula planetesimal fragments crossing the disk is taken to be 0.3. *Left lower panel*: the mass delivered within a given radius from Saturn. *The lower right panel*: the planetesimal size that gets either captured or ablated as function of radial distance

radioactive heating, accretion, transformation of amorphous to crystalline ice and melting of water ice in the formation of trans-neptunian objects (which take longer to grow than planetesimals forming closer in) with sizes ∼10 km finds that the occurrence of liquid water may be common.[8] The disruption of partially differentiated planetesimals would then lead to a population of icy/rocky fragments available to ablate through the extended Kronian gas disk.

[8]But note that the presence of liquid water is not required for partial differentiation to take place. For instance, enhanced conductivity by water vapor flow through porous media may discourage melting, but would still result in a layered internal state as the core would become water-depleted (Prialnik et al. 2008).

2 Alien Moons: The Capture of Triton and the Irregular Satellites

The origin and the subsequent evolution of the irregular satellites have profound implications for our understanding of the satellite systems of the giant planets. In this section we discuss the origins of Triton and of the irregular satellites. A separate chapter in this volume (Schubert et al. 2010, this issue) extends the discussion to the secular evolution of irregular satellites, with a focus on Phoebe.

2.1 The Irregular Satellites

Since the time of their discovery, irregular satellites have been known to belong to a separate population of objects. While the regular satellites orbit their parent planets on inner, equatorial and almost circular orbits, the orbits of the irregular satellites are about an order of magnitude larger, and are characterised by high eccentricities and inclinations. Moreover, a significant fraction of irregular satellites move on retrograde orbits (i.e., with inclinations in the range [90°–180°]). These dynamical features are incompatible with formation in prograde circumplanetary disks, which strongly suggests that irregular satellites are captured objects that originated elsewhere in the Solar System.

Depending on the formation regions of the parent bodies, irregular satellites would be characterised by different compositions. Therefore, once captured, irregular satellites would introduce exogenous materials into their host systems. Moreover, both theoretical studies and observational results suggest that mass transfer between regular and irregular satellites takes place, introducing contaminants on the surfaces of the inner moons. The identification of the irregular satellite capture mechanism (or mechanisms), and of the formation regions of the parent bodies, are thus key to our understanding of such contamination processes.

2.1.1 Capture Mechanisms and Formation Regions

To permanently capture the parent bodies of the irregular satellites within the Hill radii of the giant planets, a fraction of their kinetic energy must be removed during passage. Historically, three dissipation mechanisms have been proposed to explain the capture of the irregular satellites: gas-drag (Pollack et al. 1979), *Pull-Down* (Heppenheimer and Porco 1977), and collisional capture (Colombo and Franklin 1971). The first two scenarios are based on the presence of nebular gas and thus place the capture of the irregular satellites at the time of the formation of the giant planets. In the Pollack et al. (1979) scenario gas is needed to slow the bodies crossing the Hill's spheres of the giant planets through friction. In the Pull-Down scenario by Heppenheimer and Porco (1977) gas is needed to rapidly increase the mass of the giant planets and expand their Hill's spheres, thus trapping the irregular satellites in their enlarged gravitational fields. Heppenheimer and Porco (1977) suggest that the gas-drag and Pull-Down mechanisms can be combined into a single scenario, where the interplay of the different processes can enhance the capture efficiency. We refer the reader to recent reviews (see e.g. Jewitt and Haghighipour 2007 and references therein) for more detailed discussions of the viability of gas-based scenarios; nevertheless, we point out that, while such scenarios can be applied to gas-rich Jupiter and Saturn, their applicability to Uranus and Neptune is uncertain. The collisional capture scenario by Colombo and Franklin (1971) has been somewhat neglected, yet it poses the least amount of constraints on the time of capture, has in principle the same capture efficiency when it comes to the gaseous and ice/rock giant planets, and may provide an explanation for collisional families among the irregular satellites of Jupiter. The Pull-Down scenario by Heppenheimer and Porco (1977) explains the existence of these families by allowing for the post-capture disruption of more massive parent bodies through collisional events as postulated by Colombo and Franklin (1971). The gas-drag model of Pollack et al. (1979) accounts for the existence of collisional families through the break-up of captured bodies due to friction. However, owing to the size-dependence of gas drag, one might expect smaller bodies on inner orbits and larger ones on outer orbits, which is not observed in the present satellite systems.

More recently, a new class of dynamical models has been proposed to explain the capture of the irregular satellites. These new scenarios fit within the so-called *Nice Model*, wherein it is argued that the structure of the outer Solar System is the result of a phase of chaotic rearrangement of the orbits of the giant planets moderated by the presence of a residual disk of planetesimals (for details we refer the reader to the original series of articles by Gomes et al. 2005; Morbidelli et al. 2005; Tsiganis et al. 2005). Since the chaotic rearrangement of the orbits of the giant planets would destroy any pre-existing system of irregular satellites (see Nesvorny et al. 2007, and references within), gas-based models would not be viable, with Jupiter as the only possible exception due to its limited participation in the orbital rearrangement. Proposed mechanisms to resupply the populations of irregular satellites rely on exchange reactions (i.e., the disruption of a binary system of planetesimals while crossing the Hill's sphere of a giant planet, see Agnor and Hamilton 2006; Vokrouhlicky et al. 2008), and three-body effects due to the mutual gravitational perturbations of the giant planets during close encounters in a Nice Model-like scenario (Nesvorny et al. 2007). In addition, collisional capture has been studied in more detail for Saturn's system in light of the observational data supplied by the Cassini–Huygens mission (Turrini et al. 2009).

A detailed discussion of capture mechanisms is beyond the scope of this chapter. In the framework of the processes shaping the moons of the giant planets, the most important influence of the irregular satellites is that of contaminating the surfaces of the inner satellites, introducing exogenous elements with potentially different compositional features. The nature of the contaminants delivered depends on the composition of the irregular satellites, which is linked to the formation regions of the parent bodies. The underlying assumption of early studies is that the source of the parent bodies of the irregular satellites is located near the orbital regions of the host planets. This would imply that one should expect a compositional gradient with heliocentric distance between the families of irregular satellites of the different planets, and that the overall composition of the irregular satellites should be compatible with that of the regular ones. To the contrary, the investigation of the collisional capture scenario performed by Turrini et al. (2009) shows that the possible orbits of the putative parent bodies of the irregular satellites span over a wide range of semimajor axes and eccentricity values, covering the whole outer Solar System. This means that irregular satellites orbiting a given planet could have formed in different regions of the Solar System, and thus have significant differences in their compositions. The same conclusion also holds true in the Nice Model, due to the orbital rearrangement suffered by the early Solar System.

2.1.2 Implications of the Capture Origins

While our data on the irregular satellites of Uranus and Neptune are still limited, both dynamical and observational evidence suggests that collisions between members of the populations of irregular satellites of Jupiter and Saturn played an important role in the past. The orbital structure of the Jovian irregular satellites bears a clear imprint of the existence of collisional families (Nesvorny et al. 2003, 2004). The orbital structure of the Kronian irregular satellites is more difficult to interpret, due to Jupiter's strong gravitational perturbations, yet there are clues of the possible existence of collisional families (Turrini et al.

2008). Moreover, there are strong dynamical indications that Phoebe underwent a period of intense collisional activity in the past with now-extinct irregular satellites located near its orbital region (Turrini et al. 2008), and such a hypothesis is supported by the observations of the instruments on-board Cassini of Phoebe's deeply cratered surface (Porco et al. 2005). The collisional activity of the irregular satellites would produce fragments, which could be ejected on orbits that intersect those of the regular satellites. Depending on their size, dust grains would be expelled into heliocentric orbits due to radiation pressure, or migrate inward due to Poynting–Robertson drag. Dust migrating inward would intersect the orbits of the regular satellites and, depending on their sweeping efficiency, reach different depths in the satellite systems. Not all dust grains are collected with the same efficiency: bigger grains migrate more slowly, enhancing the probability for the inner satellites to capture them. Also, dust particles in retrograde orbits would experience more close encounters with the satellites, and are more likely to be collected than grains in prograde orbits. Since the parent bodies of the irregular satellites originated in very different environments, this dust production/transfer leads to the contamination of the surfaces of the icy satellites with exogenous material characterised by a different chemical composition than the local material, and possibly coming from different sources. The origin of the dark material coating the leading hemisphere of Iapetus is probably the best example of this process. Still, whether the cause of the leading-trailing side dichotomy on Iapetus is endogenous, as opposed to exogenous, continues to be debated.

2.2 The Case of Triton

Another case in point is Triton, where the transfer process acts in the opposite way, i.e., the regular satellites are the source of contamination on Triton. Like other irregular satellites, Triton's retrograde orbit points to a capture origin (Goldreich et al. 1989; Agnor and Hamilton 2006). Yet, Triton's semimajor axis is over an order of magnitude smaller than that of the other irregular satellites, and Triton's size makes it one of the most massive moons in the Solar System. The small orbital distance of the satellite from Neptune suggests that its close-in, circular orbit is the result of tidal dissipation (Goldreich et al. 1989); it has been pointed out, however, that Triton's high inclination ($i \approx 157°$) implies a timescale of the order of the age of the Solar System for tidal dissipation to shrink the orbit of the satellite from probable post-capture values of the semimajor axis to its present location (Cuk and Gladman 2005). Triton's gravitational perturbations from its initial post-capture orbit would have caused pre-existing regular satellites to collide with each other and shatter. Taking the mass of the pre-existing satellites to be comparable to that of Triton, which these authors justify by analogy with the Uranian satellite system, the resulting debris disk could have shrunk Triton's orbit through scatterings and collisional debris-drag on a timescale of about 10^5 years, possibly allowing for the survival of the outermost regular and irregular satellites. Such a hypothesis might be in agreement with the results obtained by Schenk and Zahnle (2007) in reprocessing the Voyager's data on Triton, suggesting that the asymmetric distribution of craters on the surface of the satellite could be best explained by assuming a planetocentric source for the impactors, i.e., that the craters on Triton's leading hemisphere are due to impacts with a population of prograde small satellites or collisional shards. The paucity of craters on Triton could be an indication that the surface of the satellite is extremely young, with its age varying in the range 10^7–10^8 years, diagnostic of an active geology possibly linked to the tidal evolution of the satellite. This young surface age could indicate that the planetocentric impactors responsible for the craters could be the surviving members of the original debris disk. While the possibility of linking Triton's origin and evolution to its present state is intriguing, it should be noted that at the moment no quantitative study of the collisional lifetime of such a debris disk exists. Moreover, the data available on Triton and on Neptune's system in general are quite limited, making it difficult to derive better observational constrains for the theoretical models. Nevertheless, it is clear that one needs to take into account Triton's history and its exogenous origin to properly interpret the data currently available and those to be supplied by future space missions. While the Voyager's data indicate that Triton's surface is quite young and possibly uncontaminated by external material, the craters on the leading hemisphere could contain residual material from the impactors, perhaps more similar in composition to the present regular satellites of Neptune than to the uncontaminated surface of Triton. At any rate, Triton's capture disrupted Neptune's pre-existing satellite system, implying that care needs to be exercised when attempting to extract information about proto-Neptune's subnebula from the structure of the present satellite system of the giant planet.

3 Evolution: The Path to Present

As planetary formation sets the boundary conditions for satellite formation, so satellite formation sets the initial conditions for satellite evolution. Dynamical and geochemical evolution, in fact, secularly reshapes the satellites, erasing or modifying their primordial features. Therefore, in order to enhance our understanding of the present state of the moons of the outer Solar System, we need to be able to discriminate the effects of their secular evolution from those properties inherited from their time of formation. Nevertheless, thanks to the recent Cassini mission to Saturn and the efforts of those who made it possible, we currently have in hand a sufficient number of reliable observational constraints to give us justifiable hope that it is indeed possible to make significant progress in our understanding of the origin and evolution of the satellite systems of the giant planets.

References

I. Adachi, C. Hayashi, K. Nakazawa, Prog. Theor. Phys. **56**, 1756–1771 (1976)

C.B. Agnor, D.P. Hamilton, Nature **441**, 192 (2006)

Y. Alibert, O. Mousis, W. Benz, Astron. Astrophys. **439**, 1205–1213 (2005)

J.D. Anderson, R.A. Jacobsen, T.P. McElrath, W.B. Moore, G. Schubert, P.C. Thomas, Icarus **153**, 157–161 (2001)

S.K. Atreya, M.H. Wong, T.C. Owen, P.R. Mahaffy, H.B. Niemann, I. de Pater, P. Drossart, T. Encrenaz, Planet. Space Sci. **47**, 1243–1262 (1999)

B. Ayliffe, M.R. Bate, Mon. Not. R. Astron. Soc. **397**(2), 657–665 (2009)

M.R. Bate, S.H. Lubow, G.I. Ogilvie, K.A. Miller, Mon. Not. R. Astron. Soc. **341**, 213–229 (2003)

P. Bodenheimer, J.B. Pollack, Icarus **67**, 391–408 (1986)

V.A. Bronshten, Meteo. Iss. **8**, 38–50 (1983) (in Russian)

G. Bryden, X. Chen, D.N.C. Lin, R.P. Nelson, C.B. Papaloizou, Astrophys. J. **514**, 344–367 (1999)

R.M. Canup, W.R. Ward, Astron. J. **124**, 3404–3423 (2002)

R.M. Canup, W.R. Ward, in *Europa Book*, ed. by W. McKinnon, R. Pappalardo, K. Khurana (University of Arizona Press, Tucson, 2009, in press)

P. Cassen, D. Pettibone, Astrophys. J. **208**, 500–511 (1976)

S. Charnoz, A. Morbidelli, Icarus **166**, 141–156 (2003)

E.I. Chiang, M.K. Joung, M.J. Creech-Eakman, C. Qi, J.E. Kessler, G.A. Blake, E.F. van Dishoeck, Astrophys. J. **547**, 1077–1089 (2001)

G. Colombo, F.A. Franklin, Icarus **15**, 186–189 (1971)

A. Coradini, P. Cerroni, G. Magni, C. Federico, in *Origin and Evolution of Planetary and Satellite Atmospheres*, ed. by S.K. Atreya, J.B. Pollack, M.S. Matthews (University of Arizona Press, Tucson, 1989), pp. 723–762

M. Cuk, B.J. Gladman, Astrophys. J. **636**, L113–L116 (2005)

J.N. Cuzzi, S. Weidenschilling, in *Meteors and the Early Solar System II*, ed. by D.S. Lauretta, H.Y. McSween (University of Arizona Press, Tucson, 2006), pp. 353–381
G. D'Angelo, W. Kley, T. Henning, Astrophys. J. **586**, 540–561 (2003)
P.R. Estrada, I. Mosqueira, J.J. Lissauer, G. D'Angelo, D.P. Cruikshank, in *Europa Book*, ed. by W. McKinnon, R. Pappalardo, K. Khurana (University of Arizona Press, Tuscon, 2009, in press)
P. Goldreich, Y. Lithwick, R. Sari, Astrophys. J. **614**, 1024–1037 (2004)
P. Goldreich, N. Murray, Y. Longaretti, D. Banfield, Science **245**, 500–504 (1989)
R. Gomes, H.F. Levison, K. Tsiganis, A. Morbidelli, Nature **435**, 466–469 (2005)
J. Goodman, R.R. Rafikov, Astrophys. J. **552**, 793–802 (2001)
T.A. Heppenheimer, C. Porco, Icarus **30**, 385–401 (1977)
O. Hubickyj, P. Bodenheimer, J.J. Lissauer, Icarus **179**, 415–431 (2005)
R.A. Jacobson, P.G. Antreasian, J.J. Bordi, K.E. Criddle, R. Ionasescu, J.B. Jones, R.A. Mackenzie, M.C. Meek, D. Parcher, F.J. Pelletier, W.M. Owen, D.C. Roth, I.M. Roundhill, J.R. Stauch, Astron. J. **132**, 2520–2526 (2006)
D. Jewitt, N. Haghighipour, Annu. Rev. Astron. Astrophys. **45**, 261–295 (2007)
T.V. Johnson, J.I. Lunine, Nature **435**, 67–71 (2005)
T.V. Johnson, J.C. Castillo-Rogez, D.L. Matson, P.C. Thomas, Phoebes shape: possible constraints on internal structure and origin. 40th Lunar and Planetary Science Conference (2009)
D.M. Kary, J.J. Lissauer, Y. Greenzweig, Icarus **106**, 288–307 (1993)
M.H. Lee, S.J. Peale, 31st DDA meeting. Bull. Am. Astron. Soc. **32**, 860 (2000)
H. Li, S.H. Lubow, S. Li, D.N.C. Lin, Astrophys. J. **690**, L52–L55 (2009)
D.N.C. Lin, J. Papaloizou, in *Protostars and Planets III*, ed. by E.H. Levy, J.I. Lunine (University of Arizona Press, Tucson, 1993), pp. 749–836
J.I. Lunine, D.J. Stevenson, Icarus **52**, 14–39 (1982)
A.B. Makalkin, V.A. Dorofeeva, E.L. Ruskol, Sol. Syst. Res. **33**, 456 (1999)
R. Malhotra, Icarus **106**, 264–273 (1993)
W.B. McKinnon, Icarus **130**, 540–543 (1997)
W.B. McKinnon, Formation time of the Galilean satellites form Callisto's state of partial differentiation. 37th Lunar and Planetary Science Conference (2006)
R. Merk, D. Prialnik, Icarus **183**, 283–295 (2006)
A. Morbidelli, H.F. Levison, K. Tsiganis, R. Gomes, Nature **435**, 462–465 (2005)
I. Mosqueira, P.R. Estrada, Icarus **163**, 198–231 (2003a)
I. Mosqueira, P.R. Estrada, Icarus **163**, 232–255 (2003b)
I. Mosqueira, P.R. Estrada, On the origin of the Saturnian satellite system: did Iapetus form in-situ? 36th Lunar and Planetary Science Conference, League City, Texas, abstract no. 1951 (2005)
I. Mosqueira, P.R. Estrada, S. Charnoz, Decoding the origin of the regular satellites of gaseous giants—Iapetus: the Rosetta Ice-moon. Icarus (2009)
I. Mosqueira, P.R. Estrada, J.N. Cuzzi, S.W. Squyres, Circumjovian disk clearing after gap-opening and the formation of a partially differentiated Callisto. Proceedings of the 32nd LPSC, 12–16 March 2001, Houston, Texas, abstract no. 1989 (2001)
D. Nesvorny, J.L.A. Alvarellos, L. Dones, H. Levison, Astron. J. **126**, 398–429 (2003)
D. Nesvorny, C. Beaugé, L. Dones, Astron. J. **127**, 1768–1783 (2004)
D. Nesvorny, D. Vokrouhlicky, A. Morbidelli, Astron. J. **133**, 1962–1976 (2007)
E.C. Ostriker, Astrophys. J. **513**, 252–258 (1999)
J.B. Pollack, J.A. Burns, M.E. Tauber, Icarus **37**, 587–611 (1979)
J.B. Pollack, A.S. Grossman, R. Moore, H.C. Graboske, Icarus **29**, 35–48 (1976)
J.B. Pollack, M. Podolak, P. Bodenheimer, B. Christofferson, Icarus **67**, 409–443 (1986)
J.B. Pollack, O. Hubickyj, P. Bodenheimer, J.J. Lissauer, M. Podolak, Y. Greenzweig, Icarus **124**, 62–85 (1996)
J.B. Pollack, R.T. Reynolds, Icarus **21**, 248–253 (1974)
C.C. Porco et al., Science **307**, 1237–1242 (2005)
D. Prialnik, M. Podolak, Icarus **117**, 420–430 (1995)
D. Prialnik, G. Sarid, E.D. Rosenberg, R. Merk, Space Sci. Rev. **138**, 147–164 (2008)
R.G. Prinn, B. Fegley, Astrophys. J. **249**, 308–317 (1981)
R.R. Rafikov, Astrophys. J. **572**, 566–579 (2002)
V.S. Safronov, G.V. Pechernikova, E.L. Ruskol, A.V. Vitiazev, in *Satellites*, ed. by J.A. Burns, M.S. Matthews (University of Arizona Press, Tucson, 1986)
P.M. Schenk, K. Zahnle, Icarus **192**, 135–149 (2007)
G. Schubert et al., Space Sci. Rev. (2010, this issue)
A.P. Showman, R. Malhotra, Icarus **127**, 93–111 (1997)
S.A. Stern, P.R. Weissman, Nature **409**, 589–591 (2001)

D.J. Stevenson, W.B. McKinnon, R. Canup, G. Schubert, M. Zuber, The power of JIMO for determining Galilean satellite internal structure and origin. American Geophysical Union Fall Meeting, abstract P11C-05 (2003)
H. Tanaka, T. Takeuchi, W.R. Ward, Astrophys. J. **565**, 1257–1274 (2002)
P.C. Thomas, J.W. Armstrong, S.W. Asmar, J.A. Burns, T. Denk, B. Giese, P. Helfenstein, L. Iess, T.V. Johnson, A. McEwen, L. Nicolaisen, C. Porco, N. Rappaport, J. Richardson, L. Somenzi, P. Tortora, E.P. Turtle, J. Veverka, Nature **448**, 50–56 (2007)
K. Tsiganis, R. Gomes, A. Morbidelli, H.F. Levison, Nature **435**, 459–461 (2005)
D. Turrini, F. Marzari, H. Beust, Mon. Not. R. Astron. Soc. **391**, 1029–1051 (2008)
D. Turrini, F. Marzari, F. Tosi, Mon. Not. R. Astron. Soc. **392**, 455–474 (2009)
D. Vokrouhlicky, D. Nesvorny, H.F. Levison, Astron. J. **136**, 1463–1476 (2008)
S.J. Weidenschilling, D.R. Davis, Icarus **62**, 16–29 (1985)
G.W. Wetherill, G.R. Stewart, Icarus **106**, 190 (1993)
A.N. Youdin, F.H. Shu, Astrophys. J. **580**, 494–505 (2002)

Space Sci Rev (2010) 153: 447–484
DOI 10.1007/s11214-010-9635-1

Evolution of Icy Satellites

G. Schubert · H. Hussmann · V. Lainey · D.L. Matson · W.B. McKinnon · F. Sohl · C. Sotin · G. Tobie · D. Turrini · T. Van Hoolst

Received: 10 July 2009 / Accepted: 8 February 2010 / Published online: 8 May 2010

Abstract Evolutionary scenarios for the major satellites of Jupiter, Saturn, Neptune, and Pluto-Charon are discussed. In the Jovian system the challenge is to understand how the

G. Schubert (✉)
Department of Earth and Space Sciences and Institute of Geophysics and Planetary Physics, University of California, Los Angeles, CA 90095, USA
e-mail: schubert@ucla.edu

H. Hussmann · F. Sohl
German Aerospace Center (DLR), Institute of Planetary Research, 12489 Berlin, Germany

H. Hussmann
e-mail: Hauke.Hussmann@dlr.de

F. Sohl
e-mail: Frank.Sohl@dlr.de

V. Lainey
IMCCE-Observatoire de Paris, UMR 8028 du CNRS, 77 Avenue Denfert-Rochereau, 75014 Paris, France
e-mail: Valery.Lainey@imcce.fr

D.L. Matson
JPL 183-335, 4800 Oak Grove Drive, Pasadena, CA 91109, USA
e-mail: dennis.l.matson@jpl.nasa.gov

W.B. McKinnon
Department of Earth and Planetary Sciences and McDonnell Center for the Space Sciences, Washington University, One Brookings Drive, Saint Louis, MO 63130, USA
e-mail: mckinnon@wustl.edu

C. Sotin
JPL/Caltech, 4800 Oak Grove Drive, Pasadena, CA 91109, USA
e-mail: christophe.sotin@jpl.nasa.gov

G. Tobie
University of Nante, Nantes, France
e-mail: gabriel.tobie@univ-nantes.fr

present Laplace resonance of Io, Europa, and Ganymede was established and to determine whether the heat being radiated by Io is in balance with the present tidal dissipation in the moon. In the Saturnian system, Enceladus and Titan are the centers of attention. Tidal heating is the likely source of activity at the south pole of Enceladus, although the details of how the heating occurs are not understood. An evolutionary scenario based on accretion and internal differentiation is presented for Titan, whose present substantial orbital eccentricity is not associated with any dynamical resonance. The source and maintenance of methane in Titan's present atmosphere remain uncertain. Though most attention on the Saturnian moons focuses on Titan and Enceladus, the mid-size satellites Iapetus, Rhea, Tethys, and the irregular satellite Phoebe also draw our interest. An evolutionary scenario for Iapetus is presented in which spin down from an early rapidly rotating state is called upon to explain the satellite's present oblate shape. The prominent equatorial ridge on Iapetus is unexplained by the spin down scenario. A buckling instability provides another possible explanation for the oblateness and equatorial ridge of Iapetus. Rhea is the only medium-size Saturnian satellite for which there are gravity data at present. The interpretation of these data are uncertain, however, since it is not known if Rhea is in hydrostatic equilibrium. Pluto and Charon are representative of the icy dwarf planets of the Kuiper belt. Did they differentiate as they evolved, and do either of them have a subsurface liquid water ocean? New Horizons might provide some answers when it arrives at these bodies.

Keywords Outer planet moons · Io · Europa · Enceladus · Dione · Titan · Iapetus · Rhea · Tethys · Phoebe · Pluto · Charon · Satellite evolution

1 Introduction

The satellites of the outer planets provide important clues about the formation and evolution of the solar system as a whole. In this chapter we discuss the evolution of the major satellites of Jupiter, Saturn and Neptune. Pluto and Charon are also included since they represent the same population of objects as Neptune's captured moon Triton. The outer planet moons are striking in their diversity and evolutionary paths. They have experienced physical processes that are unfamiliar in the bodies of the inner solar system, and, accordingly, they comprise a fascinating collection of objects for study.

2 Satellites in Resonance: Strong Thermal/Orbital-Dynamical Coupling

Resonances can play an important role in the evolution of satellites. Orbital periods of satellites in resonance are commensurable and their mutual gravitational perturbations at conjunction (where perturbations are near maximum) occur periodically at the same orbital phase. Perturbations to the satellites' orbital evolution are therefore significantly stronger

D. Turrini
INAF-IFSI, Via del Fosso del Cavaliere 100, 00133 Rome, Italy
e-mail: diego.turrini@ifsi-roma.inaf.it

T. Van Hoolst
Royal Observatory of Belgium, Avenue Circulaire 3, Uccle, 1180 Brussels, Belgium
e-mail: tim.vanhoolst@oma.be

Fig. 1 Two examples of resonances in the outer solar system involving internal heating of satellites. Rotational energy of the primary planet and angular momentum are transferred to the innermost satellite due to tidal torques. Because of the resonance coupling, energy and angular momentum are distributed among the satellites locked in resonance. Part of the energy is dissipated as heat in the satellites' interiors due to tidal flexing. This affects mainly the inner satellites close to the primary, in the first case Io and, to a lesser extent, Europa, and in the second example Enceladus (sizes and distances not to scale)

compared with the non-resonant (stochastic) case (see Greenberg 1982 and Peale 1986 for a general description). The main implication for the evolution of satellites is that the orbital eccentricities are forced and maintained as long as a resonance is stable, i.e., on geological timescales if the coupling is strong. Because the global tidal heating rate (due to tidal interaction with the primary) depends on the eccentricity squared, long-term internal heat production is strongly linked to the occurrence of resonances. In some cases when the primary is close—for instance, Io and Jupiter—this type of tidal heating can significantly exceed the heat production due to radiogenic heating. In addition to the high eccentricities associated with stable resonant equilibrium configurations, orbital eccentricities can vary considerably when the satellites pass through resonances or when oscillatory states (strongly varying orbital and thermal states) occur. Oscillatory behavior results from the disequilibrium of eccentricity forcing in a resonance and involves eccentricity damping due to tidal dissipation in the satellite.

Mean motion resonances and tidal heating play an important role in the Jupiter system, mainly for Io and Europa, and in the Saturn system in the case of Enceladus (Fig. 1). Both examples are discussed below.

2.1 Io, Europa, Ganymede

2.1.1 The Laplace Resonance

The three inner Galilean satellites are locked in various resonances. For the thermal-orbital evolution the 2:1 Io–Europa mean-motion resonance and the 2:1 Europa–Ganymede mean-motion resonance are the most important ones. Conjunctions of Io and Europa are locked to Io's perijove (resonance angle librating about 0°) and to Europa's apojove (resonance angle librating about 180°). Conjunctions of Europa and Ganymede are locked to Europa's perijove (resonance angle librating about 0°) but to neither apsis of Ganymede (resonance angle circulating through 360°). The combination of the two 2:1 resonances yields the libration of the Laplace angle $l_1 - 3l_2 + 2l_3$ about 180°. The l_i $(i = 1, 2, 3)$ are the mean longitudes of Io, Europa, and Ganymede, respectively. This implies that whenever Europa and Ganymede are in conjunction, Io is on the opposite side of Jupiter. The Laplace configuration is stable and, after differentiating the mean longitudes with respect to time $\mathrm{d}l_i/\mathrm{d}t = n_i$, is usually expressed by

$$n_1 - 3n_2 + 2n_3 = 0 \tag{1}$$

where the n_i are the mean motions of Io, Europa and Ganymede, respectively. This three-body coupling is called the Laplace resonance, named after Pierre Simon de Laplace, who first demonstrated the stability of the orbital commensurabilities on theoretical grounds. Detailed reviews of the dynamics of the Galilean satellite system are given by Greenberg (1982) and Peale (1986).

The forced eccentricities associated with the above mentioned 2:1 resonances are 0.0041, 0.0101 and 0.0006 for Io, Europa and Ganymede, respectively. In the case of Europa the eccentricity is forced by both the 2:1 resonance with its inner neighbor Io and the 2:1 resonance with its outer neighbor Ganymede. Whereas the free eccentricities are negligible for Io and Europa (order of 10^{-5}), the free eccentricity of 0.0015 is the major contribution to the eccentricity of Ganymede. The free eccentricity is the remnant of the initial eccentricity after satellite formation (or after an unusual event, e.g., a major impact or a former resonance passage), which decreases with time due to tidal dissipation in the satellite. However, Showman and Malhotra (1997) have shown that an impactor capable of creating Ganymede's free eccentricity would have to have had a mass 10^2 to 10^3 times greater than the mass of the impactor that formed Gilgamesh, the largest impact basin on Ganymede. The free eccentricity is not associated with any resonance and can be regarded as the eccentricity that would persist if all the other satellites in the system were removed. Because of their forced eccentricities and their vicinity to Jupiter, Io and Europa are tidally heated on geological timescales up to the present.

2.1.2 Origin and Evolution of the Laplace Resonance

At present it is unclear how the three-body resonance has formed. There are two conceivable scenarios: (a) a primordial origin by migration of the newly formed satellites due to interactions with the circumjovian disk (Peale and Lee 2002) and (b) a subsequent formation of the resonances by differential expansion of the orbits due to tidal torques from Jupiter (Yoder 1979; Yoder and Peale 1981). The latter is the 'classical scenario' in which Io spirals outwards more rapidly than Europa because Io is closer to Jupiter and experiences larger tidal torques. After reaching the 2:1 mean-motion commensurability the two satellites get locked in resonance (probability ~1 for a wide range of initial conditions) and from then on they spiral outwards together (still due to torques exerted on Io by Jupiter) with the 2:1 ratio maintained. Upon catching up with Ganymede, the second 2:1 resonance is established, and eventually the Laplace resonance sets up (probability of 0.9). This scenario requires that dissipation in Jupiter—usually parameterized by the Jovian quality factor Q_J (Goldreich and Peale 1968)—must be efficient enough to exert significant torques on the moons. The amount of rotational energy dissipated in Jupiter, which determines the rate of energy transferred to the satellite system and the timescale of orbital evolution, is highly uncertain. Q_J must be of the order of 10^4 to 10^5 for this scenario to work. This roughly corresponds to the estimates of Q_J from orbital analysis in Lainey et al. (2009).

An alternative scenario, suggested by Malhotra (1991) and later re-examined by Showman and Malhotra (1997), involves passage of the three satellites through Laplace-like three-body resonances characterized by $(2n_2 - n_1)/(2n_3 - n_2) = 1/2$ (Malhotra 1991), 3/2 or 2 (Showman and Malhotra 1997). The corresponding ratio (see (1)) for the Laplace resonance is 1, which would then later be established by tidal torques. These scenarios are attractive because Ganymede would be significantly tidally heated during a high-eccentricity phase before the formation of the present Laplace resonance. This would yield a natural explanation for the extensive resurfacing on Ganymede involving tectonism and presumably high internal temperatures which obviously did not occur on Callisto. Constraints on Q_J in these models (a few times 10^5) are not as severe as compared to the scenario proposed by Yoder (1979) and Yoder and Peale (1981). However, in the numerical calculations the ratio of the tidal dissipation rate of Io and Jupiter Q_{Io}/Q_J has to be assumed constant including an 'ad hoc' step to a higher value to disturb the system from the primordial resonance and to allow for an evolution into the current Laplace configuration.

The primordial origin, originally proposed by Greenberg (1987), also does not place severe constraints on the Jovian quality factor. However, the model did not provide a mechanism for placing the satellites in the resonance configuration at the time of formation of the system. Peale and Lee (2002) showed how such a scenario could be established by taking into account interactions with the Jovian gas-disk based on the accretion model of Canup and Ward (2002). In this model. Ganymede's orbit converges on the orbits of Europa and Io by its faster type-I migration in the disk because of its greater mass. In this case the satellites are evolving out of resonance implying a decrease of forced eccentricities. However, later in the history, tidal torques from Jupiter become more important eventually leading to the current configuration.

2.1.3 Thermal States

Besides the orbital configuration, the most important observational constraint on the evolution of the satellites in resonance is Io's output of thermal energy. Heat flow measurements of Io yield values in the range of several watts per meter squared (Veeder et al. 1994, 2004; Matson et al. 2001; Marchis et al. 2005). If we assume a mean value of 2 Wm^{-2}, the corresponding total heat flow of about 8×10^{13} W significantly exceeds the present radiogenic heating rate (based on a chondritic composition) of the order of a few 10^{11} W. We can estimate the rheological parameters $\mathrm{Im}(k_2)$ (Segatz et al. 1988) or k_2/Q_{Io} (Peale 1986) for an equilibrium between tidal heat production and surface heat flux ($\mathrm{Im}(k_2)$ is the imaginary part of the complex second degree Love number of a satellite. In the expression k_2/Q_{Io}, k_2 is the purely elastic (real) second degree Love number and Q_{Io} is the dissipation factor of Io). The value of $\mathrm{Im}(k_2)$ obtained, of the order of 0.01, would be consistent only with high temperatures in Io's silicate shell. Figure 2 (see also Segatz et al. 1988 and Moore 2003) shows Io's heat production as a function of the temperature of the viscoelastic silicate mantle assuming

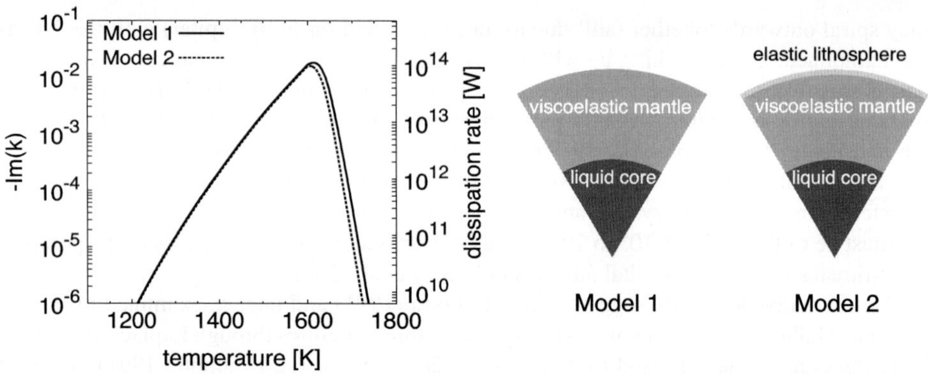

Fig. 2 Io's dissipation rate and $\mathrm{Im}(k_2)$ as a function of mantle temperature for the present orbital configuration assuming a viscoelastic Maxwell model for the tidal response of Io's silicate mantle. Maximum viscoelastic dissipation (consistent with Io's present heat flow) occurs if the Maxwell time, the ratio of the mantle viscosity to the rigidity, is near the period of the forcing, i.e., close to Io's orbital period. In the low-temperature range the response is elastic, in the mid-range viscoelastic and in the high-range fluid. The interior structure models are consistent with Io's moment of inertia. In model 2 an elastic shell is additionally taken into account which, however, does not affect the tidal heating rates significantly

a Maxwell rheology. According to this model Io's mantle temperature (if heat flow and heat production were in equilibrium) must be around 1600 K corresponding to near-maximum dissipation rates. Io is thus presently in a high-dissipation state with temperatures sufficient to produce partial melting in its silicate layer, as originally predicted by Peale et al. (1979) shortly before the Voyager encounter. Lainey et al. (2009) have recently shown that Io's tidal heating and emitted heat are in approximate balance.

There are no thermal constraints for the other two satellites. The heat flows of Europa and Ganymede are significantly smaller and have not yet been measured. In principle, values up to 10^{13} W could be reached in Europa's silicate mantle (Moore and Hussmann 2009) if a state similar to Io's high-dissipation state is assumed. Dissipation in the ice shell could reach maximum values of 10^{12} W. However, whether (and when) such high-dissipation states were obtained during Europa's evolution remains unclear. Thermal-orbital evolution scenarios including tidal heating are therefore constrained by the orbital parameters (e.g., mean motions and eccentricities) the resonance configuration (e.g., libration amplitudes) and the high-dissipation state (heat flux) of Io.

2.1.4 Coupled Evolution Scenarios

Depending on both the internal temperature and the orbital state, the tidal response of the ice and/or rock and the corresponding tidal heating rate is the link between the thermal and orbital parts of coupled evolution models. The temperature dependence of $\mathrm{Im}(k_2)$ or equivalently k_2/Q_{Sat} opens up the possibility of runaway heating (Peale et al. 1979), cooling or oscillations of thermal-orbital states (Greenberg 1982). Possible oscillatory thermal states of Io were first quantitatively investigated by Ojakangas and Stevenson (1986).

Feedback mechanisms between the orbital and thermal energy of the satellite can possibly lead to periodic variations in the surface heatflow of Io and in the orbital eccentricities of Io, Europa, and Ganymede, the latter due to the resonance locking (see Fig. 3). In this model k_2/Q_{Io} has a strong inverse temperature dependence, implying strong dissipation near the solidus of Io rock (hot state in Fig. 3). An equilibrium between the decrease in eccentricity

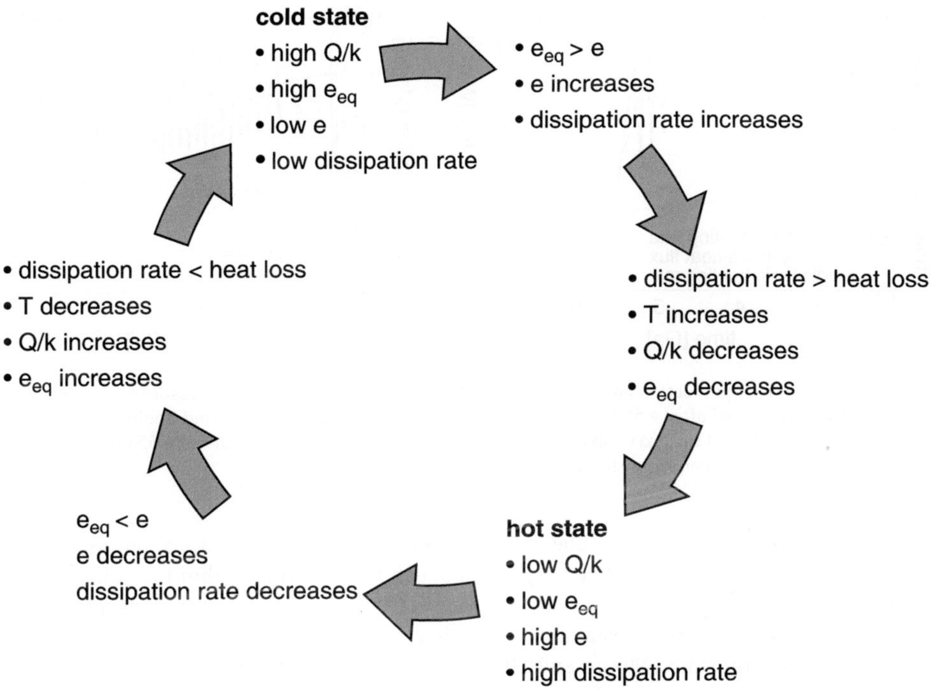

Fig. 3 Schematic view of episodic tidal heating. Io might oscillate between the two states at the upper left and the lower right. The latter implies partial melting and thus allows for volcanic activity. Io can run through this cycle several times. T is the satellite's internal temperature, Q/k is the ratio of the quality factor and the Love number of the satellite, e_{eq} is the eccentricity at which a decrease of eccentricity due to dissipation in the satellite compensates the increase in eccentricity due to torques exerted by the central planet and e is the actual eccentricity (after Ojakangas and Stevenson 1986)

due to dissipation in the satellite and an increase in eccentricity due to tidal torques exerted by the planet is not assumed. This implies that the satellites evolve deeper into resonance when Io runs from the cold to the hot state and evolve out of resonance when Io evolves from the hot state to the cold state (Fig. 3). Io could run through such a cycle several times, implying periodic occurrence of partial melt and probably episodic volcanism (super-solidus temperatures) with an active period on the order of 10 to 30 Myr separated by quiescent periods of about 60 to 200 Myr (Ojakangas and Stevenson 1986).

The model discussed so far was extended by Fischer and Spohn (1990), who identified three possible stages in Io's evolution (Fig. 4, left): (a) A stable phase where Io is in the hot equilibrium state with tidal heat production and surface heat flow being equal. In this phase the mean motion and the eccentricity decrease continuously. (b) An oscillatory phase, similar to the one described above, characterized by disequilibrium between tidal heat production and heat transport. (c) Runaway cooling at the end of the evolution where tidal dissipation plays a minor role and Io would be heated only by radiogenic elements. In this model, where the satellites evolve in resonance, with (1) being assumed throughout, Io might presently be in an equilibrium or an oscillatory state. In both cases it is possible to obtain the present orbital configuration along with a surface heat flux of a few watts per meter squared. The model was extended by Hussmann and Spohn (2004) to include Europa's thermal evolution. Figure 4 (right) shows an example where the heat production oscillates at present. In this

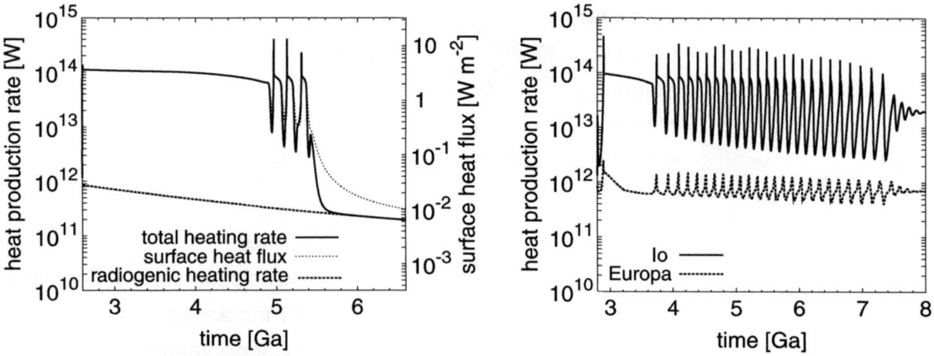

Fig. 4 *Left*: Tidal heat production and surface heat flux of Io evolving in the Laplace resonance. In this case the present state, obtained after 4.55 Gyr, is in the equilibrium phase (after Fischer and Spohn 1990). *Right*: Tidal heating rate of Io and Europa evolving in resonance. In this case the present state at 4.55 Gyr is obtained in the oscillation phase (taken from Hussmann and Spohn 2004)

model Europa is heated by 30% due to radiogenic heating from the deep interior and by 70% due to tidal heating in the ice shell. How much tidal heating within the silicate mantle contributes to Europa's energy budget at present is unclear. In principle, depending on mantle temperature, Europa's heating can reach values up to 10^{13} W in a state similar to Io's high-dissipation state (Moore and Hussmann 2009). Variations in heat production on Europa can have significant consequences for the thickness of its ice shell, which may thus have varied significantly during Europa's evolution (Hussmann and Spohn 2004). However, a consistent evolution relating the geological record of Europa's surface to variations in heat production and ice thickness has not yet been derived.

Because of its larger distance from Jupiter, tidal heating probably plays a minor role for Ganymede after capture into resonance. However, it may have significantly contributed to the thermal energy budget if the three satellites passed through a Laplace-like resonance before being locked in the present configuration (Showman et al. 1997). How this may have affected the generation of Ganymede's intrinsic magnetic field is still an open question.

2.2 Enceladus, Dione

Enceladus (radius 252 km) and Dione (radius 562 km), two icy satellites of Saturn, are locked in a stable 2:1 mean motion resonance (Hussmann et al. 2010, this issue). The corresponding forced eccentricities are 0.0044 and 0.0022 for Enceladus and Dione, respectively. Conjunctions are locked to Enceladus' perijove with a libration amplitude near 1° (Peale 1986). Capture into resonance from tidal torques exerted by Saturn is certain and it could be a viable scenario for the formation of the resonance. However, while Enceladus is moving away from Saturn it may have passed through resonances with other Saturnian satellites before the current resonance locking with Dione was established (Meyer and Wisdom 2008b).

Recent and even ongoing geologic activity on Enceladus, already suspected from Voyager imaging data ((Prockter et al. 2010), this issue; Smith et al. 1982), was spectacularly revealed by the Cassini mission. Enceladus' surface is very heterogeneous, displaying young, tectonically modified terrain, as well as old, heavily cratered surface areas. The youngest features, including the tiger stripes, roughly parallel lineaments about 500 m deep, 2 km wide, ~130 km in length and flanked by about 100 m-high ridges, are found in the south-pole region (Porco et al. 2006). An outstanding discovery of the Cassini mission was the detection of venting plumes mainly composed of water vapor and ice particles, emanating from the south polar region of Enceladus. The sources are highly correlated with the location of the tiger stripes. Furthermore, the vigorous activity near the south-pole is associated with strong thermal activity. Temperatures of about 90 K at Baghdad Sulcus—almost exactly located at the south-pole—exceed the expected equilibrium temperature of 60 K due to solar insolation at the poles (Spencer et al. 2006). The intrinsic power derived from these temperature anomalies is estimated to be $5.8 \pm 1.9 \times 10^9$ W (Spencer et al. 2006), an extremely large value for such a small icy satellite. It significantly exceeds the output of heat expected from radioactive decay of long-lived isotopes in Enceladus' silicate component, which would be only $\sim 0.3 \times 10^9$ W.

Tidal heating in Enceladus' ice and silicates, already suggested by Voyager images revealing Enceladus' strongly modified surface (Poirier et al. 1983; Squyres et al. 1983; Ross and Schubert 1989; Peale 2003), is a possible efficient energy source. Due to its proximity to Saturn and its forced eccentricity, Enceladus is subject to periodic tidal deformation which will lead to internal frictional heating. There are two major problems connected with this possibility:

1. Tidal forces act in a symmetric way. If a body is not heterogeneous with regard to its internal density, rheology and chemical composition, one would expect the same amount of heating at the north-pole and at the south-pole. However, there is no evidence for intense heating at the north-pole.
2. Under the same conditions (i.e., same rheology) the tidal heating rate in Mimas would be several orders of magnitude greater than in Enceladus (e.g., Squyres et al. 1983), simply because of Mimas' larger eccentricity of 0.02 and smaller distance to Saturn. Schubert et al. (2007) suggested that the higher rock content of about 50 wt.% in Enceladus compared to 20 wt% in Mimas and the resulting higher radiogenic heating rate in Enceladus, could have triggered tidal heating early in the satellite's history. However, continuous tidal heating over several Gyr might be difficult to maintain up to the present time. Therefore, the question remains: Why did Enceladus evolve along a hot branch while Mimas stayed geologically inactive and heavily cratered?

Whether the activity of Enceladus is recent, persisting, or episodic and how it is connected with the 2:1 mean motion resonance locking with Dione are still open questions. Meyer and Wisdom (2007) investigated a scenario in which the eccentricity damping due to tidal dissipation in Enceladus exactly cancels the increase of eccentricity due to torques exerted by Saturn ($de/dt = 0$). They found that the equilibrium state is not capable of providing sufficient heat to account for Enceladus' thermal activity. Equilibrium heating rates for the 2:1 Enceladus-Dione resonance can reach about 2.4 GW, still about a factor of 2 too small. Other low-order resonances (e.g., 3:2 Mimas- Enceladus) provide even less heat.

An alternative scenario is the oscillatory one shown in Fig. 3 for Io. However, stability analysis and numerical simulations both show that, in contrast to Io, oscillations about the equilibrium state are not obtained for Enceladus (Meyer and Wisdom 2008a). A better understanding of the effect of radial and lateral variations of the rheologic behavior inside Enceladus and better knowledge of the temperature and frequency dependence of the planetary material properties on tidal timescales is required to further investigate the oscillation scenario. The heterogeneity of both the surface features and the thermal activity furthermore suggest that a simple spherical model is insufficient. Alternative heat sources, e.g., exothermic chemical reactions (Matson et al. 2007), might be another way to achieve the high temperatures that may trigger subsequent tidal heating because of the lower viscosity involved with high temperatures.

To understand the mechanism that drives Enceladus' intense activity including the linkage between interior and surface and the connection to the thermal-orbital evolution is still a challenge. It involves investigations of Enceladus' interior structure, surface geology, composition and texture, in-situ measurements of plume composition and dynamics, and precise characterization of the satellite's orbital and rotational state. The extended Cassini mission may provide some answers to these open questions.

3 Titan

The composition of Titan's atmosphere and the low density of impact craters suggest that Titan has undergone geologically recent exchange between its interior and atmosphere (Prockter et al. 2010, this issue; Coustenis et al. 2010, this issue). This exchange would have reshaped Titan's surface, erasing previous structures such as impact craters. The presence of ^{40}Ar and the very low $^{36}Ar/N_2$ ratio suggest that argon formed by the decay of ^{40}K has been released in the atmosphere whereas primordial argon has escaped. In addition, nitrogen must have been released recently. The presence of methane, which has a lifetime of a few tens of million years in Titan's atmosphere (Yung et al. 1984) also suggests recent degassing from the interior (Tobie et al. 2010, this issue). Finally, the $^{15}N/^{14}N$ in N_2 and $^{13}C/^{12}C$ in CH_4 also suggest recent outgassing of these species.

The interior structure of Titan is not well constrained. Preliminary values of the degree 2 gravity coefficients suggest that Titan is at least partially differentiated (Rappaport et al. 2008; Sotin et al. 2009). Models coupling orbital characteristics and internal evolution suggest that an ammonia-rich ocean is present within Titan (Tobie et al. 2005; Bills and Nimmo 2008). However, no direct evidence of an internal ocean has yet been found by the Cassini spacecraft. Lorenz et al. (2008b) claimed that the Cassini radar data showed evidence of non-synchronous rotation and accordingly inferred that Titan's crust was decoupled from its deeper interior by a liquid water ocean. A reanalysis of the radar data has not supported the assertion of non-synchronous rotation (Stiles et al. 2010). During its descent through Titan's atmosphere, the Huygens probe made one measurement that is consistent with the presence of an ammonia-rich liquid water ocean below a 45 km-thick, icy crust. The probe recorded an electric field at 36 Hz that can be explained by the propagation of electric waves (Schumann resonance) between the ionosphere and an electrically conductive layer at a depth of 45 ± 15 km (Béghin et al. 2009).

These few data suggest that Titan has undergone at least some internal differentiation and interior-atmosphere exchange that extends until the present.

3.1 Accretion and Atmosphere Formation

During the accretion process, part of the impact energy is converted into heat through two primary processes: shock heating and ejecta blanket deposition (Senshu et al. 2002). The impact creates a shock wave that compresses the satellite beneath the impact site. Peak pressure is almost uniform in a quasi-spherical region below the impact site called the isobaric core (Senshu et al. 2002). Shock compression below the impact site increases the temperature. The temperature increase can be determined from the peak pressure within the isobaric core, which is directly related to the impact velocity, and from the intrinsic properties of the target materials (density, specific heat, thermal expansion, etc.) (Senshu et al. 2002; Monteux et al. 2007). Furthermore, material near the impact is excavated and ejected, forming the crater and the surrounding ejecta blanket. When the ejecta fall back onto the satellite's surface, kinetic energy is deposited in a heated ejecta blanket. Excavation and post-impact rebound of the crater transfer thermal energy from the vicinity of the impact to the

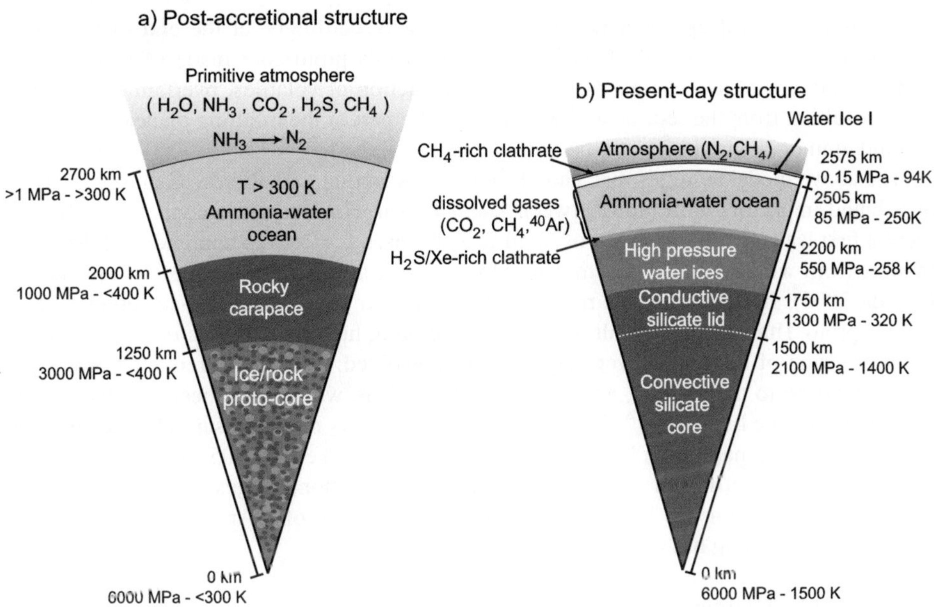

Fig. 5 Possible structure and composition of Titan after accretion (**a**) and at present (**b**). For the post-accretional interior model, the radii of the internal interfaces and the corresponding pressures and temperatures are approximate. The present structure is from the evolution model of Tobie et al. (2006), assuming an initial ammonia concentration of 5% in the primordial water ocean and a silicate mass fraction of 55% (Tobie et al. 2009)

surroundings. This reduces the temperature rise just beneath the impact, but it induces an increase of temperature at shallow depths around the impact, in particular in the ejecta deposit zone. Since the accretion timescale (10^4–10^6 years) is not long enough for most of the accretion energy to escape into space by radiation, the surface temperature increases.

Despite important progress in our understanding of impacts in the last decades, the amount of energy that can be retained within a growing satellite is still poorly constrained. Most studies of icy satellites use a simplified approach based on a factor, h, the fraction of accretional energy retained at depth and that progressively heats the surface of the growing satellite (Kaula 1979; Schubert et al. 1981; Lunine and Stevenson 1987; Mueller and McKinnon 1988; Grasset and Sotin 1996). Assuming that h is between 0.1 and 0.4 (Grasset and Sotin 1996), it can be deduced that melting and vaporization of surface materials occur when the growing proto-satellite reaches a radius of roughly 1000–1500 km (Lunine and Stevenson 1987; Kuramoto and Matsui 1994; Grasset and Sotin 1996). If contaminants such as ammonia are present, melting could occur at lower temperature and therefore at an even smaller proto-satellite radius.

As the proto-satellite grows larger than this threshold radius, infalling materials separate. The rocky part sediments to the base of a liquid layer formed from the melting of the ice phase. A proto-atmosphere, whose composition is determined by the composition of the infalling materials and the stability of the volatile-rich solid phases, is also generated. Once formed, the proto-atmosphere limits the radiation into space, resulting in a further increase of the surface temperature. Owing to the blanketing effect of the proto-atmosphere, the surface temperature can reach values higher than 300 K, and as high as 500 K (Kuramoto and Matsui 1994). This leads to the formation of a very massive and hot steam atmosphere in equilibrium with a deep water-rich ocean (Fig. 5). Accordingly, at the end of accretion, Titan's interior is comprised of an inner homogeneous proto-core made of a mixture of rocky and icy materials containing a significant fraction of volatiles, overlain by a silicate layer, resulting from the sedimentation of the silicate part of the infalling planetesimals, and a liquid ammonia-enriched water layer (Fig. 5).

The above model assumes that Titan formed by accreting a multitude of small uniformly distributed impactors with radii between 1 and 10 km. However, it is conceivable that accretion first leads to the formation of satellite embryos, a few tens to hundreds of kilometers in size, from which Titan is assembled (Mosqueira et al. 2010, this issue; Mosqueira and Estrada 2003). In this case, it is much more difficult to predict the post-accretional structure. If proto-Titan was the result of slowly assembling hundreds of volatile-rich, cold embryos, melting of the outer layer may have been avoided. The undifferentiated proto-core may extend up to the surface, and the proto-atmosphere would have been relatively tenuous. However, the high $^{15}N/^{14}N$ ratio and the low $^{36}Ar/N_2$ ratio measured by the Huygens GCMS (Gas Chromatograph Mass Spectrometer) (Niemann et al. 2005) implies that a massive ammonia-rich atmosphere was generated during accretion, suggesting that a significant portion of the interior has been melted, while the presence of ^{40}Ar suggests as well that substantial outgassing has occurred.

3.2 Internal Differentiation and Evolution

The bulk density of Titan (1882 $kg\,m^{-3}$ based on the mass from Jacobson et al. 2006 and the global shape of Zebker et al. 2009) indicates that the body is roughly 0.5–0.7 by mass silicate (for a silicate density between 3000 and 4000 $kg\,m^{-3}$), with the remainder mostly water ice. The proto-core was relatively cold and complete separation of ice and rock occurred only after decay of radiogenic isotopes contained in the silicates raised the temperature above the melting point of water ice. Melting induced a net volume change of the proto-core and a rupture of the overlying silicate layer. This leads to a rapid segregation of the different materials present in the interior (Lunine and Stevenson 1987). According to thermal evolution models, this core overturn would have occurred between 0.1 and 0.5 billion years after accretion (Kirk and Stevenson 1987; Lunine and Stevenson 1987).

The internal differentiation is mainly driven by the density contrast between the different materials composing the interior (silicate, ice, liquid water, hydrate, etc.). Silicate, which is the densest material, migrates toward the center forming a discrete rock core, and the H_2O materials migrate toward the outer regions, forming a very thick mantle mixture of ammonia-water liquid, water ices in different pressure-induced phases, and gas hydrates. The stabilities of the different phases and their densities determine the structure of the thick H_2O mantle. Depending on the pressure-temperature conditions, H_2O can be in the form of ice I (at $P < 207$ MPa), a high-pressure phase of ice (at $P > 207$ MPa) or a liquid. Accordingly, the cooling and solidification of the liquid water layer results in the formation of an ice I layer at the top of the ocean and a dense high-pressure ice layer at the bottom. Ammonia, which reduces the temperature of solidification of the ocean, does not enter the ice phase during solidification and therefore its concentration in the ocean increases as it solidifies (Grasset and Sotin 1996; Grasset and Pargamin 2005; Tobie et al. 2005; Mitri et al. 2008).

The solidification rate of the ammonia-water ocean is determined by a balance between the heat flow out of the silicate core and the heat flow transferred through the outer ice I layer. The energy source within the silicate core is mostly provided by the radioactive decay of long-lived isotopes of K, U, Th with tidal dissipation being negligible there (Sohl et al. 2003; Tobie et al. 2005). Models of thermal evolution indicate that the silicate core should be convective at present (Grasset et al. 2000; Sohl et al. 2003; Tobie et al. 2005, 2006). However, because of the strong temperature dependence of the silicate viscosity, the convective motions are confined below a thick rigid lid (estimated to be 250 km in Fig. 5b). Temperature increases quasi-linearly through the lid since heat is transferred by thermal diffusion, and then it almost follows an adiabat down to the center (Fig. 5b). If a sufficient amount of native iron is present, an iron core may eventually form about one billion years after differentiation (Grasset et al. 2000). In this case, the deep interior would be comprised of an iron core, probably liquid (not shown here, see Grasset et al. 2000), and a silicate convective mantle. The gravity measurements performed by the Radio Science System on board Cassini suggest that the deep interior is not fully differentiated (Sotin et al. 2009). However, the presence of an iron core cannot be definitely ruled out, and additional measurements are required to determine whether Titan possesses an iron core like its Jovian cousin Ganymede.

During the differentiation process, interactions between rock and liquid water might have promoted some chemical reactions (Sohl et al. 2010, this issue). The presence of a few percent of ammonia in solution with liquid water might have facilitated ionic exchanges between the liquid and the silicate minerals (Engel et al. 1994). In particular, NH_3 might have reacted with sulfate-rich brines leached from the silicate core during its hydration and might have led to the formation of a liquid layer of aqueous ammonium sulfate (Fortes et al. 2007). It is also possible that aqueous alteration of olivine-rich rocks might have produced a significant amount of H_2. This process, known as serpentinization is commonly observed on Earth in peridotites dredged from the seafloor and in ophiolites on Earth (Lowell and Rona 2002, and references therein). If carbon monoxide or carbon dioxide is present, then H_2 would reduce those gases to form methane (Atreya et al. 2006). However, it is uncertain how sulfur or ammonia present in the proto-core may have affected the thermochemical equilibrium of the serpentinization reactions. Additional laboratory experiments and theoretical models of Titan's interior are needed to determine the kinetics of reactions in the presence of NH_3, H_2S and SO_2 and under high pressure conditions and to better assess the possibility of methane production.

3.3 Primitive Crust Formation

After the accretion epoch, efficient clathration (incorporation of gases into cages in the lattice structure of ice) of gaseous species is expected to occur as the primitive atmosphere cooled down, for as long as the atmosphere and liquid water ocean were in contact; methane was preferentially sequestered in clathrate and nitrogen was preferentially retained in the atmosphere (Lunine et al. 1989, 2009). However, depending on the clathrate density, which is mainly determined by composition, clathrate grains accumulate at the surface or sink down to the bottom of the primordial ammonia-water ocean (Lunine et al. 2009; Tobie et al. 2009). Clathrates containing a large fraction of methane have a density lower than that of dilute liquid ammonia-water solutions, so as surface temperatures dropped below 250 K, clathrate accumulation and water freezing would make a solid crust, isolating the water ocean from the atmosphere and terminating clathrate formation (see also Lunine et al. 1989). Other volatile species thought to be present in the primitive atmosphere, such as ammonia or carbon dioxide, will also condense at the surface as the surface temperature drops below ~200 K, so the primitive crust is probably composed of a mixture of different ices (H_2O, CO_2), hydrates and gas clathrates (Choukroun et al. 2010).

Even though a fraction of the atmospheric methane could be incorporated in the primitive crust during the clathration and freezing processes, this fraction is small. In contrast,

Osegovic and Max (2005) showed that during the formation of clathrate from a gas mixture, preferred hydrate-forming materials can be completely consumed, and thus removed from the surrounding environment, as long as water molecules are present in sufficient amount, as was the case when the ocean was in direct contact with the proto-atmosphere. Before a solid crust formed, xenon, which is the preferred hydrate forming material, could have been completely removed from the primitive atmosphere, and would be stored in a heavy compound clathrate (mainly composed of H_2S and Xe) layer at the base of the ocean (Tobie et al. 2009).

3.4 Internal Reservoir of Volatiles

At the end of accretion, only the inner undifferentiated portion of Titan's interior was able to hold volatiles in significant amounts. Most of the region outward of this proto-core was probably warm liquid water ($T \geq 300$ K) in which many gaseous compounds have very low solubility. Large amounts of gaseous compounds, notably methane, could therefore have ended up in the primitive atmosphere and on the surface (Tobie et al. 2009). By contrast, ammonia has a high solubility in water, so that most of the available ammonia remains in the liquid phase and only a relatively small fraction is extracted from the liquid phase and converted into nitrogen. Most of the methane initially present in the building blocks ended up in the primitive atmosphere and was lost shortly after accretion owing to the combined effect of strong atmospheric escape and enhanced solar UV photolysis.

After the early epoch, most of the remaining methane is stored in the undifferentiated proto-core. For an undifferentiated proto-core representing 15 to 25% of Titan's total mass ($R_{\text{proto-core}} = 1400$–$1600$ km), a mass fraction of methane relative to water in Titan's building blocks equal to 1%, and a silicate mass fraction of 50%, the total available mass of methane in Titan's interior would range from 1 to 1.7×10^{20} kg, i.e., 360 to 610 times the present mass of atmospheric methane (estimated to be 2.8×10^{17} kg, Niemann et al. 2005). Methane in the cold proto-core is stable in the form of clathrate hydrate (Loveday et al. 2001), but it would be released after proto-core overturn (0.5 Gyr after accretion Lunine and Stevenson 1987).

The density of clathrate hydrates depends strongly on the gas molecules trapped in their cages. As an example, clathrates of pure methane have a density of 920 kg m^{-3} at ambient pressure, whereas clathrates of carbon dioxide have a density of 1130 kg m^{-3}. Therefore, segregation of clathrates with different composition is expected during differentiation. Only clathrate hydrates whose composition is dominated by methane have a smaller density than liquid water and would be able to rise through the thick primordial ammonia-water layer during differentiation. At core overturn, the huge amount of methane clathrate released is expected to accumulate at the top of the ocean just below the primordial crust (Tobie et al. 2006).

3.5 Long-Term Evolution Models of Titan's Interior

Subsequent to core overturn, methane outgassing would occur only when the conditions required to dissociate methane-rich clathrate were reached within the icy mantle of Titan's interior. Because of the low thermal conductivity and high viscosity of methane clathrate as compared with water ice, incorporation of methane clathrate in Titan's crust is expected to strongly influence the cooling rate of Titan's interior. The subsequent thermal evolution of Titan's interior can be constrained by applying the coupled orbital-thermal model of Tobie et al. (2005, 2006). This model includes heat transfer through the outer layer, clathrate dissociation, crystallization of the liquid layer, thermal evolution of the silicate core, and tidal

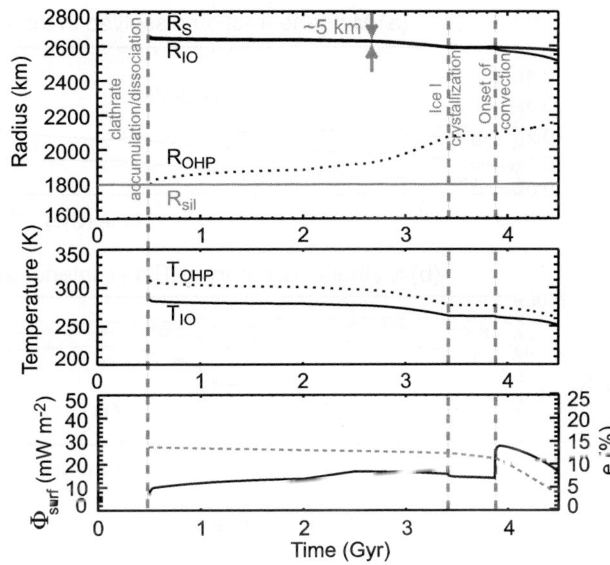

Fig. 6 Surface radius R_S (*solid line*), and upper (*solid line*) and lower (*dotted line*) ocean interface radii, R_{IO}(IO ice-ocean) and R_{OHP} (OHP ocean high pressure ice), corresponding interface temperature, T_{IO} and T_{OHP}, surface heat flux Φ_{surf} (*solid line*) and eccentricity *e* (*dotted line*), as functions of time. R_{sil} is the radius of the silicate core. After Tobie et al. (2009)

dissipation and its effect on orbital eccentricity decay. Titan has an orbital eccentricity of 0.03 that cannot be explained by orbital resonance. Titan's current free eccentricity is probably the fossil of a higher primordial eccentricity possibly ranging between 0.1 and 0.2 (Tobie et al. 2005). Figure 6 illustrates a typical simulation with initial eccentricity 0.135, mass fraction of NH_3 relative to H_2O equal to 5%, and silicate mass fraction of 55%. The simulation shows that after differentiation the internal ammonia-water layer cools very slowly owing to the insulating effect of the overlying clathrate-rich layer. This results in a slow crystallization of the water layer from the bottom, leading to a slow thickening of the high-pressure layer. Since the outer layer remains relatively thin during the first 3.5 billion years, tidal dissipation remains small and the eccentricity decays only slowly. After 3.5 Gyr, the ocean starts to crystallize from the top (Fig. 6), forming an ice I layer below the clathrate-rich crust. When the ice I layer reaches a critical thickness of 15 km at approximately 3.9 Gyr, thermal convective instabilities initiate in the ice layer. Since thermal convection is more efficient than thermal diffusion in transporting internal heat, crystallization of the ocean accelerates and the convective ice shell rapidly thickens. In addition, since the total energy dissipated by tidal friction is proportional to the thickness of the convective ice I layer (Tobie et al. 2005), tidal dissipation increases as the ice shell thickens and the orbital eccentricity rapidly decays to its present value. The surface heat flow strongly increases when thermal convection starts owing to the combined effect of enhanced tidal dissipation and ocean solidification. This releases the energies of accretion and differentiation stored in the form of latent heat in the internal water ocean. The presence of CH_4-rich clathrates in the crust delays the crystallization of the primordial ocean and reduces the dissipation rate during the first 3 Gyr. These effects allow the maintenance of a large eccentricity over approximately 4 Gyr and the triggering of thermal activity in the outer layer very late in Titan's history. As a result, the subsurface clathrate reservoir can be destabilized several billion years after accretion.

3.6 Implications for the Methane Cycle

Differentiation and subsequent evolution of Titan's interior should lead to the accumulation of CH_4-rich clathrates in Titan's crust and to their subsequent dissociation during three main

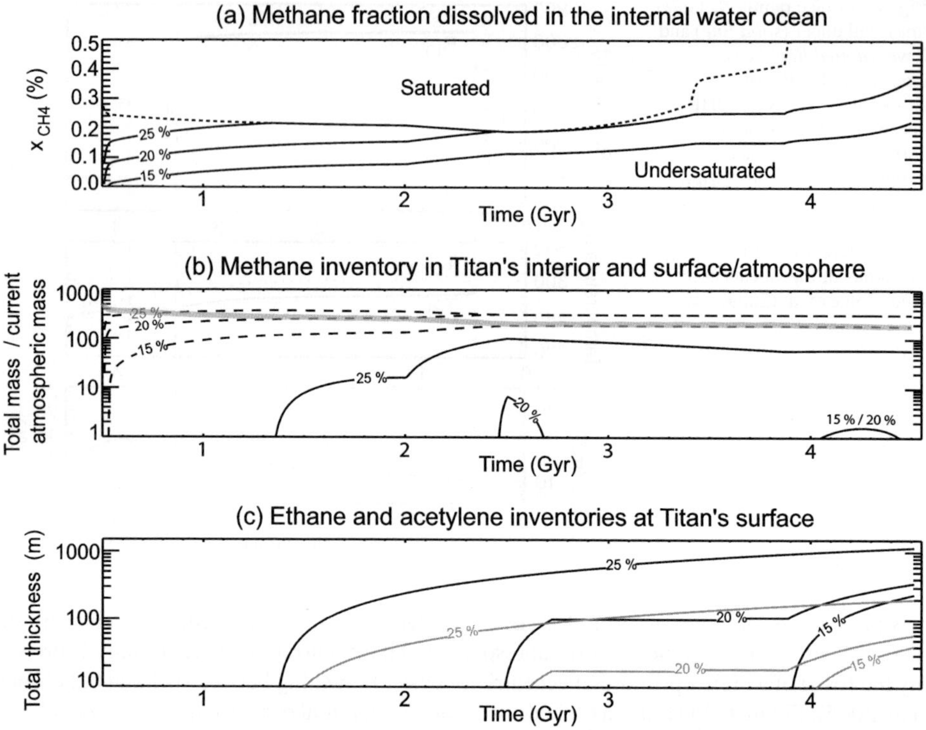

Fig. 7 **a** Evolution of methane content in the internal ocean for three different initial mass fractions of the proto-core (15, 20 and 25%) corresponding to the simulation of Fig. 6. The *dotted curve* indicates the maximum content of methane that can be dissolved in the ocean as a function of time. **b** Mass of methane, normalized to the current atmospheric mass, *dashed line*, contained in the ocean, *grey line*, in the crust and in the atmosphere/surface (*solid*) as a function of time for different mass fractions of the proto-core. **c** Total thickness of ethane (*black*, $\rho_{C_2H_6} = 550\ \mathrm{kg\,m^{-3}}$) and acetylene (*grey*, $\rho_{C_2H_2} = 745\ \mathrm{kg\,m^{-3}}$) produced by CH_4-based photochemistry, accumulated at the surface as a function of time, for different mass fractions of the proto-core and assuming the conversion rate of Toublanc et al. (1995). After Tobie et al. (2009)

epochs: (i) during and just after differentiation between 0.5 and 1 Gyr, (ii) at the onset of convection in the silicate core between 2 and 2.5 Gyr and (iii) when thermal instabilities in the crystallizing outer ice layer occur between 4 and 4.5 Gyr. However, the two first episodes have to be distinguished from the last one. Before 2.5 Gyr, the dissociation of methane clathrate occurs at the crust-ocean interface, whereas after 3.9 Gyr the dissociation is induced by upwellings of warm ice at the base of the remaining crustal reservoir (Tobie et al. 2006). Figure 7a shows the methane content in the internal water ocean as a function of time, for different mass fractions of the undifferentiated proto-core relative to the total mass of Titan (15, 20 and 25%). The accumulation of free gas at the crust-ocean interface and its subsequent outgassing to the surface are possible only once the methane content in the liquid phase exceeds its saturation solubility (Tobie et al. 2009).

According to the model of Tobie et al. (2006), during the early stage, most of the methane is stored as clathrate hydrate in the crust (Fig. 7b) and the decomposition of clathrate progressively enriches the water ocean in methane. For a 25% mass fraction of the proto-core, the ocean gets saturated in gaseous methane at 1.5 Gyr and all the insoluble gaseous methane reaches the surface and rapidly replenishes the atmosphere with methane.

Even if one uses an upper limit for methane loss rate taken from Toublanc et al. (1995) ($dM_{loss}/dt = 288\ kg\,s^{-1}$), the very high outgassing rate leads to methane accumulation on the surface. At 2.5 Gyr, the total mass of methane on the surface is 100 times larger than the present atmospheric mass. At 4.55 Gyr, a huge mass of methane (60 $\times$ observed mass) should still be present on Titan's surface, inconsistent with the present inventory of methane (Lorenz et al. 2008a).

By contrast, for a proto-core mass fraction of 15%, the water ocean never gets saturated, and methane outgassing occurs only during the last epoch between 4 and 4.5 Gyr. Tobie et al. (2006) assume that the outgassing rate during the last episode is proportional to the surface heat flow, leading to the decomposition of 10 per cent of the available crustal clathrate reservoir. For a proto-core mass fraction of 20 per cent, in addition to the late outgassing period, a short outgassing event occurs at approximately 2.5 Gyr, and the atmospheric methane totally vanishes after 200 Myr. In this model, when the last outgassing episode starts, the clathrate-rich crust still contains approximately 200 times the current mass of atmospheric methane, largely sufficient to replenish the atmosphere in methane. However, the outgassing rate strongly depends on the vigor of thermal convection in the underlying ice shell and in the ability of thermal plumes to penetrate the overlying clathrate-rich crust. Efficient dissociation could occur only if upwelling icy plumes with temperatures of 200–250 K penetrate the upper crust to relatively shallow depths (1–2 km below the surface). Recently, Choukroun et al. (2010) showed that the presence of ammonia hydrate in the crust could facilitate the dissociation of methane clathrate and the formation and rise of low viscosity cryomagmas enriched in methane. Even if ammonia hydrates can help, thermal convective instabilities in the ice shell are still required to trigger the decomposition process. If a 5 km diameter plume triggers a cryovolcanic event that dissociates all the overlaying clathrate-rich crust, the maximum amount of methane outgassed is 10^{12} kg, which is less than 10^{-5} of the total atmospheric CH_4 mass. Such a small amount would be photochemically destroyed within 100–1000 years, requiring the occurrence of similar cryovolcanic events at least every thousand years to sustain a few percent of methane in the atmosphere. If such processes are really operating on Titan, future mapping of the surface should reveal more and more cryovolcanic features.

4 Iapetus, Rhea, Tethys, Phoebe

4.1 Iapetus

Iapetus is the most distant of the regular satellites from Saturn. It is renowned for its strange 'black and white' surface markings that were discovered by Giovanni Domenico Cassini in 1671 when it was on the western side of Saturn. He was not able to see it on the eastern side until 1705 when he had a better telescope. Cassini correctly surmised that the satellite had a bright hemisphere and a dark hemisphere, and that it was tidally locked, always keeping the same face towards Saturn. The surface markings were seen in detail by Cassini-Huygens which also measured the shape of Iapetus.

Today, Iapetus is believed to be cold and inactive. However, in its past there are two types of activities that may have involved interactions between the surface and the interior. Possible venting of volatile gases is one such process and the formation of the equatorial ridge is the other. Both of these occurred a long time ago. So, in order to investigate them we must look at Iapetus' evolutionary history.

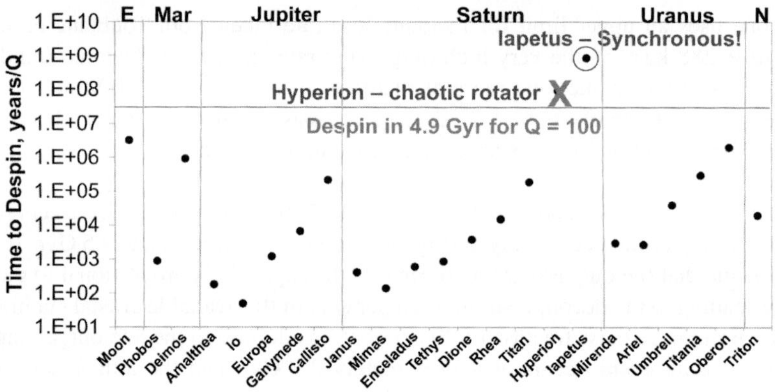

Fig. 8 Despinning times for the satellites of the solar system (after Peale 1977)

4.1.1 Iapetus' Evolution

The geophysical evolution of this satellite involves two related themes: dynamical (i.e., despinning) and thermophysical.

The fact that Iapetus is synchronously rotating with its orbital period is surprising. In his 1977 review, Peale (1977) noted that despinning is a rapid event for most satellites. The despinning rate (in rad/s) as a function of time t is given by

$$\frac{\mathrm{d}\omega}{\mathrm{d}t} = -\frac{[3k_2(t)GM_p^2R_{\mathrm{eq}}^5(t)]}{[C(t)D^6(t)Q(t)]} \tag{2}$$

where M_p is Saturn's mass, R_{eq} is the equatorial radius of the satellite, C is the polar moment of inertia of the satellite, D is the semi-major axis of the orbit, Q is the quality factor of the satellite, and k_2 is the Love number of the satellite. The rate of despinning is inversely proportional to the sixth power of the distance between the planet and the satellite. Iapetus is some 3.6×10^6 km from Saturn, much farther than any other regular satellite is from its planet. The despinning times of all the satellites in the solar system are summarized in Fig. 8 (Peale 1977). They are the times required for tidal dissipation to reduce the satellites' spin from an initial period of 2.3 h to their present orbital periods. The horizontal line, for a satellite with $Q = 100$, corresponds to a despinning time equal to the age of the solar system. While most of the satellites despin rapidly, Iapetus, mainly because of its large distance from Saturn, requires longer than the age of the solar system to despin to synchronous rotation (Fig. 8). Nevertheless, Iapetus has despun, a fact that we will be able to understand from evolutionary considerations not accounted for in Peale's (1977) despin time estimates.

Data returned by the Cassini-Huygens mission have reinvigorated the study of Iapetus' evolution. The data have shown that the bulk shape of Iapetus is that of a hydrostatic body with a 16-hour rotation period. Since Iapetus' synchronous rotation period is 79.33 days, this suggests an earlier, faster-spinning Iapetus and subsequent despinning.

Just how fast was Iapetus spinning when it finished accreting? Unfortunately, the present spin rates of other satellites cannot be used for guidance. They also have undergone despinning, so that their present rotation rates are not representative of the spin they had when they were formed. Objects not in orbit about a planet, for example, the asteroids and the transneptunian objects, have rotation rates that may be more representative of the initial rate for Iapetus. The distribution of asteroid rotation periods (Dermott and Murray 1982)

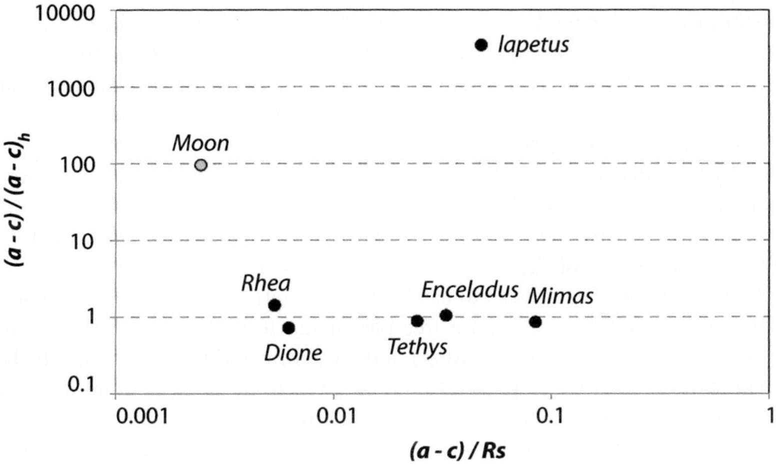

Fig. 9 Satellite shapes (Thomas et al. 2007b) compared with the shapes of uniform-density bodies in hydrostatic equilibrium. The ordinate is the difference between the equatorial axis (a) and the polar axis (c) scaled by the corresponding difference for a hydrostatic, uniform-density body. The abscissa is $a - c$ scaled by the mean radius of the satellite. The Moon is shown for reference because, until the Iapetus data became available, it was the extreme example

suggests that periods ranging from 5 to 10 hours should be reasonable estimates. As long as the body is not spinning much faster, a Maclaurin spheroid can be used to approximate the theoretical shape as a function of spin period for a uniform density, hydrostatic body. The oblateness can be computed using Chandrasekhar's (1969) formulation. As long as a body remains hydrostatic, its shape will change with time as it despins. Castillo-Rogez et al. (2007) suggested that this was the evolution for Iapetus, except that over time it became less hydrostatic as it lost heat and its interior cooled. By the time it had slowed down to a spin period of about 16 hours its lithosphere had become strong enough to maintain this nonhydrostatic figure. Thus, the formation and long-term preservation of the 16-hour figure strongly constrains models for Iapetus' geophysical evolution.

Satellites with radii greater than several hundred kilometers have relatively high internal pressures and tend to be nearly spherical in shape. Shapes can be taken as an indication of radial mass distribution, provided that the bodies are hydrostatic with respect to the tidal and rotational forces that act on them (see Schubert et al. 2004). Cassini-Huygens images indicate that the shapes of most satellites (Thomas et al. 2007b) deviate by less than a few kilometers from those of uniform density bodies in hydrostatic equilibrium. The shape data are compared with hydrostatic, uniform density shapes in Fig. 9. The comparison suggests the special nature of Iapetus.

The obvious trend in Fig. 9 is for satellites to cluster about an $(a - c)/(a - c)_{\text{hydrostatic}}$ value of 1, meaning that they are either in hydrostatic equilibrium or close to it. The outliers are Iapetus and the Moon which used to be the most non-hydrostatic satellite known. Iapetus supports a 33 km shape anomaly. This is huge compared to a 10 m bulge expected for such a body in hydrostatic equilibrium with a rotation period of 79.33 days.

Iapetus made adjustments other than its shape as it despun from its initial rotation rate. Substantial amounts of mass were repositioned and during this process there were likely opportunities for the venting of gaseous volatiles otherwise trapped deep in the interior. In order to explore these consequences of despinning, it is necessary to model Iapetus' internal thermophysical evolution.

The main problem for models of Iapetus is the paucity of heat available to drive any evolution of the interior. The key and most difficult to meet constraint for models is the shape of Iapetus. The 16 hour hydrostatic figure means that at the time when the rotation rate had slowed down to this value, the mechanical lithosphere had just become strong enough to preserve that shape against any further hydrostatic response. This synchronizes the model for despinning by tidal dissipation with the model for the growing strength of the lithosphere. In addition, after this constraint has been met, the despinning model must deliver Iapetus to its current 79.33 day rotation period. In order to meet these constraints, Castillo-Rogez et al. (2007) found that all of the available sources of heat for the interior were insufficient and that it was necessary in addition to use heat from short-lived radioisotopes (SLRI). The amount of available SLRI heat is not a free parameter. It is constrained by the amount of rock (i.e., ordinary chondritic composition) in the satellite and the time when the body was formed. This time is usually referenced to the time when the calcium-aluminum inclusions (CAI, as seen in meteorites) were formed. The currently accepted absolute date for this event is 4567.2 ± 0.6 Myr (Amelin et al. 2002). Castillo-Rogez et al. (2007, 2009) found models with formation ages between 3.4 and 5.4 Myr after CAI that satisfied the 16-hour shape constraint. For formation ages less than this range the models were too hot and the lithosphere was too weak to retain its shape when the rotation rate was 16 hours (i.e., the models ended up with figures corresponding to longer rotation periods). For formation ages greater than this interval, the models were too cool and despinning did not go to completion, i.e., they were unable to match the present synchronous rotation.

As an example, we discuss one model with a formation age in the range of 3.4 to 5.4 Myr after CAIs. The formation age of this model is ~4 Myr after CAI; the model is shown in Fig. 10.

The color coding in frame (a) shows that most of Iapetus remained frozen for its entire history. The brownish orange color shows an episode of melting at the center that started at ~800 Myr and extended outward by only ~200 km. The change in tidal dissipation associated with this event marks the major episode of despinning as shown by the decrease in the radius at this time. The other period of significant radius decrease occurred earlier, between ~2 and 20 Myr due to collapsing porosity. The material at depth was warming above the temperature at which ice grains could glide at their points of contact and as a result void space was closed. Volatiles present in the voids will be forced out and might migrate to the surface where they will encounter colder temperatures and possibly become trapped in a permafrost layer or vent to space.

The despinning at ~800 Myr is a feature of this particular model. Depending upon the starting time and other model parameters, despinning can take place between ~200 and ~900 Myr. Despinning is a highly nonlinear process because it depends on the evolving dissipation properties of the interior. Once conditions are optimum for despinning, the process itself takes only a few million years (Castillo-Rogez et al. 2007).

4.1.2 Iapetus' Equatorial Ridge

The ridge at Iapetus' equator is unique in the solar system (Figs. 11 and 12). It runs most of the way around the equator. Where visible, it sits on top of the much broader equatorial bulge. In some places the ridge deviates only slightly from lying exactly on the equator. The ridge is abundantly cratered. To the eye, there is no obvious difference between the crater densities on the ridge and on the surrounding terrain. This indicates that it is very old, comparable in age to the other terrains on Iapetus. It is clearly older than Iapetus' large basins.

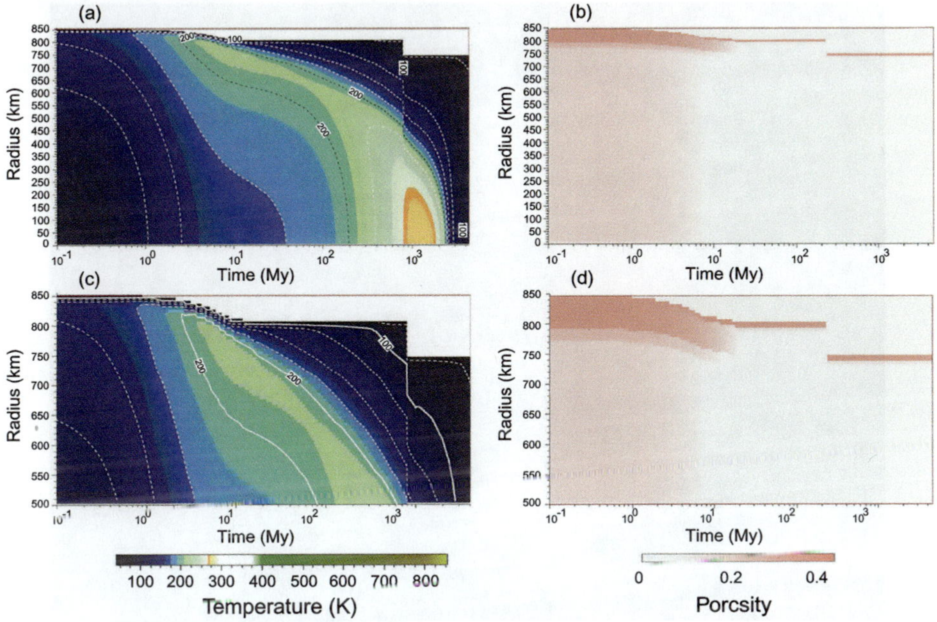

Fig. 10 Evolution of Iapetus' interior. One of the models by Castillo-Rogez et al. (2007, 2009) is shown as an example. Temperature is shown on the left and porosity on the right. The ordinate is equatorial radius and the abscissa is time (on a log scale). At the extreme left is the start of the model (time of accretion); at the extreme right is the present. The lower frames are enlargements of the upper 350 km. The temperature contour interval is 25 K. The temperature color scheme highlights geophysically significant temperature regions. This model was started with an initial complement of short-lived radioactive isotopes appropriate for accretion 4 Myr after the formation of the calcium aluminum inclusions (CAI). (The surface steps between 4 and 10 Myr are due to the granularity of the calculation)

Data for estimating the volume of the ridge are sparse. Cassini-Huygens images (Porco et al. 2005) cover only one side of the satellite. Denk et al. (2000) observed illuminated tops of equatorial peaks on the night side of the terminator in Voyager images. They cite this as evidence suggesting the continuation of the ridge into the other hemisphere. For the most part the ridge appears to be a single feature, but in some places it is a double or a triple ridge (Fig. 12). The flanks are steep; some slopes are greater than 30°. The volume of material involved in the ridge is large. It spans more than half of the equator and is well developed over a length of $\sim$1600 km. A typical (assumed to be triangular) cross-section has a base of $\sim$200 km and a height of $\sim$18 km, giving a cross-sectional area of $\sim$1800 km^2 and a volume of $\sim 3 \times 10^6$ km^3. This estimate does not include any allowance for "roots" or any low-amplitude folds that may extend outward for some distance.

A variety of mechanisms have been suggested to explain the creation of the ridge. Porco et al. (2005) and Castillo-Rogez et al. (2007) proposed that it was related to despinning. Denk et al. (2005) suggested that it was the result of volcanic activity. Giese et al. (2005, 2008) proposed upwarping of the surface due to a tectonic (rather than a volcanic) event. Ip (2006) suggested the collapse of a ring in orbit about Iapetus. Czechowski and Leliwa-Kopystyński (2008) and Roberts and Nimmo (2009) argued that it resulted from convection within Iapetus. Melosh and Nimmo (2009) suggested that dikes might be involved in producing the ridge.

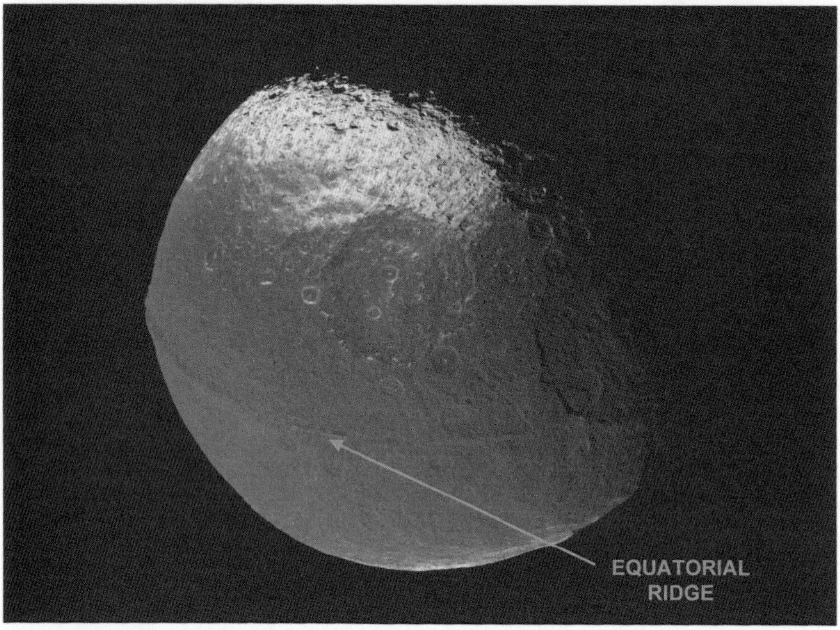

Fig. 11 Iapetus' relatively narrow equatorial ridge reaches heights of up to 20 km in some places

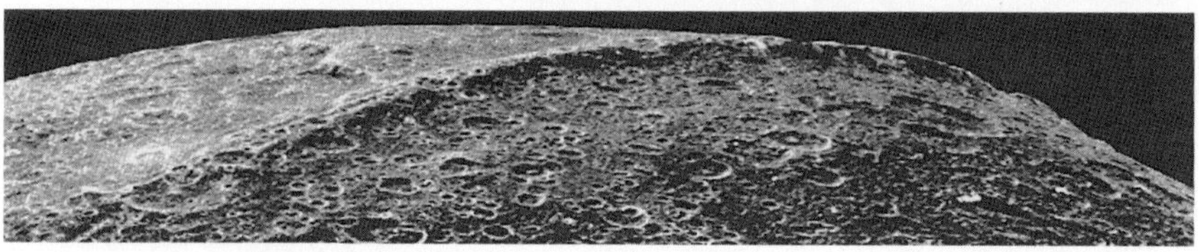

Fig. 12 Iapetus equatorial ridge as observed by Cassini ISS (Porco et al. 2005). Toward the left the ridge appears to have splayed into several components. The available images do not have enough detail to reveal how the ridge was formed

If the ridge was created by folding, as suggested by part of its length (Fig. 12) then its width of ~200 km implies that the lithospheric thickness was ~15–20 km, according to the relationship between the spatial wavelength of folds and the thickness of the lithospheric shell that formed them (Turcotte and Schubert 2002). Thus, the minimum lithospheric thickness required to preserve a 20-km high feature is about 15–20 km. However, the ridge could have been somewhat higher and subsequently sank, yielding its present height.

The ridge is small compared to the much larger, global, equatorial bulge. Removal of the ridge from limb images changes the equatorial radius of the ellipsoidal fit to the shape by only ~1 km. By way of another comparison, the volume of the ridge is less than 5% of the volume of the equatorial bulge, and clearly appears to be a superimposed form. Although the formation of the ridge and the bulge could be related, the significance of the oblate spheroid shape is independent of the formation mechanism and morphology of the ridge.

Iapetus' thermal evolution models (as previously discussed) show two periods when the surface area changed significantly. The earliest episode was when porosity decreased as a result of internal warming. Later, a large change occurred when Iapetus' oblate shape evolved

Table 1 Published values for Rhea's gravitational coefficients

	Anderson and Schubert (2007)	Iess et al. (2007)	Mackenzie et al. (2008)
GM (km^3 s^{-1})	153.9372 ± 0.0013	153.9395 ± 0.0018	153.9398 ± 0.0008
J_2 (10^{-6})	889.0 ± 25.0	794.7 ± 89.2	931.0 ± 12.0
C_{22} (10^{-6})	266.6 ± 7.5	235.3 ± 4.8	237.2 ± 4.5
J_2/C_{22}	10/3 (assumed)	3.377	3.925
C/MR^2	0.3911 ± 0.0045	0.3721 ± 0.0036	?

to a more spherical figure as a result of despinning. Since despinning occurred relatively rapidly, changes in equatorial radius and associated surface area reduction also take place quickly. Times when large decreases in surface area occur are favorable for tectonic activity because of the stress applied to the lithosphere. The amount of material available for redistribution depends on the initial rotation period. For an initial period of 5 h, the change in surface area upon despinning to a 16 h period is $\sim 0.23 \times 10^6$ km^2. If the lithosphere is ~15 km thick, then $\sim 3.5 \times 10^6$ km^3 of material would have been available for redistribution. This is comparable with the estimated volume of the ridge. Thus, it appears that this process could produce enough material for building the ridge.

There are several conditions that favor building a ridge at the equator. During despinning this is the locus and time of the maximum area change and hence it is here that the largest volume of material is available for ridge construction. This location is also favored from an energy point of view. Due to centrifugal force and distance from the center of Iapetus, it is the location at which mass could most easily be raised. Here, the restraining forces on the upward buckling of the lithosphere would be less than elsewhere on Iapetus.

Nevertheless, the elucidation of the mechanism by which the ridge came into being has continued to elude researchers. Particularly vexing is that the ridge is so precisely aligned with the equator and that the various proposed processes either produce features that are too large (i.e., characterized by low spatial frequencies, rather than high frequencies, particularly in the north-south direction), do not explain the equatorial alignment, or need interior temperatures much warmer than in the models of Castillo-Rogez et al. (2007). However, most (but not all) agree that the raising of the ridge at the equator is a surface expression of deeper, internal processes.

4.2 Rhea

On the basis of its shape Rhea appears to be in hydrostatic equilibrium although uncertainties in the shape data leave open the question of the satellite's homogeneity (Thomas et al. 2007b). However, Rhea is especially important among the medium size icy satellites of Saturn because it is the only such satellite for which there are data on the quadrupole gravitational field. The data are limited (so far) to one near-equatorial Cassini-Huygens flyby. The inference of J_2 and C_{22} from the radio Doppler data has been controversial. Table 1 lists three interpretations. The mass of Rhea (the GM values in Table 1) is not in contention. The mean radius of Rhea (764.3 ± 2.2 km) and its mass give a density of 1233 ± 11 kg m^{-3} (Thomas et al. 2007b). If Rhea is composed of pure ice with density 1000 kg m^{-3} and rock with density 3527.5 kg m^{-3} (the density of Io) then its silicate mass fraction is 0.26. (If instead the rock density is 2500 kg m^{-3}, the silicate mass fraction would be 0.31.)

The different values of the gravitational coefficients of Rhea in Table 1 are due to different approaches to the interpretation of the radio Doppler data.

Anderson and Schubert (2007) assumed that Rhea is in hydrostatic equilibrium. That assumption links J_2 and C_{22} by $J_2/C_{22} = 10/3$. Iess et al. (2007) did not *a priori* assume hydrostatic equilibrium. However their values of J_2 and C_{22} are consistent with it. If Rhea is in hydrostatic equilibrium then C_{22} can be used to infer the satellite's moment of inertia factor C/MR^2, where C is the axial moment of inertia. Anderson and Schubert (2007) find $C/MR^2 = 0.391$. This is consistent with no differentiation and a phase change from ice I to ice II at depth in Rhea's ice/rock interior. Iess et al. (2007) obtain a slightly smaller value of C/MR^2 that implies a partial separation of ice and rock inside Rhea. Since the gravitational coefficients of Mackenzie et al. (2008) are not consistent with hydrostatic equilibrium, there is no further interpretation that can be applied to them. The bottom line is that the single near equatorial flyby of Rhea cannot distinguish between hydrostatic and non-hydrostatic gravitational fields. A hydrostatic Rhea is consistent with the Doppler flyby data despite the claim of Mackenzie et al. (2008) to the contrary (Anderson and Schubert 2010). Until we can determine with some degree of certainty whether Rhea is truly hydrostatic, inferences about Rhea's moment of inertia based on C_{22} alone should be treated with caution.

Rhea's bright, clean ice surface suggests a past period of resurfacing by water. The production of such clean ice suggests that some differentiation took place. How can this be done if Rhea is largely undifferentiated? This question was addressed by Barr and Canup (2008), who modeled the accretion of Rhea. They found that their model depended upon three key parameters: how long it took accretion to occur, the starting time after the formation of CAI, and the temperature of the disk from which the satellite accreted. For relatively short accretion times they found models in which the surface is warmer than the deep interior. In the case of Rhea, there exists such a region in this parameter 3-space for which the temperatures are above the melting point of water-ice in the upper layer. In particular, they found that the parameter space could accommodate melt layer thicknesses of 46 km (Anderson and Schubert 2007) and 158 km (Iess et al. 2007). Thus it appears that the clean icy surface of Rhea is consistent with the shape and gravity interpretations because the accretion model indicates that it can be produced while keeping the vast bulk of the satellite undifferentiated.

4.3 Tethys

There was only one dedicated flyby of Tethys during the Cassini-Huygens primary mission. High-resolution imaging led to better mapping of the cratering record (Dones et al. 2009), but few constraints on its internal structure were obtained. From its physical and geological properties, it is thought that Tethys has had some endogenic activity (Schenk and Moore 1998, 2009). Tethys' relatively low density indicates a small silicate mass fraction (~6%) and possibly significant porosity. The most striking feature observed on Tethys is Ithaca Chasma, a huge rift that has a length of over 2000 km and is up to 100 km wide (Giese et al. 2007). Based upon an interpretation that explained the uplifted flanks of the chasma in terms of elastic flexure, Giese et al. (2007) inferred constraints on the thickness of the lithosphere and thus the corresponding heat flux at the time the ridge formed. The inferred heat flux was large, 18–30 m W m^{-2}! This value depends upon assumed thermodynamic properties of the surface material and its porosity. This heat flow is surprisingly large for such a small satellite with such a small silicate mass fraction. About an order of magnitude less heat flow would be expected, even 4 billion years ago. Nevertheless, this heat flux estimate led Chen and Nimmo (2008) to conclude that at the time Ithaca Chasma formed Tethys must have been subject to significant tidal heating. They suggested that the interior was warm. Possibly there was a deep ocean, allowing for substantial tidal heating during Tethys' passage through an orbital resonance with Dione (the present inclination resonance between Mimas and Tethys does not excite the eccentricity of either body). The conditions leading to such internal melting and the conjectured deep ocean in Tethys remain to be investigated.

4.4 Phoebe

Phoebe, the only irregular satellite so far observed at high resolution, is an interesting object not only in itself, but also as a probe of the connections between the evolution of the outer orbital regions of giant planets populated by the irregular satellites and the inner orbital regions inhabited by the regular satellites. The dynamical and collisional evolution of Phoebe and of the irregular satellites in general appears to be linked to the presence of dark material coating the leading hemispheres of the outermost regular satellites (i.e., Callisto and Iapetus). The data supplied by Galileo and Cassini-Huygens missions shed new light on this intriguing possibility, while theoretical works and Earth-based observations contributed to unveil the complexity of the underlying evolving scenario.

The radial distribution of the irregular satellites in Saturn's system is characterized by a gap (named *Phoebe's gap* by Turrini et al. 2008) centered on Phoebe and extending from 11.22×10^6 km to about 14.96×10^6 km from Saturn, i.e., between Ijiraq's and Paaliaq's orbits. While the absence of retrograde satellites in this orbital region can be partially explained by the stricter constraints on the conditions needed to capture the satellites (see Turrini et al. 2009 and Mosqueira et al. 2010, this issue), the same is not true of the absence of prograde satellites. The most plausible explanation is that the sweeping action of Phoebe collisionally removed the previously existing prograde satellites and the collisional shards generated by the impacts against Phoebe itself. This hypothesis is supported by statistical analyses showing that Phoebe is the most collisionally active irregular satellite in the Saturnian system (Nesvorný et al. 2003; Turrini et al. 2008). Moreover, the orbital region where Phoebe is most active in removing retrograde bodies is the one that would be populated by collisional shards generated by Phoebe itself for realistic ejection velocities. These hypotheses on Phoebe's past collisional activity are strongly supported by Cassini images of the satellite that show an intensely cratered surface (Porco et al. 2005).

Phoebe's collisional activity has likely been the source of a vast amount of dusty material ejected into the outer Saturnian system: the excavation of a medium-sized crater like Hylas (about 30 km in diameter, Giese et al. 2006) would in fact supply enough material to cover the dark hemisphere of Iapetus (Tosi et al. 2009). Spectral comparisons of Phoebe, Iapetus and Hyperion based on Cassini infrared data show a strong similarity between Phoebe's composition and that of the dark hemisphere of Iapetus (Tosi et al. 2009), in agreement with this scenario. However, other irregular satellites likely participated in the dust production process, especially if collisional families exist in the Saturnian system (Turrini et al. 2009) similar to those identified by Nesvorný et al. (2003) in the Jovian system. This hypothesis finds support in the work of Buratti et al. (2005), who reported a similarity between the color indexes of the Saturnian irregular satellites (with the exception of Phoebe) and the dark hemisphere of Iapetus, suggesting multiple sources for the dark material.

Micrometer-sized dust particles on planetocentric orbits located outside the orbits of the regular satellites would spiral inward due to Poynting-Robertson drag and intersect the orbits of the regular satellites (Soter 1974; Burns et al. 1996). Dust transfer processes have possibly been observed in the Jovian system by the Galileo spacecraft: its dust sampling data, indicate a dust population in the outer region of the system an order of magnitude higher than that of interplanetary space (Krivov et al. 2002). Moreover, Krivov et al. (2002) point out that the dynamical features of a significant fraction of the collected dust grains are compatible with particles on retrograde planetocentric orbits. The recent discovery of an outer disk of particles around Saturn (Verbiscer et al. 2009), spanning the orbital region of the irregular satellites and likely linked to Phoebe (Tosi et al. 2009; Verbiscer et al. 2009), argues for a similar scenario applying to the Saturnian system, since it proves that collisional, dust-generating processes acted and are likely still acting in its outer region. This interpretation is also supported by Hendrix and Hansen (2008), who analyzed Cassini data and found water ice in Iapetus' dark material at the warmest latitudes of the satellite consistent with a recent or ongoing emplacement of the dark material itself.

Due to their inward motion, prograde particles would impact Iapetus' leading side when on outermost orbits and Iapetus' trailing side while on innermost orbits. Dust grains moving on retrograde orbits would impact Iapetus on its leading hemisphere with impact velocities higher than those of prograde dust grains. Retrograde dust particles would experience a higher frequency of close encounters with the satellite, suggesting that regular satellites should be more efficient in collecting retrograde particles compared with prograde particles, thereby explaining the asymmetry in the distribution of dark material covering their surfaces. Tosi et al. (2009) evaluated the efficiency of Iapetus in capturing dust grains of different sizes as a function of their orbital parameters and drift time-scales due to Poynting-Robertson drag. Their results show that Iapetus would be extremely active in collecting dust grains ejected by Phoebe and the retrograde satellites. In the case of prograde orbits Iapetus' efficiency would be lower, especially in those regions of phase space that would be populated by dust grains ejected by the present prograde satellites.

The above scenario provides a plausible explanation for the origin of the dark material even though there are open issues still to be investigated, e.g., a possible higher spectral similarity in the UV between the dark material on Iapetus and Hyperion with respect to Phoebe (Hendrix and Hansen 2008). A correlation between Hyperion and Iapetus, however, is difficult to justify from a dynamical point of view. Marchi et al. (2002) show that, to guarantee an adequate delivery of material to Iapetus, fragments from Hyperion should be ejected with high velocities (greater than 100 $m\,s^{-1}$) and into collimated streams. Tosi et al. (2009) observed that the present porosity of Hyperion, estimated to be greater than 40% (Thomas et al. 2007a), would significantly affect the range of ejection velocities from the satellite, thus posing strict constraints on this mechanism and on the plausibility of a Hyperion-Iapetus connection. Moreover, Dobrovolskis and Lissauer (2009) showed that the lifetime of collisional fragments ejected by Hyperion is quite small, i.e., of the order of 10^6 years or less. Therefore, such a Hyperion-based scenario would require that the event responsible for the production of the dark material be quite recent, i.e., no older than a few million years.

Our understanding of the evolution of Phoebe and the irregular satellites has greatly improved in the last few years emphasizing how the evolution of the outer regions of the satellite systems of the giant planets is linked to that of the inner regions populated by the regular satellites. It also emphasizes that this link can still be active in the present solar system. The evolution of the satellite systems of the giant planets should therefore be studied as a whole, in order to discriminate among those morphological and compositional features on the surfaces of the regular satellites due to internal processes and those due to the global evolution of the satellite systems themselves.

5 Triton, Pluto, and Charon

The icy dwarf planets of the Kuiper belt represent a relatively newly recognized class of Solar System body. This section focuses on the best known or explored of these, Triton,

Pluto, and its major moon Charon, as long as one accepts the dominant view that Triton is a captured Kuiper belt object (KBO). Reviews of their structure and evolution, from a mid-1990s perspective, can be found in McKinnon et al. (1995, 1997), while McKinnon et al. (2008) provide a more recent overview of large KBOs. Other chapters in this volume address surface composition, internal structure, heat transfer mechanisms, and likelihood of oceans within these and similar bodies.

Triton, Pluto, and Charon are representative in that they are relatively rock-rich (as much as 70% by mass), presumably have a major bulk water-ice content, a non-negligible carbonaceous component (not directly observed), as well as one or more exotic and/or volatile ices identified on their surfaces (e.g., ammonia hydrate, N_2, CH_4). For Triton at least, the expectation is that it has differentiated into a rocky core and an icy mantle given its likely history of strong tidal heating. Pluto may also be differentiated, but in this case the evidence is based on observations of other KBOs and theoretical arguments. Compared with Rhea, Pluto is larger and more rock rich. Callisto and Titan are also less rock rich than Pluto and both are partially differentiated. These comparisons support the possibility that Pluto is differentiated.

5.1 Triton

Triton's surface is geologically young, and betrays a variety of cryovolcanic, tectonic, and atmospheric features or processes (Prockter et al. 2010, this issue). These include cantaloupe terrain, argued to be due to diapirism by Schenk and Jackson (1993), volcano-tectonic structures consistent with rifting and caldera formation (Fig. 13), and nitrogen plumes, modeled as solar- powered but conceivably endogenically driven (Kirk et al. 1995; Duxbury and Brown 1997). This geological vigor may be tied to tidal heating, as with other active icy satellites (e.g., Europa and Enceladus), but Triton is not being tidally heated now (its eccentricity is as close to zero as can be measured). Rather, any tidal heating would have occurred as its post-capture orbit circularized (McKinnon 1984). The degree to which tides circularized Triton's post-capture orbit and when this occurred are major questions. Early proposals had Triton captured from heliocentric orbit by energy loss through collision with a regular satellite of Neptune (now destroyed) (Goldreich et al. 1989) or via gas drag in a precursor protosatellite disk around Neptune (McKinnon and Leith 1995). The most likely scenario being considered at present, however, is "exchange capture," in which a binary KBO encounters Neptune deep in the planet's Hill sphere (gravitational sphere of influence), which disrupts the binary, with one half of the pair ending up in an eccentric orbit around Neptune and the other half escaping back to heliocentric orbit (Agnor and Hamilton 2006). Although seemingly improbable, binaries are now known to be common in the Kuiper belt, comprising probably $>10\%$ of the population (Pluto-Charon, the first discovered KBO, is one), so such binary encounters (as opposed to collisions, or encounters at a special time) are likely.

The total amount of tidal heat potentially available to Triton is prodigious (more than enough to completely melt the satellite; Ross and Schubert 1990). Ćuk and Gladman (2005) have argued, however, that regardless of how Triton may have been captured, it would have destabilized any pre-existing regular Neptunian satellite system, causing the satellites to collisionally self-destruct. Subsequent orbital evolution by Triton would sweep up this prograde orbiting debris, drastically reducing Triton's post-capture apoapse in 10^5–10^6 yr. The fate of Neptune's regular satellites was always in Triton's hands, though, and given that the total orbital angular momentum of Uranus' regular satellites (as an analogue) is only 45% of that of Triton, it seems unlikely that such sweep up would completely circularize Triton's orbit. A geophysically significant heating event for Triton remains likely.

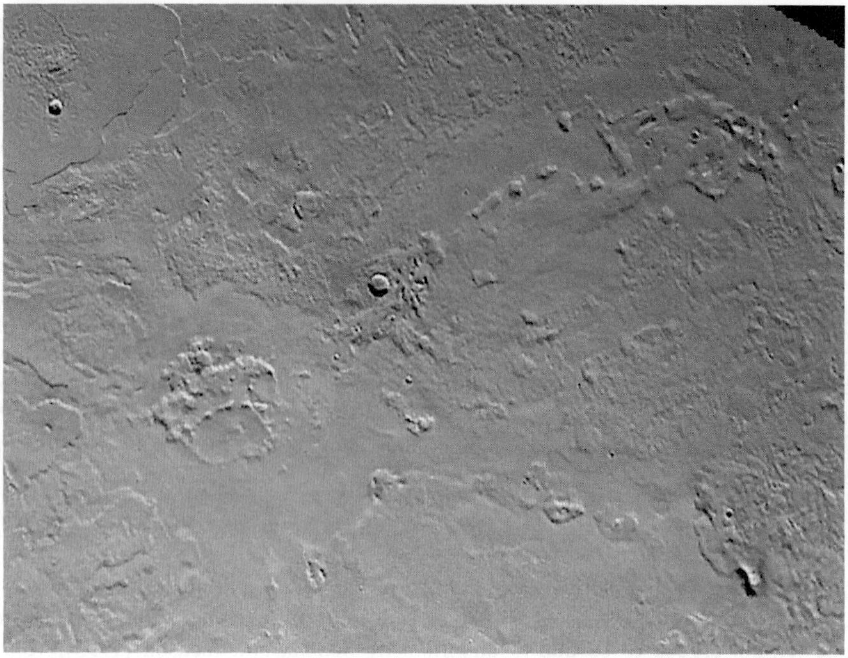

Fig. 13 Young, sparsely cratered, "cyrovolcanic" region on Triton. Smooth undulating flows apparently emanate from complex caldera-like depressions and linear alignments of volcanic pits and vents, burying pre-existing topography. Leviathan Patera, ~85-km across, sits at the vertex of two linear eruption trends (*lower left of center*)

Ćuk and Gladman (2005) also argued that Triton's post-capture orbit must have evolved quickly so as to not scatter and eliminate Neptune's distant irregular satellites. However, if these satellites are themselves not primordial, then this is only a constraint on relative capture times. For example, in the Nice model of Solar System evolution, Neptune nominally acquires its present irregular satellites during a period of strong planetary chaos (Nesvorný et al. 2007), possibly coincident with the Late Heavy Bombardment of the inner Solar System; any previously captured distant irregular satellites are lost. All that is required of Triton in this case is that it be captured, and have its orbit largely circularized, earlier than this. Vokrouhlický et al. (2008), in a detailed numerical study, further find that exchange reaction capture of Triton is *not likely* during the "planet-crossing" phase of the Nice model (encounter velocities are too high), and offer that the most opportune time for capture is within ~5–10 Myr of Solar System formation, when solar nebula gas was still present and planetesimal encounter velocities were low.

Recent work thus places Triton's capture back toward the beginning of Solar System history. Triton's surface, in contrast, is geologically quite young, and is likely undergoing active resurfacing. Stern and McKinnon (2000) inferred, from the estimated flux of small Kuiper Belt objects (comets) onto Triton, that its crater retention age does not exceed 100–300 Myr. Schenk and Zahnle (2007), in their latest calibration of the impact flux, lower this age to ~50 Myr, which would put Triton in the same activity class as Europa. Moreover, both Marchi et al. (2004) and Schenk and Zahnle (2007) argue that Triton's craters are so concentrated on its leading hemisphere of retrograde orbital motion that a predominantly prograde, planetocentric source is required. Schenk and Zahnle (2007) estimate that no more than ~20% of Triton's craters may be due to heliocentric impactors, lowering its surface age to <10 Myr, which would make it *more* active than Europa.

If the age of Triton's surface is pushed toward zero, then the flux of planetocentric debris to account for its craters must rise dramatically. Where is the reservoir of small prograde bodies? The Neptune system outside of 5 Neptune radii (R_N) is notoriously empty (as far as we can tell), and Triton's post-capture orbital evolution is usually held responsible. Both Marchi et al. (2004) and Schenk and Zahnle (2007) are forced to postulate that Triton may have been bombarded by secondary, or "sesquinary," debris from a large comet impact on Proteus, a 420-km-diameter prograde, regular satellite that orbits at 4.75 R_N (Triton is at 14.3 R_N). But can this small, distant target truly be responsible, given the much lower relative impact rate on Proteus (Zahnle et al. 2003)? We offer no solution to this puzzle, but suspect that the limitations of Voyager 2 image coverage and resolution are at fault, and that future spacecraft data may tell a very different story.

Prockter et al. (2005) have noted that ridges on Triton, while sparsely distributed, bear resemblance to their more ubiquitous cousins on Europa, and attempt to relate ridge formation on Triton to tidal stresses during orbital circularization. Ridges appear to post-date cantaloupe terrain formation, and are thus a late event even in the context of Triton's youthful geologic history. Regardless, ridge formation during tidal circularization requires either a late capture, or if capture occurred early in Solar System history, delayed circularization (Ross and Schubert 1990). Late capture is highly improbable (Vokrouhlický et al. 2008), but early capture and stalled circularization violate the constraint provided by the existence of Neptune's irregular satellites (Ćuk and Gladman 2005). We are thus led to conclude that Triton is probably presently tectonically active (the survival of tectonic structures from the epoch of extreme tidal heating was never likely to begin with).

Triton is the single icy dwarf planet for which we have close-up spacecraft imagery and other data that provide a context in which to interpret ground-based data and theoretical models. Its size makes it more akin to a Galilean satellite than to the midsized icy satellites of Saturn or Uranus. It may be that its overall size, presumed relatively large rocky core (radiogenic heat), and especially its "exotic" surface composition (low-melting-point ices) have allowed it to remain geologically active up to the present. Its deeper water-ice mantle should be convecting, depending on ice grain size (McKinnon 2006), with an internal ocean anchored at the 0.2 GPa pressure level (corresponding to the minimum melting temperature for pure ice), or ~250 km deep. The ocean could be thin, but may extend all the way to Triton's rocky core, either because ocean temperatures are high enough, or because the ocean contains sufficient dissolved salts, ammonia, and/or methanol (Hussmann et al. 2006).

5.2 Pluto

Pluto has for some time been recognized as remarkably similar in surface composition and bulk density to its larger cousin, Triton. Whether its geology has remained as active is uncertain. If Triton's activity is tied in some manner to its past history of extreme tidal heating, as is often assumed, then Pluto would not be expected to exhibit the same diversity of youthful surface features. As the above discussion makes plain, however, it is difficult to link any specific, ongoing geologic activity on Triton to what is apparently an ancient capture event. Perhaps we should not be too surprised in July 2015 if, when New Horizons flies through the Pluto system, Pluto turns out to be similarly geologically active (see Stern 2008; Weaver et al. 2008; Young et al. 2008 and related papers for a description of that mission's capabilities).

It is true that Pluto went through its own "extreme" geophysical event early in its history, i.e., the "giant" impact between two precursor bodies that created the Pluto-Charon binary and logically, the coplanar satellite system that includes Nix and Hydra as well (Stern et al. 2006). The impact and its tidal evolution aftermath are two energy sources that may have served to cause rock to differentiate from ice within Pluto. Accretional energy and the radiogenic heat from the decay of short- and long-lived radionuclides may also have contributed. All but short-lived radioisotope heating were discussed in McKinnon et al. (1997) in the context of Pluto, but those authors equivocated on the subject of Pluto's state of differentiation.

Two important pieces of circumstantial evidence have now bolstered the case for Pluto being a differentiated world. First, the dwarf planet Haumea is now recognized as having undergone a catastrophic impact that spun it up and resulted in two relatively large satellites (Ragozzine and Brown 2009) and a group of dynamically associated, heliocentric bodies, the first recognized KBO family (Brown et al. 2007). All of these bodies are spectroscopically linked in that their surfaces are dominated by water ice (Schaller and Brown 2008; Fraser and Brown 2009). This surface composition is unusual for the Kuiper belt and is the fundamental inference that Haumea (present mean radius ~700 km) was differentiated prior to the collision, and what we see now on the various surfaces is exposed water-ice "mantle" and mantle fragments. Second, it has been established from stellar occultations and determination of the Pluto-Charon barycenter that Charon ($1630 \pm 70\ \mathrm{kg\,m^{-3}}$) is less dense than Pluto (Buie et al. 2006; Person et al. 2006); in the context of an impact origin, this implies that one or both precursor bodies were differentiated, if both were initially of similar density (Canup 2005).

Pluto's diameter remains imprecisely known, but recent stellar occultations have put firmer constraints on its size and on the thickness of any deep troposphere, which translate into a radius for its solid surface between 1160 and 1184 km (see Elliot et al. 2007; Lellouch et al. 2009). With the mass from Buie et al. (2006), a density of $1935 \pm 60\ \mathrm{kg\,m^{-3}}$ and an (anhydrous) rock mass fraction close to 0.68 are implied. If differentiated, Pluto should have a relatively large rocky core (approaching 900 km radius), and a heat flow today sufficient to maintain an ocean below an ice I shell, if the ice shell is conductive. If the lower part of the ice shell is convecting, then an "antifreeze" may be required to maintain the ocean, if the convective adiabat in the ice above is too cold. There is no shortage of plausible candidate substances that would lower the freezing point of ocean water (e.g., salts, ammonia, methanol). With sufficient antifreeze, Pluto's ice mantle–water ocean system may also oscillate, on a geological timescale, between a conductive, low-heat-flow, thick shell state, and a convective, high-heat-flow, thicker shell state (Mitri and Showman 2008). Corresponding epochs of surface compression and extension may have consequences observable by New Horizons.

We can also approach the question of whether Pluto is differentiated by examining energy transport across its putative ice mantle (Hussmann et al. 2010, this issue). The conductive heat flow q across such an ice mantle, measured at the surface, is given by McKinnon (2006):

$$q = \frac{621}{D\ (\mathrm{m})} \ln\left(\frac{T_\mathrm{b}}{T_\mathrm{s}}\right)\left(1 - \frac{D}{R_\mathrm{P}}\right)\ (\mathrm{W\,m^{-2}}) \approx 2.5\ \mathrm{mW\,m^{-2}} \tag{3}$$

where D is mantle thickness (≈300 km), R_P is Pluto's radius, and T_b and T_s are the basal and surface ice temperatures (251 K, the melting threshold, and 50 K, respectively). Given the heat from long-lived radionuclide decay in Pluto's rock fraction, such a mantle would need to transport about 5 times this amount of heat 4 billion years ago in order to prevent melting. In terms of convective transport (Schubert et al. 2001), this implies a Nusselt number, the dimensionless measure of heat flow, $Nu \approx 5$. The Nusselt number in turn depends on the Rayleigh number (Ra_ad), the dimensionless measure of convective vigor, defined using the viscosity of the actively convecting ice, through $Nu = 0.52\theta^{-4/3} Ra_\mathrm{ad}^{1/3}$, where θ is the number of viscosity e-foldings across Pluto's mantle (see McKinnon 2006 for details). For diffusion creep, which is strongly temperature dependent, and a temperature drop of $\approx$200 K, $\theta \approx 20$. Thus, to prevent melting early on, Pluto must have been vigorously convecting with a minimum Ra_{ad} of order 2×10^8, which in turn implies a convecting ice-rock viscosity no larger than $\sim 2 \times 10^{15}$ Pa s. Such a viscosity is plausible for ice. In detail, however, the *effective* θ, due to mantle (shell) curvature and the strong temperature dependence of ice thermal conductivity, is probably doubled, and the viscosity of the ice-rock is probably boosted by an order of magnitude by all the admixed rock particles. Taken together, these imply that the viscosity of the ice itself can likely be no larger than $\sim 10^{13}$ Pa s if melting is not to occur. Such a low viscosity implies very fine-grained ice ($\sim$100 μm) (see Fig. 3 in McKinnon 2006).

In sum, if the short-lived heat sources listed earlier are themselves insufficient to trigger differentiation, then for Pluto to avoid differentiation due to long-lived radiogenic heat release, once thermal steady state is reached, requires ice rheological properties that are just marginally plausible. The likelihood of Pluto remaining undifferentiated is thus more precarious than originally discussed in McKinnon et al. (1997).

5.3 Charon

Charon is notable for its fairly rock-rick nature ($\sim$60% by mass), with a size and density that put it in the same class as the midsize icy satellites Dione, Ariel, and Umbriel. Dione and Ariel in turn show widespread evidence for geological activity later in their geological histories, though they are not obviously active now (Schenk and Moore 1998). The differences between Dione and Ariel on one hand and inactive Umbriel on the other are usually ascribed to present or past epochs of tidal heating. Charon has had no access to comparable amounts of tidal heat, however, so it is all the more remarkable that it is the first solid Solar System body for which there has been a definitive detection of ammonia-hydrate ice on its surface (Brown and Calvin 2000; Cook et al. 2007). This has led in turn to speculation that Charon is itself both differentiated and cryovolcanically active (Cook et al. 2007; Desch et al. 2009).

Given the evidence for Triton's ongoing activity, is it plausible that Charon be active as well? Several factors indicate caution. First, Charon is smaller and less rock-rich by comparison, and its surface heat flow is likely lower by a factor of three. Second, its surface is spectroscopically dominated by water ice, which is involatile and rigid at Charon's surface temperature of $\sim$60 K, and obviously different from Triton. Third, Charon is remote from the Sun, in a more benign UV and charged particle radiation environment than the icy satellites of the giant planets, and moreover, bathed in the escaping N_2 atmosphere of Pluto. Now, it is plausible that ammonia hydrate was erupted to Charon's surface earlier in its geological history, but the lifetime of ammonia hydrate ice exposed by ongoing cratering and regolith overturn requires careful evaluation.

Ultimately, for bodies such as Charon to be cryovolcanically active today, they must overcome the physics of conductive cooling. At low temperatures crystalline ice and rock minerals are highly conductive and possess low heat capacities, so bulk thermal diffusivities should be high (several $\times 10^{-6}$ m^2 s^{-1}), and the thermal conduction (cooling) time scale over 300 km depth (the bulk of Charon) is only $\sim$ 500 Myr. In order to sustain aqueous ammonia or methanol at late times requires something special, such as a deep, porous, low-conductivity regolith. Desch et al. (2009) require that Charon differentiate early and form a very-low-conductivity (1 W m^{-1} K^{-1}) rock core, to boost stored radiogenic heat at late times, as well as place their liquid reservoir deep and well away from Charon's cold surface. Their conclusion, that Charon is cryovolcanically active today, is not robust to more geophysically justifiable parameter or structural choices, but such modeling only increases interest in the results to come from New Horizons.

6 Summary and Conclusions

The satellites of the outer solar system are striking in their diversity and uniqueness. Their sizes range from bodies of about 100 km across and smaller to moons as large and even larger than the planet Mercury. Some moons are rocky while most are composed of both rock and ice. Some moons are differentiated (rock separated from ice) while others are nearly uniform rock/ice mixtures and a few even have metal cores. Some of the moons are geologically active while others are not. Some icy moons have internal liquid water oceans while the water in other moons is completely frozen. Size does not always matter as moons of similar size have sharply contrasting characteristics and one of the smaller moons is geologically active. Such uniqueness and diversity in the moons requires a corresponding uniqueness and diversity in their evolutions. One scenario does not fit all the evolutionary paths followed by the outer planet moons. Several of the evolutionary tracks of the outer planet satellites have been delineated in this chapter.

Tidal heating has been of extraordinary importance in the histories of outer solar system moons. In contrast, the phenomenon has hardly been visible in the inner solar system. Tidal heating closely couples the orbital-dynamical and thermal evolutions of a body and additionally couples the fates of several satellites together. It is not surprising that we have had to go to the multiple satellite systems of the outer planets to observe the significance of tidal heating and its consequences. The inner solar system provides the Earth-Moon pair, bodies little affected by tidal heating, at least at present, and the Mars-Phobos-Deimos group in which tidal effects might have influenced the evolution of the moons. We discussed the tightly coupled Io, Europa, Ganymede trio of the Jovian system and the coupled Enceladus, Dione pair of the Saturnian system. Though not all the details are completely understood, it is generally agreed that tidal heating explains Io's silicate volcanism and the preservation through geologic time of Europa's subsurface liquid water ocean.

Tidal heating is important in the outer solar system because the gravitational interactions between and among satellites maintain orbital resonances that force moons into relatively high eccentricity orbits. The 3-body Laplace resonance involving Io, Europa and Ganymede, and the 2-body resonance involving Enceladus and Dione result in the tidal heating of Io, Europa, and Enceladus. How these moons came to be in the resonances is not well understood nor is the future of the resonance configurations predictable. Theoretical considerations have raised the possibilities that Io's present heat loss and tidal heat production might not be in balance and that the satellite could be moving either toward or away from Jupiter. A recent analysis of the orbital motions of the Galilean satellites by Lainey et al. (2009) concludes that Io is in thermal steady state and is evolving inwards, towards Jupiter, while Io, Europa and Ganymede are evolving away from the exact Laplace resonance.

Tidal heating of Enceladus is difficult to understand mainly because the surface expression of the heating appears confined to the south polar region of the moon. Heterogeneity of Enceladus' interior might provide an explanation. However, the internal processes driving the strong thermal activity of such a small moon are currently not fully understood.

It is intriguing that the three largest satellites in the solar system Titan, Ganymede, and Callisto, though similar in size and mean density, are so different in most other ways. Titan is likely partially differentiated, but more thoroughly differentiated than Callisto and less completely differentiated than Ganymede (Titan might not have a metal core and this would easily explain why it has no magnetic field). Titan stands apart in having a dense atmosphere, a methane hydrologic cycle, and a surface heavily modified by liquid hydrocarbons. Titan might have an internal liquid water ocean but the limited evidence for it is far less convincing than the magnetic induction observations indicating oceans in the icy Galilean satellites of Jupiter. Tidal heating does not seem to have been especially important in Titan's evolution. Indeed, if it were, Titan's present free eccentricity would be difficult to explain. Titan's eccentricity is too large to be forced and too large to be the remnant of an early larger eccentricity if Titan were strongly tidally heated during its evolution. Models with modest tidal heating can yield the present eccentricity as a residual from early higher eccentricity. The formation and eventual dissociation of clathrate hydrates appears to have played a unique and significant role in Titan's evolution, especially with regard to the early storage and later release of methane into Titan's atmosphere. The amount of methane in Titan's atmosphere at present requires that methane be replenished from the moon's interior to balance the geologically rapid photochemical destruction of the gas. Storage of methane in crustal clathrates and eventual breakdown of the clathrates to release methane into the atmosphere is a plausible though not proven explanation of the atmospheric methane abundance.

Among the medium size Saturnian satellites Iapetus is certainly the most unusual. It is distinctive because of the bright/dark dichotomy of its trailing/leading hemispheres, its oblate shape, and its equatorial ridge, a feature so far unique among the satellites of our solar system. Our discussion of Iapetus' evolution focused on a scenario involving despinning of the moon and its early heating by short-lived radioactivity. This scenario can explain the oblateness of Iapetus but not its equatorial ridge. Another scenario, involving buckling instability of the lithosphere upon contraction of the satellite due to internal porosity reduction, may provide an explanation of both the equatorial ridge and the flattened shape (Sandwell and Schubert 2010). This scenario does not require despinning or early heating by short-lived radioactivity. Beuthe (2010) has proposed that equatorial thinning of the lithosphere might play a role in focusing the deformation of the lithosphere at the equator.

The bright/dark hemispheric asymmetry of Iapetus is likely explained by the exogenic deposition of dark material on the leading hemisphere of the satellite coupled with insolation-controlled sublimation of ice form the leading hemisphere (Denk et al. 2009; Spencer and Denk 2009). Saturn's irregular satellite Phoebe and other irregular Saturnian satellites likely played an important role in this process.

Tidal heating has played another significant role in the outer solar system in the evolution of Neptune's moon Triton. While there is general consensus that Triton is a captured Kuiper belt object similar to but larger than Pluto and Charon, the actual mechanisms responsible for its capture and orbital circularization are still debated (Mosqueira et al. 2010, this issue). Even though tidal heating is currently unimportant in Triton, it was likely involved in the post-capture circularization of Triton's orbit and easily could have released more than enough energy to completely differentiate the captured satellite.

We conclude by emphasizing once again the uniqueness and diversity of the present states and evolutions of the moons of the outer solar system. Tidal heating, largely unimportant in the inner solar system, has been a predominant heat source in many of the moons of the outer solar system, a consequence of the massive outer planets and the orbital resonances of their satellites.

Acknowledgements GS acknowledges support from the NASA Outer Planets Research and Planetary Geology and Geophysics (NASA NNG06GG70G) programs. WBM thanks the New Horizons project for support and S.A. Stern for comments.

References

C.B. Agnor, D.P. Hamilton, Nature **441**, 192–194 (2006). doi:10.1038/nature04792

Y. Amelin, A.N. Krot, I.D. Hutcheon, A.A. Ulyanov, Science **297**, 1678–1683 (2002). doi:10.1126/science.1073950

J.D. Anderson, G. Schubert, Geophys. Res. Lett. **34**, L02202 (2007). doi:10.1029/2006GL028100

J.D. Anderson, G. Schubert, Phys. Earth Planet. Inter. **178**, 176–182 (2010)

S.K. Atreya, E.Y. Adams, H.B. Niemann, J.E. Demick-Montelara, T.C. Owen, M. Fulchignoni, F. Ferri, E.H. Wilson, Planet. Space Sci. **54**, 1177–1187 (2006). doi:10.1016/j.pss.2006.05.028

A.C. Barr, R.M. Canup, Icarus **198**, 163–177 (2008). doi:10.1016/j.icarus.2008.07.004

C. Béghin, P. Canu, E. Karkoschka, C. Sotin, C. Bertucci, W.S. Kurth, J.J. Berthelier, R. Grad, M. Hamelin, K. Schwingenschuh, F. Simões, Planet. Space Sci. **57**, 1872–1888 (2009). doi:10.1016/j.pss.2009.04.006

M. Beuthe, Icarus (2010, submitted)

B.G. Bills, F. Nimmo, Icarus **196**, 293–297 (2008). doi:10.1016/j.icarus.2008.03.002

M.E. Brown, W.M. Calvin, Science **287**, 107–109 (2000). doi:10.1126/science.287.5450.107

M.E. Brown, K.M. Barkune, D. Ragozzine, E.L. Schaller, Nature **446**, 294–296 (2007). doi:10.1038/nature05619

M.W. Buie, W.M. Grundy, E.F. Young, L.A. Young, S.A. Stern, Astron. J. **132**, 290–298 (2006)

B.J. Buratti, M.D. Hicks, A. Davies, Icarus **175**, 490–495 (2005). doi:10.1016/j.icarus.2004.11.024

J.A. Burns, D.P. Hamilton, F. Mignard, S. Soter, in *Physics, Chemistry, and Dynamics of Interplanetary Dust*, ed. by B.A.S. Gustafson, M.S. Hanner. ASP Conference Series, vol. 104 (University of Chicago Press, Chicago, 1996), pp. 179–182

R. Canup, Science **307**, 546–550 (2005). doi:10.1126/science.1106818

R.M. Canup, W.R. Ward, Astron. J. **124**, 3404–3423 (2002)

J.C. Castillo-Rogez, D.L. Matson, C. Sotin, T.V. Johnson, J.I. Lunine, P.C. Thomas, Icarus **190**, 179–202 (2007). doi:10.1016/j.icarus.2007.02.018

J. Castillo-Rogez, T.V. Johnson, M.H. Lee, N.J. Turner, D.L. Matson, J. Lunine, Icarus **204**, 658–662 (2009). doi:10.1016/j.icarus.2009.07.025

S. Chandrasekhar, *Ellipsoidal Figures of Equilibrium* (Yale University Press, New Haven, 1969)

E.M.A. Chen, F. Nimmo, Geophys. Res. Lett. **35**, L19203 (2008). doi:10.1029/2008GL035402

M. Choukroun, O. Grasset, G. Tobie, C. Sotin, Icarus **205**, 581–593 (2010)

J.C. Cook, S. Desch, T.L. Roush, C.A. Trujillo, T.R. Geballe, Astrophys. J. **663**, 1406–1419 (2007)

A. Coustenis et al., Space Sci. Rev. (2010, this issue)

M. Ćuk, B.R. Gladman, Astrophys. J. Lett. **626**, L113–L116 (2005)

L. Czechowski, J. Leliwa-Kopystyński, Adv. Space Res. **42**, 61–69 (2008). doi:10.1016/j.asr.2007.08.008

T. Denk, K.D. Matz, T. Roatsch, U. Wolf, R.J. Wagner, G. Neukum, R. Jaumann, Lunar Planet. Sci. **31**, 1596 (2000)

T. Denk, G. Neukum, P. Helfenstein, P.C. Thomas, E.P. Turtle, A.S. McEwen, T. Roatsch, J. Veverka, T.V. Johnson, J.E. Perry, W.M. Owen, R.J. Wagner, C.C. Porco, the Cassini ISS Team, Lunar Planet. Sci. **36**, 2262 (2005)

T. Denk, G. Neukum, T. Roatsch, C.C. Porco, J.A. Burns, G.G. Galuba, N. Schmedemann, P. Helfenstein, P.C. Thomas, R.J. Wagner, R.A. West, Science (2009). doi:10.1126/science.1177008

S.F. Dermott, C.D. Murray, Nature **296**, 418–421 (1982)

S.J. Desch, J.C. Cook, T.C. Doggett, S.B. Porter, Icarus **202**, 694–714 (2009). doi:10.1016/j.icarus.2009.03.009

A.R. Dobrovolskis, J.J. Lissauer, Icarus **169**, 462–473 (2009). doi:10.1016/j.icarus.2004.01.006

L. Dones, C.R. Chapman, W.B. McKinnon, H.J. Melosh, M.R. Kirchoff, G. Neukum, K.J. Zahnle, in *Saturn from Cassini-Huygens*, ed. by M.K. Dougherty, L.W. Esposito, S.M. Krimigis (Springer, Dordrecht, 2009), pp. 613–635. doi:10.1007/978-1-4020-9217-6_19

N.S. Duxbury, R.H. Brown, Icarus **125**, 83–93 (1997). doi:10.1006/icar.1996.5554

J.L. Elliot, M.J. Person, A.A.S. Gulbis, S.P. Souza, E.R. Adams, B.A. Babcock, J.W. Gangestad, A.E. Jaskot, E.A. Kramer, J.M. Pasachoff, R.E. Pike, C.A. Zuluaga, A.S. Bosh, S.W. Dieters, P.J. Francis, A.B. Giles, J.G. Greenhill, B. Lade, R. Lucas, D.J. Ramm, Astron. J. **134**, 1–13 (2007)

S. Engel, J.I. Lunine, D.L. Norton, J. Geophys. Res. **99**, 3745–3752 (1994)

H.J. Fischer, T. Spohn, Icarus **83**, 39–65 (1990)

A.D. Fortes, P.M. Grindrod, S.K. Trickett, L. Vocadlo, Icarus **188**, L1–L3 (2007)

W.C. Fraser, M.E. Brown, Astron. J. **695**, 139–153 (2009). doi:10.1016/j.icarus.2006.11.002

B. Giese, T. Denk, G. Neukum, C.C. Porco, T. Roatsch, R. Wagner, Bull. Am. Astron. Soc. **37**, 728 (2005)

B. Giese, G. Neukum, T. Roatsch, T. Denk, C.C. Porco, Planet. Space Sci. **54**, 1156–1166 (2006). doi:10.1016/j.pss.2006.05.027

B. Giese, R. Wagner, G. Neukum, P. Helfenstein, P.C. Thomas, Geophys. Res. Lett. **34**, L21203 (2007). doi:10.1029/2007GL031467
B. Giese, T. Denk, G. Neukum, T. Roatsch, P. Helfenstein, P.C. Thomas, E.P. Turtle, A. McEwen, C.C. Porco, Icarus **193**, 359–371 (2008). doi:10.1016/j.icarus.2007.06.005
P. Goldreich, S.J. Peale, Annu. Rev. Astron. Astrophys. **6**, 287–320 (1968)
P. Goldreich, N. Murray, P.Y. Longaretti, D. Banfield, Science **245**, 500–504 (1989). doi:10.1126/science.245.4917.500
O. Grasset, J. Pargamin, Planet. Space Sci. **53**, 371–384 (2005). doi:10.1016/j.pss.2004.09.062
O. Grasset, C. Sotin, Icarus **123**, 101–112 (1996)
O. Grasset, C. Sotin, F. Deschamps, Planet. Space Sci. **48**, 617–636 (2000). doi:10.1016/S0032-0633(00)00039-8
R. Greenberg, in *Satellites of Jupiter*, ed. by D. Morrison, M.S. Matthews (The University of Arizona Press, Tucson, 1982), pp. 65–92
R. Greenberg, Icarus **70**, 334–347 (1987)
A.R. Hendrix, C.J. Hansen, Icarus **193**, 344–351 (2008)
H. Hussmann, T. Spohn, Icarus **171**, 391–410 (2004). doi:10.1016/j.icarus.2004.05.020
H. Hussmann, F. Sohl, T. Spohn, Icarus **185**, 258–273 (2006). doi:10.1016/j.icarus.2006.06.005
H. Hussmann, G. Choblet, D.L. Matson, C. Sotin, G. Tobie, T. Van Hoolst, Space Sci. Rev. (2010, this issue)
L. Iess, N.J. Rappaport, P. Tortora, J. Lunine, J.W. Armstrong, S.W. Asmar, L. Somenzi, F. Zingoni, Icarus **190**, 585–593 (2007). doi:10.1016/j.icarus.2007.03.027
W.H. Ip, Geophys. Res. Lett. **33**, L16203 (2006). doi:10.1029/2005GL025386
R.A. Jacobson, P.G. Antreasian, J.J. Bordi, K.E. Criddle, R. Ionasescu, J.B. Jones, R.A. Mackenzie, M.C. Meek, D. Parcher, F.J. Pelletier, W.M. Owen, D.C. Roth, I.M. Roundhill, J.R. Stauch, Astron. J. **132**, 2520–2526 (2006)
W.M. Kaula, J. Geophys. Res. **84**, 999–1008 (1979)
R.L. Kirk, D.J. Stevenson, Icarus **69**, 91–134 (1987). doi:10.1016/0019-1035(87)90009-1
R.L. Kirk, L.A. Soderblom, R.H. Brown, S.W. Kieffer, J.S. Kargel, in *Neptune and Triton*, ed. by D.P. Cruikshank (The University of Arizona Press, Tucson, 1995), pp. 949–989
A.V. Krivov, I. Wardinski, F. Spahn, H. Krüger, E. Grün, Icarus **157**, 436–455 (2002). doi:10.1006/icar.2002.6848
K. Kuramoto, T. Matsui, J. Geophys. Res. **99**, 21183–21200 (1994)
V. Lainey, J.E. Arlot, Ö. Karatekin, T. Van Hoolst, Nature **459**, 957–959 (2009). doi:10.1038/nature08108
E. Lellouch, B. Sicardy, C. de Bergh, S. Kassi, H.U. Käufl, A. Campargue, Astron. Astrophys. **495**, L17–L21 (2009)
R.D. Lorenz, K.L. Mitchell, R.L. Kirk, A.G. Hayes, O. Aharonson, H.A. Zebker, P. Paillou, J. Radebaugh, J.I. Lunine, M.A. Janssen, S.D. Wall, R.M. Lopes, B. Stiles, S. Ostro, G. Mitri, E.R. Stofan, Geophys. Res. Lett. **35**, L02206 (2008a). doi:10.1029/2007GL032118
R.D. Lorenz, B.W. Stiles, R.L. Kirk, M.D. Allison, P.P. del Marmo, L. Iess, J.I. Lunine, S.J. Ostro, S. Hensley, Science **319**, 1649–1651 (2008b). doi:10.1126/science.1151639
J.S. Loveday, R.J. Nelmes, M. Guthrie, S.A. Delmonte, D.R. Allan, D.D. Klug, J.S. Tan, Y.P. Handa, Nature **410**, 661–663 (2001)
R.P. Lowell, P.A. Rona, Geophys. Res. Lett. **29** (2002). doi:10.1029/2001GL014411
J. Lunine, M. Choukroun, D. Stevenson, G. Tobie, in *Titan from Cassini-Huygens*, ed. by R.H. Brown, J.P. Lebreton, J.H. Waite (Springer, Dordrecht, 2009), pp. 35–59. doi:10.1007/978-1-4020-9215-2_3
J.I. Lunine, D.J. Stevenson, Icarus **70**, 61–77 (1987). doi:10.1016/0019-1035(87)90075-3
J.I. Lunine, S.K. Atreya, J.B. Pollack, in *Origin and Evolution of Planetary and Satellite Atmospheres*, vol. 130 (University of Arizona Press, Tucson, 1989), pp. 605–665
R.A. Mackenzie, L. Iess, P. Tortora, N.J. Rappaport, Geophys. Res. Lett. **35**, L05204 (2008). doi:10.1029/2007GL032898
R. Malhotra, Icarus **94**, 399–412 (1991)
S. Marchi, C. Barbieri, A. Dell'Oro, P. Paolicchi, Astron. Astrophys. **381**, 1059–1065 (2002). doi:10.1051/0004-6361:20011587
S. Marchi, C. Barbieri, M. Lazzarin, Planet. Space Sci. **52**, 671–677 (2004). doi:10.1016/j.pss.2004.02.002
F. Marchis, D.L. Mignant, F.H. Chaffee, A.G. Davies, S.H. Kwokb, R. Prangë, I. dePatera, P. Amico, R. Campbell, T. Fusco, R.W. Goodrich, A. Conrad, Icarus **176**, 96–122 (2005)
D.L. Matson, T.V. Johnson, G.J. Veeder, D.L. Blaney, A.G. Davies, J. Geophys. Res. **106**, 33021–33024 (2001)
D.L. Matson, J.C. Castillo, J. Lunine, T.V. Johnson, Icarus **187**, 569–573 (2007). doi:10.1016/j.icarus.2006.10.016
W.B. McKinnon, Nature **311**, 355–358 (1984)
W.B. McKinnon, Icarus **183**, 435–450 (2006). doi:10.1016/j.icarus.2006.03.004

W.B. McKinnon, A.C. Leith, Icarus **118**, 392–413 (1995)
W.B. McKinnon, J.I. Lunine, D. Banfield, in *Neptune and Triton*, ed. by D.P. Cruikshank (The University of Arizona Press, Tucson, 1995), pp. 807–877
W.B. McKinnon, D.P. Simonelli, G. Schubert, in *Pluto and Charon*, ed. by S.A. Stern, D.J. Tholen (The University of Arizona Press, Tucson, 1997), pp. 295–343
W.B. McKinnon, D. Prialnik, S.A. Stern, A. Coradini, in *Solar System Beyond Neptune*, ed. by M.A. Barucci, H. Boehnhardt, D. Cruikshank, A. Morbidelli (The University of Arizona Press, Tucson, 2008), pp. 213–241
H.J. Melosh, F. Nimmo, Lunar Planet. Sci. **40**, 2478 (2009)
J. Meyer, J. Wisdom, Icarus **188**, 535–539 (2007). doi:10.1016/j.icarus.2007.03.001
J. Meyer, J. Wisdom, Icarus **198**, 178–180 (2008a). doi:10.1016/j.icarus.2008.06.012
J. Meyer, J. Wisdom, Icarus **193**, 213–223 (2008b). doi:10.1016/j.icarus.2007.09.008
G. Mitri, A.P. Showman, Icarus **193**, 387–396 (2008). doi:10.1016/j.icarus.2007.07.016
G. Mitri, A.P. Showman, J.I. Lunine, R.M.C. Lopes, Icarus **196**, 216–224 (2008). doi:10.1016/j.icarus.2008.02.024
J. Monteux, N. Coltice, F. Dubuffet, Y. Ricard, Geophys. Res. Lett. **34**, L24201 (2007). doi:10.1029/2007GL031635
W.B. Moore, J. Geophys. Res. **108**, 5096 (2003). doi:10.1029/2002JE001943
W.B. Moore, H. Hussmann, in *Europa*, ed. by R.T. Pappalardo, W.B. McKinnon, K. Khurana (The University of Arizona Press, Tucson, 2009), pp. 369–380
I. Mosqueira, P.R. Estrada, Icarus **163**, 198–231 (2003). doi:10.1016/S0019-1035(03)00076-9
I. Mosqueira, P. Estrada, D. Turrini, Space Sci. Rev. (2010, this issue). doi:10.1007/S11214-009-9614-6
S. Mueller, W.B. McKinnon, Icarus **76**, 437–464 (1988)
D. Nesvorný, J.L.A. Alvarellos, L. Dones, H.F. Levison, Astron. J. **126**, 398–429 (2003)
D. Nesvorný, D. Vokrouhlický, A. Morbidelli, Astron. J. **133**, 1962–1976 (2007)
H.B. Niemann, S.K. Atreya, S.J. Bauer, G.R. Carignan, J.E. Demick, R.L. Frost, D. Gautier, J.A. Haberman, D.N. Harpold, D.M. Hunten, G. Israel, J.I. Lunine, W.T. Kasprzak, T.C. Owen, M. Paulkovich, F. Raulin, E. Raaen, S.H. Way, Nature **438**, 779–784 (2005). doi:10.1038/nature04122
G.W. Ojakangas, D.J. Stevenson, Icarus **66**, 341–358 (1986)
J.P. Osegovic, M.D. Max, J. Geophys. Res. **110**, E08004 (2005). doi:10.1029/2005JE002435
S.J. Peale, in *Planetary Satellites*, ed. by J.A. Burns (The University of Arizona Press, Tucson, 1977), pp. 87–112
S.J. Peale, in *Satellites*, ed. by J.A. Burns, M.S. Matthews (The University of Arizona Press, Tucson, 1986), pp. 159–223
S.J. Peale, Celest. Mech. Dyn. Astron. **87**, 129–155 (2003)
S.J. Peale, M.H. Lee, Science **298**, 593–597 (2002)
S.J. Peale, P. Cassen, R.T. Reynolds, Science **20**, 892–894 (1979)
M.J. Person, J.L. Elliott, A.A.S. Gulbis, J.M. Pasachoff, B.A. Babcock, S.P. Souza, J. Gangestad, Astron. J. **132**, 1575–1580 (2006)
J.P. Poirier, L. Boloh, P. Chambon, Icarus **55**, 218–230 (1983)
C.C. Porco, E. Baker, J. Barbara, K. Beurle, A. Brahic, J.A. Burns, S. Charnoz, N. Cooper, D.D. Dawson, A.D.D. Genio, T. Denk, L. Dones, U. Dyudina, M.W. Evans, B. Giese, K. Grazier, P. Helfenstein, A.P. Ingersoll, R.A. Jacobson, T.V. Johnson, A. McEwen, C.D. Murray, G. Neukum, W.M. Owen, J. Perry, T. Roatsch, J. Spitale, S. Squyres, P.C. Thomas, M. Tiscareno, E. Turtle, A.R. Vasavada, J. Veverka, R. Wagner, R. West, Science **307**, 1237–1242 (2005). doi:10.1126/science.1107981
C.C. Porco, P. Helfenstein, P.C. Thomas, A.P. Ingersoll, J. Wisdom, R. West, G. Neukum, T. Denk, R. Wagner, T. Roatsch, S. Kieffer, E. Turtle, A. McEwen, T.V. Johnson, J. Rathbun, J. Veverka, D. Wilson, J. Perry, J. Spitale, A. Brahic, J.A. Burns, A.D. DelGenio, L. Dones, C.D. Murray, S. Squyres, Science **311**, 1393–1401 (2006). doi:10.1126/science.1123013
L.M. Prockter, F. Nimmo, R.T. Pappalardo, Geophys. Res. Lett. **32**, L14202 (2005). doi:10.1029/2005GL022832
M. Prockter et al., Space Sci. Rev. (2010, this issue)
D. Ragozzine, M.E. Brown, Astron. J. **137**, 4766–4776 (2009). doi:10.1088/0004-6256/137/6/4766
N.J. Rappaport, L. Iess, J. Wahr, J.I. Lunine, J.W. Armstrong, S.W. Asmar, P. Tortora, M. Di Benedetto, P. Racioppa, Icarus **194**, 711–720 (2008). doi:10.1016/j.icarus.2007.11.024
J.H. Roberts, F. Nimmo, Lunar Planet. Sci. **40**, 1927 (2009)
M.N. Ross, G. Schubert, Icarus **78**, 90–101 (1989)
M.N. Ross, G. Schubert, Geophys. Res. Lett. **17**, 1749–1752 (1990)
D. Sandwell, G. Schubert, Icarus (2010, submitted)
E.L. Schaller, M.E. Brown, Astrophys. J. Lett. **684**, L107–L109 (2008)
P. Schenk, M.P.A. Jackson, Geology **21**, 299–302 (1993)

P.M. Schenk, J.M. Moore, in *Solar System Ices*, ed. by B. Schmitt, C. de Bergh, M. Festou (Kluwer, Dordrecht, 1998), pp. 551–578
P.M. Schenk, J.M. Moore, Lunar Planet. Sci. **40**, 2465 (2009)
P.M. Schenk, K. Zahnle, Icarus **192**, 135–149 (2007). doi:10.1016/j.icarus.2007.07.004
G. Schubert, D.J. Stevenson, K. Ellsworth, Icarus **47**, 46–59 (1981). doi:10.1016/0019-1035(81)90090-7
G. Schubert, D.L. Turcotte, P. Olson, *Mantle Convection in the Earth and Planets* (Cambridge University Press, Cambridge, 2001)
G. Schubert, J.D. Anderson, T. Spohn, W.B. McKinnon, in *Jupiter: The Planet, Satellites and Magnetosphere*, ed. by F. Bagenal, T.E. Dowling, W.B. McKinnon (Cambridge University Press, Cambridge, 2004), pp. 281–306
G. Schubert, J.D. Anderson, B.J. Travis, J. Palguta, Icarus **188**, 345–355 (2007). doi:10.1016/j.icarus.2006.12.012
M. Segatz, T. Spohn, M.N. Ross, G. Schubert, Icarus **75**, 187–206 (1988)
H. Senshu, K. Kuramoto, T. Matsui, J. Geophys. Res. **107**, E125118 (2002). doi:10.1029/2001JE001819
A.P. Showman, R. Malhotra, Icarus **127**, 93–111 (1997)
A.P. Showman, D.J. Stevenson, R. Malhotra, Icarus **129**, 367–383 (1997). doi:10.1006/icar.1997.5778
B.A. Smith, L. Soderblum, R. Batson, P. Bridges, J. Inge, H. Masursky, E. Shoemaker, R. Beebe, J. Boyce, G. Briggs, A. Bunker, S.A. Collins, C.J. Hansen, T.V. Johnson, J.L. Mitchell, R.J. Terrile, A.F. Cook, J. Cuzzi, J.B. Pollack, G.E. Danielson, A. Ingersoll, M.E. Davies, G.E. Hunt, D. Morrison, T. Owen, C. Sagan, J. Veverka, R. Strom, V.E. Suomi, Science **215**, 505–537 (1982)
F. Sohl, H. Hussmann, B. Schwentker, T. Spohn, R.D. Lorenz, J. Geophys. Res. **108**, 5130 (2003). doi:10.1029/2003JE002044
F. Sohl et al., Space Sci. Rev. (2010, this issue)
S. Soter, in *IAU Colloquium 28*, Cornell University, 1974
C. Sotin, G. Mitri, N. Rappaport, G. Schubert, D. Stevenson, in *Titan from Cassini/Huygens*, ed. by R.H. Brown, J.P. Lebreton, J.H. Waite (Springer, Dordrecht, 2009), pp. 61–73. doi:10.1007/978-1-4020-9215-2_4
J.R. Spencer, T. Denk, Science **10** (2009). doi:10.1126/science.1177132
J.R. Spencer, J.C. Pearl, M. Segura, F.M. Flasar, A. Mamoutkine, P. Romani, B.J. Buratti, A.R. Hendrix, L.J. Spilker, R.M.C. Lopes, Science **311**, 1401–1405 (2006)
S.W. Squyres, R.T. Reynolds, P.M. Cassen, S.J. Peale, Icarus **53**, 319–331 (1983)
S.A. Stern, Space Sci. Rev. **140**, 3–21 (2008). doi:10.1007/s11214-007-9295-y
S.A. Stern, W.B. McKinnon, Astron. J. **119**, 945–952 (2000). doi:10.1086/301207
S.A. Stern, H.A. Weaver, A.J. Steffl, M.J. Mutchler, W.J. Merline, M.W. Buie, E.F. Young, L.A. Young, J.R. Spencer, Nature **439**, 946–948 (2006). doi:10.1038/nature04548
B.W. Stiles, R.L. Kirk, R.D. Lorenz, S. Hensley, E. Lee, S.J. Ostro, M.D. Allison, Callahan Ps, Y. Gim, L. Iess, P. Perci del Marmo, G. Hamilton, W.T.K. Johnson, R.D. West, The Cassini RADAR Team, Astron. J. **139**, 311 (2010). doi:10.1088/0004-6256/139/1/311
P.C. Thomas, J.W. Armstrong, S.W. Asmar, J.A. Burns, T. Denk, B. Giese, P. Helfenstein, L. Iess, T.V. Johnson, A. McEwen, L. Nicolaisen, C. Porco, N. Rappaport, J. Richardson, L. Somenza, P. Tortora, E.P. Turtle, J. Veverka, Nature **448**, 50–53 (2007a). doi:10.1038/nature05779
P.C. Thomas, J.A. Burns, P. Helfenstein, S. Squyres, J. Veverka, C. Porco, E.P. Turtle, A. McEwen, T. Denk, B. Giese, T. Roatsch, T.V. Johnson, R.A. Jacobson, Icarus **190**, 573–584 (2007b). doi:10.1016/j.icarus.2007.03.012
G. Tobie, O. Grasset, J.I. Lunine, A. Mocquet, C. Sotin, Icarus **175**, 496–502 (2005). doi:10.1016/j.icarus.2004.12.007
G. Tobie, J.I. Lunine, C. Sotin, Nature **440**, 61–64 (2006). doi:10.1038/nature04497
G. Tobie, M. Choukroun, O. Grasset, S. Le Mouélic, J.I. Lunine, C. Sotin, O. Bourgeois, D. Gautier, M. Hirtzig, S. Lebonnois, L. Le Corre, Philos. Trans. R. Soc. A **367**, 617–631 (2009). doi:10.1098/rsta.2008.0246
G. Tobie et al., Space Sci. Rev. (2010, this issue)
F. Tosi, D. Turrini, A. Coradini, the VIMS team, Mon. Not. R. Astron. Soc. (2009, in press). arXiv:0902.3591
D. Toublanc, J.P. Parisot, J. Brillet, D. Gautier, F. Raulin, C.P. McKay, Icarus **113**, 2–26 (1995). doi:10.1006/icar.1995.1002
D.L. Turcotte, G. Schubert, *Geodynamics*, 2nd edn. (Cambridge University Press, Cambridge, 2002)
D. Turrini, F. Marzari, H. Beust, Mon. Not. R. Astron. Soc. **391**, 1029–1051 (2008)
D. Turrini, F. Marzari, F. Tosi, Mon. Not. R. Astron. Soc. **392**, 455–474 (2009)
G.J. Veeder, D.L. Matson, T.V. Johnson, D.L. Blaney, J.D. Goguen, J. Geophys. Res. **99**, 17095–17162 (1994)
G.J. Veeder, D.L. Matson, T.V. Johnson, A.G. Davies, D.L. Blaney, Icarus **169**, 264–270 (2004). doi:10.1016/j.icarus.2003.11.016

A.J. Verbiscer, M.F. Skrutskie, D.P. Hamilton, Nature **461**, 1098–1100 (2009). doi:10.1038/nature08515

D. Vokrouhlický, D. Nesvorný, H.F. Levison, Astron. J. **136**, 1463–1476 (2008). doi:10.1088/0004-6256/136/4/1463

H.A. Weaver, W.C. Gibson, M.B. Tapley, L.A. Young, S.A. Stern, Space Sci. Rev. **140**, 75–91 (2008). doi:10.1007/s11214-008-9376-6

C.F. Yoder, Nature **279**, 767–770 (1979)

C.F. Yoder, S.J. Peale, Icarus **47**, 1–35 (1981)

L.A. Young, S.A. Stern, H.A. Weaver, F. Bagenal, R.P. Binzel, B. Buratti, A.F. Cheng, D. Cruikshank, G.R. Gladstone, W.M. Grundy, D.P. Hinson, M. Horanyi, D.E. Jennings, I.R. Linscott, D.J. McComas, W.B. McKinnon, R. McNutt, J.M. Moore, S. Murchie, C.B. Olkin, C.C. Porco, H. Reitsema, D.C. Reuter, J.R. Spencer, D.C. Slater, D. Strobel, M.E. Summers, G.L. Tyler, Space Sci. Rev. **140**, 93–127 (2008). doi:10.1007/s11214-008-9462-9

Y.L. Yung, M. Allen, J.P. Pinto, Astrophys. J. Suppl. **55**, 465–506 (1984)

K. Zahnle, P. Schenk, H. Levison, L. Dones, Icarus **163**, 263–289 (2003). doi:10.1016/S0019-1035(03)00048-4

H.A. Zebker, B. Stiles, S. Hensley, R. Lorenz, R.L. Kirk, J. Lunine, Science **324**, 921–921 (2009). doi:10.1126/science.1168905

Space Sci Rev (2010) 153: 485–510
DOI 10.1007/s11214-010-9646-y

Subsurface Water Oceans on Icy Satellites: Chemical Composition and Exchange Processes

Frank Sohl · Mathieu Choukroun · Jeffrey Kargel · Jun Kimura · Robert Pappalardo · Steve Vance · Mikhail Zolotov

Received: 24 August 2009 / Accepted: 1 March 2010 / Published online: 13 April 2010

Abstract The state of knowledge about the structure and composition of icy satellite interiors has been significantly extended by combining direct measurements from spacecraft, laboratory experiments, and theoretical modeling. The existence of potentially habitable liquid water reservoirs on icy satellites is dependent on the radiogenic heating of the rock component, additional contributions such as the dissipation of tidal energy, the efficiency of heat transfer to the surface, and the presence of substances that deplete the freezing point of liquid water. This review summarizes the chemical evolution of subsurface liquid water oceans, taking into account a number of chemical processes occuring in aqueous environments and

F. Sohl (✉)
Institute of Planetary Research, German Aerospace Center (DLR), Rutherfordstr. 2, 12489 Berlin, Germany
e-mail: frank.sohl@dlr.de

M. Choukroun · R. Pappalardo · S. Vance
Jet Propulsion Laboratory, Caltech, MS 79-24, 4800 Oak Grove Drive, Pasadena, CA 91106, USA

M. Choukroun
e-mail: mathieu.choukroun@jpl.nasa.gov

R. Pappalardo
e-mail: robert.pappalardo@jpl.nasa.gov

S. Vance
e-mail: svance@jpl.nasa.gov

J. Kargel
Department of Hydrology, University of Arizona, Tuscon, AZ 85721, USA
e-mail: kargel@hwr.arizona.edu

J. Kimura
Department of Cosmosciences, Hokkaido University, Nishi 8, Kita 10, Kita-Ku, Sapporo 060-0810, Japan
e-mail: junkim@ep.sci.hokudai.ac.jp

M. Zolotov
School of Earth and Space Exploration, Arizona State University, Tempe, AZ 85287-1404, USA
e-mail: zolotov@asu.edu

partly related to material exchange with the deep interior. Of interest are processes occuring at the transitions from the liquid water layer to the ice layers above and below, involving the possible formation of clathrate hydrates and high-pressure ices on large icy satellites. In contrast, water-rock exchange is important for the chemical evolution of the liquid water layer if the latter is in contact with ocean floor rock on small satellites. The composition of oceanic floor deposits depends on ambient physical conditions and ocean chemistry, and their evolutions through time. In turn, physical properties of the ocean floor affect the circulation of oceanic water and related thermal effects due to tidally-induced porous flow and aqueous alteration of ocean floor rock.

Keywords Outer solar system · Water oceans · Oceanic composition · Chemical evolution · Organic compounds · High-pressure ices · Clathrate hydrates · Hydrothermal activity · Porous rock · Serpentinization · Europa · Ganymede · Callisto · Titan · Enceladus

1 Introduction

Based on the notion that water ice is one of the major constituents in the outer solar system, early models of condensation and icy satellite formation predicted that incorporation of abundant water ice and silicate rock containing radionuclides could have led to ice melting and the formation of extended liquid water mantles beneath conductive outer ice shells (Lewis 1971, 1972, 1973). The maintenance of these global subsurface liquid water oceans was found to be closely related to a satellite's internal structure, chemical composition, and thermal history (Consolmagno and Lewis 1976, 1977, 1978). Later, it was argued that temperature increase with depth would affect the rheological properties of the outer ice shell and the efficiency of subsolidus convection controlling the thermal history of the deep interior (Schubert et al. 1981, 1986; Spohn and Schubert 2003). As a consequence of efficient convective cooling, the radial extent of subsurface water oceans would be strongly diminished and could eventually lead to freezing. Contrarily, laboratory experiments indicated that the melting temperature of ice will be significantly reduced by small amounts of salts and/or incorporated polar solutes such as methanol and ammonia (e.g. Croft et al. 1988; Grasset and Pargamin 2005). The latter species are present in comets and may be present in other icy bodies in the outer solar system.

The state of knowledge about the structure and composition of icy satellite interiors has improved considerably in the late 70s and mid 80s, when the *Voyager* spacecraft flew by Jupiter, Saturn, Uranus, and Neptune, followed by the advent of the *Galileo* spacecraft in the Jovian system (1995–2003) and the *Cassini* spacecraft in the Saturnian system (since 2004). *Voyager* observations of the Jovian moons Io and Europa and, to a lesser extent, Ganymede suggested tidal heating to be a major heat source, possibly resulting in near-surface viscous deformation of warmed ice and cryovolcanic resurfacing (e.g. Johnson 2005, and references therein). An outstanding finding of the *Galileo* mission, based on the interpretation of magnetometer data (e.g. Khurana et al. 1998; Kivelson et al. 1999, 2000), was the detection of induced magnetic fields in the vicinity of Europa, Callisto, and possibly Ganymede (Kivelson et al. 2004). The observed magnetic signatures suggest the presence of electrically conducting reservoirs of salty liquid water beneath the satellites' outer icy shells that may contain more water than the terrestrial oceans combined. *Galileo* imagery of Europa has been interpreted in terms of endogenic morphological features (e.g. Greeley et al. 2004) and compositional heterogeneity (e.g. McCord et al. 1999; Pappalardo et al. 1999; Carlson et al. 1999; Prockter et al. 2010, this issue; Dalton et al. 2010, this issue). The orientation of the satellite-wide, symmetric fracture pattern has been attributed to nonsynchronous rotation of the outer icy shell across Europa's permanent tidal bulge (Geissler et al. 1998; Kattenhorn 2002). Particularly strike-slip behavior and plate rotation near the anti-Jovian point (Schenk and McKinnon 1989) and episodic plate separation and fracture infill (Sullivan et al. 1998) is taken as geologic evidence that the outer ice shell is decoupled from Europa's deep interior due to the existence of a subsurface water ocean (Greenberg et al. 1998, 2002). Schenk et al. (2008) have attributed current locations of broad arcuate troughs and depressions to the episodic reorientation of Europa's floating ice shell by true polar wander of $\sim 80°$. The detection of non-icy, sulfur-bearing, and hydrated colored materials in endogenic surface features on Europa and Ganymede suggests an oceanic origin of these surface compounds (e.g. McCord et al. 1999, 2001). Cassini synthetic aperture radar (SAR) imagery of Titan has been interpreted in terms of non-synchronous rotation. Provided the satellite's outer ice shell were decoupled from the deep interior by a liquid water layer, Titan's supersynchronous spin could be attributed to seasonal exchange of angular momentum with the dense superrotating atmosphere (Lorenz et al. 2008). The detection of alkali salts in Saturn's E ring (Postberg et al. 2009), formed through eruptive activity on Enceladus (Schmidt et al. 2008), indicates past or present aqueous activity in the satellite's interior. Although a large global ocean may not exist on present-day Enceladus, sodium and potassium could have been leached from rocks early in history and accumulated at the ice-rock interface upon ocean freezing (Zolotov 2007). Geologically recent tidal heating below the south pole of Enceladus may involve the salts in eruptive plume activity.

In this review, we focus on the chemical development of subsurface liquid water oceans, physical constraints on aqueous circulation, and related exchange processes involving deep satellite interiors. In Sect. 2, we discuss some aspects of the chemical evolution of aqueous fluids and primordial oceans. Section 3 is devoted to water-ice interaction, involving segregation of high-pressure ice polymorphs and clathrate hydrates at ice-water boundaries. In Sect. 4, we discuss the impact of physical ocean floor properties such as permeability, porosity, and compressibility on Europa-like circulation of oceanic water and related thermal effects due to porous flow and aqueous alteration close to the rock-water boundary. Section 5 discusses possible implications for the compositional stratification of different classes of icy bodies. Discussion and conclusions are presented in Sect. 6.

2 Chemical Evolution

Pathways of oceanic chemical evolution are mainly affected by the bulk composition and thermal history of a satellite. The bulk composition constitutes the ratios between rocks, water ice, non-water volatiles, and organic compounds. Throughout history, the chemical evolution of subsurface water oceans has been affected by the degree of ice melting, the extent and duration of chemical interactions between rocks and liquid water, degassing of the deep interior and oceanic water, freezing and thawing cycles, and secondary inorganic and organic precipitation. A great variety of these factors and processes among icy satellites implies diverse pathways of their aqueous evolution.

2.1 Formation and Early Chemical Evolution of Aqueous Fluids

The satellites of the giant planets are thought to have originated from a compositionally diverse mixture of grains that slowly accumulated from the surrounding solar accretion disk (Canup and Ward 2002, 2009; Mosqueira et al. 2010, this issue; Coradini et al. 2010, this issue). Contrary to chondrites, these rocky materials would not have been thermally reprocessed in the accretion disk and would have consisted of dust particles ($\lesssim 1$ μm in diameter) without chondrules. A deficiency of chondrules suggests that accreted mineral dust was closer to solar composition than materials that formed asteroids. The predominant accretion of solids as dust also implies that minerals would not have been aqueously altered prior to accretion. Despite some chemical differences, icy satellites might have accreted rocky materials similar to those present on asteroids, together with abundant amorphous silicate phases, ices, and organic compounds. The rocky crystalline solids consisted of reduced and anhydrous products of solar nebula processes (mainly Fe-Ni metal, FeS, Mg- and Ca-silicates, and Ca-Al-oxides), organic matter, and minor presolar grains. By analogy with carbonaceous chondrites (e.g. Sephton 2002), the majority of accreted organic matter was in a polymer in which polyaromatic groups are linked by O-, N-, and S-bearing aliphatic units. If accreted, cometary-type materials (e.g. Bockelée-Morvan et al. 2004) also delivered CO_2, CO, methanol, ethane, acetylene, ammonia, H_2S, formaldehyde, and condensed C-H-O-N rich organic species. The initial composition of icy satellites may range between those of carbonaceous (C, D, P) asteroids (McKinnon and Zolensky 2003) and comets with increasing heliocentric distance. An increase in the abundance of C and N species with heliocentric distance reflects the increasing similarity of outer solar system bodies with the composition of the solar photosphere, which is significantly enriched in these elements compared to CI carbonaceous chondrites (Palme and Jones 2003).

Some satellites may have formed lately by accreting planetesimals (similar to CI/CM carbonaceous chondrites) that were aqueously altered and contained hydrated and oxidized phases and salts (phyllosilicates, magnetite, Fe-Ni sulfides, carbonates, phosphates, chlorides, and sulfates; Kargel 1991). For example, Kuskov and Kronrod (2005) consider L/LL-chondrites as the most likely analogue case for Europa's bulk iron-silicate composition. This compositional model would provide mean densities and dimensionless moment-of-inertia factors for Europa's silicate mantle and core that are similar to Io's, while carbonaceous chondrite compositional models would yield significantly higher values. Taking into account these cosmochemical constraints, a three-layer interior structure can be considered as the most plausible, with core radii ranging from 470 to 640 km, corresponding to bulk Fe/Si ratios of 0.9 to 1.3 (Kuskov and Kronrod 2005), i.e. less than that for CI carbonaceous chondrites of 1.7 ± 0.1 (Kerridge and Matthews 1988).

Deposition of impact energy during accretion, decay of radionuclides, energy release through water-ice differentiation, and heat of mineral hydration reactions are major heat sources for ice melting (see Hussmann et al. 2010; Schubert et al. 2010, this issue). Contrary to earlier scenarios (Schubert et al. 1986; Stevenson et al. 1986), gas-starved accretion (e.g. Canup and Ward 2009) does not provide much accretion energy, and impact melting may only be related to large planetesimals hitting large satellites. If satellite accretion was accomplished within ~ 8 Ma after the formation of Ca-, Al-rich inclusions (CAIs) in the solar nebula, heat production through the decay of short-lived ^{26}Al ($\tau_{1/2} \approx 0.7$ Ma) would surpass that of long-lived radionuclides (^{235}U, ^{238}U, ^{40}K, ^{232}Th). There is no clear understanding of how the amount of accreted ^{26}Al may have varied with heliocentric distance. Early differentiated asteroids in the inner asteroid belt (Scott 2007) and some Saturnian satellites such as Iapetus (Castillo-Rogez et al. 2007) may have accreted a significant amount of ^{26}Al a few Ma after CAI formation, while partial differentiation of Callisto (McKinnon 2006) may suggest late accretion without abundant ^{26}Al. Despite the lower but more persistent heat production by long-lived radionuclides, melting may occur on larger icy satellites and even on small bodies such as Enceladus (Schubert et al. 2007).

Due to the gradual increase in temperature over time, early melt formation could have occurred within icy satellite interiors (see Schubert et al. 2010, this issue). In the presence of non-water ice components detected or expected in the outer solar system (e.g. CH_3OH, NH_3, ethylene glycol, HCl, H_2S, CO, CO_2), eutectic solutions at temperatures well below 273 K would have formed. Heat-producing silicate particles then would have been surrounded by eutectic brine pockets. First solutions could have formed at temperatures as low as ~ 175 K, which corresponds to the NH_3-H_2O eutectic (e.g. Desch et al. 2009). Melting of HCl hydrates at ~ 186 K (Zolotov and Mironenko 2007a) would result in low pH, facilitating dissolution of mineral phases and amorphous silicate materials. This process enriched solutions by cations and caused some secondary precipitation (e.g. amorphous silica) and neutralization of fluids (Zolotov and Mironenko 2007b). Subsequently, the increase in temperature caused dilution of these primordial solutions and the formation of larger melt pockets and liquid reservoirs.

Heating and melting of ice is paramount to outward migration of aqueous solutions and rock particle settling toward the center, eventually leading to internal ice-rock differentiation and cryovolcanic resurfacing. Provided ice-rock differentiation had run to completion, silicate mantles (Europa, Ganymede, Enceladus) overlain by water ice/liquid shells would form (e.g. Schubert et al. 1986, 2007, 2009). Additional segregation of an Fe-rich core (Ganymede, Europa) is thought to be accompanied by some dehydration of phyllosilicates within the silicate mantles formed earlier. Depending on the degree of internal differentiation of ice and rock, primordial oceans on large ice-rich moons (Ganymede, Callisto, Titan) could be sandwiched between outer ice shells and high-pressure ice layers below. This separation would inhibit further compositional evolution of these liquid reservoirs. Contrary, liquid water was likely accumulated in the outer parts of smaller satellites (Enceladus, Europa, Triton) where the chemical evolution of subsurface water oceans would proceed in contact with rocky seafloors and/or oceanic sediments (e.g. Schubert et al. 2007, 2009).

Once initial solutions first appeared in the presence of ice, early liquid water on icy satellites participated in competitive oxidation and hydration reactions, analogous to aqueous processes on chondrite parent bodies (e.g. Brearley 2006). Reaction (1) exemplifies oxidation and reaction (2) illustrates hydration,

$$3Fe^0 \text{ (metal)} + 4H_2O \text{ (l)} \rightarrow Fe_3O_4 \text{ (magnetite)} + 4H_2 \tag{1}$$

$$Mg_2SiO_4 \text{ (forsterite)} + MgSiO_3 \text{ (enstatite)} + 2H_2O \text{ (l)} \rightarrow Mg_3Si_2O_5(OH)_4 \text{ (serpentine)}. \tag{2}$$

Note that metal was supplied by Fe-Ni alloys (kamacite) and a substantial portion of silicate reagents was amorphous. Phyllosilicates (serpentine, cronstedtite, saponite, etc.) and tochilinite (a hydrated Fe-Ni sulfide) formed through hydration. Heat released by hydration reactions contributed to the thermal balance early in a satellite's thermochemical evolution (e.g. Schubert et al. 2007). Oxidation led to formation of ferrous (serpentine, saponite) and ferric (magnetite, cronstedtite) minerals, sulfides (tochilinite, pyrrhotite, pentlandite), Ni-rich alloys, chromite, and phosphates. Note that Fe^{2+} and Fe^{3+} solutes formed through aqueous oxidation of Fe-Ni metal and were also incorporated into secondary silicates.

Hydrogen (H_2) could have formed through mineral oxidation by water, for example, via reaction (1). Separation and rapid release of H_2 to space through fractures in thin icy shells (Europa and Enceladus; Glein et al. 2008; Zolotov and Kargel 2009) would have favored further oxidation reactions. Despite H_2 escape, aqueous systems would not have re-equilibrated under more oxidizing conditions because redox reactions are kinetically inhibited at low temperatures. On large icy satellites (Ganymede, Titan, Triton), oxidation was limited because of restricted H_2 escape through thicker ice shells. Therefore, it is unlikely that highly oxidized species (e.g. sulfates) were able to form on large icy satellites.

In addition to dissolved volatiles and acids, the compositions of primordial oceans were a result of low-temperature (≤ 273 K) aqueous alteration of rocks, often in the presence of partially melted ice. Early in these water-rock interaction processes, aqueous compositions could have varied rapidly, as primary minerals alter with different rates (Zolotov and Mironenko 2007b). With time, neutralization of initial acid solutions and saturation of oceanic water with respect to secondary minerals led to stabilization of oceanic compositions. Early acidic Fe-Mg-Ca-Cl solutions evolved toward alkaline Na-OH-Cl solutions. Eventually, oceanic solutions reached acid-base equilibration with altered rocks, which were rich in secondary phyllosilicates and contained magnetite, sulfides and carbonates (e.g. Zolotov 2007, 2008). However, oxidized species (e.g. Fe^{3+}, carbonate and sulfate, if present) could have coexisted metastably with reduced compounds (Fe^{2+}, sulfide and native sulfur, organic matter). Multiple redox disequilibria (e.g. Zolotov and Shock 2004) may have supported metabolism of chemotrophic organisms, if they were present.

Chemical equilibrium models demonstrate that early oceans on icy satellites could have been cold, alkaline, rich in Na and Cl, depleted in Mg and S, and not saline (e.g. Zolotov 2007, 2008). Sulfur was in sulfide form ($HS^- + S^{2-}$) and its low concentration was controlled by the solubility of Fe sulfides (McKinnon and Zolensky 2003; Zolotov 2008). Sulfate sulfur can only be present if aqueously altered planetesimals accreted (Kargel et al. 2000). The concentration and activity of dissolved H_2 were mostly dependent on pressure and possible escape to space. On smaller moons (Enceladus, Europa) oceanic H_2 could have escaped easily compared to deeply located oceans on larger satellites (Titan, Ganymede, Triton). Extremely reduced conditions at high pressure in those oceans could have favored stability of CH_4 and its formation through hydrogenation of organic and inorganic (e.g, CO_2) species (Zolotov et al. 2005). Convection in the upper icy shell, if present, may have delivered some H_2 and CH_4 toward the surfaces and atmospheres (Titan, Triton).

On large and small satellites, aqueous C species could have been presented by methanol (CH_3OH), HCO_3^-, CO_3^{2-}, and formate ($HCOO^-$), which forms through hydrolysis of CO and HCN, and reactions among single C species (methanol, formaldehyde, and carbonate/bicarbonate ions; Seewald et al. 2006). Note that the abundance and speciation of aqueous C species were limited by inefficient low-temperature dissolution of primary organic matter and by low rates of redox reactions. NaCl, methanol, and ethylene glycol could be the major oceanic solutes, thereby favoring much longer ocean lifetime due to significant freezing point depletion. On some satellites in the Saturnian system and beyond, ammonia is thought to be present (Waite et al. 2009), thereby accounting for relatively cold subsurface water oceans because of the significant melting point depletion (Hussmann et al. 2006). However, the abundance of ammonia may be low or even non-existent; and nitrogen, e.g. present in Titan's atmosphere and at Triton's surface, may result from oxidation of organic compounds deep within each satellite.

2.2 Chemical Evolution of Primordial Oceans

Pathways of subsequent oceanic evolution were mostly dependent on the scale, timing, and locations of heat production in the satellite interiors. On large and/or early-formed satellites, continuous supply and transfer of heat within the silicate mantles could have induced hydrothermal circulation, dehydration of previously formed minerals, and further alteration of organic matter.

If occurred, thermal dehydration of early-formed minerals (phyllosilicates, tochilinite) would have caused upward migration of hydrothermal fluids, thereby affecting ocean mass and composition. As for primordial oceans, the composition of ocean-forming fluids was chiefly controlled by chemical interactions in the uppermost rock layers during and after core dehydration, when hydrothermal processes took place as oceanic water circulated through permeable rocks below the ocean-rock boundary. The release of tidal heat and/or magmatic activity (Europa?) in subjacent rocks enhanced the circulation. As a result, hot aqueous fluids with inorganic and organic species induced changes in ocean compositions. In turn, hydrothermal circulations caused deposition of some oceanic species (e.g. Zolotov 2007; Zolotov and Kargel 2009). However, on large satellites (Titan, Ganymede, Triton), hydrothermal circulation was unlikely because of impermeable rocks and separation of oceans from silicate mantles by high-pressure ices. On Europa, hydrothermal circulation could have affected only suboceanic rocks, and on smaller bodies (Enceladus) circulation may have affected the entire silicate mantle (Vance et al. 2007; Glein et al. 2008, and Sect. 4 of this paper).

Hydrothermal processes further altered the mineralogy of permeable rocks and favored oxidation of rocks, fluids, and organic compounds, and H_2 generation. High-temperature fluids were rich in sulfide species, and the mixing of hydrothermal fluids with oceanic water led to the precipitation of metal sulfides (e.g. pyrite) at ocean-rock interfaces (e.g. Zolotov 2007). The upward transport (bubbling, diffusion, convection) and escape of H_2 oxidized smaller icy moons (Europa, Enceladus; Zolotov and Shock 2004; Glein et al. 2008). High temperature and low-fH_2 conditions increased the carbonate ($CO_2 + HCO_3^- + CO_3^{2-}$) and sulfate concentrations of hydrothermal fluids. The oxidation of sulfides to sulfates favored the partial dissolution of Mg-silicates, which could have led to the formation of Mg-sulfates in severely oxidized satellites, such as Europa (Zolotov and Kargel 2009). Like in parent bodies of CI carbonaceous chondrites, the high Mg/Ca ratio and precipitation of Ca-bearing carbonates could have prevented major precipitation of sulfates in low-solubility Ca-sulfates (gypsum, anhydrite). As a result, Mg-sulfates could have become abundant among oceanic solutes (at least on Europa).

Presumably, the slow escape of H_2 allowed mineral assemblages to control the oxidation state (fH_2) of hydrothermal systems. Prolonged thermal history of Europa over which H_2 could have escaped, all support profound net oxidation of at least upper rock layers and accumulation of oxidized (carbonate and possibly sulfate) aqueous species in the ocean (Zolotov and Kargel 2009). On smaller moons (Enceladus; Glein et al. 2008), this process may have not occurred because of lower temperatures and an early cessation of thermal processes.

Hydrothermal conditions on small satellites favored further transformation of carbon species in fluids and permeable suboceanic rocks. If supplied by accretion, carbon monoxide could have converted to formic acid, carbonate species, methanol and some methane (Seewald et al. 2006). Hydrous pyrolysis of the high-molecular-weight polymer caused oxidation of aliphatic groups, partially liberation of aromatic molecules and led to the formation of oxygen-bearing compounds (carboxylic and amino acids, alcohols; Sephton 2002). Further increase in temperature favored oxidation of organic matter to carbonate species and N_2, and eventually led to graphitization of the polymer, as observed in metamorphosed chondrites (e.g. Brearley 2006). If H_2 escape was restricted, mineral-catalyzed Fisher-Tropsch like synthesis of hydrocarbons and CH_4 may have occurred. However, asteroidal samples (chondrites) do not reveal signs of Fisher-Tropsch type synthesis (Alexander et al. 2007) and the majority of organic species on small moons (Europa, Enceladus) could be presented by primordial and/or transformed primordial species. Note that some methane could have formed though thermal decomposition of primordial organic matter, especially in the presence of a hot aqueous phase. As a result of multiple organic and organic-inorganic interactions, an array of aromatic, aliphatic, and N-, O-, S-bearing organic species, and methane could have been delivered into water oceans. Highly soluble compounds (acids, alcohols) may have made multiple passes through hydrothermal systems causing further oxidation of organic matter in rocks and solutions. In contrast, hydrocarbons could have been exsolved from cold fluids (and ocean waters) and formed floating organic layers below the icy shells (e.g. Zolotov and Kargel 2009). Higher-density organic compounds (tar) could have been accumulated at oceanic floors. It is not excluded that oceanic waters could have been isolated from rock by heavy organic condensates.

Based on energy balance constraints, Hussmann and Spohn (2004) and Tobie et al. (2005a) studied the coupling between ice shell thickness variations and tidal heat production in the presence of resonant satellite orbits (e.g. Laplace resonance of Io, Europa, and Ganymede; see Hussmann et al. 2010, this issue). Oscillations of Europa's ice shell thickness would notably affect oceanic salinity, composition, temperature, and the mineralogy of suboceanic rocks (Zolotov and Kargel 2009). Low tidally-induced activity would cause some freezing of the icy shell, increasing the salinity of oceanic water, mineral precipitation and exsolution of gases. Higher salinities will depress freezing temperatures and imply colder oceanic water. In contrast, intense release of tidal heat would lead to dilution of oceanic water through ice melting and some dehydration of solids. Moderate changes in ice thickness may cause precipitation or dissolution of Ca-carbonates and sulfates, pyrite, ferric hydroxides/oxyhydroxides, amorphous silica, clay minerals and phosphates without changing concentrations of corresponding solutes. However, concentrations of Cl, SO_4^{2-} (if present), Mg, and Na may not be efficiently maintained by solution-mineral reactions, and concentrations of these solutes would vary with changes in ice shell thickness. For example, freezing of 50% of oceanic water in unsaturated ocean models (e.g. Zolotov and Shock 2001) would not lead to precipitation of chlorides and sulfates of Mg and Na, while double oceanic salinity. Precipitation of abundant salts can only be possible after freezing of a significant portion of oceanic water as shown for Europa (Kargel et al. 2000; Zolotov and Shock 2001) and Enceladus (Zolotov 2007). Note, however, that near-eutectic low-temperature brines should not be referred to as water oceans. Instead, near-eutectic solutions may represent low-volume brine pockets or be a small part of an ice-salt-brine slush. Such a slushy reservoir may be present on Enceladus and situated at the rock-ice interface below the south-pole region (Zolotov 2007).

3 Water-Ice Interaction at High Pressure

In large icy objects like Ganymede, Callisto, and Titan, high-pressure polymorphs of water ice, clathrate hydrates, and/or salts may have formed in the course of a satellite's thermochemical evolution. Seven stable phases of crystalline ice are known to exist at pressures and temperatures ranging between 0 and 2500 MPa and from 200 to 350 K, respectively. Among those, five phases melt to form liquid water upon heating and pressure release: ice Ih, ice III, ice V, ice VI, and ice VII. Ices Ic (cubic ice I) and ice II do not melt under these circumstances and are stable at lower temperature. These phases are the main focus of this section, adding to this panel the volatile-bearing clathrate hydrate phases and potential salts.

3.1 Phase Diagram of H_2O-Rich Phases Under Pressure

The phase diagram of H_2O in the relevant range of conditions, as well as available experimental data on the stability of some major clathrate hydrate species, is shown in Fig. 1. Two principal thermal states of icy satellite interiors are distinguished, labeled

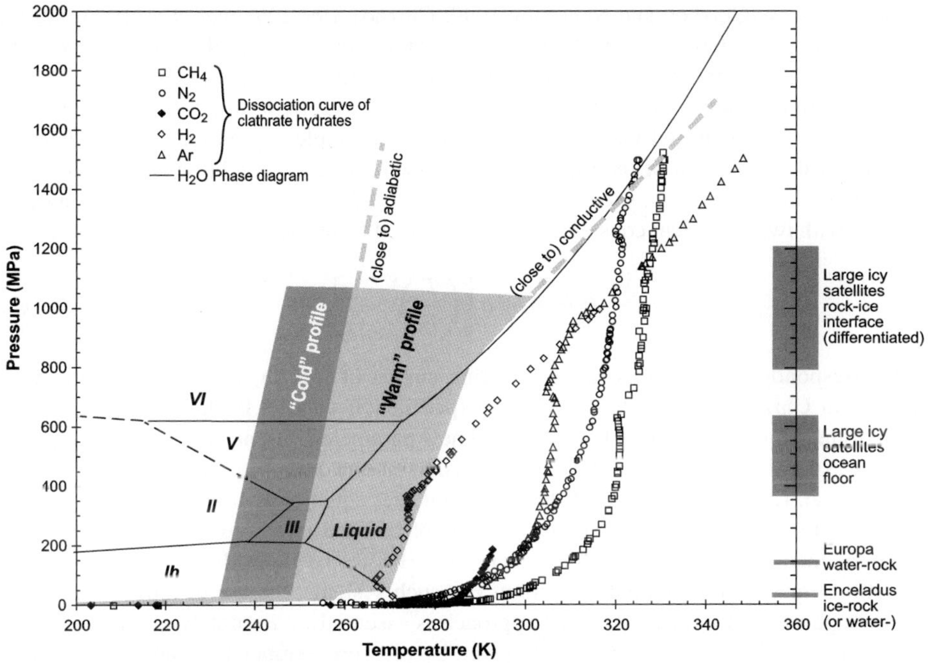

Fig. 1 P–T phase diagram showing the H_2O ices relevant to icy satellite interiors (from Chizhov 1993 and Choukroun and Grasset 2007), and the dissociation curves of major clathrate hydrates under pressure (Sloan 1998, and references therein, Dyadin et al. 1997a, 1997b, 1997c, 2001, Manakov et al. 2001, 2004). For clathrate hydrates, solid-solid phase transitions under pressure are not specified to preserve legibility of the diagram. Ranges of possible thermal profiles within icy satellites, as well as the pressure (i.e. depth) of major interfaces, are indicated in *grey*

cold and *warm*. This distinction is based upon the presence or absence of an internal liquid layer. Whereas *cold* thermal states would require vigorous convection of icy mantles to efficiently compensate for internal heating a moon's interior, *warm* thermal states would result from less efficient conductive heat transfer in the presence of low conductivity icy phases at low temperature (from 0.5 to 6 W m^{-1} K^{-1}) (see Hussmann et al. 2010; Schubert et al. 2010, this issue). For the sake of brevity, the phase diagrams of hydrated salts relevant to icy moons are not shown here (see Fortes and Choukroun 2010, this issue).

The phase diagram of Fig. 1 suggests that only the large icy satellites of Jupiter and Saturn may exhibit icy mantles deep enough to reach pressures above the ice Ih-ice II or ice Ih-ice III phase transitions. Yet, some mid-sized, ice-rich satellites of Saturn like Rhea may possess ice II or ice III in their deepest interiors, if the central part consists of an undifferentiated mixture of silicate and ice (Castillo-Rogez 2006; Anderson and Schubert 2007; Iess et al. 2007). In the case of Ganymede, Callisto, and Titan, it is likely that the present-day thermal profile intersects the high-pressure solid-liquid phase transitions, and that high-pressure ice polymorphs have allegedly formed (e.g. Lupo and Lewis 1979; Schubert et al. 1981, 1986).

A reliable description of ice melting relations as a function of pressure, i.e. depth, is a necessary tool for the prediction of the formation of high-pressure phases. The effect of impurities on melting relations is also of major importance. To that end, thermodynamic models have been developed to treat the pressure-dependent melting of planetary ices in a quantitative way (Chizhov and Nagornov 1990; Chizhov 1993; Wagner et al. 1994; Feistel and Wagner 2006; Choukroun and Grasset 2007). However, such models involve parameters that would require intense laboratory work and extrapolation to conditions beyond the range where experiments are conducted, occasionally leading to somewhat inaccurate results. On the other hand, Simon and Glatzel (1929) deduced a simple expression that facilitates a mathematical description of the melting curves. Two reference conditions are specified in the Simon-Glatzel equation, such as a triple point (P_0, T_0) in the water ice phase diagram, together with two empirical constants, a and c, derived from experimental data according to

$$P = P_0 + a\left[\left(\frac{T}{T_0}\right)^c - 1\right] \tag{3}$$

The corresponding parameters for the melting curves of ices Ih, III, V, and VI are summarized in Chizhov (1993), based on the experimental data of Bridgman (1912, 1937). Choukroun and Grasset (2007) have deduced new parameterizations of (3), also taking into account experimental work conducted by Henderson and Speedy (1987), Grasset (2001), Grasset et al. (2005). The usage of the Simon-Glatzel equation is limited in that it is only applicable to one-component systems, and cannot be used in a convenient manner to describe the melting of ices in a mixed system, like H_2O-NH_3, H_2O-$MgSO_4$, or other binary (and more) mixtures with a eutectic point. For such systems, the use of a thermodynamic and/or sophisticated numerical model approach is essential to accurately predict the melting surfaces in three-dimensional space given by pressure, temperature, and composition. The model developed by Grasset and Pargamin (2005) computes the effect of ammonia on the melting curves of ices, with application to Titan. Parameterization of the effect of salts containing major elements has been achieved by using a Pitzer method in the FREZCHEM software (Marion et al. 1999, 2010), with applications to Europa and Enceladus (Kargel et al. 2000; Zolotov 2007).

Another major type of icy solid that may be formed upon cooling of icy satellites is clathrate hydrates, in which a hydrogen-bonded H_2O skeleton forms cages that trap guest molecules individually by weak van der Waals interactions (e.g. Sloan 1998, and references therein). Clathrate hydrates are thermodynamically more stable than water ice at pressures above a few MPa because of relatively steep gradients dT/dP of the related phase stability curves. As shown in Fig. 1, most planetary-relevant gases with molecular diameters of less than 6 Å can form clathrate hydrates. In terms of absolute pressure, the phase diagram suggests that most clathrate hydrates of these gases are thermodynamically stable within icy satellites. However, one misleading property of gas hydrates is that their stability is governed by the partial pressure of the guest gas, rather than the absolute pressure. Furthermore, their ability to form from a mixture of liquid water and dissolved gas, or to dissociate, is also dependent upon the amounts of gas that can be present, i.e. it is closely related to the solubility of the gas at given ambient conditions.

One must bear in mind these peculiarities, as it implies that some clathrates would not be present, though otherwise suggested by their phase diagrams. Let us use the example of CH_4 clathrates to illustrate this issue. Numerous data have been obtained on the H_2O-CH_4 system (see Sloan 1998, and references therein), which clearly establish the high stability of methane clathrates under pressure. In all these studies, the amounts of methane were somewhat close to the ideal formula of methane clathrates, that is ($CH_4 \cdot 5.75H_2O$). Choukroun et al. (2010) have shown experimentally that, in an aqueous system that contains 0.1 wt.% of CH_4 or less, the dissociation curve under pressure reaches a maximum temperature of 275–280 K, which is 45–50 K lower than the dissociation temperature suggested by the previous

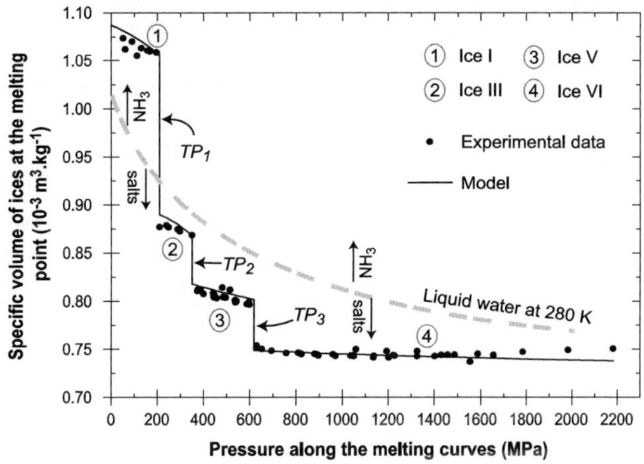

Fig. 2 Specific volume of melted phases of H_2O as a function of pressure. Available experimental data on the specific volume of ices are derived from the measurements of Bridgman (1912, 1937) along the melting curves. TP_i $(i = 1, 2, 3)$ refer to the triple points in the H_2O phase diagram: 1 for Ih-III-L (Liquid), 2 for III-V-L, and 3 for V-VI-L. The specific volume of H_2O has been calculated at 280 K, following the model of Choukroun and Grasset (2007). Reprinted with permission from Choukroun and Grasset (2007). Copyright 2007, American Institute of Physics

experimental data. This result is consistent with thermodynamic predictions using the model of Sloan (1998) with the fugacity laws from Duan et al. (1992), and is valid for any given clathrate hydrate. Therefore, for abundant gaseous species on icy bodies such as CH_4, CO_2, and N_2, the formation of clathrate hydrates seems to be a process that needs accounting for in geophysical models. However, CO_2 may escape clathration due to conversion to bicarbonate and carbonate ions in oceanic water. If formed, N_2 and H_2 are poorly soluble and could have escaped from satellites (Zolotov and Shock 2004). Further, it is still uncertain if minor species like noble gases are able to form clathrates in icy bodies.

3.2 Density and Segregation of Materials

Using phase diagrams, we are able to predict the type of solid phases that may form at the bottom of the ocean, either by solidification upon cooling (ices, clathrates) or by chemical reaction with the silicate interior (minerals). The implications of these phases at the level of a whole icy satellite then depends on: (i) where they will be preferentially located, and (ii) how they will influence heat transfer, hence the dynamics. The buoyancy of the solid phases relative to the liquid layer is the main property controlling the distribution of these phases. Buoyancy thus depends on the respective density of the phases, or their specific volume (which is the inverse of density).

The density of pure liquid water is well known in the temperature and pressure ranges of 298–373 K and 0–800 MPa, respectively (e.g. Grindley and Lind 1971). Several equations of state reproduce accurately these data (e.g. Minassian et al. 1981; Choukroun and Grasset 2007). The work of Bridgman (1912, 1937) provides phase transition data on water ices as well as associated variations in thermodynamic properties, including sudden changes in specific volume. Their experimental data are reproduced with variable accuracy by a number of numerical simulations (Chizhov and Nagornov 1990; Chizhov 1993; Tchijov et al. 2004; Choukroun and Grasset 2007). Figure 2 shows the specific volume of the solid phases of H_2O along their melting curves, as a function of pressure, and compares it to the specific volume of pure liquid water at ~ 280 K.

High-pressure phases of ice are ~ 5–15% denser than liquid water close to the melting point, therefore they would sink to the bottom of the liquid layer upon crystallization. Given both the large density contrast between consecutive ice phases and the likely slow cooling

Table 1 Typical densities of some clathrate hydrates at ~5–10 MPa and ~280 K for partial and full cage occupancy. Partial occupancy considered is 0.7 and 0.9 for small and large cages, respectively. After Sloan (1998)

Guest gas	Structure	Density, $kg\,m^{-3}$ partial occupancy	Density, $kg\,m^{-3}$ full occupancy
CH_4	sI	901	920
N_2	sII	951	1001
CO_2	sI	1084	1134
Ar	sII	1021	1093
Kr	sII	1280	1430
Xe	sI	1654	1805

rate of the liquid layer, it is expected that the layer of high-pressure ices is stratified, with ice VI overlaying the silicate core (if the transition to the rocky core is at pressures above 620 MPa), and ice V above it. Finally, on the cold side of the range of thermal profiles in Fig. 1, ice III may segregate into a thin layer as well. During this sequence of crystallization/segregation of high-pressure ices, the liquid layer is expected to remain located right beneath the outer ice Ih layer (ice Ih being the least dense crystalline H_2O phase).

This scenario, though based on the quantitative comparison shown in Fig. 2, does not take into account the effect of dissolved species on the density of the liquid layer. Dissolved species may affect the density in two ways, since they involve not only ions and neutral species (including organic species), which would increase the density of aqueous solutions, but also gases, whose solubility in liquid water generally increases with pressure, leading to a decrease in the bulk density. Overall, the changes in density associated to the dissolution of gaseous compounds are negligible compared to the large difference in specific volume between solid and liquid water phases. Therefore, we consider the effect of crystallization of clathrate hydrates of these gases. Depending on the type of clathrate hydrate that forms, and the occupancy of the cages, the density can vary significantly as shown in Table 1 for some major clathrate hydrates expected within icy satellites. Since the bulk moduli of water ice phases and those of clathrate hydrates are similar (~ 9 GPa for ice, ~ 7–8 GPa for clathrate hydrates, e.g. Sloan 1998), the density comparison at low pressure is allegedly valid at high pressures as well, despite the lack of experimental data. The density of ice Ih is of 920 $kg\,m^{-3}$ at that pressure, that of liquid water ~ 1050 $kg\,m^{-3}$. In comparison, only CH_4 and N_2 clathrate hydrates are less dense if the cages are fully occupied. Pure Ar clathrates may be less dense in the case of partial occupancies, but unless large amounts of Ar are present, these clathrate hydrates should not be stable within icy satellites. If one considers a mixed gas-water system, it appears that clathrate hydrates containing large amounts of methane will tend to be less dense than the liquid layer and thus float beneath the outer ice layer, as may have happened on Titan (Tobie et al. 2006; Lunine et al. 2009), whereas CH_4-poor clathrates will tend to sink to the bottom of the liquid layer and thicken the solid layer (Prieto-Ballesteros et al. 2005).

The effect of other compounds on the density of the liquid is much more complex, and it is shown qualitatively by the arrows in Fig. 2. For example, the density of a saturated $MgSO_4$ aqueous solution is around 1200 $kg\,m^{-3}$, i.e. about as dense as the ice III polymorph. Aqueous ammonia, on top of its notorious anti-freeze effect, decreases the density of the solution: a near-eutectic (34% NH_3) composition is less dense than water ice, 920 $kg\,m^{-3}$, at temperatures higher than 230 K (Croft et al. 1988). One must bear in mind that the concentration of solutes likely remains below 5 to 10% (see Sect. 2 in application to Europa), and therefore the possible shift of the liquid density curve shown in Fig. 2 should not exceed ~50 $kg\,m^{-3}$, i.e. $2 \times 10^{-5}\ m^3\,kg^{-1}$ if expressed in units of specific volume. This corresponds roughly to the length of the arrows indicating the effect of salts and of ammonia on liquid density. For this reason, ammonia-water solutions have long been inferred as best candidates for driving cryovolcanic eruptions in the outer solar system. As to the high-pressure phases, the presence of NH_3 will tend to decrease the thickness of the icy layers at depth due to its anti-freeze action (e.g. Sohl et al. 2003), but it is not expected to modify qualitatively the organization of these layers since NH_3 does not enter the crystalline structure of ices. Conversely, solutes (Cl^-, HCO_3^-, SO_4^{2-}, Na^+, Ca^{2+}, Mg^{2+}, etc.) will increase the density of the liquid phase. These ions also deplete the melting point of ices, though to a lesser extent than ammonia or methanol (see Fortes and Choukroun 2010, this issue, and references therein).

This is also likely to apply for sodium chloride and putative ammonium sulfate solutions. One might expect this peculiarity to result in a different segregation of the ices, with ice III tending to float to the top of the liquid rather than towards the bottom. Eventually, upon freezing of a substantial amount of oceanic water, numerous hydrated salts would precipitate from the aqueous solution. Overall, at standard temperature and pressure conditions (i.e. $\sim 10^5$ Pa and 298 K), these salts have densities that fall within the range from 1400 $kg\,m^{-3}$ ($Na_2SO_4 \cdot 10H_2O$) to 1750 $kg\,m^{-3}$ ($(NH_4)_2SO_4$). Upon crystallization from a saturated or supersaturated solution that may represent an icy satellites ocean in a late stage of freezing, they would be expected to sink to the bottom and enrich the solid layer on the ocean floor (Zolotov 2007). As several phase transitions of hydrates have been reported in the H_2O-$MgSO_4$ and the H_2O-NH_3 systems (see Fortes and Choukroun 2010, this issue), some high-pressure phases of those systems may form as well.

4 Physical Aspects of Water-Rock Interaction

On small ocean-bearing icy satellites where the ocean is in contact with silicate rock below, water-rock interactions would affect heat dissipation, oceanic composition, and the mineralogy of rocks and oceanic sediments. Earth hosts a diversity of water-rock systems, broadly referred to as hydrothermal, which advectively transport geothermal heat concentrated by magma emplacement in direct proximity (km-scale; i.e. dikes) or farther away (e.g. hotspots). Whereas seafloor spreading caused by mantle convection is the main driver for terrestrial hydrothermal activity on Earth (Sclater et al. 1980), hotspot volcanism might be sufficient to advect heat from deep rocky interiors of icy satellites if magmatic activity prevails. Note, however, that hydrothermal activity can also occur in the absence of magmatic activity.

4.1 Permeability of Ocean Floor Rock

The abundance and distribution of hydrothermal systems in an icy satellite seafloor can be estimated by applying estimates for Earth to heat flow, temperature and gravity appropriate. The exercise, as applied to Europa, highlights the importance and uncertainty of permeability. Estimates of permeability in hydrothermal systems may span a range from 10^{-17} to 10^{-9} m^2 (Fisher 1998). At mid-ocean ridges on Earth and in other places where tectonic activity opens large flow channels in the seafloor, models of flow through hydrothermal systems constrain permeability to around 10^{-12} m^2 (Fisher 1998; Fisher and Becker 2000; Becker 2000). Moving away from mid-ocean ridges, permeability decreases to 10^{-17} m^2 in older (age ∼65–200 Myr) seafloor basalt (Stein et al. 1995; Sinha and Evans 2004).

The permeability of Earth's seafloor may diminish over time due to filling of pore spaces, either by silica and other secondary minerals that precipitates from ambient fluid (Stein et al.

1995), or by the accumulation of sediments (Martin and Lowell 2000). Laboratory (Martin and Fyfe 1970; Zimmerman et al. 1986) and bore-hole investigations (Anderson et al. 1985) to ~1 km depth in Earth's lithosphere indicate high temperatures and pressures tend to close pore spaces. See Wash and Brice (1984) for a discussion of the effect of pressure on porosity and transport properties of rock. In small bodies like Enceladus, where central pressures approach a mere 20 MPa, and where the presence of ammonia may create temperatures of aqueous solutions as low as ~180 K, rock permeabilities could be much higher than those in Earth's oceanic crust. The greater depth extent of permeable rocks on small, less massive satellites is due to low internal pressure as a consequence of their lower gravitational acceleration at the surface. In deep oceans like Europa's, where average pressure at the ocean floor is ~2–7 times the average in Earth's oceans (Anderson et al. 1998). Rocks become impermeable when subjected to pressures exceeding about 300 MPa (Wash and Brice 1984). This critical pressure is about twice as high as that expected at Europa's ocean floor and suggests that a ~ 20 to 30 km thick rock layer of non-vanishing permeability may exist below the seafloor of Europa's ocean.

In regions of ocean-ridge systems where fracture-induced large-scale permeability is absent, water is expected to percolate through microfracture channels to greater depth (Lister 1974; Sinha and Evans 2004). The depth to which microfractures form may serve as an upper limit for the depth of water-rock interaction. Vance et al. (2007) adopt a model for the thermally-induced formation of microfractures controlled by the development of residual stresses at grain boundaries. In a cooling rock, anisotropic thermal stresses will increase along grain boundaries. At high temperatures, residual stresses dissipate via viscous creep processes and cracking is inhibited. At low temperatures, where dissipative processes are sluggish, residual stresses accumulate along grain boundaries. Two-layer conductive thermal models for candidate ocean worlds are used to estimate fracture formation depth from the calculated pressure and temperature at which microfracture formation ceases. For the smallest objects considered, Ceres and Enceladus, the water percolation depth would be greater than 100 km, nearly four times that expected for Europa's ocean floor. In all objects considered, the rate of exposure of new rock is one to ten thousand times slower than present rates of seafloor spreading on Earth (45 mm yr^{-1}). For differentiated ocean-bearing bodies exceeding Triton in size, cracking is inhibited since ocean floor pressure would be much too high. For Enceladus, fluid access is predicted to occur to the deepest interior in the earliest period. Water-rock interactions may continue to the present in Enceladus where low-temperatures limit reaction rates, or if tidal processes cause episodic aqueous alteration of the interior. Also notable is the effect of diminished internal heating on the depth of cracking. Where cooling rate, grain size and composition are identical to those for the regions of Earth's ocean floor under consideration, fluid circulation is predicted to occur to orders of magnitude greater depth. Revised depths of fluid circulation would change previous estimates of the biopotential of extraterrestrial lithospheres (Vance et al. 2007).

4.2 Tidally-Induced Porous Flow and Hydrothermalism

Tidal interaction between a satellite and its primary (see Hussmann et al. 2010, this issue) is relevant to the satellite's ocean-rock interface in two ways: (i) dissipation of tidal energy due to frictional losses in the body tide and ocean bottom friction (see Sohl et al. 1995, and references therein), (ii) the ice shell and the ocean below follow the equilibrium tide by adjusting to the time-variable tidal bulge, thereby inducing pressure changes at the seafloor because of the related mass loading (see Vance et al. 2007, and references therein). The first of these can in some cases contribute significantly to the satellite's energy budget (see

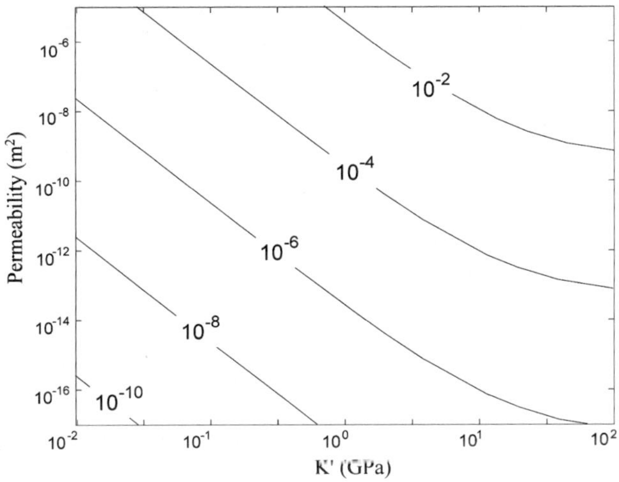

Fig. 3 Contoured rates of tidal energy release (in units of $W\,m^{-2}$) due to frictional losses in the near subsurface of Europa's ocean floor. Contours are shown as a function of permeability and confined bulk modulus of the rock matrix K' for an ocean floor porosity of $n = 0.1$, consistent with drill hole measurements in the uppermost kilometer of Earth's oceanic crust (Becker et al. 1982). The tidal height is fixed at $u_r = 27.4$ m, corresponding to locales near the sub- and anti-Jovian point, respectively. Reprinted with permission from Vance et al. (2007). Copyright 2007, Mary Ann Liebert, Inc

below). The latter process generates only modest heating for generous estimates of rock porosity. However, tidally enhanced fluid flows may be important where thermal gradients are not steep enough to drive vigorous convection.

The local height of the tidal bulge is given by the ratio between the primary tidal potential and the gravitational acceleration at the surface. In-situ tide gauge data and deep-ocean bottom pressure gauge data are routinely collected on Earth and support the view that tidally-induced ocean bottom pressure variations are caused by the tidal load of the ocean and deformation of the seafloor due to body and the ocean tides. As a consequence, the tidal forcing of the ice shell covering the water ocean below may act as a time-variable hydrostatic head. The incremental stress induced by the corresponding ocean bottom pressure variations would be partitioned between the rock matrix and interstitial pore water. Provided the seafloor rock were fractured, ocean water could readily move in response to the tidally-induced loading, thereby causing dissipation of tidal energy due to frictional losses. As detailed in Vance et al. (2007), the corresponding tidal energy release would depend on the mechanical properties of the fluid and the rock matrix, the tidal period, and the magnitude of tidal distortion of an ocean-bearing body.

The tidal energy dissipated is proportional to the product of the Darcy flow velocity through a porous matrix times the pressure gradient. The pore pressure change below the seafloor is governed by the periodic loading due to the tide and the elastic properties of the ocean floor material. Details of models and parameter that are taken to be commensurate with Earth's values are described in Wang et al. (1999) and Vance et al. (2007). Figure 3 illustrates how tidally-induced dissipation rates per unit area on Europa will depend on ocean floor permeability and bulk modulus of the rock matrix. Heating due to frictional losses is on the order of 0.001–1 $mW\,m^{-2}$, if the bulk modulus is fixed at 10 GPa and permeability would range from 10^{-17} to 10^{-9} m^2. Additionally, heat generation would vary by roughly one order of magnitude in the likely range of porosities between 0.1 to 0.5. Finally, assum-

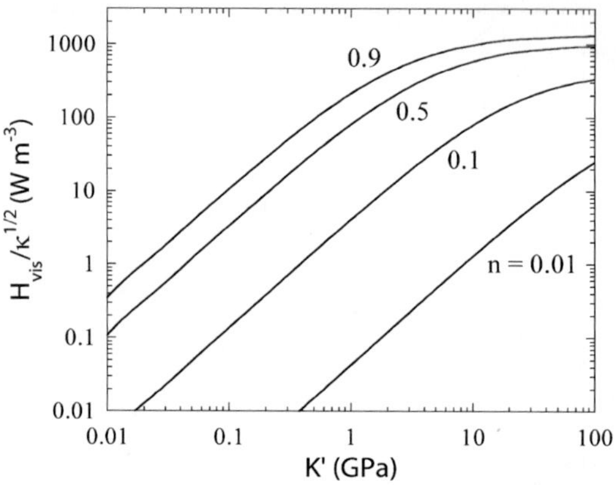

Fig. 4 Rate of tidal energy release normalized to permeability H_{vis} (in units of $\mathrm{W\,m^{-3}}$) as a function of confined bulk modulus of the rock matrix K' in the near subsurface of Europa's ocean floor. Curve labels denote ocean floor porosities ranging between $0.01 < n < 0.9$. The tidal height is fixed at $u_r = 27.4$ m, corresponding to locales near the sub- and anti-Jovian point, respectively. Reprinted with permission from Vance et al. (2007). Copyright 2007, Mary Ann Liebert, Inc

ing these material properties for potential ocean-bearing bodies, this heating mechanism is deemed significant only on Europa (Fig. 4) because tidal bulge heights on other satellites are relatively small compared to Europa's (see Hussmann et al. 2010, this issue). Where interfaces separate vertical seafloor layers with different bulk moduli, additional porous flow is induced by the difference in instantaneous response of the pore pressure across the interfaces. If the seafloor has multiple-layered structure with different bulk moduli, the amplitude of generated heat may become several times larger (Wang et al. 1999). The present discussion illustrates the potential importance of dissipation of tidal energy as a localized heat source in the near subsurface of Europa's ocean floor.

4.3 Serpentinization of Ocean Floor Rock

At the off-axis mid-Atlantic hydrothermal systems Trans-Atlantic Geotraverse (TAG), Rainbow, and Lost City, the geological process of serpentinization provides a significant source of thermal energy. In the Lost City system, it is likely that serpentinization constitutes the dominant thermal energy source (Kelley et al. 2005). Water-rock interactions on icy bodies imply the potential for serpentinization processes in extraterrestrial oceans lacking mid-ocean-ridge tectonics, with implications for astrobiology.

Ferromagnesian silicate minerals are dominant in unfractionated igneous rocks (e.g. peridotites) that comprise Earth's upper mantle, and may typify subseafloor lithologies on small ocean planets. The abundance of peridotite in a planetary lithosphere is largely governed by the temperature of formation, which depends on the timing of formation relative to the body's ability to dissipate heat (Schubert et al. 1986; McKinnon and Zolensky 2003). Even later, slower radiogenic heating can be intense enough to form peridotite, but eventually cooling takes over and the system turns retrograde—at lower temperatures, thermodynamic disequilibrium favors a transformation to the hydrated mineral form, in this case serpentine. In the event of an early period of peridotite-forming intense heating, the subsequent formation of ocean(s) could give rise to an epoch of serpentinization-driven hydrothermal activity.

An estimate of the heat generated by serpentinization can be obtained by considering the energy released in hydrating available material in the volume of rock to depth z below the seafloor. Estimates of heating for presently known and candidate ocean planets are reported by Vance et al. (2007), considering the hydration of the Mg-end member of olivine,

$$\begin{gathered} 2Mg_2SiO_4 + 3H_2O \leftrightarrow Mg(OH)_2 + Mg_3Si_2O_5(OH)_4, \\ \text{forsterite} + \text{water} \leftrightarrow \text{brucite} + \text{serpentine}. \end{gathered} \tag{4}$$

For simplicity an average enthalpy of reaction of $\Delta H = -83$ kJ per mol serpentine is assumed. Bulk composition is assumed to be lherzolitic, consistent with Earth's primordial mantle: 70% olivine, with an Fe/Mg ratio of 0.1. All of the olivine is assumed to hydrate into serpentine and brucite. Under these assumptions, present-day heating due to serpentinization would comprise less than a percent of radiogenic heating for all bodies considered. However, the slow release of this form of energy allows it to sustain gradients in chemical potential over long periods of time. These factors make even minor amounts of serpentinization important when assessing the habitable potential of icy satellites.

5 Comparison Between Icy Bodies

Henceforth, a thick layer of hydrated compounds can form beneath the putative liquid layer of icy satellites, with a wide range of compositions: from high-pressure phases of ices, to salts, and clathrate hydrates. The composition depends on the physical conditions (pressure, temperature) of the ocean floor and their evolution through time, and on the chemical evolution. As a consequence, a number of icy satellites of Jupiter or Saturn likely has an ocean floor with characteristics of its own. This means that, despite similar bulk characteristics, large icy satellites such as Titan, Ganymede, and Callisto may exhibit different internal structures.

5.1 Europa: Sulfates, Sulfides, and Prolonged Water-Rock Interaction

Europa's interior is thought to be fully differentiated into a metallic core, silicate mantle, and outer water ice/liquid shell, as inferred from measurements of the *Galileo* spacecraft of the satellite's density, low-degree gravitational field, and shape (e.g. Schubert et al. 2009). The satellite's induced magnetic field further suggests the presence of a global sub-surface liquid water ocean within Europa's outer ice shell, probably enriched in dissolved salts. The thickness of the ice Ih shell has been estimated between 1 and a few tens of kilometers from several types of measurements and models (Billings and Kattenhorn 2005, and references therein). The *Galileo* mission has also supplied numerous pieces of evidence for a geologically young surface, with tectonics processes, and compositional heterogeneities such as hydrated sulfate-bearing compounds, some of which appear to originate from the internal liquid layer (e.g. McCord et al. 1999; Pappalardo et al. 1999; Carlson et al. 1999).

In a differentiated Europa model, the internal ocean layer is approximately 100 km thick (Anderson et al. 1998; Sohl et al. 2002; Schubert et al. 2004; Kuskov and Kronrod 2005), owing to a high mean density (3010 $kg\,m^{-3}$) that implies a rocky or silicate-metal-sulfide interior of up to 90% of the satellites mass. The transition pressure between the ocean and the rocky interior is around 130–150 MPa, given the ocean thickness and density, and a gravity of 1.314 $m\,s^{-2}$. This pressure is lower than those of the ice Ih-ice II or ice Ih-ice III transition, so no high-pressure phases of ice should form. Water-rock interaction could have led to hydration (e.g. serpentinization) and oxidation of rocks and organic materials leading to the production of H_2, N_2, and CO_2 (e.g. Zolotov and Kargel 2009, and Sect. 2). The formation of sulfates could have involved elevated temperatures ($\gtrsim 200°C$) and H_2 escape (Zolotov and Shock 2004). Most clathrate hydrate species are thermodynamically stable, but the amount of gas available are likely a limiting factor in their formation. Dense hydrated salts and CH_4-poor clathrate hydrates (see Sect. 3) may form at those conditions and gather as a solid layer at the interface (Kargel et al. 2000; Prieto-Ballesteros et al. 2005). Such a barrier would have an insulating effect on the silicate portion of the interior, because of the low thermal conductivity of these materials (ranging between 0.5 and 2.5 $W\,m^{-1}\,K^{-1}$ for clathrates and salt hydrates, respectively).

The composition of Europa's tenuous atmosphere and non-ice surface materials may imply the presence of $MgSO_4$, Na_2SO_4, and H_2SO_4 hydrates among other species (H_2O_2, O_2, CO_2, SO_3, S compounds), hydrated salts and gases trapped in the satellite's icy shell (Carlson et al. 2009). The seafloor rock and sediments could contain phyllosilicates, carbonates, Ca sulfates, Fe sulfides, and organic compounds (Zolotov and Kargel 2009). As detailed in Sect. 2, the chemical evolution of Europa's subsurface ocean and alteration of the seafloor are governed by water-rock interactions and hydrogen escape.

Kargel et al. (2000) proposed a model of a near-peritectic H_2O-$MgSO_4$ composition of the Europan ocean, which would explain the presence of hydrated sulfates in the icy crust. Spaun and Head (2001) modeled the evolution of this ocean as a follow-up, and suggested that at present the 130 km-thick icy crust may consist of ice Ih +/− meridianiite ($MgSO_4 \cdot 11H_2O$), overlying a meridianiite layer. That layer would be separated from the silicate portion by a 35-km thick layer of epsomite ($MgSO_4 \cdot 7H_2O$). In their model, no ocean would have been sustained until present, but they used a stationary model without accounting for potentially large tidal heating (e.g. Sotin et al. 2002; Tobie et al. 2003). Other chemical models such as those of Zolotov and Shock (2001) and of McKinnon and Zolensky (2003) suggest lower amounts of dissolved magnesium and sodium sulfates, which would result in a thinner layer of salts at the bottom of the ocean, if any. McKinnon and Zolensky (2003), based on the reduced state of sulfur in some chondrites and the low abundance of sulfates, argued that a km-thick sulfide-rich layer at the ocean seafloor is likely to be present while sulfates would remain in solution.

Depending on the redox conditions, clathrates containing a mixture of CO_2 and SO_2 or CH_4 may have formed at the water-rock interface (e.g. Kargel et al. 2000; Prieto-Ballesteros et al. 2005), and might have migrated upwards if the liquid layer was enriched in dissolved salts and thus denser than the clathrates. Conversely, in a dilute ocean model, i.e. < 10 wt.% sulfates, clathrates would have formed an insulating layer at the bottom of the ocean because of a negative buoyancy contrast. In terms of planetary evolution, the mere existence of this insulating layer would have a dramatic influence: (i) it would make a mechanically decoupling barrier, precluding further ocean-rock interactions unless cracking occurs; (ii) it would allow a hotter silicate interior and a colder liquid layer, which would then tend to solidify further. Clay minerals, sulfur, and organic compounds could also play a role in insolating silicate mantle rocks from oceanic water. Finally, significant volume expansion of suboceanic rocks through hydration (e.g. serpentinization) would considerably decrease their porosity and permeability. The extent of these influences would depend on the nature and properties (density, thermal conductivity) of the phases that crystallize, as well as on the thickness of the insulating layer.

Europa, as an intermediate-sized object undergoing tidal heating, may constitute a rare example of an icy world with extensive and prolonged water-rock interaction. In the twenty-first century, a continuing program of missions to the outer solar system, with increasingly

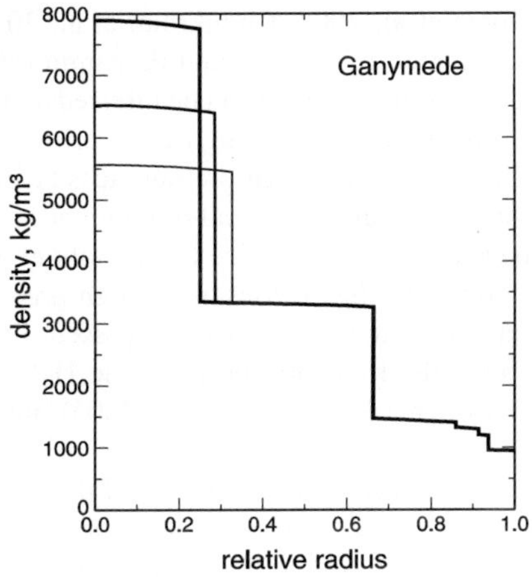

Fig. 5 Radial distribution of density for three core compositions of Ganymede. From top to bottom: pure Fe; Fe-FeS (50:50 by wt.); pure FeS core. Modified from Sohl et al. (2002)

global participation, promises to shed light on water-rock interactions in ocean-bearing icy bodies.

5.2 Ganymede, Callisto, and Titan: Ammonia and Icy Seafloors

The outer two Galilean satellites and the largest moon of Saturn have characteristics in common: (i) their mean density, 1881 $kg\,m^{-3}$ for Titan, 1940 $kg\,m^{-3}$ for Ganymede, and 1851 $kg\,m^{-3}$ for Callisto, and (ii) their radius, 2575, 2634 and 2403 km, respectively, are in the same order. This implies that they bear a similar rock mass fraction on the order of 50 wt.%. Gravity data from the Galileo mission (e.g. Schubert et al. 2004) and new results from the Radio Science experiment onboard the Cassini spacecraft (Iess et al. 2010) have shown that Ganymede is fully differentiated (Fig. 5), whereas the deep interiors of Titan and Callisto may partly contain an undifferentiated mixture of rock and ice. Ganymede's central mass concentration and self-sustained magnetic field strongly suggest the presence of an iron core (e.g. Anderson et al. 1996; Kivelson et al. 2004), whereas circumstantial evidence for the existence of a metallic core is lacking in the case of Titan. Therefore, some variability on the temperature and density stratification affects their interior. The thickness of the H_2O mantle has been estimated to 600–1000 km for differentiated Ganymede (Kuskov and Kronrod 2001; Sohl et al. 2002) and tentatively differentiated Titan (Grasset et al. 2000; Sohl et al. 2003; Grasset and Pargamin 2005; Tobie et al. 2005b), corresponding to a pressure at the ice-rock transition of ~ 0.8 to 1.2 GPa. In the event of partial differentiation of Callisto, the transition from the H_2O-dominated outer layer to the rock-ice core below must be expected to occur at a depth of 300 km (Kuskov and Kronrod 2005) to 600 km (Sohl et al. 2002). The results obtained by Iess et al. (2010) suggest that Titan's state of internal differentiation is an intermediate between Callisto and Ganymede, indicating that satellites of similar size and composition in terms of their primordial ice-rock mixtures may display distinct degrees of internal differentiation (e.g. Sotin et al. 2009).

Thicknesses of the liquid and of the high-pressure ice layers at present would depend on the internal heat production rate, and on the presence of anti-freeze constituents in the liquid layer. Of particular relevance to Titan is NH_3, which could be the source of the current atmospheric N_2 (Atreya et al. 1978; Niemann et al. 2005), and also a cryomagma component candidate that would dissociate crustal methane clathrates and replenish the atmospheric methane (Tobie et al. 2006; Choukroun et al. 2010), continuously lost in photochemistry. From the calculation by Grasset and Pargamin (2005) of the thicknesses of the outer ice Ih and the internal liquid layers, it can be inferred that Titan's high-pressure ices layer is of 300–600 km to 400–700 km, for 5 and 0 wt.% primordial NH_3 in the H_2O mantle, respectively. A similar thickness for the high-pressure ices layer can be expected on Ganymede, although its content at depth in ammonia and/or hydrated sulfates remains poorly constrained. Callisto may also have a high-pressure ice layer (Sohl et al. 2002), above the undifferentiated deep rock-ice interior. The thicknesses of the high-pressure ices layer within Ganymede and Titan are such that ice V and ice VI may be present therein. Within Callisto, the possible depth of 1700 km for the transition between the H_2O outer shell and the undifferentiated deep rock-ice interior (Kuskov and Kronrod 2005) imply that ices V, VI, and possibly even VII, may exist.

It is possible to infer the thicknesses of the ice V and the ice VI sublayers on Ganymede and Titan, although this is a mere estimation because of: (i) the uncertainty in pressure at the interfaces; (ii) the difficulty to constrain a definite thermal profile, although the solid-solid phase transitions are roughly isothermal; (iii) the poor constraints on the presence and concentration of anti-freeze constituents; (iv) the imprecise knowledge of the effect of anti-freeze constituents on the solid-solid phase transitions in H_2O ice under pressure. Depending on the models, the pressure at the ocean bottom is ranging between 400 and 700 MPa, with 800–1200 MPa pressure at the bottom of the high-pressure ice layer. This pressure range corresponds to the 300–600 km thickness inferred above for this layer. A basic scaling thus suggests that the ice V sublayer is up to $\sim$ 150–200 km thick in a pure H_2O model, while it may be less than 50 km thick in the presence of 5 wt.% NH_3. The ice VI sublayer would occupy the remaining portion of the high-pressure solid layer, i.e. 200 to 400 km, depending on the pressure of the interface and on the abundance of NH_3 or another anti-freeze constituent.

Fortes et al. (2007) have suggested that conversion of NH_3 into NH_4^+ could have resulted in an $(NH_4)_2SO_4$-rich early ocean on Titan, which would have important implications for cryovolcanic processes. Grindrod et al. (2008) also studied the impact of a concentrated ammonium sulfate ocean on the interior structure of Titan. A corollary of this model is that these dense (1780 $kg\,m^{-3}$) compounds may, upon precipitation from the liquid during cooling, contribute to some extent to the high-pressure phases layer. However, it should be noted that sulfates may not have formed on large satellites because of reducing conditions in the deep interior (see Sect. 2 of this paper). On Titan, the presence of CH_4 suggests a relatively high H_2 fugacity that would inhibit sulfate formation under those conditions, albeit the existence of sulfate salts at greater depth cannot be ruled out entirely.

6 Discussion and Conclusions

We have addressed lines of circumstantial evidence for the existence of present-day subsurface liquid water oceans on icy satellites. The ocean floor region on these moons may be the place of a wide range of exchange processes. Similarly to the Earth's oceans, where the liquid layer is directly in contact with the silicate interior, hydrothermal activity may occur, causing aqueous alteration of rocks. This is the expected situation within Europa (e.g. Zolotov and Kargel 2009) and Enceladus (e.g. Matson et al. 2007; Zolotov 2007). In large icy objects like Ganymede, Callisto, and Titan, high-pressure polymorphs of water ice, clathrate hydrates, and/or salts could have formed upon cooling.

Spacecraft observations, laboratory experiments, and theoretical modeling combined suggest that the existence and persistence of global subsurface liquid water oceans is strongly dependent on the subsurface structure, bulk composition, and coupled orbital and thermal history of an individual satellite (see Schubert et al. 2010, this issue). Four major sources of internal heat production governing the long-term thermal and chemical evolution following satellite formation can be identified: (i) radiogenic heating due to the silicate component by decay of long-lived radioactive isotopes; (ii) resonant orbital interaction between neighboring satellites and their primaries, resulting in the dissipation of tidal energy; (iii) heat of chemical reactions (e.g. hydration); (iv) latent heat release due to the crystallization of solids upon cooling. These internal and external heat sources will contribute in various extents to the total energy budget of icy moons with important implications for internal dynamics, surface tectonics and cryovolcanic resurfacing (see Hussmann et al. 2010, this issue).

Water ocean composition will be governed by the bulk ratios between rocks, water ice, non-water volatiles, and organic compounds, as well as thermal history. The silicate rock fraction incorporating long-lived radionuclides and the dissipation of tidal energy provide the dominant heat sources; the concentration of mass toward the center also affects the thermal state of the interior. Throughout history, the chemical evolution of subsurface liquid water oceans has been affected by the degree of ice melting, the extent and duration of chemical interactions between rocks and liquid water, degassing of the deep interior and oceanic water, freezing and re-melting, and secondary inorganic and organic precipitation. Pathways of subsequent oceanic evolution were mostly dependent on the scale, timing, and locations of heat production in the satellite interiors. Continuing heat production in rocky cores may have induced hydrothermal circulation in the rocky interiors, caused dehydration of previously formed minerals, and led to further alteration of organic matter. These processes suggest special conditions for water-rock interaction based on the size of an icy satellite, in the case of percolation depth, and its orbital configuration, in the case of tidally-driven circulation. Furthermore, bulk composition, degree and rate of internal differentiation of ocean-bearing bodies determine the extent of aqueous alteration (e.g. serpentinization). Subsurface water oceans are expected to have survived on larger icy bodies, e.g. the Jovian icy moons Europa, Ganymede, and Callisto, as well as Titan and Triton, due to diminished cooling and enhanced radiogenic heating compared to smaller icy bodies. However, depending on the individual amount of anti-freeze constituents, subsurface liquid water oceans in the largest mid-sized satellites of Saturn and Uranus cannot be entirely ruled out and may serve as analogues for the bulk structure of trans-Neptunian objects. On some of the latter, liquid water reservoirs beneath thermally isolating ice layers could have been maintained throughout their entire evolution (e.g. Hussmann et al. 2006; McKinnon et al. 2008; Desch et al. 2009).

Two extreme cases that may correspond to the ultimate stage of evolution of an internal liquid layer within icy satellites are compared in Fig. 6. These two scenarios are based on the distinction between *cold* and *warm* thermal profiles. The higher the salt/ammonia content of an icy satellite is, the colder the thermal profile across its interior should be.

First, consider an icy satellite subject to intense oxidation and hydration and enriched in salts, forming a thick bottom layer (from precipitation upon cooling). The latter would be thermally insulating and the predominant mode of heat transfer would be conduction (rather than advection of upwelling hot materials). This layer implies colder conditions above it, therefore a cooler ocean, enriched in salts, which would crystallize faster to balance the heat loss, forming an allegedly thick outer ice-Ih layer above a close-to-saturated ocean. This *cold* scenario may correspond to a class of models of Europa's interior involving ice and salt layers (Kargel et al. 2000; Spaun and Head 2001; Zolotov and Kargel 2009).

On the other hand, a dilute internal liquid layer containing salts and/or ammonia would be easily convective, transferring more heat to the outer regions, and releasing more heat as

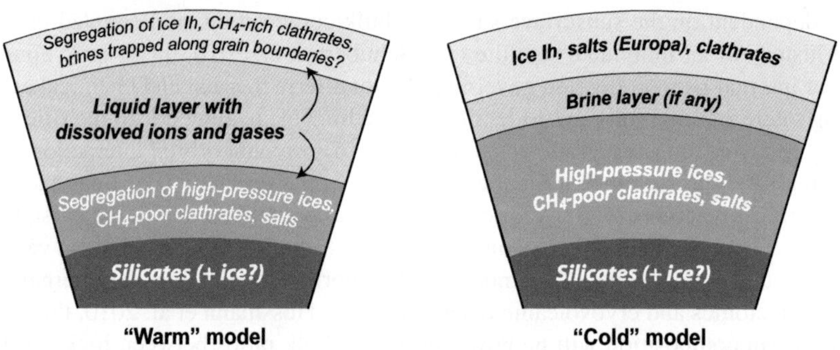

Fig. 6 Two extreme interior structure models of icy satellites, depending on the amount of heat generated and non-H_2O icy components. High-pressure ices can only be generated upon cooling of icy bodies with an H_2O mantle larger than that of Europa. Beyond an increased thickness in layers of H_2O ice, a cold model implies a higher concentration of soluble compounds in the liquid layer (if any remains), and a thicker layer of precipitated salts

ices crystallize. This more intense heat production and more efficient heat transfer would help maintain a warmer thermal profile. Furthermore, some hydrated salts and/or ammonia hydrates may crystallize lately in the evolution, or may form upon solidification of brine pockets trapped along grain boundaries in the outer ice-Ih layer. In the case of a large icy satellite, where the pressure at the bottom of the H_2O mantle is above 200 MPa, high-pressure ices would also have formed upon solidification of the ocean above. This *warm* scenario may represent the situation within Ganymede (e.g. Kuskov and Kronrod 2001; Sohl et al. 2002) and Titan (e.g. Grasset et al. 2000; Sohl et al. 2003; Tobie et al. 2005b). Provided the ice-Ih shell is subject to convection, trapped brine/ammonia pockets, representing only a small fraction of the outer ice shell, may be the driving source of resurfacing and outgassing by dissociation of clathrate hydrates (Choukroun et al. 2010). Given the current state of knowledge of the phase diagram of $MgSO_4$ and other compounds, it is difficult to assess the temperature at which the hydrated salts would form. Therefore, whether the salts would form earlier or later than the high-pressure ices remains unconstrained. Further experimental data are required to address these issues.

In an object with an initially dilute internal liquid layer, the bulk composition would remain on the water-rich side of the eutectic in sulfate-, ammonia-, or chloride-bearing systems. If sulfates are present, a compound such as epsomite ($MgSO_4 \cdot 7H_2O$) is unlikely to exist at the bottom of the liquid layer, but the heavily hydrated phases such as meridianiite ($MgSO_4 \cdot 11H_2O$) and mirabilite ($Na_2SO_4 \cdot 10H_2O$) might crystallize and form a layer on the ocean floor upon advanced freezing. A liquid layer with this kind of composition would either stay warmer than in a brine-rich case, or undergo complete freezing over time. The location of clathrate-rich layers is difficult to assess given the wide range of possible densities for these compounds and for the liquid. Overall, CH_4-rich clathrates would be buoyant compared to most liquid solutions, whereas CH_4-poor clathrates will segregate as a layer at the bottom of the ocean. Thus, floating CH_4-rich clathrates may supply methane to the outer ice shell and the surface/atmosphere of icy satellites.

Acknowledgements The authors are grateful to the organizers of the EuroPlanet-ISSI Workshop *Exchange Processes on Icy Satellites*, held November 2008 in the inspiring environment of the International Space Science Institute, Bern, Switzerland; two anonymous reviewers for their constructive comments; and O. Grasset for the manuscript handling. We would like to thank G. Tobie for providing the clathrate hydrates densities calculations. This research has been supported by the Helmholtz Association through the research alliance

Planetary Evolution and Life. A portion of this work was supported by the Jet Propulsion Laboratory, California Institute of Technology, under a contract with the National Aeronautics and Space Administration. Doctors Choukroun, Pappalardo and Vance acknowledge support from the NASA Astrobiology Institute's *Astrobiology of Icy Worlds* program at the Jet Propulsion Laboratory. Doctors Choukroun and Vance acknowledge support from the NASA Postdoctoral Program administered by Oak Ridge Associated Universities. S. Vance acknowledges support from the Caltech Postdoctoral Program. F. Sohl acknowledges discussions with H. Hussmann, V. Stamenković, and F. W. Wagner.

References

C.M.O.D. Alexander, M. Fogel, H. Yabuta, G.D. Cody, Geochim. Cosmochim. Acta **71**, 4380–4403 (2007)
J.D. Anderson, G. Schubert, Geophys. Res. Lett. **34**, L02202 (2007). doi:10.1029/2006GL028100
R.N. Anderson, M.D. Zoback, S.H. Hickman, R.L. Newmark, J. Geophys. Res. **90**(B5), 3659–3669 (1985)
J.D. Anderson, E.L. Lau, W.L. Sjogren, G. Schubert, W.B. Moore, Nature **384**, 541–543 (1996)
J.D. Anderson, G. Schubert, R.A. Jacobson, E.L. Lau, W.B. Moore, W.L. Sjogren, Science **281**, 2019–2022 (1998)
S. Atreya, T. Donahue, W. Kuhn, Science **201**(4356), 611–613 (1978)
K. Becker, Oceanus **42**(1), 2–5 (2000)
K. Becker, R.P.V. Herzen, T.J.G. Francis, R.N. Anderson, J. Honnorez, A.C. Adamson, J.C. Alt, R. Emmermann, P.D. Kempton, H. Kinoshita, C. Laverne, M.J. Mottl, R.L. Newmark, Nature **300**(5893), 594–598 (1982)
S. Billings, S. Kattenhorn, Icarus **177**, 397–412 (2005)
D. Bockelée-Morvan, J. Crovisier, M.J. Mumma, H.A. Weaver, in *Comets II*, ed. by M.C. Festou, H.U. Keller, H.A. Weaver (University of Arizona Press, Tucson, 2004), pp. 391–423
A.J. Brearley, in *Meteorites and the Early Solar System II*, ed. by D.S. Lauretta, H.Y. McSween (University of Arizona Press, Tucson, 2006), pp. 587–624
P. Bridgman, Proc. Am. Assoc. Sci. **47**(13), 441–458 (1912)
P. Bridgman, J. Chem. Phys. **5**, 964–966 (1937)
R.M. Canup, W.R. Ward, Astron. J. **124**, 3404–3423 (2002)
R.M. Canup, W.R. Ward, in *Europa*, ed. by R.T. Pappalardo, W.B. McKinnon, K. Khurana (University of Arizona Press in collaboration with Lunar and Planetary Institute, Tucson, 2009), pp. 59–83
R.W. Carlson, R.E. Johnson, M.S. Anderson, Science **286**(5437), 97–99 (1999)
R.W. Carlson, W.M. Calvin, J.B. Dalton, G.B. Hansen, R.L. Hudson, R.E. Johnson, T.B. McCord, M.H. Moore, in *Europa*, ed. by R.T. Pappalardo, W.B. McKinnon, K. Khurana (University of Arizona Press in collaboration with Lunar and Planetary Institute, Tucson, 2009), pp. 283–327
J. Castillo-Rogez, J. Geophys. Res. **111**, E11005 (2006). doi:10.1029/2004JE002379
J.C. Castillo-Rogez, D.L. Matson, C. Sotin, T.V. Johnson, J.I. Lunine, P.C. Thomas, Icarus **190**, 179–202 (2007)
V. Chizhov, J. Appl. Mech. Tech. Phys. **34**, 253–263 (1993). Translated from Russian Prikl. Mekh. i Tekhn. Fiz. 2, 113–123
V. Chizhov, O. Nagornov, J. Appl. Mech. Tech. Phys. **31**, 378–385 (1990). Translated from Russian Prikl. Mekh. i Tekhn. Fiz. 3, 41–49
M. Choukroun, O. Grasset, J. Chem. Phys. **127**, 124506 (2007)
M. Choukroun, O. Grasset, G. Tobie, C. Sotin, Icarus **205**, 581–593 (2010)
G.J. Consolmagno, J.S. Lewis, in *Jupiter*, ed. by T. Gehrels (University of Arizona Press, Tucson, 1976), pp. 1035–1051
G.J. Consolmagno, J.S. Lewis, in *Planetary Satellites*, ed. by J.A. Burns (University of Arizona Press, Tucson, 1977), pp. 492–500
G.J. Consolmagno, J.S. Lewis, Icarus **34**, 280–293 (1978)
A. Coradini, G. Magni, D. Turrini, Space Sci. Rev. (2010, this issue). doi:10.1007/s11214-009-9611-9
S.K. Croft, J.I. Lunine, J. Kargel, Icarus **73**, 279–293 (1988)
Dalton et al., Space Sci. Rev. (2010, this issue)
S.J. Desch, J.C. Cook, T.C. Doggett, S.B. Porter, Icarus **202**, 694–714 (2009)
Z. Duan, N. Moller, J.H. Weare, Geochim. Cosmochim. Acta **56**(7), 2605–2617 (1992)
Y.A. Dyadin, Y.E. Aladko, E.G. Larionov, Mendeleev Commun. **1**, 34–35 (1997a)
Y.A. Dyadin, E. Larionov, D. Mirinski, T. Mikina, L. Starostina, Mendeleev Commun. **1**, 32–34 (1997b)
Y.A. Dyadin, E. Larionov, Y. Aladko, A. Manakov, F. Zhurko, T. Mikina, V. Kumarov, E. Grachev, J. Struct. Chem. **40**, 790–795 (1997c)
Y.A. Dyadin, E. Larionov, Y. Aladko, F. Zhurko, Dokl. Phys. Chem. **378**(4–6), 159–161 (2001)

R. Feistel, W. Wagner, J. Chem. Ref. Data **35**(2), 1021–1047 (2006)
A. Fisher, Rev. Geophys. **36**(2), 143–182 (1998)
A.T. Fisher, K. Becker, Nature **403**(6765), 71–74 (2000)
A.D. Fortes, M. Choukroun, Space Sci. Rev. (2010, this issue)
A.D. Fortes, P.M. Grindrod, S.K. Trickett, L. Vočadlo, Icarus **188**(1), 139–153 (2007)
P.E. Geissler, R. Greenberg, G. Hoppa, P. Helfenstein, A. McEwen, R. Pappalardo, R. Tufts, M. Ockert-Bell, R. Sullivan, R. Greeley, M.J.S. Belton, T. Denk, B. Clark, J. Burns, J. Veverka, the Galileo Imaging Team, Nature **391**, 368–370 (1998)
C.R. Glein, M.Y. Zolotov, E.L. Shock, Icarus **197**, 157–163 (2008)
O. Grasset, High Press. Res. **21**, 139–157 (2001)
O. Grasset, J. Pargamin, Planet. Space Sci. **53**, 371–384 (2005)
O. Grasset, C. Sotin, F. Deschamps, Planet. Space Sci. **48**, 617–636 (2000)
O. Grasset, E. Amiguet, M. Choukroun, High Press. Res. **25**(4), 255–265 (2005)
R. Greeley, C.F. Chyba, J.W. Head III, T.B. McCord, W.B. McKinnon, R.T. Pappalardo, P.H. Figueredo, in *Jupiter: The Planet, Satellites and Magnetosphere*, ed. by F. Bagenal, T.E. Dowling, W.B. McKinnon (Cambridge University Press, Cambridge, 2004), pp. 329–362
R. Greenberg, P. Geissler, G. Hoppa, B.R. Tufts, D.D. Durda, R. Pappalardo, J.W. Head, R. Greeley, R. Sullivan, M.H. Carr, Icarus **135**, 64–78 (1998)
R. Greenberg, P. Geissler, G. Hoppa, B.R. Tufts, Rev. Geophys. **40**(2), 1004 (2002). doi:10.1029/2000RG000096
T. Grindley, J. Lind, J. Chem. Phys. **54**(9), 3983–3989 (1971)
P.M. Grindrod, A.D. Fortes, F. Nimmo, D.L. Feltham, J.P. Brodholt, L. Vočadlo, Icarus **197**, 137–151 (2008)
S. Henderson, R. Speedy, J. Chem. Phys. **91**, 3069–3072 (1987)
H. Hussmann, T. Spohn, Icarus **171**, 391–410 (2004)
H. Hussmann, F. Sohl, T. Spohn, Icarus **185**, 258–273 (2006)
H. Hussmann, G. Choblet, D.L. Matson, C. Sotin, G. Tobie, T. Van Hoolst, Space Sci. Rev. (2010, this issue)
L. Iess, N.J. Rappaport, P. Tortora, J. Lunine, J.W. Armstrong, S.W. Asmar, L. Somenzi, F. Zingoni, Icarus **190**, 585–593 (2007)
L. Iess, N.J. Rappaport, R.A. Jacobson, P. Racioppa, D.J. Stevenson, P. Tortora, J.W. Armstrong, S.W. Asmar, Science **327**, 1367–1369 (2010)
T.V. Johnson, Space Sci. Rev. **116**, 401–420 (2005)
J.S. Kargel, Icarus **94**, 368–390 (1991)
J.S. Kargel, J.Z. Kaye, J.W. Head, G.M. Marion, R. Sassen, J.K. Crowley, O. Prieto-Ballesteros, S.A. Grant, D.L. Hogenboom, Icarus **148**, 226–265 (2000)
S.A. Kattenhorn, Icarus **157**, 490–506 (2002)
D.S. Kelley, J.A. Karson, G.L. Fruh-Green, D.R. Yoerger, T.M. Shank, D.A. Butterfield, J.M. Hayes, M.O. Schrenk, E.J. Olson, G. Proskurowski, M. Jakuba, A. Bradley, B. Larson, K. Ludwig, D. Glickson, K. Buckman, A.S. Bradley, W.J. Brazelton, K. Roe, M.J. Elend, A. Delacour, S.M. Bernasconi, M.D. Lilley, J.A. Baross, R.T. Summons, S.P. Sylva, Science **307**(5714), 1428–1434 (2005)
J.F. Kerridge, M.S. Matthews, *Meteorites and the Early Solar System* (University of Arizona Press, Tucson, 1988)
K.K. Khurana, M.G. Kivelson, D.J. Stevenson, G. Schubert, C.T. Russell, R.J. Walker, C. Polanskey, Nature **395**, 777–780 (1998)
M.G. Kivelson, K.K. Khurana, D.J. Stevenson, L. Bennett, S. Joy, C.T. Russell, R.J. Walker, C. Zimmer, C. Polanskey, J. Geophys. Res. **104**(A3), 4609–4626 (1999)
M.G. Kivelson, K.K. Khurana, C.T. Russell, M. Volwerk, R.J. Walker, C. Zimmer, Science **289**, 1340–1343 (2000)
M.G. Kivelson, F. Bagenal, W.S. Kurth, F.M. Neubauer, C. Paranicas, J. Saur, in *Jupiter, The Planet, Satellites, and Magnetosphere*, ed. by F. Bagenal, T. Dowling, W. McKinnon (Cambridge Univ. Press, Cambridge, 2004), pp. 513–536
O.L. Kuskov, V.A. Kronrod, Icarus **151**, 204–227 (2001)
O.L. Kuskov, V.A. Kronrod, Icarus **177**, 550–569 (2005)
J.S. Lewis, Icarus **15**, 174–185 (1971)
J.S. Lewis, Icarus **16**, 241–252 (1972)
J.S. Lewis, Space Sci. Rev. **14**, 401–411 (1973)
C.R.B. Lister, Geophys. J. R. Astron. Soc. **39**(3), 465–509 (1974)
R.D. Lorenz, B.W. Stiles, R.L. Kirk, M.D. Allison, P.P. del Marmo, L. Iess, J.I. Lunine, S.J. Ostro, S. Hensley, Science **319**, 1649–1651 (2008)
J.I. Lunine, M. Choukroun, D.J. Stevenson, G. Tobie, in *Titan from Cassini-Huygens*, ed. by R.H. Brown, J.P. Lebreton, H. Waite (Springer, New York, 2009), pp. 35–59
M.J. Lupo, J.S. Lewis, Icarus **40**, 157–170 (1979)

A. Manakov, V. Voronin, A. Kurnosov, A. Teplykh, E. Larionov, Y. Dyadin, Dokl. Phys. Chem. **378**, 148–151 (2001)
A. Manakov, V. Voronin, A. Kurnosov, A. Teplykh, V. Kumarov, Y. Dyadin, J. Inclus. Phenom. Macro. Chem. **48**, 11–18 (2004)
G. Marion, R. Farren, A. Komrowski, Cold Regions Sci. Technol. **29**, 259–266 (1999)
G.M. Marion, M.V. Mironenko, M.W. Roberts, Comput. Geosci. **36**, 10–15 (2010)
B. Martin, W.S. Fyfe, Chem. Geol. **6**(3), 185 (1970)
J. Martin, R. Lowell, J. Geophys. Res. **105**(B1), 869–882 (2000)
D.L. Matson, J.C. Castillo, J. Lunine, T.V. Johnson, Icarus **187**, 569–573 (2007)
T. McCord, G. Hansen, D. Matson, T. Johnson, J. Crowley, F. Fanale, R. Carlson, W. Smythe, P. Martin, C. Hibbitts, J. Granahan, A. Ocampo, J. Geophys. Res. **104**, 11827–11852 (1999)
T.B. McCord, G.B. Hansen, C.A. Hibbitts, Science **292**, 1523–1525 (2001). doi:10.1126/science.1059916
W.B. McKinnon, Lunar Planet. Sci. Conf. **37**, 2444 (2006). Lunar and Planetary Institute
W.B. McKinnon, M.E. Zolensky, Astrobiology **3**(4), 879–897 (2003)
W.B. McKinnon, D. Prialnik, S.A. Stern, A. Coradini, in *Structure and Evolution of Kuiper Belt Objects and Dwarf Planets*, ed. by M.A. Barucci, H. Boehnhardt, D.P. Cruikshank, A. Morbidelli (2008), pp. 213–241
L. Minassian, P. Pruzan, A. Soulard, J. Chem. Phys. **75**(6), 3064–3072 (1981)
I. Mosqueira, P. Estrada, D. Turrini, Space Sci. Rev. (2010, this issue). doi:10.1007/s11214-009-9614-6
H. Niemann, S. Atreya, S. Bauer, G. Carignan, J. Demick, R. Frost, D. Gautier, J. Haberman, D. Harpold, D. Hunten, G. Israel, J. Lunine, W. Kasprzak, T. Owen, M. Paulkovich, F. Raulin, E. Raaen, S. Way, Nature **438**, 779–784 (2005)
H. Palme, A. Jones, in *Treatise on Geochemistry*, vol. 1, ed. by H.D. Holland, K.K. Turekian (Elsevier, Amsterdam, 2003), pp. 41–61
R. Pappalardo, M. Belton, H. Breneman, M. Carr, C. Chapman, G. Collins, T. Denk, S. Fagents, P. Geissler, B. Giese, R. Greeley, R. Greenberg, J. Head, P. Helfenstein, G. Hoppa, S. Kadel, K. Klaasen, J. Klemaszewski, K. Magee, A. McEwen, J. Moore, W. Moore, G. Neukum, C. Phillips, L. Prockter, G. Schubert, D. Senske, R. Sullivan, B. Tufts, E. Turtle, R. Wagner, K. Williams, J. Geophys. Res. **104**, 24015–24056 (1999)
F. Postberg, S. Kempf, J. Schmidt, N. Brilliantov, A. Beinsen, B. Abel, U. Buck, R. Srama, Nature **459**, 1098–1101 (2009)
O. Prieto-Ballesteros, J.S. Kargel, M. Fernández-Sampedro, F. Selsis, E.S. Martínez, D.L. Hogenboom, Icarus **177**, 491–505 (2005)
L. Prockter et al., Space Sci. Rev. (2010, this issue)
P.M. Schenk, W.B. McKinnon, Icarus **79**, 75–100 (1989)
P. Schenk, I. Matsuyama, F. Nimmo, Nature **453**, 368–371 (2008)
J. Schmidt, N. Brilliantov, F. Spahn, S. Kempf, Nature **451**, 685–688 (2008)
G. Schubert, D.J. Stevenson, K. Ellsworth, Icarus **47**, 46–59 (1981)
G. Schubert, T. Spohn, R.T. Reynolds, in *Satellites*, ed. by J.A. Burns, M.S. Matthews (University of Arizona Press, Tucson, 1986), pp. 224–292
G. Schubert, J.D. Anderson, T. Spohn, W.B. McKinnon, in *Jupiter: The Planet, Satellites and Magnetosphere*, ed. by F. Bagenal, T.E. Dowling, W.B. McKinnon (Cambridge University Press, Cambridge, 2004), pp. 281–306
G. Schubert, J.D. Anderson, B.J. Travis, J. Palguta, Icarus **188**, 345–355 (2007)
G. Schubert, F. Sohl, H. Hussmann, in *Europa*, ed. by R.T. Pappalardo, W.B. McKinnon, K. Khurana (University of Arizona Press in collaboration with Lunar and Planetary Institute, Tucson, 2009), pp. 353–367
G. Schubert, H. Hussmann, V. Lainey, D.L. Matson, W.B. McKinnon, F. Sohl, C. Sotin, G. Tobie, D. Turrini, T. Van Hoolst, Space Sci. Rev. (2010, this issue)
J. Sclater, C. Jaupart, D. Galson, Rev. Geophys. Space Phys. **18**, 269–311 (1980)
E.R.D. Scott, Annu. Rev. Earth Planet. Sci. **35**, 577–620 (2007)
J. Seewald, M.Y. Zolotov, T. McCollom, Geochim. Cosmochim. Acta **70**, 446–460 (2006)
M.A. Sephton, Nat. Prod. Rep. **19**, 292–311 (2002)
F. Simon, G. Glatzel, Z. Anorg. Allg. Chem. **178**, 309–316 (1929)
M. Sinha, R. Evans, *Mid-Ocean Ridges: Hydrothermal Interactions Between the Lithosphere and Oceans* (AGU Press, Washington, 2004)
E.D. Sloan, *Clathrate Hydrates of Natural Gases* (Marcel Dekker, New York, 1998)
F. Sohl, W.D. Sears, R.D. Lorenz, Icarus **115**, 278–294 (1995)
F. Sohl, T. Spohn, D. Breuer, K. Nagel, Icarus **157**, 104–119 (2002)
F. Sohl, H. Hussmann, B. Schwentker, T. Spohn, R.D. Lorenz, J. Geophys. Res. **108**(E12), 5130 (2003). doi:10.1029/2003JE002044
C. Sotin, J.W. Head III, G. Tobie, Geophys. Res. Lett. **29**(8) (2002). doi:1029/2001GL013844

C. Sotin, G. Mitri, N. Rappaport, G. Schubert, D.J. Stevenson, in *Titan from Cassini-Huygens*, ed. by R.H. Brown, J.P. Lebreton, H.W. Jr. (Springer, New York, 2009), pp. 61–73
N.A. Spaun, J.W. Head, J. Geophys. Res. **106**, 7567–7576 (2001)
T. Spohn, G. Schubert, Icarus **161**, 456–467 (2003)
C. Stein, S. Stein, A. Pelayo, in *Seafloor Hydrothermal Systems*. Geophysical Monograph Series, vol. 91 (American Geophysical Union, Washington, 1995), pp. 425–445
D.J. Stevenson, A.W. Harris, J.I. Lunine, in *Satellites*, ed. by J.A. Burns, M.S. Matthews (University of Arizona Press, Tucson, 1986), pp. 39–88
R. Sullivan, R. Greeley, K. Homan, J. Klemaszewski, M.J.S. Belton, M.H. Carr, C.R. Chapman, R. Tufts, J.W. Head, R.T. Pappalardo, J. Moore, P. Thomas, the Galileo Imaging Team, Nature **391**, 371–373 (1998)
V. Tchijov, R. Baltazar Ayala, G. Cruz León, O. Nagorno, J. Phys. Chem. Solid **65**(7), 1277–1283 (2004)
G. Tobie, G. Choblet, C. Sotin, J. Geophys. Res. **108**(E11), 5124 (2003). doi:10.1029/2003JE002099
G. Tobie, A. Mocquet, C. Sotin, Icarus **177**, 534–549 (2005a)
G. Tobie, O. Grasset, J.I. Lunine, A. Mocquet, C. Sotin, Icarus **175**, 496–502 (2005b)
G. Tobie, J.I. Lunine, C. Sotin, Nature **440**, 61–64 (2006)
S. Vance, J. Harnmeijer, J. Kimura, H. Hussmann, B. DeMartin, J.M. Brown, Astrobiology **7**(6), 987–1005 (2007)
W. Wagner, A. Saul, A. Pruß, J. Phys. Chem. Ref. Data **23**(3), 515–525 (1994)
J.H. Waite Jr., W.S. Lewis, B.A. Magee, J.I. Lunine, W.B. McKinnon, C.R. Glein, O. Mousis, D.T. Young, T. Brockwell, J. Westlake, M.J. Nguyen, B.D. Teolis, H.B. Niemann, R.L. McNutt Jr., M. Perry, W.H. Ip, Nature **460**, 487–490 (2009)
K. Wang, G. van der Kamp, E.E. Davis, Geophys. J. Int. **139**, 763–768 (1999)
J.B. Wash, W.E. Brice, J. Geophys. Res. **89**, 9425–9432 (1984)
R. Zimmerman, W. Somerton, M. King, J. Geophys. Res. **91**, 12765–12777 (1986)
M.Y. Zolotov, Geophys. Res. Lett. **34**, L23203 (2007). doi:10.1029/2007GL031234
M.Y. Zolotov, Lunar Planet. Sci. Conf. **39**, 2349 (2008). Lunar and Planetary Institute
M.Y. Zolotov, J. Kargel, in *Europa*, ed. by R.T. Pappalardo, W.B. McKinnon, K. Khurana (University of Arizona Press in collaboration with Lunar and Planetary Institute, Tucson, 2009), pp. 431–457
M.Y. Zolotov, M.V. Mironenko, Lunar Planet. Sci. Conf. **38**, 2340 (2007a). Lunar and Planetary Institute
M.Y. Zolotov, M.V. Mironenko, in *Workshop on Ices, Oceans, and Fire: Satellites of the Outer Solar System* (Lunar and Planetary Institute, Houston, 2007b), pp. 157–158
M.Y. Zolotov, E.L. Shock, J. Geophys. Res. **106**, 32815–32827 (2001)
M.Y. Zolotov, E.L. Shock, J. Geophys. Res. **109**, E06003 (2004). doi:10.1029/2003JE002194
M.Y. Zolotov, T. Owen, S. Atreya, H. Niemann, E.L. Shock, EOS Trans. AGU **86**(52), 43–44 (2005). American Geophysical Union

Space Sci Rev (2010) 153: 511–535
DOI 10.1007/s11214-009-9610-x

Exobiology and Planetary Protection of icy moons

François Raulin · Kevin P. Hand · Christopher P. McKay · Michel Viso

Received: 14 April 2009 / Accepted: 19 November 2009 / Published online: 16 February 2010

Abstract The outer solar system is an important area of investigation for exobiology, the study of life in the universe. Several moons of the outer planets involve processes and structures comparable to those thought to have played an important role in the emergence of life on Earth, such as the formation and exchange of organic materials between different reservoirs. The study of these prebiotic processes on, and in, outer solar system moons is a key goal for exobiology, together with the question of habitability and the search for evidence of past or even present life. This chapter reviews the aspects of prebiotic chemistry and potential presence of life on Europa, Enceladus and Titan, based on the most recent data obtained from space missions as well as theoretical and experimental laboratory models. The habitability of these extraterrestrial environments, which are likely to include large reservoirs of liquid water in their internal structure, is discussed as well as the particular case of Titan's hydrocarbon lakes. The question of planetary protection, especially in the case of Europa, is also presented.

F. Raulin (✉)
LISA CNRS & University Paris 7 and Paris 12, Créteil, France
e-mail: Francois.Raulin@lisa.univ-paris12.fr
url: http://www.lisa.univ-paris12.fr/

K.P. Hand
Jet Propulsion Laboratory, California Institute of Technology, Pasadena, CA, USA
e-mail: khand@jpl.nasa.gov
url: http://science.jpl.nasa.gov/people/hand/

C.P. McKay
NASA Ames Research Center, M.S. 245-3, Moffett Field, CA, USA
e-mail: chris.mckay@nasa.gov
url: http://spacescience.arc.nasa.gov/people/mckay/

M. Viso
CNES/DSP/EU, Paris, France
e-mail: michel.viso@cnes.fr
url: http://www.cnes.fr/web/CNES-fr/6919-cnes-tout-sur-l-espace.php

Keywords Exobiology · Prebiotic chemistry · Extraterrestrial life · Europa's ocean · Enceladus liquid water reservoir · Habitability · Planetary protection

1 Exobiology and Habitability

Since the emergence of exobiology in the 1960s, the definition of this new and continuously increasing field has evolved from the search for and study of life outside the Earth to the study of life in the whole universe. Exobiology (~astrobiology) thus includes today the search for extraterrestrial life, but also the study of the origin and limits of life on Earth, as well as the study of processes and structures related to life. Extraterrestrial organic chemistry, and all related processes, including the formation and exchange of organic materials between different reservoirs of planetary significance is an important part of exobiology. The later are present in many planetary objects of the solar system, especially in the outer solar system, on and in many of its moons.

Several of these extraterrestrial organic processes may be similar to those which allowed the origin of life on Earth, about 4 billion years ago. They involve low molecular weight reactive compounds, such as nitriles, HCN and HC_3CN, and hydrocarbons such as C_2H_2 and C_2H_4, and their oligomers or polymers. The chemical evolution of these compounds in liquid water yielded macromolecules capable of self-replication on the primitive Earth, and the first living systems. Studying some of the chemical steps which are occurring today in these extraterrestrial environments and in particular the exchange processes involved can provide some insights into the chemical evolution which occurred on the early Earth. Studying these processes is of paramount importance since they are not present anymore on our planet and cannot be reproduced experimentally in the laboratory because of unrealistic time requirements. Those processes are also of great interest by themselves since they allow complexification of matter in extraterrestrial environments, providing a possible step toward the emergence of life, either for life as we know it or perhaps even for alternative biochemistries. Now these processes are based on many exchanges with the environment where they occur, involving transport of matter from different parts of the planetary environment, and fluxes of energy, inducing the chemical transformation of simple organic compounds toward complex macromolecular structures. The possible internal oceans of Europa, Titan and Enceladus are potential places for such chemical exchanges and evolution. The atmosphere of Titan with chemical homogeneous reactions, as well as heterogeneous reactions involving the atmospheric aerosol particles, but also interaction with the surface is another intriguing locale. Titan's surface by itself is another location were organic exchange processes can also occur, in particular in the hydrocarbon large lakes present in the polar regions.

Another aspect related to life and exobiology is habitability. One should differentiate the question of origin of life from that of habitability. One of the crucial parameters required for habitability is the availability of "usable" liquid water. This can be quantitatively represented by the water activity. A limit of water activity frequently quoted for habitability—at least of life as we know it on Earth—is >0.6. Another key parameter is temperature. It must not be too high and incompatible with liquid water and the preservation of macromolecular biological structures. It must be not too low, to allow biochemical reactions to occur at a noticeable rate. Environments which are suitable for the origin of life may evolve in such a way that the biological evolution cannot follow the geological one. For example, liquid water may totally disappear from a planetary surface. Water activity may become smaller than 0.5, with the environment fully incompatible with life, such as the surface of Venus, for example. In such cases, life may disappear if the planet drastically evolves and becomes too hostile and not habitable. We may thus have planetary environments that have harbored life, but are not habitable anymore today. Such environments may have kept traces of this past life, which could be searched for with future missions. There are environments where life may have originated and that have maintained conditions compatible with habitability. Such environments may still harbor life today. Finally, there are planetary environments where conditions are compatible with habitability, but which have never been compatible with the development of a complex chemical evolution to life. However, the possibility of the presence of life in such environment cannot be fully ruled out if this place is reachable from materials coming from another planetary object where life may have appeared and can be exported elsewhere.

Thus there are many possibilities of environments that may be, or may have been suitable for life. Exchange processes play a major role in such environments for the development of life. The use of nutrients by living systems, necessary for metabolism, release compounds in the environment that can be used as biosignature. Carbon assimilation (like CO_2 as an example) by living organisms is followed by an increase of $^{12}C/^{13}C$ in the biomolecular structures. Homochirality, like the use of L-amino acids in protein and D-sugar in nucleotides on Earth, is another example of biosignature. Interaction and exchanges between microorganisms and their mineral environments can also generate specific biomineral structures that have specific properties and can be used as another sign of life.

If we consider the availability of liquid water as the key parameter for habitability, there are several places that can be considered as potential targets for searching for extraterrestrial life in the outer solar system. Europa, Titan and Enceladus are among the most interesting sites. What kind of life? The terrestrial scenario of the origin of life assumes that before "LUCA", our last universal common ancestor, there was an RNA-world, and before, a pre-RNA world. In this first living word the self-replicating molecules were probably similar to the current nucleic acids, but were using much simpler and more robust nucleotides analogues. Our search for extraterrestrial life must be open and should consider the possibility of such replicating systems. But extraterrestrial living beings may be even more exotic than what we know, and the possibility of another life, even without liquid water, cannot be fully ruled out and could be also taken into account, for instance in the case of Titan and its hydrocarbon lakes.

Regardless of the nature of this hypothetical extraterrestrial life, the search and exploration of potentially inhabited worlds introduces the question of planetary protection. These aspects are also presented below, after a description of the prebiotic chemical aspects and the potentiality for life aspects, as related to the exchange processes, of moons of the outer solar system.

2 Prebiotic Chemistry and Life in Other Environmental Conditions and the Role of Exchange Processes

2.1 The Case of Europa and Enceladus

Introduction The high likelihood that Europa harbors a global subsurface liquid water ocean, and that Enceladus may harbor a subsurface South Polar liquid water sea (Sohl et al. 2010, this issue), positions these two icy worlds atop the list of prime targets for future exobiological investigations.

If oceans and seas do exist on these worlds, then the habitability of their liquid water environments may be critically tied to exchange processes between the seafloor and ocean water, and between the ice shell and ocean water. Here we review these processes in the context of prebiotic chemistry and habitability. For the most part, we do not strongly distinguish between the cases of Europa and Enceladus since the primary differences between these worlds, and their putative liquid water environments, are treated elsewhere in this volume (Schubert et al. 2010, this issue; Sohl et al. 2010, this issue). Of key concern for exobiology is the presence of a liquid water reservoir beneath a relatively thin ice shell (<20 km) and liquid water in contact with a rocky seafloor. For Europa these conditions seem likely (Greeley et al. 2004), while for Enceladus the data are intriguing but not yet fully understood on either empirical or theoretical grounds.

Our study of life on Earth has revealed three broad requirements for habitability (Chyba and Hand 2005; Nealson 1997). Liquid water, a suite of $\sim$50 elements essential for life, and some form of chemical energy or radiation must be present. Europa and Enceladus may have an abundance of liquid water, but if the latter two conditions are not met then such oceans might not be habitable for life as we know it. Conditions for the origin of life are poorly constrained but include the conditions for habitability along with several additional constraints (Kasting 2000; Nisbet and Sleep 2001; Orgel 1998). Thus, both habitability and life's origin are in part contingent on the availability of key elements and energy in the environment. Exchange processes on icy worlds may provide the key to satisfying these conditions.

On Earth, exchange of material between the rocky seafloor and ocean serves to remove oxidized compounds from solution (e.g. CO_2, SO_4^{2-}) while replacing them with reduced species (e.g. H_2S, CH_4) (Elderfield and Schultz 1996). Similarly, communication between the atmosphere and ocean on Earth permits dissolution of oxidants (e.g. O_2 and CO_2) into the ocean water. This recharge and exchange of material on Earth is largely powered by radiogenic decay (responsible for $\sim$80% of the energy driving mantle convection) and solar insulation (responsible for heating of the surface and atmosphere). On icy worlds around giant planets, radiogenic decay is significant but tidal energy dissipation likely dominates geological and geochemical exchange processes.

The partitioning of tidal energy in the mantle and hydrosphere (ice + ocean) of Europa determines the extent to which the ocean is enriched with oxidants and reductants. If most of the tidal energy is dissipated in the mantle, then the ocean will be enriched with reductants. Conversely, if most of the tidal energy is dissipated in the ice shell and ocean, then the ocean will be rich in oxidants. The ideal condition for habitability is likely a balance between these two scenarios, i.e. enough tidal energy dissipated in the mantle to maintain a flux of reductants to the ocean and enough dissipation in the ice shell to maintain a flux of oxidants to the ocean below. Understanding the tidal interaction and dissipation at Europa and Enceladus is therefore important not just to our geophysical understanding of these worlds, but is also critical to our astrobiological understanding.

Elements for Life

The basic elemental building blocks for life (C, H, N, O, P, and S) are well known and appreciated in the field of exobiology. Lesser known, but yet of great importance to biological structures and mechanisms are a host of other key elements. Wackett et al. (2004) surveyed the role and importance of naturally occurring elements for life as we know it. Figure 1 shows a periodic table based on that of Wackett et al. Along with CHNOPS, the ions of Na, Mg, K, Ca and Cl are all required. Another 10 elements serve as important transition metals or key trace elements. Beyond these 21 elements, another 37 are important metabolically or for specialized uses. The remaining 34 naturally occurring elements of the periodic table apparently serve no critical purpose.

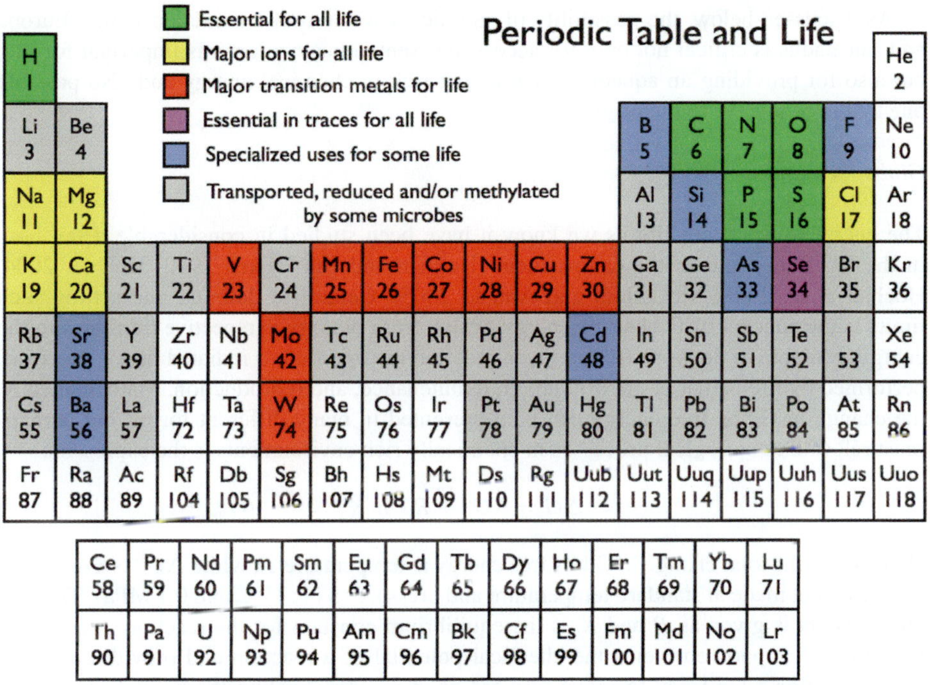

Fig. 1 The Periodic Table and elemental requirements for life. Shown in this representation of the elements are those that are essential to life, ions and heavy metals that are of major but perhaps essential importance to life, and elements that have trace, specialized, and/or key metabolic significance to life

In the context of the habitability of icy worlds, and in particular of Europa and Enceladus, it is important to consider the availability of these biologically essential elements. The initial conditions under which such elements became available on these worlds can, in large part, be attributed to exchange processes occurring within the primordial solar system. During solar system formation, volatile species such as H_2O, CH_4, H_2S, CO_2, N_2, and NH_3 would have been preferentially sequestered into bodies of the outer solar system where temperatures permitted condensation (Lewis 1971, 1974, 2004). Thus while the terrestrial planets may be rich in heavy elements, the outer planets and their moons are rich in the lighter elements that are so essential to life. McKay (1991) and later Hand et al. (2009) have shown this relationship for C, N, S, and P within our solar system. While Wackett et al. do show the importance of heavier elements, the expected chondritic composition for Europa and Enceladus (Kargel et al. 2000; Zolotov 2007; Zolotov and Shock 2001) would appear to satisfy the trace abundance levels needed for life.

On the issue of heavier elements, however, herein we note important differences between the various icy worlds. Europa is believed to have a silicate mantle, perhaps even a basaltic seafloor, in contact with the ocean water (Anderson et al. 1998; Greeley et al. 2004). Ganymede and Callisto may harbor subsurface oceans but their seafloors are likely made of higher polymorphs of ice, e.g. ice III and V (Kuskov and Kronrod 2005). The same is true of Titan, though there the ocean and seafloor ice are expected to have much higher ammonia content (Lorenz et al. 2008; Sohl et al. 2003). For Enceladus, the case of differentiation and water–rock contact is highly uncertain; however, recent Cassini spacecraft results from the particle detector do appear consistent with such activity (Burger et al. 2010, this issue).

As discussed below, the possibility of significant water–rock interaction within Europa and Enceladus is critical not only for access to essential heavy elements important for life, but also for providing an aqueous environment rich in chemical energy and also possible one that is conducive to the origin of life.

Gibbs Free Energy

The energetic limits for life as we know it have been studied in considerable detail both in the laboratory and in the environment (Heijnen and van Dijken 1992; Hoehler 2004; Hoehler et al. 2001; McCollom and Amend 2005). Central to defining and setting such limits is the concept of Gibbs free energy, which from biology perspective translates to the energy available to do work from a given reaction (as opposed to radiated heat). The work performed by life is that of growth, repair, maintenance, and reproduction. Mathematically, the change in Gibbs free energy can be represented in many different forms, but perhaps most relevant to biological processes in chemically rich environments is the expression

$$\Delta G = \Delta G^0 + RT \ln K.$$

Here ΔG is the total change in Gibbs free energy at temperature T, while ΔG^0 is a reference measurement made at standard temperature and pressure (STP, 25°C and 0.1 MPa). The gas-law constant is given as R and K is the equilibrium constant for the reaction in question. Roughly, K in part represents the chemical potential of a reaction and provides a metric of the reaction's tendency to move forward or in reverse. Scaling the natural log of that potential by the temperature of the system and the gas-constant then yields a contribution, given in Joules per mole, to the change in Gibbs free energy. For a reaction proceeding from left-to-right (forward) the natural log of K will be negative, and if ΔG^0 is negative or has an absolute value smaller than this, the total change in Gibbs free energy will be negative, indicating that work can be harnessed from the reaction. Hoehler et al. (2001) measured the smallest possible negative change in Gibbs free energy that still permitted biological maintenance (no repair, growth, or reproduction was possible) and found that near STP for methanogenesis,

$$CO_2 + 4H_2 \rightarrow CH_4 + 2H_2O,$$

the minimum is $\Delta G = -10.6 \pm 0.7$ kJ per mole of methane. For the case of sulfate reducing bacteria,

$$SO_4^{2-} + 4H_2 \rightarrow S^{2-} + 4H_2O,$$

the limit is $\Delta G = -19.1 \pm 1.7$ kJ per mole of sulfate. These reactions and metabolic pathways serve as our current baseline for the energetic requirements for life as we know it. In other words, changing the reaction or conditions such that the yield in ΔG is any closer to zero leads to an uninhabitable environment. On Earth, most chemical systems near the surface are relatively rich in energy and surpass this limit (i.e. have a great negative change in G). On other worlds that are perhaps less dynamic than the Earth, the available chemical energy may be limiting for life.

Aside from photosynthesis, the metabolic pathways that yield energy involve pairing an oxidant (electron acceptor, e.g. CO_2 and SO_4^{2-} as in above reactions) with a reductant (electron donor, e.g. H_2 as above). These serve as the positive and negative 'terminals' of the biochemical 'battery' needed to power life and are often referred to as 'redox' pairs or coupling. Interestingly, two elements essential to life, carbon and sulfur, serve in many different

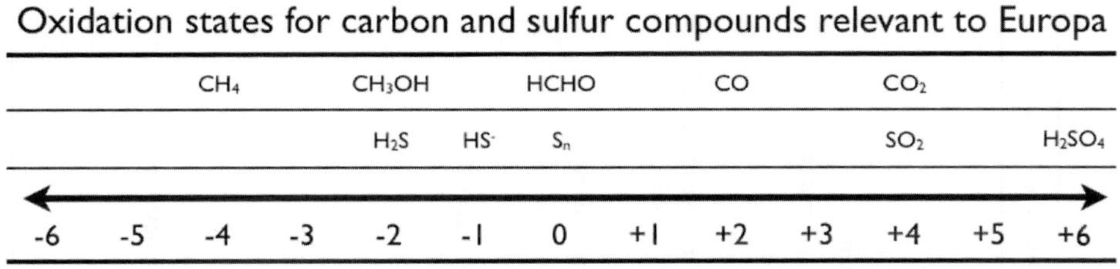

Fig. 2 Oxidation states of carbon and sulfur. The diversity of oxidation states for these compounds is in part responsible for their importance to life and biological processes (from Carlson et al. 1999b)

biological functions and also share the characteristic of having a variety of metabolically useful oxidation states. Figure 2 shows the spectrum of states and associated compounds for these two elements. On Europa it has been proposed that radiolytic chemistry drives a sulfur cycle producing reservoirs of many of the compounds listed in Fig. 2 (Carlson et al. 1999b). Seafloor chemistry could also be responsible for providing both carbon and sulfur in a variety of oxidation states (Zolotov and Shock 2004). Table 1, adapted from Zolotov and Shock (2004), illustrates some of the metabolic pathways known to function on Earth and which could be applicable to Europa (and possibly Enceladus) based on our limited knowledge of the ocean chemistry.

With energy and access to essential elements as key limiting factors for life, we now consider the role of exchange processes between different geophysical layers on Europa and Enceladus, working first from the space environment down to the interiors.

Space to Ice

The exchange of energy between Europa and the Jovian magnetic field results in an ice surface rich with oxidants. Similarly, though weaker by a few orders of magnitude (Cooper et al. 2006), the surface of Enceladus may be modified through interactions with the Saturnian magnetic field.

This interaction between a gas giant's magnetic field and the icy surfaces of its moons serves to convert energy ultimately derived from planetary angular momentum and the internal dynamo into the chemical energy released and stored in bonds of the radiolytically processed compounds on the icy surfaces. The magnetospheres are rich in ions and electrons from several sources (e.g. solar wind, ionospheres, volcanism) which then bombard the surfaces of Europa and Enceladus, splitting apart water and other compounds and generating products such as H_2O_2 and O_2 (Carlson et al. 1999a; Spencer and Calvin 2002). For the surface of Europa, this interaction results in a approximately 125 $\mathrm{mW\,m^{-2}}$ of incident particle energy deposition, over 80% of which is due to the energetic electrons (Cooper et al. 2001). Globally, this energy source accounts for $\sim 4 \times 10^{12}$ W, or approximately 3% of the absorbed solar energy (less than $\sim 1\%$ of which is capable of photolyzing water).

Important to exobiology, the radiolytic chemistry derived from the magnetic field interaction serves to 'charge' the surfaces of icy worlds with oxidants that could provide the necessary electron acceptors for biochemical processes in the subsurface. Gaidos et al. (1999) were concerned that Europa's ocean might be rich in reductants, but lacking oxidants it might be a thermodynamically, and potentially biologically, 'dead' ocean. The discovery of hydrogen peroxide on the Europan surface (Carlson et al. 1999a) inspired Chyba (2000) to hypothesize that radiolytic chemistry on the surface of Europa could provide the key for powering life in the subsurface. Based on the 0.13% by number abundance of H_2O_2 relative

Table 1 Example metabolic pathways for terrestrial microbes that may serve as models for energetically useful pathways in the putative subsurface oceans of Europa and Enceladus. Adapted from Zolotov and Shock (2004)

Metabolic process	Example reaction	Representative organism or Genus
Anaerobic Reactions		
Methanogenesis	(a) $HCO_3^- + H^+ + 4H_2(aq) \rightarrow CH_4(aq) + 3H_2O$	(a) Methanobacterium
	(b) $CO(aq) + 3H_2(aq) \rightarrow CH_4(aq) + H_2O$	(b) Methanobacterium
	(c) $4HCOO^- + H^+ \rightarrow 3CO_2(aq) + CH_4(aq) + 2H_2O$	(c) Methanomicrobium
	(d) $CH_3OH(aq) + H_2(aq) \rightarrow CH_4(aq) + H_2O$	(d) Methanosphaera stadtmanae
Homoacetogenesis	$2HCO_3^- + H^+ + 4H_2(aq) \rightarrow CH_3COO^- + 4H_2O$	Acetobacterium woodii
Coupled sulfate reduction–H_2 oxidation	$SO_4^{2-} + H^+ + 4H_2(aq) \rightarrow HS^- + 4H_2O$	Desulfosarcina
Coupled sulfate reduction–methane oxidation	$SO_4^{2-} + CH_4(aq) \rightarrow HS^- + HCO_3^- + H_2O$	Unknown, but likely mediated by coexisting methanogens and sulfate-reducers
Coupled sulfate reduction–organic carbon oxidation	$SO_4^{2-} + 4HCOO^- + H^+ \rightarrow HS^- + 4HCO_3^-$	Desulfobacter
Coupled oxidation of native sulfur and iron reduction	$S^0 + 6Fe^{3+} + 4H_2O \rightarrow SO_4^{2-} + 6Fe^{2+} + 8H^+$	Acidithiobacillus ferrooxidans
Coupled reduction of native sulfur–hydrogen oxidation	$S^0 + H_2(aq) \rightarrow H_2S(aq)$	Acidianus
Coupled reduction of native sulfur–methane oxidation	$4S^0 + CH_4(aq) + 2H_2O \rightarrow 4H_2S(aq) + CO_2(aq)$	Halobacterium
Coupled reduction of native sulfur–organic carbon oxidation	$4S^0 + CH_3COO^- + 2H_2O \rightarrow 4HS^- + 3H^+ + 2CO_2(aq)$	Desulfuromonas acetoxidans
Aerobic Reactions		
Oxidation of hydrogen	$H_2(aq) + 0.5O_2(aq) \rightarrow H_2O$	Acidovorax facilus
Oxidation of methane	$CH_4(aq) + 2O_2(aq) \rightarrow CO_2(aq) + 2H_2O$	Methylomonas
Oxidation of organics	$CH_3COO^- + H^+ + 2O_2(aq) \rightarrow 2CO_2(aq) + 2H_2O$	Pseudomonas
Oxidation of aqueous sulfides	$HS^- + 2O_2(aq) \rightarrow SO_4^{2-} + H^+$	(acido)Thiobacillus
Oxidation of native sulfur	$S^0 + 2O_2(aq) \rightarrow SO_4^{2-} + 2H^+$	Sulfolobus acidocaldarius
Oxidation of ferrous silicates, oxides, and sulfides	(a) $1.5Fe_2SiO_4 + 0.5O_2(aq) \rightarrow Fe_3O_4 + 1.5SiO_2(aq)$	(a–d) Thiobacillus ferrooxidans
	(b) $3FeS + 2O_2(aq) \rightarrow 3S^0 + Fe_3O_4$	
	(c) $FeS_2 + 3.5O_2(aq) + H_2O \rightarrow 2SO_4^{2-} + Fe^{2+} + 2H^+$	
	(d) $Fe_3O_4 + 0.5O_2(aq) + 1.5H_2O \rightarrow 3FeOOH$	

to water, as measured by Carlson et al. (1999a), Chyba and Phillips (2001) calculated that an H_2O_2 (oxidant) and HCHO (reductant) powered ecosystem could support a steady-state biomass of 10^{23}–10^{24} cells. The terrestrial soil bacterium *Hyphomicrobium* serves as a model for such a metabolic pathway, though it is important to note that O_2 serves as the oxidant and in the Chyba and Phillips model for Europa H_2O_2 rapidly decays to O_2 in the ocean and then serves as the oxidant. Also significant in this model is that the reductant, HCHO (formaldehyde), is radiolytically produced via processing of CO_2 on the surface of Europa. The order of magnitude estimates for HCHO abundance calculated by Chyba and Phillips (2001) are consistent with laboratory results (Hand et al. 2007).

Though the detection of hydrogen peroxide sparked much interest in surface oxidants, it may be one of the less abundant oxidants on the Europan surface. Measurements of condensed phase O_2 on Europa indicate concentrations possibly reaching several percent by number abundance (Hand et al. 2006; Spencer and Calvin 2002). Sulfate, also a very useful oxidant for metabolic pathways, reaches 10–90% of the surface abundance (Carlson et al. 2005). Carbon dioxide and sulfur dioxide are also observed on the surface of Europa and contribute to the oxidant inventory. As shown later in this chapter, if the ice shell of Europa is active and permits delivery of the surface oxidants to the subsurface, then the ocean of Europa could be highly enriched, even saturated, with species like oxygen, sulfate, and carbonate.

The role of radiation chemistry and the production of oxidants on Enceladus is at present uncertain and debated. Newman et al. (2007) reported the detection of hydrogen peroxide via the 3.5 μm feature in VIMS data, but Loeffler and Baragiola (2009) rejected that identification as inconsistent with laboratory measurements of the band width and position for a variety of H_2O_2+H_2O mixtures. Newman et al. (2007) argued for an abundance upper limit of 0.5% H_2O_2 by number relative to H_2O, which is several times that observed on Europa (Carlson et al. 1999a). If the 3.5 μm were correctly identified as H_2O_2 then Enceladus' surface might be rich with oxidants.

Radiation chemistry aside, the observed surface chemistry of Enceladus is remarkable in the context of exobiology. Organics, methane, carbon dioxide, carbon monoxide, and possibly several nitrogen species (e.g. N_2 and NH_3^+) have been observed on the surface or in the plume material. Though the infrared results are somewhat tenuous (Brown et al. 2006), the ion and neutral mass spectrometer (INMS) provide conclusive detection of all of these species, save for ambiguities pertaining to nitrogen species and the complexity of the organics (Waite et al. 2006).

Clearly it is impossible to know at this point whether the plume material truly reflects internal aqueous solution chemistry or if it reflects ice shell chemistry. The identification of the above compounds, combined with the observed plume activity, nevertheless provides some intriguing clues that if a subsurface sea exists, then it may have chemistry conducive to habitability and the origin of life. McKay et al. (2008) explored this possibility and concluded that several methanogenic subsurface ecosystems on Earth serve as interesting models for Europa and Enceladus, and that heterotrophic and chemosynthetic theories for the origin of life could apply given the known chemistry of Enceladus.

The issue of life's origin is particularly interesting in the context of cold, ice-covered worlds with subsurface oceans. The production and stability of long polymeric compounds is difficult in a rapidly changing, energetic, and hot environments such as the early Earth (Bada and Lazcano 2002; Miller and Orgel 1974). Cold environments may limit reactivity but freezing and the resulting eutectic solutions may promote polymerization. On both Europa and Enceladus melt regions of liquid water within the ice shell could serve to concentrate compounds in the ocean water or products from radiolytic chemistry on the surface. Hand (2007) showed via laboratory radiolysis experiments that the electron irradiation environment on Europa could lead to amino acids, formamide, HCN, and other compounds relevant to prebiotic chemistry. Such results are in agreement with those performed using UV-photolysis in ices at colder, astrophysical temperatures (Bernstein et al. 1995; Dworkin et al. 2001). Along with radiolytically driven prebiotic chemistry, Europa and Enceladus will receive a flux of organics from cometary and meteoritic bombardment. Pierazzo and Chyba (2002) estimate this flux to be 10^{12} to 10^{13} kg of carbon over Europa's history. The survival fraction of amino acids contained in the original impactor was calculated to be a few percent of the starting concentration.

Energy and material exchange processes occurring on the surface of Europa and Enceladus are significant both in the context of habitability and of life's origin. Radiolysis of water ice yields biochemically useful oxidants while radiolysis of hydrocarbons and ammonia yields prebiotic compounds. Exogenous delivery of organics from cometary and meteoric infall, and sulfur from Io volcanism (for the case of Europa), also serve to generate a chemically rich surface. The surface reservoirs of compounds on Europa and Enceladus are interesting but of limited relevance to exobiology unless they are cycled into, and have some connection to, subsurface aqueous chemistry.

Ice to Ocean

The past and present activity of the ice shells of Europa and Enceladus determine the extent to which radiolytically processed surface material reaches the liquid water ocean, and the extent to which observed surface materials represent subsurface composition. Though our understanding of exactly what geological mechanisms have, or are, taking place on both worlds remains poor, several points are unequivocal. The surface of Europa is globally young, constrained to be 10^7–10^8 years in age (Zahnle et al. 2003). Enceladus, however, shows a strong dichotomy in surface age. The northern hemisphere is >3 Gyr in age while much of the southern hemisphere is very young (Squyres et al. 1983b). The region of the South Polar Terrain and the Tiger Stripes is essentially new and fresh material (Porco et al. 2006; Spencer et al. 2006).

Enceladus is geologically active today (Porco et al. 2006) while Europa, based on very limited data, shows no current activity (Phillips et al. 2000). Europa, however, is globally active on geologically short timescales ($<10^8$ years). The geological context for life is a common theme in our efforts to understand what makes a world habitable; the rock cycle, for instance, is critical to life on Earth, maintaining the carbon cycle that is so central to terrestrial habitability. On Mars, a cooling core and mantle resulted in a stagnant lithosphere and the demise of the internal magnetic field. These conditions may have then lead to Mars becoming a much less habitable world. On icy worlds like Europa and Enceladus tidal interactions may be essential to driving rock cycles of ice. Ultimately, rock cycles—be they with silicates or ices—facilitate chemical cycles that can then be exploited by biological processes.

On Europa differing interpretations of surface features have lead to models in which the ice shell is thin (<10 km) and those in which the ice shell is thick (> 10 km; Tobie et al. 2010, this issue; Sohl et al. 2010, this issue for more detail on these models). In the context of exobiology, the critical difference between these models is that the thin-shell model permits direct communication of the ocean material with the surface, whereas convecting thick-ice models yield an ocean largely isolated from the surface. A thick shell may not provide a sufficiently dynamic rock cycle for maintaining a chemically rich and habitable ocean.

Allowing for the thin- or thick-shell model, Hand et al. (2007) considered the delivery of several radiolytically produced surface oxidants (O_2, CO_2, SO_4^{2-}) and the reductant HCHO to Europa's ocean. Even if cycling of surface material into the ocean takes 0.5 Gyr, the low-end estimates for sulfate concentration (~10%) still yield a net annual flux of 2×10^9 moles of SO_4^{2-}. For oxygen and hydrogen peroxide delivery, Hand et al. find that even with a 0.5 Gyr delivery period the molarity of O_2 in the Europan ocean still surpasses levels found in Earth's oceanic O_2-minima zones (~80 μM). If abiotic sinks consume a flux of $\sim 10^9$ moles per year of oxidants (McCollom 1999), delivery periods of $\sim 10^8$ years or less are needed to surpass O_2-minima levels. Thus, even for very long and slow geological processing and delivery of surface oxidants, the ocean of Europa achieves considerable oxidant delivery, especially when considering sulfate. Though geological mechanisms may be uncertain, and the possibility of a completely isolated brittle surface shell remains, the exchange of surface material with the subsurface ocean on even very long timescales still achieves a relatively oxidized ocean.

Interestingly, the process of entraining surface material into the ice and eventual delivery to the ocean could serve to enhance prebiotic chemistry in the ocean. Levy et al. (2000) showed that solutions with ammonium cyanide (NH_4CN) frozen at −20°C and −78°C and left for 25 years could produce the bases adenine and guanine, along with glycine, alanine, and several other amino acids. More recently, Trinks et al. (2005) have demonstrated the polymerization of ~400 nucleotides to form strands of RNA in eutectic solutions cycled between −7°C and −24°C over the course of a year. Such experiments are intriguing when considering the radiolytic production of simple compounds on the surface of icy worlds that might lead to the synthesis of prebiotically significant compounds after burial and processing within the ice shell. Along with questions of life's origin, the ice–water interface and microenvironments within the ice shell also present interesting question pertaining to the co-evolution of life and its environment (Price 2007, 2009).

Finally, we note that while the delivery of surface compounds to the subsurface may be important for habitability, the delivery of subsurface material to the surface is important for our ability to detect life. The ice shells of icy worlds serve as our window to the subsurface, and the extent to which the ice shells reflect the oceans below will determine our ability to assess the presence of any subsurface life. Hand et al. (2009) have examined this point in considerable detail for Europa and McKay et al. (2008) explored the case of biosignatures for Enceladus.

Ocean to Seafloor

Circulation between the oceans of icy worlds and rocky seafloors could provide an important flux of reductants, i.e. electron acceptors, to life in the ocean. Hydrothermal interactions may also be an important locale for prebiotic chemistry and the origin of life on such worlds. Estimates for energy flow at the seafloor of Europa range from 6 to a few hundred $\mathrm{mW\,m^{-2}}$ (Squyres et al. 1983a; Thomson and Delaney 2001). The lower limit is set by radiogenic models while upper estimates depend on how much tidal energy is dissipated in the mantle. Estimates for Enceladus are much less certain, but the observed heat flux of the South Polar Terrain is ~6 GW and in localized hotspots along the Tiger Stripes the heat flux reaches 13 $\mathrm{W\,m^{-2}}$ or greater (Abramov and Spencer 2008; Spencer et al. 2006). The heat flux at the seafloor of the Earth, by comparison, is 80 $\mathrm{mW\,m^{-2}}$ on average and reaches >200 $\mathrm{mW\,m^{-2}}$ in mid-ocean ridge and along axis regions (Elderfield and Schultz 1996). Clearly both models and observations of Europa and Enceladus are consistent with possibly active interiors. Models for the orbital evolution of Europa indicate that they might have had periods of much greater tidal energy dissipation in the past (Hussmann and Spohn 2004).

McCollom (1999) modeled Europa's hydrothermal activity as being $1/1000$ that of the Earth's and found that lithoautotrophic methanogenesis might be able to support 10^5–10^6 kg yr^{-1}, or roughly 10^{-5} that of terrestrial biomass productivity at hydrothermal vents. McCollom considered both a reduced and oxidized ocean conditions and found that the CO_2 and H_2 needed for methanogenesis will be available in both cases.

Zolotov and Shock (2004) examined possible biogeochemical pathways utilizing sulfur, iron, and carbon on Europa. They consider only low-temperature reactions and dismiss the possibility of a tectonically active seafloor. Nevertheless, their models indicate that Europa's ocean was, and still may be, an alkaline sulfate and carbonate-rich ocean in which a host of metabolic pathways might have been available for life (see Table 1). Without continued seafloor cycling and exposure of fresh rock, however, the pathways available after the formation and differentiation of Europa's ocean would have become exhausted. Considering the case of Enceladus, Zolotov (2007) found that a primordial ocean on Enceladus would likely have had alkaline and carbonate-rich chemistry, perhaps providing for similar biochemical pathways, though the issue of habitability was not specifically explored in that work. Parkinson et al. (2007, 2008) hypothesized that the observed surface chemistry of Enceladus could serve to support a subsurface ecosystem, were there to be sufficient water–rock interaction beneath the ice.

Vance et al. (2007) maintain that even if tectonics were minimal on Europa and Enceladus, seawater interaction with reduced mantle material could drive serpentinization reactions yielding H_2 that could be useful for life. The basic reaction is

$$2Mg_2SiO_4 + 3H_2O \Rightarrow Mg(OH)_2 + Mg_3Si_2O_5(OH)_4,$$

and translates to olivine (specifically shown as forsterite above) and water reacting to yield brucite and serpentine. Though the above reaction drives serpentinization, it is the oxidation of iron in olivine (now fayalite or ferrosilite) that ultimately leads to the H_2 production of astrobiological interest:

$$3Fe_2SiO_4 + 2H_2O \Rightarrow 2Fe_3O_4 + 3SiO_2(aq) + 2H_2(aq)$$

and

$$3FeSiO_3 + H_2O \Rightarrow Fe_3O_4 + 3SiO_2(aq) + H_2(aq).$$

The resulting H_2 can then be metabolized with CO_2 to yield CH_4 during methanogenesis.

Interestingly, tectonics as classically envisioned need not be a prerequisite for such interaction. Rather, the permeability of the mantle—i.e. the fracturing and micro-fracturing that occurs as a result of tidal dissipation and variations in temperature and pressure—may allow for sufficient water penetration to drive such reactions. On Europa the thermal cracking depth for an olivine-rich seafloor is calculated to be 25 km below the seafloor. For Enceladus this depth is 172 km, or essentially the entire radius. The resulting serpentinization heat flux is 0.05% and 6.56% of the radiogenic flux for Europa and Enceladus respectively. A limiting factor for such reactions is that the process is self-sealing, i.e. serpentinization products fill the cracks through which the water is permeating the subsurface. Nevertheless, Vance et al. argue that Europa likely experiences a net flux of H_2 to the ocean of $\sim 5 \times 10^9$ moles per year, while on Enceladus this may be an order of magnitude higher. For Europa this flux is comparable to, and a factor of $\sim$2–3 larger than that estimated for the reductant flux given by McCollom (McCollom 1999). The combined annual biomass may then be closer to 10^6–10^7 kg yr^{-1}, still small by terrestrial standards but significant nonetheless.

Considering both end-members of seafloor activity, the nearly inactive Vance et al. model based on serpentinization versus the McCollom model in which active hydrothermal vents are present, we find that reductants are likely in sufficient supply to power at least a marginal ecosystem. Interestingly, net fluxes of reductants in either case are not found to differ greatly. As detailed in the section on ice shell and ocean exchange, the availability of oxidants to pair with the reductants could greatly enhance net biomass productivity because of the increased energy yield from metabolic pathways made possible by the introduction of oxidants such as O_2 and sulfate.

The seafloor reactions and exchange processes discussed above also have important consequences for the origin of life. Serpentinization and reactions with alkaline fluids have been proposed as a possible pathway to early autotrophic acetogenesis (Martin and Russell 2007; Russell and Martin 2004). Similarly, the more active iron and sulfur-rich hydrothermal vents on Earth have been proposed as possible sites for the emergence of life (Cody et al. 2000; Wächtershäuser 1988, 1990). In both cases, the rocky seafloor minerals serve as catalytic surfaces on which organic synthesis can proceed. Much laboratory work remains to explore the efficacy of these hypotheses, but the active exploration of both Europa and Enceladus may serve to greatly inform our understanding of the origin of life, both here on Earth and elsewhere in the universe. Lastly, it must be emphasized that Enceladus is the only body in the solar system apart from the Earth that seems to have liquid water accessible to the surface.

2.2 The Case of Titan

Prebiotic Chemistry in the Atmosphere, on the Surface, in the Subsurface

Atmospheric chemistry in Titan allows the continuous transformation of methane and dinitrogen into complex organic compounds from the high atmosphere to the surface (see Coustenis et al. 2010, this issue; Cassidy et al. 2010, this issue).The recent Cassini data, especially from the INMS and CAPS instruments, indicates that the processes start in the ionosphere (Waite et al. 2005, 2007). The resulting organics which are produced through high energy photon and electron chemistry, include low molecular organic species, several of which are of prebiotic interest, such as HCN and HC_3N, and higher molecular compounds, tholin-like products (cf. Cassidy et al. 2010, this issue). The later probably form the aerosol particles present in the lower parts of the atmosphere. The particles induce the condensation of the volatile atmospheric species in the low stratosphere and in the troposphere. Simultaneously, methane must be regenerated in the atmosphere from a subsurface source, produced from methane clathrates or/and serpentinization processes.

This methane "cycle" (Lunine and Atreya 2008) is a good illustration of the role of processes in the chemical evolution of a planetary body toward complex organic structures. Methane trapped or formed in the internal structure of Titan feeds the atmosphere. Its gas phase and multiphase chemistry coupled to that of nitrogen produces low molecular weight volatile and high molecular weight refractory organics. The later form hazes, which have a direct impact on the chemical evolution of the atmosphere, generate heterogeneous reactions, as well as an indirect impact via the change of the radiative transfer of the atmosphere.

The most complex product of this evolution in the atmosphere is likely to be a macromolecular organic material made of C, H and N atoms, as detected by the ACP experiment on Huygens (Israël et al. 2005), and similar to some of the Titan's tholins produced in the laboratory (see Cassidy et al. 2010, this issue). Once the aerosols reach Titan's surface, they may also follow a chemical evolution, especially if in contact with water. The episodic presence of liquid water bodies formed from the melting of surface water ice by big impacts can be considered (O'Brien et al. 2005; Artemevia and Lunine 2003, 2005). With a life time of several thousands years (O'Brien et al. 2005) it could allow an efficient hydrolysis of the macromolecular organics of the aerosols into compounds of biological interest, such as amino acids (Raulin et al. 2007; Neish et al. 2008, 2009). Such an evolution may also occur, although at a much lower rate, when the tholin-like material is in contact with water ice. This could also allow the formation of O-containing organics, including compounds of biological interest on Titan's surface.

Another location where an efficient prebiotic chemistry can occur is the possible internal water ocean of Titan (Fortes 2000). Some of the organic materials which have been accreted during the formation of the satellite may have been in contact with the water of the ocean allowing prebiotic reactions to occur. This could have been even more efficient at the beginning of Titan's history, since the bottom of the early ocean was probably in contact with the bedrock, offering conditions similar to the terrestrial hydrothermal deep-sea vents, favored location for prebiotic chemistry and the emergence of life.

More exotic prebiotic-like processes, involving a non-aqueous solvent, are occurring today on Titan. The most abundant volatile organic product of Titan's atmospheric chemistry is ethane which accumulates at the surface, in particular in the polar lakes discovered by Cassini (Stofan et al. 2007; Brown et al. 2008). These surface liquid bodies provide an environment where organics can accumulate and chemically differentiate (Raulin 2008a).

The analysis of Titan's atmosphere in the low troposphere and near the surface by GC-MS (Niemann et al. 2005) indicates a methane mole fraction of 0.049, consistent with the value obtained by the Descent Imager & Spectral Radiometer (DISR) instrument on Huygens (Tomasko et al. 2005). The main composition of the atmosphere near the surface, in the Huygens landing site area, is thus $N_2 \sim 95\%$–$CH_4 \sim 5\%$ with only a very small fraction of argon. In the same area, the Huygens Atmospheric Structure Instrument (HASI) has determined a surface temperature of 93.7 K (Fulchignoni et al. 2005). In the polar regions, the surface temperature is likely to be slightly lower (Tokano 2005). The recent work by Tokano (2009) on the limnological structure of Titan's lake suggests surface temperatures of 90–94 K, for a lake in equilibrium with the atmosphere, depending on the season.

The main composition of a $CH_4{-}N_2{-}C_2H_6{-}Ar$ ocean in thermodynamic equilibrium with the atmosphere was previously computed as a function of the atmospheric methane abundance for an argon-free system at 92.5 K and 101 K and for an atmosphere containing 17% of argon at 101 K (Lellouch et al. 1989; Dubouloz et al. 1989). The conditions used by the authors at 92.5 K are close to those of Titan's polar lakes. The data (Fig. 1 in Dubouloz et al. 1989) show that a methane mole fraction of 0.05 in the atmosphere corresponds to an ocean with mole fractions of 0.65 for ethane, 0.31 for methane and 0.04 for nitrogen. The three-dimensional model of the methane cycle of Titan's troposphere by Tokano et al. (2001) indicates that the advection of methane by the atmospheric global circulation should be important and the methane mole fraction should not exhibit noticeable horizontal variations below the condensation layer. Thus it seems reasonable to assume that the methane mole fraction in the polar regions is close to the value of 4.9% measured by Huygens in equatorial region. This assumption is also used by Tokano (2009) in his recent work on Titan's lakes. Thus this indicates that, if the liquid of Titan's lakes is in equilibrium with the atmosphere, the liquid of these lakes should be richer in ethane than in methane, with a few percent of dissolved dinitrogen. This result, which is in agreement with the detection of liquid ethane in Ontario Lacus (Brown et al. 2008), has also interesting consequences on the possible concentration of minor species (Raulin 2008a).

Most of the minor organic species present in Titan's atmosphere—essentially hydrocarbons and nitriles—must condense in the low atmosphere, especially on the haze particles, and accumulate on the surface. If they reach the liquid surface of a lake, they partly or totally dissolve in the liquid, depending on their solubility, and the insoluble part sediment down to the bottom of the lake or float at its surface, or even form a suspension (Tokano 2009), depending on the density of the material relative to that of the liquid.

Knowing the flux of minor atmospheric species relative to ethane, and assuming as a very first approximation that this relative flux has been constant during Titan's history, one can deduce the total abundance of these compounds relative to ethane on Titan's surface, in particular in Titan's lake. Assuming that compound (i) is totally dissolved in the liquid, one can easily calculate from its relative flux to ethane $n(i)/n(C_2H_6)$, its mole fraction, $X(i)$, in the liquid:

$$X(i) = n(i)/n(C_2H_6) \times X(C_2H_6)$$

knowing the mole fraction of ethane, $X(C_2H_6)$, in the lake. Its concentration, Ci, can also be calculated knowing the molar volume of the liquid (Raulin 1987).

In parallel it is also possible to determine by calculation or even experimentally the solubility of minor species in the liquid of the lake. If the solubility of (i) is higher than its concentration calculated from its relative flux, then the later provides a value of Ci. On the contrary, if the solubility is lower than its concentration calculated from its relative flux, then Ci is equal to the solubility.

Raulin (1987) and Dubouloz et al. (1989) have calculated the solubility of many species of interest for Titan's atmosphere in a CH_4–N_2–C_2H_6–Ar ocean, as a function of the methane atmospheric mole fraction and have estimated the flux of these species relative to ethane, using the data of photochemical models and experiments simulating the evolution of Titan's atmosphere. The results suggest that for all studied alkanes and alkenes the solubility is too high to reach the saturation and the concentration in the liquid is imposed by the atmospheric flux. On the contrary, the more polar organic compounds, such as alkynes and nitriles, and benzene, are generally not very soluble in this non-polar solvent and their concentration is imposed by the solubility. The data also show that the solubility, except for CO_2 and butane, strongly decreases when the mole fraction of methane in the solvent increases. Again, it is possible to apply today these data to Titan's lake, taking into account the temperature and main composition of the lake discussed above. The corresponding results are presented in Table 2. For compounds which have already been detected in the low atmosphere of Titan (mainly in the stratosphere, where their concentration is higher than in the troposphere), the atmospheric concentration is also given, for comparison (Raulin 2008b). These data show that for almost all expected organic compounds the solubility is high enough to allow a concentration in the lake of the order of 1–100 ppm for the more polar compounds to several thousands ppm for the non-polar ones. For all organics already detected in Titan, the concentration in the lake is much higher than that in the stratosphere. Ethane, ethylene and propane concentration is multiplied by $\sim 10^5$, acetylene by $\sim$200, diacetylene by $\sim$500, benzene by $\sim 2 \times 10^4$ and propyne by $\sim 4 \times 10^3$. The same tendency is observed for nitriles, the concentration of which is in the range of 100–1 ppb in the stratosphere and becomes 50–$\sim$1 ppm in the lake. Only the detected O-containing compounds, CO (decrease by factor of 10) and water (decrease by factor of 100) exhibit a much lower concentration in the lake, than in the atmosphere, at the exception of CO_2(strong increase, by $\sim 10^4$). However, the solubility calculated for highly polar compounds, especially H_2O, is probably largely underestimated. However it is interesting to note that ammonia, which is likely to be present on Titan's surface but has not been detected so far, could be present at 5 ppm level.

Table 2 Estimated concentration (order of magnitude) of minor constituents in Titan's lakes (T: 92,5 K; 65% ethane), adapted from Raulin (1987). If no other indication, concentration is given in ppm = part per million in volume. (s): Indicates that the limiting factor is the solubility, otherwise it is the atmospheric flux or (?) it is not known. The atmospheric concentration of minor species is a stratospheric value (adapted from Raulin 2008b and references therein)

Solutes	Concentration (ppm if not %)			Solutes	Concentration (ppm if not %)		
	Lake		Atmosphere		Lake		Atmosphere
Hydrocarbons				Nitriles			
Alkanes				*Alkanitriles*			
Ethane C_2H_6	65%		10	Methanenitrile HCN	3	s	0.1
Propane C_3H_8	2%		0.5	Ethanenitrile CH_3CN	30	s	0.02
Butane C_4H_{10}	4000			Ethanedinitrile (cyanogen) C_2N_2	0.6	s	1×10^{-3}
2-Methylpropane $(CH_3)_3CH$	4000			Propanenitrile C_2H_5CN	50	s	
Pentane C_5H_{12}	400			Butanenitrile C_3H_7CN	1		
2-Methylbutane $(CH_3)_2CC_2H_5$	400			2-Methylpropanenitrile $(CH_3)_2CHCN$	1	s	
Dimethylpropane $(CH_3)_4C$	400			2-Methylbutanenitrile$(CH_3)_2CHCH_2CN$	1		
Hexane C_6H_{14}	40			*Alkenenitriles*			
2-Methylpentane $(CH_3)_2CHC_3H_7$	40			Propenenitrile (acrylonitrile) CH_2CHCN	10	s	
2,2-Dimethylbutane $(CH_3)_3C_3H_7$	40			2-Butenenitrile $CH_3CHCHCN$	1	s	
2,3-Dimethylbutane $(CH_3)_2CC(CH_3)_2$	40			2-Methylpropenenitrile $CH_2C(CH_3)CN$	1	s	
3-Methylhexane $C_2H_5C_H(CH_3)C_3H_7$	4			2-Butenenitrile CH_2CHCH_2CN	1	s	
Alkenes, Benzene				*Alkynenitriles*			
Ethene (ethylene) C_2H_4	4%		0.4	Propynenitrile (cyanoacetylene) HC_3N	3	s	1×10^{-3}
Propene C_3H_6	50			2-Butynenitrile CH_3CCCN	2	s	
2-Methylpropene $CH_2C(CH_3)_2$	5						
1-Butene $CH_2CHC_2H_5$	5			Other compounds			
2-Butene (cis & trans) $CH_3CHCHCH_3$	5						
1,3-Butadiene $CH_2CHCHCH_2$	5			*N-Heterocycles*			
Benzene C_6H_6	5	s	3×10^{-4}	Pyrimidine $C_4H_4N_2$	2	?	

Table 2 (*Continued*)

Solutes	Concentration (ppm if not %)			Solutes	Concentration (ppm if not %)		
	Lake		Atmosphere		Lake		Atmosphere
Alkynes, Allene				Adenine $C_5H_5N_5$	0.01	?	
Ethyne (acetylene) C_2H_2	400	s	2	*Inorganics*			
Propyne (methylacetylene) CH_3CCH	30	s	8×10^{-3}	Carbon dioxide CO_2	10	s	1.6×10^{-3}
1-Butyne $CHCC_2H_5$	400	s		Carbon monoxide CO	4	s	45
1,3-Butadiyne (diacetylene) C_4H_2	0.5	s	1×10^{-3}	Water H_2O	2×10^{-7}	s	4×10^{-4}
Allene CH_2CCH_2	100	s		Ammonia NH_3	5	?	

The relative fluxes to ethane, which were used to calculate the concentration of the minor species when saturation is not reached (Raulin 1987), only give an order of magnitude of the concentration since they are estimated from 1-D photochemical models or simulation experiments. 3-D models providing more reliable fluxes of the different products in the high latitudinal regions of Titan, which would be needed for a better estimate of their concentration in the lake, have not been published yet.

Titan's aerosols when reaching the lake, are also feeding the cryogenic solvent with macromolecular compounds, which are present in the nucleus of the particle, likely to be made of C, N and H atoms (Israël et al. 2005) and similar to tholins (cf. Cassidy et al. 2010, this issue). The solubility of polymeric materials in a CH_4–C_2H_6 cryogenic solvent has been previously studied theoretically (Raulin 1987) as a function of the methane mole fraction. The results indicate that an ethane-rich mixture is much more favorable to the dissolution of organic polymers than a methane-rich mixture. However, the solubility remains very low, and the only soluble material (in a mixture with $> 90\%$ ethane) is polyethylene, a polymer not supposed to be representative of the macromolecular constituents of Titan's haze particles. HCN polymers, which can be considered, like tholins, as a laboratory analogue of Titan's aerosols refractory organics, are very insoluble in this cryogenic mixture, whatever the percent of ethane is. Some experimental studies have also been carried out on that question. McKay (1996) has studied the solubility of Titan's tholins in various solvents and has noticed that it is insoluble in liquid ethane. Coll (1997) has obtained similar results with other Titan's tholins using pentane and hexane at room temperature. Consequently, it seems very likely that the complex organic material included in Titan's aerosol will not dissolve in Titan's lakes.

The volumic mass of a CH_4-C_2H_6 liquid mixture made of 65% ethane at temperature close to 92.5 K is 0.62 $g\,cm^{-3}$ (Raulin 1987). All organics of interest for Titan, as well as water and ammonia, when in a solid phase, have a much higher volumic mass than the cryogenic solvent, assuming those solids are not fluffy. The non-soluble part of these compounds, including insoluble refractory organic materials, should thus sink and settle down to the bottom of the lake. However, the lake may be well mixed and the material accumulating at the bottom of the lake may form a suspension (Tokano 2009).

Titan's lake thus appears as a very interesting chemical reactor, were many organics have a concentration high enough to allow their detection by in situ analysis, as planned on a planetary mission like TSSM or TandEM (Coustenis et al. 2008).

Habitability and the Possible Presence of Life in the Internal Liquid Water

In spite of the very low temperature of its surface, the possibility of the presence of life on Titan has been considered, due to the complex organic chemistry of this environment. Huygens and the many pictures taken by DISR of the same location on Titan during more than one hour after landing did not show any sign of (macroscopic) life on Titan's surface. However, this does not exclude the possibility of the presence of a microscopic life, on Titan's surface or even below. If a biota is localized far from the surface, in the deep internal structure of Titan its metabolic activity may produce chemical species which diffuse through the ice mantle covering the hypothetical internal ocean and feed the atmosphere. It has even been speculated in several publications that the methane present in the atmosphere today is the product of biological activity (Simakov 2001). If this was the case, the atmospheric methane would be notably enriched in light carbon. Indeed, on Earth, biological processes induce an isotopic fragmentation producing enrichment in light carbon, ^{12}C. Consequently, the $^{12}C/^{13}C$ ratio increases from 89 (the reference value, in the Belemnite of the Pee Dee

Formation) to about 91–94 depending on the terrestrial biosynthesis processes. The methane $^{12}C/^{13}C$ ratio in the low atmosphere of Titan, as determined by the GC-MS instrument on Huygens, is 82 (Niemann et al. 2005). Although no abiotic reference for $^{12}C/^{13}C$ on Titan is available, this low value strongly suggests that the origin of atmospheric methane is likely to be abiotic.

However this does not exclude that life may be present in Titan's internal ocean. Models of formation and evolution of Titan indicate that this ocean was in contact with the bedrock on one side and with the atmosphere on the other side, at the beginning of Titan's history, a situation similar to that on the primitive Earth. With the potential availability of hydrothermal vents in this hypothetical liquid water ocean, together with organic compounds (coming from accreted chondritic materials) and energy sources, minimum conditions for the emergence of life were thus available on Titan.

However, if life emerged in this (still hypothetical) Titan's subsurface ocean, one should wonder if the conditions in the ocean are compatible with the sustaining of life. Fortes (2000) has shown that there are no insurmountable obstacles. With a possible temperature of this ocean as high as about 260 K and the possible occurrence of cryovolcanic hotspots allowing 300 K, the temperature conditions in Titan's subsurface oceans could allow for the development of living systems. Even at depth of 200 km, the expected pressure of about 5 kbar is not incompatible with life, as shown by terrestrial examples. The expected pH of an aqueous medium made of 15% by weight of NH_3 is equivalent to a pH of 11.5. Some bacteria can grow on Earth at pH 12. Even the limited energy resources do not exclude the sustaining of life.

Taking into account only the potential radiogenic heat flow ($\sim 5 \times 10^{11}$ W) and assuming that 1% of that is used for volcanic activity and 10% of the later is available for living system metabolism, Fortes (2000) estimated an energy flux available in the subsurface oceans of about 5×10^{8} W. Considering the terrestrial biosphere, such a flux corresponds to the production of about 4×10^{11} mol of ATP per year and about 2×10^{13} g of biomass per year. If we assume an average turn over for the living systems in the order of one year, the biomass density would be 1 g/m^2. This is very small compared to the lower limit of the value of the biomass for the Earth (about 1000 to 10000 g/m^2). Nevertheless, this indicates the possible presence of a limited but not negligible bioactivity on the satellite. The biota on Titan, if any, assuming that the living systems are similar to the ones we know on Earth, and based on the chemistry of carbon and the use of liquid water as solvent, would thus be localized in the subsurface deep ocean. Several possible metabolic processes such as nitrate/nitrite reduction or nitrate/dinitrogen reduction, sulfate reduction and methanogenesis have been postulated (Simakov 2001) as well as the catalytic hydrogenation of acetylene (Abbas and Schulze-Makuch 2002; McKay and Smith 2005).

The possibility of another life based on a pseudo-biochemistry, evolving in the absence of a noticeable amount of O atoms, such as a nitrogen-chemistry based on "ammono" analogues replacing the oxygen-chemistry, has previously been considered (Raulin and Owen 2002). Different alternatives of terrestrial biochemistry have also been re-examined by Schulze-Makuch and Irwin (2004), and by the Space Studies Board and Board on Life Sciences committees (SSB-BLS 2007). They consider, in particular, the possible replacement of liquid water as a solvent by liquid ammonia or other liquid N-compounds. Thus several ways can be envisaged to drive chemistry to prebiotic chemistry and even to biotic systems in Titan's environment.

An Exotic Life in Titan's Lakes?

A particularly relevant alternative liquid for life in Titan is liquid methane and ethane. It is now established that Titan has polar lakes that contain liquid methane and ethane. There are clouds—both thin cloud and thunderheads—composed of liquid methane, and there is direct evidence that the surface is moist with liquid methane/ethane (Niemann et al. 2005). Thus, there is an active surface hydrological cycle on Titan where methane and ethane play the same role that liquid water plays on the Earth. Titan is the only world in the solar system other than Earth with a hydrological-like cycle operating at its surface.

The possibility that this active fluid of methane and ethane on Titan could be a fluid in which life operates, has been suggested by several authors (Benner et al. 2004; Schulze-Makuch and Grinspoon 2005; McKay and Smith 2005). Two possible objections to life in liquid methane solutions are: (1) the low solubility of organics in such a non-polar solvent, and (2) the low temperatures and associated low reaction rates. However these objections are not fundamental. The low solubility of organic material may be compensated for by organisms with large surface areas and with active transport mechanisms. Life on Earth demonstrates the ability to take up nutrients at very low concentrations. Low temperatures may inhibit chemical reactions but that may be good for biology. Life survives by catalyzing reactions that are thermodynamically favorable but kinetically inhibited. An environment in which reactions proceeded quickly would have no energy available for life. On Titan the reactions are slow due to the low temperatures so there is considerable chemical potential energy. By the use of catalyst, organism could release this energy at rates that are not dependent on the temperature.

A possible energy source for life on the surface of Titan is the hydrogenation of reduced organic materials, especially C_2H_2 (Schulze-Makuch and Grinspoon 2005; McKay and Smith 2005). McKay and Smith (2005) show that the energy available from this reaction is comparable to the energy required to support methanogens on Earth.

If life on Titan can occur in liquid methane and consume the hydrocarbons produced in the atmosphere, then we would expect that life would be widespread on the surface of Titan. If life is widespread on Titan we would expect it to have a global effect on the environment. This is seen, for example, on Earth where the presence of widespread life results in obvious changes in the composition of the atmosphere (O_2 and CH_4 both present in the atmosphere). McKay and Smith (2005) point out that life on Titan could consume enough H_2 so that it would affect the concentration in the lower atmosphere. Thus H_2 depletion in the lower atmosphere may be a "biosignature" for liquid methane based methanogenic life on Titan's surface.

The hypothesis of life in liquid methane on Titan is quite speculative and may not warrant a targeted mission or instrument. However, we should be looking for evidence that unexplained processes are occurring at the surface and considering if these processes might be biological. Clearly such life on Titan would represent an independent origin of life and would demonstrate that life might exist in a myriad of forms in the universe.

3 Planetary Protection and the Icy Moons of the Giant Planets

Exploratory missions to the Jovian and Saturnian systems have revealed the diversity of the icy moons of these giant planets. From the data gathered, it also appears that some of them have active surface and some could have, beneath an icy crust, oceans or pocket reservoirs containing mixtures of liquid water. Based on an increasing body of knowledge about the microbial diversity of terrestrial microorganisms, it appears that the presence of indigenous life forms in the liquid water of these worlds cannot be ruled out—nor can we discard the potential for some terrestrial organisms to survive or proliferate there.

Since several space agencies are committing themselves to explore those icy moons, it is possible that future missions could inadvertently export portions of the terrestrial biosphere to these worlds. Accordingly, these agencies and other parties to the United Nations Outer Space Treaty (*Treaty on Principles Governing the Activities of States in the Exploration and Use of Outer Space, Including the Moon and Other Celestial Bodies;* U.N. document 6347 January 1967) have a responsibility to protect these moons against possible terrestrial contamination.

To meet the provisions of Article IX of the Treaty, which states that "harmful contamination" should be avoided and that the Earth's biosphere should be protected, the Committee for Space Research (COSPAR) maintains a planetary protection policy which applies to activities on or around these bodies. This policy is regularly updated, taking into account the most recent scientific knowledge and technologies. This policy might be considered as a soft law implemented in the regulations of the countries or the instructions of the agencies.

3.1 Categorization

From the last update of the planetary protection policy promulgated by COSPAR, five categories of space missions have been defined and are summarized in Table 3. Beyond these recommendations, COSPAR is proposing some figures for contamination probability and methods to reach them.

Note that there is no general categorization of the icy moons of the giant planets leaving room for a case-by-case assessment depending on the actual knowledge about the structure of each moon as stated in the planetary protection policy.

3.2 The Case for Europa

A specific appendix was edited for Europa in 2002 taking into account the last results of the US mission Galileo. This document is acknowledging a proposal coming from two publications issued by the US Space studies board in 1999 (SSB-NRC 1999) and 2000 (SSB-NRC 2000). The probability of inadvertent contamination of a possible Europa's ocean must be less that 10^{-4} for each mission to or around Europa. The calculation of this probability includes conservative estimates of several factors like number of microorganisms on the spacecraft by the time of the launch, the survival rate of these microorganisms along the cruise and within the Jovian environment, the probability of the spacecraft to hit Europa's surface (this probability is 1 for a Europa lander or a Europa orbiter) as well as the timescale and then the probability of the organisms to reach the Europan ocean. Listing all these factors, the space studies board proposed a formula to be used to evaluate the probability taking into account 5 types of microorganisms depending on their capability to form spores and to be radiation-resistant. From the model displayed in the example from the SSB, the probability of contamination for Europa is coming from the rare highly resistant microorganisms displaying a resisting profile like *Deinococcus radiodurans;* even such bacteria are not forming spores.

Several agencies are already sponsoring studies on the microbiological load of the spacecraft and the microbial fauna of the assembly rooms. Studies are also performed on the effect of the spaceflight on the survival of various species.

Table 3 Categorization of the missions according with the planetary protection policy promulgated by COSPAR

Cat.	Type of Mission	Target Bodies	Recommendation
1	All	Sun, Mercury, Venus, Metamorphosed asteroids	Impact probability Contamination passive control
2	All	Venus, Saturn, Uranus, Neptune, Comets, Jupiter, Chondritic Asteroids, some moons of the external planets, Pluto, Charon, Kuiper-Belt objects...	Crash probability minimization Passive control of the contamination
3	Flyby and orbiters	Mars, Europa...	Minimization of crash probability Active control of the contamination
4	Contact probes ground or atmosphere	Europa... No astrobiology in the payload Subcategories applies only for Mars	Minimization of probability of off-nominal entry or landing Active control of the contamination
4 A	Contact probes ground or atmosphere	Subcategories applies only for Mars No astrobiology in the payload	Minimization of probability of off-nominal entry or landing Active control of the contamination
4 B		Astrobiology in the payload	Same + Stringent active control of the contamination
4 C		Penetrating a special region	Same plus more stringent active control of the contamination
5 no R	Earth return	All	No restriction
5 R	Earth return	Mars, Europa, others to be defined on a case by case basis	Restricted Earth crash avoidance Contact-chain breaking Sample quarantine

3.3 Conclusion

The present requirements for planetary protection of the icy moons of the giant planets are not specific but for Europa. Such requirements will be updated for the icy moons as well as for Europa in the next years as the knowledge of their structure and subsurface environment is increasing. The exploration of these bodies of the solar system will involve, in most cases, a large international cooperation. COSPAR is probably the most appropriate structure to organize the review of the ongoing studies and to propose the required updates of the planetary protection policy.

References

O. Abbas, D. Schulze-Makuch, ESA SP **518**, 345–348 (2002)
O. Abramov, J.R. Spencer, Icarus **199**, 189–196 (2008)
J.D. Anderson, G. Schubert, R.A. Jacobson, E.L. Lau, W.B. Moore, W.L. Sjogren, Science **281**, 2019–2022 (1998)

N. Artemevia, J.I. Lunine, Icarus **164**, 471–480 (2003)
N. Artemevia, J.I. Lunine, Icarus **175**, 522–533 (2005)
J.L. Bada, A. Lazcano, Science **296**, 1982–1983 (2002)
S.A. Benner, A. Ricardo, M.A. Carrigan, Curr. Opin. Chem. Biol. **8**, 672–689 (2004)
M.P. Bernstein, S.A. Sandford, L.J. Allamandola, S. Chang, M.A. Scharberg, Astrophys. J. **454**, 327–344 (1995)
R.H. Brown, L.A. Soderblom, J.M. Soderblom, R.N. Clark, R. Jaumann, J.W. Barnes, C. Sotin, B. Buratti, K.H. Baines, P.D. Nicholson, Nature **454**, 607–610 (2008)
R.H. Brown, R.N. Clark, B.J. Buratti, D.P. Cruikshank, J.W. Barnes, R.M.E. Mastrapa, J. Bauer, S. Newman, T. Momary, K.H. Baines, G. Bellucci, F. Capaccioni, P. Cerroni, M. Combes, A. Coradini, P. Drossart, V. Formisano, R. Jaumann, Y. Langevin, D.L. Matson, T.B. McCord, R.M. Nelson, P.D. Nicholson, B. Sicardy, C. Sotin, Science **311**, 1425–1428 (2006)
M.H. Burger et al., Space Sci. Rev. (2010, this issue)
R.W. Carlson, M.S. Anderson, R.E. Johnson, W.D. Smythe, A.R. Hendrix, C.A. Barth, L.A. Soderblom, G.B. Hansen, T.B. McCord, J.B. Dalton, R.N. Clark, J.H. Shirley, A.C. Ocampo, D.L. Matson, Science **283**, 2062–2064 (1999a)
R.W. Carlson, R.E. Johnson, M.S. Anderson, Science **286**, 97–99 (1999b)
R.W. Carlson, M.S. Anderson, R. Mehlman, R.E. Johnson, Icarus **177**, 461–471 (2005)
T.A. Cassidy et al., Space Sci. Rev. (2010, this issue). doi:10.1007/s11214-009-9625-3
C.F. Chyba, Nature **403**, 381–382 (2000)
C.F. Chyba, K.P. Hand, Annu. Rev. Astron. Astrophys. **43**, 31–74 (2005)
C.F. Chyba, C.B. Phillips, Proc. Natl. Acad. Sci. USA **98**, 801–804 (2001)
G.D. Cody, N.Z. Boctor, T.R. Filley, R.M. Hazen, J.H. Scott, A. Sharma, H.S. Yoder Jr., Science **289**, 1337–1340 (2000)
P. Coll, PhD thesis, Université Paris 12, France, 1997
J.F. Cooper, R.E. Johnson, B.H. Mauk, H.B. Garrett, N. Gehrels, Icarus **149**, 133–159 (2001)
J.F. Cooper, E.C. Sittler Jr., P.D. Cooper, M.H. Burger, S.J. Sturner, A.M. Rymer, M.E. Hill, R.E. Johnson, Bull. Am. Astron. Soc. **38**, 509 (2006). Abstract 38.15.06
A. Coustenis et al., Exp. Astron. (2008). doi:10.1007/s10686-008-9103-z
A. Coustenis et al., Space Sci. Rev. (2010, this issue). doi:10.1007/s11214-009-9615-5
N. Dubouloz, F. Raulin, E. Lellouch, D. Gautier, Icarus **82**, 81–96 (1989)
J.P. Dworkin, D.W. Deamer, S.A. Sandford, L.J. Allamandola, Proc. Natl. Acad. Sci. USA **98**, 815–819 (2001)
H. Elderfield, A. Schultz, Annu. Rev. Earth Planet Sci. **24**, 191–224 (1996)
A.D. Fortes, Icarus **146**, 444–452 (2000)
M. Fulchignoni, F. Ferri, F. Angrilli, A.J. Ball, A. Bar-Nun, M.A. Barucci, C. Bettanini, G. Bianchini, W. Borucki, G. Colombatti, M. Coradini, A. Coustenis, S. Debei, P. Falkner, G. Fanti, E. Flamini, V. Gaborit, R. Grard, M. Hamelin, A.M. Harri, B. Hathi, I. Jernej, M.R. Leese, A. Lehto, P.F. Lion Stoppato, J.J. Lopez-Moreno, T. Makinen, J.A.M. McDonnell, C.P. McKay, G. Molina-Cuberos, F.M. Neubauer, V. Pirronello, R. Rodrigo, B. Saggin, K. Schwingenschuh, A. Seiff, F. Simoes, H. Svedhem, T. Tokano, M.C. Towner, R. Trautner, P. Withers, J.C. Zarnecki, Nature **438**, 785–791 (2005)
E.J. Gaidos, K.H. Nealson, J.L. Kirschvink, Science **284**, 1631–1633 (1999)
R. Greeley, C.F. Chyba, J.W. Head III, T.B. Mccord, W.B. Mckinnon, R.T. Pappalardo, P. Figueredo, in *Jupiter: The Planet, Satellites and Magnetosphere*, ed. by F. Bagenal, T.E. Dowling, W.B. McKinnon (Cambridge University Press, Cambridge, 2004), pp. 329–362
K.P. Hand, 2007. PhD Thesis Stanford University, 2007
K.P. Hand, R.W. Carlson, R.W. Cooper, C.F. Chyba, Astrobiology **6**, 463–482 (2006)
K.P. Hand, R.W. Carlson, C.F. Chyba, Astrobiology **7**, 1–18 (2007)
K.P. Hand, C.F. Chyba, J.P. Priscu, R.W. Carlson, K.H. Nealson, in *Europa*, ed. by R.T. Pappalardo, W.B. McKinnon, K. Khurana (Univ. of Arizona Press, 2009 in press)
J.J. Heijnen, J.P. van Dijken, Biotechnol. Bioeng. **39**, 834–857 (1992)
T.M. Hoehler, Geobiology **2**, 205–215 (2004)
T.M. Hoehler, M.J. Alperin, D.B. Albert, C.S. Martens, FEMS Microbiol. Ecol. **38**, 33–41 (2001)
H. Hussmann, T. Spohn, Icarus **171**, 391–410 (2004)
G. Israël, C. Szopa, F. Raulin, M. Cabane, H.B. Niemann, S.K. Atreya, S.J. Bauer, J.-F. Brun, E. Chassefière, P. Coll, E. Condé, D. Coscia, A. Hauchecorne, P. Millian, M.J. Nguyen, T. Owen, W. Riedler, R.E. Samuelson, J.-M. Siguier, M. Steller, R. Sternberg, C. Vidal-Madjar, Nature **438**, 796–799 (2005)
J. Kargel, J.Z. Kaye, J.W. Head III, G.M. Marion, R. Sassen, J.K. Crowley, O. Prieto Ballesteros, S.A. Grant, D.L. Hogenboom, Icarus **148**, 226–265 (2000)
J.F. Kasting, in *The Origins of Life*, ed. by D. Baltimore et al. (Academic Press, New York, 2000), pp. 193–200

O.L. Kuskov, V.A. Kronrod, Icarus **177**, 550–569 (2005)
E. Lellouch, A. Coustenis, D. Gautier, F. Raulin, N. Dubouloz, C. Frère, Icarus **79**, 328–349 (1989)
M. Levy, S.L. Miller, K. Brinton, J.L. Bada, Icarus **145**, 609–613 (2000)
J. Lewis, Science **186**, 440–443 (1974)
J.S. Lewis, Icarus **15**, 174–185 (1971)
J.S. Lewis, *Physics and Chemistry of the Solar System* (Academic Press, San Diego, 2004)
M.J. Loeffler, R.A. Baragiola, Astrophys. J. Lett. **694**, L92–L94 (2009)
R.D. Lorenz, B.W. Stiles, R.L. Kirk, M.D. Allison, P. Persi del Marmo, L. Iess, J.I. Lunine, S.J. Ostro, S. Hensley, Science **319**, 1649–1651 (2008)
J.I. Lunine, S.K. Atreya, Nat. Geosci. **1**, 159–164 (2008)
W. Martin, M.J. Russell, Philos. Trans. R. Soc. Lond. B, Biol. Sci. **362**, 1887–1926 (2007)
T.M. McCollom, J. Geophys. Res. **104**, 30,729–30,742 (1999)
T.M. McCollom, J.P. Amend, Geobiology **3**, 135–144 (2005)
C.P. McKay, Icarus **91**, 93–100 (1991)
C.P. McKay, Planet. Space Sci. **44**, 741–747 (1996)
C.P. McKay, H.D. Smith, Icarus **178**, 274 (2005)
C.P. McKay et al., Astrobiology **8**, 909–919 (2008)
S.L. Miller, L.E. Orgel, *The Origins of Life on the Earth* (Prentice-Hall, Englewood Cliffs, 1974)
K.H. Nealson, J. Geophys. Res. **102**, 23675–23686 (1997)
C.D. Neish, A. Somogyi, J.I. Lunine, M.A. Smith, Astrobiology **8**(2), 273–287 (2008)
C.D. Neish, J.I. Lunine, M.A. Smith, Icarus **201**, 412–421 (2009)
S.F. Newman et al., Astrophys. J. Lett. **670**, 143–146 (2007)
H.B. Niemann, S.K. Atreya, S.J. Bauer, G.R. Carignan, J.E. Demick, R.L. Frost, D. Gautier, J.A. Haberman, D.N. Harpold, D.M. Hunten, G. Israel, J.I. Lunine, W.T. Kasprzak, T.C. Owen, M. Paulkovich, F. Raulin, E. Raaen, S.H. Way, Nature **438**, 779–784 (2005)
E.G. Nisbet, N.H. Sleep, Nature **409**, 1083–1091 (2001)
D.P. O'Brien, R.D. Lorenz, J.I. Lunine, Icarus **173**, 243–253 (2005)
L.E. Orgel, Trends Biochem. Sci. **23**, 491–495 (1998)
C.D. Parkinson, M.-C. Liang, H. Hartman, C.J. Hansen, G. Tinetti, V. Meadows, J.L. Kirschvink, Y.L. Yung, Astron. Astrophys. **463**, 353–357 (2007)
C.D. Parkinson, M.-C. Liang, Y.L. Yung, J.L. Kirschvink, Orig. Life Evol. Biosph. **38**, 355–369 (2008)
C.B. Phillips, S. McEwen, G.V. Hoppa, S.A. Fagents, R. Greeley, J.E. Klemaszewski, J. Geophys. Res. **105**(E9), 22579–22597 (2000)
E. Pierazzo, C.F. Chyba, Icarus **157**, 120–127 (2002)
C.C. Porco, P. Helfenstein, P.C. Thomas, A.P. Ingersoll, J. Wisdom, R. West, G. Neukum, T. Denk, R. Wagner, T. Roatsch, S. Kieffer, E. Turtle, A. McEwen, T.V. Johnson, J. Rathbun, J. Veverka, D. Wilson, J. Perry, J. Spitale, A. Brahic, J.A. Burns, A.D. DelGenio, L. Dones, C.D. Murray, S. Squyres, Science **311**, 1393–1401 (2006)
P.B. Price, FEMS Microbiol. Ecol. **59**, 217–231 (2007)
P.B. Price, Can. J. Microbiol. **55**, 1–11 (2009)
F. Raulin, Adv. Space Res. **7**(5), 71–81 (1987)
F. Raulin, Nature **454**, 587–589 (2008a)
F. Raulin, Space Sci. Rev. **135**(1–4), 37–48 (2008b)
F. Raulin, T. Owen, Space Sci. Rev. **104**(1–2), 379 (2002)
F. Raulin, M.J. Nguyen, P. Coll, in *Instruments, Methods, and Missions for Astrobiology*, ed. by X. Richard, B. Hoover, G.V. Levin, A.Y. Rozanov, P.C.W. Davies, Proc. SPIE, vol. 6694 (2007). doi:10.1117/12.732883
M.J. Russell, W. Martin, Trends Biochem. Sci. **29**, 358–363 (2004)
G. Schubert et al., Space Sci. Rev. (2010, this issue)
D. Schulze-Makuch, D.H. Grinspoon, Astrobiology **5**, 560–567 (2005)
D. Schulze-Makuch, L.N. Irwin, *Life in the Universe. Expectations and Constraints* (Springer, New York, 2004)
M.B. Simakov, ESA SP **496**, 211–214 (2001) and refs included
F. Sohl, H. Hussmann, B. Schwentker, T. Spohn, R.D. Lorenz, J. Geophys. Res. **108**(E12), 101029 (2003). CiteID 5130
F. Sohl et al., Space Sci. Rev. (2010, this issue)
J.R. Spencer, W.M. Calvin, Astron. J. **124**, 3400–3403 (2002)
J.R. Spencer, J.C. Pearl, M. Segura, F.M. Flasar, A. Mamoutkine, P. Romani, B.J. Buratti, A.R. Hendrix, L.J. Spilker, R.M.C. Lopes, Science **311**, 1401–1405 (2006)
S.W. Squyres, R.T. Reynolds, P.M. Cassen, S.J. Peale, Nature. **301**, 225–226 (1983a)
S.W. Squyres, R.T. Reynolds, P.M. Cassen, S.J. Peale, Icarus **53**, 319–331 (1983b)

SSB-BLS, Space Studies Board and Board on Life Sciences, *The Limits of Organic Life in Planetary Systems* (The National Academies Press, Washington, 2007)

SSB-NRC, Space Studies Board, National Research Council, *A Science Strategy for the Exploration of Europa* (National Academy Press, Washington, 1999)

SSB-NRC, Space Studies Board, National Research Council, *Preventing the Forward Contamination of Europa* (National Academy Press, Washington, 2000)

E.R. Stofan, C. Elachi, J.I. Lunine, R.D. Lorenz, B. Stiles, K.L. Mitchell, S. Ostro, L. Soderblom, C. Wood, H. Zebker, S. Wall, M. Janssen, R. Kirk, R. Lopes, F. Paganelli, J. Radebaugh, L. Wye, Y. Anderson, M. Allison, R. Boehmer, P. Callahan, P. Encrenaz, E. Flamini, G. Francescetti, Y. Gim, G. Hamilton, S. Hensley, W.T.K. Johnson, K. Kelleher, D. Muhleman, P. Paillou, G. Picardi, F. Posa, L. Roth, R. Seu, S. Shaffer, S. Vetrella, Nature **445**, 61–64 (2007)

R.E. Thomson, J.R. Delaney, J. Geophys. Res. **106**, 12355–12365 (2001)

G. Tobie et al., Space Sci. Rev. (2010, this issue)

T. Tokano, F.M. Neubauer, M. Laube, C.P. McKay, Icarus **153**, 130–147 (2001)

T. Tokano, Icarus **173**, 222–242 (2005)

T. Tokano, Astrobiology **9**, 147–164 (2009)

M.G. Tomasko, B. Archinal, T. Becker, B. Bezard, M. Bushroe, M. Combes, D. Cook, A. Coustenis, C. de Bergh, L.E. Dafoe, L. Doose, S. Doute, A. Eibl, S. Engel, F. Gliem, B. Grieger, K. Holso, E. Howington-Kraus, E. Karkoschka, H.U. Keller, R. Kirk, R. Kramm, M. Küppers, P. Lanagan, E. Lellouch, M. Lemmon, J. Lunine, E. McFarlane, J. Moores, G.M. Prout, B. Rizk, M. Rosiek, P. Rueffer, S.E. Schröder, B. Schmitt, C. See, P. Smith, L. Soderblom, N. Thomas, R. West, Nature **438**, 765–778 (2005)

H. Trinks, W. Schröder, C. Biebricher, Orig. Life Evol. Biosph. **35**, 429–445 (2005)

S. Vance, J. Harnmeijer, J. Kimura, H. Hussmann, B. deMartin, J.M. Brown, Astrobiology **7**, 987–1005 (2007)

G. Wächtershäuser, PNAS **87**, 200–204 (1990)

G. Wächtershäuser, Syst. Appl. Microbiol. **10**, 207–210 (1988)

L.P. Wackett, A.G. Dodge, L.B.M. Ellis, Appl. Environ. Microbiol. **70**, 647–655 (2004)

J.H. Waite Jr., H. Niemann, R.V. Yelle, W.T. Kasprzak, T.E. Cravens, J.G. Luhmann, R.L. McNutt, W.-H. Ip, D. Gell, V. De La Haye, I. Müller-Wordag, B. Magee, N. Borggren, S. Ledvina, G. Fletcher, E. Walter, R. Miller, S. Scherer, R. Thorpe, J. Xu, B. Block, K. Arnett, Science **308**, 982–986 (2005)

J.H. Waite Jr., M.R. Combi, W.-H. Ip, T.E. Cravens, R.L. McNutt Jr., W. Kasprzak, R. Yelle, J. Luhmann, H. Niemann, D. Gell, B. Magee, G. Fletcher, J. Lunine, W.-L. Tseng, Science **311**, 1419–1422 (2006)

J.H. Waite Jr., D.T. Young, T.E. Cravens, A.J. Coates, F.J. Crary, B. Magee, J. Westlake, Science **316**, 870–875 (2007)

K. Zahnle, P. Schenk, H. Levison, L. Dones, Icarus **163**, 263–289 (2003)

M.Y. Zolotov, Geophys. Res. Lett. **34**, L23203 (2007)

M.Y. Zolotov, E.L. Shock, J. Geophys. Res. **106**, 32815–32827 (2001)

M.Y. Zolotov, E.L. Shock, J. Geophys. Res. **109**, E06003 (2004). doi:10.1029/2003JE002194

Space Science Series of ISSI

1. R. von Steiger, R. Lallement and M.A. Lee (eds.): *The Heliosphere in the Local Interstellar Medium*. 1996 ISBN 0-7923-4320-4
2. B. Hultqvist and M. Øieroset (eds.): *Transport Across the Boundaries of the Magnetosphere*. 1997 ISBN 0-7923-4788-9
3. L.A. Fisk, J.R. Jokipii, G.M. Simnett, R. von Steiger and K.-P. Wenzel (eds.): *Cosmic Rays in the Heliosphere*. 1998 ISBN 0-7923-5069-3
4. N. Prantzos, M. Tosi and R. von Steiger (eds.): *Primordial Nuclei and Their Galactic Evolution*. 1998 ISBN 0-7923-5114-2
5. C. Fröhlich, M.C.E. Huber, S.K. Solanki and R. von Steiger (eds.): *Solar Composition and its Evolution – From Core to Corona*. 1998 ISBN 0-7923-5496-6
6. B. Hultqvist, M. Øieroset, Goetz Paschmann and R. Treumann (eds.): *Magnetospheric Plasma Sources and Losses*. 1999 ISBN 0-7923-5846-5
7. A. Balogh, J.T. Gosling, J.R. Jokipii, R. Kallenbach and H. Kunow (eds.): *Co-rotating Interaction Regions*. 1999 ISBN 0-7923-6080-X
8. K. Altwegg, P. Ehrenfreund, J. Geiss and W. Huebner (eds.): *Composition and Origin of Cometary Materials*. 1999 ISBN 0-7923-6154-7
9. W. Benz, R. Kallenbach and G.W Lugmair (eds.): *From Dust to Terrestrial Planets*. 2000 ISBN 0-7923-6467-8
10. J.W. Bieber, E. Eroshenko, P. Evenson, E.O. Flückiger and R. Kallenbach (eds.): *Cosmic Rays and Earth*. 2000 ISBN 0-7923-6712-X
11. E. Friis-Christensen, C. Fröhlich, J.D. Haigh, M. Schüssler and R. von Steiger (eds.): *Solar Variability and Climate*. 2000 ISBN 0-7923-6741-3
12. R. Kallenbach, J. Geiss and W.K. Hartmann (eds.): *Chronology and Evolution of Mars*. 2001 ISBN 0-7923-7051-1
13. R. Diehl, E. Parizot, R. Kallenbach and R. von Steiger (eds.): *The Astrophysics of Galactic Cosmic Rays*. 2001 ISBN 0-7923-7051-1
14. Ph. Jetzer, K. Pretzl and R. von Steiger (eds.): *Matter in the Universe*. 2001 ISBN 1-4020-0666-7
15. G. Paschmann, S. Haaland and R. Treumann (eds.): *Auroral Plasma Physics*. 2002 ISBN 1-4020-0963-1
16. R. Kallenbach, T. Encrenaz, J. Geiss, K. Mauersberger, T.C. Owen and F. Robert (eds.): *Solar System History from Isotopic Signatures of Volatile Elements*. 2003 ISBN 1-4020-1177-6
17. G. Beutler, M.R. Drinkwater, R. Rummel and R. von Steiger (eds.): *Earth Gravity Field from Space – from Sensors to Earth Sciences*. 2003 ISBN 1-4020-1408-2
18. D. Winterhalter, M. Acuña and A. Zakharov (eds.): *"Mars" Magnetism and its Interaction with the Solar Wind*. 2004 ISBN 1-4020-2048-1
19. T. Encrenaz, R. Kallenbach, T.C. Owen and C. Sotin: *The Outer Planets and their Moons* ISBN 1-4020-3362-1
20. G. Paschmann, S.J. Schwartz, C.P. Escoubet and S. Haaland (eds.): *Outer Magnetospheric Boundaries: Cluster Results* ISBN 1-4020-3488-1
21. H. Kunow, N.U. Crooker, J.A. Linker, R. Schwenn and R. von Steiger (eds.): *Coronal Mass Ejections* ISBN 978-0-387-45086-5

22. D.N. Baker, B. Klecker, S.J. Schwartz, R. Schwenn and R. von Steiger (eds.): *Solar Dynamics and its Effects on the Heliosphere and Earth* ISBN 978-0-387-69531-0
23. Y. Calisesi, R.-M. Bonnet, L. Gray, J. Langen and M. Lockwood (eds.): *Solar Variability and Planetary Climates* ISBN 978-0-387-48339-9
24. K.E. Fishbaugh, P. Lognonné, F. Raulin, D.J. Des Marais and O. Korablev (eds.): *Geology and Habitability of Terrestrial Planets* ISBN 978-0-387-74287-8
25. O. Botta, J.L. Bada, J. Gomez-Elvira, E. Javaux, F. Selsis and R. Summons (eds.): *Strategies of Life Detection* ISBN 978-0-387-77515-9
26. A. Balogh, L. Ksanfomality and R. von Steiger (eds.): *Mercury* ISBN 978-0-387-77538-8
27. R. von Steiger, G. Gloeckler and G.M. Mason (eds.): *The Composition of Matter* ISBN 978-0-387-74183-3
28. H. Balsiger, K. Altwegg, W. Huebner, T.C. Owen and R. Schulz (eds.): *Origin and Early Evolution of Comet Nuclei, Workshop honouring Johannes Geiss on the occasion of his 80th birthday* ISBN 978-0-387-85454-0
29. A.F. Nagy, A. Balogh, T.E. Cravens, M. Mendillo and I. Mueller-Wodarg (eds.): *Comparative Aeronomy* ISBN 978-0-387-87824-9
30. F. Leblanc, K.L. Aplin, Y. Yair, R.G. Harrison, J.P. Lebreton and M. Blanc (eds.): *Planetary Atmospheric Electricity* ISBN 987-0-387-87663-4
31. J.L. Linsky, V. Izmodenov, E. Möbius and R. von Steiger (eds.): *From the Outer Heliosphere to the Local Bubble: Comparison of New Observations with Theory* ISBN 978-1-4419-0246-7
32. M.J. Thompson, A. Balogh, J.L. Culhane, Å. Nordlund, S.K. Solanki and J.-P. Zahn (eds.): *The Origin and Dynamics of Solar Magnetism* ISBN 978-1-4419-0238-2
33. U.R. Christensen, A. Balogh, D. Breuer and K.-H. Glaßmeier (eds.): *Planetary Magnetism* ISBN 978-1-4419-5900-3
34. C.W.F. Everitt, M.C.E. Huber, R. Kallenbach, G. Schäfer, B.F. Schutz and R.A. Treumann (eds.): *Probing The Nature of Gravity: Confronting Theory and Experiment in Space* ISBN 978-1-4419-1361-6
35. O. Grasset, M. Blanc, A. Coustenis, W.B. Durham, H. Hussmann, R.T. Pappalardo and D. Turrini (eds.): *Satellites of the Outer Solar System: Exchange Processes Involving the Interiors* ISBN 978-1-4419-7438-9